# POLLUTANT REMOVAL HANDBOOK

# POLLUTANT REMOVAL HANDBOOK

Marshall Sittig

**NOYES DATA CORPORATION**

Park Ridge, New Jersey          London, England

1973

Published in the United States of America by
Noyes Data Corporation
Noyes Building, Park Ridge, New Jersey 07656

# FOREWORD

This Pollutant Removal Handbook attempts to clarify the ways and means open to the alert industrial processor who must keep his polluting wastes down to a minimum.

The book contains a total of 128 subject entries arranged in an alphabetical and encyclopedic fashion. Because of its encyclopedic arrangement, it can also serve as a textbook and professional aid to eager students in the fields of antipollution engineering and pollution abatement.

During the past few years, the words pollution, environment and ecology have come into frequent usage and the cleanliness of the world we live in has become the concern of all people. Pollution, for example, is no longer just a local problem involving litter in the streets or the condition of a nearby beach. Areas of the oceans, far-reaching rivers and the largest lakes are now classified as polluted or subject to polluting conditions. In addition, very surprisingly, lakes and streams remote from industry and population centers have been found to be contaminated.

For these reasons, this handbook gives pertinent and concise information on such widely divergent topics as the removal of oil slicks in oceans to the containment of odors and particulates from paper mills.

Aside from practical considerations such as indicating by its extensive bibliographical lists, further sources to consult for additional information, this book is also helpful in explaining the new lingo of pollution abatement. This field is developing new concepts and a new terminology all its own, for example, particulates, polyelectrolytes, flocculation, recycling, activated sludge, gas incineration, catalytic conversion, industrial ecology, etc.

Advanced composition and production methods developed by Noyes Data are employed to bring our new durably bound books to you in a minimum of time. Special techniques are used to close the gap between "manuscript" and "completed book." Industrial technology is progressing so rapidly that time-honored, conventional typesetting, binding and shipping methods are no longer suitable. We have bypassed the delays in the conventional book publishing cycle and provide the user with an effective and convenient means of reviewing up-to-date information in depth.

# CONTENTS

# Contents

# Contents

# INTRODUCTION

The purpose of this handbook is to provide a one-volume ready reference for the handling of pollutants, and particularly those pollutants emanating from industrial processes.

The major sources of the information contained herein are U.S. patents and U.S. Government reports, primarily from the Environmental Protection Agency. Through citations in these and other sources, however, hundreds of references to books and periodical literature are given here. Thus, this book provides a ready reference to the entire spectrum of published literature on pollutant removal. While all of this material is presumably available and in the public domain, in actual fact its location and abstracting may be a tedious, time-consuming and expensive process. Thus, this volume is specifically designed to save the concerned reader time and money in his or her search for pertinent information on the control of specific pollutants.

This book is addressed to the industrialist, to the local air pollution control officer, to legislators who are contemplating control measures, to the conservationist who is interested in exactly what can be done about the effluents of local factories, as well as to the student and concerned citizen.

As one specific set of references to complement this volume, the reader is referred to the series on pollution control, industry by industry, by H.R. Jones. The titles to date include:

*Environmental Control in the Organic and Petrochemical Industries,* 1971
*Environmental Control in the Inorganic Chemical Industry,* 1972
*Pollution Control in the Nonferrous Metals Industry,* 1972
*Waste Disposal Control in the Fruit and Vegetable Industry,* 1973
*Pollution Control in the Textile Industry,* 1973
*Pollution Control and Chemical Recovery in the Pulp and Paper Industry,* 1973.

All of these volumes have also been published by Noyes Data Corporation, Park Ridge, New Jersey.

## ACIDS

Industrial wastes of many types, and from a variety of processes, exhibit extreme pH values which can greatly influence the quality and aquatic life of receiving waters. Table 1 lists some representative industries, and gives waste pH characteristics associated with each industry. Wastes may be released on a batch basis as acid processing vats are dumped, or released as a continuous waste stream resulting from a specific industrial process (e.g., rinsing). The volume and composition of a waste stream containing acidic compounds can be quite variable. Dickerson and Brooks (1) have described the waste stream from a cellulose production line of Hercules Powder Co. as varying in volume from 3,000 to 9,000 gpm with acidity, measured as free sulfuric acid, ranging from 30 to 1,100 mg./l.

The problem of acid waste disposal has also been addressed by Cooper (2).

The reader of this handbook is referred to specific sections which follow on Hydrogen Chloride, Phosphoric Acid and Sulfuric Acid.

## TABLE 1: pH CHARACTERISTICS OF INDUSTRIAL WASTES (3)

| Industrial Process | Waste pH Characteristics |
|---|---|
| Food and Drugs: | |
| Pickling | Acidic or alkaline |
| Soft Drinks | High pH |
| Apparel: | |
| Textile | Highly alkaline |
| Leather Processing | Variable |
| Laundry | Alkaline |
| Chemicals: | |
| Acids | Low pH |
| Phosphate and Phosphorus | Low pH |
| Materials: | |
| Pulp and Paper | High or low pH |
| Photographic Products | Alkaline |
| Steel | Mainly acid, some alkaline |
| Metal Plating | Acids |
| Oil | Acids |
| Rubber Stores | Variable pH |
| Naval Stores | Acid |
| Energy: | |
| Coal Processing | Low pH |

Source:   Report PB 204,521

### Removal of Acids from Air

See under removal of specific acids from air, for example:  Aromatic Acids and Anhydrides, Hydrogen Chloride, Hydrogen Fluoride, Phosphoric Acid, Sulfuric Acid.

### Removal of Acids from Water

Methods include:  [1] mixing acid and alkaline wastes so that the net effect is a near-neutral pH; [2] passing acid wastewaters through beds of limestone; [3] mixing acid wastes with lime slurries or dolomite lime slurries; [4] adding the proper amounts of concentrated caustic soda (NaOH) or soda ash ($Na_2CO_3$) to acid wastewaters; [6] bubbling waste boiler-flue gas through alkaline wastes; [7] adding strong acid to alkaline wastes, according to Patterson et al (4).

The selection of a caustic agent to neutralize an acid waste is usually between sodium hydroxide (caustic soda, NaOH), sodium carbonate (soda ash, $Na_2CO_3$) and various limes (calcium oxide compounds). The important factors in selection of a caustic reagent include purchase price, neutralization capacity, storage and equipment costs, and neutralization end products. The costs of the various caustic reagents have been summarized by Ross (5). Although sodium hydroxide is far more expensive than the other materials, it is frequently selected due to composition uniformity, ease of storage and feedings, rapid reaction rate, and solubility of end products. Sodium carbonate is not as reactive as sodium hydroxide and can produce foaming problems due to release of carbon dioxide.

Hansen (6) has reported the use of both sulfuric acid and hydrated lime, $Ca(OH)_2$, in treating a metal finishing waste. Both acid and alkaline wastes were produced in the industrial process, and an effluent pH of approximately 9.0 was routinely achieved. It has been pointed out that acid wastes which contain iron salts, as most metal plating and ferrous industry wastes do, may required 3 to 10 times the amount of neutralizing caustic agent required by acidic solutions of equivalent pH which do not contain the iron salts as discussed by Omya (7). MacDougall (8) has also discussed the problem associated with the presence of acidic iron salts in acid wastes. He reports that waste sulfuric acid, used in cleaning rust from steel products, typically contains 2 to 7% (or 2,000 to 7,000 mg./l.) free acid, and 15 to 22% ferrous sulfate (an acid salt).

As discussed above, salts of weak acids or bases influence the pH of effluents and receiving waters, and exert a chemical demand for additional neutralizing agent.

K.S. Watson (9) has summarized the treatment efficiency and neutralizing agent demand for an acid waste originating from cleaning, plating and other metal-finishing operations of a General Electric Company plant. The acid waste was over-neutralized to an alkaline pH, in order to allow later precipitation of waste heavy metals. Thus the large quantities of lime used resulted from four requirements: [1] neutralization of the free acids; [2] neutralization of the combined acids (i.e., acid salts); [3] over-neutralization to an alkaline pH, in order to achieve conditions suitable for heavy metal-hydroxide precipitation; and [4] the lime required to react with the heavy metals, forming insoluble hydroxides.

Because of the high lime demand, operating costs of this acid waste treatment process were extremely high at $26.20 per 1,000 gallons, in contrast to the usual 10 to 20¢ per 1,000 gallons required for simple acid/alkaline neutralization, described by Zievers et al (10). Total treatment costs, including equipment depreciation, were $65.50 per thousand gallons. The author points out that acid treatment costs are high because the acid waste is comparatively concentrated, and considerable lime is required for the neutralization. Capital cost of the acid waste treatment facility was $106,387, exclusive of the waste collection system, according to Watson (9).

#### References

(1) Dickerson, B.W. and Brooks, R.M., "Neutralization of Acid Wastes," *Ind. Eng. Chem.* 42, 599-605 (1950).
(2) Cooper, J.E., "How to Dispose of Acid Wastes," *Chemistry and Industry* 1950, 684-685.
(3) Nemerow, N.L., *Theories and Practices of Industrial Waste Treatment*, Reading, Mass., Addison-Wesley Publishing Co. (1963).
(4) Patterson, J.W. and Minear, R.A., "Wastewater Treatment Technology," *Report PB 204,521*, Springfield, Va., National Technical Information Service (Aug. 1971).
(5) Ross, R.D., *Industrial Waste Disposal*, New York, Reinhold Book Corp. (1968).
(6) Hanson, N.H., "Design and Operation Problems of a Continuous Automatic Plating Waste Treatment Plant at the Data Processing Division, IBM, Rochester, Minn.," *Proc. 14th Purdue Industrial Waste Conf.*, pp. 227-249, 1959.
(7) Omya, S.A., "Treatment of Residual Acid Effluent," *Wat. Waste Treatment* 12, 27-28, 1968.
(8) MacDougall, H., "Waste Disposal at a Steel Plant: Treatment of Sheet and Tin Metal Wastes," *ASCE Separate No. 493*, September, 1954.
(9) Watson, K.S., "Treatment of Complex Metal-Finishing Wastes," *Sew. Industr. Wastes* 26, 182-194, 1954.
(10) Zievers, J.F., Crain, R.W. and Barclay, F.G., "Waste Treatment in Metal Finishing: U.S. and European Practices," *Plating* 55, 1171-1179 (1968).

## ADIPIC ACID

**Adipic Acid Removal from Wastewater**

A process developed by C.R. Campbell, D.E. Danly and M.J. Mathews III (1) is a process for the removal of monobasic and dibasic acids, mineral acids, and other organic and inorganic material from aqueous mixtures prior to the discharge thereof to waste.

An example of a commercially important chemical process which gives rise to a waste disposal problem is the production of pure dibasic acids by the oxidation of cycloparaffins. This process is of considerable commercial importance because of the extensive and expanding use of such acids in the preparation of polyamide resins and because of the ready availability of cyclic hydrocarbons as starting materials from which the acids may be made. Much art is known and available which describes processes for the direct oxidation of cycloparaffins to a variety of products including a mixture of monobasic and dibasic acids. Improvements in the processes known in the art have shown that yields of individual dibasic acids can be increased markedly when the oxidation is carried out in stages in which the cycloparaffins are air oxidized first to corresponding cyclic alcohols and ketones, and then the mixture of alcohols and ketones is nitric acid oxidized to a mixture of corresponding dibasic acids.

In the more sophisticated oxidation processes only selective portions of the air oxidation product are subjected to nitric acid oxidation, and mixtures of monobasic and dibasic acids are obtained which contain predominate amounts of a single dibasic acid and only minor amounts of other monobasic and dibasic acids. In accordance with this two-step oxidation process, cycloparaffins are first air oxidized and a selective part of the air oxidation product is nitric acid oxidized to give mixture of monobasic and dibasic acids predominant in one dibasic acid in an aqueous nitric acid solution. Either this aqueous mixture or the mother liquor remaining after crystallization of a portion of the predominant dibasic acid therefrom is subjected to evaporating conditions to reduce the nitric acid concentration of the liquor, and further crystallization of the predominant dibasic acid is effected.

The mother liquor from this second crystallization contains additional dibasic acids which cannot be crystallized successfully, and therefore, the mixture of dibasic acids contained in this second mother liquor or the mixture remaining after a second evaporation to reduce the nitric acid concentration may be separated, if desired, from each other in excellent yields by treatment with a suitable immiscible solvent or other means to permit recovery of substantially all the dibasic acids contained therein.

An example of such a process is the manufacture of adipic acid by the air oxidation of cyclohexane followed by the nitric acid oxidation of a selective portion of the air oxidation product. In this process the aqueous nitric acid mixture resulting from nitric acid oxidation is predominantly adipic acid and contains other dibasic acids such as glutaric and succinic as well as monobasic acids.

An example of a waste aqueous stream which is high in chemical oxygen demand and which cannot be discharged safely without treatment is the aqueous mixture which results from the overhead or make stream from the evaporation, either before or after crystallization of adipic acid, of the aqueous product of the nitric acid oxidation described above.

In this process, the aqueous mixture described above, or any similar aqueous mixture, is contacted either in single or multiple stages with a liquid ion exchange material dissolved in a suitable solvent to remove the chemical oxygen demand causing components in the solvent phase, thereby leaving purified water suitable for safe, nonchemical polluting waste discharge to rivers or other natural waste areas. The resulting solvent phase may be contacted with an inorganic base, anhydrous ammonia, an aqueous solution of an inorganic base or ammonia, or mixtures thereof to regenerate the solvent solution of ion exchange material for recycle or other use, and the remaining concentrated aqueous solution of salts of the organic acids and other chemical oxygen demand causing components which remain in much reduced volume may be disposed of by incineration or other suitable nonpolluting means or recovered for profitable use. Figure 1 is a block flow diagram showing the essentials of this process.

The aqueous solution from which the high chemical oxygen demand material must be removed prior to the discharge of the water to waste enters the process at **10**. In the example of the aqueous mixture for which the process will be described, the nitric acid concentration of this aqueous mixture may be from 0 to 1% by weight of the water therein; however, generally the nitric acid concentration may be between 0.05 and 0.5% by weight. The water-soluble organic acid concentration of this aqueous mixture may be from 0 to 5.0% by weight of the water therein, however, generally may be between 0.1 and 3.0%.

Nonacidic organic liquids which may be entrained in the aqueous mixture to be purified and which may operate as a solvent or ion exchange resin contaminant, both of which will be described later, may be flashed overhead in flash tank **11**, with heat supplied by a source, not shown, or these organic liquids may be removed by any other means such as decanting or other. The overhead stream, as indicated at **12**, may be condensed in condenser **13** and discharged to a sump or other as indicated by line **14**. Cooling water for condenser **13** enters at **16** and exits at **15**.

FIGURE 1:  BLOCK FLOW DIAGRAM FOR PURIFICATION OF WASTEWATERS
CONTAINING ADIPIC ACID

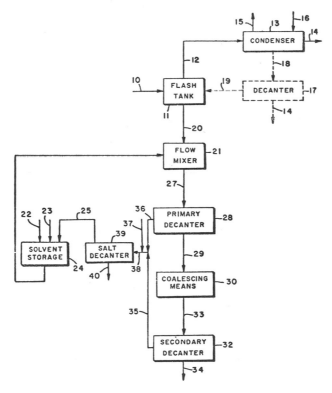

Source:  C.R. Campbell, D.E. Danly and M.J. Mathews III, U.S. Patent 3,267,029;
August 16, 1966

As an example, cyclohexyl nitrate may be present in the described aqueous mixture to be purified in a concentration up to 4,000 parts per million, and since this material will concentrate in the immiscible solvent, causing continuing dilution of the ion exchange resin, the cyclohexyl nitrate may be flashed overhead with a portion of the water by introducing waste process steam into feed line **10** or into flash tank **11**.

If it is desired to reduce the volume of liquid flowing in line **14**, decanter **17** may be provided. If the process is operated in this manner all or a part of the liquid condensed in condenser **13** may proceed, as indicated by line **18** to decanter **17** where the liquid is permitted to separate into an aqueous layer and a water-immiscible layer with the aqueous layer being returned to the process, as indicated by line **19**, and the water-immiscible layer being discharged to a sump or other suitable waste disposal means, as indicated by **14**.

Ion exchange resin and solvent are introduced to the process, as indicated by lines **22** and **23** respectively, and a solvent solution of the ion exchange resin may be prepared by mixing or other means and maintained for use in the process in solvent storage **24**. In operation or after a sufficient quantity of solvent solution of ion exchange resin has been prepared, only minor quantities of solvent and ion exchange resin may be added to the process by their respective lines to provide makeup with the major quantity of solvent for the process being provided by recycled regenerated solvent solution of ion exchange resin as indicated by line **25**.

The ion exchange resin may be liquid or a solid capable of forming a solution, and may be any one of a mixture of more than one of any of a number of commercially available mixtures of high molecular weight primary, secondary or tertiary amines. Several of such ion exchange resins are marketed under the trademark Amberlite by Rohm and Haas Company with the designation LA-1 and LA-2.

The solvent used as carrier for the ion exchange resin may be any organic liquid material or mixture of organic liquid materials miscible with the ion exchange resin but substantially immiscible with water, such as xylene, benzene, mixed aromatic organic liquids having more than 9 carbon atoms, and kerosene.

The ion exchange resin and the solvent may be mixed in any manner such as mechanical stirring or flow mixing, and the concentration by weight of the ion exchange resin in the solvent may be between 1 and 100% with the preferred concentration range being between 10 and 30% by weight.

In the example of the process being described, the solvent may be either kerosene or xylene with kerosene being preferred, and the concentration of the Amberlite LA-2 in the kerosene or xylene may be 25% by weight. The solvent solution of ion exchange resin proceeds to flow mixer **21**, as indicated by line **26**, where it is mixed with the aqueous mixture, indicated entering flow mixer **21**, by line **20**.

The flow mixing of the solvent solution and the aqueous mixture may be accomplished by any suitable means which will provide good contacting of the solvent solution of ion exchange resin and the aqueous mixture for good extraction efficiency, but will not cause a dispersion of solvent solution in the aqueous mixture that is extremely slow in separating in subsequent steps of the process. As an example, in the embodiment being described, pumping of the solvent solution and the aqueous mixture may introduce sufficient energy to form a tight emulsion which is extremely difficult to separate and may require excessive holdup periods to separate into two phases. Proper mixing may be obtained by using a ball-type mixing valve and controlling the pressure drop across the valve. The amount of the pressure drop which is necessary and suitable is dependent upon the solvent used and the aqueous mixture to be purified, as well as other factors; and in the embodiment of the process being described, the pressure drop may be controlled at 1 lb./in².

The solvent solution to aqueous mixture ratio which is necessary is dependent upon many factors such as the chemical oxygen demand of the aqueous mixture or its acidity, the concentration of the ion exchange resin in the solvent, and others. It has been found in the embodiment being described that when the acidity of the aqueous mixture is not in excess of 1.0 normal, the ratio of equivalents of ion exchange resin in the solvent to equivalents of acidity in aqueous mixture may be as low as 0.75 with good purification results being obtained. As is clear to those skilled in the art, the ratio of solvent solution of ion exchange resin to the aqueous mixture may be adjusted accordingly depending upon the ion exchange resin concentration in the solvent and the acidity of the aqueous mixture.

After mixing, the combined mixture of ion exchange resin solution and aqueous mixture proceeds, as indicated by line 27, to primary decanter 28 where the combined mixture is permitted to separate into an immiscible solvent phase and an aqueous phase. Primary decanter 28 may be of any design well known and suitable for the separation of two-phase systems and should be of sufficient size to provide adequate holdup time in the process to permit good separation of the immiscible solvent phase and the aqueous phase. In the described example, it has been found that the holdup time for good separation of the kerosene or xylene phase from the aqueous phase should be no less than five minutes to prevent entrainment of the aqueous phase in the solvent phase or excessive carryover of the solvent phase in the aqueous phase.

The aqueous phase containing a small amount of nonseparated solvent proceeds, as indicated by line 29, to coalescing means 30 where the solvent phase which did not separate in primary decanter is coalesced to assist in its separation in secondary decanter 32. Any suitable coalescing media such as sand or other may be used in coalescing means 30, and in the process of the example, 10 to 20 mesh sand in a bed 6 ft. high may be used successfully as the coalescing media at mass flow rates at least as high as 4,500 lbs./hr./sq. ft.

After passing through coalescing means 30, the aqueous phase proceeds, as indicated by line 33, to secondary decanter 32 where the final separation of the aqueous and immiscible solvent phases is permitted to take place. The secondary decanter should be sufficiently large to provide adequate holdup time for good separation or the inlet to the vessel may be sufficiently large to provide sufficient deceleration of the velocity of the material prior to its entry into the main body of the decanter to provide good separation in a smaller vessel. In the process of the example, it has been found that a vessel sufficiently large to provide a separating holdup time of 2 to 4 minutes for the liquid is adequate provided the inlet to the vessel is such that fluid velocities are about 0.3 to 2.5 ft./sec.

After separation of the phases in secondary decanter 32, the resulting aqueous phase which is now substantially free of chemical oxygen demand material may be discharged safely to any river, pond or other without danger of pollution, as indicated by line 34. The resulting immiscible solvent phase may be joined, as indicated by line 35, with the immiscible solvent phase leaving primary decanter 28, as indicated by line 36, for recovery and regeneration of the solvent solution of ion exchange resin.

Regenerating material is added to the immiscible solvent phase, as indicated by line 37, and this mixture proceeds, as shown by line 38, to decanter 39 where the regenerated solvent solution of ion exchange resin is separated either from an aqueous phase or a precipitate and returned to solvent storage 24, as indicated by line 25. The aqueous phase or the precipitate formed in the regeneration is removed from the process as indicated by line 40.

The regenerating material added as indicated at 37 may be an inorganic base, anhydrous ammonia, an aqueous solution of an inorganic base or ammonia, or mixtures thereof. If an aqueous solution of an inorganic base or ammonia is used, an aqueous phase is separated from the solvent solution of the ion exchange resin in decanter 39 and removed at 40; and if anhydrous ammonia or an inorganic base is used, a precipitate is separated and removed.

Contacting of the spent immiscible solvent phase of lines 35 and 36 with the regenerating material of line 37 may be accomplished by any suitable means that is convenient and efficient, and in the example of the process being described, contacting of the aqueous ammonia, which may be the regenerating material, and the spent immiscible solvent solution may be effected in a recirculating centrifugal pump.

As is clear to those skilled in the art, care must be taken in controlling the amount of regenerating material added to the process. If insufficient regenerating material is added, the ion exchange resin solution will not be regenerated completely and the efficiency of the process will be reduced; and if an excessive amount of regenerating material is added, it will be returned to the process along with the solvent solution where it will prevent preferentially the ion exchange material from effecting efficient removal of the chemical oxygen demand material from the aqueous stream to be treated. In the example being described, a slight excess of the stoichiometric amount of ammonia needed to neutralize the chemical oxygen demand material in the solvent solution may be used.

The results of a typical operation of a single-stage process on a continuous basis according to the preferred embodiment are summarized below; all numbers are in pounds per hour unless otherwise specified. The column identified as "Feed" is line **10** in Figure 1, and the column identified as "Product" is line **34**.

|  | Feed | Product |
|---|---|---|
| Water | 100,000 | 97,137 |
| Nitric Acid | 287 | 0 |
| Organic Acids | 729 | 176 |
| Ammonium Salts | 0 | 22 |
| Solvent | 0 | 9 |
| Cyclohexyl nitrate | 49 | 5 |

As can be seen clearly from the above results, all of the inorganic acid and approximately 76% of the organic acids can be removed simply by the single-stage process. It is clear, also, that multistage purification by the use of this process can be accomplished easily if it is desired to reduce the chemical oxygen demand of the product below that which is possible with the use of a single stage.

References

(1) Campbell, C.R., Danly, D.E. and Mathews, M.J. III, U.S. Patent 3,267,029; August 16, 1966; assigned to Monsanto Company.

# ALDEHYDES

The most characteristic and important effect of aldehydes, particularly of low molecular weight aldehydes, for both humans and animals is primary irritation of the eyes, upper respiratory tract, and skin. The observed symptoms in humans from inhalation of low concentrations of aldehydes include lacrimation, coughing, sneezing, headache, weakness, dyspnea, laryngitis, pharyngitis, bronchitis, and dermatitis. In most cases, the general and parenteral toxicities of these aldehydes appear to be related mainly to these irritant effects. The unsaturated aldehydes are several times more toxic than the corresponding aliphatic aldehydes. Also, the toxicity generally decreases with increasing molecular weight within the unsaturated and aliphatic aldehyde series. Sensitization has occurred from contact with formaldehyde solutions and other aldehydes, but sensitization of the pulmonary tract rarely is produced by inhalation of aldehydes. The anesthetic properties of aldehydes are generally overshadowed by the stronger irritant effects. Furthermore, concentrations that can be tolerated via inhalation can usually be metabolized so rapidly that systemic symptoms do not occur.

Formaldehyde concentrations as low as 600 $\mu$g./m.$^3$ have been shown to cause cessation of the ciliary beat in rats. Animal experiments have shown that aldehydes can affect the responses of the respiratory system, causing such effects as an increase in flow resistance and in tidal volume and a decrease in the respiratory rate. Exposure of rates to 150$\mu$g. per m.[3] of acrolein for two months caused a rise in the number of luminescent leukocytes in the blood. Exposure of animals to high concentrations of aldehydes has been shown by several investigators to produce edema and hemorrhages of the lungs and fluid in the pleural and peritoneal cavities. In a Russian study, formaldehyde was found to prolong the mean duration of pregnancy in rats and decrease the number of offspring. In addition, the weight of the lungs and liver of the offspring was less than that of the controls' offspring, but other organs exhibited an increase in weight.

Animal experiments also indicate possible synergistic effects between aldehydes and aerosols. Thus, acrolein and formaldehyde in the presence of certain inert aerosols appeared to be more toxic to mice than the pure compounds. Experiments with guinea pigs showed that formaldehyde with sodium chloride aerosols produced significant increases in the "respiratory work" compared with the effect of the pure vapor.

In addition, to the toxic effects, aldehydes may contribute to the annoyances of odor and eye irritation caused by polluted air. Aldehyde concentrations have been shown to correlate with the intensity of odor of diesel exhaust and the intensity of eye irritation during natural and chemically produced smog. Data indicate that as little as 12$\mu$g/m.[3] of formaldehyde can cause human eye response.

Aldehyde air pollution may result in oxidant-type damage to plants, although atmospheric photochemically produced products from the aldehydes may actually cause the damage rather than direct attack by aldehydes. There are no data available to indicate the effect of aldehyde air pollution on materials.

Vehicle exhaust, particularly from automobiles, appears to be the major emission source of aldehydes. However, significant amounts may also be produced from other combustion sources such as open burning and incineration of solid waste materials, and the burning of fuels (gas, fuel oil, coal). Another source of aldehyde emission is the thermal decomposition of hydrocarbons by pyrolysis in the presence of air or oxygen. Sources of these emissions include chemical manufacturing plants and industries that use drying or baking ovens to remove organic solvents in such processes as automobile painting and the manufacture of coated paper and metals.

Air sampling data indicate that plants manufacturing formaldehyde may be local sources of aldehyde pollution; over 4 billion pounds of formaldehyde were manufactured in the United States in 1968. However, the major amount of aldehyde pollution in some areas of the United States is from the photochemical reaction between nitrogen oxides and hydrocarbons. Hydrocarbons that yield formaldehyde are olefins, and to a lesser degree, other aldehydes and aromatic hydrocarbons. Diolefins produce most of the atmospheric acrolein. Some data indicate that in certain areas over two-thirds of the atmospheric aldehydes may have resulted from photochemical reactions. Of course, the sources that emit aldehyde pollutants are generally the same as those emitting hydrocarbons and nitrogen oxides.

In addition, aldehydes themselves may undergo photochemical reactions. They may produce, at low partial pressure in the presence of nitrogen oxides, other products such as carbon monoxide, lower aldehydes, nitrates, and oxidants. The oxidants produced include ozone, peroxyacyl nitrates, and alkyl hydroperoxides (hydrogen peroxide in the case of formaldehyde). No peroxyacids or diacetyl peroxides are found at low partial pressure of aldehydes.

In 1967 the National Air Sampling Network began to report data for aliphatic aldehydes. The data for 1967 for several cities show that the average concentrations of aldehydes ranged from 3 to 79$\mu$g./m.[3] and that the maximum values ranged from 5 to 161$\mu$g./m.[3]. A Los Angeles area report indicates that the maximum values for two "smog" days in 1968 were 208$\mu$g./m.[3] for aliphatic aldehydes, 163$\mu$g./m.[3] for formaldehyde, and 27$\mu$g./m.[3] for acrolein. Generally, formaldehyde accounts for 50% or more of the total aldehydes, while acrolein accounts for about 5%, according to Q.R. Stahl (1).

### Removal of Aldehydes from Air

Control methods for the removal of aldehydes from air streams, as reported by Stahl (1) include more effective combustion methods and the use of direct-flame and catalytic after-burners. Paradoxically, although these methods generally decrease the amount of emissions, they may actually produce greater amounts of aldehydes and other oxygenated hydrocarbons from the burning of streams containing a variety of organic compounds.

Table 2 shows reported aldehyde emissions from various sources. As noted there, the amount of aldehydes increases tenfold by the use of afterburners in some drying oven processes.

### TABLE 2: REPORTED ALDEHYDE EMISSIONS FROM VARIOUS SOURCES

| Source | Aldehyde Emissions (as formaldehyde) | Reference |
|---|---|---|
| Amberglass Manufacture | | |
| Regenerative furnace, gas fired | $8,400\mu g./m.^3$ | 2 |
| Brakeshoe Debonding | | |
| (single-chamber oven) | 0.10 lb./hr. | 2 |
| Core Ovens | | |
| Direct gas fired (phenolic resin core | | |
| binder from oven) | $62,400\mu g./m.^3$ | 2 |
| Direct gas fired (linseed oil core | | |
| binder from afterburner) | $<12,000\mu g./m.^3$ | 2 |
| Indirect electric (linseed oil core | | |
| binder from oven) | $189,600\mu g./m.^3$ | 2 |
| (from afterburner) | $<22,800\mu g./m.^3$ | |
| Insulated Wire Reclaiming, covering | | |
| Rubber $^5/_8''$ o.d. | $126,000\mu g./m.^3$ | 2 |
| Secondary burner off | $6,000\mu g./m.^3$ | 2 |
| Secondary burner on | | |
| Cotton-Rubber-Plastic $^3/_8$-$^5/_8''$ o.d. | | |
| Secondary burner off | 10,800 to $43,200\mu g./m.^3$ | 2 |
| Secondary burner on | $4,800\mu g./m.^3$ | 2 |
| Meat Smokehouses | | |
| Pressure mixing burner | | |
| Afterburner inlet | 0.04 lb./hr. | 2 |
| Afterburner outlet | 0.22 lb./hr. | 2 |
| Multijet burner | | |
| Afterburner inlet | 0.49 lb./hr. | 2 |
| Afterburner outlet | 0.22 lb./hr. | 2 |
| Meat Smokehouse Effluent, gas fired | | |
| boiler-firebox as "afterburner" | | |
| Water-tube, 426 hp. | | |
| Afterburner inlet | 0.22 lb./hr. | 2 |
| Afterburner outlet | 0.09 lb./hr. | 2 |
| Water-tube, 268 hp. | | |
| Afterburner inlet | 0.39 lb./hr. | 2 |
| Afterburner outlet | 0.40 lb./hr. | 2 |
| Water-tube, 200 hp. | | |
| Afterburner inlet | 0.39 lb./hr. | 2 |
| Afterburner outlet | 0.30 lb./hr. | 2 |
| Locomotive, 113 hp. | | |
| Afterburner inlet | 0.03 lb./hr. | 2 |
| Afterburner outlet | 0.0 lb./hr. | 2 |
| HRT, 150 hp. | | |
| Afterburner inlet | 0.03 lb./hr. | 2 |
| Afterburner outlet | 0.18 lb./hr. | 2 |

(continued)

TABLE 2: (continued)

| Source | Aldehyde Emissions (as formaldehyde) | Reference |
|---|---|---|
| Meat Smokehouse Exhaust | | |
|     Gas fired afterburner inlet | 104,400$\mu$g./m.$^3$ | 2 |
|     Gas fired afterburner outlet | 40,200$\mu$g./m.$^3$ | 2 |
|     Electrical precipitation system inlet | 88,800$\mu$g./m.$^3$ | 2 |
|     Electrical precipitation system outlet | 56,400$\mu$g./m.$^3$ | 2 |
| Mineral Wood Production | | |
|     Blow chambers | 109$\mu$g./m.$^3$ | 2 |
|     Curing ovens | | |
|         Catalytic afterburner inlet | 1.90 lb./hr. | 2 |
|         Catalytic afterburner outlet | 0.90 lb./hr. | 2 |
|         Direct-flame afterburner inlet | 2.20 lb./hr. | 2 |
|         Direct-flame afterburner outlet | 0.94 lb./hr. | 2 |
|     Wool coolers | 32$\mu$g./m.$^3$ | 2 |
| Litho Oven Inlet | 120$\mu$g./m.$^3$ | 3 |
| Litho Oven Outlet | 32,880$\mu$g./m.$^3$ | 3 |
| Litho Oven Outlet | 4,680$\mu$g./m.$^3$ | 3 |
| Paint Bake Oven | | |
|     Nozzle mixing burner | | |
|         Afterburner inlet | 0.19 lb./hr. | 2 |
|         Afterburner outlet | 0.03 lb./hr. | 2 |
|     Atmospheric burner | | |
|         Catalytic afterburner inlet | 0.07 lb./hr. | 2 |
|         Catalytic afterburner outlet | 0.31 lb./hr. | 2 |
|     Premix burner | | |
|         Catalytic afterburner inlet | 0.3 to 0.4 lb./hr. | 2 |
|         Catalytic afterburner outlet | 0.2 to 0.5 lb./hr. | 2 |
| Phthalic Acid Plant | 135,600$\mu$g./m.$^3$ | 3 |
| Phthalic Anhydride Production Unit | | |
|     (multijet burner) | | |
|     Afterburner inlet | 1.75 lb./hr. | 2 |
|     Afterburner outlet | 0.43 lb./hr. | 2 |
| Reclaiming of Electrical Windings | | |
|     (single-chamber incinerator) | | |
|     100 hp. generator starter | 0.08 lb./hr. | 2 |
|     14 pole pieces | 0.08 lb./hr. | 2 |
|     Auto armatures | 0.13 to 0.29 lb./hr. | 2 |
|     Auto field coils (multiple chamber) | 0.49 lb./hr. | 2 |
|     Auto field coils afterburner | 0.08 lb./hr. | 2 |
|     14 generator pole pieces | 0.08 lb./hr. | 2 |
| Varnish Cooking Kettles | | |
|     Four-nozzle mixing burner | | |
|         Afterburner inlet | 0.30 lb./hr. | 2 |
|         Afterburner outlet | 0.11 lb./hr. | 2 |
|     Inspirator burner | | |
|         Afterburner inlet | 0.29 lb./hr. | 2 |
|         Afterburner outlet | 0.02 lb./hr. | 2 |
| Webb Press | 480$\mu$g./m.$^3$ | 3 |
| | 360$\mu$g./m.$^3$ | 3 |
| | 480$\mu$g./m.$^3$ | 3 |
| | 1,920$\mu$g./m.$^3$ | 3 |

Source: Report PB 188,081

## Removal of Aldehydes from Water

See discussion of removal of formaldehyde from water under Formaldehyde.

References

(1) Stahl, Q.R., "Air Pollution Aspects of Aldehydes," *Report PB 188,081*, Springfield, Va., National
    Technical Information Service (September 1969).
(2) Danielson, J.A., *Air Pollution Engineering Manual*, U.S. Department of Health, Education & Welfare,
    Cincinnati, Ohio (1967).
(3) Levaggi, D.A. and Feldstein, M., "The Collection and Analysis of Low Molecular Weight Carbonyl
    Compounds from Source Effluents," *J. Air Pollution Control Assoc.* 19, 43 (1969).

# ALKALIS

## Removal of Alkalis from Air

A patent by T.S. Dean (1) is directed specifically to the recovery of substantially pure
alkali metal salts from exhaust gases and dust from the kilns of a cement-producing operation. As is well known in the art, most cement raw materials contain alkali metal compounds in some amounts. As these raw materials enter the clinkering zone of a cement
kiln, a portion of the alkalis are volatilized and are carried by the exhaust gases toward
the exit door of the kiln and removed from the kiln itself.

In many of these installations, these exhaust gases and dust are vented into the atmosphere
where they may cause a pollution problem. During the clinkering or burning of the cement
raw materials, these alkali metals are converted into alkali metal sulfates or chlorides and
these compounds if vented into the atmosphere carry along with them the sulfate or chloride radicals themselves. Both of these materials will cause air pollution problems. Additionally, it is well known to dissolve these alkali metal sulfates or chlorides in water after
they leave the kiln or blast furnace and then once in solution they are discharged into any
open stream. This also causes pollution problems in the water of that area.

The pH of the stream may be increased to the point where fish and water life will be killed.
In addition, the volatilized parts of the raw material used in producing cement which has
been carried out of the kiln with discrete particles will form dissolved solids in the water
which, if allowed in too concentrated form are toxic to and will kill fish and water life.
Also, the fine discrete particles of the raw materials themselves are not easily removed from
the discharge water and therefore there is an increase in the suspended solids which may
kill fish and other water life if allowed to increase in concentration.

These exhaust gases from a cement manufacturing plant kiln will carry with it finely divided
particles of raw materials themselves. Thus, there is an economic loss of the raw materials
if these discrete particles are exhausted into the atmosphere or discharged into an open
stream or lake. This is costly in that there is a loss of part of the raw materials used in the
production of concrete clinkers.

Along with the discrete particles of the raw materials carried from the kiln as previously
pointed out, alkali metal salts, usually sulfates, are also moved from the kiln. It is of importance that these alkali metal sulfates which are soluble in water, be removed from the
water which may be discharged into an open stream or sewer system. If they can be recovered in a substantially pure condition, they will bring a reasonable price on the open
market and thus the cost of producing one barrel of cement clinkers will be substantially
reduced. Further, if an attempt is made to return the discrete particles of raw materials to
the kiln itself, without removal of these alkali metal salts there will be a buildup of these
salts in the cement clinkers themselves which is most objectionable.

As these salts are water-soluble, and if this cement is used in construction of roads or buildings, when exposed to the weathering effects of rain, this will result in the alkali metal salts
being leached from the cement and results in it cracking which is most objectionable, especially in the surfacing of highways. In addition, with an increase in the alkali metal salts

content of the cement clinkers, the cement thus produced would not meet Federal, State or ASTM specifications as to alkali metal content and therefore there would be a very limited sales market.

It is therefore the primary object of this process to provide a method for removing and recovering soluble alkali metal salts, primarily sulfates, from kiln exhaust gases and dust. Figure 2 shows the essential features of the process.

The discrete particles of the raw materials and the particles of the alkali metal salts, which have condensed from exhaust gases of a blast furnace or kiln, the dust from a kiln not shown in the drawing, are carried through pipe 10 into mixing chamber 11 where these particles are thoroughly mixed with water. The ratio of water to solids should be at least 7 parts of water by weight to 1 part of solid by weight, with the preferred ratio being 10 to 1. It has been found that if the water to solids ratio is much below 7 to 1, a gel is formed which, of course, will be very difficult to pump as a practical matter.

Once a slurry is produced in the mixing tank, it is carried through pipe 12 into separating device 14 where the slurry is maintained at a temperature of just below boiling or between 140° and 200°F. This separating device may be any mechanical equipment used to separate a slurry into two fractions, one fraction with a low specific gravity and one fraction with a higher specific gravity. These devices may include such equipment as settling tanks, filters, centrifuges or hydrocyclones. The separating device 14 separates the slurry into two fractions, clarified water and dewatered slurry.

FIGURE 2: PROCESS FLOW DIAGRAM FOR REMOVAL OF ALKALI METAL SALTS FROM FLUE DUST

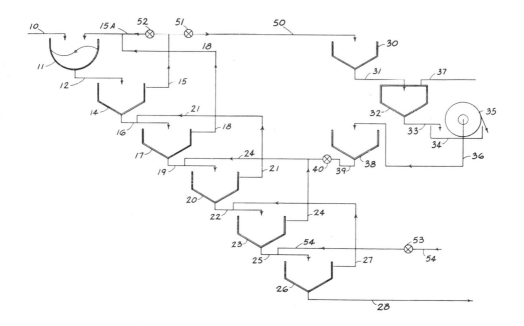

Source: T.S. Dean; U.S. Patent 3,647,395; March 7, 1972

The clarified water is maintained with a concentration of 150 g. of the dissolved alkali metal salts per liter of solution. This clarified water is then fed through pipe **15**, valve **52** and pipe **15A** into the mixing chamber **11** to be used with incoming dust to form additional slurry. Some of this clarified water is fed through valve **51** and pipe **50** to crystallizer holding tank **30**. This occurs when the equilibrium of the entire separating system has been reached and is maintained at 150 g. of alkali metal salts per liter of water. The above dewatered slurry is intimately mixed in pipe **16** with the clarified water from settling device **20** to form a slurry and is maintained at a temperature of 100° to 175°F. It is noticed that the temperature may be maintained at a somewhat lower temperature than the clarified water from settling device **14**. This slurry is fed into separating device **17** through pipe **16**.

The separation device **17** separates the slurry into two fractions, clarified water and dewatered slurry. The concentration of the alkali metal salts in this clarified water is maintained at about 80 to 140 g. per liter of solution and is fed through pipe **18** and pipe **15A** back to the original mixing tank **11** where it is used as mixing water to form additional slurry. Thus the concentration of the alkali metal salts and the clarified water can be maintained somewhat lower than the clarified water concentration from separating device **14**. It is of importance that the total clarified water leaving separating devices **14** and **17** and which is flowing into the mixing chamber **11** should equal the total amount of water flowing out of the mixing chamber **11** through pipe **12**.

The dewatered slurry from separating device **17** is intimately mixed in pipe **19** with the clarified water from separating device **23**,and the water, that has the excess alkali metal salts removed from it, from the crystallizer **32**, to form a slurry and the total is fed into separating device **20**. This slurry is maintained at a temperature of around 60° to 130°F. Separating device **20** separates the slurry into two fractions, clarified water and dewatered slurry. This clarified water is maintained at a concentration of approximately 40 to 100 g. per liter of the alkali metal salts.

The dewatered slurry from separating device **20** is intimately mixed in pipe **22** with the clarified water from separating device **26** to form an additional slurry. This slurry is fed into separating device **23** and maintained at a temperature of about 50° to 100°F. This separating device separates the slurry into two fractions, clarified water containing dissolved alkali metal salts and the dewatered slurry.

The clarified water from separating device **23** is fed through pipe **24** and is intimately and thoroughly mixed with the dewatered slurry from separating device **17** in pipe **19** along with the water from the crystallizer. The concentration of the alkali metal salts in this clarified water is maintained at from 20 to 60 g. per liter of solution. The dewatered slurry leaving separating device **23** is intimately and thoroughly mixed in pipe **25** with fresh water entering the system through pipe **54** and is fed into separating pipe **26**. This separating device separates the slurry into two fractions, clarified water and dewatered slurry. The clarified water is fed through pipe **27** and is mixed intimately and thoroughly in pipe **22** with the dewatered slurry leaving separating device **20**. The concentration of the alkali metal salts in this clarified water is maintained at approximately 10 to 40 g. per liter of solution and above a temperature of 50°F. The dewatered slurry from separating device **26** is fed back into the kiln or blast furnace to be processed again, through pipe **28**.

The word slurry used hereinbefore is a mixture of flue dust and water. The phrase clarified water as used hereinbefore is the part having the lower specific gravity after the slurry has passed through a separating device. It will contain all three elements involved in this process, a solvent, dissolved solids and suspended solids. The phrase dewatered slurry as used is a mixture of dust and water after the mixture has passed through a separating device. This fraction will have the higher specific gravity of the two parts remaining after separation. It will also contain a solvent, dissolved solids and suspended solids. The main difference between the dewatered slurry and clarified water at any given stage will be the percent of solvent and the percent of solids.

The amount of moisture in the dewatered slurry leaving a separating device is not of importance. However, a practical range will be 30 to 40% moisture. The crystallizer, described hereinafter, will return water at a concentration of between 20 to 60 g. per liter of the alkali metal salts. The concentration of these salts in the clarified water when it leaves the first separating device is arbitrarily held at 150 g. of alkali metal salts per liter of solution.

The saturated or supersaturated solution of the alkali metal salts are sent into crystallizer holding tank 30 from separating device 14 through pipe 15, pump 51 and pipe 50. When holding tank 30 is full, this solution is directed to crystallizer 32 through pipe 31. Once this solution is in the crystallizer, it is cooled by any of the well known methods such as running chilled water around the outside of the crystallizer in pipes or by vacuum cooling. The exact process used in cooling the solution is not of importance.

After cooling has taken place, most of the alkali metal salts have precipitated out as crystals and are mechanically separated from the solution by any of several well known methods of separation. The solids or crystals may be withdrawn from container 34 by way of drum 35 where the crystals are then sent to a dryer. Once the alkali metal salts crystals and the solution have been separated, the solution containing a very small amount of unprecipitated alkali metal salts is fed back into the system through pipe 36 to storage tank 38, through pipe 39, valve 40 and through pipe 24. The crystals withdrawn from the container 34 are then prepared for commercial use such as sodium sulfate and potassium sulfate, have uses as a fertilizer component, in the manufacture of soap, etc.

### Removal of Alkalis from Water

The treatment of alkaline wastes has been discussed by J.W. Patterson et al (2). The choice of an acidic reagent for neutralization of an alkaline wastewater is generally between sulfuric acid and hydrochloric acid, as discussed by Ross (3). Sulfuric acid is usually selected because of its lower cost. Zievers et al (4) have reported that sulfuric acid for neutralization costs (1968 price) $3.75/lb. in Chicago; more on the East and West Coasts. Hydrochloric acid has the advantage of soluble reaction end products. However, the concentration of chloride resulting from hydrochloric acid neutralization may cause the waste to exceed effluent chloride standards. Carbon dioxide has been employed much less frequently to neutralize alkaline wastes than have sulfuric or hydrochloric acids.

The disposal of spent caustic petroleum refinery wastes by treatment with spent pickle liquor has been described by Knowlton (5). As he points out, in many petroleum refineries, gasoline blending stocks are treated with a strong caustic solution, which in some cases may contain free sulfur dissolved therein. During this treatment the caustic solution extracts undesirable sulfur compounds such as corrosive mercaptans, and cresylic acids from the gasoline. Cresylic acids is a general term used in the petroleum industry to refer to the entire range of phenolic and related compounds found in crude petroleum or formed during catalytic cracking. The caustic will contain considerable amounts of these cresylic compounds usually as the sodium salts thereof and, since the gasoline stocks commonly contain hydrogen sulfide, the caustic solution will also contain sulfide compounds after the treatment.

Disposal of waste caustic refinery liquor obtained by this treatment creates a serious problem because it is not possible to pump it into wells or adjacent streams and waterways because of violation of local health and pollution laws. A stream is generally considered polluted if its phenolic content exceeds a maximum value set by law, if the combined biochemical oxidation demand (BOD) and chemical oxidation demand (COD) exceed the available dissolved oxygen in the stream, and if the odor is bad. Because the spent caustic leaves the refinery at temperatures of the order of 115°F., and because of the heat of dilution of the caustic with water, streams into which it is léd are warmed considerably and this reduces the normal content of absorbed oxygen. If the temperature of the stream is, for example, 80°F., it contains about 7 parts of oxygen per million parts of water at equilibrium. The combined BOD (bacterial activity is high at this temperature) and COD (due to sulfur compounds in the caustic) thereof far exceed this available oxygen value, so that the stream acquires an odor which is exceedingly objectionable. Thus the sulfides and cresylic acids in the spent caustic become a problem and represent, in the case of the cresylic compounds at least a substance to eliminate from the disposal wastes and to recover for other uses where the recovery can be accomplished economically. The disposal of spent refinery caustic waste is an important problem.

The steel industry likewise is faced with a serious waste liquor disposal problem. The removal of oxide scale from certain steel products is an essential operation prior to further processing. Usually this is accomplished by immersing the steel in a bath of dilute sulfuric acid for a relatively brief period. The acid dissolves the scale together with some of the base metal, resulting in an accumulation of ferrous sulfate in the bath. This treatment is called pickling and the liquor which results is called pickle liquor. Eventually the pickle liquor becomes ineffective and must be disposed of. Its disposal has been a problem for many years.

A variety of acids (sulfuric, hydrochloric, nitric, hydrofluoric and phosphoric, individually and in combination) is employed, depending on the kind of product being treated, but sulfuric acid accounts for more than 90% of the tonnage pickled, and therefore disposal of sulfate pickle liquors presents the most serious problem. Spent sulfate pickle liquor normally has a composition ranging from 0.57 to 7% free sulfuric acid and from 15 to 30% ferrous sulfate when a batch pickler is used, while waste pickle liquor from a continuous pickler has from 2 to 7% free sulfuric acid and from 14 to 16% ferrous sulfate.

In accordance with this process, pickle liquor is combined with spent caustic refinery waste liquors in order to separate cresylic acids and sulfur compounds therefrom and to render both wastes innocuous. In the first stage of the process, spent pickle liquor is added to the spent caustic liquor in an amount to adjust the pH thereof to between about 6 and about 11.5, preferably between about 6.5 and 10.5. In this pH range, iron sulfides, iron hydroxides, and insoluble iron salts of cresylic acids precipitate and may be separated mechanically from the liquor such as by settling, filtering or centrifuging.

In an alternative embodiment of the process, either the pickle liquor or the spent refinery caustic may be oxidized before they are mixed or alternatively the mixture may be oxidized before it is filtered. The oxidation may be accomplished by blowing oxygen through the liquid or it may be accomplished by an oxidizing agent such as chlorine, chlorates, hypochlorates, chlorine dioxide, hydrogen peroxide and other peroxides and other compounds well known in the art as oxidizing agents. The oxidizing step may be carried out at a pressure of from atmospheric to 200 lbs./in.$^2$ and at a temperature of about 35°F. up to a boiling point of the liquor at the pressure of the oxidation.

After the precipitate is separated, cresylic acids may be recovered by adding additional spent pickle liquor, or preferably a mineral acid such as sulfuric or hydrochloric acids, to the residual liquor to adjust the pH to a value in the acid range. The exact pH will depend on the particular caustic liquor being treated and the cresylic acids to be separated, but it will generally be between about 3 and 5 or 6, preferably about 5. The cresylic acids, depending on their specific gravity, may separate as a water-insoluble upper layer and may be skimmed from the top or separated centrifugally.

If any soluble cresylic acids remain, they may be removed by extraction with a water-immiscible solvent therefor, such as petroleum naphtha. A final residual liquor may be run to waste through the sewers. If any solids remain before discharge they may be settled or the liquid filtered before discharge. Before the solvent extraction or before discharge the acidity may be neutralized in part or completely. The following is one specific example of the conduct of the Knowlton process. A spent caustic refinery waste liquor was taken having the following analysis:

| | |
|---|---|
| Gravity | 8.8°Bé. (7% NaOH by weight) |
| pH | 12.6 |

(continued)

| | |
|---|---|
| S= | 1.91% by weight as $Na_2S$ |
| HS- | 0.40% by weight as NaHS |
| $S_2O_3$= | 0.93% by weight as $Na_2S_2O_3$ |
| Cresylic (phenolic compounds) | 0.93% by weight, as phenol |
| Total sulfur | 3.41% by weight, as S |
| Sodium polysulfide ($Na_2S_{2.5}$) | 2.98% by weight |

To a quantity of the abovedescribed spent refinery caustic waste was added an amount of spent pickle liquor containing 29.3 weight percent ferrous sulfate and 3.5 weight percent sulfuric acid so that the mixture had a pH of 6.5. A precipitate formed which was separated by filtration. The filtrate was acidified with sulfuric acid to a pH of 4.0 and the oily layer of the cresylic acids was skimmed from the top.

### References

(1) Dean, T.S., U.S. Patent 3,647,395, March 7, 1972.
(2) Patterson, J.W. and Minear, R.A., "Wastewater Treatment Technology," *Report PB 204,521*, Springfield, Va., National Technical Information Service (August 1971).
(3) Ross, R.D., *Industrial Waste Disposal*, New York, Reinhold Book Corp. (1968).
(4) Zievers, J.F., Crain, R.W. and Barclay, F.G., "Waste Treatment in Metal Finishing: U.S. and European Practices," *Plating* 55, 1171-1179 (1968).
(5) Knowlton, R.E., U.S. Patent 2,679,537, May 25, 1954, assigned to The Standard Oil Co. (Ohio).

# ALKALI CYANIDES

## Removal of Alkali Cyanides from Air

A process developed by J.S. Mackay (1) relates to the removal of potassium cyanide and other solid cyanides present in the gases from ferromanganese blast furnaces.

In the process for producing ferromanganese in a conventional blast furnace higher temperatures are required than in iron ore reduction. At the high temperatures present in the ferromanganese blast furnace cyanides are synthesized from the nitrogen gas and the alkali metal carbides which are also present. The cyanides formed, particularly potassium cyanides are condensed higher in the furnace and carried out in the gas stream as a fine sublimate.

In the past, the solid cyanides have conventionally been disposed of in two ways. First, the solids are removed from the gas by a Cottrell precipitator and then the powders are either burned to destroy the cyanides or alternatively where contamination is not a problem, the solids are washed out with water and disposed of in streams. Both of these procedures are relatively expensive and in areas where the streams must contain potable water the second procedure cannot be used at all. It is possible to destroy the cyanides in aqueous solution by chlorination in an alkali medium and the effluent then passed into the stream, but the cost of such a procedure is very appreciable.

It has been found that a large part of the cyanide can be removed by recycling the wash water liquor from the thickener back to the gas washer. Ideally, 100% of the water is recycled. As a practical matter, normally only about 95% of the water from the overflow is recycled. Normally up to 3% of the total water is removed with the sludge and it has been found desirable to recycle 80 to 97% of the total water or to recycle 80 to 99% of the clarified water after sludge removal.

The success of the process is due at least in part to the fact that the potassium cyanide and similar cyanides are dissolved in the wash water and are then hydrolyzed during the process as the water goes to the settling tank for the removal of solids and the solution of cyanide and other soluble salts is recycled to the washer. The period of contact with the water is important in the hydrolysis product. Of importance also is the lowering in pH accomplished by the recycling. Generally, the lower the pH the greater is the amount of potassium cyanide hydrolyzed to HCN. In a process wherein the wash water is not recycled, the pH of the wash water as it is removed from the system, e.g., by dumping into a stream, is about 9.5. At a pH of 9, potassium cyanide is hydrolyzed to the extent of 65% to HCN, whereas at a pH of 8, the hydrolysis is 95%.

The ferromanganese blast furnace gases which meet the wash water in the gas washer are acidic since they contain $CO_2$. Recycling the wash water permits it to become saturated with $CO_2$. The high $CO_2$ pressure consequently reduces the pH to well below 9 at the top and gas exit of the washer and further increases the hydrolysis of potassium cyanide to HCN. Unlike potassium cyanide which is a solid and quite soluble in water, HCN is a gas and the gas stream moving countercurrently to the water in the gas washer purges the water of the HCN. In addition, the accumulation of $CO_2$ in the water tends to further volatilize the HCN. The above hydrolysis cycle is continuously repeated.

Unlike the problems presented by dumping wastewater containing cyanides into streams, there are no problems in getting rid of the volatilized HCN in the exit gases since these gases are burned in the boilers for fuel. In addition, the gas is already dangerous due to CO and HCN does not change the toxicity of the gas in contrast to the dangerous concentration of cyanides normally present in the unrecycled wash water dumped into the stream in other processes. The essentials of this process are shown in the flow diagram in Figure 3.

Referring to the drawing, 2 is a ferromanganese blast furnace having a gas take-off pipe 4. The gas is passed through a conventional dry dust catcher 6 and then leaves through pipe 7 and enters gas washer and disintegrator 5 near the bottom thereof. The gas in washer 8 is met by a countercurrent flow of water entering the washer near the top thereof through spray units 10. The washed gas emerges from the top of the washer into pipe 12 from which it passes through gas cleaning equipment 14 and out through pipe 16 to the boilers and stoves.

The water descends from the sprays 10, picks up cyanides, salts, $CO_2$ and other soluble material from the gases and exits from the gas washer 8 through pipe 18 from which it enters settling tank 20 which preferably is a Dorr thickener. The solids are removed as a sludge through line 22 and are recharged to the furnace. The aqueous liquor flows out through a pipe 24 near the top of the settling tank and then goes to water collection sump 26. From the sump the liquor is pumped through line 28 with the aid of recirculating water pump 30. Instead of recirculating all of the aqueous liquor, a small fraction thereof can be bled off through pipe 32 and disposed of in the sewer or is utilized as feed to the blast furnace.

A portion of the recirculating aqueous liquor in pipe 28 is bled off through pipe 34 to the gas cleaning equipment 14 and returns to the settling tank 20 via line 36. Generally, 10 to 25% of the aqueous liquor in pipe 28 is bled off through pipe 34.

As previously stated, it is desirable to remove a small portion, e.g., 5 to 20% of the recirculating liquor through pipe 32. Sufficient fresh water is added to the recirculating aqueous liquor through pipe 38 to make up for the water lost through waste disposal pipe 32, as well as for the small amount of water lost through sludge line 22. It has been found desirable to limit the amount of water lost through lines 22 and 32 to not more than 20% of the recirculating water.

If the liquor to be disposed of is to be dumped into the sewer, it may be necessary to chlorinate the liquor or otherwise treat the liquor to destroy the hydrogen cyanide therein in the event the sewer enters into a source of potable water or water containing fish. However, only a relatively small amount of chlorination agent is required compared to that employed in conventional processes. A specific example follows.

## FIGURE 3: PROCESS OF REMOVING ENTRAINED ALKALI METAL CYANIDES FROM FERROMANGANESE FURNACE GASES

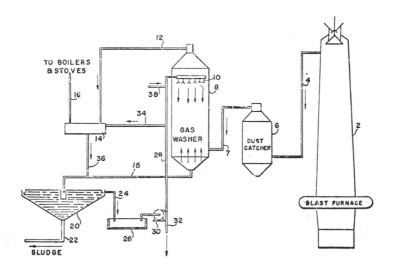

Source:  J.S. Mackay; U.S. Patent 2,877,086; March 10, 1959

The ferromanganese blast furnace was set to produce about 300 tons a day of ferromanganese.  The furnace exit gas entering gas take-off pipe **4** was 60,000 scfm.  This gas contained 5.1% $CO_2$, 34.3% CO, 58.7% $N_2$ and 1.9% $H_2$.  Entrained with the gas were solids in an amount of 77 lbs./min.  These solids contained 3.5 lbs./min. of cyanide measured as HCN but actually as alkali metal cyanides, chiefly KCN.  The solids were washed out of the gas in the gas washer and disintegrator **8** by countercurrent flow of water.  A total of 2,500 gallons of water per minute flowed through the system.  Of this amount, about 2,000 gallons per minute were continuously recycled to the gas washer and 400 gal./min. to gas cleaning equipment **14** or a total of 2,400 gal./min. were recycled in all.

The water at the end of the first cycle through the system contained 168 ppm of cyanide at the Dorr thickener.  Upon recycling, the cyanide content rose to an amount which varied between 200 and 300 ppm in the feed to the Dorr thickener.

The solids which settled out in the Dorr thickener and were removed as sludge through line **22** contained 20 to 40 gal./min. of water.  Approximately 70 gal./min. of water were continuously removed from the system through pipe **32**.  (The water thus removed can vary from 50 to 500 gal./min.)  The water thus removed was fed to the furnace for cooling purposes.  The contact of the water with the cyanide at these high temperatures resulted in the cyanide being largely hydrolyzed to ammonia.

The discharge on occasion was also fed to the Ohio River.  The acceptable maximum cyanide content in this river is about 2 ppm.  To insure sufficient breakdown of the cyanide to nontoxic materials, the discharge was chlorinated before being fed to the river.  The chlorine consumption is 5 lbs./lb. of HCN and this amounted to less than 5% of the chlorine required if all of the recycled liquor were so treated.

Makeup water is added through pipe **38** at the same rate at which water is lost from the system and is normally between 50 and 500 gal./min.

References

(1) Mackay, J.S., U.S. Patent 2,877,086; March 10, 1959; assigned to Pittsburgh Coke & Chemical Co.

## ALUMINUM

### Aluminum Recovery from Scrap

A process has been described by E.A. Mathis (1) whereby aluminum is recovered from scrap by volatilization of organics from the scrap in a nonoxidizing furnace having multiple hearths, and thereafter separating the aluminum from the ash by mechanical means. A considerable amount of such scrap is generated at plants where printing is done on aluminum foil. The foil is usually laminated with an organic backing material such as wax, plastic, lacquer, paper or paperboard. The metallic aluminum may constitute anywhere from 7 to 55% of the total weight of the scrap.

Scrap aluminum is generally collected and baled for storage or for shipment. It is processed according to prior art techniques by breaking open the bales and shredding the material, and thereafter blowing the shredded material through a continuous combustion unit. The material becomes ignited in the combustion unit and then passes into one end of an elongated combustion trough. The bottom of the trough is made of a flat steel plate on which rides a drag chain conveyor with steel flights, so as to move the burning scrap continuously through the trough. An insulated hood having several smoke stacks is positioned over the trough and combustion air is admitted into the trough just under the hood. At the other end of the trough the burned material is milled or hammered and subjected to mechanical separation. The aluminum portion is then passed through a rotary kiln at about 950°F. where the carbon is mostly burned off.

The continuous combustion technique described above is subject to several serious defects. The most serious of these defects involves the occurrence of thermite type reactions whereby some of the aluminum reacts very violently with oxygen to produce temperatures in the region of 2300°F. These high temperatures melt the surrounding aluminum and cause it to adhere to and clog the interior of the trough. As a result, the operation of the system is intermittent rather than continuous. Also the system becomes very expensive to operate and maintain.

An additional difficulty associated with the continuous combustion trough is that is produces an exorbitant amount of heavy black smoke which presents a nuisance. This smoke has been measured consistently at No. 4 on the Ringleman scale, and very rarely at No. 3. The Ringleman scale extends from No. 0 to No. 5, where No. 5 is smoke of maximum blackness and maximum density.

This process overcomes the abovedescribed problems of the continuous combustion system for removal of organic materials from aluminum scrap. With this process, aluminum scrap may be continually processed in relatively compact and inexpensive equipment. Further, the aluminum values are recovered at a high degree of purity, and no danger of thermite type reaction is presented. The equipment is never subject to temperatures higher than about 1100°F. so that melting does not occur; nor is there any destructive effect produced on the insulative materials in the equipment.

In general the process achieves its superior results by heating the dispersed aluminum scrap in a nonoxidizing (e.g., carbonaceous) atmosphere up to a temperature (e.g., 700° to 1100°F.) sufficient to evaporate or to drive off the volatile components of the backing material. This leaves a product comprising aluminum, carbon and ash, from which the aluminum may be readily separated by mechanical means.

As will be described more fully hereinafter aluminum foil scrap which is shredded and dispersed, is processed, according to one example, in a multiple hearth furnace in which there is maintained a nonoxidizing atmosphere. The scrap material is moved through the furnace along each of its hearths in succession. This movement of scrap is accomplished by means of rabble arms which rotate over each hearth. The rabble arms are provided with canted rabble teeth which plow through the scrap material; and by virtue of the cant angle of each set of teeth, they urge the material in a radially inward or outward direction along the associated hearth toward drop holes through which the material eventually passes to the next lower hearth.

During its passage through the furnace, the scrap material is relieved of its volatile components which proceed to the top of the furnace and are subjected to controlled burning to supplement the heat input. The output of the furnace is in the form of metallic aluminum, carbon and ash. The carbon and ash, now relieved of the volatile components which bound them to the aluminum are then easily separated from the aluminum by mechanical means. Figure 4 shows the essential components involved in the conduct of this process.

As shown there, bales **10** of aluminum scrap are moved along a belt conveyor **12** and are dropped into a hopper **14** of a shredding machine **16**. The shredding machine separates the baled material into individual fragments which then pass out through a conveyor tube **18**. The conveyor tube opens into an air conveyor **20** which is powered by a blower **22**. The blower serves to convey the shredded aluminum scrap to a cyclone **24** where the scrap fragments are separated from the air stream. The scrap passes downwardly from the cyclone where the scrap fragments are separated from the air stream. The scrap passes downwardly from the cyclone and through an airtight lock **26** into a controlled atmosphere furnace **28**.

The interior of the furnace **28** is maintained at a temperature above 700°F., and preferably between 900° and 1100°F. Also the composition of the furnace atmosphere is kept under close control so that no oxygen will come into contact with the metallic aluminum in the scrap material. As a result of this, the volatile organic materials forming the major portion of the nonaluminum values are vaporized and separate from the solid scrap. These materials are then discharged or burned and their products of combustion are discharged out through a stack **29**. The remainder of the scrap material, i.e., the solid residue composed chiefly of metallic aluminum values and solid, loosely adherent carbon and ash are passed through a mechanical separator where these materials are crushed and separated to provide high grade aluminum.

FIGURE 4: APPARATUS FOR ALUMINUM RECOVERY FROM SOLID WASTE

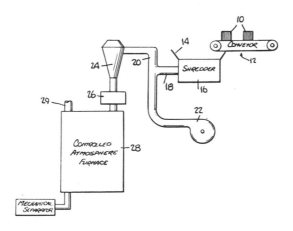

Source:  E.A. Mathis; U.S. Patent 3,650,830; March 21, 1972

References

(1) Mathis, E.A.; U.S. Patent 3,650,830; March 21, 1972; assigned to Nichols Engineering & Research.

## ALUMINUM CELL EXIT GASES

The emissions from electrolytic cells for aluminum production have been described by the Washington State Department of Health (1) and cited by H.R. Jones (2) (3). See also Removal of Fluorine Compounds from Air under Fluorine Compounds in this volume.

### Removal from Air

A process developed by A.W. Kielback (4) involves scrubbing the gases from aluminum reduction cells (of the electrolytic type using molten baths), or the gases from the buildings, i.e., potrooms, where such cells or pots are operated. The scrubbing is effected with water, preferably water having a suitable chemical content to promote the desired absorption of gaseous or other material from the waste gases. That is to say, in such scrubbing operations, it is usually desired to recover fluorine values, present as gaseous hydrogen fluoride and particles of solid fluorine compounds (e.g., fluorides). The gases withdrawn from the pots or potrooms, as by suitable pumping, contain fluorine in various amounts, as for example from 2 to 40 mg./cu. ft., a relatively minor proportion of the total gas. These gases also contain other solid particles, such as alumina, tars, carbon and the like, which it is desired to remove.

As stated, the fluorine in such gases is present in part as hydrogen fluoride, but also in part as solid particulate fluorine compounds, which may account for as much as 30 to 50% of the total fluorine loss from the reduction cells. At least a major portion of this solid material in the gas results from volatilization of electrolyte constituents, which condense to extremely small fume particles. Removal of these very fine particles. understood to have sizes of a fraction of a micron, has created difficulty in getting a high efficiency of recovery. Whereas hydrogen fluoride gas alone can be removed efficiently with equipment having a low pressure drop (e.g., 0.5 to 1 inch of water), such as open spray towers, satisfactory removal of fluoride fumes requires expenditure of more energy, as in higher pressure drop apparatus. This process is highly effective for such purpose, and is specially advantageous in that it does not become clogged with accumulations of solid material. This process utilizes a device known as a floating-bed scrubber as shown in Figure 5.

The apparatus comprises a tower or vessel 10 having provision for countercurrent flow of fluids. The tower may be rectangular (as shown), circular or of other suitable shape in horizontal cross section and is of sufficient vertical extent to provide an upper discharge chamber 11 and a lower or contact chamber 12, the latter being bounded at its top by transverse grid means 13, e.g., a grid extending horizontally across the vessel 10. While it is conceived that other shapes of such grids or perforate structures, e.g., sloping or upwardly or downwardly convex, may be employed, a simple horizontal grid is found to be highly effective and indeed specially advantageous in most cases.

The openings 14 in the grid are preferably as large as possible, their function being essentially only to prevent upward escape of the spherical elements, while the solid structure of the grid should be relatively thin, i.e., in a vertical direction (as by having a vertical dimension, say, less than half the side of a rectangular grid opening), to avoid a tendency, in some cases, to accelerate the gas so that the upper layer of spheres is held firmly against the grid. A second or lower grid means 16 may be provided, if desired, being spaced below the grid means 13 by a distance substantially greater than the vertical thickness of the floating bed, for the purpose of supporting the latter when the apparatus is shut down or otherwise not in full operation.

# FIGURE 5: FLOATING-BED SCRUBBER FOR ALUMINUM CELL EXIT GAS TREATMENT

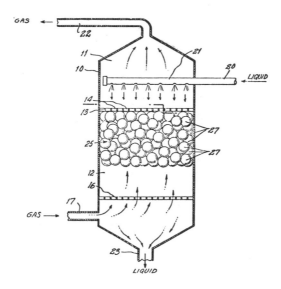

Source: A.W. Kielback; U.S. Patent 3,122,594; February 25, 1964

Gas is introduced to or near the bottom of lower part of the vessel **10**, i.e., at the foot of the contact chamber below the grid **16**, as by a pipe **17** opening through the wall of the vessel. Liquid is similarly introduced at or near the top of the outlet chamber **11**, or otherwise above the grid means **14**, through a pipe **20** and preferably through a suitable distributor **21** so that a multiplicity of streams are descending or are projected downwardly toward the grid **13** and the packing, to assure distribution of the liquid over the surface of the bed. Gas and liquid are withdrawn from the vessel at appropriate localities, for instance through a gas discharge pipe **22** at the top and a liquid discharge pipe **23** at the bottom.

The packing or contact-promoting assembly, shown here as functioning in the form of a floating bed **25**, consists of a large multiplicity of spheres or balls **27**, sufficient in number to provide, in approximate effect, a plurality of layers, and very preferably at least several layers (such as five or six, as shown) extending entirely across the contact chamber **12**. Thus the bed **25** transversely fills the chamber and has a vertical depth equaling a plurality and, with special advantage, a multiplicity of diameters of the individual elements. While in some cases it is conceived that the spheres of a given bed may differ among themselves in size or density or both, it is of special convenience and advantage, and indeed thus a particular feature, to provide a packing **25** wherein the elements **27** are all of the same diameter and density.

The elements may be made of various materials and are conveniently hollow, thin-walled balls of plastic or synthetic resin or the like. Alternatively, other low density constructions may be employed, such as various foamed plastics or other porous materials, with an impervious external surface. In some cases hollow or other low density spheres of metal may be used, or conceivably inflated plastic balls, with a very thin wall, containing gas under pressure, to provide elements of extremely low density.

In operation, the scrubbing or other gas-liquid contact is effected in the illustrated apparatus by continuously supplying liquid through the pipe **20** and gas through the pipe **17**.

The liquid falls in sprays or advantageously in streams in force of gravity toward and through the grid means **13** and into and through the bed **25** in the contact chamber **12**, finally falling or draining into the outlet pipe **23**. The gas is projected from the inlet pipe **17**, at a lower part of the contact chamber, so as to rise through the liquid and between the balls. Assuming that in starting up, for example, the bed of balls is resting on the lower grid **16** and the supply of liquid is initiated first, it will pass freely through the interstices between balls and drain into the outlet. When the gas is now passed upward, it crosses the falling streams of liquid (below the grid **16**) and forces its way up between the balls, coming into closely effective contact with the descending liquid. More specifically, it there interferes with the downward flow of liquid, increasing the amount of liquid hold-up within the bed and correspondingly increasing the buoyancy acting on the balls.

Thus a gas-liquid mixture is in effect established in the spaces between the packing elements, causing them to float and assemble in a mobile, multilayered manner as shown at **25**, beneath the grid **13**. In such position, the individual spheres **27** have the following forces acting downwardly upon them, viz. gravity (the weight of the sphere), the impact of the falling liquid, and the downward frictional drag of the liquid. Upward forces on each ball comprise the frictional lift or upward drag of the passing gas, and the buoyancy of the gas-liquid mixture which in effect surrounds the ball. It appears that in operation, the chief opposing forces are gravity and the buoyancy of the fluid mixture, or that in a general sense the forces of drag and impact may be deemed to cancel each other, whereby the floating tendency of the balls is essentially a function of their buoyancy.

Under the desired floating conditions of the bed **25**, the region of the chamber **12** below the bed contains the gas in continuous phase (and under pressure relative to the upper space **11**), with the liquid falling through it in more or less turbulent or broken streams, to drain through the outlet **23**. As stated, there is mutual dispersion and distribution of the liquid and gas in the floating bed **25**, and there is ordinarily no continuous level of liquid above the grid **13**, the area **11** thus containing the departing gas in continuous phase, with liquid descending through it.

The bed **25** is unrestrained at its lower boundary and as stated involves a relatively loose assembly; whereas the floating action tends to bring all of the balls into contact with each other, the downward impact and flow of the liquid and the pressure of the rising gas keep sufficient interstices between the balls for the desired travel of both fluids. The liquid in effect provides a thin film flowing over at least a large part of each ball surface, with corresponding layers or bodies of gas, mixed with liquid, between the film spheres; the gas thus turbulently ascends through all voids of the packing. An exceptionally thorough interphase contact of the fluids is achieved, over a relatively large area (substantially the total of the sphere surfaces) and with both fluids passing as thin or small bodies.

At the same time, because the assembly can be regarded as truly floating in the gas-liquid mixture which fills its interstices, most or all of the balls have distinct freedom of movement, not only to a slight extent in various directions of translation, but especially in a rotating sense. This loose and mobile character of the bed effectively prevents channeling, and especially with the rotating or turning movements of the balls, inhibits any tendency of solid particles, in either the liquid or the gas, to accumulate in the interstices.

It will be understood that the operation should be controlled to avoid so-called flooding. For example, if the gas velocity is allowed to rise, in any given operation, a point may be reached where the liquid no longer flows down through the bed. In such case the upper part **11** of the tower may tend to fill with liquid, through which some of the gas may bubble upward, or indeed in some such cases the gas may build up to a certain high pressure in the chamber **12** while the accumulated liquid in the chamber above may afford an effective lock or the like, substantially arresting further passage of gas or indeed of both liquids. Where the entire region below the grid **13** becomes filled essentially with gas alone, the balls, or most of them, will tend to fall down to or toward the lower grid **16**.

In other words, the desired floating action is absent and the bed fails to be or remain elevated as a complete unit. Although the passing gas may have some tendency to elevate the balls partially, in a random manner (if there is any travel of gas into and through the upper chamber), the desired floating effect and distributed interphase contact of the fluids are not achieved. When such condition occurs, it can be readily corrected, e.g., by reducing the velocity or amount of gas flow until a liquid-gas mixture reappears in and substantially throughout the chamber **12** so that the balls assume their desired, floating action.

A process developed by L.L. Knapp and C.C. Cook of Aluminum Co. of America (5) is one in which hydrogen fluoride and finely divided solids are removed from gases evolved in the electrolytic production of aluminum, by means of a bed of alumina particles fluidized by a stream of such gases. Alumina particles combined with the hydrogen fluoride, and finely divided solids entrapped in the bed, discharge from the bed. This design is reported to be the basis of the proprietary process known as Alcoa Process A-398. The process is shown schematically in Figure 6.

In the elongated chamber **1** a horizontally disposed, perforated plate **2** supports a dense, turbulent fluidized bed **3** consisting primarily of finely divided alumina particles. At the base of the chamber **1** is a plenum chamber **4** extending substantially the entire length of the chamber **1** and having a blower unit **5** connected thereto. A conduit **6** connects the blower unit with gas collecting hoods **7** over conventional electrolytic cells **8** (indicated diagrammatically) for the production of aluminum by reduction of alumina dissolved in molten cryolite. Gas distribution chambers **9** below the plate **2** communicate with the chamber **4** through spaced openings **10** in the top of chamber **4**.

Conventional bag filters **11** made of porous fabric extend across the upper portion of chamber **1**, above the bed **3**. At the top of chamber **1** is a gas discharge conduit **12**. Air jets **13** are provided for periodically dislodging deposited solids from the exterior of the filter bags by blowing a reverse air stream through the bags.

**FIGURE 6:  ALCOA PROCESS FOR FLUORIDE REMOVAL IN FLUIDIZED
ALUMINA BED**

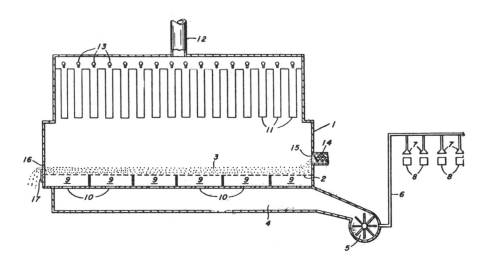

Source:   L.L. Knapp and C.C. Cook; U.S. Patent 3,503,184; March 31, 1970

At one end of chamber **1** is a conveyor **14** for feeding finely divided alumina into the bed **3** through an inlet opening **15** in an end wall of the chamber. At the opposite end of the chamber **1** is an outlet opening **16** through which solids from the bed **3** can overflow to a suitable conveyor (not shown). An adjustable weir **17** at the opening **16** permits control of the depth of the bed **3** to assure adequate contact of the gas stream with the bed for the desired purposes.

In carrying out the process in such apparatus, a stream of gases from the electrolytic cells **8** is blown through the plenum chamber **4** and the plate **2** by the blower **5**, at a sufficient velocity to keep the bed **3** fluidized. That bed has a depth of only 2 to 12·inches, whereby the pressure drop through the bed is low (e.g., 3 to 6 inches of water) with attendant low cost of operation of the blower **5**. However, 50 to 150 pounds of alumina is kept in the bed per pound of gaseous hydrogen fluoride present in the gas stream, thereby providing a large ratio of alumina to hydrogen fluoride.

The gas stream (which is at a temperature of 150° to 280°F.) passes through the bed **3** in from 0.25 to 1.5 seconds, depending on the bed depth and the velocity of the gas stream. Solids in the gas stream are entrapped in the bed **3**. The gas stream leaving the bed **3** flows through the bag filters **11**, depositing on the exterior of the bags remaining finely divided solids from the electrolytic cells, as well as alumina particles small enough to be carried upward out of the bed. Periodically sharp jets of air are blown into the bags **11** by the air jets **13**, dislodging the solids that had been deposited on the bags' exterior surfaces, whereupon the dislodged solids fall back into the bed **3**.

As additional alumina particles are fed into the bed **3** by the conveyor **14**, solids adjacent the outlet **16** flow automatically out of the bed through that outlet. As a result of the substantial distance between the inlet **15** and the outlet **16**, and the heterogeneous, irregular path of movement of particles in a fluidized bed, the average time of passage of a particle of alumina across and out of the bed is 2 to 14 hours, thereby insuring ample opportunity for contact of the particles with gaseous hydrogen fluoride in the bed. Such average time of passage of the particles through the bed is referred to herein as "average residence time."

The alumina particles fed into the bed **3** are of the same grade of alumina as is normally used for production of aluminum in electrolytic cells. That grade of alumina is of low porosity, but nevertheless gaseous hydrogen fluoride in the bed is taken up by the alumina particles, apparently by adsorption.

The materials leaving the bed consist of solids which have been removed from the gas stream and entrapped in the bed, as well as alumina particles carrying hydrogen fluoride. The latter particles are referred to herein as alumina particles containing hydrogen fluoride, although it may be that hydrogen fluoride initially sorbed on the alumina has reacted, at least in part, with the alumina by the time the alumina particles leave the bed. The mixture leaving the bed can then be fed into the fluoride bath in one or more of the cells **8**, thus returning the recovered fluorine values to the cells and thereby providing a portion of the fluoride values needed in the fluoride bath of the cell, as well as supplying alumina to the cell for conversion to aluminum.

By procedures described above, 99 to 100% of the total hydrogen fluoride and other fluorides in the gas stream can be removed, over a wide range of total fluoride content in the gas stream, and even though the percentage of fluorides present in the gas stream is quite low, e.g., less than 35 mg./cu. ft. Moreover, the process is highly economical in view of the relatively inexpensive equipment and operating conditions employed, and the very high efficiency of recovery of fluoride.

A process developed by H.H. Predikant (6) involves a multistage method for scrubbing a waste gas stream containing HF and $SO_2$ such as from the aluminum production industry where aluminum dust can also be present. Control of wash streams is maintained to preclude too much HF reaching the second stage washing zone causing sodium fluoride to precipitate and then block valves and piping. A combined effluent from each wash stage is passed to a neutralization zone where an alkaline stream such as caustic soda will cause precipitation of sodium fluoride and sodium sulfate along with entrained dust or aluminum particulates.

**References**

(1) Washington State Department of Health, *Report on the Primary Aluminum Industry*, Seattle, October 1969.
(2) Jones, H.R., *Fine Dust and Particulates Removal*, Park Ridge, N.J., Noyes Data Corp. (1972).
(3) Jones, H.R., *Pollution Control in the Nonferrous Metals Industry*, Park Ridge, N.J., Noyes Data Corp. (1972).
(4) Kielback, A.W., U.S. Patent 3,122,594; February 25, 1964; assigned to Aluminum Laboratories, Ltd.
(5) Knapp, L.L. and Cook, C.C., U.S. Patent 3,503,184; March 31, 1970; assigned to Aluminum Co. of America.
(6) Predikant, H.H., U.S. Patent 3,709,978; January 9, 1973; assigned to Universal Oil Products Company.

# ALUMINUM CHLORIDE

In carrying out secondary aluminum purification operations, it is customary to periodically introduce gaseous chlorine into molten batches of aluminum so as to effect a degasification thereof and the removal of resulting slag-like impurities. For example, chlorine will combine with the magnesium content of molten metal, where such metal may enter into the furnace along with aluminum scrap, to effect the formation of magnesium chloride which will in turn form a slag-like material that may be removed. However, the addition of chlorine gas to the molten metal may also form aluminum chloride which leaves the furnace in gaseous form. As the gas cools and sublimes, the aluminum chloride is converted to finely divided particulate matter which is in the size range of 0.1 to 1.0 micron so that as such material is released to the atmosphere there is a resulting dense opaque effluent. The effluent stream also contains air, chlorine, gaseous aluminum chloride and a small amount of magnesium chloride, all of which can be particularly obnoxious and damaging to the surrounding area.

## Removal of AlCl$_3$ from Air

Various systems have been used, or tried out, to effect the neutralization and removal of entrained particles from effluent streams of secondary smelting furnaces, as for example, the use of spray towers, packed towers, submerged inlet tanks, etc.; however, none of the previous types of systems have been entirely satisfactory. Packed bed towers are avoided as being particularly troublesome inasmuch as fixed bed packing becomes rapidly clogged with aluminum chloride fines and its efficiency impaired.

Tomany (1) has described a method of removing chlorine and entrained aluminum chloride particles from a hot gaseous stream comprising cooling and agglomerating at least a portion of the particles in the gaseous stream by contacting it with a liquid spray which simultaneously absorbs chlorine and thereafter passing the gas upwardly through a loose mobile floating bed of contact elements in countercurrent contact with a descending stream of an alkaline absorbent solution. Figure 7 shows a suitable form of apparatus for the conduct of the Tomany process.

There is indicated a vertically elongated tower 1 with a plurality of internal horizontally disposed perforated plate members 2 which in turn divide the tower into a plurality of superposed contact zones 3 for accommodating beds of contact elements 4. A multiplicity of low density contact elements 4 are, of course, utilized in each of the separate zones 3, such that they can operate as a fluidized moving bed of elements for each of the independent zones or stages of contact. The contact elements 4 may be thin walled spheres of metal or plastic having a smooth outer surface or, alternatively, they may be of a foamlike

FIGURE 7:  SCRUBBER FOR AlCl₃ AND Cl₂ REMOVAL FROM WASTE GASES

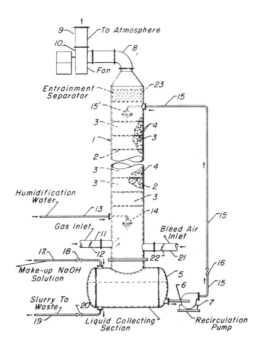

Source:  J.P. Tomany; U.S. Patent 3,445,182; May 20, 1969

nature formed of foamed plastics such as polyurethane, polystyrene, or the like, whereby fluidization and random movement may be effected by an up-flowing gaseous stream to be scrubbed, substantially in accordance with the teachings of U.S. Patent 3,122,594.

The lower end of tower 1 is indicated as being provided with a liquid collecting reservoir section 5 having outline line 6 connect with pump means 7.  The upper end of the tower 1 is provided with conduit or duct means 8 which in turn connects with a stack 9 having fan means 10 to maintain a constant suction on the upper portion of the tower.  The gas stream inlet line 11, with flow control damper means 12, connects with the lower portion of the scrubber tower 1 at a level above the upper portion of the collection section 5 such that the gas stream may be subsequently admitted to rise upwardly through each of the contact zones 3.  A humidification liquor inlet line 13 is provided to connect with an internal spray head 14 at a level just below the lowermost contact zone 3 such that water, or a slightly alkaline liquid stream, may effect an initial spraying and contacting of the gas inlet flow prior to the latter entering the upper contact zones 3.

At an upper level of the tower, above the uppermost contact zone 3, there is a line 15 connective with an internal spray head 15' that serves to introduce the principal quantity of alkaline scrubbing liquid which continuously carries downwardly through each of the successive contact zones 3 countercurrently to the upward gas flow of the waste gas stream entering by way of conduit 11.  In the actual operation of the unit, the gas and liquid flows are adjusted such that the light weight contact elements 4 in each of the superposed zones are caused to float above a lower perforate plate 2 and effect a random rotational floating bed action for each of the individual elements **4**. There is thus insured an inter-phase contact between the highly laden gas stream and the liquid stream with a substantial prevention of the channeling of either stream through the unit. Each of the spaced perfor-ate plate members **2**, of course, assists in effecting a redistribution of the gas and liquid streams at each of the vertically stacked separate zones **3**.

The diagram indicates a pump **7** discharging into the line **15**, having control valve **16**, so as to recycle scrubbing liquor to the upper contact zones **3**. Makeup sodium hydroxide solution is periodically, or continuously, introduced into the lower collection section **5** by way of inlet line **17** and control valve **18**, so as to insure an adequate quantity and proper pH solution within the system. A waste slurry stream from the lower portion of section **5** may be discharged by way of line **19** and control valve **20**.

A scrubber or contact tower of preferred design utilizes one or more mist extraction means or de-entrainment means at the upper end of the tower ahead of the discharge line such as indicated diagrammatically in the drawing as **23**. Thus, a minimum quantity of liquid is permitted to be entrained and carried to the atmosphere by way of outlet conduit **8** and stack means **9**.

For purposes of controlling and stabilizing gas flow through the unit, a bleed-in air inlet duct **21** with control damper means **22** is provided at the lower end of the scrubber tower below the first contact zone **3**. Such bleed-in air stream may be controlled manually or automatically and, as indicated hereinbefore, the introduction of air shall be used primarily to compensate for fluctuations in effluent stream flow to the unit and stabilize upward flow through contact zones **3**. The bleed-in air stream may also be used to assist in tem-pering a high temperature gas stream being introduced to the scrubber tower. The follow-ing are specific examples of operation of the Tomany process.

*Example 1:* In one example of the employment of a multistage floating-bed contactor or scrubber, using multiplicity of hollow plastic spheres of approximately 1½ inches in diame-ter for contact elements, there is provided for the countercurrent liquid treating of some 2,000 scfm of effluent gas stream containing free chlorine, aluminum chloride vapor and aluminum chloride particles. With the effluent originating from a molten aluminum puri-fication or smelting system, the stream is at a high temperature of the order of 800°F. and is heavily laden with aluminum chloride particles to provide 150 grains per standard cubic foot of effluent gas.

Utilizing an initial humidifying and agglomerating water spray of 2 gal./min. across the area of the contactor and below the zones of contact elements, as well as the use of some 75 gallons per minute of a 5% sodium hydroxide solution for the primary scrubbing liquor through the upper part of the contactor, there is a reduction in solids loading from 150 to 0.75 grains per cubic foot in the effluent stream. The resulting efficiency of removal thus is 99.5%. The effluent gas stream temperature leaving the contactor is of the order of 90°F. and the averaged quantity of aluminum chloride solids leaving the scrubber with a liquor effluent stream is about 2.0 lbs./hr.

*Example 2:* In another example of gas stream scrubbing, there is again considered the treatment of an effluent gas stream from an aluminum purification system such that there is free chlorine in the gas and a resulting relatively heavy aluminum chloride particle load-ing of some 60 grains per standard cubic foot. With the employment of the same type of multistage floating-bed scrubber as considered for Example 1 except for the elimination of the humidification-agglomeration spray stream, there is received 1,600 scfm of effluent gas at a temperature of the order of 750°F.

Again, this effluent stream is treated with a principal scrubbing liquor stream, comprising 5% sodium hydroxide, at a rate providing 75 gallons per minute down through the column to effect the desired countercurrent contacting in the presence of the floating and moving plastic spheres. In this instance, there is a reduction of particle loading in the gas stream leaving the contactor to provide of the order of 0.24 grains of aluminum chloride/scf with a resulting scrubbing efficiency of 99.6%. The gas stream temperature from the contactor is of the order of 70°F; and the solids content of the effluent liquor from the bottom of the unit is of the order of 80.0 lbs./hr.

Another process developed by J.P. Tomany (2) involves multiple stage system for removing chlorine and aluminum chloride particles from a hot gaseous stream. It embodies an initial condensation step, preferably of the nature of a centrifuging-cooling zone, so as to provide a partial cooling of the stream and effect some collection of sublimed aluminum chloride particles and then, subsequently, there is a countercurrent scrubbing of the partially cooled laden gaseous stream in the presence of at least one bed of loose mobile contact elements. A preferred operation may also incorporate a cooling-humidification stage directly ahead of the scrubbing stage.

A process developed by E.E. Lindenmaier, E. Jackel and L.G. Mathys (3) is one in which the waste gas from a chlorine treated aluminum melt is passed with a temperature above the sublimation temperature of the gaseous aluminum chloride contained therein into a falling film condenser in which walls, completely covered with a film of a cooling liquid which dissolves condensed aluminum chloride, enclose the waste gas stream.

**References**

(1) Tomany, J.P.; U.S. Patent 3,445,182; May 20, 1969; assigned to Universal Oil Products Company.
(2) Tomany, J.P.; U.S. Patent 3,582,262; June 1, 1971; assigned to Universal Oil Products Company.
(3) Lindenmaier, E.E., Jackel, E., and Mathys, L.G.; U.S. Patent 3,435,592; April 1, 1969; assigned to Krebs & Co. Ltd., Switzerland.

# ALUMINUM SILICATE PIGMENT

Increasingly stringent requirements on the level of contaminants, particularly of fines allowable in industrial gases vented to the atmosphere necessitate development of more efficient and effective devices and of machines able to wash greater quantities of such gases. A process developed by N.J. Panzica, R.M. Jamison and E. Umbricht (1) utilizes an improved high capacity gas washer for the removal of particulates from waste gases.

Such a device consists of a generally cylindrical housing through which gas or air to be washed and cleaned is passed. The gas is passed through a zone of spray droplets of a washing liquid within the housing which is generated by a rotary generator. A blower in the housing moves the gas. Pump means in the housing base feeds liquid for the spray to the generator. In the housing between the spray zone and the gas exhaust are baffles, set in horizontal tiers, commonly called eliminators, for deposition and removal of liquid droplets caught in the gas flow.

Washers of that type have gas washing capacities of from several thousand to about 30,000 standard cubic feet per minute; their size varies according to capacity from about 4 to 10 ft. in diameter, and they are generally about half again as high as their width. The generators within such washers have diameters from about 1 to 4 feet and varying heights, usually greater or lesser than the diameter for small or large, respectively, diameter rotors. Rotational speeds of generators commonly range from less than one thousand up to several thousand rpm. About 100 to 500 gallons per minute of liquid are circulated and generated into cleansing spray.

A number of tests have been conducted to compare the efficiency of such gas washers with those previously known. One model, having a capacity for washing 40,000 standard cubic feet per minute of gas was compared to an older type gas washer having a 32,000 standard cubic foot per minute capacity. Since there is a difference in the gas capacities of the two washers, the test values set forth below are based on concentration terms and percentages for purposes of comparing the two machines.

A paint pigment of very fine aluminum silicate, designated ASP-100, approximately 49% of whose particles are less than ½ micron in size, was used in some tests. Using the older washer, and feeding an average of 962 grains of the ASP-100 pigment per thousand standard cubic feet of gas into the washer, resulted in removal of all but 6.83 grains per thousand cubic feet. The overall efficiency for that washer was about 99.3%. When using this type of washer, and an inlet contaminant loading average of 842 grains per thousand standard cubic feet, the outlet loading concentration was found to be 2.47 grains per thousand standard cubic feet, an overall washing efficiency of 99.7%.

While in relative terms the improvement may not appear significant, in absolute terms there is a remarkable increase in efficiency, since the contaminant is composed of ultrafine particles. Specifically, the reduction from 6.83 to 2.47 in the outlet loading represents a reduction of 63.8% in the amount of contaminants remaining in the outlet.

A second test using smaller concentrations of ASP-100 verified the results. The older washer reduced an inlet loading of 333 grains to 5.48 grains per 1,000 standard cubic feet, an efficiency of 98.3%. However, this type of washer, fed an average of 328 grains of ASP-100, removed all but 1.85 grains per 1,000 standard cubic feet, an efficiency of 99.44%. Thus, the new washer reduced the outlet loading 66.2% compared to the older washer.

### References

(1) Panzica, N.J., Jamison, R.M. and Umbricht, E.; U.S. Patent 3,444,669; May 20, 1969; assigned to Ajem Laboratories, Inc.

## AMMONIA

The main source of atmospheric ammonia is naturally-produced ammonia which is released from land and ocean areas. In terms of total air content of ammonia, urban-produced ammonia is of lesser importance, though it may be important from the air pollution standpoint in localized situations, as noted by Miner (1).

Atmospheric concentrations of ammonia are generally below the level at which health hazards and deleterious effects to humans, animals, plants, and materials are known to occur. High concentrations, usually caused by accidental spillage, can result in corrosive action to mucous membranes, permanent injury to the cornea, damage to the throat and upper respiratory tract, chronic bronchial catarrh, and edema. In sufficiently high doses (1,700,000 $\mu$g./m.$^3$), ammonia acts as an asphyxiant. Ammonia combines with sulfur oxides and metallic materials in the atmosphere to form ammonium sulfate and zinc ammonium sulfate aerosols. These aerosols are thought to be in part responsible for the irritant effects of the air during the Donora Smog Episode in 1948. Severe irritation was produced by these salts in guinea pigs.

Ammonia has been shown to be toxic to most plant life, producing injury to leaf and stem tissue, and reducing or delaying germination of seeds at concentrations of 700,000 $\mu$g./m.$^3$. It also has a corrosive effect on certain materials.

The bulk of the ammonia found throughout the world in the atmosphere is produced by natural biological processes which release this substance to the air. The background concentration of ammonia is 6 $\mu$g./m.$^3$ in the mid-latitudes and 140 $\mu$g./m.$^3$ near the equator. A secondary source of atmospheric ammonia is urban-produced ammonia, which may be an important source in localized situations. The average concentration of ammonium in the urban atmosphere is 20 $\mu$g./m.$^3$, although measurements as high as 7,200 $\mu$g./m.$^3$ have been recorded in areas adjacent to chemical manufacturing complexes.

### Removal of Ammonia from Air

The primary source of ammonia air pollution in cities is the combustion process involved in the combustion of fuels, incineration of wastes, and use of the internal combustion engine. Ammonia may be released to the air from a variety of process sources, including: coke ovens; fertilizer manufacture; organic synthesis; oil refineries. Other sources are stockyards and similar installations, where ammonia is formed by biological degradation. Finally, ammonia is found in the off-gases from diazo copying machines. See also the section of this book on Ammonium Phosphate Fertilizer Plant Effluents.

Methods used to abate other pollutants with which it is associated also reduce the quantity of ammonia that reaches the atmosphere. For example, in smokeless charging of coke ovens (that is, collecting the bulk of escaping coke oven gas, coal dust, and tar by vacuum during coke oven charging), the ammonia emissions to the atmosphere are cut in half (2).

In incineration systems where wet scrubbers are used to remove fly ash, the ammonia in the gas stream leaving the incinerator should also be reduced. However, no specific information was found on this subject by Miner (1).

In the chemical industry, where ammonia is used as a raw material, its recovery is a matter of fundamental economic importance; methods have therefore been designed to minimize its loss. For high concentrations of ammonia, gas wet scrubbers can be used. For ammonia concentrations in air between approximately 16 to 27% (flammable range) the gas can be flared. Impregnated activated charcoal has been used to remove ammonia from the air in laboratories that use animals in research and in other places where animals are kept in large numbers (3). Where the ammonia occurs as a solid (as ammonium sulfate in the fertilizer industry for instance) conventional methods for solids removal can be used such as bag filters, electrostatic precipitators, and wet scrubbers.

A process developed by Mohr (4) involves the separation of ammonia from off-gas obtained in the synthesis of melamine from urea. The off-gas, after melamine has been separated, is treated with a melt which contains ammonium nitrate and/or ammonium thiocyanate and/or urea. Temperatures which are between the boiling point of ammonia and the decomposition temperature of ammonium carbamate are maintained in the treatment.

It is known that urea can be converted into melamine in the presence of ammonia and catalysts at atmospheric or superatmospheric pressure at temperatures of from 320° to 450°C. As a rule about 2 to 5 m.$^3$ (STP) of ammonia has to be used per kilogram of urea. An off-gas is formed according to the equation:

$$6NH_2-CO-NH_2 \rightarrow C_3N_6H_6 + 6NH_3 + 3CO_2.$$

The off-gas contains 1 mol of melamine, 6 mols of ammonia and 3 mols of carbon dioxide as well as the ammonia added. Melamine is condensed from this off-gas by cooling to a temperature of from 150° to 250°C. Unreacted urea which has been entrained from the reaction zone with the off-gas can be condensed out in a further cooling stage, generally by cooling to a temperature of from 120° to 140°C. The off-gas obtained consists mainly of ammonia together with carbon dioxide and small amounts of inert gases; its usability is considerably impaired by the carbon dioxide content.

In principle it is possible to supply this off-gas after it has been reheated to the reactor as fluidizing gas. A high content of carbon dioxide however impairs the conversion of urea to melamine and is also unfavorable from the point of view of energy consumption because of the necessity of cooling and reheating large quantities of gas.

It is known that in order to separate the excess ammonia from this off-gas the latter can be scrubbed with water in a column so that almost all the carbon dioxide and some of the ammonia are absorbed. Ammonia containing water vapor is thus obtained and this has to be dried before being reused for example as fluidizing gas in the melamine reactor. This may be done for example by cooling the gas mixture in the condenser to temperatures as low as –6°C. For this reason the method is very expensive as regards energy consumption.

It is further known from U.S. Patent 2,950,173 that ammonia can be separated from melamine synthesis off-gas by treatment with an anhydrous solvent, for example dimethyl formamide, ethylene glycol or diethylene glycol, so that at the same time a suspension of ammonium carbamate in the solvent is obtained. The temperature of the solvent is kept at from 0° to 20°C. The ammonium carbamate is separated from the solvent and split, by heating at about 100°C., into ammonia and carbon dioxide from which solvent which has been discharged together with the moist carbamate must be removed by condensation. Moreover, cooling brine has to be used for cooling the liquid. This improved process is shown in Figure 8.

311 kg. per hr. of molten urea containing biuret is supplied through line **1** to a fluidized bed reactor **2**. During the same period 603 m.³ (STP) of ammonia which contains 0.5% by volume of carbon dioxide is supplied through line **3**. The ammonia serves to fluidize the catalyst consisting of aluminum dioxide which is situated in the reactor. The urea introduced is converted to the extent of 98% into melamine at a temperature of 380°C. and at atmospheric pressure.

The off-gas containing melamine, which is at about 320°C., is passed through line **4** to a separator **5** where it is cooled by being mixed with 920 m.³ (STP) of ammonia (supplied through line **6**) which is at 40°C. and which contains 0.5% by volume of carbon dioxide and 0.9% by volume of inert gases, and thus brought to a temperature of 210°C. so that 100 kg. per hr. of 99.5% melamine is desublimed.

The off-gas freed to the extent of about 98% from melamine is supplied through line **7** to a column **8** in which it is scrubbed with a urea melt containing biuret at a temperature of from 130° to 140°C. for removal of residual melamine and unreacted urea. The melt is recycled and the heat absorbed is withdrawn in a cooler **9**. Urea can be discharged from the system through line **10**

### FIGURE 8: PROCESS FOR AMMONIA REMOVAL FROM MELAMINE PROCESS OFF-GAS

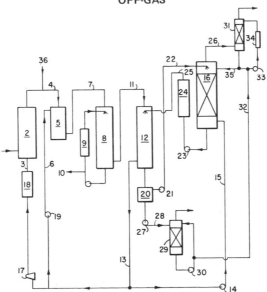

Source: R. Mohr; U.S. Patent 3,555,784; January 19, 1971

The off-gas, which has a temperature of 138°C. and contains 3.6% by volume of carbon dioxide and 0.9% by volume of inert gas (nitrogen) in addition to 95.0% by volume of ammonia, is supplied through line **11** to a carbamate separator **12**, where the off-gas is scrubbed at about atmospheric pressure and a temperature of 40°C. with 47 m.$^3$ (STP) per hour of a melt which is at a temperature of 45°C. and contains 46.0% by weight of ammonium thiocyanate, 30.5% by weight of urea, 18.6% by weight of ammonia and 4.8% by weight of carbon dioxide. In relation to the temperature of 40°C. and the pressure, this melt contains about 1.65% more ammonia than corresponds to the equilibrium.

The carbamate which separates is suspended in the melt and the heat liberated is removed by ammonia evaporating from the melt.

2,518 m.$^3$ (STP) per hour of gaseous ammonia having a residual content of 0.2 to 0.5% by volume of carbon dioxide and 0.4 to 1.0% by volume of inert gas is withdrawn through line **13**. Of this, about 995 m.$^3$ (STP) is compressed by means of a blower **14** to 0.6 to 0.7 atmosphere gauge and supplied through line **15** to an ammonia absorber **16**. 603 m.$^3$ (STP) of ammonia is returned to the melamine reactor through a compressor **17** and a heat exchanger **18**. The remaining 920 m.$^3$ (STP) of ammonia is supplied (as described above) through blower **19** and line **6** to the melamine separator **5**.

The melt containing carbamate is withdrawn at the base of the separator **12** and passes into a settling tank **20**. From this, 46 m.$^3$ per hour of the supernatant and clarified melt is returned continuously by means of a pump **21** through line **22** to the abovementioned ammonia absorber **16** and saturated there at 45°C. and the pressure prevailing therein (which is higher than that prevailing in the carbamate separator **12**) with ammonia supplied through line **15**. The melt treated with ammonia is withdrawn by means of a pump **23** and passed into a cooler **24** where the heat of absorption is withdrawn. The melt saturated with ammonia (which has the abovementioned composition) is supplied through line **25** to the carbamate separator **12**.

A suspension containing 40% by weight of ammonium carbamate is withdrawn at the bottom of the settling tank **20** by means of a pump **27** and passed through line **28** into the carbamate decomposer **29**. The ammonium carbamate is decomposed therein at a temperature of 75°C. and the melt freed from the carbamate is recycled by means of a pump **30**. At the top of the decomposer, 194 m.$^3$ (STP) per hour of off-gas containing 72% by volume of ammonia and 28% by volume of carbon dioxide is removed from the process through an air lock and may be processed for example into fertilizers, for example ammonium sulfate, or may be used as starting material for the synthesis of urea.

The gas (80% by volume of ammonia and 20% by volume of inert gas) removed through an air lock from the top of the ammonia absorber **16** through line **26** in an amount of 20 m.$^3$ (STP) per hour is supplied (to utilize its ammonia content) to an absorber **31** and treated therein at 0.6 to 0.7 atmosphere gauge and 30°C. with a melt of urea and ammonium thiocyanate supplied through line **32** from the decomposer **29** in an amount of 320 kg./hr. The melt is recycled through a pump **33** and a cooler **34** where its heat is withdrawn.

5.5 m.$^3$ (STP) per hour of a gas containing 73% by volume of inert gas and 27% by volume of ammonia is removed at the top of the absorber **31**. The melt laden with ammonia is returned to the process through line **35**. About 35 m.$^3$ (STP) of ammonia is supplied through line **36** to make up for the ammonia lost in the plant, mainly through the off-gas.

A process developed by H.J. Clausen (5) is one whereby ammonia can be recovered from the exit gases of an ammonium polyphosphate plant by scrubbing the reactor exit gases with partially ammoniated superphosphoric acid having a pH above about 3, and recycling the product obtained thereby to the reactor. The essentials of this process are shown in Figure 9.

## FIGURE 9: SCHEME FOR AMMONIA RECOVERY FROM AMMONIUM POLYPHOSPHATE PLANT OFF-GASES

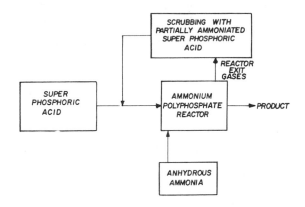

Source: H.J. Clausen; U.S. Patent 3,687,618; August 29, 1972

The process starting material is concentrated superphosphoric acid containing 60 to 85% by weight phosphorus, expressed as $P_2O_5$. The superphosphoric acid can be obtained in any convenient way such as by concentration of wet process acid or from furnace acid. The term superphosphoric acid as used defines a phosphoric acid containing substantial quantities of polyphosphoric and orthophosphoric acid. The polyphosphoric acids include pyrophosphoric acid, and the various polymeric acids varying from triple phosphoric acids to metaphosphoric acid and higher polyphosphates. It is desired that the concentrated superphosphoric acid contain as little orthophosphoric acid as possible in order to maximize the yield of ammonium polyphosphate.

To provide sufficient water so that the heat evolved during the ammoniation process is removed by the heat required for the evaporation of water, the superphosphoric acid can be diluted prior to ammoniation. Superphosphoric acid, however, tends to hydrolyze in water to the ortho form and it is therefore necessary to provide a method of preventing or inhibiting hydrolysis and consequent formation of orthophosphate. To this end, superphosphoric acid is conveniently diluted with water just prior to ammoniation and under conditions of rate and temperature to inhibit hydrolysis of a polyphosphate.

In the preparation of ammonium polyphosphates a suitable range for dilution of the concentrated superphosphoric acid is in the range of from about 49 to 60% (percent $P_2O_5$ by weight) superphosphoric acid. Undiluted superphosphoric acid having a concentration (expressed as percent by weight $P_2O_5$) from about 60 to 80% can be used without dilution if desired. The superphosphoric acid, in convenient concentration, is continually cooled by conventional means to maintain a feed acid temperature of about 135°F.

The diluted superphosphoric acid is passed into the ammonium polyphosphate reactor; desirably, it is continuously sprayed therein.

Anhydrous ammonia is continuously passed into the ammonium polyphosphate reactor. Here, the ammonia gas is continuously and thoroughly mixed with the diluted superphosphoric acid to obtain the ammonium polyphosphate product. In a preferred embodiment, the quantity of water in the diluted superphosphoric acid can be such that it is sufficient to maintain the temperature of the reactor bed at a desired level, for example below about 180°F. while also being vaporized, thereby removing the heat of neutralization so as to yield a satisfactory dry ammonium polyphosphate product. The product is continuously removed while excess ammonia is vented along with the reactor exit gases.

As indicated, the reactor exit gases are scrubbed with partially ammoniated superphosphoric acid. The exit gases are desirably absorbed and scrubbed with the partially ammoniated superphosphoric acid under such conditions as to minimize the hydrolysis of the polyphosphate to orthophosphate. Thus, it has been found that when the partially ammoniated superphosphoric acid has a pH above about 3, and preferably between about 5 and 6, the hydrolysis of the polyphosphate to orthophosphate is significantly reduced.

The scrubbing temperature is significant and is desirably maintained in the range of from about 100° to 175°F. At lower temperatures there is a possibility of precipitating some polyphosphate material while if the temperature is increased significantly above 175°F. there are attendant corrosion problems and the hydrolysis of polyphosphate is increased. In a preferred embodiment, the temperature is maintained in the range of from about 140° to 170°F., with superior results being obtained between 150° and 160°F.

The scrubbing solution can have any convenient concentration, but in a preferred embodiment the solution is diluted. Dilution is required to overcome problems of viscosity, concentrated superphosphoric acid being extremely viscous. In a preferred embodiment the acid is diluted to about 50 to 60% ($P_2O_5$ by weight) to obtain a solution having a viscosity which is convenient to work with.

As indicated, the diluted superphosphoric acid is partially ammoniated. This can be done in any convenient way, the ammoniation advantageously being performed so as to provide a solution containing from about 5 to 15% nitrogen by weight. In a preferred embodiment the acid is ammoniated to provide from about 5 to 10% by weight of nitrogen. Typical solutions will have a composition such as 7-24-0 (N-P-K) and 13-44-0. At higher degrees of ammoniation, excess orthophosphate formation is obtained.

In a preferred embodiment the solution obtained upon scrubbing the reactor exit gases, which solution is characterized by a low orthophosphate content is recycled for reaction in the ammonium polyphosphate reactor.

A process developed by H. Hashimoto, N. Kasahara and T. Yoshino (6) utilizes a deodorization device particularly adapted for use with ammonia process diazo copying machines which includes an odor removing conduit having two odor removing sections, through which ammonia containing exhausts are caused to flow. A filler is provided in the first odor removing section, a deodorizing chemical being applied to the filler, while an ammonia adsorbent is provided in the second odor removing section, the ammonia containing exhausts flowing first through the first section and then through the second section.

As is well known, ammonia gas is used with ammonia process diazo copying machines for developing exposed sheets. All the ammonia gas supplied to the developing section is not used for developing purposes but part of the gas leaks out of the developing section into other sections of the machine and out of the machine. Ammonia gas gives of an offensive smell and has deleterious effects on human beings. If ammonia gas leaks out of the copying machine into the operation chamber, it will be injurious to the health of the operator and markedly reduce operation efficiency.

In order to obviate this problem means are provided in ammonia process diazo copying machines for removing ammonia gas in the machine and forcibly venting the same to atmosphere outside the operation chamber. However, the provision of such venting system has disadvantages in that an additional cost is involved in providing the venting system and an extra space is required for mounting such system, thereby the place for installing the copying machine being strictly restricted.

The process has as its objective the provision of a deodorization device which obviates the aforementioned disadvantages of the prior art deodorization system for ammonia process diazo copying machines and permits to effect deodorization economically and positively without requiring installation of a particular venting system. Figure 10 shows the apparatus employed.

An odor removing conduit 2 is disposed perpendicularly in a casing 1 of the deodorizing device. A fan 3 is connected on its delivery side to the inlet of the odor removing conduit at the lower end thereof and on its suction side 4 to a copying machine through a connecting line. 7 is a motor for operating the fan.

The odor removing conduit consists of two portions or a lower minor dimension portion and an upper major dimension portion. A filler 8 arranged on a support, for example a net 9, is mounted in the lower minor dimension portion, and an adsorbent 11 arranged on a support, for example a net 10, is mounted in the upper major dimension portion. Arranged above the filler in the odor removing conduit is a nozzle 12 which is connected by a line 13 to a chemical supply pump 14 which in turn is connected by a line 15 to a chemical supply tank 16. Disposed at the lower end of the odor removing conduit is a drain port 17 which is connected by a line 18 to a waste chemical discharge pump 19 which in turn is connected by a line 20 to a waste chemical tank 21 disposed adjacent a chemical supply tank 16.

The motor 7 for operating the fan comprises a shaft which is connected, at an end opposite to the end at which the shaft is connected to the fan, to a speed reducing means 22 for operating the chemical supply pump 14 and waste chemical discharge pump 19. Provided on the top of the casing 1 is a cover 23 on which a warning device 24 for measuring the degree of saturation of the adsorbent and giving warning is provided. The warning device is connected, on one hand, to the adsorbent by a line and, on the other hand, to the suction side of the fan by a line. The operation of the deodorization device constructed as aforementioned will be explained.

FIGURE 10: APPARATUS FOR AMMONIA REMOVAL FROM DIAZO COPYING
MACHINE VENT GASES

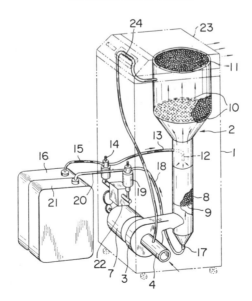

Source: H. Hashimoto, N. Kasahara and T. Yoshino; U.S. Patent 3,679,369; July 25, 1972

Ammonia-containing exhausts are vented from the copying machine **6** and sent through the connecting line **5** to the suction side **4** of the fan **3** which delivers the exhausts to the lower portion of the odor removing conduit **2**. The exhausts move upwardly from the lower portion and pass through the filler **8** which consists of glass beads 3 mm. in diameter, particles of pumice stone 5 to 7 mm. in diameter or other material having a large surface area and high chemical resistance.

The chemical contained in the supply tank **16** is delivered by the supply pump **14** and applied uniformly to the filler through the nozzle **12**. The chemical applied to the filler in this way moves downwardly through the filler in an orderly manner. The chemical used for this purpose may be a solution of one or more substances which readily react with ammonia gas and are highly soluble in water, such as a 5 to 10% aqueous solution of mineral acids, organic acids and the like, such as a 5 to 20% aqueous solution of acetic acid.

Accordingly, the major portion of ammonia gas in the exhausts moving upwardly through the odor removing conduit reacts with the chemical in the filler and has its odor removed, so that the exhausts reaching the adsorbent in the upper portion of the conduit have had their odor substantially removed. The chemical that has reacted with ammonia gas is led through the drain port **17**, line **18**, discharge pump **19** and line **20** to the waste chemical tank **21**.

The exhausts that have had the major portion of ammonia gas content removed by the chemical in the filler are led through the adsorbent where the rest of the ammonia is completely removed from the exhausts. The exhausts that have had their odor completely removed are vented out of the casing of the deodorization device through the sides and backside of the cover **23**. The adsorbent is an acidic or basic deodorant capable of adsorbing odor components. For example, a porous amphoteric ion exchange resin in which acidic and basic group are exchangeable groups, such as one of ion exchangers made by condensing aromatic amines, for example m-phenylenediamine, and phenols with formaldehyde, may be used in pellet, particle or sphere form with a diameter of about 3 to 5 mm. If the adsorbent is placed in a nonsealing bag such as a gauze bag, the replacing of the old adsorbent with a fresh one can be effected with ease merely by opening the cover **23**.

It should be noted that the chemical applied through the nozzle **12** and the ammonia gas contained in the exhausts form countercurrents in the filler. By this arrangement, the portion of chemical that has already reacted with ammonia gas to a certain degree comes into contact with ammonia gas of high concentration in the lower portion of the filler and further reacts with it and then the ammonia gas which has now lower concentration is brought into contact with a fresh supply of chemical in the upper portion of the filler and further reacts with it. Thus, the chemical reacts with the ammonia gas in the exhausts uniformly in the entire area of the filler so that odor can be removed satisfactorily from the exhausts.

### Removal of Ammonia from Water

A process developed by H. Siewers, K. Flasche, A. Stetter and S. Pfeiff (7) involves the destruction of ammonia contained in waters resulting from the operation of coke ovens based on the principle of expelling the gas from the water, which comprises heating the water to be treated, with lean gas at an elevated temperatures of such a degree that the temperature of the water after treatment thereof is at least 60°C., conveying the lean gas charged with ammonia to regenerators of industrial furnaces, and heating the same therein to such a temperature that the ammonia will become dissociated to nitrogen and hydrogen. By this treatment, ammonia and other harmful gases will be completely eliminated from the waters which are then safe for discharge into the sewers.

The removal of ammonia from coke oven gas is necessary in order to prevent the gas pipes from becoming corroded and, furthermore, in order to avoid formation of nitrogen oxides in the smoke upon combustion of coke gas. The condensates formed in the gas cooling units of coke oven plants likewise contain ammonia and other harmful substances; they must, however, not be discharged into the drains, since according to existing regulations wastewaters must be practically free of ammonia, hydrogen sulfide and other harmful products.

As a rule, coke oven plants dispose of the ammonia contained in the gas and in the condensates (called coal water) by converting them into ammonium sulfate. However, that is no longer profitable, so that the plants sustain greater or smaller losses in the production of ammonium sulfate. The production of other ammonium salts or concentrated ammonia is likewise not economical for various reasons, and it is therefore no longer in use.

As a consequence, many attempts have been made over the years to find a process for converting the ammonia from this source in a more profitable manner than by the conventional methods. In the indirect by-product recovery process, both the condensates containing ammonia as well as the wash waters obtained in the ammonia washing process are further treated; in the semidirect recovery process, only the condensates have to be worked up.

Thus, a process is known according to which clouds of ammonia resulting from injecting steam into ammonia-containing waters obtained in coke oven plants in the conventional manner, are passed with combustion air below a gas generator, where ammonia is dissociated into nitrogen and hydrogen within the glowing coke. This process, however, depends on the presence of a gas generator plant, which has to be in permanent operation the year round. There exists the further drawback that a certain amount of nitrogen oxides may be formed.

A further process is known for burning ammonia formed in coke oven plants; in that process, ammonia-containing waters are made to release ammonia in cloud form by a known expelling process, whereupon the ammonia is heated by the combustion of a fuel, the hot mixture being then passed through a zone of dissociation charged with a catalyst, and the hot gases leaving the dissociation zone being then completely burned by addition of more air. The last-mentioned process is dependent on the presence of a catalyst and requires operating above a certain limit temperature which is quite high, in order to avoid formation of nitrogen oxides.

Another known method operates by introducing ammonia clouds liberated from ammonia-containing waters by an expelling process as mentioned above into the heating gas for the coke ovens. This method, however, has not found practical application in coke oven plants because it is quite expensive and leads to considerable operational difficulties in the heating of coke ovens due to the entrained steam. The condensates formed contain ammonia and have to be treated again. Sometimes damage is incurred to the heating equipment and the brick lining of the regenerators, or deposits of condensate may destroy the masonry in the chimney.

High expenses for steam are a disadvantage inherent in all known processes, since they operate by expelling ammonia from the ammonia-containing water. The expenses are further increased by the costs for the continuous operation of the expelling process.

In order to overcome these shortcomings, it has been suggested to remove ammonia and other harmful substances from the waters by treatment with gases. It has, for instance. been tried to inject air or waste gas into the ammonia-containing waters. However, in that manner, ammonia is only transported from the water into the air and that is contrary to existing regulations on air pollution.

Another known process is the injection of ammonia-containing water into smokestacks. There, too, the only effect is the transfer of ammonia from the water into the air, whereby the latter is polluted so that the regulations are violated. In that case, too, the masonry will often be destroyed by the deposit of condensates in the chimney.

It is finally known to introduce ammonia-containing water mixed with cooling water, to which phosphates have been added, into a combined scrubbing and cooling unit for hot generator gas. In that case, however, only part of the ammonia is transmitted from the water into the gas, which gives off again part of the absorbed ammonia in a dry desulfuring process. The balance is finally burned with the generator gas below the ovens. However, the wastewater discharged from the scrubber still contains considerable amounts of ammonia and other harmful ingredients. This can only be considered a coarse cleaning of the ammonia-containing water. In accordance with existing regulations, such water cannot be discharged into the sewers.

It is the object of this process to provide a method for eliminating ammonia from the water accumulating in coke oven plants, which is free of the drawbacks of the known processes. The arrangement of apparatus which may be used for the conduct of this process is shown in Figure 11.

FIGURE 11: PROCESS FOR REMOVAL OF AMMONIA FROM WATER

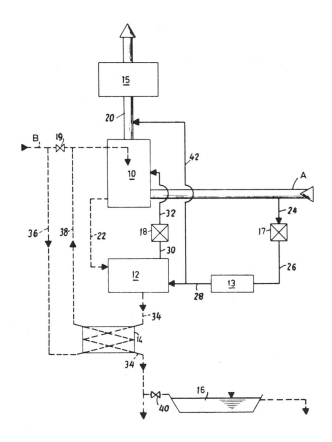

Source: H. Siewers, K. Flasche, A. Stetter and S. Pfeiff; U.S. Patent 3,540,189; November 17, 1970

In the drawing, a scrubber is designated by **10**; a pipe **B** empties into the scrubber for admitting water to be treated over a valve **19**, while it is in open position; a pipe **A** is provided for admitting lean gas to the scrubber (from a source not shown).   A regenerator **15** which may, for instance, form part of a coke oven plant, is connected to the scrubber by a pipe **20**.   Lean ammonia-containing gas passes through pipe **20** into the regenerator, where it is heated to a high temperature at which dissociation of ammonia into nitrogen and hydrogen will occur.   The heated gases escaping from the regenerator may pass, for instance, into the heating system of a coke battery (not shown).

The water is passed from scrubber **10**, where it has been freed from ammonia to a large extent over pipe **22** to a high power washer **12**, where it is treated with lean gas of elevated temperature.   For that purpose, a part of the lean gas arriving through pipe **A** branched off, and is conveyed by pipe **24** over a regulator **17**, and from there by pipe **26** and through a heating device **13**, arriving at the high power washer by way of pipe **28** at a temperature of 400°C., to which it was raised in heater **13**.   The lean ammonia-containing gas leaving the high power washer **12** is fed over pipe **30** to a regulator **18** and from there through pipe **32** to scrubber **10**.   Water discharged from the high power washer is made to pass through a heat exchanger **14** by way of a pipe **34**, through which it leaves the heat exchanger in order to discharge directly into the sewers, with a valve **40** closed; or, when further purification is desired, it will be fed, with valve **40** open, to a wastewater purification pool **16**.

While it is possible to feed water directly to scrubber **10**, from pipe **B** through valve **19** in open position, it is preferable to lead the water with valve **19** closed, over pipe **36** to the heat exchanger **14**, where it is preheated and then passed over pipe **38** to scrubber **10**. This mode of operation is more economical since it utilizes to full extent the heat generated in heater **13**.

In order to avoid corrosion in the gas pipes arranged between the scrubber and regenerator, the lean gas charged with ammonia is heated.   For this purpose, a partial current of heating gas is conveyed to line **20** by way of a line **42** branched off from line **28**.

The removal of ammonia from wastewater by ion exchange has been described by the staff of Battelle-Northwest (8).

**References**

(1)  Miner, S., "Air Pollution Aspects of Ammonia," *Report PB 188,082,* Springfield, Va., National Technical Information Service (September 1969).
(2)  Kapitulskii, E.H., "A Comparison of the Hygiene Characteristics of the Smokeless and Ordinary Methods of Charging Coke Ovens," *Coke Chem. USSR*, No. 8 (1966).
(3)  Lee, D., "Removal of Reactive Light Gases with Impregnated Activated Charcoal," *Fourth Annual Technical Meeting and Exhibit of the American Association for Contamination Control,* Miami Beach, Florida (May 1965).
(4)  Mohr, R., U.S. Patent 3,555,784; January 19, 1971; assigned to Badische Anilin- & Soda Fabrik AG, Germany.
(5)  Clausen, H.J., U.S. Patent 3,687,618; August 29, 1972; assigned to Cities Service Co.
(6)  Hashimoto, H., Kasahara, N. and Yoshino, T.; U.S. Patent 3,679,369; July 25, 1972.
(7)  Siewers, H., Flasche, K., Stetter, A., and Pfeiff, S.; U.S. Patent 3,540,189; November 17, 1970
(8)  Battelle-Northwest, "Wastewater Ammonia Removal by Ion Exchange," *Report PB 209,934,* Springfield, Va., National Technical Information Service (February 1971).

# AMMONIUM PHOSPHATE FERTILIZER PLANT EFFLUENTS

## Removal from Air

A process developed by W.J. Sackett, Sr. (1) embodies a fully implemented control system for preventing the discharge of obnoxious wastes from plants such as those of the ammonium phosphate fertilizer types. The problem that the wastes are of different vapors and dusts in differing combinations is solved by the use of regenerating combinations of dry and wet cyclones and scrubbers taking into account that certain effluents are at elevated temperature. Figure 12 illustrates the arrangement of apparatus which may be employed in the conduct of this process.

In the drawing, reference numeral 10 indicates generally an ammonium-phosphate-potassium processing plant of improved design. This plant includes an ammoniator-granulator system 12 of the type well known in the art and consisting of a rotating ammoniator-granulator 14, a rotating dryer 20 and a rotating cooler 26. The previously blended solid portion of the material formula introduced by delivery chute 16 to the rotating ammoniator-granulator is reactively treated with ammonia, sulfuric and/or phosphoric acids, steam and water. A chute 18 conveys the resulting hot plasticized mass into the dryer. Here lifting flights shower the material thus introduced through the hot bases from the combustion chamber 22, thus drying the material as it progresses to the discharge end.

At the end of the dryer, a chute 24 discharges the material into the cooler where through a showering action it is exposed to an oppositely directed air stream and delivered through chute 28 to classifying screens then to the product storage area of the plant.

It should be understood the foregoing process results in wastes of three different overlapping classes which in the past were openly disposed of or only partially treated. These classes are (a) ammonia, water vapor, acid fumes and finely divided ammonium and potassium chloride particles from the ammoniator-granulator; (b) ammonia and combustion dust products from the dryer, and (c) ammonia and product dust from the cooler. The corresponding retrieval systems are indicated in the drawing by reference numerals 30, 32, and 34 respectively.

System 30 requires that the duct 66 which exhausts the ammoniator-granulator 14 be provided with an exhaust fan 68 to feed the duct 70. System 32 requires that the duct 36 which exhausts the combustion gases from the dryer be provided with an exhaust fan 46 to feed the duct 44. System 34 requires that the duct 52 which exhausts the air from the cooler 26 be provided with an exhaust fan 62 to feed the duct 60. The two systems 32 and 34 which handle dust, as related, each include first a dry cyclone 38 and 54, respectively. Here "fines" are collected for ultimate recycling in the manufacturing process, for example as disclosed in U.S. Patent 3,272,596.

The outlet ducts 40 and 56 from the cyclones 38 and 54, respectively, carry ammonia gas as well as some vapor-borne residual dust to associated wet cyclones 42 and 58. Ammonia gas is taken up as hydroxide by the water sprays therein and the particulate dust is taken up in liquid suspension and drawn off as a hydroxide-charged slurry through vane feeders 116 as will be related. Wet vapors emerging from the wet cyclones 42 and 58 are transferred by ducts 44 and 60. The flow is aided by the aforementioned exhaust fans 46 and 62 in the respective lines leading to ducts 48 and 64. These ducts are downwardly directed at their ends, 48 within a dryer scrubber 50, and 64 within a cooler scrubber 65.

The scrubbers 50 and 65 are identical each consisting of a vertical tank with spray heads 84 and 110. Each further has a demisting bed 86 such as fiberglass filters or the like in the top and a discharge stack 51 thereabove to the atmosphere, for discharging the air from the scrubbers 50 and 65 as clean air after being treated in the systems.

A system 72 comprising an input line 74, a water meter 76, a valve 78, and lines 80 and 82 introduces fresh water to the scrubbers 50 and 65 through their spray head manifolds 84. The water wets the beds 86 above and the surplus collects in a pool 85 at the bottom. These pools 85 are also augmented by the recirculating scrubbing liquids through the return lines 118 of the vane feeders 116 previously mentioned.

The wet vapors from the discharge ducts 48 (and 64) impinge on the surface of the pool and rise within the scrubber. Any remaining dust, water or air suspended is driven forcefully

## FIGURE 12:  AIR POLLUTION CONTROL SYSTEM FOR AMMONIUM PHOSPHATE FERTILIZER PLANT

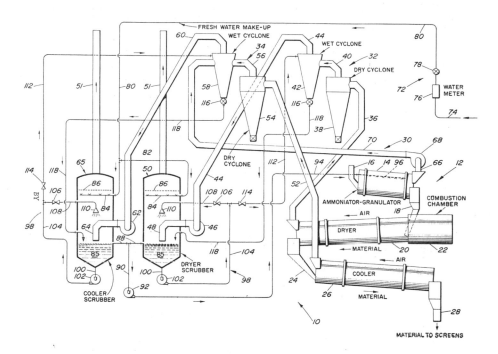

Source:  W.J. Sackett, Sr.; U.S. Patent 3,499,731; March 10, 1970

into the pool **85**.  A pump **102** in the line **100** withdraws the recirculating scrubbing waters from pool **85** recirculating it through a system **98** comprising a line **104**, a valve **106**, and a line **108** to the discharge head **110**.  The liquid spraying therefrom scrubs the rising wet vapors.  A valve **114** leading to branch line **112** taps the recirculating system **98** for liquid supply to the wet cyclones **42** and **58**.

Overflow pipes **88** and a line **90** lead from the pool **85** to a pump **92** which removes excess slurry collected in the scrubbers **50** and **65** and transfers it over a line **94** to the ammoniator-granulator **14** to be introduced therein from a serrated launder **96**.  The chemicals contained in the overflow liquids are in this manner reintroduced into the process. Where the proportion of airborne dust component is low and the dust is mainly vapor wetted particles as in the case of the ammoniator-granulator **14**, the dry cyclone is bypassed, the vapor being carried by duct **70** directly to wet cyclone **58**.

### References

(1)  Sackett, W.J., Sr., U.S. Patent 3,499,731; March 10, 1970.

## AMMONIUM SULFATE

**Removal from Air**

A process developed by W. Dieters (1) involves the removal of solid polar particles, particularly of ammonium salts, from streaming gases in which they are suspended, by making the particles descend while rotating without eddy formation, whereby the particles agglomerate with each other and are then thrust out continuously and collected, while the gases escape in a pure state. A suitable form of apparatus for use in the conduct of this process is shown in Figure 13.

As shown there, an elongated cylindrical vessel is generally designated at **10**; it consists of a narrower upper part **11** and a wider lower part **15**. Centrally arranged in the vessel is a shaft **12**, the shaft being mounted for rotation by drive means **17**. Disposed on the shaft is a number of horizontal discs **13**. The uppermost portion of cylinder **11** is free of such discs, since it serves for receiving gas through an inlet pipe **18** in such a manner that separation of solid particles from the gas is at first avoided. Shaft **12** also carries blades **14** provided for rotating the gas in the upper portion of the cylinder.

The bottom of the vessel designated by **15** is to serve as collecting and settling chamber for the separated particles. It is much wider, about 2 to 2.5 times the diameter, of the upper portion **11**. The gas, substantially free of suspended particles leaves the apparatus by a discharge tube **16**. The tube is fitted at **19** into the chamber **15** to serve as a stopper while the separated solid particles are collecting in the chamber. When removal of the particles and cleaning of chamber **15** becomes desirable, tube **16** may be withdrawn to open a discharge outlet **20**. The following is a specific example of the operation of this process.

A very fine aerosol of ammonium sulfate which is difficult to separate from the gas, is obtained when completely anhydrous ammonium bisulfate, evaporated at 400° to 450°C., is reacted with ammonia in a current of air. An aerosol, as mentioned, which contains per liter about 0.3 g. ammonium sulfate, is passed from inlet **18** through the cylinder **11** at a rate of flow of about 100 l./hr. and a temperature of 50°C. The cylinder which consists of glass has a height of about 300 mm. Shaft **12** with discs **13** is rotated at 5,000 rpm.

**FIGURE 13: AGGLOMERATOR FOR REMOVAL OF AMMONIUM SULFATE PARTICLES FROM GASES**

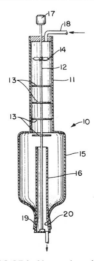

Source: W. Deiters; U.S. Patent 3,410,054; November 12, 1968

In the uppermost portion of the cylinder, the blades **14** effect a stirring of the gas, in order to prevent precipitation of some of the solid particles on the glass wall. The blades are slightly inclined downwardly, exerting a light pressure on the aerosol while it descends. The aerosol then passes through the several zones of agglomeration, where the solid particles are concentrated and finally, upon reaching the settling chamber **15**, are deposited therein, while the pure gas free of particles leaves the apparatus through discharge tube **16**. The tube has a diameter of about 35 mm. and its top opening is spaced at 45 mm. from the inlet to chamber **15**.

The rotating discs have a diameter of 58 mm. and are spaced by 50 mm. from each other. The glass cylinder **11** has an inner diameter of 65 mm. The chamber **15** has a diameter of 140 mm. and a height of about 300 mm. The lowermost of the discs **13** extends 8 mm. into the chamber.

In operation, aerosol rings are formed between the discs, as mentioned before, and the mean grain size of the suspended particles increases downwardly from ring to ring due to advancing agglomeration. When the agglomerates reach a certain size, they are thrust out in spirals and arrive finally in the settling chamber **15** into which the agglomerates from the bottom discs **13** are also thrown. In the case of the ammonium sulfate aerosol here described, four rotating discs **13** are required. In the settling chamber, about 95% by weight of the ammonium sulfate will have collected; the balance is adhering to parts of the apparatus. The gas is completely free of sulfate.

### References

(1) Dieters, W., U.S. Patent 3,410,054; November 12, 1968; assigned to Inventa AG für Forschung und Patentverwertung, Switzerland.

# AMMONIUM SULFIDE

## Recovery from Wastewater

A process developed by K.B. Brown, W.K.T. Gleim and P. Urban (1) is one in which wastewater is treated in such a manner as to convert the sulfide impurities to a form having an oxygen demand which is considerably reduced and in some cases is practically nil.

In various industrial applications, water containing sulfur impurities is collected prior to disposal. For example, in a petroleum refinery, large quantities of water are used in refining operations such as purifying hydrocarbon fractions, steam distillation, heat transfer, diluting corrosive materials, etc. When used as a purifying medium, the water becomes contaminated with the impurities removed from the petroleum. When used otherwise in contact with petroleum, the water will contain at least an equilibrium distribution of the impurities contained in the petroleum. The more abundant of these impurities are hydrogen sulfide and ammonia, although other impurities are present as, for example, aliphatic mercaptans, thiophenols, phenols, etc.

With the increase in the size of refineries and in the number of processing steps in a refining operation, the amount of impurities in the water is increased to an extent that may be harmful to marine life when the wastewater is disposed of in the neighboring streams.

The impurities in wastewater from petroleum refineries include ammonium sulfide, sodium sulfide, potassium sulfide, and in some cases hydrogen sulfide, as well as mercaptans, phenols, etc. Although these impurities comprise a minute portion of a large volume of water, the sulfides, for example, consume oxygen when disposed in neighboring streams and rob aquatic life of necessary oxygen. This method of treating water containing such a sulfur impurity in a concentration of less than 5% by weight of the water, comprises reacting the sulfur impurity with an oxidizing agent in the presence of a phthalocyanic catalyst. The following is one specific example of the operation of this process.

A composite of cobalt phthalocyanine sulfonate on activated carbon was prepared by dissolving cobalt phthalocyanine sulfonate in water to which a trace of ammonium hydroxide (28%) solution was added. Activated carbon granules of 30 to 40 mesh were added to the solution with stirring. The mixture was allowed to stand overnight and then was filtered to separate excess water. The catalyst was then dried and was calculated to contain 1% by weight of the phthalocyanine catalyst.

10 cc of the composite catalyst prepared in the above manner were mixed in a separatory funnel with 100 ml. of water containing 0.0112% by weight of ammonium sulfide. The mixture was shaken at room temperature and analyzed periodically by titration with silver nitrate to determine the disappearance of the sulfide ions. The air contained in the separatory funnel was sufficient for the desired purpose. After 13 minutes of contact in the above manner, the sulfide concentration was reduced to 0.00032% by weight. From the above data, it will be seen that the ammonium sulfide was reduced from 0.0112% by weight to 0.00032% by weight within 13 minutes.

A process developed by K.M. Brown (2) involves treating wastewater containing a sulfur impurity in a concentration of less than 2% by weight of the water, which comprises reacting the sulfur impurity with ascending air in contact with a phthalocyanine sulfonate catalyst during descent of the water in a cooling tower. The following is one specific example of the operation of this process.

In a petroleum refinery processing approximately 20,000 barrels per day of crude oil, there will be approximately 20,000 gallons per minute of wastewater to be cooled. Approximately two parts per million of cobalt phthalocyanine disulfonate, based upon the water to be cooled, is commingled with the wastewater at a temperature of about 120°F. and then passed downwardly through a conventional cooling tower, in countercurrent contact with ascending air. This treatment serves to cool the water to a temperature of about 90°F. and to convert sulfur impurities contained in the wastewater.

A process developed by R.J.J. Hamblin (3) involves the production of elemental sulfur and ammonia by the oxidation of wastewater containing an ammonium sulfide salt in the presence of an oxidation catalyst, such as solid phthalocyanine catalyst, followed by decomposition of the resultant ammonium polysulfide, thereby forming aqueous ammonium thiosulfate. The continuous recycle to the oxidation step of an aqueous stream containing $(NH_4)_2S_2O_3$ is carried out in order to suppress side reactions leading to this product in the oxidation step. Figure 14 is a flow diagram showing the essentials of the process.

Referring now to the drawing, a water stream containing about 5 weight percent sulfur as ammonium hydrosulfide enters the process through line **1** and is commingled with an air stream at the junction of line **2** with line **1**. The resulting mixture is heated in a suitable heating means, not shown, to a temperature of about 140°F. and passed into treatment zone **3**. The amount of oxygen contained in the air stream entering the process through line **2** is sufficient to react about 0.4 mol of oxygen per mol of sulfide charged to treatment zone **3**. This includes not only the sulfide present in the water stream entering the treatment via line **1**, but also the sulfide that is contained in the aqueous recycle stream entering the treatment zone through lines **18** and **19**.

The total amount of ammonium hydrosulfide contained in the water stream plus the recycle stream is equivalent to about 6 weight percent sulfur as ammonium hydrosulfide on a combined stream basis. Moreover, the aqueous recycle stream, entering treatment zone **3** via line **18** and **19**, is at a temperature of about 120°F. and is injected at two separate injection points spaced along the axis of flow of the water stream through treatment zone **3**, thereby providing quench streams for the exothermic reaction taking place within treatment zone **3**. Despite the fact that only two injection points are shown in the drawing, in many cases it is advantageous to use a plurality of injection points spaced along this axis by methods well known to those skilled in the art to carefully control the temperature

FIGURE 14: PROCESS FLOW DIAGRAM FOR AMMONIUM SULFIDE REMOVAL
FROM WASTEWATER

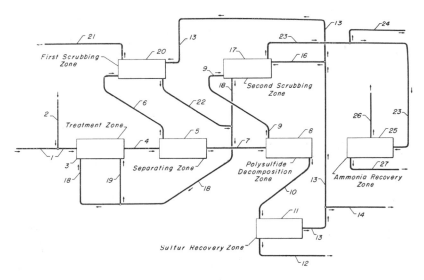

Source: R.J.J. Hamblin; U.S. Patent 3,627,465; December 14, 1971

rise observed across treatment zone 3; or the recycle stream, in whole or part, may be commingled with air and fed to the front of the treatment zone, in which case, the fresh feed may
be fed at some intermediate point in the treatment zone. Also, the aqueous recycle stream
injected via lines 18 and 19 contains ammonium thiosulfate which serves to control thiosulfate
production within treatment zone 3.

The amount of ammonium thiosulfate entering the treatment zone via this recycle stream
is equivalent to about 0.1 to 5 weight percent or more sulfur on a combined stream basis;
the exact value within this range being dependent upon the exact conditions utilized within
the treatment zone coupled with the amount of a drag stream withdrawn via line 14 from
this aqueous recycle stream, as will be explained below. The aqueous recycle stream will
also contain $NH_4OH$ resulting from the operation of scrubbing zone 17 as hereinafter explained.

The treatment zone contains a fixed bed of a solid catalyst comprising cobalt phthalocyanine monosulfonate combined with an activated carbon carrier material in an amount such
that the catalyst contains 1.0 weight percent of the phthalocyanine component. The activated carbon granules used as the carrier material are in a size of 10 to 30 mesh. The mixture of the fresh feed stream, the aqueous recycle stream, and the gas stream flow through
the bed of catalyst in downflow fashion. The conditions utilized in the zone 3 are: a
temperature of about 140°F. at the inlet to this zone, an outlet temperature of about
185°F., a pressure of about 5 psig and a liquid hourly space velocity based on the total
volume of the combined space velocity based on the total volume of the combined water
stream feed and aqueous recycle stream of about 2.0 hr.$^{-1}$.

Following the oxidation step, an effluent stream is withdrawn from zone 3 via line 4 and
passed to separating zone 5. This effluent stream contains ammonium polysulfide, $NH_4OH$,
$(NH_4)_2S_2O_3$, and $H_2O$, $N_2$ and unreacted $NH_4HS$. In a separating zone 5 a gas stream comprising nitrogen, $H_2O'$ $NH_3$, and $H_2S$, is separated from a liquid water stream containing
ammonium polysulfide, $NH_4HS$, $NH_4OH$, and $(NH_4)_2S_2O_3$. The gas stream leaves separating
zone 5 via line 6, and the liquid stream leaves via line 7.

The liquid stream from separating zone 5 is charged via line 7 to polysulfide decomposition zone 8. In this case polysulfide decomposition zone 8 is a stripping column containing suitable gas liquid contacting means such as fractionating plates, baffles, or other suitable contacting means. Heat is supplied to the bottom of this stripper column by means such as a steam coil or reboiler near the bottom of the tower. In zone 8, the liquid stream is heated to a temperature of about 280°F. at a pressure of about 40 psig which is sufficient to produce an overhead vapor stream containing $NH_3$, $H_2S$, and $H_2O$ which is withdrawn via line 9, and a bottom stream containing liquid sulfur dispersed in an aqueous solution of ammonium thiosulfate which is withdrawn from zone 8 via line 10. Essentially all of the $NH_3$ contained in the input stream to zone 8 is recovered in the overhead vapor stream.

The bottom stream from the polysulfide decomposition zone is withdrawn via line 10 at a rate which is sufficient to prevent the liquid sulfur from collecting at the bottom of zone 8. The resulting stream is passed to sulfur recovery zone 11, which in this case is a settling zone wherein the liquid sulfur separates from an aqueous phase. In many cases, the flow parameter within zone 8 can be adjusted such that the separation of the liquid sulfur can occur in the bottom regions of this zone, if desired; however, here the separation is performed in a separate settling zone. The liquid sulfur that separates in sulfur recovery zone 11 is withdrawn from the system via line 12. The aqueous phase containing ammonium thiosulfate is withdrawn from the system via line 14 in order to remove at least a portion of the net water charged to the system, as previously explained, if the amount of thiosulfate salt contained therein can be tolerated in the treated water product stream from this process.

In addition, the amount of water stream withdrawn via line 14 can serve to regulate the amount of ammonium thiosulfate recycled to the treatment zone 3, if desired. In another mode of operation, the aqueous stream withdrawn via line 14 can be subjected to a suitable evaporation step to produce a substantially pure water stream as overhead and an enriched ammonium thiosulfate bottom stream which can then be recycled to treatment zone 3, if desired, or evaporated to dryness to produce ammonium thiosulfate crystals which then can be utilized in any manner known to those skilled in the art.

The remaining portion of the aqueous stream withdrawn from the sulfur recovery zone 11 is passed via line 13 to the junction of line 16 with line 13 where it, in turn, is divided into two additional portions, the first portion continuing on via line 13 to first scrubbing zone 20 and the second portion being charged via line 16 to second scrubbing zone 17. In zone 20, a portion of the aqueous stream withdrawn from zone 11 is countercurrently contacted with the gas stream withdrawn from zone 5 via line 6 in a suitable scrubbing column containing contacting means. Zone 20 is preferably operated at a relatively low temperature and a pressure corresponding to the pressure in the separating zone 5.

Normally, intimate contact between the gas stream and the liquid stream is effected in a vertically positioned tower at a gas-to-liquid loading which is sufficient to produce a nitrogen-rich gas stream which exits from the tower near the top thereof via line 21, and an aqueous effluent stream containing ammonium thiosulfate and ammonium hydrosulfide which is withdrawn near the bottom of the tower via line 22. The aqueous stream withdrawn via line 22 contains substantially all of the hydrogen sulfide and ammonia which was flashed off in separating zone 5.

In the second scrubbing zone, zone 17, another portion of the aqueous stream recovered from zone 11 is countercurrently contacted with the overhead vapor stream from polysulfide decomposition zone 8 which is charged to the lower region of zone 17 via line 9. Zone 17 is operated at a pressure of about 35 psig, and a temperature of about 270°F., and a liquid-to-gas loading sufficient to produce an overhead stream containing $NH_3$ and $H_2O$ which is substantially free of sulfide, and an aqueous bottom stream containing $(NH_4)_2S_2O_3$, $NH_4OH$, and $NH_4HS$. The overhead vapor stream is withdrawn from zone 17 via line 23, condensed to form an ammoniacal aqueous stream, and the resulting liquid stream passed to ammonia recovery zone 25. If desired, an ammoniacal aqueous product stream can be withdrawn from the system via line 24. Ammonia recovery zone 25 is a stripping zone wherein an ammonia-rich stream is stripped from an aqueous stream to produce an ammonia concentrate which is recovered via line **26** and a substantially sulfide-free and ammonia-free treated water product stream which is recovered via line **27**. The conditions utilized within ammonia recovery zone **25** are well known to those skilled in the art and will not be repeated here.

Returning to the bottom streams from zone **20** and zone **17**, these streams are combined at the junction of line **22** with line **18** and the resulting mixture cooled by a suitable cooling means not shown. The resulting cooled mixture is passed via line **18** and line **19** back to treatment zone **3** as previously explained. The purpose of the two scrubbing operations is to recapture the unreacted sulfide contained in the effluent from the oxidation step in order to prevent pollution problems that could be caused by the disposal of this unreacted sulfide, and to increase the yield of elemental sulfur recovered via line **12**.

Accordingly, the principal effect of the aqueous recycle stream being charged to treatment zone **3** is to control the production of thiosulfate in the system, and, the process is operated in the manner indicated for a substantial period of time, and it is determined that this procedure is an effective and economical way to solve the problem of production of ammonium thiosulfate as a side product in the oxidation step.

Another process developed by R.J.J. Hamblin (4) relates to the treatment of a water stream containing ammonium sulfide salts. The key feature in this particular process variation is the use of a scrubbing step on the vent gases from the treating step operated at a relatively low pressure and a relatively high liquid gas loading, coupled with a combination scrubbing and fractionating step on the bottom stream from the scrubbing step and on the vapor stream from the polysulfide decomposition step. This combination scrubbing and fractionating step is operated at a relatively high pressure, thereby increasing the amount of $H_2S$ that can be absorbed. The principal advantages of the resulting process are significant improvements in the amount of ammonia and sulfur recovered and a substantial simplification of the overall process.

A process developed by P. Urban and R.H. Rosenwald (5) also relates to the treatment of aqueous waste streams containing $NH_4HS$. The key feature of this particular treatment method is the use of a carbon monoxide reduction step to enable the continuous recycle of the treated water stream back to the hydrocarbon conversion process with consequential abatement of water pollution problems and substantial reduction of requirements for makeup water.

A process developed by K.M. Brown (6) is still another variation of the basic Universal Oil Products Company process for treatment of an aqueous stream containing water-soluble inorganic sulfide compounds. The key features of this process variation involve the use of a first catalytic oxidation step which is run at relatively low temperatures and pressures to produce polysulfide, coupled with a second catalytic oxidation step which is run at relatively high temperatures and pressures to selectively oxidize the polysulfide to elemental sulfur, thereby preventing the deposition of elemental sulfur on the catalyst used during these oxidation steps while simultaneously minimizing the amount of oxygen which must be supplied at the relatively high pressure.

**References**

(1) Brown, K.M., Gleim, W.K.T. and Urban, P., U.S. Patent 3,029,201; April 10, 1962; assigned to Universal Oil Products Company.
(2) Brown, K.M., U.S. Patent 3,029,202; April 10, 1962; assigned to Universal Oil Products Company.
(3) Hamblin, R.J.J., U.S. Patent 3,627,465; December 14, 1971; assigned to Universal Oil Products Company.
(4) Hamblin, R.J.J., U.S. Patent 3,634,037; January 11, 1972; assigned to Universal Oil Products Company.
(5) Urban, P. and Rosenwald, R.H., U.S. Patent 3,672,835; June 27, 1972; assigned to Universal Oil Products Company.
(6) Brown, K.M., U.S. Patent 3,672,836; June 27, 1972; assigned to Universal Oil Products Company.

## AMMONIUM SULFITE

**Removal from Water**

A process developed by P. Urban (1) is one in which a water stream containing a water-soluble sulfite compound is treated in order to reduce its total sulfur content while minimizing the formation of sulfate by-products by the steps of: (a) converting the sulfite compound contained in the water stream to the corresponding thiosulfate compounds; (b) reacting the resulting thiosulfate compound with carbon monoxide at reduction conditions selected to produce the corresponding sulfide compound; and thereafter (c) stripping hydrogen sulfide from the effluent stream from step (b) to form a substantially sulfate-free treated water stream which is substantially reduced in total sulfur content relative to the input water stream.

Principal utility of this treatment procedure is associated with the regeneration of a sulfite-containing absorbent stream which is commonly produced by contacting a flue gas stream containing sulfur dioxide with a suitable aqueous absorbent stream containing an alkaline reagent. The treated water stream produced by this method can then be reused in the absorption process or discharged into a suitable sewer without causing pollution problems. Key features of this method involve the selective conversion of the sulfite compound to the corresponding thiosulfate compound, the subsequent reduction of the thiosulfate compound to the corresponding sulfide compound in a highly efficient, economic and selective manner, and the minimization of undesired sulfate by-products during both of these conversion steps. Figure 15 is a flow diagram of this process.

Referring now to the drawing, an aqueous waste stream containing ammonium sulfite in an amount of about 10 weight percent thereof is continuously introduced into the system via line 1 and passed into the upper region of the first reaction zone, zone 2. Zone 2 is a conventional liquid-gas contacting zone designed to effect intimate contact between a downflowing liquid stream and an upflowing gas stream.

**FIGURE 15: FLOW SCHEME OF PROCESS FOR AMMONIUM SULFITE REMOVAL FROM WASTEWATER**

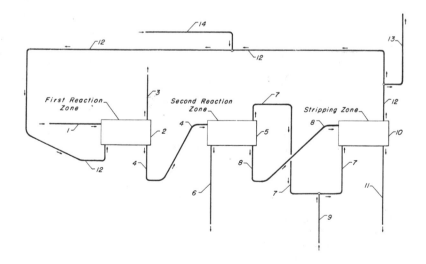

Source: P. Urban; U.S. Patent 3,574,097; April 6, 1971

Also introduced into zone 2 via line 12 is a gas stream containing hydrogen sulfide. During startup of zone 2 sufficient $H_2S$ is introduced therein via line 14 and 12 to initiate the desired reduction reaction. Thereafter, a portion of a hydrogen sulfide-containing gas stream which is produced in a subsequently described stripping step is passed from zone 10 to zone 2 via line 12. In either case, the amount of hydrogen sulfide supplied to zone 2 is sufficient to react about 0.5 mol of hydrogen sulfide per mol of ammonium sulfite charged to this zone. By conventional means, zone 2 is maintained at a temperature of about 100°C., and a pressure of about 200 psig, the residence time of the reactants in zone 2 is about 0.5 hour. In addition, the pH of the input water stream entering zone 2 via line 1 is maintained in the range of about 4 to about 7.

An overhead gaseous stream containing unreacted hydrogen sulfide, carbon dioxide and carbon monoxide is then withdrawn from zone 2 via line 3 and vented from the system. Additionally, an aqueous effluent stream is withdrawn from zone 2 via line 4 and charged to the second reaction zone, zone 5. This aqueous effluent stream contains ammonium thiosulfate in an amount corresponding to a conversion in zone 2 of greater than 90% of the input ammonium sulfite to ammonium thiosulfate. Furthermore, the amount of undesired ammonium sulfate contained in this stream is less than 3% of the input ammonium sulfite. Accordingly, the aqueous effluent stream from zone 2 principally contains ammonium thiosulfate, with minor amounts of unreacted ammonium sulfite and ammonium hydroxide and with only a trace amount of ammonium sulfate.

The second reaction zone, zone 5, is another liquid-gas reaction zone designed to effect intimate contact between an ascending gas stream and a descending liquid stream. The aqueous effluent stream from zone 2 is introduced into the upper region of zone 5. Likewise, a carbon monoxide stream is introduced into the lower region of zone 5 by means of line 6. Zone 5 contains a catalyst comprising 10 to 12 mesh particles of activated carbon having a cobalt sulfide component combined therewith in an amount of about 5 weight percent, calculated as cobalt sulfide. The amount of carbon monoxide introduced thereto via line 6 corresponds to a carbon monoxide to ammonium thiosulfate mol ratio of 5.46:1. The reduction conditions maintained in zone 5 are a temperature of 200°C., a pressure of 500 psig, and a liquid hourly space velocity of 1 hr.$^{-1}$.

An aqueous effluent stream is then withdrawn from the lower region of zone 5 via line 8 and passed to stripping zone 10. Likewise, an overhead gaseous stream is withdrawn from the upper region of zone 5 via line 7 and passed to the lower region of stripping zone 10. An analysis of the stream flowing throughout line 8 indicates that 99% of the ammonium thiosulfate charged to zone 5 is converted to ammonium hydrosulfide. An analysis of the overhead stream withdrawn from zone 5 via line 7 indicates that it contains relatively large amounts of unreacted carbon monoxide and carbon dioxide with minor amounts of hydrogen sulfide, ammonia and water. At the junction of line 9 with line 7 additional quantities of $CO_2$ may be added in some cases to this gas stream in order to increase the efficiency of the stripping operation conditions within zone 10. In most cases the amount of $CO_2$ contained in the overhead stream from zone 5 is sufficient for the stripping step, and the addition of $CO_2$ via line 9 is not necessary.

In stripping zone 10, the aqueous effluent stream from zone 5 is countercurrently contacted with an ascending gaseous stream which essentially comprises the overhead gaseous stream from zone 5. Zone 10 is operated at a relatively low temperature and pressure as compared to zone 5. In fact excellent results are obtained at a temperature of about 70°C., and at atmospheric pressure. Once again zone 10 typically contains suitable means for effecting intimate contact between a descending water stream and an ascending gaseous stream.

An overhead gaseous stream is then withdrawn from zone 10 via line 12. It contains relatively large amounts of hydrogen sulfide, unreacted carbon monoxide, carbon dioxide and small amounts of ammonia and water. A portion of this gaseous stream is withdrawn from the system via line 13. It contains the net sulfide product of this method, and it can be charged to any suitable process for the recovery of sulfur therefrom; for example, an indirect oxidation procedure like a conventional Claus unit. Another portion of this over-head gas stream is passed via line **12** to zone **2** in order to supply hydrogen sulfide reactant thereto. A treated water stream is also withdrawn from zone **10** via line **11**. It contains substantial amounts of ammonium carbonate with minor amounts of unreacted ammonium thiosulfate, ammonium sulfite, and ammonium hydrosulfide. The total sulfur content of this treated water stream is substantially less than 10% of the total sulfur content of the input water stream which enters the system via line **1**. Moreover, this treated water stream contains only a trace amount of undesired ammonium sulfate. An overall sulfur balance on this system indicated that approximately 85% of the sulfur entering the process in the form of ammonium sulfite via line **1** is converted to hydrogen sulfide, while only less than about 3% of the entering sulfur is converted to undesired ammonium sulfate. These results evidence the ability of the process to regenerate sulfite-containing water streams without producing substantial amounts of undesired ammonium sulfate by-product.

**References**

(1)  Urban, P.; U.S. Patent 3,574,097; April 6, 1971; assigned to Universal Oil Products Company.

## AROMATIC ACIDS AND ANHYDRIDES

### Removal from Air

A process developed by D.C. Ferrari and C.G. Bertram (1) is one in which chemical efflu-ent waste gases from chemical plants, particularly effluent waste gases from phthalic an-hydride and maleic anhydride plants, are effectively water washed of residual organic mat-ter (98 to 99% removal) in a wet scrubber using recycled water to concentrate the organic pollutants in the scrubber liquor. A concentrated liquid purge (blowdown) from the scrubber recycle circulating loop is directed to a thermal incinerator where the purge is vaporized and the organic pollutants are oxidized to nonpollutant products.

In the production of phthalic anhydride (hereinafter referred to as PA), switch condensers (these are condensers for condensing the PA as a solid out of the hot reaction product gas formed by the vapor phase catalytic oxidation of naphthalene or o-xylene with air, the re-action product gas stream being alternately switched from one bank of solid condensers to another to permit the solidified PA to be melted out of the off-stream condensers), have been classically treated to remove organic pollutants therefrom by washing with large quan-tities of once-through water in an appropriate scrubber prior to discharging to the atmo-sphere via a vent stack. Typical removal efficiencies of organic pollutants range upwards of 95% with once-through water scrubbers. However, with the use of once-through water at the scrubber in this way, a considerable water pollution problem has always existed due to the large quantity of polluted water which must be disposed of.

In recent years, in order to overcome the water pollution problem, the switch condenser off-gases have been thermally incinerated at 1200 to 1350°F. without the use of a scrubber. This method involves handling an enormous quantity of gas and without heat recovery, requires approximately 12,500 Btu/lb. PA produced of additional fuel. Heat recovery, via this method, can reduce the fuel requirements to 2,000 to 4,000 Btu/lb. PA; however, at a considerable increase in investment since heat is exchanged between the incoming off-gases and combusted incinerator effluent gas. Furthermore, if there is a heat recovery above 40%, then preheating the off-gases to high temperatures (850° to 1000°F.) cannot be avoided.

This must be carefully done to avoid preignition of the combustible content of the off-gases. With the use of conventional multiple solid PA switch condensers operating in cycles, the carryover of solid PA dust, which is combustible, has consistently been a problem, especi-ally during startups and shutdowns. It is this intermittent carryover of combustible material, which can preignite (explode) in a preheat section of a thermal gas incineration system, which labels this system as a dangerous safety hazard.

The problem of economically and safely removing organic pollutants from chemical waste gases to avoid air pollution and without creating a water pollution problem is found in many chemical plants other than phthalic anhydride plants.

This process eliminates both the water pollution problem of the once-through scrubber and the explosion hazard of the thermal gas incinerator while achieving between 98 and 99% removal of the organic pollutants. In addition, the economics, based on both fuel consumption and additional investment, are markedly better.

This is achieved by scrubbing or washing the effluent waste gases in a wet, preferably countercurrently operated, scrubber with an aqueous scrubbing liquor, which is recycled to build up the concentration of the organic pollutants in the liquor so that the scrubbing liquor comprises water with a relatively high concentration of the organic pollutants. A concentrated liquid purge or blowdown is removed from the recycle circulating loop of the concentrated scrubbing liquor and is passed to an incinerator where it is vaporized and the organic pollutants oxidized to harmless nonpollutant compounds, such as $CO_2$ and water. Figure 16 is a flow diagram of this process.

With reference to the drawing, **10** represents a single shell two-stage, liquid-gas, slurry-handling scrubber and **11** represents a conventional aqueous incinerator for incinerating liquids, such as an incinerator sold under the trade name Thermal Oxidizer. Switch condenser off-gas from the solid condensers of a phthalic anhydride plant containing organic pollutants, i.e., phthalic anhydride, maleic anhydride, benzoic acid and naphthalquinone, and at a temperature of 150° to 200°F. enters the scrubber **10** through line **30** and passes upwardly through the first scrubbing stage **12**, where it is contacted with a countercurrent, downwardly flowing scrubbing liquor, comprising an aqueous slurry of the organic pollutants removed from the off-gas during preceding cycles.

The aqueous scrubbing liquor is introduced into the top of the first scrubbing stage **12** through line **13** and distribution manifold **13b**. Essentially all of the phthalic anhydride, benzoic acid and naphthaquinone, along with approximately 80% of the maleic anhydride, are removed from the off-gases by the scrubbing liquor in this way in the first scrubber stage **12**. The organic anhydrides absorbed from the gas stream are hydrolyzed to the corresponding acids (one mol water per mol of organic anhydride), i.e., phthalic acid and maleic acid, in the scrubbing liquor by the water in the scrubbing liquor. It is also in the first scrubber stage **12** where the temperature of the off-gas is reduced from 150° to 200°F. to approximately 100°F. as evaporation (of the water) cooling takes place and the gas stream becomes saturated with the evaporated water.

The organic pollutants are removed from the off-gas by absorption into the scrubbing liquor and also by precipitation thereof when the liquor becomes saturated therewith since they exist in a solid state at the scrubbing temperature. Accordingly, an aqueous slurry of solid organic pollutants in a saturated solution thereof in water is formed. The concentrated slurry at the bottom of first stage **12** is continuously recycled through line **13** to the upper portion of the first stage **12** by pump **13a**. The scrubbed gas stream from the first stage then proceeds upwardly through the second scrubbing stage **15**, then through the conventional mist eliminator **16** and then out of the stack **17** to the atmosphere at **35**.

In the second stage **15**, the gas is scrubbed with a countercurrently and downwardly flowing scrubbing liquor, comprising a dilute aqueous solution of organic pollutants (chiefly maleic acid) which is introduced into the top part of the upper stage **15** from line **14** and spray nozzles **14b** and which is isolated from the first stage **12** and the first stage scrubbing liquor by a conical shaped deflector plate **18** and collection tray **19** in the intermediate portion of the scrubber. The collected liquor on tray **19** supplies hold up through the line **14** for the second stage liquor recycle pump **14a** which pumps it to the upper end portion of the second stage **15** through line **14**. This recycle liquor **14**, together with fresh makeup

## FIGURE 16:  APPARATUS FOR TREATMENT OF PHTHALIC ANHYDRIDE PLANT WASTE GASES

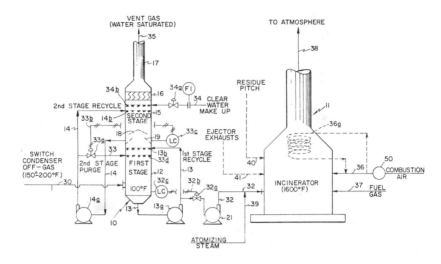

Source:  D.C. Ferrari and C.G. Bertram; U.S. Patent 3,624,984; December 7, 1971

water **34**, which is introduced into the top of the second stage **15** at a point below the mist eliminator **16** and above the spray nozzles **14b** and through the line **34**, valve **34a** and spray nozzles **34b**, form the makeup scrubbing liquor for the second stage scrubber.

The scrubbed off-gas from the first scrubber stage **12** flows upwardly past the deflector **18** and past the second scrubber stage hold up liquor on tray **19** into the bottom of the second stage scrubber **15**.

A portion of the recycle liquor **14** is continuously purged from the second stage recycle circuit **14** through the second stage purge or blowdown line **33** to control the concentration of organic pollutants in the second stage scrubbing liquor, the purge liquor **33** being flowed to the upper end portion of the first scrubber stage **12** through line **33** and spray nozzles **33d** to combine with the first stage recycle liquor **13** to make up the first stage scrubbing liquor.  The rate of second stage purge **33** removed from the recycle line **14** is controlled by a level controller device **33a**, **33b** and **33c**. It is in the second stage where the remainder of the maleic acid is removed from the gas stream, thereby providing an exit gas stream **35** essentially free of organic pollutants.

A controlled portion of the first stage concentrated recycle scrubbing liquor **13** is continuously purged from the recycle circuit **13** through first stage purge line **32** to control the concentration of organic pollutants in the scrubbing liquor of the first stage.

The rate of makeup water introduced at **34b** is adjusted by valve **34a** to replace water removed from the scrubber system (a) in the saturated scrubbed gas **35** (this is the water which is evaporated in the scrubber and saturates the gas stream) by evaporation and (b) in the purge line **32**, these being the only exits from the system.

Each of the scrubber stages contains a fluid bed of packing on a supporting grid or grids to achieve intimate liquid-gas contact.  The packing is made up of relatively large but light smooth-surfaced pieces such as spheres or rings or cylinders, e.g., ping-pong balls or Pall Rings, which are large in bulk but very light in weight so that the bed can be easily fluidized but yet does not hold up the solid organic pollutant particles in the slurry. This type of packing and scrubber are conventional. One such scrubber construction using ping-pong balls is sold under the trade name Turbulent Contact Absorber.

Purge stream 32 from the scrubber first stage 12 is fed from the recycle line 13 through line 32 to the liquid incinerator 11 via pressure booster pump 21. The rate of flow of purge 32, which controls the concentration of organic pollutants in the first stage scrubbing liquor is controlled by a level controller device 32a, 32b, 32c.

Atomizing steam 39 is introduced into the liquid purge stream 32 to finely divide it immediately before it enters the combustion area of the incinerator 11. Preferably the atomizing guns discharge the atomized liquid directly into the combustion zone. It is preferred to atomize the purge immediately before incineration to eliminate the danger of converting maleic acid to the more insoluble fumaric acid before burning occurs. The purge can be atomized with pressurized air if desired.

Combustion air (through air pump 50) and fuel gas, e.g., methane, are introduced into the incinerator at 36 and 37, respectively, in conventional manner, to provide a temperature of 1400° to 1600°F. The rate of flow of combustion air is in excess of the stoichiometric amounts required to completely burn the fuel gas to $CO_2$ and $H_2O$ and to completely oxidize the organic pollutants in the purge stream to $CO_2$ and water. A part of the air can also be introduced with the atomizing steam.

Additional organic pollutant-containing liquid waste streams from the PA plant, including the residue pitch 40 from the purification area or areas of the PA plant and PA distillation column ejector exhausts 41, may also be fed to the incinerator, where, at 1400° to 1600°F., the organic pollutants therein are 99.9% oxidized to carbon dioxide and water.

If desired, the combustion air 36 can be preheated by the hot combustion gases by leading it through heat exchange coil 36a located at the top of the incinerator, as shown in broken lines, before passing it into the incinerator at 36, to thereby provide recovery of heat. Also, the coil 36a can be used to generate steam for export to increase heat recovery. Preheating of the aqueous scrubber or blowdown stream 32 is inadvisable since the maleic acid may be converted to fumaric acid.

## Removal from Water

A process has been developed by E.L. Cole and H.V. Hess (2) for treating wastewater containing very stable and resistant contaminants such as maleic acid, fumaric acid, phthalic acid, terephthalic acid and the like. Currently, these are handled by bacterial oxidation in ponds but this requires considerable land to hold the waste stream and is slow.

In this process, a waste feed stream containing essentially water-soluble organic wastes is continuously subjected to noncatalytic air oxidation at a temperature in the range of 400° to 700°F. at a pressure within the range of 300 to 3,100 psi under turbulent conditions for a contact time ranging from 0.1 minute to 2 hours whereby substantially all the organic wastes are oxidized to carbon dioxide and water.

### References

(1) Ferrari, D.C. and Bertram, C.G.; U.S. Patent 3,624,984; December 7, 1971; assigned to The Badger Company, Inc.
(2) Cole, E.L. and Hess, H.V.; U.S. Patent 3,642,620; February 15, 1972; assigned to Texaco, Inc.

## ARSENIC

Arsenic is toxic to some degree in most chemical forms. Arsenical compounds may be ingested, inhaled, or absorbed through the skin. Industrial exposure to arsenic has shown that it can produce dermatitis, mild bronchitis, and other upper respiratory tract irritations including perforation of the nasal septum. However, because of the irritant qualities of arsenic, it is doubtful that one could inhale sufficient amounts to produce systemic poisoning.

Skin cancer can result from prolonged therapeutic administration of arsenic. Similar cancers have not been observed among industrial workers. Moreover, lung tumors which resulted from inhaling mixed industrial dusts were often thought to be the result of inhaling arsenic. Recently, this relationship has been questioned because animal experiments have failed to demonstrate that arsenic is a carcinogen. Therefore, the causal relationship between cancer and arsenic is disputed. Arsenic is poisonous to both animals and plants, but no damage to materials was found.

Air pollution caused by arsenical compounds has been observed near gold and copper smelters as well as in the areas where arsenic is used for agricultural purposes. A small amount of arsenic can be measured in the air in most cities, according to Sullivan (1).

Arsenic is a common industrial nuisance wherever arsenical ores are smeltered. Before the advent of organic insecticides (e.g., DDT) the use of arsenicals as pesticides was increasing. However, since then, their use has leveled off and declined as the organic pesticides have taken their place. As a result, the supply of arsenic is greater than the demand, and the only economic incentive to remove arsenic from the exhaust fumes of smelters is the presence of other trace metals, such as tellurium, selenium, tin, zinc, and antimony.

Two air pollution episodes in the United States have shown that there is an arsenical air pollution potential at every smelter which refines arsenical ores.

Arsenical compounds are used as insecticides and herbicides. Although the use of arsenical pesticides declined sharply after the appearance of DDT and 2,4-D, arsenical compounds are still used as desiccants, herbicides, and sterilants. Some undetermined amounts of air pollution take place during spraying and dusting operations with arsenical pesticides. Pollution from cotton gins and cotton trash burning has been cited as an important source of agricultural pollution. While the emission rates from cotton trash burning have not been determined, as much as 1,258,000 $\mu$g./m.$^3$ of exhaust air (580,000 $\mu$g./min.) may be emitted during the ginning operation. This produced concentrations of only 0.14 $\mu$g./m.$^3$ of arsenic in the air 150 feet from the gin.

Arsenic is found to the extent of approximately 5 $\mu$g./g. in coal. Therefore, the air of cities which burn coal contains some arsenic. Air quality data from 133 sites monitored by the National Air Sampling Network showed an average daily arsenic concentration of 0.02 $\mu$g./m.$^3$ in 1964.

Control of arsenic emissions requires special attention to the temperature of exhaust gases since arsenic trioxide sublimes at 192°C. For this reason exhaust fumes must be cooled to approximately 100°C. prior to removing them as particulates.

Arsenic and arsenical compounds have been reported in aqueous waste products of the metallurgical industry, glassware and ceramic production, tannery operations, dye, and pesticide manufacture. The manufacture of Paris green and calcium meta-arsenate, both insecticides, was reported to produce wastewaters containing 362 mg./l. of arsenious oxide (2). Arsenical wastes from these sources would be expected to assume less importance in the future, however, as noted above.

High arsenic levels have been encountered in raw municipal water supplies, necessitating arsenic removal by the water treatment industry. Ground water supplies in Central and

South America are frequently reported to contain excessive arsenic levels (3)(4). Shen and Chen (5) have described deep well waters in Taiwan which contain up to 1.1 mg./l. arsenic, and are believed to be responsible for an endemic illness in the area called "black-foot" disease.

## Removal of Arsenic from Air

In general, the removal of particulate material will control arsenic emissions if the control equipment operates at a temperature low enough (~100°C.) to condense the arsenic fumes. An electrostatic precipitator has been reported to reduce the arsenic from 5 to 17 ppb to 0 to 4 ppb (6). Cooling flues, bag houses, and electrostatic precipitators have been used in the smelting industry. No data have been reported in the United States on their removal efficiency for arsenic. However, at a chemical plant in the U.S.S.R., the efficiency for arsenic removal was greatly improved by using wet vacuum pumps instead of fabric filters. When the fabric filters were used, the arsenic content in the air frequently reached several hundred thousand micrograms per cubic meter. After the wet scrubbing vacuum pumps were installed, the removal is reportedly 100% effective (7).

In the cotton industry, removal of particulate material emitted from cotton gins should control the arsenic emissions (8). However, methods need to be developed to control the arsenical emissions produced by the burning of cotton trash.

## Removal of Arsenic from Water

Only limited information was encountered on current levels of arsenic in industrial wastes, and on current treatment processes and removals obtained according to J.W. Patterson and R.A. Minear (9). Most of the literature describing industrial sources and treatment of arsenic is 30 years or more old. More recent industrial literature, while referring to arsenical wastes and to the severe pollution resulting from their discharge, presents no specific treatment processes or industrial waste values. More up-to-date information is available on the removal of arsenic from drinking water, and in fact the methods for arsenic treatment of both drinking water and industrial wastes are similar. The treatment methods and arsenic removal efficiencies are summarized in Table 3. No treatment costs directly related to wastewaters were encountered in the literature search reported by Patterson and Minear (9).

## TABLE 3: SUMMARY OF ARSENIC TREATMENT METHODS AND REMOVALS ACHIEVED

| Treatment | Initial Arsenic (mg./l.) | Final Arsenic (mg./l.) | Percent Removal | Reference |
|---|---|---|---|---|
| Lime Softening | 0.2 | 0.03 | 85 | (10) |
| Charcoal Filtration | 0.2 | 0.06 | 70 | (10) |
| Ferric Sulfide Filter Bed | 0.8 | 0.05 | 94 | (5) |
| Coagulation with Ferric Sulfate | 25.0 | 5 or less | 80 or more | (11) |
| Coagulation with Ferric Chloride | 3.0 | 0.05 | 98 | (12) |
| Precipitation with Ferric Hydroxide | - - - | 0.6 | - - | (13) |
| Precipitation with Ferric Hydroxide | 362.0* | 15-20* | 94-96 | (2) |

* As arsenious oxide.

Source: Report PB 204,521

### References

(1) Sullivan, R.J., "Air Pollution Aspects of Arsenic and Its Compounds," *Report PB 188,071*, Springfield, Va., National Technical Information Service (September 1969).
(2) Cherkinski, S.N. and Genzburg, F.I., "Purification of Arsenious Wastewaters," *Water Poll. Abstr.* 14, 315-316 (1941).

(3)  Trelles, R.A. and Amato, F.D., "Treatment of Arsenical Waters with Lime," *Water Poll. Abstr.* 23, 125 (1950).
(4)  Viniegra, G. and Marquez, R.E., "Chronic Arsenic Poisoning in The Lake Region: Section 4, Treatment of Drinking Water," *Water Poll. Abstr.* 38, 430-431 (1965).
(5)  Shen, Y.S. and Chen, C.S., "Relation Between Black-foot Disease and the Pollution of Drinking Water by Arsenic in Taiwan," *Proc. 2nd Internat. Conf. Wat. Poll. Research, Tokyo,* 1, 173-190, New York, Pergamon Press (1964).
(6)  Chambers, L.A., Foter, M.J. and Cholak, J., "A Comparison of Particulate Loadings in the Atmosphere of Certain American Cities," *Proc. Third Natl. Air Pollution Symposium,* pp. 24-32, Pasadena, Calif. (1955).
(7)  Matsak, V.G., "The Utilization of Air Dust and Smoke Purification Equipment," *Survey of the U.S.S.R. Literature on Air Pollution and Related Occupational Diseases,* 3, 141 (1960).
(8)  "Control and Disposal of Cotton-ginning Wastes," A Symposium, *Public Health Service Publication* 999-AP-31 (1966).
(9)  Patterson, J.W. and Minear, R.A., "Wastewater Treatment Technology," *Report PB 204,521,* Springfield, Va., National Technical Information Service (August 1971).
(10) Magnuson, L.M., Waugh, T.C., Galle, O.K. and Bredfeldt, J., "Arsenic in Detergents: Possible Danger and Pollution Hazard," *Science* 169, 389-390 (1970).
(11) Buswell, A.M., Gore, R.C., Hudson, H.E., Wiese, H.C. and Larson, T.E., "War Problems in Analysis and Treatment," *Jour. Amer. Water Works Assoc.* 35, 1303-1311 (1943).
(12) Irukayama, K., "Discussion-Relation between Black-foot Disease and the Pollution of Drinking Water by Arsenic in Taiwan," *Proc. 2nd Internat. Conf. Wat. Poll. Research, Tokyo* 1, 185-187, New York, Pergamon Press (1964).
(13) Berezman, R.I., "Removal of Inorganic Arsenic from Drinking Water under Field Conditions," *Water Poll. Abstr.* 29, 185 (1965).

## ASBESTOS

Inhalation of asbestos may cause asbestosis, pleural or peritoneal mesothelioma, or lung cancer. Mesothelioma is a rare form of cancer which occurs frequently in asbestos workers. All three of these diseases are fatal once they become established. The dose necessary to produce asbestosis has been estimated to be 50 to 60 million particles per cubic foot-years. No information is available on the dose necessary to induce cancer. Random autopsies of lungs have shown "asbestos bodies" in the lungs of one-fourth to one-half of samples from urban populations. Thus, the apparent air pollution by asbestos reaches a large number of people. Animals have been shown to develop asbestosis and cancer after exposure to asbestos. No information has been found on the effects of asbestos air pollution on plants or materials.

The likely sources of asbestos air pollution are uses of the asbestos products in the construction industry and asbestos mines and factories. Observations in Finland and Russia indicate that asbestos does pollute air near mines and factories. However, no measurements were reported of the concentration of asbestos near likely sources in the United States. A concentration in urban air of 600 to 6,000 particles per cubic meter has been estimated. No satisfactory analytical method is available to determine asbestos in the atmosphere, however, according to Sullivan and Athanassiadis (1).

### Removal of Asbestos from Air

The control of asbestos fiber emissions from industrial and commercial sources has been described by W.B. Reitze, D.A. Holaday, E.M. Fenner and H. Romer (2). Those authors have summarized the operations involved in asbestos mining, milling and manufacturing and have tabulated the control methods involved in each case.

**References**

(1)  Sullivan, R.J. and Athanassiadis, Y.C., "Air Pollution Aspects of Asbestos," *Report PB 188,080,* Springfield, Va., National Technical Information Service (September 1969).
(2)  Reitze, W.B., Holaday, D.A. Fenner, E.M. and Romer, H., paper presented before Second International Clean Air Congress, Washington, D.C. (December 1970).

## AUTOMOTIVE EXHAUST EFFLUENTS

### Removal from Air

Since this is such a wide-ranging topic in itself and since there are so many references available to proprietary devices for treating complex automotive exhaust mixtures, no detailed treatment of this topic will be given here. Instead, the reader is referred to comprehensive published works by McDermott (1), Post (2) and Sittig (3).

#### References

(1) McDermott, J., *Catalytic Conversion of Automobile Exhaust*, Park Ridge, N.J., Noyes Data Corp. (1971).
(2) Post, D., *Noncatalytic Auto Exhaust Reduction*, Park Ridge, N.J., Noyes Data Corp. (1972).
(3) Sittig, M., *Air Pollution Control Processes and Equipment*, Park Ridge, N.J., Noyes Development Corp. (1968).

## BARIUM

Very little information is available on the air pollution aspects of barium and its compounds; however, the introduction of barium compounds into diesel fuels as a means of reducing black smoke emissions has focused interest on the effects of barium in the environmental air. While it is generally accepted that the insoluble compounds, such as barium sulfate, are nontoxic, the soluble compounds are known to be highly toxic when ingested, and inhalation of barium compounds can produce a benign pneumoconiosis, known as baritosis. However, the effects of barium air pollution cannot be stated with certainty because of insufficient knowledge of the effects of atmospheric concentrations of barium compounds, particularly in the micron-particle size emitted from exhausts of diesel engines fueled with smoke-suppressant additives. Data on sources and emissions of barium have been presented in systematic form by W.E. Davis & Associates (1).

### Removal of Barium from Air

No information has been found on the abatement of barium or its compounds in the environmental atmosphere. Since most of the barium processed by industry is in solid form, the conventional methods for removal of solids, such as bag filters, electrostatic precipitators, and wet scrubbers, should prove effective in preventing their escape to the atmosphere. No information has been found on the control of emission of barium-containing exhaust solids from diesel engines operating with fuel containing barium-base smoke-suppressant additives, according to Miner (2).

### Removal of Barium from Water

Treatment of barium contained in an explosives manufacturing wastewater has been described in the literature (3). Barium was removed by precipitation as barium sulfate, upon addition of sodium sulfate. Barium sulfate is extremely insoluble, having a maximum theoretical solubility of approximately 1.4 mg./l. as barium, and the treatment process is presumably quite effective. Coagulation of suspended solids in the wastewater (including the finer nonsettleable barium sulfate particles) with ferric sulfate provided solids removal. It has been reported that the above process of precipitation of barium sulfate by addition of sodium sulfate is used by barite producers to recover barium from solution (6).

The second barium removal treatment process discussed in the literature is ion exchange. This is discussed in specific relation to barium removal from nuclear wastes (4)(5). Ayres (4) has reported extremely high removals of 99.999% for radioactive barium, 140 by ion exchange, while Pressman, Lindsten and Schmitt (5) report 99% removal of nonradioactive barium in a U.S. Army mobile ion exchange unit. These authors also found 99.9% removal of barium by an Aquamite – 30B electrodialysis demineralization unit.  No costs were reported for sulfate precipitation, ion exchange, or electrodialysis.

Based upon relative economics, however, and the very low solubility of barium sulfate, the precipitation method of barium removal would probably be the method of choice by industry.  The effectiveness of this method could presumably be enhanced by use of coagulation and filtration to remove fine, nonsettleable particles of barium sulfate.

References

(1)  Davis, W.E. and Associates, "National Inventory of Sources & Emissions: Section 1, Barium," *Report No. PB 210,676,* Springfield, Va., National Technical Information Service (May 1972).
(2)  Miner, S., "Air Pollution Aspects of Barium & Its Compounds," *Report PB 188,083,* Springfield, Va., National Technical Information Service (September 1969).
(3)  "Taming Explosives," *Indust. Engng. Chem.* 46 (12), 20A (1954).
(4)  Ayres, J.A., "Treatment of Radioactive Waste by Ion Exchange," *Indust. Engng. Chem.* 43, 1526–1531 (1951).
(5)  Pressmen, M., Lindsten, D.C. and Schmitt, R.P., "Removal of Nuclear Bomb Debris; Strontium – 90 – Yttrium – 90, and Caesium – 137 – Barium – 137 From Water with Corps of Engineers Mobile Water Treatment Equipment," *Wat. Poll. Abst.,* 36, 42–43 (1963).
(6)  *The Economics of Clean Water, Vol. III, Inorganic Chemicals Industry Profile,* U.S. Dept. Interior, Washington (1970).

# BERYLLIUM

Beryllium and its compounds, when present in the environmental air, are of concern because of their effect on the health of humans and animals, since beryllium is among the most toxic and hazardous of the nonradioactive substances being used in industry.  Beryllium is commonly found as an atmospheric pollutant within the confines and in the proximity of industrial plants producing or using beryllium substances.

Almost all the presently known beryllium compounds are acknowledged to be toxic in both the soluble and insoluble forms, depending on the amount of material inhaled and the length of exposure according to Durocher (1).  Soluble beryllium compounds, such as beryllium sulfate and beryllium chloride, commonly produce acute pneumonitis; insoluble compounds, such as metallic beryllium and beryllium oxide, can produce chronic pulmonary disease (berylliosis).  However, it should be noted that the toxic beryllium effect is not limited to berylliosis but instead is a body-wide systemic disease.

Expanded industrial use of beryllium in the 1930's, particularly the large-scale production of fluorescent lamps using phosphors of beryllium oxide, produced a number of cases of pulmonary diseases which were initially attributed to causes other than exposure to beryllium.  Recognition of the toxic effects of beryllium, however, beginning in 1943, resulted in establishment of the first Community Air Limit for metal substances, as well as in health safety procedures which were exceptionally effective in controlling further occurrences of the disease.

However, increased use of berryllium in the metallurgical industry, along with its proposed use as a high-energy fuel for rocket motors, suggests that further study should be made of the air pollution aspects of this highly toxic material.  An annotated bibliography has been published dealing with beryllium and air pollution (2).

## Removal of Beryllium from Air

Efforts to reduce concentrations of beryllium in the atmosphere have generally centered on the problem of reducing industrial concentrations of beryllium, where exceptional success has been achieved.  Control activity has mainly surrounded significant beryllium plants, where concentrations of beryllium dust, produced within the plant, are discharged to the outside atmosphere through intent or by accident. Recognition of beryllium as a cause of industrial disease in the early 1940's has led to sweeping changes in manufacturing processes and controls designed to reduce the concentration of atmospheric beryllium, both in-plant and out-of-plant (1).

Conventional air-cleaning devices have been used. For wet chemical processes, scrubbers, venturi scrubbers, packed towers, organic wet collectors, and wet cyclones have been found to be effective. For dry processes, conventional bag collectors, reverse-jet bag collectors, electrostatic precipitators, cyclones and unit filters have been used. Silverman (3) has published a listing of air-cleaning devices and their expected efficiency in handling effluents from various processes handling beryllium and its compounds.

The effectiveness of abatement activities can be appreciated by comparing present-day industrial exposure concentrations with concentrations existing prior to recognition of the significance of beryllium air pollution to industrial health. Breslin (4) states that current practices limit employees' exposure to beryllium to about 2 $\mu$g./m.$^3$ or less, with out-of-plant neighborhood concentrations limited to 0.01 $\mu$g./m.$^3$. By comparison, University of Rochester samples of beryllium concentration in a Cleveland factory as reported by Laskin et al (5) revealed concentrations as high as 4,710 $\mu$g./m.$^3$.

Another form of control is exemplified by discontinuance of the use of beryllium in fluorescent lamp tubes in 1949; the action eliminated an industrial exposure to beryllium which had been responsible for approximately one-third of all beryllium respiratory illnesses (6).

Schulte and Hyatt (7) reported one particularly detailed study on the economics of industrial hygiene control in a beryllium machinery plant in the Los Alamos Scientific Laboratory (New Mexico) in 1958. According to Tepper (8), in many instances the recovery of valuable materials will significantly offset the costs of abatement of beryllium air pollution.

### References

(1) Durocher, N.L., "Air Pollution Aspects of Beryllium & Its Compounds," *Report PB 188,078,* Springfield, Va., National Technical Information Service (September 1969).

(2) "Beryllium & Air Pollution: An Annotated Bibliography," *Publ. No. AP-83,* Research Triangle Park, North Carolina, Air Pollution Control Office, U.S. Environmental Protection Agency (Feb. 1971).

(3) Silverman, L., "Control of Neighborhood Contamination Near Beryllium – Using Plants," *A.M.A. Arch. Inc. Health* 19, 172 (1959).

(4) Breslin, A.J., "Exposures & Patterns of Disease in the Beryllium Industry," Chapter 3 in *Beryllium – Its Industrial Hygiene Aspects,* H.E. Stokinger, ed., New York, Academic Press (1966).

(5) Laskin, S., Turner, R.A.N. and Stokinger, H.E., "An Analysis of Dust & Fume Hazards in a Beryllium Plant," 6th Saranac Symposium (1947) in *Pneumoconiosis: Beryllium, Beryllium Fumes, Compensation,* A.J. Vorwald, ed., New York, Paul B. Hoebler, Inc. (1950).

(6) "Fluorescent Lamp Makers Stop Use," *Beryllium Industry Hygiene Newsletter* 9 (1949).

(7) Schulte and Hyatt, *Workshop on Beryllium,* The Kettering Laboratory, University of Cincinnati (January 1961).

(8) Tepper, L.B., Hardy, H.H. and Chamberlin, R.I., *The Toxicity of Beryllium Compounds,* London, Elseviar (1961).

# BLAST FURNACE EMISSIONS

## Removal from Air

A process developed by R. Kemmetmueller (1) deals with dust removal from a blast furnace in conjunction with an oxygen steel making process. The blast furnace exhaust gas containing entrained iron and iron oxide particles is lead to a smelting chamber where the iron oxide is reduced to iron and the iron is melted.

The particulate pollutants are thus recovered in the form of molten iron.

## Removal from Water

The reader is referred to the section of this book on "Cyanides" where U.S. Patent 2,989,147 refers to the removal of cyanides from blast furnace wash water by oxidation.

### References

(1) Kemmetmueller, R.; U.S. Patent 3,364,009; January 16, 1968.

# BORON

Boron and its compounds have presented hazards to the health of humans and animals, and therefore should logically be investigated as a cause of environmental contamination. The most common health hazards have been the accidental ingestion of household chemicals, such as boric acid or borax, and absorption of boric acid from wounds or burns; these hazards, of course, are not directly relatable to environmental pollution. A less common hazard is the contamination of the atmosphere by boron dusts, mainly produced during the manufacture of boron compounds and products.

The most serious hazard is the danger of atmospheric contamination by boron hydrides, or boranes, highly-toxic compounds used as high-energy fuels for rocket motors and jet engines. The use of boron as a fuel additive in the petroleum industry undoubtedly contributes to boron air pollution, however, its impact is unknown.

No data are available on the concentrations of boron in the atmospheric environment, and no measurements of this element or its compounds are known to be included in current air-monitoring programs.

Information on industrial waste levels of boron, and on treatment processes for removal of boron from industrial wastewaters is almost nonexistent in the literature surveyed. Industrial waste sources of boron include the following:

Detergent manufacture                    Production of leather and carpets
Weather-proofing wood processes          Cosmetics
Fire-proofing fabric processes           Photography supplies
Manufacture of glassware and procelain   Rocket fuels (boron hydrides)

## Removal of Boron from Air

Normal entrapment and precipitation procedures applicable to most particulate pollutants should be effective in controlling or reducing the amount of boron compound dusts emitted into the atmosphere. No specific methods for abatement of atmospheric borane compounds have been identified, but strict regulatory practices are used to prevent accidental emissions. No information has been found on the economic costs of boron air pollution or on the costs of its abatement, according to Durocher (1). Methods of analysis are available, but these are not sufficiently sensitive or selective for determining atmospheric concentrations of boron and its compounds.

## Removal of Boron from Water

A process has been developed by R.W. Goeldner (2) for the removal of boron from water. It involves evaporation, and recondensing the vapor. The distillation process appears not particularly effective, as it is reported that upon evaporation of a waste containing 21,000 to 22,000 mg./l. of boron, the recondensed vapor still contained 50 to 80 mg./l. of boron.

Upon passing this water through a 6 ft. column containing ceramic Raschig contact rings, the condensed vapor had a boron content of 2 to 3 mg./l. No costs were reported for the process.

### References

(1) Durocher, N.L., "Air Pollution Aspects of Boron & Its Compounds," *Report No. PB 188,085,* Springfield, Va., National Technical Information Service (September 1969).
(2) Goeldner, R.W.; U.S. Patent 3,480,515; November 25, 1969; assigned to Aqua-Chem, Inc.

# BREWERY WASTES

## Removal from Water

A process developed by E.H. Pavia (1) involves treating high protein liquid wastes under anaerobic conditions to effect a substantial reduction in the biochemical oxygen demand thereof, in a relatively short treatment time.

Figure 17 shows a suitable form of apparatus for the conduct of the process and the process will now be described in further detail with reference to that figure. As shown, **10** designates a line carrying the raw liquid waste of high protein content, which is to be treated in a sealed chamber generally indicated at **11**. Chamber **11** may take the form of a tank having a conical lower portion **12**, a domed top portion **13** and an intermediate cylindrical portion **14**.

The incoming waste line **10** enters tank portion **14** at inlet **15** with valves **16, 17** in the line. The tank **11** is sealed and therefore the treatment is conducted under anaerobic conditions. A water seal line **18** extends from an outlet **19** in crowned tank portion **13**. A drainage line **20** extends downwardly from the bottom outlet **21** in conical tank portion **12**; with valves **22, 23** in the line.

Mixing in tank **11** is accomplished by continuous circulation, via a line **25** which extends from drainage line **20** at a point between valves **22, 23** therein, to an inlet **26** at the top of tank portion **14**, with valves **27, 28** and a pump **29** interposed in line **25**. Selective mixing in tank **11** is also accomplished by continuous circulation via a line **30** extending from an outlet **31** located at a level depending on the characteristics of the waste being treated, to line **10** at a point between valves **16, 17**; the line **30** having valves **32, 33** and a pump **34** interposed therein.

In operating the system, with tank **11** empty, all valves except valves **16, 17** are closed and the raw waste is supplied to tank **11** from line **10** until a level **35** therein is reached. Valves **16** and **17** are then closed, while valves **22, 27** and **28** are opened and pump **29** is started. Pump **29** operates at a relatively high rate of speed to provide a very fast mixing of the entire contents of tank **11** and to achieve a proper mixture of liquid and suspended solids at a selected ratio thereof. During this mixing period, some gas is generated which will drive some of the water out of seal **18**, a relief chamber, not shown, being provided to receive such water.

When the proper mixed liquor-suspended solids ratio is reached, valves **22, 27** and **28** are closed; pumpage is stopped and valves **32, 33** and **17** are opened. Pump **34** is started and operates at a relatively low rate of speed to continue mixing. This mixing operation is effected to provide a gentle mixing of the contents of tank **11** at the lower portions thereof and the formation of a blanket of colloidal form as at **40** which is slightly above the level of inlet **15**.

The speed rate of mixing at both the high and low rate will depend on the characteristics of the waste being treated and the hydraulic characteristics of the tank. In tests, the ratio between the high rate of mixing and the low rate has varied from 2:1 to 6:1 depending on the individual waste treated.

Blanket **40** is effective to prevent solids from rising into the liquid zone lying between level **35** and a level at inlet **15**. During the low rate mixing operation, the pH of the recirculated liquid is checked carefully as it drops from an initial value of the order of 3.5 to 6.0 to a value of the order of 2.0 to 3.5. At this time, gas generated is reabsorbed by the liquid waste through oxidation and the reduction in pH causes soluble protein to precipitate out from the liquid.

**FIGURE 17:  APPARATUS FOR REDUCTION OF BOD OF HIGH PROTEIN BREWERY EFFLUENTS**

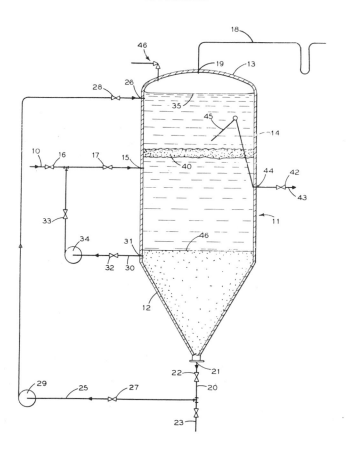

Source:   E.H. Pavia; U.S. Patent 3,520,802; July 21, 1970

When the pH has been suitably reduced, valves **32, 17** and **33** are closed.  A valve **42** in a line **43** connected to a draw off outlet **44** in an upper portion of tank portion **14** is opened.  Internally of tank portion **14** is a jointed draw off conduit **45**, connected to outlet **44**, which is used to draw off the treated liquid to a selected lower level.  Draw off conduit **45** may be manipulated by suitable means, not shown.

Valve **42** is then closed and valves **22, 23** are opened to draw the sludge accumulated in the lower portions of tank **11** to a level **46**, immediately above outlet **31**. This provides seed material for the second mixing phase as described above. The two phase mixing operation described above, is then repeated. The high protein content sludge drawn off may be recovered and utilized commercially. A pressure-vacuum relief valve is indicated at **46**.

It has been found that the initial mixing phase may extend over a period of from 5 to 20 minutes while the second mixing phase has a time period of from 50 to 70 minutes. Accordingly, the treatment has a total time period not exceeding 55 to 90 minutes.

The mixing intervals will vary with the portion content of the waste, the suspended solids of the waste and the temperature maintained in the tank. In tests it has been found that utilizing the same waste, the time for the completed reaction will be decreased approximately 1 minute per each 3°F. increase in temperature maintained. By way of example, liquid waste samples from spent brewers grain were taken before and after treatment in accordance with the process disclosed herein, with the following test results:

| Sample | | BOD, mg./l. | Total Suspended Matter, mg./l. | Total Protein, g./l. | Total Carbohydrate, g./l. |
|---|---|---|---|---|---|
| 7/19 | Raw | 23,850 | 21,000 | 12.50 | 15.80 |
| 7/19 | Treated | 13,500 | 2,150 | 3.56 | 13.60 |
| 7/21 | Raw | 19,440 | 27,000 | 18.80 | 8.90 |
| 7/21 | Treated | 11,490 | 1,350 | 2.38 | 7.20 |
| 7/21 | Sludge | 38,250 | 210,000 | 44.50 | 14.50 |

These results are based on composite samples of several runs each day. The five day 20°C. Biochemical Oxygen Demand (BOD) tests and the total suspended determinations were conducted in accordance with standard test practice. Total protein content was determined by the method of Lowry et al (J. Biol. Chem., 193, 265–275, 1951). Total carbohydrate content was determined by the anthrone method of Morris (Science 107,254,1948).

It will be apparent that with a relatively short treatment period, the BOD is substantially reduced, so that the effluent is in a more manageable condition for passage to municipal sewage disposal lines or to further treatment or to a stream. Further, the sludge represents a substantial buildup in protein, which may be recovered by suitable procedures known in the art, for commercial usage.

A process developed by E. Krabbe (2) is a two stage biological process for purification of brewery effluent where fermentation of sugars by means of aeration with yeast precedes treatment with activated sludge, thereby enhancing flocculation and separation of biota from the purified effluent. The yeast may be recovered.

As used here, the term "brewery effluent" refers to aqueous waste from the production of beer. Brewing is normally a batch operation, but it includes aqueous discharges from processing: such as the last sparges or tailings from lautering; press-water from spent grains and hop sparge; as well as various process precipitates such as trub, settlings and yeast. These discharges are flushed into the brewery sewers during tank cleaning.

The original sources of the organic load in brewery effluent are wort, beer and yeast, each having a BOD of 100,000 ppm or more; but, as a result of blending with rinse-water, cleaning solutions and cooling water, they are diluted 50 to 100 times as they are discharged into the sewers in the brewery.

The BOD or biological oxygen demand of the untreated brewery effluent may vary from 600 to 4,000 ppm. The average BOD load of the raw effluent from a modern brewery is frequently 2,000 ppm as compared to 200 to 250 ppm for domestic sewage. The discharge of waste from a brewery is about 10 times greater than the volume of the beer production. The total discharge of wastewater from all the breweries in the United States is of the order of 100,000,000 gal./day. For the treatment process which will be described later, it is important to note that the organic load in brewery waste is of two types: one fraction originating from the wort production section of the brewery contains fermentable sugars; the other fraction originating from the fermentation and aging section contains live yeast cells but only traces of fermentable sugars. The effluent from the packaging plant is of the second type but contains few yeast cells.

The traditional and common methods for purifying brewery effluent are adaptations of the methods used for treating domestic sewage. Primary treatment, namely gravity settling, is followed by biological treatments: in activated sludge systems, or on trickling filters on which biota grow. These continuous biological processes are referred to as secondary treatment. In order to accommodate brewery waste, activated sludge systems have been designed with a very low BOD loading; and for trickling filters a high rate of recycle has been provided. But neither of these processes has been free from operational problems.

Although brewery effluent is readily digested by the microorganisms in biota from activated sludge or trickling filters, it frequently upsets and severely impairs the operational efficiency of conventional sewage treatment plants. Such disturbances result in bad odors and the discharge of foam and slime. The effects are offensive to the public and injurious to the environment.

The adverse impact of brewery effluent on the conventional effluent treatment systems entails a shift in the microbial population away from that normally encountered with domestic sewage. This disturbance is so pronounced that the effluent from a new brewery may upset the operation of a sewage treatment plant in a medium-sized municipality when the brewery is discharging directly to the sewer system.

The shift in the microbial population entails a proliferation of slime bacteria such as *Sphaerotilus nanans*. This profuse growth of slime bacteria induced by brewery effluent causes these operational upsets: bulking, clogging and general impairment of the operational efficiency in conventional sewage treatment systems.

Slime-forming bacteria are always present in small numbers in the natural biota of the activated sludge process and on trickling filters. But carbohydrates, and particularly fermentable sugars in brewery waste, intensely stimulate the growth of the disturbing slime bacteria.

According to one aspect of the purification process for brewery effluent, there is provided a continuous biological process with two separate and consecutive stages of aerated microbial treatments. In the first stage, fermentable sugars from the wort production are metabolized by means of live yeast cells from the fermentation and aging section of the brewery to form new yeast cell material. In the second stage the effluent, now free of fermentable sugars, can be treated in an activated sludge system or on trickling filters, without operational disturbances, to reduce the BOD level of the effluent to a satisfactorily low level.

The yeast aeration tank also performs a useful purpose as a buffer tank. This is because brewing is a batch operation which normally is discontinued during the weekend; whereas the biological treatment with sludge is a continuous process requiring uninterrupted feed to maintain its biological flora during the weekend. The holdup in the yeast aeration tank serves as an equalization storage for the biological sludge treatment in the second stage.

The contents of the yeast aeration tank can be substantially reduced and be used for feeding the activated sludge in the second stage during the weekend interruption of the brewing operation. Because fresh yeast issues forth when the brewing production commences, the operation of the first stage of microbial treatment is not endangered by depletion during the weekend.

Because the organic discharges from the brewing process are wholesome in nature, it is desirable to provide for recovery of potential food for humans or animals. For this purpose a centrifugal recovery of solids is provided between the first and the second stage of microbial treatment. This recovery step also serves to reduce the ultimate BOD load which otherwise ends up as excess sludge and is disposed of by burning, burying or disposal at sea; thus contributing to the pollution load on the environment.

By fermenting the soluble sugars with yeast during the first step, the sugar fraction of the BOD load is converted into solid recoverable yeast cell material. Some BOD-laden process wastes, such as drippings from the beer filling machines in the packaging plant and the rinse water from the cleaning of kegs and returnable bottles, contain contaminations from the outside and should not be used for recovery purposes. Because the waste beer is low in fermentable sugars, it may be added directly to the activated sludge in the second stage, without causing growth of slime bacteria or impairment of flocculation therein.

Figure 18 is a schematic outline of the two-stage continuous biological purification process. Sanitary piping system **1** serves for collecting and transferring the first internal washings from brewery kettles and tanks to aeration tank **2**. These washings contain the major BOD load in as clean and wholesome a condition as the original sources, wort yeast and beer.

## FIGURE 18: BIOLOGICAL PROCESS FOR PURIFICATION OF BREWERY EFFLUENT

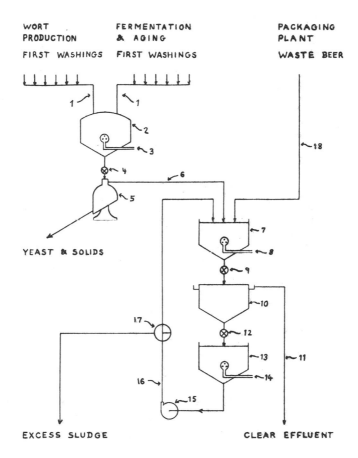

Source: E. Krabbe; U.S. Patent 3,650,947; March 21, 1972

Tank 2, designed for sanitary operations, provides a liquid residence-time equivalent to 16 to 24 hours of average effluent flow, so as to provide at least 4 hours residence at all times. During this period, the yeast contained in the washings from the fermentation section metabolizes the sugars contained in the washings from the wort production section of the brewery, thereby generating new yeast cell material. In this way the sugars are removed before the effluents come in contact with the natural biota in the second stage.

Clean compressed air from an external source is injected into the combined washings in tank 2 by means of air dispersion nozzle 3 and at a rate which maintains at least 1 ppm of dissolved oxygen in the aqueous mixture. The outflow from tank 2 is regulated by means of valve 4 as it flows to centrifuge 5 which is of the automatic desludging or self-opening type.

The separated solids from centrifuge 5 are discharged in a wet form and may be dried or processed separately to provide an edible by-product; or they may be blended into other by-products from the brewing process, such as spent grains. If this mixture is used as is for animal food, it is preferable to pasteurize the wet solids from the centrifuge in order to inactivate the yeast cells.

The clarified liquor is discharged through pipe 6 to aeration tank 7 of the second stage. The clarified liquor is now free of fermentable sugar and the remaining soluble BOD material can be digested by the biota of the activated sludge without interference from profuse growth of slime bacteria. Air is supplied from an external source through dispersion nozzle 8 at a rate which maintains at least 1 ppm of dissolved oxygen in mixture of liquor and biota in tank 7.

The air sparge also provides agitation and blending in tank 7. The volume of tank 7 provides a liquid residence time which accommodates the desired degree of BOD removal. Depending on the nature of the particular brewery waste and on the digestion rate of the natural biota, the residence time in tank 7 may vary from 5 to 10 hours.

Valve 9 serves to regulate the outflow from tank 7 to clarifier 10 which provides gravity separation of the biota. The clarified overflow of treated effluent is discharged through pipe 11 for final disposal. The withdrawal of the underflow of concentrated biota is regulated by means of valve 12, as the biota flows into postaeration tank 13. Air from an external source is injected through dispersing nozzle 14 to supply oxygen for the aerobic digestion of bioadsorbed BOD material in the biota.

Pump 15 recycles the biota through pipe 16 to a second-stage aeration tank 7 for the purpose of maintaining a suitable concentration of biota therein. Threeway valve 17 in recycle line 16 serves to withdraw the excess sludge in order to maintain a proper sludge age of the biota in the second stage of biotreatment. Waste beer from the packaging plant is unsuitable for recovery purposes, but is free of fermentable sugars and is therefore conducted directly to activated sludge aeration tank 7 by means of conduit 18.

The process may be practiced with or without the feature of recovery of yeast solids. In the latter case all effluent may be introduced into aeration tank 2 of the first stage, and after treatment therein, it may be transferred directly to aeration tank 7 of the second stage without centrifugation.

When effluents contain only wort, or sugary substances such as corn syrup, then yeast from an external source may be added to remove the fermentable sugars by means of aeration. The object of this process is to improve flocculation of biota in biological treatment of brewery effluent. Therefore, the following arrangements are useful.

If the brewery is connected to a public sewer then fermentable sugar may be removed from the brewery effluent by means of aeration with yeast at the brewery site, and the resulting pretreated effluent, free of fermentable sugars, may be discharged to the public sewer for aerobic treatment with biota of the activated sludge type at a municipal sewage treatment plant. In that case pipe 6 will represent the public sewer and the second stage, 7 through 17, will represent a municipal sewage treatment plant.

Another useful modification of the process serves to reduce the disturbing impact of brewery waste on municipal treatment plants. According to this modification, the process as described in the drawing is used for removing fermentable sugars and for reducing the BOD of pretreated effluent to about 250 ppm, equivalent to that of domestic sewage, and then discharging this effluent together with stabilized excess sludge to a public sewer for final treatment at a municipal sewage treatment plant.

References

(1) Pavia, E.H.; U.S. Patent 3,520,802; July 21, 1970.
(2) Krabbe, E.; U.S. Patent 3,650,947; March 21, 1972.

## CADMIUM

Cadmium and cadmium compounds are toxic substances by all means of administration, producing acute or chronic symptoms varying in intensity from irritations to extensive disturbances resulting in death. However, despite increasing use of this metal and increasing attention to it toxic nature, the exact manner in which it affects human or animal organisms is not yet known. Cadmium is toxic to practically all systems and functions of the body, and is absorbed without regard to the levels of cadmium already present, thereby indicating the lack of a natural homeostatic mechanism for the control of organic concentrations of cadmium.

Inhalation of cadmium fumes, oxides, and salts often produces emphysema, which may be followed by bronchitis. Prolonged exposures to airborne cadmium frequently cause kidney damage resulting in proteinuria. Cadmium also affects the heart and liver. Statistical studies of people living in 28 U.S. cities have shown a positive correlation between heart diseases and the concentration of cadmium in the urban air.

Cadmium may also be a carcinogen. While there is little evidence to support this conclusion from studies of industrial workers, animal experiments have shown cadmium may be carcinogenic. No data were found that indicated deleterious effects produced by airborne cadmium on commercial or domestic animals. However, experiments with laboratory animals have shown that cadmium affects the kidneys, lungs, heart, liver and gastrointestinal organs, and the nervous and reproductive systems, according to Athanassiadis (1).

The metals industry is the major source of emissions of cadmium into the atmosphere. Cadmium dusts and fumes are produced in the extraction, refining, and processing of metallic cadmium. Since cadmium is generally produced as a by-product in the refining of other metals, such as zinc, lead, and copper, plants refining these materials are sources of cadmium emissions as well as of the basic metal. Also, because cadmium is present in small quantities in the ores of these metals, cadmium emissions may occur inadvertently in the refining of the basic metal.

Common sources of cadmium air pollution occur during the use of cadmium. Electroplating, alloying, and use of cadmium in pigments can produce local contaminations of the atmosphere. Also, since cadmium is added to pesticides and fertilizers, the use of these materials can cause local air pollution.

In 1964, the average concentration of cadmium in the ambient air was 0.002 $\mu$g./m.$^3$, and the maximum concentration was 0.350 $\mu$g./m.$^3$.

Potential industrial sources of dissolved cadmium are reported by McKee and Wolf (2) and Santaniello (3) to be: metallurgical alloying, ceramics, electroplating, photography, pigment works, textile printing, chemical industries and lead mine drainage.

Of the industries listed above, only the electroplating industry is discussed to any extent in the literature with regard to cadmium bearing wastes and the treatment thereof. High concentrations of cadmium in lead mine drainage have been reported, however (2). Cadmium wastewater concentrations reported in the literature are summarized in Table 4. A report on groundwater contamination from plating wastes gives a concentration of 1.2 milligram per liter cadmium in a groundwater recharge basin receiving the plant wastes. Private wells in the area reported 0.3 to 0.6 mg./l., while one test well contained 3.2 milligrams per liter of cadmium (4).

## TABLE 4: CADMIUM CONCENTRATIONS REPORTED FOR INDUSTRIAL WASTEWATERS

| Process | Cadmium Concentration, mg./l. | Reference |
|---|---|---|
| Automobile heating control manufacturing | 14 – 22* | 5 |
| Plating rinse waters (large installations) | 15 av., 50 max. | 6 |
| Plating rinse waters | | |
| 0.5 gph dragout | 48 | 7 |
| 2.5 gph dragout | 240 | 7 |
| Plating bath | 23,000 | 7 |
| Lead mine acid drainage | 1,000 | 2 |

*24 hour composite on days when plating baths were dumped. Cadmium levels much reduced due to dilution when solution dragout was the only contributing source.

Source: Report PB 204,521

## Removal of Cadmium from Air

Air pollution control procedures are employed at some metal refinery plants in order to recover the valuable cadmium that would otherwise escape into the atmosphere. Electrostatic precipitators, baghouses, and cyclones are effectively used for abatement. However, little information has been found on the specific application of these procedures for the purpose of controlling cadmium air pollution, according to Athanassiadis (1). The procedures for recovering cadmium from exhaust in a copper extraction plant collected significant quantities of valuable cadmium, at the same time reducing local air pollution levels.

## Removal of Cadmium from Water

Methods of cadmium wastewater treatment currently in use are essentially those employed for most heavy metal wastes encountered in the plating and metal processing industries. These methods include precipitation (generally considered a destructive treatment), and ion exchange. Although not mentioned specifically in the cadmium literature, electrolytic recovery and evaporative recovery processes as used for copper, silver, and nickel waste recovery are technically feasible, but only for a concentrated waste or after a concentration step such as ion exchange.

The lack of specific cadmium data in the literature may be related to in-house dilution through central waste collection systems. Cadmium plating rinse streams are likely only a small part of total waste flow in a large plating operation, and resultant dilution can be assumed to reduce cadmium to very low levels.

*Chemical Precipitation:* Cadmium forms an insoluble and highly stable hydroxide at alkaline pH. Jenkins et al (8) report that freshly precipitated cadmium hydroxide leaves approximately 1 mg./l. of residual cadmium ion in solution at pH 8, but that this is reduced to 0.1 mg./l. at pH 10. Weiner ⑨ substantiates this behavior by reporting that cadmium is relatively soluble at pH 8 and requires higher pH for effective removal in the absence of iron. However, he indicates that coprecipitation with iron hydroxide at pH 8.5 effects complete removal.

Weiner (9) further reports that in the presence of complexing agents (e.g., cyanide) it is impossible to precipitate out the cadmium ion. In the event that appreciable complexing agents are in the waste solution, pretreatment to destroy the agents would be required. Cadmium plating wastes normally contain cyanide. Removal of cadmium from spent plating baths or the dilute plating rinse waters is therefore dependent upon prior cyanide removal. Fortunately, cyanide breakdown is relatively rapid and easy, as has been demonstrated for zinc and copper cyanide baths (10)(11). Costs would be expected to be essentially those of cyanide oxidation.

In the absence of cyanide, unit operations involved in cadmium removal are similar to those of the conventional tertiary treatment sequence of coagulation settling, and sand filtration. The latter step has become of increasing importance for complete removal of non-settling and slow settling metal hydroxide flocs. DiGregorio has provided approximate cost data for this sequence of operations (12).

Hanson and Zabban ⑬ have reported on lime precipitation of cadmium from a waste stream at the Rochester, Minnesota IBM Machine manufacturing plant. An effluent pH of 9.0 and cadmium concentration of 0.54 mg./l. was reported for the lime precipitation process. Although sand filtration was not used, the process incorporated coagulant aid addition (Separan NP-10) at 1 to 2 mg./l. The effluent contained only 2 mg./l. of suspended solids, indicating extremely good solids removal by sedimentation.

Linstedt et al (14) have reported extremely high cadmium removal for a pilot plant lime coagulation, settling process. Treating combined sewage effluent from a secondary treatment plant, these workers achieved 94.5% cadmium removal for the trace quantities present (14).

A proprietary hydrogen peroxide oxidation-precipitation system called the Kastone process has been developed, which simultaneously oxidizes cyanides and forms the oxide of cadmium (rather than the hydroxide). Cadmium oxide is claimed to be much easier to remove from solution than cadmium hydroxide ⑮. This process is reported as suitable for small plating operations, as solids removal is readily accomplished with slight modification to simple filtration apparatus already present in most shops.

*Ion Exchange:* Ion exchange can be used as a polishing treatment or recovery process. The concentrated solution obtained upon regeneration of the ion exchange resin is more suitable to economical recovery procedures than the dilute waste streams from which the cadmium was removed. The recovery value of cadmium is estimated as $1.20 to $6.00 per 1,000 gallons, at a solution concentration of 50 to 250 mg./l. of cadmium (12).

Mattock (16) states that the extra costs of ion exchange can be offset in ½ to 2 years when product recovery value is substantial, as is the case with cadmium. Odland and Hesler (17) on the other hand, indicate that ion exchange is unsuitable for recovery of cyanide solutions. No specific applications of ion exchange to cadmium cyanide solutions were found. Linstedt et al (14) have reported 99% reduction of trace quantities of cadmium by a pilot plant cation exchanger, and 99.9% removal by cation plus anion exchange. The waste stream originated from a secondary sewage treatment plant.

*Summary:* General removal of cadmium from a waste stream can be accomplished by precipitation as the hydroxide at elevated pH, but values of pH 10 or greater are necessary for effective removal. Removal solely on a solubility basis is restricted to 0.1 mg./l. residual cadmium at pH 10. Evidence was presented which indicated that coprecipitation with or adsorption on iron floc may enhance removal. Organic or inorganic complexing agents in the waste stream will reduce the effectiveness of precipitative removal.

A scarcity of operating data and costs specifically related to cadmium treatment exists in the literature. However, costs associated with the various treatment procedures for other nonferrous heavy metals should apply to cadmium solution of similar general characteristics, due to the similarity of treatment methods. Recovery processes would be expected to be more attractive than with other nonprecious metal systems, due to the high value of cadmium metal.

### References

(1) Athanassiadis, Y.C., "Air Pollution Aspects of Cadmium & Its Compounds," *Report No. PB 188,086,* Springfield, Va., National Technical Information Service (September 1969).

(2) McKee, J.E. and Wolf, H.W., "Water Quality Criteria," 2nd ed., *California State Water Quality Control Board, Pub. No. 3-A* (1963).

(3) Santaniello, R.M., "Air & Water Pollution Quality Standards, Part 2: Water Quality Criteria & Standards for Industrial Effluents," in *Industrial Pollution Control Handbook,* H.F. Lund, ed., New York, McGraw-Hill (1971).

(4) Lieber, M. & Welsch, W.F., "Contamination of Ground Water by Cadmium," *Jour. Am. Water Works Assoc.* 46, 541-547 (1954).

(5) Gard, S.M., Snavely, C.A. and Lemon, D.J., "Design and Operation of a Metal Wastes Treatment Plant," *Sew. Ind. Wastes* 23, 1429-1438 (1951).

(6) Pinkerton, H.L., "Waste Disposal. A. Inorganic Wastes," *Electroplating Engineering Handbook,* 2nd ed., A. Kenneth Graham, ed., New York, Reinhold Publishing Co. (1962).

(7) Nemerow, N.L., *Theories and Practices of Industrial Waste Treatment,* Reading, Mass., Addison Wesley (1963).

(8) Jenkins, S.H., Keight, D.G. and Humphreys, R.E., "The Solubility of Heavy Metal Hydroxides in Water, Sewage and Sewage Sludge, I. The Solubility of Some Metal Hydroxides," *Int'l. Jour. Air Wat. Poll.,* 8, 53-56 (1964).

(9) Weiner, R.F., "Acute Problems in Effluent Treatment," *Plating,* 54, 1354-1356 (1967).

(10) Eden, G.E., Hampson, B.L. and Wheatland, A.B., "Destruction of Cyanide in Waste Waters by Chlorination," *Jour. Soc. Chem. Ind.,* 69, (8), 244-249 (1950).

(11) Lancy, L.E., "An Economic Study of Metal Finishing Waste Treatment," *Plating,* 54, 157-161 (1967).

(12) DiGregorio, D., "Cost of Wastewater Treatment Processes," Robert A. Taft Water Research Center *Report No. TWRC-6,* U.S. Dept. Interior, Washington, D.C., (1968).

(13) Hansen, N.H. and Zabban, W., "Design and Operation Problems of a Continuous Automatic Plating Waste Treatment Plant at Data Processing Division, IBM, Rochester, Minnesota," *Proc. 14th Purdue Industrial Waste Conf.,* pp. 227-249 (1959).

(14) Linstedt, K.D., Houck, C.P. and O'Connor, J.T., "Trace Element Removals in Advanced Wastewater Treatment Processes," *Jour. Wat. Poll. Control Fed.,* 43, 1507-1513 (1971).

(15) "New Process Detoxifies Cyanide Wastes," *Env. Sci. Technol.,* 5, 496-497 (1971).

(16) Mattock, G., "Modern Trends in Effluent Control," *Metal Finishing Jour.,* 14, 168-175 (1968).

(17) Odlan, K. and Hesler, T.C., "Profitable Recovery of Plating Wastes by Reconcentration of Reduced Volume Rinses," *Plating,* 43, 1022-1025 (1956).

# CARBON

## Carbon Black Removal from Air

Carbon black is ultrafine soot manufactured by the burning of hydrocarbons in a limited supply of air. This finely divided material (10 to 400$\mu$ in diameter) is of industrial importance as a reinforcing agent for rubber and as a colorant for printing ink, paint, paper and plastics.

Three basic processes currently exist in the United States for producing this compound. They are: the furnace process, accounting for about 83% of production; the older channel process, which accounts for about 6% of production; and the thermal process (1970 figures). Atmospheric pollutants from the thermal process are negligible since the exit gases which are rich in hydrogen are used as fuel in the process. In contrast, the pollutants emitted from the channel process are excessive and characterized by copious amounts of highly visible black smoke. Emissions from the furnace process consist of carbon dioxide, nitrogen, carbon monoxide, hydrogen, hydrocarbons, particulate matter, and some sulfur compounds. For the furnace process, collection equipment is an integral part of the process for collection of the product as described in more detail by Vandegrift et al (1).

The channel process emits large quantities or carbon because no way has yet been developed to separate the escaping black and avoid upsetting the burning conditions, which in turn would drastically affect yield and quality (2). Production by the channel process has declined over the years, and at present only 3 plants are known to be in operation (1). These three plants produce a total of approximately 71,500 tons/yr. representing about 6.0% of the total carbon black produced in the United States.

For the furnace process, collection equipment is an integral part of the process for collection of the product. The types of equipment commonly used for effectively separating and collecting finely divided black from a gas stream are agglomerators, electrostatic precipitators, cyclone separators, scrubbers and baghouses (2).

In old plants, the black-laden gases are first cooled to about 450° to 550°F., and then passed through a dry electrostatic precipitator which agglomerates the black. The increased diameter and density of the agglomerated black permits it to be removed effectively from the gas stream by using several cyclone separators. Together these have a recovery efficiency of 85 to 90%, leaving about 10% of the initial carbon in the off-gases.

Collection or removal of nearly all the remaining 10% is accomplished when the off-gases are passed through either a bag filter to give 99% recovery or a wet-scrubber system with 97 to 98% collection overall. The scrubber system may comprise water scrubbing and wet electrostatic precipitation or washing in a slot scrubber followed by a wet cyclone scrubber. In order to recover the black, the slurry is circulated back to the reactors where it is used for quench. The black is thus reentrained in the smoke and recovered.

The electrostatic precipitator is rapidly disappearing from use in the carbon-black industry. Even at some older plants they are no longer in operation, since they have been shut off to save operating and maintenance expenses. They are not energized, and no carbon black is removed from them. The current trend is to do without electrostatic precipitators in new furnace-black plants. The trend is to a mechanical agglomeration device (i.e., cyclones) placed ahead of bag filters. When the older plants were built, suitable filters had not yet been developed, and the use of cyclones by themselves was inadequate.

In recent years, the carbon-black industry turned to agglomeration apparatus of lower first cost and of equivalent or superior performance compared with the electrostatic precipitator. Accordingly, in most plants, the black-laden gases, cooled to about 550°F., enter a series of large diameter cyclones (usually four) which separate about 70% of the carbon. The gases with the remaining black may be cooled further to about 360°F. They are then passed to a bag filter which separates the remaining black from combustion gases for an overall recovery of 99+%.

In some plants, cyclones (as well as electrical precipitators) have been eliminated and the design for carbon-black separation calls for a system with an agglomerating device and a single bag filter connected in series (2).

A process developed by D.C. Williams (3) relates to a bag-type filter means for separating carbon black particles from a carbon black-laden smoke. In the art of producing carbon black by pyrolysis a hot gaseous effluent or smoke at a temperature of about 1200° to 3000°F. emanates from the pyrolysis reactor. Generally, this carbon black-laden smoke is cooled to a temperature in the order of 400° to 600°F. by quenching with water in the reactor and thereafter cooling in an external cooler.

Generally, the quenching operation consists of contacting the effluent near the exit of the reactor with a spray of water. The quenched effluent is then conveyed to the cooler and thereafter to a recovery system, where the carbon black is separated from the smoke. In order to recover the greatest amount of carbon black from the smoke and, additionally, in order to obviate creating a nuisance by discharging carbon black-containing gas to the atmosphere, it is customary to filter the smoke through cloth or fabric bag-type filters.

By and large the fabric bag-type filters used in the recovery of carbon black are glass cloth bags which are comparatively resistant to high temperatures and chemical attack. The latter is an important consideration since there are significant amounts of acidic components in carbon black smoke, particularly when employing feedstocks of petroleum origin. However, one prominent disadvantage of glass cloth bags is that they are subject to wear and deterioration by mechanical stress when subjected to conventional shaking operations which are normally practiced to maintain the bag filters clean and free of carbon black deposits adhering to the surface thereof.

The usual carbon black filter bag is closed at its top and this closed top is generally attached to the bag shaking means. The open lower end of the suspended bag is attached to tubular nipple or flange mounted over an aperture in a cell plate spanning the lower portion of the filter unit. The carbon black-laden smoke is introduced below the cell plate and passes upwardly to the interior of a plurality of such filter bags.

Clarified gas or smoke passes through the bags to the exterior thereof in a filtering zone and then to a clarified gas outlet. Most of the carbon black is retained on the surface of the bag. In a cleaning, or what is termed conventionally as a "repressurizing" cycle, clarified gas obtained from one or more filter zones, which are in their filtering cycle, is introduced into the filter zone on the clarified gas side of the filter bags. The reverse flow of repressuring gas through the bags, coupled with the bag-shaking action, effectively cleans the bags of the accumulated layer of carbon black. Most of the carbon black drops into the bottom hopper of the filter unit and is suitably retrieved therefrom.

It has normally been proposed to discharge the carbon black from the filter unit hopper by means of a positive gas lock valve, such as a star valve or screw conveyor, in order to prevent passage of any significant amount of gas into the solids transport or recovery system. The solids transport system may include a screw conveyor or other appropriate transport means. It has previously been suggested that a part of the repressure gas be used as a transport medium for the separated carbon black. However, it is obvious that such transport gases are subject to serious fluctuations in pressure even when all of the repressure gas from one of a multiplicity of filter units is utilized.

Where a multiplicity of parallel-connected filter units is employed there is a problem of maintaining an even and continuous flow of carbon black-laden smoke to the filter units and also in maintaining an even and continuous flow of separated carbon black from the units.

Where a multiplicity of filter cells are utilized in the separation of carbon black from carbon black-laden smoke, there are also serious problems of corrosion which occur in the repressuring system. Since only one or less than all of the units undergoes repressuring at any given time, it is obvious that the manifold line supplying repressuring gas to the system will have intermittent flow therethrough and flow at one end in the manifold may be stagnant for considerable periods of time. As a result, this portion of the manifold cools, condensation occurs in the manifold and, subsequently, corrosion of the system occurs.

In accordance with the process, it has been discovered that solids may be effectively separated from solids-laden gas in a down-flow filter having tubular open ended bags; and that, contrary to the prior art, the separated solids can be continuously discharged from the filter along with a small portion of the solids-laden gas. It has also been found that a portion of the solids-laden gas discharged along with separated solids can be used as a transport medium along with a part of the solids-laden gas taken from the gas inlet to the filter system. A further discovery is that a filter system can be provided with a closed loop repressuring system and thereby eliminate, to a great extent, uneven flow and corrosion in the repressure system.

FIGURE 19:  BAG FILTER DESIGN FOR CARBON BLACK REMOVAL FROM AIR

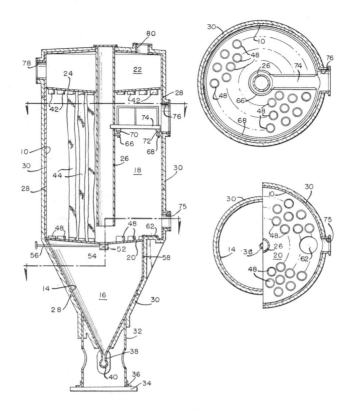

Source:  D.C. Williams; U.S. Patent 3,491,518; January 27, 1970

In the process, as shown in Figure 19, the main body of the filter unit is comprised of a cylindrical steel shell **10**, terminating at its upper end in slightly inclined top closure **12** and at its lower end in a frustoconical member whose base is above its apex or hopper **14**. The lower hopper zone **16** is adapted to receive separated solids and is separated from the central filter zone **18** by downwardly inclined lower cell plate **20**.

In like manner, central filter zone **18** is separated from upper deceleration and distribution zone **22** by upper cell plate **24**, which is also inclined downwardly toward its center.  Passing along the central axis of filter zone **18** from a point just above lower cell plate **20** and then through distribution zone **22** and out through top closure **12** of the unit is a clarified gas outlet duct **26**.

Circling the exterior of shell **10** and hopper **14** at spaced vertical points are insulation support ribs **28**.  These support ribs **28** support a layer of insulation **30**, which surrounds the entire unit.  Inasmuch as any metal structure which comes in contact with the filter unit must be well insulated beyond the point of contact and also must be periodically sand blasted and painted, the insulation and maintenance of such support structures is a considerable problem.

Accordingly, the filtering unit has no superstructure supporting the unit but is almost totally supported by cylindrical skirt **32**.  Skirt **32** is appropriately welded to the exterior of cone **14** at a point approximately midway between the top and bottom of the cone. Skirt **32** is attached at its bottom end to a base **34** by means of a rolled angle **36** welded to the skirt. Because of this support structure and means of supporting the filtering unit, it is possible to attach all necessary platforms and the like to the shell itself by means of lugs extending from the exterior of the unit.

Located at the outlet of cone **114** is an orifice-type solids discharge means **38**. Orifice **38**, as will be pointed out in more detail hereinafter, is open at all times but is sized in a manner to permit discharge of solids while at the same time providing a continuous leak of gas to horizontally-disposed solids transport duct **40**.

Upper cell plate **24** has formed therein a plurality of circular apertures. These circular apertures are extended downwardly by means of short tubular thimbles or bag-connector means **42**. Thimbles **42** vary in length so that their lower ends terminate at the same vertical plane. Thus, the thimbles form an upper terminus for filter bags **44**. The filter bags are cylindrical, tubular bags having both ends open.

The upper end of the filter bags are slipped over the upper thimbles and held in place by banding **46**. The filter bags terminate at their lower ends of lower thimbles **48** which extend upwardly from a plurality of circular apertures formed in lower cell plate **20**. Thimbles **48** are also of different lengths so that their ends terminate at a single horizontal plane. The filter bags have their lower ends attached to thimbles **48** by means of appropriate banding **50**.

Lower cell plate **20** also has formed therein a central opening which is extended by downwardly projecting tube **52**. Mounted in the tube is floor cleanup valve **54**, whose valve handle **56** extends through hopper zone **16** to the exterior thereof. Valve **54** is, of course opened for the purpose of cleaning up the upper surface of lower cell plate **20**.

For such cleanup operations and for the insertion of bags in filter chamber **18** a semicylindrical, vertically-disposed depression **58** is formed in the outer surface and adjacent the top of cone **14** and extends upwardly through lower cell plate **20**. The semicylindrical depression terminates at its upper end in an annular flange or step **60** above the level of lower cell plate **20**. Normally, the opening formed by semicylindrical depression **58** is closed by plate **62** held to step **60** by means of bolts.

A man-way **75** can also be provided for access to the cell. Provision is also made for the servicing of bags adjacent the top of filter zone **18**. For this purpose, track **66** is formed annularly about outlet duct **26**. At the same horizontal plane, track **68** is formed about the interior of shell **10**. Mounted on tracks **66** and **68** through rollers or wheels **70** and **72**, respectively, is bag service platform **74**.

Bag service platform **74** serves as a platform for a man to attach or remove the upper ends of bags **44** on or from thimbles **42**. Access to this platform and an opening for removing the platform, if desired, is provided through man-way **76** through the side of shell **10**.

For like access to the upper surface of upper cell plate **24**, an appropriate man-way **78** passes through shell **10** at an appropriate point along the height of the shell. A solids-laden gas inlet connection **80** is formed in top closure **12** for the introduction of solids-laden gas to deceleration and distribution zone **22**.

Figure 20 shows such a filter installed in a multiunit filter system equipped with a solids transport system. The system, as shown, comprises ten filter units **110** arranged in two parallel rows. By way of example, to show the proportions and arrangement, a typical cell **110** will be about 15 feet in diameter, each pair of cells will be spaced on center lines about 17 feet apart and the center lines through the two rows will be spaced about 19½ feet from one another.

## FIGURE 20: PLANT DESIGN FOR CARBON BLACK REMOVAL FROM AIR

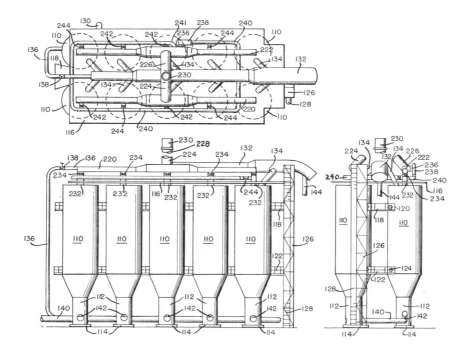

Source: D.C. Williams; U.S. Patent 3,491,518; January 27, 1970

The cells are substantially completely supported by individual skirt units 112 into which a portion of the cone or hopper section of the cell 110 passes and to which the hopper portion is welded. Skirts 112 are in turn attached to base elements 114. Extending across the top of the entire system of filters 110 is a valve service platform 116.

Lower down the filters 110 and attached to the interior sides of the units is an upper cell service platform 118 which provides access to the upper portion of the filter unit for the removal and attachment of the upper ends of filter bags. Such access is provided through man-ways 120. Adjacent the lower end of filter units 110 and attached to the interior sides of the units is lower cell plate service platform 122 which permits access to the lower portion of the filter through man holes 124, for purposes of attaching and detaching the lower ends of the bag elements. Extending from ground level to valve service platform 116 is man-lift 126. Appropriate caged ladders 128 and 130 also provide access to platforms 116, 118 and 122.

Extending between the rows of filters 110 above valve service platform 116 is solids-laden gas inlet manifold or header 132 for the introduction to the filter units of solids-laden gas. Such introduction is provided by side branches 134 which extend from manifold 132 downwardly into filter units 110.

Manifold 132 forms a portion of the separated solids transport system. It is also to be noted that manifold 132 is of diminishing cross section from its inlet end to its terminus at the fifth pair of filter units. For example, the drawings show a stepped design where the bottom of the manifold is straight, the initial section is about 60 inches in diameter adjacent the first two filters. The diameter then diminishes to about 54 inches adjacent the second and third pair of filters and, finally, diminishes to a 36-inch diameter adjacent the fourth and fifth pairs of filters. The diminishing cross section can also be continuous rather than stepped. In any event, the diminishing cross section of header **132** accommodates the decrease in volume of solids-laden gas caused by the withdrawal by the filters and provides a constant and even flow through the manifold with no low or high pressure spots at any point.

Obviously, with such continuous, even flow, filtering will be much more effective and consistent from one unit to another. Extending from the terminal end of inlet manifold **132**, adjacent the fifth pair of filters **110** is withdrawal line **136**. The withdrawal line provides a continuous, small flow of solids-laden gas from the inlet manifold **132** to act as a transport gas stream for separated solids, as hereinafter pointed out.

The withdrawal line is provided with valve **138** adapted to adjust the amount of gas withdrawn. The withdrawal line passes to the end of solids transport manifold **140** adjacent the rearmost one of the fifth pair of filter units **110**. Transport manifold **140** passes down the rearward row of filters **110** and then loops back below the front row of filter units **110** and is connected to and in open communication with the solids discharge means **142** of each of filter units **110**.

The transport manifold then passes to appropriate apparatus (not shown) for further processing of the solids. In the preferred embodiment of the process the transport manifold gradually increases in diameter as it passes from the first to the last filter unit to accommodate the solids plus gas added to the manifold by each successive unit.

In the exemplified system it will initially be the same size as line **136**, that is, 8 inches in diameter, adjacent the rearward filter of the fifth row and then increase in diameter to about 15 inches adjacent the solids discharge means of the forward filter of the fifth row. Thus, an even, continuous flow of solids through the transport manifold can be provided with no hangup of solids, no low or high pressure spots and dead spots in the manifold. Also, in the preferred embodiment, the solids discharge means of the filter units are orifices which provide a continuous leakage of hot gas into the transport manifold.

Orifices **142** are of decreasing size from the rearward one of the fifth pair of filter units **110** to the forward one of the fifth pair of the filter units. Accordingly, the orifices provide a portion of the gas necessary to convey solids in transport manifold. For example, about half of the transport gas may be supplied by the withdrawal line, while the other half is supplied from the filter units through the orifices along with the discharge of the solids through the orifices.

It is also obvious that by sizing the orifices as indicated the flow through the transport manifold will be continuous and constant. After solids-laden gas from the transport manifold has passed through appropriate collection and processing equipment, the gas passes back to inlet manifold **132** by means of the recycle line **144**. Thus, a closed loop system may be provided. The recycle line will normally be the same size as the terminal end of the transport manifold, or specifically, about 15 inches in diameter.

Clarified gas manifolds **220** and **222** are mounted above the front and rear rows of the filter units, respectively. In the preferred embodiment manifolds are of varying or stepped cross section, as shown in the drawings. If, for example, the filters are 15 feet in diameter and there are ten units, as shown, the manifolds are preferably about 48 inches in diameter where they serve the third or middle pair of filters and about 30 inches in diameter on either end where they serve the first and second and the fourth and fifth pairs, respectively.

From the top of the manifolds branch ducts **224** and **226**, respectively, connect the manifolds to a common discharge stack **228**. At the top of discharge stack is mounted a conventional burner **230**, which further clarifies the exhaust gases before discharging the same to the atmosphere. Exhaust gas from the filters discharges through standpipes 232 which are connected at their lower ends to the output means of filters 110 and at the upper ends to the manifolds, respectively. Mounted in the upper portion of each standpipe is exhaust or clarified gas outlet valve 234 which is opened and closed for purposes of repressuring the filter units.

Leading from one of the manifolds, in this case 222, is duct 236. This duct withdraws a portion of the clarified gas from manifold 222 and feeds it to repressure blower 238. The repressure blower discharges into looped repressure manifold 240. The repressure manifold forms a complete loop passing each of the standpipes of the filters and then returns to manifold 222 and is controlled by valve 241. From loop 240 branch lines 242 pass to each standpipe just below the valves 234. Each of the branch lines also has mounted therein a repressure valve 244.

A process developed by E.K. Caskey (4) involves an improved method and apparatus for collecting carbon black or separating finely divided solids that are entrained in a gaseous stream. More specifically, after hot gaseous effluent containing suspended carbon black is passed through a series of bag filters, the resultant clean gas continues along a generally parallel path to completely surround and heat the collection hopper for the previously entrained carbon black.

The use of the high temperature clean gas to maintain the temperature of the collection hopper above the dew point of the atmosphere within the collection hopper prevents the formation of condensate therein, and the resultant caking and corrosion associated therewith.

## Carbon Black Removal from Water

A process developed by R.M. Dille et al (5) is concerned with improvements in a method for the separation and recovery of free carbon particles from a carbon-water slurry formed in a synthesis gas generation process.

It is known that in the production of synthesis gas by the partial oxidation of a solid hydrocarbonaceous material there is obtained as a by-product a certain amount of finely-divided carbon in the resulting product gases, carbon monoxide and hydrogen. It is also known that the presence of these solid carbon by-products in the gaseous products obtained from a synthesis gas generator interfere with the main reactions in subsequent processes for conversion of the synthesis gas products into hydrocarbons, ammonia, and oxygenated hydrocarbons, e.g., Fisher-Tropsch type synthesis, ammonia synthesis, or methanol etc. synthesis.

Many methods have been proposed to effect separation of the free carbon particles from the synthesis gas products. These known carbon separation methods in general comprise treatment of the effluent products from the synthesis gas generator with water or water containing a low molecular weight oxygenated hydrocarbon in a scrubbing tower to effect removal of the entrained carbon particles therefrom and the removed carbon particles are then withdrawn from the scrubbing zone in the form of a carbon-water slurry.

In general the carbon content of the slurry varies from 0.1 to 3% by weight dependent on the generator feed stocks and reaction conditions in the generator. The carbon particles are thereafter separated from the carbon-water slurry in a conventional manner such as by filtration or by evaporation. When the carbon is separated from the slurry by a filtration operation the resulting water filtrate is commonly recycled to the scrubbing tower for reuse as the scrubbing liquid by suitable piping, pumps, etc.

The carbon particles separated from the aqueous slurry are not suited for immediate use as fuel feed stock or in carbon black manufacture because of the high water content of such particles. The carbon particles recovered from the slurry in a filtration operation contain from 80 to 90 or 95 weight percent water, the balance carbon. In order to utilize the carbon particles recovered from a filtration operation in an efficient and economical manner it is generally necessary to decrease the water content of the filter cake somewhat.

particularly when the carbon is to be used as a fuel feed stock. One known method of decreasing the water content of the carbon filter cake is by an evaporation operation but this
method is not completely satisfactory in large scale industrial operations due to the considerable amount of heat required and also the equipment needed for such an operation.

Solvent extraction has been employed previously in the separation of undesirable liquids
including water from materials but such methods likewise have certain objectionable features such as, solvent-handling and solvent loss which make them unattractive to industry.
These disadvantages can be overcome or avoided and the water content of the separated
carbon particles can be substantially decreased by the process, which comprises separating
the carbon particles from the carbon water slurry in a separation zone in the presence
of a water-soluble nonionic wetting agent.

The recovered carbon particles of reduced water content may be utilized as a fuel feed
stock for a fuel-consuming operation or employed in the form of carbon black in the manufacture of rubber products without further treatment for reduction of its water content.

As shown in the flow diagram in Figure 21, in the reaction zone of a synthesis gas generator **10** a mixture of methane and oxygen introduced therein by feed lines **12** and **14**, respectively, is subjected to controlled partial combustion at elevated temperatures and superatmospheric pressures whereby hot gaseous products of partial combustion including carbon monoxide, hydrogen and free carbon particles are produced.

**FIGURE 21: PLANT DESIGN FOR CARBON BLACK REMOVAL FROM SYNTHESIS
GAS PLANT WASTEWATER**

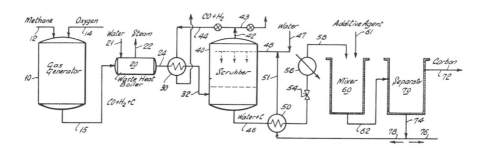

Source:  R.M. Dille and R.W. Chapman; U.S. Patent 2,980,521; April 18, 1961

The generator effluent comprising the hot gaseous combustion products CO and $H_2$ and
the free carbon particles is withdrawn through line **15** from the generator **10** and passed
to a waste heat boiler **20** whereby the generator effluent is cooled to a temperature not
substantially above 800°F., and preferably not above about 450°F.

A cooling liquid such as water is passed through line **21** to boiler **20** and by-product steam
is withdrawn therefrom by line **22** for utilization as a source of heat in the system. The
resulting partially cooled generator effluent is then fed through line **24**, through indirect
heat exchanger **30** where further cooling is effected, then passed through line **32** to a
scrubbing tower or zone **40** where the free carbon particles (in the generator effluent)
are separated from the generator effluent gases.

Carbon separation is accomplished in the tower **40** by countercurrent contacting of the
generator effluent gases with an aqueous scrubbing liquid that is introduced into the upper portion of scrubber **40** through line **48**. Scrubbing liquid is fed through line **47** to
feed line **48** from a water source not shown. Optionally, makeup fresh water can be

passed through lines **47** and **48** and mixed in line **48** with water recovered at a subsequent point in the process, i.e., from separator **70**; the recovered water being passed to line **48** through feed lines **74** and **78**, heat exchanger **50** and feed line **51**. The scrubbed effluent synthesis gases are removed from scrubber **40** by way of line **42** and passed to storage facilities not shown or to a synthesis gas reaction zone through line **43**, for example, an oxide ore reduction zone or a hydrocarbon synthesis zone. If desired, the scrubbed synthesis gases can be passed from scrubber **40** through line **42** and line **44** to heat exchanger **30** to serve as the cooling agent therefor, before being passed to storage facilities or utilized in a synthesis gas reaction zone.

The aqueous wash liquid or slurry containing the carbon particles removed from the generator gas products is withdrawn from scrubber **40** through line **46** and passed through a heat exchanger **50** to reduce the temperature of the slurry to about 150°F., then passed through line **52** and pressure-reducing valve **54**, whereby the pressure on the slurry is decreased to substantially atmospheric pressure, then passed to cooling zone **56** to further reduce the temperature of the slurry to about 70° to 115°F.

The cooled slurry is then passed through line **58** to a mixing tank **60** where the carbon-water slurry is blended with the nonionic additive agent introduced into vessel **60** through line **61**. If desired, vessel **60** can be provided with suitable mechanical mixing means, not shown, to assist in the mixing of the slurry and the additive agent. The resulting aqueous mixture is withdrawn through line **62** from vessel **60** and passed to a separation zone **70** where the carbon is separated from the wash liquid by filtration or by centrifugation. The recovered carbon is withdrawn through line **72** from separation zone **70**.

The substantially carbon-free aqueous liquid is withdrawn from separation zone **70** through line **74** and may be recycled through line **78** to heat exchanger **50**, serving as the cooling agent therefor, then passed through line **51** to scrubbing water feed line **48**. Optionally, a portion of the filtrate can be withdrawn from line **74** and passed through line **76** to storage facilities not shown, or can be discarded.

In the case when the filtrate from separation zone **70** is recycled to the scrubbing zone **40** by way of lines **74**, **78**, heat exchanger **50**, lines **51** and **48**, it is desirable to maintain the temperature within the scrubbing zone **40** below about 350°F. for the most satisfactory results.

A process developed by P. Visser et al (6) is one in which solid particles such as soot are removed from aqueous suspensions by a two-step process, the first step being to agglomerate the particles by contacting them with gentle agitation with a water-immiscible liquid, and then to contact the agglomerate-containing aqueous phase with a continuous phase that is water immiscible under conditions such that the agglomerates enter the nonaqueous phase.

References

(1) Vandegrift, A.E. et al in "Particulate Pollutant System Study, III, Handbook of Emission Properties," *Report PB 203,522*, Springfield, Va., National Technical Information Services (May 1971).
(2) Drogin, I., "Carbon Black," *Journal of the Air Pollution Control Association*, 18, (4), 216–228 (1968).
(3) Williams, D.C.; U.S. Patent 3,491,518; January 27, 1970; assigned to Ashland Oil & Refining Co.
(4) Caskey, E.K.; U.S. Patent 3,520,109; July 14, 1970; assigned to Ashland Oil & Refining Co.
(5) Dille, R.M. and Chapman, R.W.; U.S. Patent 2,980,521; April 18, 1961; assigned to Texaco, Inc.
(6) Visser, P. and Ter Haar, L.W.; U.S. Patent 3,694,355; September 26, 1972; assigned to Shell Oil Co.

# CARBON MONOXIDE

Air quality criteria for carbon monoxide have been reviewed by the National Air Pollution Control Administration (1).

## Removal from Air

See also the section on Foundry Effluents, Removal from Air.

The control techniques for carbon monoxide emissions from stationary sources have been reviewed in a publication of the U.S. Department of Health, Education of Welfare (2).

Table 5 summarizes methods for controlling CO emissions from stationary combustion sources.

### TABLE 5: SUMMARY OF METHODS FOR CONTROLLING CARBON MONOXIDE EMISSIONS FROM STATIONARY COMBUSTION SOURCES

| Control Method | Remarks |
|---|---|
| **Change of fuel or energy source:** | |
| Change to gas from oil and coal | Accepted emission factors for burning of coal, oil, and gas show decreasing CO emissions for these three fuels, in the order given. But CO emissions from boilers and furnaces are so low a fraction of total CO emissions that fuel change is not justified. |
| Change to nuclear power or hydro-electric generation | Use of nuclear power is expected to grow; hydroelectric generation will grow slowly. Nuclear power involves generation of some CO due to the periodic test operation of standby power-generating units employing conventional fuels. |
| Replace industrial, commercial, and household thermal requirements with central power | Generation of electric power is increasing. CO emissions are easier to control at a central power plant than at small installations and households. Efficiency is lower for indirect use of fuel through electricity than for direct burning. Reduction in local CO concentrations may result in increased oxides of nitrogen ($NO_x$) emissions at distant power plants. |
| **Combustion control:** | |
| Air supply | A well-adjusted gas-fired boiler may emit less than 1 ppm of CO, but may emit more than 50,000 ppm if insufficient combustion air is supplied. Insufficient air always causes CO formation; too much air may do the same by flame quench. |
| Residence time | Short residence times tend to cause more CO in exit gases. Proper residence time allows the use of less excess air. |
| Temperature | High temperature is desirable, but dissociation of $CO_2$ into CO becomes noticeable at 2800°F. Rapid cooling and low oxygen concentration tend to hinder recombination of $CO_2$. Flame temperatures above 3000°F. are conducive to formation of oxides of nitrogen ($NO_x$). |
| Mixing | Good mixing is very important for burning of CO; appliance and burner design should facilitate mixing. |
| Flame contact | Contact of flame with cold surfaces tends to form CO by quenching, i.e., it reduces residence time at effective oxidation temperature. |
| **Change of waste disposal method:** | |
| Sanitary landfill | Replaces open-burning and incineration. |
| Various treatments for coal-waste piles | These are not deliberately burned, but ignite by spontaneous combustion or accident. See AP-52, *Control Techniques for Sulfur Oxide Air Pollutants.* |

Source: National Air Pollution Control Administration Publ. No. AP-65 (1970)

Table 6 summarizes methods for controlling CO emissions from stationary process sources.

TABLE 6:  SUMMARY OF METHODS FOR CONTROLLING CARBON MONOXIDE
EMISSIONS FROM STATIONARY PROCESS SOURCES

| Source | Control Method | Remarks |
|---|---|---|
| Iron and steel industry: | | |
| Blast furnace | CO generated is burned as fuel. | Emissions can be produced by faulty equipment and accidents. |
| Grey iron cupola | Flame afterburner. | Not all controlled. |
| Basic oxygen steel furnace | Burned inside hood and dispersed by stack. | Collection for use as fuel not common in United States. |
| Sintering furnace | None | |
| Coke oven | Proper design, scheduling, operation and maintenance. | Controls are same as those to control particulates and $SO_2$. See AP-52, *Control Techniques for Sulfur Oxide Air Pollutants.* |
| Petroleum industry: | | |
| Petroleum catalytic cracking unit | Burned as fuel in CO boiler. | CO produced during regeneration of catalyst. Burning as fuel usually requires supplementary fuel for stability. |
| Petroleum fluid coker | Burned as fuel in CO boiler. | Gas produced in coker burning section of coking unit is rich in CO. |
| Chemical industry: | Most commonly burned as waste. | Moderate amounts generated in chemical industry as a whole, but this actually occurs only in specified segments of the industry. Emissions are from gas purging, leaks, abnormal operations such as startup, upsets and shutdown, or relief of overpressure. |

Source:  National Air Pollution Control Administration Publ. No. AP-65 (1970)

Determining the costs involved in control of CO emissions is seldom straightforward, and is often impossible (2).

Enormous amounts of CO are generated in a blast furnace, but this gas is cleaned and used as fuel.  Cleaning entails removing particulate matter; and if costs were to be allocated to air pollution control, it would be logical to allocate them to particulate removal rather than to CO removal.  Particulates also constitute the real air pollution problem in the operation of the basic oxygen furnace.

The CO generated is usually burned, or it can be collected for use as fuel.  If the CO is collected for fuel, the cost of the gasholder and associated piping could be allocated to utilities rather than to CO air pollution control.  Total costs are, of course, not necessarily recovered in the heating value of the CO collected.

The chemical industry generates a moderate amount of CO in reforming operations that usually has to be removed by suitable processes in order to make the desired product-hydrogen, or a mixture of hydrogen and nitrogen.  If CO is burned in a waste-gas flare, the costs of flare operation could be allocated to CO control unless the flare is used to burn various waste gases from other chemical processes.

The economics of a CO boiler serving a petroleum catalytic cracking unit are separable from those of any equipment required to clean the boiler feed gas.  In this case, the boiler handles only clean CO-rich gas, and abates only CO emissions.  Costs of such a boiler and its auxiliaries should, however, be based on engineering study and cost quotations from CO boiler suppliers.  Cost estimates for CO control, when applicable, may be made by the general methods described in AP-51, *Control Techniques for Particulate Air Pollutants.*

A process developed by R.J. Frundl et al (3) involves an afterburner system and arrangement for cupola furnaces where a series of burners are arranged near the charging opening of the cupola stack and ignite and combust the carbon monoxide gas emitted during furnace operation as it leaves the furnace bed. The afterburning raises the stack gas temperature thereby permitting complete combustion of high molecular weight organic compositions and minimizing the amount of free air pulled in through the charging opening.

Until a short time ago, cupola furnace stacks were allowed to discharge their effluents directly into the atmosphere and the resultant discharge was characterized by a distinctive objectionable odor and heavy black, brown or reddish smoke from which dust deposited on nearby areas. The characteristic smoke and odor were the result of iron oxide particles plus organics and hydrocarbons that were only partially combusted within the furnace itself. With the increasing furor over air pollution many devices have been tried in conjunction with cupola furnace stacks in an attempt to eliminate the air pollution hazard.

Among some of these air pollution control devices are dust collectors, to remove ash dust, and air washers to clean and scrub particulate matter from the stack exhaust. While these devices have been helpful in removing some of the objectionable contaminants in the cupola exhaust, they have not been completely successful. As more stringent air pollution control codes were enacted limiting the amount of contaminant that can be emitted from industrial furnaces, it has become acutely necessary to devise more efficient systems and apparatus to remove a greater proportion of contaminants from cupola furnace exhausts.

Afterburners positioned high up in the cupola furnace stack have also been tried but these have not been found to be entirely satisfactory. The positioning of some of the early designed afterburners high up in the stack was believed necessary in order to sustain continuous combustion since the composition of the gaseous products of combustion vary greatly and caused blowout of the early afterburner flame at lower elevations.

Further, the afterburners positioned above the charging opening were burning the stack gases mixed with the unmetered air drawn in through the charging opening. Then, the efficiency of the afterburner combustion varied constantly. Also, the air drawn in through the charging opening had a cooling effect on the stack gases and the afterburner combustion temperatures were not high enough to burn carbon monoxide and high molecular weight organic residues.

The most seriously objectionable constituents in furnace exhaust gases, which cause unwanted air pollution, are carbon monoxide (resulting from incomplete oxidation within the furnace itself) and various high molecular weight organic residues. These organic residues often result from the use of scrap metal in the furnace along with materials such as coke, flux and iron. On a great deal of this scrap metal, oil and grease and other organic material including paint and like surface coatings are present in appreciable quantities.

These grease and paint resin components have a high molecular weight and are difficult to burn completely. As differences in temperature occur in different regions of the furnace itself, the burning process is not simple; and many of these organic residues are thermally cracked, decomposed and semioxidized to form numerous organic derivatives. Since most cupola furnaces utilize a rapid combustion process the materials do not remain in the furnace for a long period and even if the temperature is high, there is often an insufficient oxygen supply within the furnace itself to effect the transformation and complete oxidation of these high molecular weight organic products.

As a result, the oils and greases are distilled and thermally cracked but do not burn to carbon monoxide and/or carbon dioxide and water vapor. Also, due to the variation in the air supply to the furnace, the amount of oxygen available for assisting in the oxidation of these materials may be limited and consumed by the high volume of other more readily oxidizable materials present in the furnace mix.

As a result, the organic residues become cracked with formations of low molecular weight derivatives which leave the furnace in an unburned state. This results in the formation of submicron liquid aerosol products which are difficult to remove in any dust collection or air washing system and thus pass through these systems to contaminate the air as they pass out of the stack.

In addition, since the combustion process within the furnace is rapid, and often there is insufficient oxygen for complete combustion, an unduly large residue of carbon monoxide passes out through the stack. This carbon monoxide can pass through the air washers and dust collection systems and, while it may be cleaned and scrubbed free of solid particulate material it nevertheless is ultimately exhausted to the atmosphere.

In accordance with the process, a plurality of afterburner jets are spaced at varying elevations around the rear and side portions of a cupola furnace stack in the area opposite the charging opening. The burner jets are inclined at a downward angle to direct the afterburner flame into the area immediately below the charging opening. The jets include a gas fuel supply, a furnace gas supply, and an air or oxygen supply line with appropriate valving to control the combustible mixture in the afterburner according to the character of combustion products encountered in the cupola furnace.

Figure 22 shows such an arrangement. There is shown a stack 10 for a cupola furnace including a row of tuyeres 12, to admit the oxygen which is required for the formation of the melting zone in the cupola furnace, and a charging opening 14 to allow for the introduction of the coke, iron and flux into the furnace itself.

The charge for the cupola furnace is admitted through the charging opening 14 in any suitable manner, i.e., by a mechanical charging bucket, and drops down into the stack through the melting zone. The gaseous products of combustion rise in the stack 10 from the combustion zone, adjacent the tuyere zone near the bottom of the cupola, and up through the charge and to the top of the stack, from which they escape through air washing and dust collecting systems (not shown).

Due to the incomplete combustion and the high molecular weight organic products present with the gaseous products of combustion, the gases discharged from the stack contain an undesirably high content of carbon monoxide gas as well as submicron organic particulate matter.

Therefore, to produce a more complete combustion of the gaseous products, a plurality of afterburner elements 16 are provided. The afterburners 16 are disposed at three elevations 18, 20 and 22 respectively, around the portion 24 of the cupola stack opposite the charging opening 14 and, illustratively, are downwardly inclined at a 30° to 60° angle in order to effectively combust the carbon monoxide and the accompanying organic residues within a short distance above the cupola furnace bed.

The afterburners may, of course, be installed at lower elevations and at different angles to the stack, as desired, and at varying numbers of elevations, the important consideration being to have them effectively oriented so that the afterburner flame jets penetrate the area immediately below the charging opening. At this point the carbon monoxide is richest since it has not been diluted by air entering through the charging opening and the temperature of the gases is at their highest level in respect to the metallurgical operation of the cupola furnace.

Additionally, the combined effect of the downwardly directed afterburner jets and the combustion zone formed below the charging opening forms a screen which shields the zone below the charging opening from the diluting and cooling effect of the unmetered air being drawn through the charging opening 14. Then, the air drawn in through the charging opening passes up the stack to assist in drawing off the gaseous products of combustion.

In a normal cupola furnace operation, there is a wide deviation in temperatures of the gases above the bed which is partly due to the manner in which the furnace is operated, particularly in respect to the rate at which air is forced into the bottom of the cupola furnace through the tuyeres. By placing the afterburners near the charging opening level and angled downward so that the flame from the afterburner penetrates to levels below the opening, but still above the furnace bed, the temperature in this zone can be more closely controlled by adjusting the intensity and number of afterburners that are operating.

### FIGURE 22: ELEVATION SHOWING AFTERBURNER SYSTEM FOR CUPOLA FURNACE

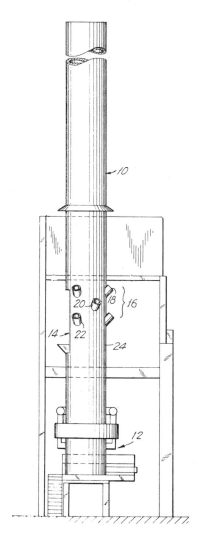

Source:   R.J. Frundl, V.W. Hanson, R.M. Jamison and O.M. Arnold; U.S. Patent 3,545,918; December 8, 1970

The afterburners in this zone advantageously provide for a more complete combustion of the carbon monoxide. This in turn elevates the stack gas temperature, due to the heat from the afterburners and from the combustion of the carbon monoxide, which results in a more complete thermal cracking of the high molecular weight organic residues and in the organic residues burning more completely. Additionally, the higher temperatures in the afterburner region preheats the charge as it descends to the furnace bed, which improves the efficiency of the metallurgical reactions in the furnace itself.

Another process developed by R.J. Frundl et al (4) involves an in-bed burner system and arrangement for stack furnaces where a series of burners are mounted on the stack below the charge door opening to direct a jet of burning gases into the stack below the top surface of the charge bed. Supplementary air is supplied to the stack from a position adjacent the burners to support relatively complete combustion of the products of combustion rising in the stack.

Oxygen can augment or replace the supplementary air. The in-bed burners and supplementary air effect burnout of carbon monoxide in the stack exhaust gases and maintain a relatively high temperature in the top portions of the charged bed to thermally crack and burn out organic materials present in the charge.

### References

(1)  U.S. Department of Health, Education & Welfare, "Public Health Service, National Air Pollution Control Administration," *Air Quality Criteria for Carbon Monoxide,* Wash., D.C. (1970).
(2)  U.S. Dept. of Health, Education & Welfare, "Control Techniques for Carbon Monoxide Emissions from Stationary Sources," *National Air Pollution Control Administration Publication No. AP-65,* Wash., D.C., U.S. Government Printing Office (March 1970).
(3)  Frundl, R.J., Hanson, V.W., Jamison, R.M. and Arnold, O.M.; U.S. Patent 3,545,918; December 8, 1970; assigned to Ajem Laboratories, Inc.
(4)  Frundl, R.J., Hanson, V.W., Jamison, R.M. and Arnold, O.M.; U.S. Patent 3,666,248; May 30, 1972; assigned to Ajem Laboratories, Inc.

# CARBON OXYSULFIDE

## Removal from Air

A process developed by T. Nicklin et al (1) is one in which removal of organic sulfur compounds, carbon disulfide and carbon oxysulfide, from inorganic gases, such as air, or exhaust gases, polluted therewith, is effected by contacting a mixture of steam and the gases to be treated with a catalyst comprising uranium oxide, and preferably a carrier such as alumina.

### References

(1)  Nicklin, T. and Farrington, F.; U.S. Patent 3,649,169; March 14, 1972; assigned to The Gas Council.

# CEMENT KILN DUSTS

Atmosphere emissions from cement manufacture have been discussed in some detail by T.E. Kreichelt et al (1).

A bibliography with abstracts dealing with the air pollution aspects of cement manufacturing has been published by the Environmental Protection Agency (2).

The economic impact of pollution control costs on the cement industry has been reviewed by the Boston Consulting Group (3).

## Removal from Air

See also the section on Alkalis, Removal from Air.

A process developed by G.W.Barr (4) relates to baghouses and more particularly to an improved construction designed to operate continuously at high efficiency and capable of filtering large quantities of undesirable contaminants from the gases discharged by industrial processing equipment as for example, a continuously operating cement kiln.

A typical cement kiln discharges approximately 6,000,000 ft.$^3$ of high temperature, dust-laden gases per hour, there being many tons of dust carried in these gases during each 24 hours of daily operation. Formerly the problem of attempting to separate such quantities of extremely fine dust from this huge volume of gas presented grave problems. To safeguard against the escape of this dust during recent years, there has been proposed a great variety of dust collectors including baghouses, dust chambers, cyclones and electrostatic precipitators.

Owing to the huge volume of hot gases with variable moisture content and of dust to be handled these various types of collectors, of necessity, are bulky and costly. An ever present problem encountered in all types is the provision of effective means for removing separated dust in a manner maintaining as nearly uniform as possible the efficiency of the separating operation. The problems are aggravated if the wet process of cement making is employed since the gases discharging from the kiln contain large quantities of water vapor which condense and cake dust filtering equipment if the gas temperature falls below the dew point.

Inasmuch as the process concerns the baghouse type separator, particular mention will be made of the problems encountered by designers of this type collector. As previously designed, baghouses characteristically comprise a large vertical enclosure for a number of tubular filter bags held suspended in various ways and opening at their lower ends into a distributing chamber for the dust-laden stream issuing from a dust source such as an operating cement kiln.

As this dirty stream passes axially through these bags and through the porous walls thereof, the dust separates out and falls into an underlying hopper or accumulates as a layer on the interior surface of the bags. After a short period of operation the accumulated layer increases the back pressure to such an extent that continued filtering operation is impractical and it is necessary to interrupt filtering operations in that baghouse until the dust layer can be removed and the filtering capability restored. Discontinuing the operation of the baghouse for cleaning purposes necessitates the provision of an additional or standby baghouse to which the dust-laden stream can be shunted during the reconditioning operation.

In order to reduce the investment represented by the provision of duplicate baghouses, various attempts have been made to shut off one portion of the baghouse for cleaning purposes while the remainder continues in regular operation. Although such designs have been successful to the extent that they avoided the need for full size duplicate baghouses, they have been subject to many disadvantages and shortcomings which are obviated by the process. Among the features characteristic of prior baghouses designed for continuous operation is the elimination of the shaker mechanism, or scraper suction cleaning mechanisms of any kind, having for their purpose the forced removal of the accumulated dust layer from the inner surface of the filter bags.

For example, it has been essential previously to provide some type of mechanical equipment for forcibly removing the accumulated dust layer. In one type of equipment, use of the filter is discontinued and the bags are subjected to vigorous shaking by mechanical shakers. Such shakers, though effective as dust dislodgers, are expensive to construct and operate and result in very short bag life. This is particularly true where the bags are formed of temperature resistant synthetic materials such as fiber glass which cannot withstand repeated flexure. Other types of mechanical devices make use of traveling scrapers and suction devices mounted within the filter bags. Some of these permit continuous cleaning while the bag remains in operation. However, all are objectional from cost and maintenance viewpoints and interfere with the flow of gases within the bags. Periodic shutdown is also necessary to remove accumulated foreign matter from this bag cleaning equipment.

Still another disadvantageous characteristic feature of prior baghouse designs is the enclosure of the filter bags within a structure substantially closed to the surrounding atmosphere except for air inlet and outlet openings of inadequate size and location. These baghouse designs are unsuitable for use in filtering very hot gases issuing from boilers, furnaces and kilns because of the inability of the filtering fabric to withstand high temperatures and fail after a short period of service.

These and related problems are handled most satisfactorily by a baghouse design freely open to the atmosphere in a zone surrounding the baghouse immediately between the roof and the top portions of the filter bags. The admission of atmospheric cooling air to the lower portions of the baghouse is arranged to be controlled in such manner as to provide for the cooling of the bags to a temperature approaching but above the dew point.

Preferably provision is made for the circulation of large volumes of atmospheric cooling air by convection upwardly through the baghouse. Restriction on the flow is conveniently provided by covering or uncovering the grillwork catwalks at the base of the filter bags by lightweight panels of plywood or the like.

There is provided in the process an improved and simplified baghouse of the self-cleaning type so designed as to obviate the disadvantages and shortcomings of prior designs including those referred to above. More specifically, the baghouse is characterized in particular by its simplicity of design and mode of operation, and more particularly in that the bags are supported by fixed means free of movable shaker, scraper, beater or suction cleaning nozzles of any character for removing dust layers therefrom. Instead, the process provides a baghouse operating continuously at uniform efficiency for periods of months without need for shutdown.

Provision for continuous reconditioning of the filter bags relies upon a suction fan and sequentially actuated control valves for periodically cleaning each section of the filter bags without interference with the operation of the remainder. An important feature of the reconditioning operation is the use of sub- and superatmospheric pressures in mutual cooperation to impress a traveling annular wave upon the bag wall of a type highly effective in dislodging adhering dust therefrom thereby obviating the need for mechanical cleaning devices of the various types previously considered essential. The use of the traveling wave in the bag walls progressively shears the dust layer from the bag wall with minimum flexure of the fabric and contributes to the effectiveness and speed of the cleaning operation as well as to the long life of the bag fabric.

Another feature of the process is the provision of a baghouse and the associated duct means for conveying a dust-laden stream thereto from a cement kiln in such manner that the very high temperatures of the gases issuing from the kiln are reduced sufficiently to avoid injury to the fabric used for the filter bag yet sufficiently above the dew point of the moisture in the dirt stream to avoid wetting of the dust layer and the caking thereof.

These objectives are accomplished by the cooperative action of several factors including the use of the ducting for the dust-laden stream as a heat exchanger, to the number and surface area of the filter bags maintained continuously in operation, to the exposure of the filter bags throughout their lengths to the ambient atmospheric air, and to the introduction of relatively cool atmospheric air used in reconditioning a filter cell to the hot, main gas stream.

The essential equipment features of the process are shown schematically in Figure 23. There is shown a baghouse facility designated generally **10** including as an example of a source of high temperature contaminants a continuously operating rotary cement kiln of

conventional design having its axis included slightly to the horizontal. It will be understood that the kiln is supported in the usual manner with gas outlet **11** suitably supported
in framework **12**, the right-hand half of framework **12** including at its upper end suitable
facilities, not shown, for feeding the kiln with the raw materials to be processed therein.

### FIGURE 23: BAGHOUSE INSTALLATION FOR REMOVING CEMENT KILN DUSTS FROM AIR

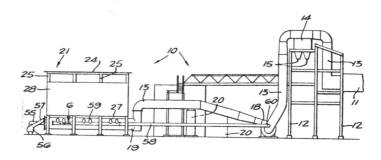

Source:   G.W. Barr; U.S. Patent 3,266,225; August 16, 1966

Gas outlet **11** leads directly into a large diameter duct **13** extending upwardly to the top
of framework **12** and through suitable cyclone separators **14** comprising centrifugal dust
separators where a considerable volume of the heavier constituents of the dust settle out
for the removal through hoppers **15** into which the cyclones discharge.

The dust-laden stream depleted of its heavier constituents continues through duct **13** which
extends downwardly into the inlet of a heavy duty blower **18** which raises the pressure of
the stream very appreciably above atmospheric pressure forcing it to flow rapidly through
following portions of duct **13** and into manifold **19** extending lengthwise of the baghouse.
As is indicated, duct **13** is suitably supported throughout its length by suitable frameworks
or bents **20, 20**.

The baghouse proper, designed generally **21**, suitable for housing the filter sections required in separating dust from approximately 6,000,000 ft.³/hr. of cement kiln gases, comprises a building having a framework measuring approximately 27 ft. in width and 65 ft.
in length and having a height of 50 ft. Preferably, and as herein shown, this building comprises a structural steel framework provided at its top with suitable roofing **24**.

The main supporting columns **25** are suitably crossbraced by cross-members **26** and by
horizontally disposed stringers **27**. It will be understood that baghouse **21** is preferably
covered along its sides and across its ends with suitable paneling or siding **28**, such paneling extending from stringers **27** to a level somewhat above the upper ends of the tubular
filter bags.

Rigidly secured to and supported by the lower row of stringers **27** are two rows of dust
hoppers extending lengthwise of the baghouse along either side thereof.

In the baghouse illustrated, each row of hoppers forms part of a filtering section or cell,
each section including a plurality of filter bags consisting of six rows of eight filter bags
each. The forty-eight bags constituting a single filtering section are in communication
with one another. Although it will be understood the number, arrangement and dimensions of the filter bags in each filtering section may vary widely, it has been found in
actual practice that twelve similar sections of filters, each having forty-eight filter bags approximately one foot in diameter and 25 ft. in height, provides high efficiency, continuous

filtering capacity adequate to handle the output from a single cement kiln 330 feet in length and discharging approximately 6,000,000 ft.³ of dust-laden gas per hour. Preferably the filter bags are made of fabric capable of resisting relatively high temperatures as is found necessary in practice to avoid the caking of dust layers on the interior surfaces due to cooling of the gases below the dew point. For this reason it is necessary to forego available exterior cooling of the filter bags using atmospheric air and advantageous in that it would permit the use of a cheaper filter fabric. Suitable fabrics for fabric operating temperatures in excess of 225°F. include the synthetic material Dacron, fiber glass and others. Fiber glass is particularly resistant to higher gas temperatures.

The filtering sections and associated dust hoppers on the far side of the baghouse utilize a dust return duct 55 opening into the intake of a suction blower 56, the outlet side of which is connected to manifold 19 by a duct 57. The filtering sections and associated hoppers on the near side of the baghouse employ an alternate return system comprising a dust return duct 58 having its inlet end connected through outlet elbows 59 with each of the dust chambers in hoppers and its discharge end 60 opening into the main stream duct 13 on the inlet side of blower 18.

A process developed by S. Masuda (5) relates to an apparatus for the electrical precipitation of cement dust in a dust-containing gas.

A process developed by H. Duessner (6) is a process for removing dust from the exhaust gases of a cement manufacturing installation in which the dust-laden exhaust gases, after leaving a raw-powder-preheater, are conducted into the lower portion of a vertically extending moistening compartment. Water in an excess amount is sprayed into the upwardly moving gas stream and the excess water loaded with dust collects in a sump at the bottom of the compartment.

A pump withdraws the muddy water from the sump and discharges it firstly into nozzles tangentially extending into the sump to circulate its content, and secondly into the upper portion of the compartment to wet the inner wall of the same. Another pump conveys a portion of the muddy water from the sump to the raw-powder-preheater at a point before the one where the raw-powder is introduced.

Figure 24 shows the essential equipment involved in the process. A rotating furnace 1 for the production of cement is traversed in direction of the arrow by the furnace gases. The gases in a rising main flow direction pass through a group of consecutively connected cyclones 2, which serve for the preheating of the raw-powder-cement. The raw-powder-cement flows through this aggregate in known manner against the main direction of flow of the gases.

A blower 3 withdraws the gases from the cyclones and conveys the gases to the lower portion of a vertically extending cooling and moistening-compartment 4, through which the gases flow in rising direction. By means of a fresh water inlet, for example, represented by a pump 5, water is fed into the compartment 4, and control members 6 adjust the fresh water supply for example in relation to the water level above the bottom of the compartment 4, the so-called muddy water-sump 8.

The fresh water is sprayed finely divided by means of nozzles 7 into the compartment 4. Insofar as the fresh water does not evaporate in the rising gases, it precipitates in the form of dust-laden drops on the floor or on the inner wall 12 of the compartment 4 and collects in the sump 8 as muddy excess water.

By means of a water pipe 9 and a pump 10, the excess muddy water is forced into the overflow ring trap 11 from where it uniformly moistens the inner wall 12 of the compartment. Furthermore spray pipes 13 are likewise connected with the muddy water pump 10 and extend tangentially into the sump 8. Through the kinetic energy of the muddy water jets, the contents of the sump is caused to perform a circular movement, so that it does not come to any stagnant deposits. With a constant number of revolutions of the

FIGURE 24: SCRUBBER APPARATUS FOR REMOVING DUST FROM CEMENT KILN GASES

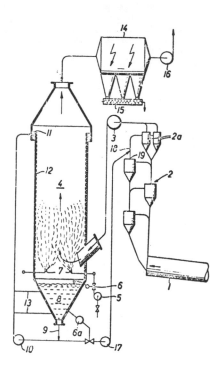

Source:  H. Deussner; U.S. Patent 3,485,012; December 23, 1969

pump **10**, the built up fluid pressure is dependent on the weight and on the viscosity of the muddy water. A manometer, not shown in the drawing, may accordingly furnish an indication, under the mentioned prerequisites, as to whether the muddy water has the desired consistency, or whether more or less fresh water should be supplied or whether more or less excess water is to be sprayed into the raw-powder preheater **2**.

Other members **6a** of a type known per se, may, however, also be employed for determination of the specific weight and/or viscosity of the excess water, for example in operative connection with an adjusting member for the feed quantity of the pump **17**. The greatly cooled and moistened gases in the compartment **4**, are conveyed to the electrofilter **14** and after the electro-dust-removal are conveyed through the blower **16** into the atmosphere. The dust dropping down in the dust discharge **15** may in known manner again be used for the cement production method.

The muddy excess water loaded with dust particles is withdrawn from the sump **8** by means of the pump **17** and by means of turbulence nozzles **18** is sprayed into the gas conduit leading to step **2a** of the cyclone group **2** mainly carried out as double-cyclone, last viewed in direction of the gas flow. At suitable distance in front of the turbulence nozzles **18**, likewise viewed relatively to the direction of gas flow, is disposed the raw-powder delivery point **19** constructed according to known art. From the gas, which in this area still has a temperature of about 500°C. a considerable quantity of heat will be removed or extracted by the introduced cool raw powder. Nevertheless, the spraying in of muddy water taking place shortly thereafter causes a further cooling off of the gases through removal of evaporation heat and heating of the dust precipitated by the evaporation. By means of the spraying in of the muddy water in the gases loaded with raw-powder, a substantial part of the muddy water agglomerates with the raw-powder particles, so that only relatively little fine-grain results, which in the last cyclone step **2a** is not separated and accordingly once more passes through the blower **3** and into the compartment **4**.

With the method and devices described, a highest possible cooling of the exhaust gases is attained at a highest possible dewpoint, without that in this way loss of heat is incurred. Owing to the quality of the exhaust gases attained, the filter may be dimensioned smaller than previously. Incrustations in the moistening compartment, in the cyclone group as well as depositions of mud are effectively prevented, furthermore no waste mud is produced whose removal is always problematic and only possible with appreciable expenditures.

Due to the fact that muddy water is sprayed into the raw-powder-preheater, not only the solids contained therein are again supplied to the manufacturing method, but they themselves effect already a cooling of the gases, which is of advantage in the interest of as small as possible supply of water to the raw-powder-preheater itself.

A process developed by G. Deynat (7) is one which extracts alkali from exhaust gases of a cement kiln and which includes a duct including a curtain of endless chain elements with chain cleaning arrangements for the chains.

**FIGURE 25: WETTED CHAIN DEVICE FOR DUST REMOVAL FROM CEMENT KILN GASES**

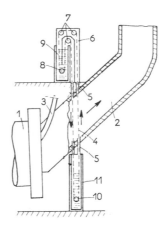

Source: G. Deynat; U.S. Patent 3,503,187; March 31, 1970

Figure 25 shows the essentials of the process. It shows the end of a kiln connected to ducting **2** for the outlet of gases to a reheater (not shown), and pulverulent materials preheated in the reheater being introduced into the kiln via channels **3**. The alkali-extracting device is disposed on the ducting **2** adjacent the kiln and is mainly formed by a plurality of endless chains **4** whose two vertically oriented runs extend through the wall of the ducting **2** in guide tubes **5** attached in sealing-tight relationship to the ducting.

Disposed above the ducting **2** is a frame **6** supporting wheels **7** for guiding each of the chains, one of the wheels being mounted on a shaft common to the chain assembly and driven by a motor (not shown). The two runs of each of the chains are wound on a return wheel **8** immersed in a tank **9** containing the water for washing and cooling the chains. Below the ducting **2** the chains are wound around return wheels **10** which are also immersed in a washing and cooling tank **11**.

The two tanks **9**, **11** have double walls to allow cooling by water circulation of the washing and cooling liquid (water) whose temperature is preferably kept at about 50° to 60°C., to accelerate the dissolving of the alkalis entrained by the chains **4**, and to reduce the effect of the thermal shocks to which the chains are subjected.

A process developed by L. Kraszewski et al (8) is one in which the alkali content of cement clinker produced in an apparatus having a suspension preheater which delivers preheated raw meal to a rotary kiln is reduced by the steps whereby:

(a)   a portion of the hot exhaust kiln gases are withdrawn from a point adjacent the gas exhaust of the rotary kiln;

(b)   the portion is passed through a wet scrubber to remove alkali and solid material therefrom and produce a thin slurry;

(c)   the gaseous portion obtained from step (b) being then passed to an electrostatic precipitator; and

(d)   the slurry is discharged from the scrubber to waste.

The apparatus, in Figure 26, includes a conventional tilted rotary kiln **10**, a vertical suspension preheater **11** whose lower end **27** communicates with the gas exhausting end of the kiln and an outlet **28**, a raw metal bin **12** and feeder **13**, an exhaustor **14** and electrostatic precipitator **15** with duct **16** connecting the preheater outlet to the exhaustor and precipitator.

**FIGURE 26:   ELECTROSTATIC PRECIPITATOR INSTALLATION FOR CEMENT KILN DUST REMOVAL**

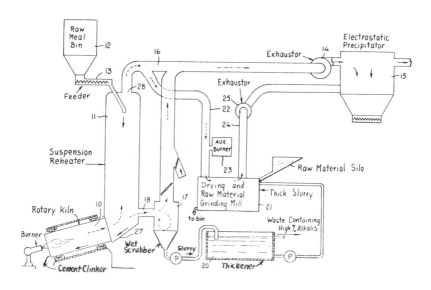

Source:   L. Kraszewski and G.A. Zulauf; U.S. Patent 3,507,482; April 21, 1970

The feature of the process comprises a wet scrubber 17 which communicates with the lower end 27 of the preheater to the precipitator through duct 18 and the duct 16 leading from the top end of the preheater to the exhaustor. When operating, a portion of the gases exhausted from the kiln pass through the wet scrubber which removes solids and alkali from gases exhausted by the kiln.

The resulting slurry which collects in the scrubber may be dumped directly to waste. However it is more economical and practical to pump the slurry to a thickener 20 where the water, which contains a high percentage of alkali, is allowed to overflow to waste leaving a thick slurry which is then pumped to the raw material and grinding mill 21 for grinding and drying raw material used in the clinker manufacturing process. The ground material (meal) is conveyed to the bin 12.

The system may include a further bypass which comprises a duct 22 connected to the duct 16 and the mill 21, the duct 22 being connected to the duct 16 between the preheater outlet 28 and the junction of the duct 18 with the duct 16. The duct 22 includes an auxiliary burner 23. Leading away from the mill 21 is a duct 24 having an exhaustor 25. The exhaustor can discharge directly to the precipitator 15 as shown. The advantages of the above described process are:

[1] Removal from the cement clinker burning system of a substantial portion of the alkali thus permitting production of higher quality cement.
[2] Reduction of dust air pollution by improving the dust-collecting efficiency of the rotary kiln circuit electrostatic precipitator by humidifying the bypass gas.
[3] Reduction of dust air pollution by improving the dust-collecting efficiency of the electrostatic precipitator by humidifying the exhaust gases of the mill by the addition of a thick slurry of leached dust to the raw material.

A process developed by J.G. Hoad (9) involves a combined wet scrubber and clarifier or settling basin for removing dust and other particulate matter from exhaust gases associated with various metallurgical and chemical processes, such as cement manufacture.

**References**

(1) Kreichelt, T.E., Kemnitz, D.A. and Cuffe, S.T., "Atmospheric Emissions from the Manufacture of Portland Cement," *Publ. No. 999-AP-17*, Cincinnati, Ohio, U.S. Public Health Service (1967).
(2) Environmental Protection Agency, Office of Air Programs, "Air Pollution Aspects of Emission Sources: Cement Manufacturing," *Publ. No. AP-94*, Research Triangle Park, N.C. (May 1971).
(3) Boston Consulting Group, "The Cement Industry: Economic Impact of Pollution Control Costs," *Reports No. 207,150 and 207,151*, Springfield, Va., National Technical Information Service (November 1971).
(4) Barr, G.W.; U.S. Patent 3,266,225; August 16, 1966; assigned to Southwestern Portland Cement Co.
(5) Masuda, S.; U.S. Patent 3,444,668; May 20, 1969; assigned to Onoda Cement Co. Ltd., Japan.
(6) Duessner, H.; U.S. Patent 3,485,012; December 23, 1969; assigned to Klockner-Humboldt-Deutz AG, Germany.
(7) Deynat, G.; U.S. Patent 3,503,187; March 31, 1970; assigned to Societe des Forges et Ateliers du Creusot, France.
(8) Krazewski, L.; and Zulauf, G.A.; U.S. Patent 3,507,482; April 21, 1970.
(9) Hoad, J.G.; U.S. Patent 3,577,709; May 4, 1971; assigned to John G. Hoad & Assoc., Inc.

# CHLORIDES

Chlorides are present in practically all waters (1). They may be of natural origin, or derived (a) from seawater contamination of underground supplies; (b) from salts spread on fields for agricultural purposes; (c) from human or animal sewage; or (d) from industrial effluents, such as those from paper works, galvanizing plants, water softening plants, oil wells, and petroleum refineries (2). Herbert and Berger (3) have reported waste chloride concentrations from a Kraft process paper mill of 350 to 1,760 mg./l.

An industrial waste study on blast furnaces and steel mills (4) reveals that average size steel mills using advanced production technology produce from their pickling line 9,700 lbs./day of chloride as the ferrous salt, and 4,500 lbs./day as hydrochloric acid. The waste stream from the pickling line contains a total chloride content of 7,000 mg./l. This is diluted to a as little as 15 mg./l. chloride upon mixing of the pickling line waste stream (approximately 0.24 MGD) with the waste flow from other steel processes (approximately 113 MGD).

Oil refineries and petrochemical processes are major producers of high chloride wastes. Crude oil desalting is a major source of chloride in petroleum refinery wastes. Berger (5) describes a refinery effluent as containing total dissolved solids of 1,163 mg./l., primarily as sodium chloride dissolved out of crude oil by desalting. Petrochemical processes, including chlorination, oxidation, polymerization and alkylation produced chloride wastes, usually associated with calcium or aluminum (6).

The Solvay process, widely used in making soda ash, yields wastewaters high in chloride. Soda ash is a major raw material of many industries, including glass, chemicals, pulp and paper, soap and detergent, aluminum and water treatment. The Solvay process uses sodium chloride brine solution, and yields a substantial amount of waste calcium chloride (7). Jones has reported that the chloride content of a typical soda ash manufacturing waste is 90,000 mg./l. (10). While some calcium chloride is recovered by distillation, this practice is not sufficiently prevalent to prevent the Solvay process having a major waste disposal problem (7).

Frequently chloride ion is added to a wastewater as a result of waste treatment required to remove some less desirable constituent. Sodium chloride and hydrochloric acid are often used to regenerate spent ion exchange resins used in water purification. Chloride ions on the regenerated resins then exchange with undesirable anions as the wastewater passes through the ion exchange bed. Another example of chloride addition is chlorination treatment of cyanide waste streams.

### Removal from Water

Two of the most frequently employed methods of controlling high chloride wastes involve techniques that are more closely ultimate disposal than chloride reduction per se. These methods are deep well injection and holding basins, including solar evaporating ponds. Each method is briefly discussed below. Methods of actually removing chloride ion from a waste stream include anion exchange and electrodialysis, in addition to other, broadly applicable, total dissolved solids reduction methods.

*Deep Well Injection:* Henkel (8) has reported the deep well disposal of chemical industry wastewaters containing 10 to 15% sodium chloride (60,000 to 90,000 mg./l. of chloride) at volumes of 500 to 600 gpm. The waste stream also contained dissolved trace metals and trace organics. Other reports of deep well disposal of high chloride wastes are also in the literature (6). Capital investment and operating costs for deep well disposal are given in a recent federal report (7). Injection pressures of 200 to 5,000 psi have been reported in the literature for chloride wastes (6).

*Ponds and Basins:* The concepts behind the use of holding ponds and solar evaporating basins are similar. Holding ponds are used to store wastewater in times of low receiving stream flow, with gradual wastewater release to the stream in high flow periods at rates necessary to maintain acceptable stream quality standards. Solar evaporating ponds utilize the slow loss of water by evaporation from the sun's heat to avoid any release of wastewater at all to receiving streams. A discussion of treatment of petroleum refinery effluents, including crude desalting wastes, emphasizes that disposal by evaporation is severely limited by geographical location, climate and land availability (9). It is obviously a very attractive method where stringent effluent regulations are in force, and the geographical, climatic and land conditions are favorable. Pollutant removals are essentially complete providing floating oils, which retard evaporation, are removed.

Janes (10) has reported the use of a holding pond and controlled discharge of a chloride waste (Pittsburgh Plate Glass Co., Barberton, Ohio), where underground disposal was prohibited. The waste stream contained approximately 90,000 mg./l. of chloride. A holding basin, having a capacity of 150,000 ft.$^3$, was constructed at a cost of approximately $1,000,000 (including land, 1958 prices). Discharge of chloride waste from this basin (into the Ohio River) was proportionate to river flow.

*Ion Exchange:* Higgins (11) has discussed the use of ion exchange to remove chlorides, and points out that removal of strong acid and acidic ions such as chloride, sulfate, nitrate and phosphate anions requires a weak base type ion exchange resin. Essentially complete removal of the undesirable ions is achieved, until the resin exchange capacity is exhausted. Exchange capacity can be regenerated with weak alkali-ammonia solution (11); cost data are presented.

*Electrodialysis:* Electrodialysis has been employed for chloride treatment, particularly with saline and brackish waters in the municipal water supply industry. Smith and Eisenmann (12) have reported electrodialysis in a wastewater recycle operation, whereby chlorides were reduced from 115 to 80.5 mg./l. Other ions in the waste stream were also reduced. Katz (13) in discussing a desalination electrodialysis plant (Dell City, Texas), reported chloride reduction from 122 mg./l. down to 58 mg./l., in addition to reduction of other ions.

In summary, there are relatively few reports in the technical literature on treatment of chloride wastes. High levels of treatment can be achieved, however, by both ion exchange and electrodialysis as discussed above. For small waste volumes, costs per 1,000 gallons treated are quite high.

However, the economy of scale resulting from larger treatment plants results in significant reduction of unit costs. In addition to ion exchange and electrodialysis, holding basins and evaporative ponds, and deep well injection have been employed to dispose of concentrated chloride wastes. As more stringent regulations are placed on subsurface disposal, the choice of industry will probably become holding basins and evaporative ponds, according to Patterson et al (1).

One specific ion exchange process for chloride and other mineral pollutant removal from water has been described by De Pree (14).

Th process is shown schematically in Figure 27. The water containing ionic pollution is introduced by line 10 to the mixed bed ion exchange 12. The deionized water is discharged through line 14. The spent mixed ion exchange material is passed via line 16 to a separator 18 where the material is segregated into cation ion exchange material and anionic ion exchange material. The former is then passed to cation ion exchange material regeneration zone 20 and the latter to anionic ion exchange material regeneration zone 22.

Cation regenerant is introduced to zone 20 via line 24 and anion regenerant is introduced to zone 22 via line 26. The spent cation regenerant is discharged from zone 20 via line 28 and the spent anion regenerant is discharged from zone 22 via line 30. The regenerated cation ion exchange material is passed through line 32 and joins with the regenerated anion ion exchange material from line 34 in line 36 for recycle to the mixed bed ion exchange 12.

Th spent cation regenerant is liberated from the pollutant cations in zone 38 and the spent anion regenerant is effectively freed of the pollutant anions in zone 40. The zone 38 is supplied with metal ion precipitant generated as a by-product in zone 40 via line 42. The metal carbonates formed in zone 38 are cycled to zone 40 via line 44 where they are

## FIGURE 27: BLOCK FLOW DIAGRAM OF ION EXCHANGE PROCESS FOR CHLORIDE REMOVAL FROM WASTEWATERS

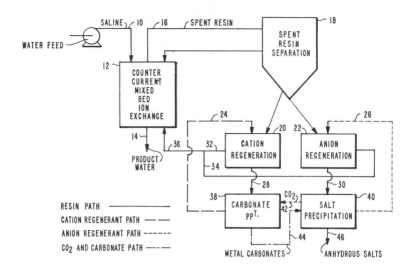

Source:  D.O. De Pree; U.S. Patent 3,700,592; October 24, 1972

treated to yield metal salt products removed at line **46**. As can be seen, the mineral pollutants are recovered as anhydrous precipitates for easy disposal rather than as a concentrated aqueous stream which must be further treated. All chemicals used in the regeneration of both the ion exchange resins and in the salt recovery are recovered and recycled, resulting in maximum economy rather than consumption as in conventional processes.

The process can be further modified to provide for the selective isolation of specific anions and cations and their recombination to recover specific salts preferentially through the use of specially selected chelating agents and amines in the regeneration steps.

### References

(1) Patterson, J.W. and Minear, R.A., "Wastewater Treatment Technology," *Report No. PB 204,521,* Springfield, Va., National Technical Information Service (August 1971).

(2) McKee, J.E. and Wolf, H.W., *Water Quality Criteria,* California State Water Quality Control Board, Publication No. 3-A (1963).

(3) Herbert, A.J. and Berger, H.F., "A Kraft Bleach Waste Color Reduction Process Integrated with the Recovery System," *Proc. 15th Purdue Industrial Waste Conf.,* pp. 49–57 (1962).

(4) *The Cost of Clean Water, Vol. III, Industrial Waste Profiles, No. I – Blast Furnaces and Steel Mills,* U.S. Dept. Interior, Washington, D.C. (1967).

(5) Berger, M., "The Disposal of Liquid and Solid Effluents from Oil Refineries," *Proc. 21st Purdue Industrial Waste Conf.,* pp. 759–767 (1966).

(6) Gloyna, E.F. and Ford, D.L., *The Characteristics and Pollutional Problems Associated with Petrochemical Wastes – Summary,* U.S. Dept. Interior, Washington, D.C. (1970).

(7) *The Economics of Clean Water, Vol. III, Inorganics Chemical Industry Profile,* U.S. Dept. Interior, Washington, D.C. (1970).

(8) Henkel, H.O., "Deep Well Disposal of Chemical Waste Water," *Proc. 5th Annual Sanitary & Water Resources Eng. Conf.,* Vanderbilt Univ., p. 26 (1966).

(9) *The Cost of Clean Water, Vol. III, Industrial Waste Profiles, No. 5 – Petroleum Refining,* U.S. Dept. Interior, Washington, D.C. (1967).

(10) Jones, M.W., "Construction and Operation of a Chloride Holding Basin," *Proc. 16th Purdue Industrial Waste Conf.,* pp. 186–192, (1961).

(11) Higgins, I.R., "A Unique Process for Demineralizing Waste Water," *Indust. Water Engrs.*, 2, 26–29 (1969).

(12) Smith, J.D. and Eisenmann, J.C., "Electrodialysis in Waste Water Recycle," *Proc. 19th Purdue Industrial Waste Conf.*, pp. 738–760 (1964).

(13) Katz, W.E., "Electrodialysis Saline Water Conversion for Municipal and Governmental Use," *Proc. Western Water and Power Symp.*, pp. 113–125, Los Angeles, California (1968).

(14) De Pree, D.O.; U.S. Patent 3,700,592; October 24, 1972; assigned to Aerojet-General Corporation.

# CHLORINATED HYDROCARBONS

## Removal from Air

A process developed by M.R. Zimmermann (1) employs a device for the recovery of solvents from air, especially of per- and trichloroethylene, by the introduction of a filter containing activated charcoal into an air supply canal for the air containing the solvent for the purpose of adsorption of the solvent from the air onto the activated charcoal.

The process utilizes subsequent switching over of the filter chamber for the purpose of desorption of the solvent from the activated charcoal by means of flushing steam by connection on one side to an individual steam generator provided for the desorption and on the other side to a condenser for the purpose of subsequent separation of flushing steam and solvent by condensation, the steam generator being directly connected, by means of its steam chamber, to the filter chamber.

Figure 28 shows a twin arrangement of two activated charcoal units operating alternately in the adsorption and desorption operation. As shown there, the two activated charcoal devices are operated alternately in the adsorption and desorption operation, especially in order to recover solvents, such as per- and trichloroethylene, from the air, such as is present, for example, in machines for cleaning textiles, plants for degreasing metals or in an extraction plant. Of the two activated charcoal chambers **1** and **2**, at any one time, one of them, for example, the left-hand chamber **1**, works in the adsorption operation and the other one, for example, the right-hand chamber **2**, works in the desorption operation.

**FIGURE 28: APPARATUS FOR THE ADSORPTION RECOVERY OF CHLORINATED SOLVENTS FROM THE AIR**

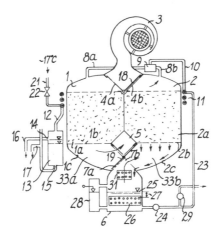

Source:  M.R. Zimmermann; U.S. Patent 3,486,305; December 30, 1969

In the case of the adsorption operation, the air to be purified and from which the solvent is to be recovered, is drawn in by means of a fan **3**, via an entry opening **4a**, and forced into the left-hand activated charcoal chamber **1** and, after purification, blown back via an exit opening **5** into the atmosphere.

In the case of the desorption operation in the right-hand chamber **2**, flushing steam is passed from a steam generator **6** via a steam entry opening **7b** into the right-hand chamber **2** and emerges, loaded with solvent, via a pipe **8b**. Via a differential pressure valve **9** and a pipe **10**, this steam-solvent vapor mixture is passed to a countercurrent condenser **11**, from which the liquid mixture of steam condensate and solvent condensate flows via a pipe **12** to a separation vessel **13** for water and solvent.

From this separation container, the steam condensate is taken off at a comparatively high point **14** and the solvent condensate, which has a higher specific gravity, is taken off from the bottom at **15**. The steam condensate flows off via a pipe **16**, while the solvent can be led off via a pipe **17** to a collecting vessel for reuse.

After a certain period of time, depending upon the operating conditions, for example, after one or two hours, in which, as is known from experience, the activated charcoal **1a** in the filter chamber **1** has become enriched with solvents to such an extent that a continuation of the adsorption would be uneconomical and in which, at the same time, in the right-hand filter chamber adsorbed solvent has been flushed out of the activated charcoal **2a** to such an extent that a continuation of the desorption operation would be uneconomical, the functions of the two filter chambers are changed by switching over the left-hand chamber from adsorption to desorption and switching over the right-hand chamber from desorption to adsorption.

For this switching over, there are provided flaps **18** and **19** which, for the switching over, are tilted by about 90° in a counterclockwise direction. By means of the flap **18**, the air entry opening **4b** of the chamber **2**, which was closed during the desorption operation in the chamber **2**, is opened to the fan **3** and the air entry opening **4a** for the filter chamber **1** is closed. By means of the flap **19**, the steam entry opening **7a** to the filter chamber **1**, which was closed during the adsorption operation in the filter chamber **1**, is opened and the air entry opening **5** of the chamber **1** is closed.

The switch-over of the steam exit canals **8a** and **8b** of the filter chambers **1** and **2** is carried out automatically by the differential pressure between the two chambers, which acts upon the valve **9** constructed as a differential pressure valve and adjusts this valve in such a manner that the connection of the canal **8b** to the pipe **10** is closed and the connection of the exit canal **8a** to the pipe **10** is opened.

The differential pressure valve **9** is, on the one hand, subjected to the pressure of the air from the fan **3** of, for example, 0.03 atmosphere, which acts via the connection **8b** on the differential pressure valve and via the connection canal **8a** from the pressure of the steam-solvent vapor mixture in the filter chamber **1**, which amounts to about 0.2 to 0.3 atmosphere.

The feed water for the steam generator **6** is preheated in the condenser **11**. For this purpose, the condenser consists of a coil of tubes concentrically enclosing one another. The inner tube coil is connected to the water pipe **21** via a closure valve **22** and the cooling water has an entry temperature of about 17°C. in countercurrent to the steam-solvent vapor mixture to be condensed, which flows through the outer pipe coil. The cooling and feed water emerges from the condenser with a temperature of up to about 100°C. and passes via a feed pipe **23**, provided with a magnetic valve **24**, into the steam generator **6**.

The magnetic valve **24** is operated in response to the water level **25** in the steam generator **6**, so that in the steam generator there is continuously maintained a head of water **27** above the immersed electrical heating body **26**. For this purpose, a water level regulator **28** is attached to the steam generator **6** which controls the magnetic valve **24** via a switch contact operated by the liquid level. If, after the desired water level in the steam generator is reached, the magnetic valve is closed, then an overflow valve 29 opens a branch pipe 30 to an overflow in order to maintain the water flow in the condenser 11. The overflowing water can be used again. The steam formed in the steam generator is superheated to 140° to 150°C. by a superheater 31 in the steam chamber of the steam generator so that desorption takes place with dry steam.

The activated charcoal fillings 1a and 2a rest upon perforated plates 1b and 2b, respectively, which thus, in the desorption operation, form the upper limit of the steam chamber of the steam generator. Particles of activated charcoal entrained with solvent and falling through the perforated plate must not, when, for example, the solvent is chlorinated or fluorinated hydrocarbons which are subject to dangerous chemical conversion reactions be heated up to the critical temperature at which the chemical change of the solvent used takes place.

These charcoal particles falling down from the filter chamber must, therefore, be prevented from coming into contact with the hot surfaces of the heating bodies 26 and 31. A direct contact with the hot surface of the heating body 26 is prevented since this heating body is continuously submerged in water 27 so that particles of charcoal reaching the steam generator fall into water heated to about 100°C. Solvent present in the charcoal particles is evaporated, together with the water, and passes into the recovery process.

A contact with the hot surfaces of the superheater 31 is, in the example illustrated, prevented in that the superheater lies in a part 32 of the steam chamber which is arranged laterally to the steam chamber region which lies vertically under the perforated plate 2b. For the particles of activated charcoal and droplets of condensate falling from the filter chamber, there is thus provided a continuous fall space which leads from the activated charcoal chamber to the water level 25 of the immersion evaporator and is free of heating body surfaces.

The fall spaces 33a and 33b below the perforated plate 1b and 2b, respectively, are defined at the bottom by guiding surface 1c and 2c, respectively, which are inclined downwardly to the steam generator 6, and guide condensate and particles of activated charcoal into the steam generator without coming into contact with the heating bodies of the steam generator.

A process developed by J.E. Panzarella (2) involves an improved method for the recovery of valuable organic compounds and deleterious air polluting chemicals from water-containing process off-gases. It comprises [1] contacting the process off-gases with a cold surface upon which a film of water-miscible liquid is maintained; [2] separating the noncondensible gases from the condensate which forms upon contact of the off-gases with the cold wall and the water-miscible liquid; [3] separating the water-miscible liquid, water and condensate one from the other; [4] returning the water-miscible liquid to the first step to form the film on the cold wall and [5] recovering the condensates free of the water-miscible liquid and water.

Figure 29 illustrates the arrangement of apparatus used in the conduct of the process. Off-gas and an ethylene glycol-H$_2$O mixture are led through a conduit 10 into the top of the cooling zone 11 so that it covers the interior walls of the zone.

The liquid-gas mixture is then removed from the zone 11 by way of a conduit 12, and led into a gas-liquid separator 13. The gases so separated are vented to the atmosphere via a conduit 14, and the liquids removed via a conduit 15 to a separator 16 which removes the condensed organics, which exit via a conduit 17.

It is to be noted that if a water-miscible liquid is used which is miscible with the condensate as well, one or more stills would be required in the place of the phase-separator 16. The glycol-H$_2$O mixture, now diluted over that introduced into zone 11, is removed by a conduit 18 to a concentrator 19, where sufficient H$_2$O is removed to raise the glycol concentration to the desired level. The excess H$_2$O is removed via a conduit 20, and the concentrated glycol-$H_2O$ is removed via a conduit **21**. It is to be understood that the process may be run in a countercurrent fashion with equal and often increased efficiency.

## FIGURE 29: APPARATUS FOR RECOVERY OF CHLORINATED SOLVENT FROM AIR BY CONDENSATION

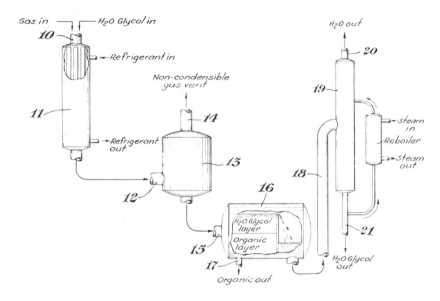

Source: J.E. Panzarella; U.S. Patent 3,589,104; June 29, 1971

The following are specific examples of the operation of the process.

*Example 1:* A vertical tube and shell condenser was utilized for the experiment. Liquid ammonia (-41°C.) was the coolant in the shell. The feed gas was 87.95 volume percent $N_2$, 3.18 volume percent $H_2O$ and 8.87 volume percent chlorinated hydrocarbons. The hydrocarbons were primarily ethylene dichloride and smaller amounts of ethyl chloride, cis- and transdichloroethylene and β-trichloroethane.

The feed gas (38°C.) entered the tube side of the condenser at a flow rate of 614 scf/hr. and a pressure of 7.8 psig. Expansion into the system dropped the pressure to 0.28 psig. A 60 weight percent aqueous solution of ethylene glycol, at a rate of 6.56 lb./hr. and a temperature of -28°C., was also fed into the tube so that it flowed down and covered the walls of the tube. The gas and glycol flow was concurrent.

The gas, at -21°C., exited the condenser at 545 scf/hr., and contained 0.576 volume percent chlorinated hydrocarbons. The exiting glycol solution, separated from the condensate by phase separation, contained 52 weight percent glycol due to $H_2O$ dilution from the feed gas. Only 400 ppm glycol was detected in the recovered chlorinated hydrocarbons.

*Example 2:* Utilizing the same condenser as in Example 1, a gas (91.33 volume percent $N_2$, 1.71 volume percent $H_2O$ and 6.96 volume percent chlorinated hydrocarbons), at a temperature of 33°C. was fed into the condenser tube at 504 scf/hr. and a pressure of 7.2 psig. Expansion into the system dropped the pressure to 0.164 psig. A 65 weight percent aqueous solution of ethylene glycol (-29°C.), at a rate of 6.60 lb./hr., entered the tube as before.

The gas, at -22°C., exited the condenser at 473 scf/hr. and contained 0.49 volume percent chlorinated hydrocarbons.

## Removal from Water

As pointed out in *Chemical Week* for December 6, 1972, p. 39, with traditional disposal routes such as open-pit burning, ocean discharge and deep-well burial being choked off by legislation, there is sharpening interest in systems to dispose of such hard-to-treat wastes as chlorinated hydrocarbons. *Chemical Week* estimates that chlorinated hydrocarbon residues are generated at a rate of 350,000 tons/year in the United States alone.

A process developed by R.G. Woodland et al (3) is an improved process capable of converting halogenated organic residue materials to hydrogen halide recoverable as a salable product. Halogenated organic residues from halogenation processes, and in particular chlorination and bromination processes, such as the residues obtained from the processes for manufacturing materials such as hexachlorocyclopentadiene, benzoyl chloride, benzyl chloride, chlorendic acid, and the like, have in the past presented a difficult disposal problem.

These halogenated organic residues are toxic to plant and animal matter so that they generally cannot be released into rivers or lakes, nor can they be dumped on land where the drainage therefrom will reach waters used for human consumption. Attempts to disposed of these halogenated materials by burning, using conventional furnace apparatus, have generally not been satisfactory, principally because the large amount of halogenated material in these residues made them difficultly combustible or resulted in a combustion gas having an appreciable free halogen content which corroded the equipment used and contaminated the surrounding atmosphere. Moreover, in such a disposal process, recovery of the free halogen from the combustion gases, in a salable form, is virtually impossible so that the halogen content of the residue materials being treated was lost.

In an effort to overcome the above difficulties, a process was developed involving the use of a self-regenerative furnace where these halogenated organic residue materials are converted into a hydrogen halide product containing substantially no free carbon or halogen and organics. Moreover, in this process, the furnace operates continuously, thus not requiring intermittent shutdowns to regenerate the needed heat of ignition, and, additionally, generally does not require auxiliary fuel during the disposal operation, thereby reducing the disposal cost.

A suitable form of apparatus for the conduct of the process is shown in Figure 30. The furnace is shown as having an outer cylindrical fire wall 1, vertically positioned on a foundation 2 at the bottom and covered at the top with refractory brick 3 and an insulated manhole cover 4 which serves as an explosion release means. The outer fire wall is formed with a refractory material 5 on the inside, which has a low heat conductivity and which is resistant to free halogen, oxygen, and hydrogen halide, such as mullite, a composition of aluminum silicate. The refractory material is backed by fire brick 6 and an air gap 7 separates the fire brick from a gas-tight steel shell 8.

The furnace has an inner cylindrical fire wall 9 concentrically aligned with the outer fire wall so that an annular space 10 is formed between the inner and outer walls. The inner fire wall 9 is positioned on the foundation and is in communication at the bottom with an insulated outlet 11. This inner fire wall is in open communication with the top inner space 12 of the outer fire wall.

Preferably, the inner fire wall is constructed of heat conductive refractory material, such as silicon carbide brick, so that there is an exchange of heat from the inside 13 of the inner wall cylinder through the refractory material to the annular space between the inner and outer fire walls.

A feed inlet 14 is positioned in the outer fire wall. While this inlet may be intermediate the top and bottom of the inner fire wall and formed at a downward angle and tangential

FIGURE 30:  INCINERATOR DESIGN FOR DISPOSAL OF LIQUID CHLORINATED
HYDROCARBON WASTES

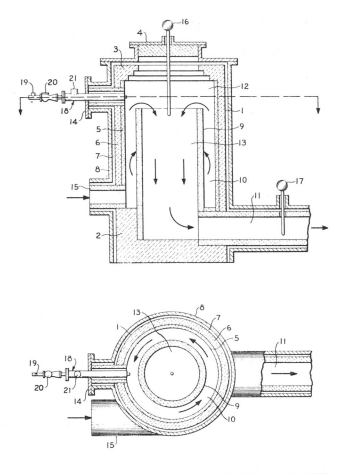

Source:   R.G. Woodland and M.C. Hall; U.S. Patent 3,445,192; May 20, 1969

to the cylindrical outer fire wall, it is preferably positioned horizontally above the inner
fire wall and is radial with respect to the outer fire wall.  A second inlet **15** is positioned
in the outer wall, preferably adjacent the bottom of the annular space between the inner
and outer walls.  This inlet is preferably located below the first inlet **14** and is tangential
to the annular space.

This second inlet means is preferably used to start up the furnace by injecting air and hydrogen or fuel gas therein, and afterward to control or regulate the composition of the
final product by injecting steam or air or a hydrogen source, etc., as required.  Temperature indicating means **16** and **17** are positioned in the top inner space **12** of the outer wall
and in the exit **11**, respectively.

Positioned within the inlet **14** is a mixer-nozzle means indicated as **18**.  This mixer-nozzle
has a steam inlet **19**, an inlet **20** for the halogenated organic material and an inlet **21** for
air or steam.  The mixer-nozzle is formed with an inner conduit portion **22** through which

the mixture of steam and halogenated organic materials is introduced into the annular combustion chamber 10 of the furnace and an outer conduit means surrounding the inner means so as to form an annular conduit 23 around the inner conduit 22. The inlet 21 is in communication with this conduit 23. Suitable flanges are provided for securing the mixer-nozzle in the inlet 14 of the furnace.

In actual operation, a furnace constructed in the manner shown and having an inside volume of about 60 cubic feet, was preheated to a temperature of about 900°C. A halogenated organic material was then metered into the mixer-nozzle 18 through the inlet 20 at a rate of about 425 lbs./hr. This halogenated organic material was comprised of substantial quantities of the following materials: $CCl_4$, $C_2Cl_4$, $C_2Cl_6$, $C_4Cl_6$, $C_5Cl_8$ and $C_6Cl_6$.

The overall average composition of this feed material was about 20% carbon and 80% chlorine. Steam, at the rate of 300 to 500 lbs./hr. and a pressure of 70 psig was also added to the mixer-nozzle 18 through the inlet 19. Within the mixer portion 24 of the mixer-nozzle, the steam and halogenated organic material was intimately admixed and the mixture vaporized as it passed through the second constricted zone 28 into the second expanded zone 29 and into the inner conduit 22 of the nozzle portion 25 of the apparatus.

Air, at the rate of about 660 lbs./hr., was also added to the combustion zone 10 of the furnace, a portion of this air being added to the inlet 21 in the second conduit 30 and the remainder being introduced through the bottom furnace inlet 15. Burning of the atomized mixture within the combustion chamber 10 of the furnace took place about 2 to 3 feet from the end of the nozzle, the burning being characterized by a bright flare.

Additionally, it was noted that there was no fall-out from the atomized combustion mixture, i.e., no unburned or partially burned particles built up on the relatively cool lower portion of the combustion chamber 10. The exit gases from the outlet 11 were comprised of about 350 lbs./hr. of hydrogen chloride, 311 lbs./hr. of carbon dioxide, 507 lbs./hr. of nitrogen, and the balance excess steam and oxygen with traces of carbon monoxide. The exit gases contained no detectable free chlorine, carbon, or organics. The hydrogen chloride in this exit gas stream may then be recovered as a salable product by any known means, such as a hydrogen chloride absorption apparatus.

A process developed by A.M. Essex et al (4) involves the recovery of organic solvents from aqueous mixtures by passing steam therethrough, condensing steam and solvent, separating water and solvent by decantation, discharging waste products and repeating the cycle, whereby solvent cleans the apparatus. The apparatus and process are useful for recovering solvents from photopolymerizable materials containing ethylenically unsaturated monomers and macromolecular polymers.

The apparatus for carrying out this process comprises a container for holding the spent solvent, a steam source, a condenser, a separation decanter, means for drawing off the polymer- and monomer-laden residue after each batch of solvent is processed, means for injecting additional spent solvent into the container, and electrical means for automating the entire apparatus.

The process and apparatus eliminates fouling of the working parts due to the buildup of residue in two ways. First, no solid heating surfaces are used, which minimizes the tendency of the polymer and monomer to solidify on those surfaces. A slight buildup occurs initially on the pipe conveying the steam into the container, but this is limited since the residue itself gradually becomes a heat insulator and no more buildup can occur.

Second, the use of a batch process determines that any residue left after a given batch of solvent is purified will be dissolved by the next batch of spent solvent introduced into the container. This process allows recovery of 80 to 98% of the spent solvent.

Figure 31 shows such an apparatus. The process is begun with the activation of a positive displacement gear pump 1 which draws in the spent solvent with dissolved polymer and

## FIGURE 31: STEAM DISTILLATION PROCESS FOR CHLORINATED HYDROCARBON REMOVAL FROM WATER

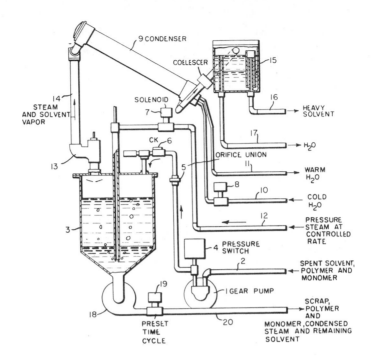

Source: A.M. Essex and R.B. Heiart; U.S. Patent 3,666,633; May 30, 1972

monomer from an outside source through pipe **2**. The pump forces the solvent into the container **3** for a preset time based on the amount of solvent which vessel or container **3** is designed to hold. For example, if the container is to be charged to 5 gallons, the pump could introduce the solvent at a rate of 2½ gal./min. for 2 minutes.

The entrance line **2** is equipped with a pressure switch **4** and an orifice union **5** which provides for automatic stopping of the system if no liquid pressure is detected for 5 seconds during the "fill cycle." The solvent passes through a check valve **6** which prevents back flow of the solvent when steam is entering the container.

At the end of the filling time, the pump automatically stops and the fill cycle is over. At that instant, solenoid valves **7** and **8** open. Water flows through valve **8** from pipe **10** through the shell- and tube-type condenser **9** and out of pipe **11** in order to cool the condenser and liquefy the vapor that will be entering it. Valve **7** admits steam from pipe **12** under pressure at a controlled flow rate.

The steam is bubbled through the solvent and after a few minutes distillation begins. A thermo-switch **13** senses the rise in temperature as the distillation proceeds. The distillation temperature is less than 100°C. and gradually climbs as the steam to solvent ratio increases during distillation.

The nonvolatile impurities are left behind in the container. The mixture of steam and solvent vapor travels up pipe **14** and into condenser **9**. There it liquefies. The liquefied mixture then travels to the decanter 15 where the heavier liquid is separated by gravity from the lighter liquid and is drained through and collected from pipe 16. The water is drained through pipe 17. It can be determined experimentally at what temperature, detected by the temperature switch 13, optimum recovery of the solvent is obtained.

For example, using methyl chloroform as the solvent to be recovered, 90% of the solvent will have been distilled and recovered when the vapor temperature at the thermo switch 13 reaches 70°C. At this point, the temperature switch 13 closes valves 7 and 8 thus ending the steam or distillation cycle, and activates a centrifugal pump 18 and opens solenoid valve 19. Pump 18 sucks out the scrap polymer and monomer left behind in the container after distillation along with the condensed steam and the remaining amount of solvent.

The waste leaves through pipe 20. The pump 18 operates on a preset time cycle, then stops. Solenoid valve 19 closes. At this point, the gear pump 1 is activated and another batch of spent solvent is introduced into the container and the process begins again as just described. The new batch of solvent dissolves any scrap polymer and monomer which the pump 18 was not able to remove. Thus all working parts remain clean throughout recovery and the apparatus may be used continuously without fear of gumming or fouling by residue buildup. When the parameters of operation have been settled upon, the apparatus works on its own normal cycle without manual intervention.

The process and apparatus can be used to purify and recover any solvent which is at least partially immiscible with water. The process is useful in recovering methyl chloroform, methylene chloride and trichloroethylene. The spent solvent generally contains 0.1% to 10% polymer composition by weight. During the steam or distillation cycle, the vapors given off are mostly solvent and a relatively small amount of steam by volume. Thus the process is extremely efficient.

Separation of the liquefied water and solvent can be made more complete by the addition of a liquid coalescer between condenser 9 and the decanter 15. This increases the size of the droplets of the internal phase going into the decanter 15 and thus aids separation. A specific example of how this process is used follows.

The gear pump 1 drew in spent methyl chloroform and charged the container at a rate of 2.5 gal./min. The desired amount of solvent in the container was 5 gal., thus time delay relay 25 was set to time out in 2 minutes, ending the fill cycle. As described above, the steam cycle automatically begins. Steam was introduced into the solvent in the container at a rate of about 30 lbs./hr. After approximately 23 minutes, the vapor temperature as detected by the thermo switch 13 had risen from 67° to 70°C., the point at which optimum recovery of solvent was obtained.

The switch 13 automatically turned off the steam and ended the steam cycle. The pump out cycle began. Pump 18 drew out the residual polymer-monomer mixture and steam condensate. Time delay relay 23 had been set for 35 seconds. When it timed out, the pump 18 stopped, ending the pump out cycle. The next batch of spent solvent was then introduced by pump 1 into container 3, dissolving any scrap that pump 18 had not emptied and thus cleaning the apparatus as well. The process went on as before. The recovery and purification of the methyl chloroform was 88%; i.e., 88% of the spent solvent was purified.

A process developed by E. Feder et al (5) is one in which steam-volatile organic solvents are removed from process wastewaters by intimately mixing the process wastewaters with steam to form an azeotropic steam mixture, withdrawing the azeotropic steam mixutre from the resultant mixture of steam and water, and condensing the azeotropic steam mixture. The major amount of process wastewater, thus freed of solvent content, is discharged as general sewage.

Figure 32 shows a suitable form of apparatus for the conduct of the process. In the drawing, the process wastewater, fed by a pump 2, is supplied, in a horizontally positioned

FIGURE 32:  STEAM STRIPPING PROCESS FOR CHLORINATED SOLVENT REMOVAL
FROM WASTEWATER

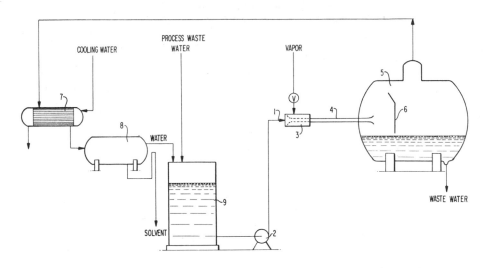

Source:  E. Feder, K. Deselaers and G. Czehovsky; U.S. Patent 3,669,847; June 13, 1972

pipe **1**, with a sufficient quantity of steam by way of an injector **3**, made of a corrosion-proof material, that a mixing temperature of 90° to 100°C. is obtained as the ambient condition at the end of the mixing zone disposed behind the injector and fashioned as a pipe **4**.  This mixing zone has a length of about 2 meters and has a turbulent flow therein.

The mixing pipe **4** terminates in a voluminous settling and separating tank **5** having a baffle **6**, from which the liquid, aqueous or sludge laden phase can be discharged at the bottom, whereas the azeotropic steam mixture is withdrawn at the head (top) and is condensed in a condenser **7**.  A relatively small amount of solvent-saturated water and anhydrous (water-free) solvent mixture is obtained in a two-phase separator **8** connected downstream of the condenser.  The residual proportion of the water is combined with the process wastewater in a collecting tank **9**.  The following are specific examples of the utilization of the process.

*Example 1:*  Wastewater having a chlorinated hydrocarbon content is fed in an amount of 5 to 6 cubic meters per hour by means of pump **2** into the apparatus of the process.  The wastewater is mixed in the injector **3** with the required amount of factory steam of 4 to 8 atmospheres gauge so that a temperature of 90° to 100°C. is present at the end of mixing pipe **4**.

The following results are obtained.  Gas chromatographic analysis indicates a content of 4,011 ppm of dichloroethane and 381 ppm of perchloroethylene in the supplied wastewater.  The water leaving the settling tank exhibits a solvent content of 62 ppm of dichloroethane and 14 ppm of perchloroethylene.

*Example 2:*  Wastewater containing solvent and sludge is fed at the rate of 7 cubic meters per hour into collecting tank **9**.  An amount of steam is admixed therewith such that a temperature of 96°C. is obtained at the end of mixing pipe **4**.  The results obtained by gas chromatography analysis are as follows.  The analysis shows, in the supplied wastewater, solvent contents of 2,733 ppm of 1,2-dichloroethane and 313 ppm of perchloroethylene.  The water discharged from the settling tank has a solvent content of 66 ppm of 1,2-dichloroethane and 70 ppm of perchloroethylene.

References

(1) Zimmerman, M.R.; U.S. Patent 3,486,305; December 30, 1969.
(2) Panzarella, J.E.; U.S. Patent 3,589,104; June 29, 1971; assigned to the Dow Chemical Co.
(3) Woodland, R.G. and Hall, M.C.; U.S. Patent 3,445,192; May 20, 1969; assigned to Hooker Chemical
    Corporation.
(4) Essex, A.M. and Heiart, R.B.; U.S. Patent 3,666,633; May 30, 1972; assigned to E.I. du Pont
    de Nemours and Company.
(5) Feder, E., Deselaers, K. and Czehovsky, G.; U.S. Patent 3,669,847; June 13, 1972; assigned to
    Dynamit Nobel AG, Germany.

## CHLORINE

Low concentrations of chlorine gas cause irritation of the eyes, nose and throat of humans (approximately 3,000 $\mu$g./m.$^3$ or 1 ppm is the threshold value). Excessively prolonged exposures to low concentrations, or exposure to higher chlorine concentrations, may lead to lung diseases such as pulmonary edema, pneumonitis, emphysema, or bronchitis. Recent studies indicate that other residual effects may occur, such as a decrease in the diffusing capacity of the lungs. Furthermore, there is evidence which suggests that continuous exposure to low concentrations may cause premature aging and increased susceptibility to lung diseases, according to Stahl (1).

Chlorine is also a phytotoxicant which is stronger than sulfur dioxide but not as strong as hydrogen fluoride. Several episodes have been reported in which chlorine emissions from accidental leaks or spillages have resulted in injury and death to humans, animals, and plants. Material damage from chlorine is possible, since chlorine has strong corrosive properties.

The major chlorine production processes involve the electrolysis of alkali chloride solution. Possible sources of chlorine pollution from these processes include the liquefication process, the filling of containers or transfer of liquid chlorine, and the emission of residual gas from the liquefication process. No information is currently available on the concentrations of chlorine gas in ambient air. However, chlorine has been classified and is presently analyzed as one of the "oxidants" of the air.

Effective methods are available for controlling chlorine emissions. In some cases the chlorine or its by-products can be recovered for further use. No information has been found on the economic costs of chlorine air pollution or on the costs of its abatement. Methods of analysis are available; however, they are not sufficiently sensitive or selective for determining atmospheric concentrations of chlorine.

### Removal from Air

In plants producing chlorine gas, the electrolytic cells, the chlorine coolers, and the drying systems are operated at a slight vacuum, which normally prevents the emission of chlorine from these systems. In spite of this precaution, an upset or malfunction of the pumps or a large leak in the system may cause a temporary discharge of chlorine to the atmosphere.

The major source of chlorine emissions, however, is the presence of residual gases remaining after the liquefication of the chlorine cell gas. These residual gases may contain from 20 to 50% chlorine by volume, depending on the conditions of liquefication. Other sources of emission include the loading and cleaning of tank cars, barges or cylinders; dechlorination of spent brine solutions and power or equipment failures (2)(3)(4).

To prevent air pollution from these sources, the emissions from these systems can often be piped to the sniff gas system or tied in directly either to a scrubber system or a high stack for dispersion. When scrubbers are used, the effluent from the scrubber is vented

to a tall stack.  Several efficient methods have been developed to control chlorine emissions
(2)(4)(5).  The most common of these methods include the use of liquid scrubbers which
employ either water (6)(7), alkali solutions (5)(8), carbon tetrachloride (7)(9)(10)(11), or
brine solution (12) as well as solid absorbents such as silica gel (13).

With the exception of the alkali solution technique, all these methods can be used to recover the chlorine.  Other control methods which have been reported use sulfur dichloride
(14), hexachlorobutadiene (15) and stannic chloride (9).  Some producers send the untreated emission gases containing chlorine to other in-plant operations, where they are used
directly, for such operations as chlorination of hydrocarbons (2).

For example, a process developed by Kraus (16) involves removing free chlorine from residual tail gases formed in a chlorination of titaniferous ore at temperatures ranging from 400°
to 1300°C. by reacting the free chlorine in the present of $CO_2$ with a stoichiometric amount
of a low molecular weight hydrocarbon having the formula $C_nH_{2n+2}$ where n is 1 to 4.

*Water Scrubbers:*  When controlled by water scrubbers, the chlorine-containing vent gas is
passed countercurrently to a water stream in a tower filled with ceramic packing.  Upon
treatment with a water scrubber, a vent gas initially containing 15% chlorine by volume
yields an effluent gas containing 15,000,000 to 30,000,000 $\mu$g./m.$^3$ (5,000 to 10,000 ppm)
of chlorine (6).

It is common practice to pass gases from the water scrubber through the more efficient
alkali scrubbers, or to tall stacks for dispersion.  The chlorine-rich scrubber solution is
heated so that the chlorine may be stripped and recovered.  An alternate method of treatment of the chlorine-rich water is to pass the water over activated charcoal or iron filings.
The result is an oxidation-reduction reaction that converts the chlorine to the noninjurious
chloride ion.

*Alkali Scrubbers:*  Contact of chlorine with alkali solutions (usually caustic or lime solutions) produces an effluent gas with a lower residual chlorine concentration (often below
the odor threshold of about 1.0 ppm) than can readily be attained by water scrubbing (1).
The reaction products are bleach, salt and water.  The main disadvantage of the method is
the cost as well as the difficulty of disposing of the hypochlorite bleach solution.

Such bleach solution is sometimes reused by local plants (such as pulp and paper mills) or
can be treated with carbon to reduce the chlorine to chloride ion.  However, some producers dispose of the solution by dumping it in rivers and streams.

One specific lime scrubbing process has been described by C.J. Howard et al (17).  It is an
improved process for treatment of gases containing small amounts of chlorine to separate
and recover therefrom chlorine gas in concentrated and purified form.

Commercially there is produced as a by-product from chlorine producing plants, dilute
waste gases referred to in the industry as tail gas or "sniff" gas.  Discharge of such dilute
chlorine gases into the atmosphere, would pollute the atmosphere and further would result
in a loss of valuable chlorine.

While it is possible and indeed often necessary due to governmental regulation to convert
the chlorine in the dilute gases to another form, e.g., sodium hypochlorite or calcium hypochlorite, such operations are generally not profitable because of the low concentration
of the chlorine in the gases.

Chlorine in purified and concentrated form is recovered from dilute chlorine gas by a
process of passing the dilute chlorine gas in contact with a calcium hydroxide slurry to
effect reaction of the chlorine and calcium hydroxide to produce an aqueous solution of
calcium hypochlorite and calcium chloride.  The relative proportions of calcium hydroxide and chloride are regulated to provide an excess of not more than 20% calcium hydroxide, preferably 1 to 6% excess calcium hydroxide in the reaction mixture.  The clear liquor

containing dissolved calcium hypochlorite and calcium chloride is separated from insoluble solids contained in the reaction mixture. The clear liquor is introduced into a reaction zone where the clear liquor is admixed and reacted with sulfuric acid to produce chlorine and calcium sulfate dihydrate.

Chlorine is discharged from the reaction zone, maintaining the concentration of calcium sulfate dihydrate solids in the reaction zone in an amount of 20 to 40% solids suspended in any part of the liquid in the reaction zone. Sufficient acid is added to the reaction mixture in the reaction zone to reduce the pH of the clear liquor to a pH 1.2 to 1.7.

The temperature of the reaction mixture is maintained within the range of 30° to 60°C. The slurry of calcium sulfate dihydrate is discharged from the reaction zone. The calcium sulfate dihydrate slurry is separated into a more dilute slurry and a more concentrated slurry of calcium sulfate dihydrate. The more concentrated slurry of calcium sulfate dihydrate is then returned to the reaction zone in an amount sufficient to maintain at least 20% concentration of solids in the slurry in the reaction zone.

A flow diagram of the process is given in Figure 33. The first step of the operation involves the absorption of chlorine from tail and "sniff" gases from a chlorine plant by passing the gas countercurrent to a falling stream of calcium hydroxide slurry. The lime slurry is chlorinated until its chlorine absorption capacity is nearly exhausted at which point it can be referred to as "spent lime slurry."

FIGURE 33: APPARATUS FOR RECOVERY OF CHLORINE FROM WASTE GASES
BY ALKALI SCRUBBING

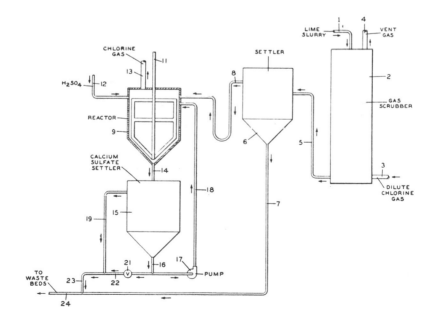

Source:  C.J. Howard and E.B. Port; U.S. Patent 3,357,796; December 12, 1967

Referring to the drawing the lime slurry in which the amount of calcium hydroxide may vary from 2% to in excess of 10% by weight of the slurry is passed through line **1** into gas scrubber **2** which may be any suitable vessel equipped with baffles or plates or filled with packing to effect intimate contact between the gas and lime slurry.

Dilute chlorine gas containing minor amounts of chlorine of the order of a few percent down to as low as a fraction of a percent chlorine is passed through line **3** into the bottom of gas scrubber **2**, then upwardly through the gas scrubber countercurrent to the down flowing lime slurry and the scrubbed gas exits into the atmosphere via vent **4**. The amount of lime slurry fed to the gas scrubber is controlled so that only a small excess of lime remains in the slurry leaving the gas scrubber via line **5** to settler **6**. The absorption of chlorine from dilute chlorine gases by passing the gases in contact with calcium hydroxide slurry in gas scrubber **2** may be illustrated by the following equation:

$$2Ca(OH)_2 \ + \ 2Cl_2 \ \longrightarrow \ Ca(OCl)_2 \ + \ CaCl_2 \ + \ 2H_2O$$

Insoluble matter settles to the bottom of settler **6** and is withdrawn via line **7**. Relatively clear liquor containing calcium hypochlorite, calcium chloride and a minor amount of calcium hydroxide overflows from settler **6** through line **8** to reactor **9**. The overflow through line **8** constitutes generally well over 90% of the liquor with the remaining few percent being discharged through line **7**.

Reactor **9** is agitated preferably with a paddle type agitator **11**, so that a fairly uniform suspension of calcium sulfate dihydrate is maintained therein. Sulfuric acid aqueous solution is fed to reactor **9** through line **12**. Desirably the entry of line **12** is placed so that it is well removed from the entry point of line **8** to avoid precipitation of poorly settling $CaSO_4 \cdot 2H_2O$ which often results when these two streams are introduced at the same point in the reactor.

The concentration of solids in the slurry is kept at a level such that at least 20% solids are present in any given part of the slurry volume, with overall solids in suspension kept in the range of 20 to 40%. Substantially lower concentrations give rise to poorer crystals and a slurry difficult to settle, while substantially higher concentrations produce slurries that are difficult to pump and agitate.

The acid addition is controlled so that free chlorine is liberated from the solution which is maintained at a pH below 2, preferably about 1.5 (measured in liquor cooled to room temperature). Chlorine leaves the reactor via line **13** which chlorine is of a sufficient purity and concentration that it generally can be combined directly with the chlorine stream produced in the plant, e.g., line **13** can be connected directly to the header drawing off chlorine from the electrolysis cells. The temperature of the reactor is kept below 90°C. and may be operated as low as about 10°C. but is preferably operated within the range of 30° to 60°C.

The slurry from reactor **9** passes through line **14** to calcium sulfate settler **15**. Here the slurry is thickened so that a slurry containing 30 to 50% suspended solid $CaSO_4 \cdot 2H_2O$ is collected at the bottom of the tank. This is withdrawn through line **16** to pump **17** and returned through line **18** to reactor **9**. The thin slurry from settler **15** overflows through lines **19**, **23** and **24** to waste beds. This slurry may be combined with the settled solids from calcium hypochlorite settler **6** which has sufficient alkali to raise the pH of the combined streams to above **7**. Excess thickened slurry from settler **15** can be drawn through valve **21** and line **22** and sent to waste beds via lines **23** and **24**.

The reaction in reactor **9** between the clear liquor containing calcium hypochlorite and calcium chloride to produce chlorine gas which is recovered and calcium sulfate precipitate may be illustrated by the following equation:

$$Ca(OCl)_2 \ + \ CaCl_2 \ + \ 2H_2SO_4 \ \longrightarrow \ 2Cl_2 \ + \ 2CaSO_4 \ + \ 2H_2O$$

In some plants there is available waste lime and waste sulfuric acid which products often present a disposal problem. The process permits the use of waste lime and waste sulfuric acid to recover a valuable product present in dilute form in a waste gas and converts sulfuric acid and waste lime into water and an insoluble solid composed primarily of calcium sulfate dihydrate. Such a procedure reduces pollution of the air and water table in the area surrounding the plant.

The reader is also referred to the earlier section of this handbook on Aluminum Chloride, Removal from Air for discussion of the treatment of gas streams containing both aluminum chloride and chlorine.

*Carbon Tetrachloride Scrubbers:* The advantage of using $CCl_4$ as an absorbent is that its absorbing capacity for chlorine gas is 10 to 12 times greater than water, and the recovery of chlorine is much more complete (4). However, losses of $CCl_4$ have been reported as high as 30 pounds per ton of chlorine recovered (10).

*Acid Scrubbers:* A process developed by M.E.J. Cathala (18) consists essentially in dissolving the chlorine, contained in a mixture of gases, in chlorosulfonic acid, maintained preferably at a low temperature, and in degasifying the solution thus obtained in a tower or in a degasifying apparatus of known type, in which a suitable temperature gradient and pressure are maintained, and then in expanding the hot chlorosulfonic acid passing but of the degasifying apparatus in an appropriate apparatus in any known manner.

**Removal from Water**

A process developed by L.A. Fabiano (19) involves the vacuum dechlorination of acidic alkali metal chloride brine effluent from a mercury cathode electrolytic cell. Chlorine is recovered and an effluent environmentally acceptable is produced using steam jet vacuum with rectification to remove chlorine to acceptable amounts in the aqueous effluent.

In the operation of an electrolytic chlorine plant using mercury cells, the weak brine effluent from the cells is partially depleted in salt content and saturated with chlorine. It is ordinarily treated for recycle to the cells by dechlorination, suitably with vacuum followed by blowing with air, resaturation by contact with solid salt, purification, particularly with respect to iron and other metals introduced as contaminants with the salt, by the addition of caustic soda, soda ash and/or barium carbonate or barium chloride followed by settling and/or filtration to remove the precipitated metal compounds. The purified alkalized brine is acidified preferably to a pH of about 3 as described in detail in U.S. Patent 2,787,591 and recycled to the cells.

The dechlorination of the effluent brine is partly for the purpose of recovering chlorine and partly to reduce the corrosiveness of the brine to equipment used in fortifying and purifying the brine for recycle to the cells. Steam jets are the cheapest and most effective vacuum producing means for this service.

However, complete dechlorination is uneconomical when the cost of dechlorination exceeds the value of the recovered chlorine plus the cost of replacing corroded equipment or the cost of avoiding corrosion by chemical means. While it is cheaper in some circumstances to discharge unrecovered chlorine to the environment, this is a currently unacceptable practice. It is therefore a principal object of this process to provide a method for reducing the chlorine content of the effluent to such levels that residual chlorine discharged from the operation is below limits acceptable to avoid environmental contamination of rivers, streams, lakes or other bodies of water or ground or air and to accomplish the above objects economically.

These and other objects are accomplished by vacuum dechlorination using steam exhaust to produce the required vacuum including the steps of:

[1] vaporizing overhead from the brine a stream of water and chlorine

containing a major proportion of the chlorine dissolved in the brine;

[2] recycling the resulting dechlorinated brine to a brine preparation system for fortification, purification and reuse in the cell;

[3] fractionating the overhead stream of water and chlorine to provide a second overhead stream of water and chlorine and an aqueous bottom product containing less than 250 ppm of chlorine;

[4] condensing the second overhead stream of chlorine and water together with the condensate from the steam exhaust and recycling the resulting mixed condensates by combining them with the acidic brine from the cells.

The objects are further accomplished in a particularly advantageous mode of vacuum dechlorination in two stages where the second stage is the method described above and the first stage is a preliminary stage where the steps are:

(a) vaporizing overhead from the brine from the cell a first stream of water and chlorine containing a major proportion of the chlorine dissolved in the brine;

(b) condensing the first overhead stream of chlorine and water together with the condensate from the steam exhaust and recycling the resulting mixed condensates by combining them with the acidic brine from the cells;

(c) charging the resulting partially dechlorinated brine to the second stage.

The single stage version of the process is shown schematically in Figure 34. In cell **11**, brine is electrolyzed to produce chlorine removed by line **12** and amalgam removed by line **13** to decomposer **14**. In the decomposer, the amalgam reacts with water introduced by line **15** to form hydrogen removed by line **16** and aqueous caustic removed by line **17**. Denuded mercury is removed by line **18** and returned to cell **11**. The chlorine-saturated brine is removed from cell **11** by line **19** to brine receiver **20**. Chlorine disengaged in the brine receiver is removed by line **21** under control of valve **22** and combined with chlorine in line **12** and transferred to the chlorine recovery operation (not shown).

Brine removed from receiver **20** is passed by line **123** to flash chamber **42**. The overhead therefrom is removed by line **43** to separator **44** and separated liquid phase is returned by line **45** to liquid line **46**. The stream of chlorine and water vapor removed from separator **44** is transferred by line **47** to rectifier column **48** in which the bottoms are heated by steam lines **48A**. Chlorine-free water is removed from rectifier column **48** by line **49** to storage **50** where it is neutralized and discharged to effluent by line **51**.

Water vapor and the remaining chlorine is removed from rectifier column **48** by line **52** to separator **53** which provides reflux to rectifier column **48** by line **54**. Vacuum is produced by steam jet **55** supplied with steam by line **64** and operatively connected to separator **53** by line **56**. Pressure is controlled by controller **57** which measures the pressure in rectifier column **48** by line **58** and operates valve **59** to admit air through line **60** to line **56** as necessary to maintain suitable pressure.

Condensate from steam jet **55** containing dissolved chlorine passes by line **61** to condenser **62**. Uncondensed gas from condenser **62** passes to steam jet **63** supplied with steam via line **65**. Condensate from steam jet **63** is returned by line **166** to brine receiver **20**. Condensate from condenser **62** is removed by line **67** to brine receiver **20**. It is advantageous in some circumstances, in order to handle the volume of gases, to provide duplicates (not shown) of steam jets **55** and **63** and to arrange the duplicates in parallel with jets **55** and **63**.

The liquid in line **46** is chlorine-free brine which is returned to weak brine tank **68**. Weak brine is removed therefrom by line **69** to brine purification system **70**. Purified brine is removed therefrom by line **71** to strong brine tank **72** for storage. Strong brine is transferred by line **73** to head tank **74**, acidified by hydrochloric acid introduced by line **75**. The acidified brine is fed to cell **11** by line **76**.

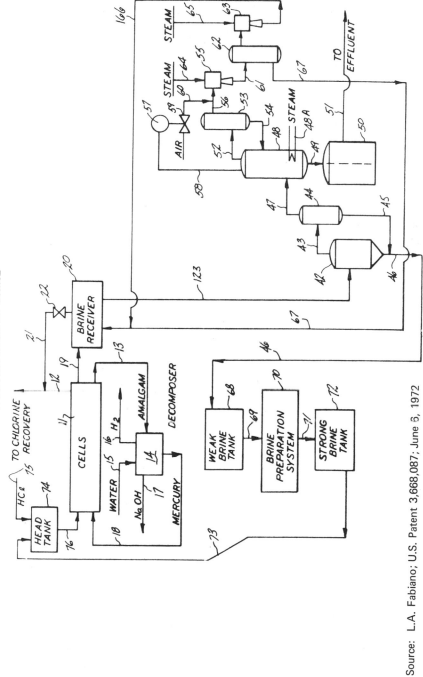

FIGURE 34:  FLOW SHEET OF PROCESS FOR DECHLORINATION OF BRINE

Source:  L.A. Fabiano; U.S. Patent 3,668,087; June 6, 1972

The following is a specific example of the operation of the process. In a single stage operation substantially as shown in the accompanying figure, chlorine-containing brine at a temperature of 185°F. under atmospheric pressure was charged at a rate of 857,800 lbs./hr. to a flash chamber operating at a pressure of 8.67 inches of mercury. The brine contained 23.6% of sodium chloride and 0.025% of dissolved chlorine. The liquid effluent from the flash chamber amounting to 844,550 lbs./hr. of completely dechlorinated brine at a temperature of 166°F. was transferred to the weak brine tanks for fortification, purification, and recycle.

The overhead from the flash chamber, containing about 220 lbs./hr. of chlorine, passed through the separator to the chlorine rectifier column with the bottom temperature maintained at 137.5°F. and the pressure at 5.86 inches of mercury. The liquid effluent from the rectifier column was substantially pure water containing about 0.015% or 150 ppm of chlorine. This concentration is acceptable for discharge into rivers. Substantially all (99.3%) of the chlorine in the brine charged was recovered and returned to the brine receiver and eventually to the chlorine recovery system.

### References

(1) Stahl, Q.R., "Air Pollution Aspects of Chlorine Gas," *Report PB 188,087,* Springfield, Va., National Technical Information Service (September 1969).

(2) "Atmospheric Emissions from Chlor-Alkali Manufacture," U.S. Dept. of Health, Education, and Welfare, Public Health Service, National Air Pollution Control Administration and Manufacturing Chemists' Association, Inc., *Report AP-80* (1971).

(3) Manufacture of Chlorine and Sodium Hydroxide, *J. Air Pollution Control Assoc.,* 14 (3), 88 (1964).

(4) Stern, A.C., (ed.), *Air Pollution,* vol. III, New York, Academic Press, pp. 234–238 (1968).

(5) "The Restriction of Chlorine Gas Emissions," *Verein Deutscher Ingineure, Kommission Reinhaltung der Luft,* Duesseldorf, Germany, VDI 2103 (1961).

(6) Bryson, H.W., *Oregon State Coll. Eng. Expt. Sta. Circ.,* 29, 147 (1963).

(7) "For Chlorine Recovery Take Your Choice, *Chem. Eng.,* 64 (6), 154 (1957).

(8) Cady, F.H., *A Kraft Mill Waste Chlorine Gas Recovery Scrubber,* presented at the Second Annual Meeting, Pacific Northwest International Section, Air Pollution Control Association, Portland, Oreg. (November 5–6, 1964).

(9) Hulme, R.E.; U.S. Patent 2,765,873; October 9, 1956; assigned to Diamond Alkali Co.

(10) Molyneux, F., "Recovery of Chlorine from By-Product Hydrochloric Acid, *Chem. & Process Eng.,* 43, 267 (1962).

(11) Sutter, R.C., "Recovery of Chlorine from Air-Chlorine Mixtures, *J. Air Pollution Control Assoc.,* 7, 30 (1957).

(12) Henegar, G.P. and Gardiner, W.C.; U.S. Patent 3,052,612; September 4, 1962; assigned to Olin Mathieson Chemical Corp.

(13) Wynkoop, R.; U.S. Patent 2,800,197; July 23, 1957; assigned to National Distillers & Chemical Corp.

(14) Karpiuk, R.S.; U.S. Patent 2,881,054; April 7, 1959; assigned to Dow Chemical Co.

(15) Sconce, J.S., *Chlorine-Its Manufacture, Properties, and Uses,* American Chemical Society Monograph Series, New York, Reinhold (1962).

(16) Kraus, P.B.; U.S. Patent 3,485,577; December 23, 1969; assigned to E.I. du Pont de Nemours & Co.

(17) Howard, C.J. and Port, E.B.; U.S. Patent 3,357,796; December 12, 1967; assigned to Allied Chemical Corp.

(18) Cathala, M.E.J.; U.S. Patent 2,868,325; January 13, 1959.

(19) Fabiano, L.A.; U.S. Patent 3,668,087; June 6, 1972; assigned to Olin Corporation.

## CHROMIUM

Chromium (Cr) is commonly known for its use as a decorative finish in chrome plating. Most of the chromium ore produced is used in the production of stainless and austenite steel. However, chromium chemicals appear to be more important than chromium or chromium ore in relation to air pollution. Chromium concentrations in the urban air average 0.015 $\mu$g./m.$^3$ and range as high as 0.350 $\mu$g./m.$^3$. Although the exact sources of chromium air pollution are not known, some possible sources are the metallurgical industry, chromate-producing industry, chrome plating, the burning of coal, and the use of chromium chemicals as fuel additives, corrosion inhibitors, pigments, tanning agents, etc.

The exposure of industrial workers to airborne chromium compounds and chromic acid mists, particularly the hexavalent chromates, has been observed to produce irritation of the skin and respiratory tract, dermatitis, perforation of the nasal septum, ulcers, and cancer of the respiratory tract. Chromium metal is thought to be nontoxic. Hexavalent compounds appear to be much more harmful than trivalent compounds, with the toxic effects depending on solubility.

Two effects that appear to be particularly important in relation to air pollution are hypersensitivity to chromium compounds and induction of cancers in the respiratory tract. Exposure of industrial workers in the chromate-producing industry has shown an incidence of deaths from cancer of the respiratory tract which is over 28 times greater than expected. Time-concentration relationships for induction of cancer are not known.

No evidence of damage by airborne chromium to animals or plants has been found. Chromic acid mists have discolored paints and building materials. In 1964, atmospheric concentrations of total chromium averaged 0.015 $\mu$g./m.$^3$ and ranged as high as 0.350 $\mu$g./m.$^3$, according to Sullivan (1).

Chromium occurs in nature as both the trivalent ($Cr^{+3}$) and the hexavalent ($Cr^{+6}$) ion. Hexavalent chromium present in industrial wastes is primarily in the form of chromate $CrO_4^=$) and dichromate ($Cr_2O_7^=$). Chromium compounds are added to cooling water to inhibit corrosion. They are employed in manufacture of paint pigments, in chrome tanning, aluminum anodizing and other metal plating and electroplating operations.

In the metal plating industry, automobile parts manufacturers are probably the largest producers of chromium-plated metal parts. Frequently, the major source of waste chromium is the chromic acid used in such metal plating operations.

Hexavalent chromium treatment frequently involves reduction to the trivalent ($Cr^{+3}$) form prior to removing the chromium from the industrial waste. Thus trivalent chromium in industrial waste may result from one step of the waste treatment process itself, that of chemical reduction of hexavalent chromium.

In addition to the hexavalent form, some few mg./l. of trivalent chromium may be encountered in raw plating wastes, even prior to chemical reduction of $Cr^{+6}$. For example, Anderson and Iobst (3) report for one process waste a hexavalent chromium concentration ranging from 0.0 to 18.0 mg./l. and a total chromium concentration of 2.9 to 31.8 milligrams per liter. By difference, their data indicate a trivalent chromium concentration in this one waste stream of 2.9 to 13.9 mg./l.

These same authors report for an acid bath waste stream a hexavalent content of 122 to 270 mg./l. and trivalent chromium levels of 37 to 282 mg./l. Thus, according to these authors, trivalent chromium represents a major component of this acid waste, even before reduction of hexavalent chromium. After reduction the trivalent content of this latter waste stream would be approximately 160 to 550 mg./l.

## Removal of Chromium from Air

Chromium air pollution usually occurs as particulate emissions, which may be controlled by the usual dust-handling equipment, such as bag filters, precipitators, and scrubbers according to Sullivan (1). Chromium poses no peculiar control problems except when it is emitted as aerosols (e.g., sprays and chromic acid mists).

Some chrome-plating facilities have installed methods of preventing air pollution. Moisture-extractor vanes in the hood-duct system have been used to break up bubbles in the exhaust gases (4). Most exhaust systems use slot hoods to capture the mists discharged from the plating solutions. The device most commonly used to remove air contaminants from exhaust gases in the hard-chrome-plating facilities is a wet collector. The scrubber water becomes contaminated with the acid; therefore, efficient mist eliminators must be used in the scrubber to prevent a contaminated mist from being discharged to the atmosphere. The mist emissions from a decorative-chrome-plating tank with lesser mist problems can be substantially eliminated by adding a suitable surface-active agent to the plating solutions. The action of the surface-active agent reduces the surface tension, which in turn reduces the size of the hydrogen bubbles. Several of these mist inhibitors are commercially available.

In a new plating facility, there will be an attempt to replace the hood system with a coating of synthetic material that will float on the electrolyte. The coating is expected to suppress the chromic acid mist to prevent it from leaving the bath (1).

### Removal of Hexavalent Chromium from Water

Reduction of hexavalent chromium from a valence state of plus six to plus three, and subsequent precipitation of the trivalent chromium ion is the most common method of hexavalent chromium disposal. To meet increasingly stringent effluent standards, some industries have turned to ion exchange to treat chromate and chromic-acid wastes. Both the reduction and ion exchange processes are discussed here with respect to the degree of treatment achievable and costs associated with treatment. Evaporative recovery of concentrated chromate and chromic acid wastes has also proven technically and economically feasible as a hexavalent chromium waste treatment alternative, and is also considered.

*Reduction:* The standard reduction treatment technique is to lower the waste stream pH to 3.0 or below with sulfuric acid and convert the hexavalent chromium to trivalent chromium with a chemical reducing agent such as sulfur dioxide, sodium bisulfite or ferrous sulfate. The trivalent chromium is then removed, usually by precipitation with lime (5)(6). The reduction of hexavalent to trivalent chromium is not 100% effective, and the amount of residual nonreduced hexavalent chromium depends upon the allowed time of reaction, pH of the reaction mixture, and concentration and type of reducing agent employed. Treatment of chromic acid waste does not normally require pH adjustment with sulfuric acid, as the pH of the chromic acid waste itself is sufficiently low for the reduction reaction to proceed.

The treatment of a metal electroplating waste containing 140 mg./l. of hexavalent chromium has been reported (7). Reduction was carried out at pH 2.5 to 2.8, employing sodium bisulfite as the reducing agent. The process was reported to reduce the hexavalent chromium to concentrations of 0.7 to 1.0 mg./l. The Electric Autolite Company (Sharonville, Ohio) also attempted to use sodium bisulfite to reduce the waste from chrome plating automobile bumpers (8). They abandoned the use of sodium bisulfite as a reductant because of odors and corrosion hazards associated with its use, and now employ sulfur dioxide in its place.

Sulfur dioxide appears to be the most popular reducing agent used in treatment of chromium wastes. Hulse et al (9) have reported its use at a Boeing plant to treat chromium in metal finishing wastes. The waste was adjusted to pH 2.5 with sulfuric acid before reduction. Schink (10) also reports the use of sulfur dioxide, to treat chromic acid plating bath and chromic acid etch bath wastes.

At an average waste flow of 40 gal./min., and for a 20 to 30 minute treatment period, Schink reported that residual hexavalent chromium was less than 1.0 mg./l. Hexavalent chromium concentration of the raw waste was not reported.

Sulfur dioxide has also been used as the reductant at an IBM plant (11), to treat a waste estimated to contain 1,300 mg./l. of hexavalent chromium. The chromate wastes were treated with sulfur dioxide at pH 2. Treatment detention time was approximately 90 minutes. Sulfuric acid was added to the waste, to maintain the proper pH. Chemical analyses by the state Department of Health have shown zero hexavalent chromium in the treated effleunt.

An unusual source of sulfur dioxide is employed by the Fisher Body Plant of Elyria, Ohio, to treat the hexavalent chromium in its plating waste (12). Sulfur dioxide is washed out of the power house stack gas, and used as the chemical reducing agent for converting hexavalent chromium to the trivalent form. The process has been in operation since 1959. On occasions when additional reduction is necessary, sodium bisulfite is employed as a supplement. No values for raw or treated waste chromium concentrations are reported by the authors, although it is claimed that the process is highly effective. Lacy and Cywin (20) report that the use of sulfur dioxide for reduction imparts an oxygen demand to the waste effluent, unless the effluent is oxygenated by passing air through it.

Yuronis (13) has provided cost data on the treatment of chromium wastes by reduction. For complete treatment, including pH adjustment, reduction and neutralization, and removal of the trivalent chromium, Yuronis indicates that treatment operating costs range from $55 to $100 per day, dependent on the cost of municipal water to an industry. The basis for his costs is a waste flow of 100 gal./min. at 120 mg./l. of chromium, and a 16-hour day. Based upon his values, and the cost of water in Chicago (25¢/1,000 gal., 1968, Ref. 14), daily operating costs would be about $75 for the waste flow and concentration given. This cost is less than $1.00/1,000 gal. of waste treated.

Zievers et al (14) have also published treatment costs for chromium wastes, which include capital and operating costs for the reduction of hexavalent chromium and subsequent pH neutralization to an acceptable level.

A process developed by R.P. Selm et al (15) involves the reduction and precipitation of chromates in aqueous industrial wastes by the steps of dissolving flue gas containing carbon dioxide into the wastes to carbonate same, passing the carbonated wastes rapidly through a bed of ferrous chips to activate the bed, recycling the wastes that pass through the ferrous chip bed at least once through the steps of dissolving flue gas containing carbon dioxide therein and again passing the wastes through the activated ferrous chip bed to maintain a rapid flow therethrough for reduction of the chromates in the flow of wastes, and withdrawing a portion of the wastes after passage thereof through the ferrous chip bed for disposal as nontoxic waste.

The essential components involved in the conduct of the process are shown in Figure 35. Inlet pipe 1 discharges the chromate-containing water into tank 2. The tank contains three sections, a separator section 3, treating section 4 and outlet section 5. However, for the purposes of this process, the sections could be separate tanks. The separator section contains a bubbler plate 7 or gas discharge nozzle which disperses bubbles of gas adjacent the bottom of the separator section for purposes which will appear below.

The separator section has a drawoff device or pipe 9 spaced upwardly from the bubbler plate and preferably positioned at the desired level of liquid in the separator section. The treating section is separated from the other sections of the tank by spaced transverse baffles 11 and 12 which are perforated to provide a plurality of small openings 13 to freely allow fluid to pass therethrough but retain ferrous metal chips or particles 14.

The baffles and the side walls of the tank define a treating bed 15. The outlet section has a drawoff or pipe 17 into which the treated effluent flows, the pipe being spaced upwardly from the bottom of the outlet section to discharge the effluent from above a desired liquid level therein.

A pump 19 withdraws fluid from the tank outlet section and pumps it under pressure into an absorption tower 20 where it passes through bubble caps or packing 21 finding its way to a bottom storage portion 22 of the absorption tower. A flow line 23 communicates the bottom storage portion with the treating section. A level control 24 operates valve 25 in the flow line 23 to prevent the liquid 26 from falling below a predetermined level

FIGURE 35:  APPARATUS FOR THE TREATMENT OF WASTE CHROMATE SOLUTIONS

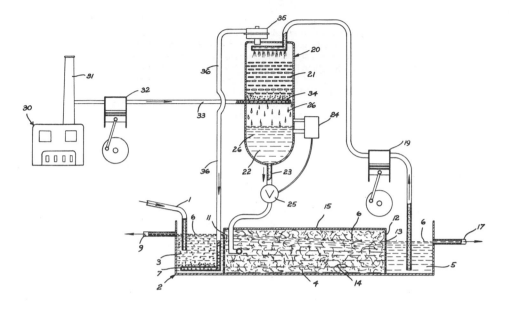

Source:  R.P. Selm and B.T. Hulse; U.S. Patent 3,027,321; March 27, 1962

and thereby prevents the absorption tower from losing gas pressure through the flow line
in case of low fluid inlet flow.  The solution leaves the bottom of the absorption tower
through the fluid line and enters directly into the ferrous metal chip treating bed adjacent
the baffle **11** in the tank.

Carbon dioxide, which, for this process, may be used in the form of waste flue gas, is
obtained by tapping boiler stack **31** of boiler **30**.  The flue gas is pumped to an appropriate pressure by a compressor **32** and is delivered under pressure through a line **33** to the
absorption tower **20** where it is discharged through nozzles **34**.

A pressure regulator **35** on the absorption tower controls the pressure therein to maintain
same substantially at a predetermined pressure.  An excess of carbon dioxide gas is normally delivered from the source of supply to the absorption tower and the pressure regulator
vents the excess through a line **36** to the bubbler plate in the separator section.

The treating section of the tank contains iron material **14** which may be in the form of
chips, particles or other forms, the only requisite being that the treating bed is porous to
the flow of liquid therethrough.  It is highly desirable to prevent short circuiting flow
through the treating bed which may be caused by an uneven distribution of chips or particles therein, or by the level of liquid **6** rising above the upper level of the bed, because
portions of fluid would then pass the treating section without coming into sufficient contact with the treating metal, reducing efficiency.

In practicing this process, wastewater containing chromates in solution enters separator
tank **3** from pipe **1**.  Such water very often contains free oil or small solid particles in
suspension which, for more efficient results, should be separated from the water prior to
the treatment for chromate removal since such materials may coat the iron particles, close
openings in the bed or otherwise interfere with chemical reactions in the treating bed.  In
order to separate the free oil and particles from the water, carbon dioxide gas is delivered

by the line 36 to the bubbler plate and dispersed as bubbles at the bottom of separator section of the tank to aid gravitational separation of the water and such materials by floating the free oil and particles to the surface where they can be drawn off by gravity through the pipe 9.

The bubbles of gas rising in separator tank 3 perform a function in addition to the flotation of foreign material which is, like the absorption tower, to dissolve carbon dioxide into the water to create an acidic condition rendering it highly reactive with ferrous materials. The solution leaves the separator section 3 and flows through the baffle 11 into the treatment bed where, it is believed, a carbonic acid-ferrous corrosion process takes place, forming a reactive ferrous bicarbonate complex which is highly efficient in reducing chromate ions. The products of the process are inert hydroxides of iron and chromium which take the form of precipitates.

As the solution passes through the treating bed, flow resistance thereof or friction loss causes dropping of the level 6 of the solution as it passes toward the outlet section. If desired, the treating bed may be tilted slightly so that its upper surface slopes at approximately the same extent as the solution surface level passing therethrough. It is to be noted that the process may be satisfactorily carried out by passing the solution vertically through the bed or at any convenient angle so long as the solution comes into intimate contact with reactive ferrous metal surface.

When the solution leaves the treating bed, it flows through the baffle 12 into outlet section where part of it is drawn off through effluent pipe 17. The major portion of the solution entering the outlet section, however, passes through a recycle process during which carbon dioxide is dissolved into the solution. In recycling, the pump 19 withdraws the solution from the outlet section and pumps it under appropriate pressure into the absorption tower 20 where it gravitates down through bubble caps or packing 21 in the presence of carbon dioxide.

The vast majority of carbon dioxide dissolution occurring in this process takes place in the absorption tower rather than in the separator section. Since impurities found in common waste flue gas do not adversely affect this process, the carbon dioxide may be pumped into the absorption tower in this form, although it is apparent that it may be obtained from other sources.

A pressure of 40 psig in the absorption tower with a solution temperature of about 78°F. has been found to be satisfactory, although variations of pressure to 150 psig and higher are usable and temperature, within wide limits, is not critical. Economic conditions, such as the cost of powering the pumps, will determine the optimum working pressure in the absorption tower.

It is preferable that the solution in the absorption tower dissolve sufficient carbon dioxide to become supersaturated at atmospheric pressure and operating temperature so that as it enters the treatment bed bubbles of carbon dioxide will form which aid in keeping the solution in a high state of turbulence for maximum contact with the reaction ferrous metal surface, as well as preventing residue from clogging the bed. Supersaturation also helps ensure that the solution pH will be low enough to promote the process reaction regardless of the pH of the raw wastewater.

The use of recycle flow, by increasing the rapid movement of fluid through the bed, permits the handling of large quantities of water in a relatively small apparatus, as well as maintaining high turbulence for efficiency and helping to prevent clogging or plugging of the bed. Recycle flow is also beneficial in that the dilution effect at the inlet of the treatment bed is greater, being directly proportional to the recycle ratio. This prevents the passivation of the bed due to high chromate concentrations at the inlet.

This process has several distinct advantages over prior processes of chromate removal in that only inexpensive treating materials are required, and the process may operate continuously without appreciable labor cost. Ferrous chips or particles are produced in large quantities by metal fabricating plants and are usually sold as waste material. Such material provides a satisfactory ferrous reactive material for charging the treatment bed. As stated above, the carbon dioxide used in this process is suitable in the form of waste flue gas which would otherwise be released to the atmosphere through a boiler stack or the like.

Within wide limts, the conditions of operation of this process need not be adjusted for varying concentrations of chromates in the wastewater since the process does not produce side reactions which result in other toxic compounds if the water is overtreated. If a significant change in chromate concentration is expected to occur over a long period of time, it is usually necessary only to vary the flow through pump **19** and correspondingly adjust the valve **25** to change the recycle rate so that maximum efficiency can be attained without undue loss of ferrous material through excess corrosion.

It has been found that the process conditions are not critical so long as sufficient carbon dioxide is forced into solution. The chromate removal efficiency is generally increased as the carbon dioxide solution concentrations and recycle flow rates are increased. With low chromate concentrations, the process may be carried out satisfactorily if flue gas is simply bubbled through a chip bed at atmospheric pressure, dispensing with the absorption tower and allied equipment. An example of a set of satisfactory conditions for 99+% removal of chromates from the wash water is:

| | |
|---|---|
| Chromate solution feed rate, gal./min. | 1.0 |
| Chromate feed concentration, ppm | 20 |
| Chromate feed pH entering chip bed | 7.0 |
| Chromate feed temperature, °F. | 78 |
| Recycle ratio | 7.0 |
| Recycle flow through absorption tower, gal./min. | 7 |
| Flue gas flow to absorption tower, scfm | 0.80 |
| Flue gas flow vented from absorption tower, scfm | 0.75 |
| Absorption tower operating pressure, psig | 40 |
| Treatment bed, loosely packed chips – | |
|   Width, 30" | |
|   Height, 30" | |
|   Length, 10'6" | |

In the above example, scfm means cubic feet per minute reduced to standard conditions.

A process developed by L.E. Lancy (16) provides a method of quickly and effectively conditioning spent cooling water having a toxic hexavalent chromium solution content to make the water innocuous. It comprises, providing a mass of hard metal sulfide granules, moving the spent cooling water through the mass in contact with surface portions of the granules and surface-reacting the granules with the hexavalent chromium solution content to fully convert it into trivalent chromium, while inhibiting the forming of a soluble sulfide in the water from the sulfide granules, and while maintaining surfaces of the metal sulfide granules reactive to the hexavalent chromium solution content of the water, and removing the water in an innocuous condition from the mass and discharging it.

*Ion Exchange:* Ross, in his text on industrial waste treatment, claims that ion exchange processes can be economically utilized for chromium recovery and elimination of waste (17). Cation exchange can be applied to remove trivalent chromium, as described in a separate report, and anion exchange employed to remove hexavalent chromium in the chromate form normally present in industrial wastes.

When the anion exchange resin is exhausted, it is regenerated (usually with sodium hydroxide), and sodium chromate is eluted from the ion exchange resin. The eluted sodium chromate can be passed through a cation exchange resin to recover purified chromic acid. If chromic acid is not recovered, the concentrated waste yielded by the resin regeneration process must be disposed of. This may be accomplished by reduction to trivalent chromium, followed by lime precipitation. Waste treatment by ion exchange produces a reusable water, which may provide an economic edge when water costs are high.

Yuronis (13) has reported that ion exchange recovery is economically and technically feasible for wastes containing chromate ion concentrations up to 200 mg./l. This study indicated that chromate wastes of 100 to 500 mg./l. were suitable for reduction and precipitation. Wastes with chromate concentration exceeding 500 mg./l. were suitable for evaporative recovery.

Rothstein (18) reported successful ion exchange treatment of a metal finishing waste from Fairchild Camera and Instrument Corp. (Syosset, L.I.) to meet the chromate effluent standard of 0.05 mg./l. This chromate level is equivalent to a hexavalent chromium concentration of 0.023 mg./l. The chromate and chromic acid wastes treated at this plant by ion exchange were segregated from other metal finishing wastes.

Recovery of chromic acid was practiced, and resulted in a five year savings of $8,342 above the cost which would have been required for chromium reduction and precipitation. Rothstein reports that standard reduction and precipitation of this waste would cost $1.25/lb. of chromate removed. This is equivalent to a treatment cost of $1.00/1,000 gallons at a hexavalent chromium concentration of approximately 50 mg./l.

Lacy et al (20) report that the cost of ion exchange treatment depends upon the industrial source of the waste. Waste treatment operating costs are given by these authors as 16 to 24¢/1,000 gallons of waste stream treated. Ross (17) has reported ion exchange costs for a waste stream of 73 gal./min. with a chromic acid concentration of approximately 5,000 milligrams per liter (equivalent to 2,300 mg./l. of hexavalent chromium).

Operating costs for the waste treatment process totaled $72.24 per day, with a credit of $57.30 per day resulting from chromic acid recovery and reuse of purified water. Net operating costs were, therefore, $12.94 per day. By contrast, reduction and precipitation would have cost $33.19 per day for that waste.

Besselievre (6) has reported ion exchange treatment costs of $17 per day for a chromium waste, versus an estimated reduction and precipitation cost of $63 per day. Keating and Duff in 1955 reported a net profit of $10 per day on an ion exchange chromium recovery system, as well as a recycle of 86,500 gallons of treated wastewater to the plant (19).

Yuronis has reported ion exchange costs of 15¢/lb. of chromate removed (13). He estimated capital costs for a plant treating 100 gal./min. and a chromium content of 50 to 100 mg./l. as approximately $40,000. Daily operating costs would be $65,000, or 68¢ per 1,000 gallons.

A process developed by A.J. Saraceno et al (21) is an improved anion exchange process for the removal and recovery of trace quantities of chromates from water. The process utilizes a bed of basic anion exchange resin and the combination of upflow exhaustion of the chromates from the water and downflow regeneration of the resin. Efficient recovery of chromates is accomplished while virtually eliminating chromate leakage during the initial portion of the exhaustion operation. Regeneration can be accomplished without the use of strong, expensive chemicals.

Water available for industrial use often is high in dissolved solids and may be corrosive. To render such water less corrosive for use as a recirculating coolant, it is common to add chromate ions to the water in the form of hexavalent chromium compounds, such as alkali metal chromates. When this water is passed through cooling towers, some of it is lost through evaporation and windage. This loss is reflected in a buildup in the concentration of dissolved solids in the system. Consequently, whenever this concentration reaches a selected level, it is customary to blow down the system by discharging a fraction of the water therein and replacing this blowdown fraction with makeup water containing a lower concentration of dissolved solids. It is desirable to process cooling system blowdown to recover the chromate therein before discharging the blowdown to holding ponds, streams, or sewers. The blowdown from a cooling tower system typically contains 15 to 50 ppm chromate and 500 to 2,000 ppm dissolved solids such as metal phosphates, sulfates, and carbonates.

In some instances, blowdown contains as much as 500 ppm chromate. (Where chromate concentrations are referred to herein they are expressed as $CrO_4$.) Thus, when appreciable quantities of blowdown are involved it is economically attractive to recover the expensive chromate for reuse as a corrosion inhibitor. The economies attending the recovery of chromates from cooling tower blowdown increase with the size of the cooling system, and as is well-known could be large in the case of plants on the scale of the United States gaseous diffusion plants for the separation of uranium isotopes.

Even if reuse of the abovementioned chromate were not a consideration, its removal from blowdown often is necessary in order to meet local regulations on the content of materials discharged to sewers and streams. Proposed Federal, State or local regulations for restricting pollution may limit the chromate content of discharged waters to as little as 2 ppm, and the general trend appears to be toward still lower permissible maximum concentrations, such as a limit of 0.1 ppm for drinking water.

Because of the incentives mentioned above, the removal of chromates from water has been studied intensively by many investigators. Through the years, anion exchange processes have been developed for such recovery. Some of these processes, for example, those described in U.S. Patents 3,223,620, 3,306,859 and 3,414,510, have been described specifically in terms of the recovery of chromates from blowdown water.

All of the known prior processes for the recovery of chromate from water by anion exchange are conducted in at least two steps, an exhaustion step and a regeneration step. In the exhaustion step, it is the practice to pass the chromate-containing water (e.g., blowdown) downwardly through a bed of a basic anion exchange resin to selectively absorb the chromate ions on the resin, where they are retained.

In the course of this exhaustion step, the total ion exchange capacity of the bed is reduced, perhaps to the point where the bed is essentially saturated with respect to chromate. In the subsequent regeneration step, an alkaline solution is passed either upwardly or downwardly through the bed to remove chromate from the resin, thus restoring the ion exchange capacity of the bed to at least a part of its original value.

Unfortunately, the prior art anion exchange processes for recovering chromate from water are subject to various disadvantages. For example, they employ costly regenerating chemicals which, because of their strongly alkaline nature, reduce the life of the anion exchanger resin. Even more important, the prior art processes do not consistently reduce the chromate content of the input water to less than about 3 ppm.

A primary object of the process is to provide an anion exchange process for consistently reducing the chromate content of cooling system wastewater to less than 2 ppm in a single pass of the water through an ion exchange column and under conditions promoting long life of the resin. It is another object to provide an anion exchange process for reducing the chromate content of chromate-treated water to 0.1 ppm or less.

In the process, these objects are achieved by utilizing the discovery that the recovery of chromate from water by anion exchange is unexpectedly and significantly improved by utilizing the combination of upflow exhaustion of the chromate from the water and downflow regeneration of the anion exchange resin. Basically, the process is conducted by adjusting the pH of the chromate-containing water to within the range of 3 to 5.5.

The pH-adjusted water is then passed upwardly through a bed of a basic anion exchange resin to selectively remove the chromates and retain them on the resin. When the chromate loading of the bed reaches a selected value or the chromate concentration in the effluent water reaches a specified maximum, the exhaustion operation is stopped. Subsequently, the chromate-retaining resin is regenerated to a selected degree by passing downwardly through the bed an alkaline aqueous solution which contains regenerant anions.

The regenerant solution leaving the column contains the eluted chromate and, being low in dissolved solids, can be returned to the process system for reuse as a corrosion inhibitor. One of the applications in which this process can be used to advantage is the removal and recovery of chromates from blowdown waters of the kind referred to above, where the chromate concentration typically exceeds 15 ppm but is small compared to the total concentration of dissolved solids.

A process developed by L. Sloan et al (22) permits recovery of chromate ions from the blowdown of a cooling system involving cooling towers. In the case of air conditioning or other cooling systems, water is used to absorb heat and is then passed to cooling towers in which cooling is effected by evaporation of a portion of the water. To minimize corrosion chromate ions are added to the recirculating water usually in the form of sodium chromate or chromic acid.

Because of the fact that evaporation takes place during operation, the salts present in the water which is used become concentrated, and to limit the concentration it is necessary to blowdown the system either continuously or intermittently, bleeding off portions of the water which are then replaced by makeup. The blowdown has been wasted to the sewer, but this has two major objections: first, the expensive chromate ions are thus lost, and second, the chromate ions represent a pollution problem and are in some cases required to be removed but in such fashion that recovery is impractical from the standpoint of cost.

Removal by ion exchange has been used, but has been carried out in expensive fashion. A typical blowdown may include twenty to several hundred parts per million of chromate ion, $CrO_4$. The chromate ions may be accompanied by, typically, 900 ppm of sulfate ions, 300 ppm of chloride ions and 15 to 20 ppm of phosphate ions.

It involves a two-zone anion exchange process, the first of which contains a high concentration of the absorbed ionic material, and a second of which contains a lower concentration of the same absorbed ionic material.

*Evaporative Recovery:* This process consists of evaporating metal plating rinse water, to drive off the water as vapor and thus concentrate the chromic acid for recovery and reuse. Evaporative recovery can be used on almost all process rinse water systems, with the exception of those which deteriorate with use. Culotta and Swanton (23) reported in a case study of plating waste recovery that rinse water containing only a few milligrams per liter of chromic acid could be concentrated to above 900 mg./l. Yuronic (13) reported evaporative recovery operating costs of $2.50 to $10.00 per 1,000 gallons. Capital equipment costs for a continuous flow evaporative recovery system were estimated at $60,000.

In a process developed by A. Yagishita (24) for reclaiming plating wastes containing chromic acid and the like, a plurality of wash tubs are provided into which plated articles are dipped successively to rinse off the plating solution. The tub water from the tub, into which the articles are first dipped, is sucked into a tower heated by steam to concentrate the tub water to plating strength; and this water is returned to the plating tank.

Water is siphoned from the other tubs successively back to the first tub. The vapor in the tower is used in a condenser to heat water flowing to the tub most remote from the first tub, to replenish the system. The tower has an inner container; and the tower proper and the space between this container and the tower proper is held under vacuum.

In another process developed by A. Yagishita (25) for reclaiming plating wastes containing chromic acid and the like, the rinse liquid from the first of a plurality of wash tubs into which plated articles are dipped successively to rinse off the plating solution, is sucked into a tower where it is heated by steam partially to vaporize and concentrate it to plating strength. The vapor generated in the tower is passed through a check valve to an ejector, where it is introduced transversely into a jet of cooling water to be condensed and con–veyed with the stream of cooling water to a reservoir. The cooling-water jet develops a vacuum in the tower; and a trap is interposed between the tower and ejector to trap any cooling water which a faulty check valve might otherwise allow to back up into the tower.

*Chemical Precipitation:* A process developed by R. Richards (26) involves chemically re-moving chromium ions from industrial waste solutions by direct precipitation with barium carbonate in such solutions acidified with nitric or hydrochloric acid or their salts.

A process developed by G.J. Nieuwenhuis (27) is one in which hexavalent chromium con-tained in aqueous metal treating wastes is substantially completely removed from the aque-ous solution by metering the acid wastes into rinse water, metering stoichiometric amounts of a lead compound such as lead nitrate into the rinse water containing the hexavalent chromium to form an insoluble lead chromate precipitate and an aqueous precipitate residue which is subsequently treated with a caustic cleaning solution to neutralize the acids and precipitate the remaining metals and metal compounds in the aqueous treating solution. The residual aqueous solution is suitable for reuse.

Figure 36 shows the essential features of the process in diagrammatic form. In such a metal treating system, the metal products are first subjected to a caustic cleaner to clean the metal surface, secondly dipped in one or more chrome treating solutions to obtain a rust inhibiting surface, and then rinsed in the dipping solution.

FIGURE 36: HEXAVALENT CHROMIUM REMOVAL BY PRECIPITATION AS LEAD
CHROMATE

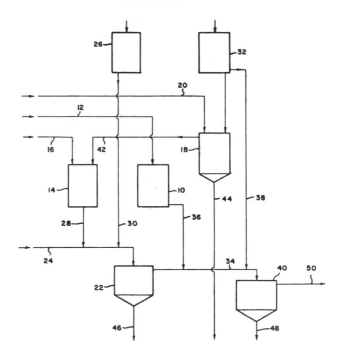

Source: G.J. Nieuwenhuis; U.S. Patent 3,493,328; February 3, 1970

The caustic cleaner is passed to a holding vessel **10** in line **12**. The chromic-sulfuric acid wastes of the dipping solution are passed to a holding vessel **18** in line **20** for a reason that is explained hereinafter. Metal treating solutions are passed to vessel **14** in line **16**. The rinse waters are passed to a first settling vessel, or clarifier, **22** in line **24**.

To briefly complete the description of the figure, the contents of holding vessel **14** and lead nitrate solution from vessel **26**, are metered into rinse water line **24**, in lines **28** and **30**, respectively, line **28** being upstream of line **30** with respect to settling vessel **22**. The contents of vessel **10**, and any required amount of calcium hydroxide from vessel **32**, are metered into the effluent line **34** leading from the first settling vessel **22** to the second settling vessel **40** in lines **36** and **38**, respectively, line **36** being upstream of line **38**, with respect to the second settling vessel, or clarifier, **40**.

Calcium hydroxide from holding vessel **32** is metered into holding vessel **18** and any hexavalent chromium in vessel **32** is passed to holding vessel **14** in line **42**. The precipitate residues from vessels **18**, **22** and **40** are drawn off in lines **44**, **46** and **48**, respectively. Clear water is drawn from the process from the second settling vessel **40** in line **50** for disposal or for reuse in the metal treating system.

The lead chromate product recovered in the process depicted in the figure should preferably precipitate as relatively fine particles. If the particles are too large, they must be ground into smaller particles after recovery before they are useful as pigment. It has been found that lead chromate particle size can be controlled by controlling the concentration of hexavalent chromium entering settling vessel **22**, with smaller concentrations yielding smaller particle sizes.

The concentration of lead and chromate reactants for this particular process should not exceed about 0.05 molar to yield a particle size that meets pigment industry specifications for a first grade lead chromate pigment. Of course if larger sized particles can be tolerated, an increase in the molarity of the lead solution is acceptable. The lead nitrate is metered into line **24** downstream of the input from vessel **14** in the proper amount to combine all of the hexavalent chromium into lead chromate with no excess of lead. Of course, a suitable rinsing and feedback control system would monitor the solution in settling vessel **22** to insure that there is no remaining hexavalent chromium or lead present to pass into the second settling vessel **40**.

The caustic cleaner is metered into line **34** to adjust the pH of the solution passing into settling vessel **40** to 7.0 to neutralize the solution and precipitate the metals leaving ionized water. This ionized water contains primarily sodium and calcium ions and is suitable for cleaning and other purposes. If deionized, the water can also be used as rinse or process water. Should the alkali content of the caustic cleaner be insufficient to neutralize the solution from settling tank **22** without use of undue quantities, calcium hydroxide from vessel **32** can be added to adjust the pH to 7.0.

As mentioned above, if any chrome-containing sulfuric acid is present, such acid wastes are passed to vessel **18** where sufficient calcium hydroxide is added to precipitate gypsum and all toxic metals except hexavalent chromium. This precipitate is useless because of the concentration of gypsum. If trivalent chromium is present in sufficient quantities sodium hydroxide can be used instead of calcium hydroxide to recover trivalent chromium. The remaining hexavalent chromium is passed in line **42** to vessel **14** for processing in the manner described above.

If chrome-containing sulfuric acid wastes were not processed in this manner but rather combined with the other acid wastes, two undesirable results would occur. First, when lead nitrate is added, both lead chromate and lead sulfate would precipitate thereby producing a composite pigment having little or no commercial utility. Second, when basic solution is added, the formation of gypsum would eliminate the opportunity to recover desirable metals such as trivalent chromium from the residue in settling vessel **40**. In the process depicted in the figure, if trivalent chromium is present in such residue in sufficient concentration, it can be economically recovered whereas it could never be economically recovered from gypsum.

*Solvent Extraction:* A process developed by R.R. Dougherty (28) is one in which spent or used aqueous chromic acid etchants are treated to separate chromic acid from by-products of the etching process which are dissolved or suspended in the spent etchants, thereby permitting the by-products of the etching process to be concentrated or isolated for waste disposal and/or recovery of the metal values contained therein.

In a typical process, chromic acid and water are separated from the by-products by acetone extraction. The extract is then stripped of the acetone to produce an aqueous solution of chromic acid which can be reused in the etching process. The raffinate is rich in the by-products of the etching process.

Figure 37 shows the essentials of the process flow scheme. Spent aqueous chromic acid etchant containing metal-containing by-products of the etching process is withdrawn from reservoir **10** and mixed with a water-miscible or water-soluble ketone (e.g., acetone) from reservoir **11**. A mixing head **12** has been shown for this purpose. This mixture is then fed to or introduced to a separation vessel **13** which include a separation zone.

Vessel **13** can take various forms, although an open tank will suffice. Desirably, separation vessel **13** will be equipped with some means (not shown) for continuously or periodically vibrating or agitating the mixture which has been introduced into vessel **13** from mixing head **12**. An ultrasonic vibrator can be used for this purpose. The use of vibrating or shaking means is preferred over vigorous agitation (i.e., the type which completely disrupts the separation of the phases).

However, vigorous agitation can be used if its use is followed by a quiescent period or a period of mild motion. The mixture introduced into separation vessel **13** will separate with time (e.g., five minutes) into an upper or extract phase **14** and a lower or raffinate phase **15**. It has been observed that the extract phase **14** will contain a significant quantity of chromic acid and substantially reduced or even insignificant quantities of the metal-containing by-products of the etching process (e.g., a reduced copper content).

**FIGURE 37: APPARATUS FOR TREATMENT OF USED CHROMIC ACID ETCHING SOLUTIONS BY EXTRACTION WITH ACETONE**

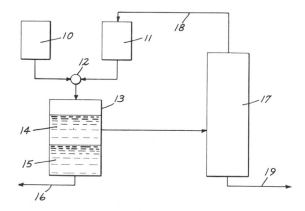

Source: R.R. Dougherty; U.S. Patent 3,531,262; September 29, 1970

The raffinate phase **15** is withdrawn from the separation vessel **13** (e.g., at the bottom) through line **16** and can be disposed of as sewage, or more preferably, subjected to a further treatment to separate the copper and chromium metal values contained therein. Such a separation process can be conducted as an integral part of the process or performed separately, as by a different business concern engaged in the salvage of waste products. The raffinate phase **15** will typically appear as a watery dark green or black sludge or precipitate and may contain some visually observable suspended or floating small solid particles or crystals.

The upper or extract phase (frequently amber or red in color) consisting essentially of acetone, chromic acid, water and probably some sulfuric acid is then removed from separation vessel **13** and delivered to separation vessel **17**. Typically, separation vessel **17** will be distillation apparatus. In vessel **17**, the acetone is removed from extract phase **14** and can be recycled through line **18** to vessel **11**.

The aqueous chromic acid which remains is withdrawn from vessel **17** through line **19**. This reclaimed or recovered chromic acid can then be reused in the etching process. Frequently, the chromic acid which has been recovered will be activated by the addition of some sulfuric acid.

*Reverse Osmosis:* A process developed by A. Geinopolos et al (29) is one in which process water containing hexavalent chromium is pretreated to bring it to a state of dichromate dominance and the water is then subjected to reverse osmosis using a loose membrane. The membrane allows passage of most of the dichromate ions with the product water while concurrently effecting a significant reduction in the hardness and dissolved solids levels. The softened product water containing dichromate ions is recyclable with the recirculated process water and the wasted fraction contains substantially lower amounts of potentially harmful chromium.

*Summary:* Removal of hexavalent chromium ions from a waste stream may be accomplished by reduction and precipitation, ion exchange, or evaporative recovery. Ion exchange and evaporative recovery have the advantage that the chromium may be recovered for reuse. Whether ion exchange is used to recover chemicals, or to concentrate them for further treatment, it always has the advantage of producing reusable water.

Water vapor from the evaporative recovery process may be recondensed, and also reused. If waste flow rates and chromium concentrations are low (less than 50 mg./l.), the total water volume reuse, and the amount of chromium involved does not warrant reclamation. In such a case reduction to trivalent chromium, and precipitation appears most feasible.

## Removal of Trivalent Chromium from Water

Trivalent chromium may be removed by precipitation as the hydroxide upon addition of lime or caustic soda, or it may be concentrated by ion exchange and recovered. The high cost of chromium as an industrial material dictates that ion exchange and recovery be seriously considered as a waste treatment process. However, the literature indicates that at this time precipitation of (trivalent) chromium hydroxide, and disposal of the resulting sludge, remains the most common practice. Both processes will be considered below, with respect to relative degrees of purification achieved and costs associated with treatment.

*Precipitation:* Trivalent chromium can be removed as insoluble chromic hydroxide, $Cr(OH)_3$, by precipitation with caustic soda or lime (6). The chromic hydroxide sludge formed by the process frequently is disposed by landfill. The precipitation process is most effective at pH 8.5 to 9.5 due to the low solubility of chromic hydroxide in that range.

Because a typical metal bearing waste will include other metal ions, as well as chromium to be removed by precipitation, and because the optimum pH of precipitation varies for different metal hydroxides, an average waste stream pH of about 8 seems to achieve the best overall results on a waste containing several metal ions (14). Thus much of chromium treatment reported in the literature occurred at pH 7 to 8. Schink (10) has reported chromium precipitation at an even higher pH of 9, in an electroplating waste. No values of raw or effluent trivalent chromium concentrations were given. An effluent containing less than 1.0 mg./l. hexavalent chromium and 2.0 mg./l. total heavy metals was achieved, via the reduction-neutralization-precipitation treatment sequence (10).

Soda ash has been used to precipitate trivalent chromium (6)(8). However, the most commonly employed chemical for precipitation appears to be lime, $Ca(OH)_2$. Avrutskii (7) has reported reducing the chromium content of an electroplating waste from 140 to 1.0 mg./l. This was achieved by reduction, neutralization and precipitation with lime at pH 7 to 8.

In a detailed discussion of waste treatment at an IBM plant, Hansen and Zabben (11) report that hexavalent chromium (at 1,300 mg./l.) in one waste stream was converted to trivalent chromium and precipitated with lime to achieve an effluent containing zero hexavalent and 0.06 mg./l. total chromium. As total chromium must equal hexavalent plus trivalent forms, the effluent thus contained 0.06 mg./l. trivalent chromium.

A coagulant aid (Separan NP-10) was employed to improve the precipitation-sedimentation removal of chromic hydroxide. This coagulant aid undoubtedly accounted for the high degree of treatment achieved, as the authors reported that while the influent to the sedimentation basin contained 1,000 to 2,000 mg./l. suspended solids, the effluent routinely contained less than 10 mg./l. Thus 99 to 99.5% suspended solids removal was achieved by employing the coagulant aid.

Filtration of effluent from the postprecipitation sedimentation basin has been employed to remove the small, nonsettleable chromic hydroxide particles not readily removable by simple sedimentation. Filtration was reported to improve the effluent quality with respect to chromium content (8). Stone (30) has also reported the use of filtration to improve the effluent quality of a chromium bearing waste from a British metal alloy plant.

Hexavalent chromium, as sodium dichromate was employed to produce a bright finish on copper alloy products. The plating was carried out in a mixture of the sodium dichromate and sulfuric acid. Waste chromate from the process was reduced with sulfur dioxide to trivalent chromium, and precipitated at pH 8.5 with lime. Hydroxides of copper, zinc and nickel also precipitated in the process.

The metal hydroxides were removed by sedimentation. During a two-week surveillance period, chromium content in the settling basin effluent was found to range from 1.3 to 4.6 mg./l. Passage of this effluent through a sand filter to remove the finely divided chromic hydroxide particles carried over reduced the effluent chromium concentrations to 0.3 to 1.3 mg./l. This residual should principally be in the soluble $Cr^{+6}$ form, rather than as trivalent chromium hydroxide.

Anderson et al (3) report the use of reduction and lime precipitation to treat chromium bearing waste from a General Electric appliance plant. Sedimentation of the precipitate was improved by addition of an anionic polyelectrolyte. Plant effluent was reported to contain no hexavalent chromium and an average of 0.75 mg./l. total chromium.

Large quantities of sludge are produced in the precipitation of metal hydroxides from industrial wastes. Table 7 summarizes the treatment of these sludges, as reported for four typical industrial processes. The filtrate from sludge dewatering is normally returned to the original sludge sedimentation basin for further treatment.

One article referenced in Table 7 reported that vacuum filtration was used at one time to dewater the chromium (and other metals) hydroxide sludge (12). However, filtering became less and less effective as presoaks, cleaners, and emulsifying agents used in industrial plating became more effective. Finally, it became impossible to develop filterable sludge, and the vacuum filter process was abandoned.

## TABLE 7:  TREATMENT OF METAL HYDROXIDE SLUDGES

| Initial Sludge Solids, mg./l. | Sludge Dewatering Process | Final Sludge Solids, mg./l. | Ultimate Sludge Disposal | Reference |
|---|---|---|---|---|
| – | Vac. filt. | – | Landfill | 1 |
| 30,000 – 60,000 | Vac. filt. | 100,000 – 120,000 | City dump | 7 |
| 100,000 | Vac. filt. | 290,000 | – | 8 |
| 30,000 – 50,000 | None | 30,000 – 50,000 | Landfill | 9 |

Source:  Report PB 204,521

Hansen and Zabben (11) report that vacuum filtration frequently is not necessary, when sludge from the sedimentation basin is held for short periods of time in sludge storage tanks. The sludge stored in the sludge holding basins thickens (or compacts) to a suspended solids concentration of 10 to 12%, equivalent to that obtainable by vacuum filtration. The thickening is thought due to the application of a coagulant aid in the sedimentation basin.

The treatment costs associated with trivalent chromium removal by precipitation are due to the lime precipitation process and the expense of sludge disposal. In the Midwest, lime costs 2.25¢/lb. in 1 ton lots (1968 price). This compares to a cost of 1.75¢/lb. on both the east and west coasts (14). Yuronis (13) has given a somewhat higher lime cost of 3.6¢ per pound in his discussion of chromate treatment. He estimates a lime cost of approximately $2.50/1,000 gal. for chromium precipitation from a waste containing 2,000 mg./l. of $Cr^{+3}$.

The cost for treatment of a stronger or weaker waste would vary directly. For example, a waste containing 1,000 mg./l. $Cr^{+3}$ would have a lime cost of $1.25/1,000 gal. Cost of complete treatment from the hexavalent chromium form would include reduction cost, precipitation cost, and sludge handling, and disposal costs. Reduction costs are approximately $1.00/1,000 gal.

Anderson et al (3) have given details of the costs associated with treatment of a metal bearing waste which contains large quantities of chromium and lesser concentrations of nickel, copper, iron and aluminum. The treatment plant had an initial installation cost of $1,090/1,000 gal. capacity per day. In 1967, the plant treated 205,400 gal. of waste per day, at a total cost of approximately 80¢/1,000 gal.

Costs include chemicals, manpower, utilities, depreciation and taxes. Cost of chromium treatment represent about 20% of total treatment costs. This results in an estimated chromium treatment installation cost of $218/1,000 gal. treated per day and 16¢/1,000 gal. for operating costs. These costs include sludge disposal.

Zievers et al (14) have also discussed sludge disposal and associated costs. Major costs associated with sludge disposal result from vacuum filtration to concentrate the sludge solids. These authors point out that there is a growing emphasis on filtration of metal hydroxide sludges, due to more stringent enforcement of effluent suspended solids standards.

*Ion Exchange:*  Relatively little information is available in the technical literature on ion exchange treatment of trivalent chromium. This is in part because most industrial waste chromium exists in the hexavalent form, and if recovery is preferred direct ion exchange of the hexavalent chromium, chromate, or dichromate ion is employed. Ion exchange is dependent upon the electron charge on an ion.

Trivalent chromium has a plus 3 charge, and hexavalent chromium in its usual form of chromate or dichromate has a minus 2 charge. Thus different ion exchange resins must be employed to recover each form of chromium. An anion type of resin is required to capture negatively charged forms such as the hexavalent chromium chromate and dichromate ions. A cation type of resin is employed for positively charged (e.g., $Cr^{+3}$) forms. Ross (17) has pointed out two reasons to recover trivalent chromium by ion exchange. The first is the passage of concentrated chromic acid baths through a cation resin to remove metallic contaminants such as iron, aluminum and trivalent chromium from the chromic acid. This process would purify the chromic acid solution, which would then be reused.

The second reason discussed by Ross is for the complete purification of a plating rinse water, in which this rather dilute waste would be passed through both an anionic and a cationic resin to remove chromium forms of both positive and negative charge. The treated rinse water could then be recycled. No capital cost figures are available for ion exchange treatment of trivalent chromium.

Lacy et al (20) have reported operating costs for ion exchange treatment of 16 to 24¢ per 1,000 gallons. Ross (17) has reported daily operating costs for ion exchange treatment of trivalent chromium as approximately $28/day for a waste flow of 73 gal./min. Assuming a 16-hour day, operating costs are estimated at 40¢/1,000 gal. If the assumption is erroneous and the process day consists of only 8 hours, operating costs double to 80¢ per 1,000 gallons.

A process developed by J.F. Zievers et al (31) for treating mixed rinse waters used to rinse workpieces removed from chromium and other metal treating baths provides for reducing hexavalent chromium ions carried by the workpieces prior to rinsing and mixing the rinse waters from a plurality of different rinses prior to passage thereof through cation and anion exchangers. The method and system also provide for the immediate removal of CN from the anion regenerant while it has a high pH value and then combining the cation and anion regenerants prior to neutralization.

Figure 38 shows a chrome plating line comprising a plurality of adjacent tanks generally arranged in the shape of a horseshoe with the work traveling in a clockwise direction as it is carried from one tank to the next in the system.

As shown, the workpieces are first immersed in an alkaline cleaning bath 10 and then rinsed in a cold water rinse tank 12 before moving into an electrolytic cleaning bath 14. The workpieces are again cleaned in a cold water rinse tank 16 and lightly dipped in an acid bath 18 before again being washed in a cold water rinse tank 20.

The parts are then electroplated with copper while immersed in a copper plating solution in the tank 22. Typically, this solution will contain $Cu(CN)_2$, KCN and KOH. After removal from the copper plating bath tank 22, the workpieces are washed in a cold water rinse tank 25 and again in a second cold water rinse tank 26.

In accordance with common plating practices, the fresh water enters the cold water rinse tank 26 and flows over a weir between the tanks 24 and 26 into the tank 24 and flows out of the tank 24 at the same rate as water enters the tank 26. The workpieces are then immersed in a nickel plating solution contained in a tank 28 where a layer of nickel is electroplated over the copper layer previously plated onto the workpieces. The nickel plating bath will typically contain Ni, $NiCl_2$, boric acid, sodium formate, cobalt formate, formaldehyde and brighteners.

After being plated with the desired thickness of nickel, the workpieces are moved from the bath 28 into a first cold water rinse tank 30, then into a second cold water rinse tank 32 and into a third cold water rinse tank 34. As shown, fresh water enters the rinse tank 34 and flows over suitable weirs first into the tank 32 and then into the tank 30 from which it exits at a rate equal to the rate at which water enters the third tank 34.

After being thus thoroughly washed, the nickel plated workpieces are now immersed in a chromium plating solution contained in the tank 36 where a layer of chromium is electroplated thereon over the nickel. The chromium plating solution typically includes $H_2CrO_4$ and $H_2SO_4$. This solution will contain a substantial number of hexavalent chromium ions which are generated during the plating operation. The workpieces after leaving the chromium plating tank 36 are immersed in a dragout bath contained in a tank 38.

# FIGURE 38: PROCESS FOR RECOVERY OF MIXED PLATING RINSES

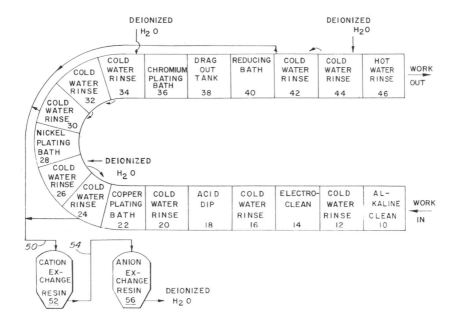

Source: J.F. Zievers and C.J. Novotny; U.S. Patent 3,681,210; August 1, 1972

The solution in the tank **38** is essentially the same as the solution in the tank **36** although at a substantially lower concentration. Accordingly, the tanks **36** and **38** both contain the chromium plating solution including both hexavalent and trivalent chromium ions.

The workpieces leaving the chromium plating solution contained in the tank **38** are immediately immersed in a reducing solution contained in tank **40**. The reducing agent may, for example, be $NaHSO_3$ and its purpose is to reduce or convert the hexavalent chromium ions to trivalent chromium ions before the workpieces are washed. When the workpieces are removed from tank **40**, they move through a pair of cold water rinse tanks **42** and **44** where the solution carried by the workpieces is substantially completely washed therefrom. As shown, fresh water is supplied to tank **44** and passes over a weir into the tank **42** before exiting at a rate substantially equal to the rate at which fresh water enters the tank **44**. After leaving tank **44**, the workpieces may be immersed in a hot water rinse contained in tank **46** to minimize dry out time.

The rinse waters from the rinse tanks **42**, **30** and **24** are combined or mixed and supplied via a line **50** to a cation exchange column **52** containing a cation exchange resin which removes the trivalent chromium ions and any other cations contained in the rinse water. For example, in the system illustrated and described above, sodium ions will also be removed from the rinse water by the cation exchange resin in tank **52**.

The solution then passes through a line **54** to an anion exchange column **56** containing an anion exchange resin. The various anions contained in the solution are removed by the anion exchange resin in the column **56** and completely deionized water flows out of the column **56** from which it is recirculated to the various cold water rinse tanks **26**, **34** and **44**. It will be realized by those skilled in the art that it is not necessary to utilize deionized water for rinsing the workpieces after the copper plating and nickel plating operations. Nevertheless, it is economically possible to use deionized water at all rinse stations in the plating system, thereby providing better washing than has been achieved with the prior art. Any make up water to be added to replace that water which has evaporated may be added just prior to the cation exchange resin or if desired it can be added at either tank 26 or tank 34.

The cation exchange resin in the column 52 will be regenerated by passing an acid such, for example, as $H_2SO_4$ therethrough and the anion exchange resin will be regenerated by passing a hydroxide such as NaOH therethrough. The regeneration solution coming off the cation exchange resin will thus contain $Na_2SO_4$, $Cr_2(SO_4)_3$ at a low value of pH e.g., approximately 3. The regenerant coming off the anion exchange resin will contain NaCN, NaCl, at a pH of approximtely 10.

This latter regenerant may be immediately treated with chlorine while at the high pH value to remove the CN ions and then blended directly with the regenerant from the cation exchange resin prior to the addition of an acid or alkali to neutralize the combined regenerant solutions. Accordingly, the expense of repeatedly altering the pH level of the regenerant solutions in order to remove the cyanide ions from the anion exchange resin and then to remove the hexavalent chromium ions therefrom is obviated with the system.

*Summary:* Most trivalent chromium in industrial waste results from reduction of hexavalent chromium as a preliminary step in waste treatment. This trivalent chromium is most often treated with lime to precipitate chromic hydroxide. The chromic hydroxide sludge is dewatered by vacuum filtration, or disposed of directly to landfill or otherwise. Treated effluent trivalent chromium concentrations to 0.06 mg./l. have been reported (11).

Under special circumstances, ion exchange removal of trivalent chromium have been employed. Ion exchange is characterized by effecting complete $Cr^{+3}$ removal, until the exchange capacity of the resin is exhausted. This exchange capacity may be recovered by regenerating the resin.

## Removal of Chromium from Waste Sludges

A process developed by C.O. Weiss (32) permits recovering the chromium content from waste sludge produced from the separation of the sludge as precipitated solids from the waste effluent streams of metal finishing processes, the chromium content of the sludge being in the form of barium chromate.

The above is achieved by forming an aqueous slurry of the sludge, adding sulfuric acid thereto, agitating the resulting admixture for a period of time at ambient temperature, separating the insolubles therefrom, adding to the filtrate from the separation an amount of a carbonate selected from the group consisting of $Ca^{+2}$, $Sr^{+2}$, $Ba^{+2}$ and $Pb^{+2}$ to precipitate out a substantial portion of the sulfate ion from the reacting environment, adding to the filtrate of the second separation step strontium carbonate to precipitate out the remaining sulfate ion, and separating the insolubles from the reacting environment to obtain a solution containing all the $CrO_3$ content from the initial extraction.

The following example illustrates this method. A representative sludge was prepared of a 55 kg. sample of a mixture of $BaCrO_4$ (40%) and $BaCO_3$ (60%). This mixture was slurried with sufficient water to give a readily stirrable mixture. To the slurry was added 25.1 kg. of $H_2SO_4$ (13.9 liters) and the admixture volume was adjusted to about 85 liters. The mixture was agitated with no external heating for 2½ hours.

The insolubles were then filtered off and the cake washed with several portions of water until the solution analyzed 5.5 g./100 ml. of $CrO_3$ and 7.8 g./100 ml. of $H_2SO_4$. The solution was then charged with 8.0 kg. of $CaCO_3$ and agitated at room temperature for 10 minutes. The insolubles were filtered off and the cake washed as described above to give a solution volume of about 115 liters. This solution still contained 2.0 kg. of $H_2SO_4$, and this solution in turn, was charged with 4.5 kg. of $SrCO_3$ and agitated at room temperature for about 15 minutes. The insolubles were filtered off and the cake washed as described above to produce a final volume of 130 liters. The resulting solution contained no $H_2SO_4$ and virtually all the $CrO_3$ obtained in the initial extraction.

### References

(1) Sullivan, R.J., "Air Pollution Aspects of Chromium & Its Compounds," *Report PB 188,075,* Springfield, Va., National Technical Information Service (September 1969).

(2) Patterson, J.W. and Minear, R.A., "Wastewater Treatment Technology," *Report PB 204,521,* Springfield, Va., National Technical Information Service (August 1971).

(3) Anderson, J.S. and Iobst, E.H., Jr., "Case History of Wastewater Treatment in a General Electric Appliance Plant," *Jour. Water Pollution Control Federation,* 10, pp. 1786-1795 (1968).

(4) Udy, M.J. (ed.) *Chromium, Vol. 2, Metallurgy of Chromium and Its Alloys,* American Chemical Soc. Monograph 132, New York, Reinhold, (1956).

(5) MacDougall, H., "Waste Disposal at a Steel Plant: Treatment of Sheet and Tin Mill Wastes," *ASCE Proc. Separate No. 493,* (1954).

(6) Besselievre, E.B., *The Treatment of Industrial Wastes,* McGraw-Hill Book Co., New York (1969).

(7) Avrutskii, P.I., "Control of Chromium (VI) Concentration in Waste Waters," *Chem. Abst.* 70, 206-207 (1969).

(8) Avrutskii, P.I., "Liquid $SO_2$ in New Reduction Role," *Indust. Eng.,* 53, 29A-30A (1961).

(9) Hulse, B.T., Selim, R.P. and Summers, G.E., "Control of Metal Finishing Wastes Using ORP," *Jour. Wat. Poll. Control Fed.,* 32, 975-981 (1960).

(10) Shink, C.A., "Plating Wastes: A Simplified Approach to Treatment," *Plating,* 55, 1302-1305 (1968).

(11) Hansen, N.H. and Zabben, W., "Design and Operation Problems of a Continuous Automatic Plating Waste Treatment Plant at the Data Processing Division, IBM, Rochester, Minnesota," *Proc. 14th Purdue Industrial Waste Conf.,* pp. 227-249 (1959).

(12) Fisco, R., "Plating and Industrial Waste Treatment at the Expanded Plant of Fisher Body, Elyria, Ohio," *Proc. 25th Purdue Industrial Waste Conf.* (1970).

(13) Yuronis, D., "Metal Finishing Waste Treatment: Comparative Economics," *Plating,* 55, 1071-1074 (1968).

(14) Zievers, J.F., Crain, R.W. and Barclay, F.G., "Waste Treatment in Metal Finishing: U.S. and European Practices," *Plating,* 55, 1171-1179 (1968).

(15) Selm, R.P. and Hulse, B.T.; U.S. Patent 3,027,321; March 27, 1962; assigned to Wilson & Co.

(16) Lancy, L.E.; U.S. Patent 3,294,680; December 27, 1966; assigned to Lancy Laboratories, Inc.

(17) Ross, R.D., (ed.), *Industrial Waste Disposal,* New York, Reinhold Book Corp. (1968).

(18) Rothstein, S., "Five Years of Ion Exchange," *Plating,* 45, 835-841 (1958).

(19) Keating, R.J. and Duff, J.H., "Plating Waste Solutions: Recovery or Disposal," *Proc. 10th Purdue Industrial Waste Conf.,* (1955).

(20) Lacy, W.J. and Cywin, A., "The Federal Water Pollution Control Administration's Research and Development Program: Industrial Pollution Control," *Plating,* 55, 1299-1301 (1968).

(21) Saraceno, A.J., Walters, R.H., Jones, D.B. and Wiehle, W.E.; U.S. Patent 3,664,950; May 23, 1972; assigned to U.S. Atomic Energy Commission.

(22) Sloan, L., Nitti, N.J. and Pratt, J.B.; U.S. Patent 3,306,859; February 28, 1967; assigned to Crane Co.

(23) Culotta, J.M. and Swanton, W.F., "Case Histories of Plating Waste Recovery Systems," presented at *56th Annual Conf. Amer. Electroplaters Soc.,* Detroit (1969).

(24) Yagishita, A.; U.S. Patent 3,542,651; November 24, 1970.

(25) Yagishita, A.; U.S. Patent 3,616,437; October 26, 1971.

(26) Richards, R.; U.S. Patent 3,371,034; February 27, 1968.

(27) Nieuwenhuis, G.J.; U.S. Patent 3,493,328; February 3, 1970.

(28) Dougherty, R.R.; U.S. Patent 3,531,262; September 29, 1970; assigned to Control Data Corp.

(29) Geinopolos, A., Gupta, M.K. and Katz, W.J.; U.S. Patent 3,625,885; December 7, 1971; assigned to Rex Chainbelt, Inc.

(30) Stone, E.H.F., "Treatment of Non-Ferrous Metal Process Waste at Kynoch Works, Birmingham, England," *Proc. 25th Purdue Industrial Waste Conf.,* pp. 848-865 (1967).

(31) Zievers, J.F. and Novotny, C.J.; U.S. Patent 3,681,210; August 1, 1972; assigned to Industrial Filter & Pump Manufacturing Co.

(32) Weiss, C.O.; U.S. Patent 3,552,917; January 5, 1971; assigned to M & T Chemicals, Inc.

## CLAY

### Removal of Clay from Water

A process developed by J.W. Ryznar (1) involves the coagulating and settling of finely divided solids which are predominantly inorganic and normally remain suspended in water, more particularly dilute suspensions in water containing concentrations within the range of about 0.003% to 3% by weight of the solids.

One of the most difficult industrial problems is to clarify industrial waste which would otherwise create a nuisance and cause pollution in lands and streams. Examples of such wastes are phosphate mine waters, coal washing waters, clay suspensions, calcium carbonate suspensions, and other suspensions of finely divided solids in water which result from industrial processes such as mining, washing, purification, and the like. These suspensions normally contain the solid materials in very finely divided form in concentrations within the range of about 0.003% to 3% by weight of the suspension. Such suspensions will remain stable for days and many of them are not affected by the addition of ordinary coagulants such as alum. If the solids are allowed to remain in suspension, the resultant suspension cannot be utilized for industrial processes and also presents a disposal problem.

In accordance with this process it has been found that it is possible to produce coagulation and settling of finely divided solids which are predominantly inorganic and normally remain suspended in water in concentrations of 0.003% to 3% by weight of the suspension by treating the suspensions with about 1 to 15 parts per million based on the weight of the total suspension of a synthetic water dispersible polymer having a weight average molecular weight of at least 10,000 and having a structure derived by the polymerization of at least one monoolefinic compound through the aliphatic unsaturated group, said strucure being substantially free of crosslinking.

The treating agents which have been found to be especially effective are water dispersible synthetic polymers having a linear hydrocarbon structure and containing in a side chain a hydrophilic group from the class consisting of carboxylic acid, carboxylic acid anhydride, carboxylic acid amide, hydroxy, pyridine, pyrrolidone, hydroxy alkyl ether, alkoxy, and carboxylic acid salt groups.

A process developed by F.W. Camp (2) is a process for separating suspended solids from a liquid comprising admixing the liquid with a fluid immiscible with the solids-containing liquid, settling the resulting mixture to form an interfacial zone in which the solids have concentrated, separately removing interfacial zone material from the settler, and treating the material by a conventional process to separate the solids from the liquid. The process is particularly applicable to the clarification of water discharged from a hot water process for the separation of bitumen from tar sands.

The following table illustrates particular systems to which the process can be applied. Column A of the table enumerates several solids which can be suspended in the volatile liquids of Column B. Column C enumerates various volatile fluids which can be added to the particular liquid of Column B to form the two-phase interface:

| A<br>Solids | B<br>Liquid | C<br>Added Fluid |
|---|---|---|
| Clay | Water | Hexane |
| Sewage sludge | Water | Mineral spirits |
| Clay-viscous oil mixtures | Water | Mineral spirits |
| Iron rust | Gasoline | Water |
| Carbon black | Lubricating oil | Water |
| Tank bottom sludge | Gasoline | |
| | Jet fuel | Water |
| | Home heating oil | |

A chemical treatment step can be added to the process to render the solids more surface active and consequently more attracted to the interface zone between the two phases. The chemical treatment step can comprise treatment with a flocculant type or surfactant type of reagent or with both a flocculant and a surfactant type. If a flocculant type, the particular reagent used, of course, depends on the particular system. Flocculation appears to cause the suspended solids to form a gel-like mass which is more readily trapped in the interface zone.

If a surfactant type of reagent is used, the surfactant should carry on one of its portions (the hydrophilic portion in the case of a system with a water phase), a charge opposite to the surface charge on the suspended solids. For example, if the treated system is a clay in water system, the hydrophilic portion of the surfactant should have a positive charge, a cationic surfactant. The other portion of the surfactant is of a molecular structure and size for easy solubility in the added immiscible fluid. In the case of a clay-water-benzene system the hydrophobic portion of the surfactant is of a molecular structure and size for easy solubility in the benzene. A good surfactant for this system is $RNH_2Cl$ where R is an aliphatic $C_{18}$ carbon chain.

### References

(1) Ryznar, J.W.; U.S. Patent 3,492,224; January 27, 1970; assigned to Nalco Chemical Co.
(2) Camp, F.W.; U.S. Patent 3,526,585; September 1, 1970; assigned to Great Canadian Oil Sands Ltd., Canada.

## COKE OVEN EFFLUENTS

### Removal from Air

A process developed by G. Nashan and J. Knappstein (1) for the controlled gas-free and dust-free discharge and quenching of coke for horizontally arranged coke furnace batteries includes a closed coke-receiving chamber which is provided with gas and dust exhaust devices. The chamber extends over the entire length of the coke battery and encloses the doors to the individual furnace chambers. In addition, the receiving chamber is connected with a quenching tower and it includes means for conducting the coke to a conveyor which leads to the quenching tower. The quenching tower itself includes either an arrangement of sprays directed over the conveyor or a quenching pool into which the coke is delivered and which includes a means for conveying the quenched coal into a receiving bin at the exterior of the quenching tower.

A process developed by E.J. Helm (2) is one in which a hood is movably mounted to a coke quenching car and cooperates with a coke guide that keeps the hood in place relative to the coke guide as the coke quenching car moves relative to the hood. Dust and smoke from pushed coke passing through the hood are collected in the hood. Gas scrubbing equipment is associated with the coke guide and the hood to remove particulate matter from the gases collected in the hood.

A process developed by F. Breitbach and G. Choulat (3) involves decomposing the ammonia of impure ammonia-containing vapor clouds, particularly ammonia-containing vapor clouds emanating from coke oven plants and gas works and contaminated by hydrogen sulfide, cyano compounds like HCN, benzene, naphthalene and the like. The vapor clouds are first heated to the decomposition temperature of the ammonia, whereafter the hot clouds are passed at this temperature through a decomposition zone to cause decomposition of the ammonia into hydrogen and nitrogen. The combustible matter in the decomposed gases is then burned in an intermediately installed combustion zone by supplying combustion air. The combustible matter including the hydrogen may be burned at a place remote from the decomposition plant or may be added to other gases.

Figure 39 is a schematic representation of the essential elements of the process.

FIGURE 39:  APPARATUS FOR THERMAL DECOMPOSITION OF AMMONIA IN
GASEOUS COKE OVEN EFFLUENTS

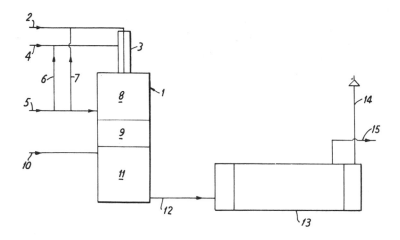

Source:  F. Breitbach and G. Choulat; U.S. Patent 3,661,507; May 9, 1972

Reference numeral **1** indicates schematically a reactor to be heated to the required decomposition temperature.  For this purpose the reactor is heated by the combustion of a gaseous or liquid fuel which is supplied to a burner **3** through line **2**.  The combustion air necessary to accomplish combustion of the fuel flowing through line **2** is supplied to the burner through line **4**.  The amount of combustion air is adjusted so that a reducing atmosphere is produced in the reactor.

The ammonia vapor clouds to be decomposed can be supplied to the reactor in different ways.  According to one embodiment, these ammonia clouds may be supplied through the line **5** so that they will directly enter the combustion chamber **8** of the reactor.  According to a second possibility, the ammonia clouds may be supplied jointly with the combustion air through the lines **5**, **6** and **4**.  Further, it is also possible to supply the ammonia clouds jointly with the fuel through the lines **5**, **7** and **2**.  In all these instances, the ammonia clouds are introduced into the chamber **8** of the reactor **1**, which constitutes the heating zone, and are heated to the reaction or decomposition temperature.  In the embodiment shown in the figure, there thus occurs actual mixing of the heating medium and the ammonia clouds.

After the ammonia clouds have been heated to the necessary decomposition temperature, they are then, jointly with the off-gases from the combusted heating gases, passed through the decomposition zone **9** for complete decomposition of the ammonia.  The zone **9** may be filled with filling bodies or with a suitable nickel catalyst.  The admixing of a part or the whole of the vapor clouds to the heating gas or to the combustion air before the burner makes it possible that during the combustion with a deficiency of air no carbon black is formed.

By supplying secondary air which, through line **10**, enters a combustion zone, combustion of the combustible matter formed in the decomposition zone may be effected close to the decomposition zone.  This means that any combustible matter which is formed in the decomposition of the ammonia is completely burned up in the combustion zone **11** due to the introduction of the secondary air through line **10**.  The hot off-gases flow subsequently through line **12** to reach the waste heat boiler **13**.  The off-gases are discharged to the atmosphere through the chimney **14**. The steam which is produced in the waste heat boiler flows through line **15** to any suitable point of utilization, i.e., it may be supplied to the steam network of the plant. The hot off-gases of the aftercombustion zone may, of course, also be utilized or cooled in any other suitable manner. The burnable hydrogen-containing gas may, however, be burned at a place remote from the plant. In other words, the burning may be delayed until the gas has been conveyed to a suitable place of utilization where the gases are burned in situ, if desired, after admixture with other gases. In this case the gas, which leaves the reactor at about 1000°C., is also conducted into a waste heat vessel for using its content as waste heat.

The following is a specific example of the operation of the process. The coking of 1,000 tons of dry coal results in the formation of 2,600 kg./day of ammonia. This ammonia may, for example, be contained in 220 m.$^3$/day of wash water and condensate. The ammonia is expelled from this liquor in known manner so that the ammonia clouds formed contain about 250 kg. of ammonia per long ton and in addition to the other customarily accompanying substances 750 kg. of steam per ton. These ammonia clouds are supplied to the reactor **1** through line **5** as shown in Figure 39. 260 Nm.$^3$/hour (atmospheric pressure) of coke oven gas are supplied through line **2** while 1,100 Nm.$^3$/hour of air are introduced through line **4**.

The coke gas amount is adjusted in such a manner that the heat resulting from its combustion is utilized for the decomposition of the ammonia into nitrogen and hydrogen and that the gas mixture which exits from the decomposition zone **9** has a temperature of about 1000°C. This hot mixture enters into the combustion zone **11** into which are supplied about 750 Nm.$^3$/hour air through line **10**.

After the combustion of the combustible gases which have been formed, the combustion gases leave the aftercombustion zone **11** through line **12** and enter the waste heat boiler **13**. 40,000 kg./day of steam (40 atm.) are obtained which are discharged from the waste heat boiler through line **15**. The off-gases which leave through the chimney **14** do not contain any significant amount of nitrogen oxides.

If, by contrast, ammonia clouds are supplied to an ordinary prior art combustion furnace, where the clouds are burned up in a single stage with the same coking gas amount and in the presence of 1,850 Nm.$^3$/hour of air, so that nitrogen and water are formed from the ammonia, an off-gas leaves the chimney which contains up to 4.8 grams of nitrogen oxides per Nm$^3$. By contrast, according to this process, only the fuel is burned at the beginning In the heating zone, and the hot mixture to be decomposed is conveyed first to a decomposition zone and subsequently and under admixture of additional combustion air to a combustion chamber where the hydrogen is completely burned up so that an off-gas is obtained which does not contain any nitrogen oxides.

A process developed by J.J. Keimar (4) involves the overall design of a nonpolluting by-product coal carbonization plant. As coal is heated to produce coke and coke oven gas, the coke is delivered to a reactor where a bed of it is maintained. Most of the tar and ammonia and oil are removed from the coke oven gas and then the gas is passed through the coke bed. A tar plant receives the tar removed from the gas and produces coal tar pitch, which is delivered with calcium carbonate to the reactor to desulfurize the coke and gas therein and enriches the gas.

The desulfurized gas is continuously removed from the reactor and cooled. Periodically a batch of desulfurized coke is withdrawn from the reactor. Steam is produced in cooling the coke and is used in heating an ammonia still. Water from the still is processed and used in an ammonia scrubber along with make-up water produced by condensing water vapor from the coke oven combustion gas. Only nonpolluting products leave the system. The rest are consumed or recirculated in it.

A process developed by W. Kubsch (5) involves a collection hood for coke-quenching cars and comprises an anti-pollution apparatus carried on a cantilever arm supported on a hood car mounted on tracks. The hood extends over a coke-quenching car mounted on the same tracks as the hood car and is connected to a suction and dust extraction device such that gases and dust which rise from the coke batch as it is pushed out of a horizontal by-product coke oven and into the quenching car will be conveyed to a gas-stream purifying device rather than being permitted to escape into the air. As the quenching car moves from the coke oven to a quenching station, the hood car and the hood carried thereby follows such that the hood continually covers the quenching car and prevents the escape of pollutants.

A process developed by R. Kemmetmueller (6) provides a method and apparatus for carrying out pollution-free charging of coke ovens. During introduction of a charge into a coke oven from a larry car gas which escapes through the larry car connections to the coke oven is collected with a suitable gas-collecting pipe. Standpipes which communicate with the interior of the coke ovens communicate with a second gas-collecting pipe so that additional gas escaping from the interior of the coke ovens is collected. The entire space over the coke ovens is enclosed and the interior of the enclosure communicates with a third gas-collecting pipe in which gas is collected from the space over the coke ovens. All of this collected gas is cleaned before being discharged to the outer atmosphere.

### Removal from Water

A process developed by P.X. Masciantonio (7) relates to the removal of contaminants from coke-works wastewaters and the like, to render them suitable for further use or discharge into streams, and to the reclamation of the contaminating substances as useful chemicals. Typical wastes advantageously treated by this process are those from the ammonia stills of by-product coke-works and coal gas plants, which contain highly obnoxious stream pollutants. Of the contaminants present in the wastewaters, the organic compounds are most troublesome of disposal. These include phenol and its homologues, all of which are commonly grouped under the term phenols.

Various expedients have been employed to dispose of the offending wastes, the most generally used being evaporation by spraying onto incandescent coke. However, serious corrosion of adjacent coke-works equipment results from this practice, and the presence of contaminants on the coke leads to subsequent problems in blast furnace wastes and impairs its value as domestic coke.

Methods for recovering the chemical constituents from these waters are generally impractical, since the contaminants are present in low concentrations and the expense of removing them outweighs their value. Removal of the phenols by adsorbents, such as charcoal, has been proposed. However, regeneration or discard of spent adsorbent is expensive, and any chemical recovery was impractical due to equipment requirements and operating expenses.

This process permits purification of the coke-works wastewaters and the like, and reclamation of their chemical constituents without necessity for special facilities, other than a bed of suitable adsorbent, which is used in combination with conventional by-product coke-works equipment and the like. Although the contaminants are removed by adsorption, the adsorbent used requires no regeneration or discard and the adsorbed chemicals are recovered relatively effortlessly, utilizing usual by-product recovery practices.

Figure 40 is a block flow diagram of the process and shows an adsorbent bed 1 in functional relationship with conventional equipment of a by-product coke-works, comprising coal chemical recovery units 2, coke quencher 3, and coke ovens 4. The process contemplates use of a carbonaceous adsorbent agent in bed 1 which is a suitable coke oven charge component either by virtue of its coke-forming qualities or its compatibility with other coke-making charges, to produce metallurgical coke or other carbonized products of desired characteristics. Specific examples of preferred adsorbents are bituminous coal and bituminous coal treated to increase its adsorptive activity for organics such as phenols.

In the drawing, the adsorbent bed 1 is shown as receiving contaminated waters of the type previously mentioned from the coal chemical recovery units **2** and the coke quencher **3**. In the further operation of the process, upon adsorption of contaminants, including the organic compounds and ammonia, the resulting purified water is separated from the adsorbent and the latter, after having spent its adsorptive capacity, is charged into the coke ovens **4**, carrying with it the adsorbed contaminants. Upon carbonization of the charge in the coke ovens **4** the contaminants are driven off, along with the usual coal chemicals, whereupon they are available for processing in the coal-chemical recovery units **2** for recovery of useful chemical products.

Since the process permits of operation in conjunction with usual coke-works practices, coking coal is shown as being charged into the coke ovens **4**. Also, incandescent coke, which includes coke-making components from the adsorbent, is shown as being fed to the coke quencher **3**, which along with the coal-chemical recovery units **2**, is shown as receiving water for the respective processes involved.

The adsorbent bed, insofar as structural arrangement is concerned, may be of any suitable type, such as would occur to persons familiar with the use of solid adsorbents for the removal of dissolved constituents from liquids. A bed arranged for continuous operation is preferred.

**FIGURE 40: METHOD OF PURIFYING COKE-WORKS WASTE LIQUOR**

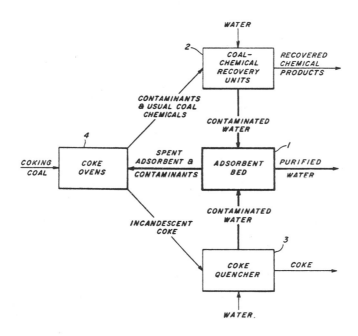

Source: P.X. Masciantonio; U.S. Patent 3,373,085; March 12, 1968

**References**

(1) Nashan, G. and Knappstein, J.; U.S. Patent 3,630,852; December 28, 1971; assigned to Firma Carl Still, Germany.
(2) Helm, E.J.; U.S. Patent 3,647,636; March 7, 1972; assigned to Koppers Co., Inc.

(3)  Breitbach, F. and Choulat, G.; U.S. Patent 3,661,507; May 9, 1972; assigned to Firma Carl Still, Germany.
(4)  Keimar, J.J.; U.S. Patent 3,661,719; May 9, 1972.
(5)  Kubsch, W.; U.S. Patent 3,675,400; July 11, 1972; assigned to Hartung, Kuhn & Co., Maschinenfabrik GmbH, Germany.
(6)  Kemmetmueller, R.; U.S. Patent 3,697,381; October 10, 1972; assigned to American Waagner-Biro Co., Inc.
(7)  Masciantonio, P.X.; U.S. Patent 3,373,085; March 12, 1968; assigned to United States Steel Corp.

# COLOR PHOTOGRAPHY EFFLUENTS

## Removal from Water

A process developed by B.A. Hutchins and R.S. Walsh (1) is one in which potential water pollutants including primary aromatic amine color developing agents are recovered, alone, or simultaneously with color-forming couplers or benzyl alcohol from developer solutions in a separable, second phase by admixture with a water-soluble salt under limited pH conditions.

Large quantities of color developing solutions are used in the processing of exposed photographic elements in order to produce color photographs. It is the color developing solutions that produce the dye images that form the color picture in the processed element. In the processing of color subtractive-type photographic elements, the final picture is usually made by the formation of a cyan dye image, a yellow dye image and a magenta dye image in superposed relationship.

The magenta dye image development, for example, is accomplished with an alkaline developer solution which contains a primary aromatic amine color developing agent, such as N-(2-amino-5-diethylaminophenethyl)methane sulfonamide hydrochloride, and a magenta-forming coupler which is usually a 2-pyrazolin-5-one or a coumarone type coupler. Where exposed silver halide in the magenta image layer is developed to silver, color developing agent is oxidized to a compound that couples with the magenta-forming coupler present to form a nondiffusing dye image. The cyan and yellow dye images are formed in a similar manner with cyan dye-forming and yellow dye-forming couplers, respectively.

In commercial color processing of color film, a continuous flow of color film is conducted through various processing solutions including the color developers in a processing machine. The concentrations of the components of the developer solutions must be maintained within certain specified ranges in order to give good sensitometric results. Thus, it is usual practice to replenish the various developer solutions by a controlled flow of the appropriate replenisher solution at the necessary rate.

The flow of color developer replenisher into a processing machine color development tank, for example, displaces "used color developer solution" which still contains substantial amounts of color developing agent, color-forming coupler and other such materials which are relatively expensive to replace. Not only is it undesirable to lose these materials, but it is highly undesirable to sewer such solutions, since some of them contain materials such as benzyl alcohol in addition to the developing agents which have a substantial biochemical oxygen demand and an adverse effect on fish and wildlife that eventually come into contact with the waters into which these materials are deposited.

It has been proposed to recover the various components of the color developing solution for reuse by means of certain techniques including solvent extraction, evaporative or freezing-out procedures, ion exchange and the like. However, none of these processes has been completely satisfactory for commercial purposes. Certain of the proposed processes are complicated and difficult to operate. Furthermore, while some of the proposed processes permit the recovery of certain of the components of the color developing solution, e.g., the color-forming coupler, the color developing agent and benzyl alcohol (when used in the developer) still remain in solution and must be discarded.

An improved process is here provided which comprises admixing a color developer solution having a pH in the range of between 9 and 14 with a water-soluble salt with the solution in an amount sufficient to result in the formation of a developing agent containing second phase that is separable from the main body of color developer solution phase. The newly formed second phase can be either in the form of a liquid or a solid phase depending upon the materials being recovered.

The developing agent phase is easily recovered or separated from the main body of the developer solution by any suitable means, including filtration, centrifuging, decanting, and the like, thus substantially reducing the biochemical oxygen demand of the main body of the solution which is then discarded. The recovered developing agent or developing agent with color-forming coupler or developing agent with benzyl alcohol is then advantageously used to make fresh color developer solution for processing.

**References**

(1) Hutchins, B.A. and Walsh, R.S.; U.S. Patent 3,502,577; March 24, 1970; assigned to Eastman Kodak Company.

## COPPER

Primary sources of copper in industrial waste streams are metal process pickling baths and plating baths. Brass and copper metal working requires periodic oxide removal by immersing the metal in strong acid baths. Solution adhering to the metal surface, referred to as "dragout" is rinsed from the metal and contaminates the waste rinse water. Similarly, metal parts undergoing copper or brass plating drag out some of the concentrated plating solution. Jewelry manufacturers employ copper plating either directly or as a base metal for silver and other precious metal surfaces. Copper is also employed in the alkaline Bemberg rayon process, as cupro-ammonium salts. Copper bearing acid mine drainage also contributes significant quantities of dissolved copper to waste streams. Table 8 summarizes values reported for copper in various industrial process wastes.

### TABLE 8: CONCENTRATIONS OF COPPER IN PROCESS WASTEWATERS

| Process | Copper Concentration, mg./l. | Reference |
|---|---|---|
| Plating wash | 20-120* | 1 |
| Plating wash | 0-7.9 | 2 |
| Brass dip | 2-6 | 2 |
| Brass mill rinse | 4.4-8.5 | 3 |
| Copper mill rinse | 19-74 | 3 |
| Metal processing | 204-370 | 4 |
| Brass mill wash | | 5 |
| Tube mill | 74 | |
| Rod and wire mill | 888 | |
| Rolling mill | 34 | |
| Brass mill bichromate pickle | | 5 |
| Tube mill | 13.1 | |
| Rod and wire mill | 27.4 | |
| Rolling mill | 12.2 | |
| Copper rinse | 13-74 | |
| Brass mill rinse | 4.5 | |
| Appliance manufacturing | | 6 |
| Spent acids | 0.6-11.0 | |

(continued)

TABLE 8: (continued)

| Process | Copper Concentration, mg./l. | Reference |
|---|---|---|
| Alkaline wastes | 0-1.0 | |
| Typical large plater | | 7 |
| Rinse waters | up to 100 (20 ave) | |
| Four plating operations | 6.4-88 | 8 |
| Automobile heater production | 24-33 (28 ave) | 9 |
| Silver plating | | 10 |
| Silver bearing | 3-900 (12 ave) | |
| Acid wastes | 30-590 (135 ave) | |
| Alkaline wastes | 3.2-19 (6.1 ave) | |
| Brass industry | | 10 |
| Pickling bath wastes | 4.0-23 | |
| Bright dip wastes | 7.0-44 | |
| Business Machine Corp. | | 10 |
| Plating wastes | 2.8-7.8 (4.5 ave) | |
| Pickling wastes | 0.4-2.2 (1.0 ave) | |
| Copper plating rinse water | 5.2-41 | 11 |
| Copper tube mill waste | 70 (ave) | 12 |
| Copper wire mill waste | 800 (ave) | 12 |

*High values from spray rinse

Source:   Report PB 204,521

The reader of this handbook is also referred to a review of pollution control in the non-ferrous metals industry by H.R. Jones (13).

## Removal of Copper from Air

A report has been prepared by W.E. Davis and Assoc. (14) regarding the nature, magnitude and extent of the emissions of copper in the United States for the year 1969. Background information concerning the basic characteristics of the copper industry has been assembled and included. Brief process descriptions are given; they are limited to the areas that are closely related to existing or potential atmospheric losses of the pollutant. The copper emissions and emission factors are based on data obtained from production and reprocessing companies. Additional information was acquired during field trips to inspect the air pollution control equipment and observe processing operations.

Emissions to the atmosphere during the year were 13,680 tons. About 64% of the emissions resulted from the metallurgical processing of primary copper, and about 20% from the production of iron and steel. The combustion of coal was the only other significant emission source.

## Removal of Copper from Water

As with most heavy metal wastes, treatment processes employed for reduction of soluble copper in wastewaters may be of a destructive nature involving precipitation and disposal of resulting solids, or recovery processes which include ion exchange, evaporation and electrolysis. The value of copper metal frequently makes recovery processes more attractive than for iron or zinc.

The importance of good internal housekeeping and process modification in reducing overall treatment requirements have been summarized by Domey and Stiefel (15) and Foulk (16). Modification of rinsing techniques for metal parts emerging from pickling baths and plating baths can greatly reduce rinse water volumes, and require treatment plant capacity. Such modifications also increase the concentration of metal in the rinse water and make recovery processes more feasible technically.

As pointed out by E.A. Tomic (17) as little as 2 to 4 micrograms of copper per liter of aqueous solution in streams flowing through aluminum processing equipment causes extensive corrosion damage. Thus, he has proposed a process for the removal of copper from nonaqueous and aqueous solutions employing a polymeric agent especially selective for copper.

The process employs a chelating polymer having a repeating thiosemicarbazide unit of the structural formula:

$$-NH-\overset{\overset{\textstyle S}{\|}}{C}-NH-N\overset{\diagup}{\diagdown}$$

*Destructive Treatment:* The standard destructive treatment for copper and most other heavy metals is precipitation of the relatively insoluble metal hydroxide at alkaline pH. As most heavy metal wastes are initially acidic, lime is the desired base for pH adjustment, due to its low cost and reaction to form the metal hydroxide.

Copper hydroxide solubility is minimal around pH 9.0 to 10.3 (18). Jenkins, et al (19) have reported a maximum solubility of 0.01 mg./l. at pH 10. This value would correspond to the theoretical minimum effluent level which could be achieved by precipitation. Theoretical levels are seldom attained, due to poor settling of colloidal precipitates, slow reaction rates, pH fluctuations, and the influence of other ions in solution. Levels of copper treatment and associated costs reported in the literature are discussed below.

Parsons and Rudolfs (1) reported that copper in a plating bath waste was reduced from 33 mg./l. to less than 1 mg./l. by precipitation with lime.

Nyquist and Carroll (4) report treatment of a waste containing from 204 to 385 mg./l. copper, reduced to an average effluent copper concentration of 0.5 mg./l. Treatment also included zinc reduction from initial levels of 55 to 120 mg./l. to less than 1 mg./l.

Stone (20) reports that copper precipitation at pH 8.5 to 9.0 provided an effluent concentration of copper ranging from 0.2 to 2.5 mg./l. prior to installation of rapid sand filters. After filter installation, more consistent results were obtained, although minimum values were not reduced. The post-filter concentrations of copper ranged from 0.2 to 0.5 mg./l. Costs for construction of the entire waste treatment plant, which treated between 1.5 and 2.0 MGD, was nearly $800,000.

Hanson and Zabban (21) report an effluent concentration of 2.2 mg./l. copper after lime treatment and coagulant addition at the Rochester, Minnesota IBM plant.

Nemerow (10), in a lab study of copper cyanide waste from a silver plating operation, was able to reduce 14 to 18 mg./l. of copper to 0.05 to 0.5 mg./l. at pH 8.0. The process involved cyanide destruction with sodium hypochlorite, pH adjustment to pH 6.5 with ferric chloride, further addition of ferric chloride to 200 mg./l. for coagulation, and addition of lime to pH 8.0 for metal precipitation.

In-plant operation of this process, also employing sand filtration, produced an effluent containing 0.16 to 0.3 mg./l. copper. Untreated waste concentration of copper was, on the average, 30 mg./l.

Watson (22) reports effluent copper concentrations of 0 to 1.2 mg./l. for metal rinse waters, with the average copper level being 0.15 to 0.19 mg./l. No initial concentrations were given, but the author stated that copper levels were high enough to warrant treatment.

Tallmadge (12) reports that the Scoville Brass Mill of New Milford, Connecticut, disposes of its rinse waters and waste pickle liquors by lime precipitation. Rinse waters containing 10 to 20 mg./l. copper are treated on a continuous basis, while the more concentrated waste pickle liquors are pumped into holding tanks and bled into the waste stream slowly, thus allowing continuous treatment of the high concentrate solutions. Lime is used to neutralize the acid and precipitate copper. The resultant effluent containing 1 to 2 mg./l. of residual copper is diluted four-fold with other process waters to yield a final concentration of 0.25 to 0.50 mg./l. of copper in the discharge effluent.

The standard domestic sewage tertiary treatment sequence of chemical coagulation, sedimentation and rapid sand filtration is equivalent to metal precipitation operations, except for neutralization of acids to near neutral pH. Treatment costs should, therefore, be similar for the treatment units.

*Recovery Processes:* Recovery of copper is more attractive than some other metals due to its higher value. A recent study (23) indicated that the recovery value for copper solutions ranging from 100 to 500 mg./l. is $0.60 to $3.00/1,000 gallons. In brass rinse waters containing 40 to 250 mg./l. of copper, the recovery values are given as $0.20 to $1.20 per 1,000 gallons. Unless a recovery process is a closed loop operation, however, the wastewater may still require final treatment for residual metal removal.

Process modification for waste stream segregation and flow reduction is usually an integral part of recovery processes. In fact, direct recovery can result from such modifications, as reported by Barnes (11). Use of countercurrent rinse copper plating baths allowed direct return of the rinse flow into the plating baths as evaporative makeup water. Rinse water flow was reduced from 480 gal./hr. to 16 gal./hr. for the copper plating lines. The final rinse tank concentration of copper was reduced to 0.9 mg./l. from a previous value of 40 mg./l.

Domey and Stiefel (15) report that copper plating bath rinse water volume could be reduced from 4,100 gal./day to 600 gal./day by rinsing modifications.

*Evaporative Recovery:* Use of evaporative recovery has been practiced for over twenty years (24). In the event that water evaporation from the plating bath is not sufficient to allow direct return of the dilute rinse water, preliminary evaporation of rinse water may be necessary in order to increase its copper concentration. Cost of evaporative recovery versus destructive treatment can be compared based upon a study by Culotta and Swanton (25). They report the cost of copper cyanide plating solution to be $0.60/gal. Their cost analysis equations allow calculation of approximate costs for destructive treatment and for evaporative recovery. An economic analysis for zinc cyanide waste demonstrated that at a plating solution dragout rate of greater than 4 gal./hr., recovery had the economic edge over destruction. Copper plating solution has a higher cost than zinc (60¢ versus 45¢ per gal.), and therefore more favorable recovery economics.

*Ion Exchange:* Bothan and Bryson (26) reported copper removal by ion exchange from 1.02 mg./l. down to less than 0.03 mg./l. at flow rates of approximately 0.80 gal./ft.$^2$/min. McGarvey (3) reports that "complete" removal of copper was accomplished from alkaline rayon process wastewaters prior to World War II by ion exchange recovery. Ion exchange is capable of achieving very high levels of copper removal, particularly from dilute wastes.

The initial costs of an ion exchange unit depend upon the required resin volume. For dilute systems, flow rate will be the limiting factor, while the resin exchange capacity will dictate the volume for more concentrated systems. Costs of various units and types of resins are given by McGarvey et al (5) as a function of required capacity. These costs are for complete units (valves, resin, etc.).

*Electrolytic Recovery:* Direct electrolytic recovery of copper metal is possible from relatively concentrated waste solutions. This process generally requires a preconcentration step such as ion exchange or evaporation. Acid leach solution from copper ore is electrolyzed directly by the Inspiration Consolidated Copper Company of Arizona to re-cover copper. The resultant regenerated sulfiric acid is also recycled (27).

For copper cyanide baths, copper recovery by electrolysis can be accomplished in conjunc-tion with cyanide destruction (7)(28)(29). The process is not feasible for dilute solutions but is appropriate for treatment of spent plating bath solutions or ion exchange concen-trates. The process economics compares favorably with other copper cyanide treatment processes. Operating data are presented by Easton (28), and an economic analysis of the process (1967 prices) is given. It has been claimed however that electrolytic recovery of copper from copper etching solutions (including pH neutralization) is too expensive.

A preferred process proposed by Lancy (30) is nonelectrolytic and simply involves adding an excess of caustic soda to the toxic solution and heating the solution, to convert the ammonium and copper content into ammonia gas and copper oxide, and to drive off the ammonia gas and precipitate out or remove the copper oxide from the solution to form a copper-free and ammonia-free supernatant liquid.

*Filtration:* A process developed by W.N. Schjerven, Jr. (31) relates to a method of re-covering solid particulate copper from a liquid, and more particularly to a method of re-covering such metallic solids from a neutralized metal processing solution, to make the solution suitable for reuse or for discharge into a sewer.

In the manufacture of rolled copper rod which is to be drawn into electrical conductor wire, it is standard practice to dip the copper rod in a pickling solution, which includes sulfuric acid, certain other process enhancing agents and water, to remove copper oxides which have formed on the surface of the rod during its passage through a rolling mill. The rod subsequently is dipped in a series of water rinse tanks to remove acid solution remaining on the rod as a result of the pickling process, and finally in a neutralizing soap dip. Eventually, the water in the rinse tanks becomes contaminated with solid particulate material from the surface of the copper rod, which must be removed from the water be-fore it can be reused in manufacturing operations or discharged into a sewer. In this re-gard, these copper solids usually are in the form of copper oxide particles (cupric and cuprous), pure copper fines and small amounts of copper hydroxide, with traces of copper sulfate in solution.

One well-known procedure for removing the copper solids from the rinse water involves introducing the contaminated rinse water into a settling tank in which the copper solids settle to the bottom of the tank and in which, after a preselected period of time, the rinse water is withdrawn from the tank. In a specific known system of this type, the water, instead of being withdrawn from the tank, is permitted to seep out of the tank through pervious vertical side walls and through a filter bed.

In other known processes the effluent from a settling tank is passed through a relatively thick filter bed of a suitable filter material, such as dolomite. The filter material may be pretreated so as to react chemically with any suspended or colloidally dissolved copper salts in the water, so as to remove these contaminants from the water. When the filter bed becomes clogged, it is backwashed to remove the copper contaminants therefrom, and the backwash water and the copper contaminants are returned to the settling tank for re-settling.

One disadvantage of these prior known systems is that the recovery process generally is relatively time consuming. Further, settling tank systems require a large amount of space and settling tank systems in which the water is withdrawn after a predetermined settling time, without further treatment, are not particularly suited for the recovery of small solid particles and fines, because of the tendency for these materials to remain in suspension. In addition, however, systems in which the water subsequently passes through a filter bed also are disadvantageous in that usually a relatively large amount of filtering material is required, there is a tendency for the filter bed to become clogged rapidly when there are a large number of fines and colloidal suspensions in the water, and it is necessary to take the filter bed out of operation periodically when backwashing is required.

Filtration processes and equipment for removing the copper solids from the rinse water also are known in which a filter aid, in the form of a 2 to 3 inch layer of a filtering material, such as diatomaceous earth, is deposited on a unitary supporting filter media, such as an automatically indexable filter paper. The rinse water, after strip acidity has been neutralized, then is passed through the diatomaceous earth and the filter paper, which cooperate to filter out the copper solids. This filtering process is assisted by providing a vacuum beneath the filter paper, and when the diatomaceous earth and the filter paper become clogged with solid particulate material such that the vacuum increases to a preselected value, a new portion of the paper is automatically indexed into filtering position and a filter aid bed of the diatomaceous earth is formed on the new portion of the paper. The clogged diatomaceous earth and filter paper subsequently may be collected in the same container for disposal, or the diatomaceous earth may be removed from the filter paper, in which case they are collected in separate containers for disposal.

This last mentioned system also is disadvantageous, however, for several reasons. For example, while the copper solids have been removed from the neutralized rinse water, whereby it can be reused in manufacturing operations or discharged into a sewer, the clogged diatomaceous earth and the paper filter media still must be disposed of in a suitable manner, such as by hauling them to a land fill area. Another disadvantage of the system is that there is no practical procedure from an economic standpoint for recovering the copper solids from the diatomaceous earth and reusing the solids in manufacturing, and they must be discarded with the diatomaceous earth as a sludge or "soft scrap."

In the process by Schjerven a filter aid in the form of a relatively thin layer of solid particulate material of the same type as the solid particulate material to be recovered from the liquid, is formed on a unitary filter medium. The liquid is then passed through this filter aid and the filter media, which cooperate to remove solid particulate material from the liquid. Subsequently, the filtered solid particulate material and the solid particulate material forming the filter aid are removed from the filter medium.

More specifically, solid particulate material of the type to be removed from the liquid initially is mixed with a volume of the same type of liquid to form a relatively thick slurry. The slurry then is passed through the filter medium and the material which is filtered out is leveled to a uniform depth on the order of $^1/_{16}$ to $^1/_8$ of an inch to form a filter aid on the filter medium. After the initial operating cycle has been completed, a portion of the solid particulate material which has been removed from the filter medium may be recycled through the process and used to make the slurry for forming the filter aid. Further, after removal of the solid particulate material from the filter medium, the medium may be incinerated to recover any particulate material embedded therein or adhering thereto.

*Reduction and Filtration:* A process developed by E.B. Saubestre (32) involves the removal of substantially all copper from a substantially cyanide-free alkaline waste solution containing ionic copper. A reducing agent, for example, formaldehyde, for the ionic copper is added to the solution if not already present therein, and usually a complexing agent for the ionic copper. The process provides for the recovery of marketable zero valent metallic copper.

The alkaline waste solution is contacted with a material or metal catalytic to the reduction of the ionic copper by the reducing agent to zero valent metallic copper, and the catalytic contacting continued until the desired amount of copper is precipitated as zero valent metallic copper. The metallic copper precipitate is then separated, for instance, by filtration, from the solution. Exemplary of the catalytic metals for use in catalyzing the reduction of the ionic copper to zero valent metallic copper are palladium, platinum, gold and silver.

*Summary:* Copper removal by destructive treatment is currently accomplished by precipitation with lime. Effluent concentrations below 0.5 mg./l. are achieved on a consistent basis only with proper pH control and either proper clarifier design accompanied by good sludge settling characteristics or final effluent treatment with sand filtration. Final copper concentrations below 0.1 mg./l. are reported but do not seem readily attainable on a consistent basis with current precipitation methods. Since copper hydroxide solubility increases at higher pH after passing through a minimum at approximately pH 9 to 10, variations in the waste stream may be a causitive factor in limiting the removal of copper.

Process modification and waste stream separation provide a means of enhanced treatment of copper bearing wastes. When such modification is introduced, recovery processes become more attractive. Recovery of one type or another is practiced in many industrial operations and appears economically attractive due to the value of recovered copper.

### Recovery of Copper from Solid Wastes

A process developed by J.E. Perry (33) involves the recovery of metal values in scrap materials by destructive distillation removal of organic nonmetal materials, at temperatures which leave the metal relatively unaffected.

Among the methods which have been suggested for recovering the metal values separately from the nonmetal associated therewith are incineration or burning of combustible material, such as plastics, rubber, resin or paper, carried on the metal, which it is desired to reclaim. Such methods require an oxidizing atmosphere and usually are carried out at relatively high temperatures. As a consequence, substantial amounts of the metal are oxidized with consequent loss of metal product salvaged as metal. Another method which has been suggested involves heating metal scrap and the nonmetallic materials found with it in a retort or closed chamber. Such a procedure requires the treatment of small batches of product and is inherently more expensive than a process in which the material is processed continuously in an orderly manner through a suitable apparatus.

These and other suggested procedures in which the materials are exposed to relatively high temperatures are objectionable for at least two reasons. First, because the metal is contaminated by oxidation, and second because the physical properties of the metal which is later recovered are usually adversely affected by the overheating, and are therefore not as good as those possessed by the metal before it was processed for removal of the undesired organic or inorganic materials which may be considered as contaminating the metal to be recovered.

An object of this process is to provide an operation in which waste materials associated with metal scrap are continuously removed from the metal by a simplified, continuous process which avoids the smoke generated and overheating accompanying combustion processes, and in which the metal is recovered in a relatively uncontaminated state while it retains its physical properties and in which the major portion of the contaminating nonmetallic material is recovered as a volatile or condensible gaseous product which is useful as a fuel or heating means.

Briefly, the method consists of heating the contaminated scrap, in the absence of oxygen, to a temperature at which all of the volatile components are distilled off, using a continuous feed and removal of residue. At the temperatures involved, some volatile components are formed by destructive distillation. The volatile components, liquid and gas fractions, may be collected for reuse or they may be ignited and burned with a smokeless, odorless flame.

Such an apparatus is shown in Figure 41. As shown, the apparatus includes a furnace 10 which may be similar to a conventional powder metallurgy furnace or a rotary furnace. The furnace has an inlet end 12 and an outlet end 14, and is provided with a conveyor 16, for example, a stainless steel belt for carrying scrap material through the furnace. Furnace 10 is provided with means to supply heat to the furnace for startup and for maintaining the temperature of the furnace substantially constant at a controlled level during operation. A flame curtain 18 is provided at the inlet end of the furnace to keep air out and to maintain a nonoxidizing atmosphere in the furnace. Gases produced by the destructive distillation in the furnace burn at the flame curtain **18** at the inlet end. From the furnace, the product discharged at the outlet end **14** falls through a chute **20**. Beneath the chute is a hopper **24** with a bottom discharge through a star valve **26** and a side discharge through a side arm **28**.

As noted above, the scrap product emerging from the outlet end **14** of furnace **10** usually carries a coke or ash-like residue which adheres to the metal. This may be detached from the metal by means located in chute **20**, such as a rolling ring crusher **30**, or by other physical means. The scrap and loosened char fall onto a vibrating screen **34** which permits the coke-like residue to fall free of the metal to a collection bin. The metal, with any remaining residue, is conveyed from the screen **34** to a heated fused salt bath **40**, such as is described in U.S. Patent 3,000,766 and in U.S. Patent 3,615,815.

**FIGURE 41: APPARATUS FOR COPPER METAL RECOVERY FROM SOLID WASTES**

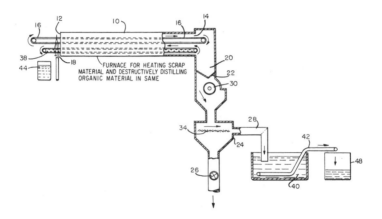

Source: J.E. Perry; U.S. Patent 3,697,257; October 10, 1972

After the furance is started and before raw material is put through on a continuous belt or continuous bucket conveyor, an initially neutral atmosphere can be provided consisting of gases such as nitrogen, hydrogen, cracked ammonia, and/or carbon dioxide. As soon as the organic material on its wire starts to pyrolyze, in substantially all cases, the gaseous products of the pyrolyzed organic material will form a protective atmosphere which is nonoxidizing in character, since the majority of organic insulation is usually composed of synthetic resin or paper, or natural resins.

The gases evolved provide the necessary controlled atmosphere. As much as 90% or more of the usual organic insulation is charred and destructively distilled in a matter of minutes in a temperature range between 900° and 1500°F., and the char formed is easily detached from the metal, as indicated above. At such temperatures the metal is not oxidized appreciably, nor is it adversely affected by the brief heating to which it is subjected.

The coke was removed from the wire with a rolling ring crusher and the mixture was separated on a vibratory screen. The screen was formed of a perforated sheet with ¼ inch holes. The product was of uniform quality independent of diameter of wire; guage sizes tested varied from No. 6 to No. 25. A slight discoloration on the wire was removed in the salt bath. Only a slight reaction was noted when the wire was placed in the salt bath and the reaction was complete in less than a minute.

It should be noted that the operating temperature is above the melting point of lead. Lead sheathed wire may be fed to the furnace and the lead removed by melting. The mechanism for removal of lead from the furance would be the belt **38** which would be designed with buckets to collect the molten lead and deposit the metal on the inlet end of the furnace. This additional feature is shown schematically in the drawing at **44**. The second conveyor **38** positioned beneath conveyor **16** in the furnace catches nongaseous products formed when the insulation and other coating materials burn off the wire and discharges them at the inlet end of the furnace.

Crushing and screening should be done while the metal scrap, e.g., wire, is hot to conserve heat at the salt bath. These operations would be done in the atmosphere of the furnace. Coke is removed through a star (rotary) valve **26** and metal or wire is conducted into the salt bath **40** into which it passes below the surface of the salt. A continuous conveyor **42** removes the wire from the salt bath and discharges the wire product into a container **48**.

The method is described more fully in the following example. Feed material was from representative samples of insulation wire. The insulation material included polyvinyl chloride, neoprene, paper, cloth, rubber, polyethylene, fiber glass, and various varnishes and waxes. The feed was cut to a size to produce a free-flowing mixture. Each of the insulation materials was treated in the above described apparatus, separately and then in mixtures; no difference was found in reactivity of the organic materials under the conditions of operation specified above, i.e., in the absence of oxygen and at temperatures at which organic materials were destructively distilled.

In a series of runs in which tin coated copper wire was processed, the tin content of the coke residue averaged about 0.32 weight percent. The tin was completely removed from the copper wire by maintaining the furnace temperature at 850°F. or higher. At higher temperatures, such as 1200°F., the tin is completely removed in as short a residence time in the furnace as 1.5 minutes. Similar results were obtained for the removal of lead solders from soldered scrap, by melting the solder.

It is also to be noted that the furnace includes two parallel conveyors, one disposed above the other. These run in opposite directions so that scrap being processed is conveyed by the upper run of the uppermost conveyor away from the entry end and towards the discharge end of the furnace, while separated material is conveyed by the upper run of the lowermost conveyor towards the entry end and away from the discharge end of the furnace. Also the evolution of gas during the distillation process results in an excess of gas in the furnace and some gas passes down into hopper **20** and out valve **22** when it is opened, further protecting the furnace from contamination by air or oxygen. The metal product may be recovered without crushing or sieving or fused salt cleanup, if desired, or it may be melted or otherwise processed.

### References

(1) Parsons, W.A. and Rudolfs, W., "Lime Treatment of Copper Pyrophosphate Plating Wastes," *Sew. Ind. Wastes Eng.*, 22, 313-315 (1951).

(2) Simpson, R.W. and Thompson, K., "Chlorine Treatment of Cyanide Wastes," *Sew. Ind. Wastes Eng.*, 21, 302-304 (1950).

(3) McGarvey, F.X., "The Application of Ion Exchange Resins to Metallurgical Waste Problems," *Proc. 7th Purdue Industrial Waste Conf.*, pp. 289-304 (1952).

(4) Nyquist, O.W. and Carroll, H.R., "Design and Treatment of Metal Processing Wastewaters," *Sew. Indust. Wastes*, 31, 941-948 (1959).

(5) McGarvey, F.X., Tenhoor, R.E. and Nevers, R.P., "Brass and Copper Industry: Cation Exchangers for Metals Concentration from Pickle Rinse Waters," *Ind. Eng. Chem.*, 44, 534-541 (1952).

(6) Anderson, J.S. and Iobst, E.H., Jr., "Case History of Wastewater Treatment in a General Electric Appliance Plant," *Jour. Wat. Poll. Control Fed.*, 40, 1786-1795 (1968).

(7) Pinkerton, H.L., "Waste Disposal" in *Electroplating Engineering Handbook*, 2nd ed., A. Kenneth Graham, ed.-in-chief, New York, Reinhold Pub. Co. (1962)

(8) Wise, W.S., "The Industrial Waste Problem: IV. Brass and Copper Electroplating and Textile Wastes," *Sew. Indust. Wastes*, 20, 96-102 (1948)

(9)  Gard, C.M., Snavely, C.A. and Lemon, D.J., "Design and Operation of a Metal Wastes Treatment Plant," *Sew. Ind. Wastes*, 23, 1429-1438 (1951).

(10) Nemerow, N.L., *Theories and Practices of Industrial Waste Treatment*, Reading, Mass., Addison Wesley (1963).

(11) Barnes, G.E., "Disposal and Recovery of Electroplating Wastes," *Jour. Wat. Poll. Control Fed.*, 40, 1459-1470 (1968).

(12) Tallmadge, J.A., "Nonferrous Metals," in *Chemical Technology, Vol. 2, Industrial Waste Water Control*, C. Fred Gurnham, ed., New York, Academic Press (1965).

(13) Jones, H.R., *Pollution Control in the Nonferrous Metals Industry*, Park Ridge, New Jersey, Noyes Data Corp. (1972).

(14) Davis, W.E. and Assoc., "National Inventory of Sources and Emissions," Section III, Copper, *Report PB 210,678*, Springfield, Va., National Technical Information Service (May 1972).

(15) Domey, W.R.,and Stiefel, R.C., "Wastewater Reduction in Metal Finishing Operations," presented at 43rd Annual Conf. Wat. Poll. Control Fed., Boston, Mass (October 5, 1970).

(16) Foulk, D.G., "Metal Finishing Products," in *Chemical Technology, Vol. 2, Industrial Waste Water Control*, C. Fred Gurnham, ed., New York, Academic Press (1965).

(17) Tomic, E.A.; U.S. Patent 3,196,107; July 20, 1965; assigned to E.I. du Pont de Nemours and Co.

(18) Stumm, W. and Morgan, J.J., *Aquatic Chemistry*, New York, Wiley-Interscience (1970).

(19) Jenkins, S.H., Keight, D.G. and Humphreys, R.E., "The Solubility of Heavy Metal Hydroxides in Water, Sewage and Sewage Sludge. I. Solubility of Some Metal Hydroxides," *Internat. Jour. Air Wat. Poll.* 8, 53-56 (1964).

(20) Stone, E.H.F., "Treatment of Non-ferrous Metal Process Waste of Kynoch Works, Birmingham, England," *Proc. 25th Purdue Industrial Waste Conf.*, pp. 848-865 (1967).

(21) Hansen, N.H. and Zabban, W., "Design and Operation Problems of a Continuous Automatic Plating Waste Treatment Plant at Data Processing Division, IBM, Rochester, Minnesota," *Proc. 14th Purdue Industrial Waste Conf.*, pp. 227-249 (1959).

(22) Watson, K.S., "Treatment of Complex Metal-Finishing Wastes," *Sew. Ind. Wastes* 26, 182-194 (1954).

(23) Resource Eng. Assoc., "State of the Art Review on Product Recovery," *Report PB 192,634*, Springfield, Va., National Technical Information Service (November 1969).

(24) Culotta, J.W. and Swanton, W.F., "Case Histories of Plating Waste Recovery Systems," *Plating*, 57, 251-255 (1970).

(25) Culotta, J.W. and Swanton, W.F., "The Role of Evaporation in the Economics of Waste Treatment for Plating Operations," *Plating*, 55, 957-961 (1968).

(26) Botham, G.H. and Bryson, W.R., "Note on the Corrosion of Aluminum Dairy Equipment by Water," *J. Dairy Res.*, 20, 154-155 (1953).

(27) Simpson, R.W. and Thompson, K., "Chlorine Treatment of Cyanide Wastes," *Sew. Ind. Wastes Eng.*, 21, 302-304 (1950).

(28) Easton, J.K., "Electrolytic Decomposition of Concentrated Cyanide Plating Wastes," *Jour. Wat. Poll. Control Fed.*, 39, 1621-1625 (1967).

(29) American Electroplating Society, "A Report on the Control of Cyanides in Plating Shop Effluents," *Plating*, 56, 1107-1112 (1969)

(30) Lancy, L.E.; U.S. Patent 3,218,254; November 16, 1965; assigned to Lancy Laboratories, Inc.

(31) Schjerven, W.N., Jr.; U.S. Patent 3,696,928; October 10, 1972; assigned to Western Electric Co.

(32) Saubestre, E.B.; U.S. Patent 3,666,447; May 30, 1972; assigned to Enthone, Inc.

(33) Perry, J.E.; U.S. Patent 3,697,257; October 10, 1972; assigned to Horizons Research Inc.

# CRACKING CATALYSTS

It is common practice in the petroleum refining industry to subject hydrocarbon streams to catalytic cracking operations to produce products having greater value. Various types of fluidized catalytic crackers have been developed for this purpose. These catalytic crackers normally comprise a reaction vessel and a catalyst regeneration vessel. In the regeneration vessel, the catalyst particles are subjected to an oxidizing atmosphere at relatively high temperatures to burn coke deposits off the catalyst particles. Unfortunately, a substantial volume of catalyst fines is produced in a fluidized catalytic cracking operation of this type, and large amounts of these fines are often entrained in the effluent flue gases from the regenerator.

## Removal from Air

A process has been developed by F.A. Zenz (1) for reducing dust emission to the atmosphere from a fluidized bed regenerator. The system employs cyclone separators one of which, in addition to returning separated particles to the bed through a dip-leg, is provided with means for bypassing a small part of the gas stream to a filtering system. The system reduces dust emission by subjecting part of the gas stream carrying nonseparable particles as an underflow stream to a filtering operation while returning separated fines to the fluid bed. The system is shown schematically in Figure 42.

**FIGURE 42: DUST CONTROL APPARATUS FOR FLUIDIZED BED REACTORS**

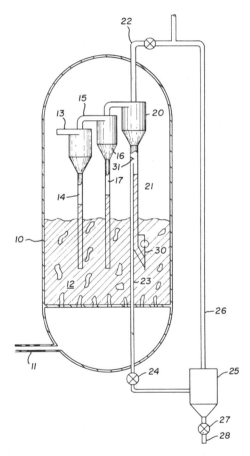

Source: F.A. Zenz; U.S. Patent 3,698,874; October 17, 1972

Referring to the drawing, a gas stream enters the regenerator vessel **10** through an inlet pipe **11** and passes up through a particulate bed **12** which is thereby fluidized. In passing through the bed the gas stream entrains small particles from the bed surface with the result that the gas enters the first stage cyclone **13** as a dust laden stream. The coarser particles are separated from the gas stream in the cyclone and are returned to the fluidized bed through dip-leg **14**.

Those particles that are not separated in cyclone 13 are carried by the gas stream lines through duct 15 to the second stage cyclone 16 where they are subject to more intensive centrifugal forces. This accomplishes separation of still finer particles which are returned to bed 12 through dip-leg 17. Those particles which are not separated in cyclone 16 are carried by the gas to a third cyclone 20 where further separation takes place. Here again those particles which are separated are returned to the fluidized bed 12 through dip-leg 21 while those particles still entrained in the gas stream are exhausted to the atmosphere through stack 22. In this example, a three stage cyclone separator is described and shown, although fewer or more stages may be employed.

In any case, the last stage of the cyclone system is provided with a separate duct 23 which exits from reactor vessel 10 and through valve 24 goes to an additional separation apparatus 25. Since, as will be described, only a small portion of the gas stream leaving the vessel will pass through duct 23, separation apparatus 25 may be an absolute filter that removes all particles carried by the gas through duct 23. The filtered gas leaves apparatus 25 through pipe 26 to merge with the dust entrained gas in stack 22. It will be clear that the merged gas will contain less dust than if a portion of the gas did not pass through apparatus 25. The fines filtered out by apparatus 25 are collected and drawn off periodically through valve 27 and pipe 28.

Duct 23 must exit from the cyclone below the cone of the cyclone so that the gas entering the duct does so as an underflow stream. Such a stream increases the efficiency of the cyclone by carrying a higher percentage of nonseparable particles than would be carried by a comparable stream exiting normally from the cyclone. In the preferred embodiment, the underflow gas stream will be from 1 to 5% of the gas stream entering the cyclone, but in no event should it be more than 10% of the gas stream, otherwise the requirements for filtering apparatus 25 might increase the costs of the system to an uneconomic level.

Duct 23 must terminate at its upper extremity at or preferably some distance above the level the separated particles would normally reach in dip-leg 21. This level is determined by the bulk density of the particles and the difference in pressure within the bed at the foot of the dip-leg 21 and the pressure at the junction of the dip-leg with the cone of the cyclone. A trickle valve 30 may be provided at the foot of dip-leg 21. It may also be necessary to provide a baffle or coarse screen 31 above duct 23 to prevent plugging of the duct by material that might flake or spall off the interior surface of cyclone 20.

While duct 23 has been shown adjacent dip-leg 21 and passing through the grid plate of vessel 10, it could exit from the vessel either through the sidewall or the top of the vessel. Also while a serial arrangement of cyclones has been shown, they could be arranged in a parallel configuration or a series parallel configuration. In certain of the parallel configurations where a number of ducts, similar in purpose to duct 23, are provided, those ducts could lead to a header from which a single duct will lead to the exterior of vessel 10 and to separation apparatus 25. The header may be located above or below the fluidized bed.

A process developed by J.B. Rush (2) is one in which catalyst fines are removed from a gaseous stream by scrubbing the stream with oil. The resulting oil, containing catalyst fines, is introduced as feed into a distillation column wherein asphalt is separated as a product stream. The catalyst fines appear in the asphalt.

In order to prevent atmospheric contamination by the catalyst fines, it is necessary to subject the flue gases to rather expensive treating steps. One system which has been used involves passing these gases through electrostatic precipitators. However, large precipitators are required and these are quite expensive. Another procedure which has been proposed for the removal of catalyst fines comprises scrubbing the flue gas with an oil to remove the fines. However, the scrubbing oil must then be treated to remove the fines. Filtering operations to perform this function are quite expensive, and there still remains the problem of disposing of substantial volumes of the fines, which are then present in the form of a paste with the scrubbing oil.

In accordance with this process, a method is provided for removing catalyst fines from gaseous streams in a relatively inexpensive manner and for disposing of the removed catalyst fines. This is accomplished by first scrubbing the flue gas with an oil to remove the catalyst fines. The resulting oil containing the catalyst fines is then introduced into a distillation unit wherein asphalt is being produced. It has been discovered that substantial volumes of catalyst fines can be incorporated into the resulting asphalt in this manner without materially affecting the desired properties of the asphalt. This permits removal of substantial quantities of fines from a refining operation without extensive disposal procedures.

### References

(1) Zenz, F.A.; U.S. Patent 3,698,874; October 17, 1972.
(2) Rush, J.B.; U.S. Patent 3,494,856; February 10, 1970; assigned to Phillips Petroleum Company.

## CYANIDES

Cyanide, as sodium cyanide (NaCN) or hydrocyanic acid (HCN) is a widely used industrial material. Cyanide waste streams result from industries such as ore extracting and mining, photographic processing, coke furnaces, and synthetics manufacturing. A major source of waste cyanide is the electroplating industry. Electroplaters use cyanide baths to hold ions such as zinc and cadmium in solution. The metal ions are then electro-deposited onto ferrous metals. "Drag-over" of the plating solution, containing cyanide ions and metal-cyanide complexes, contaminates rinsing baths.

### Removal of Cyanides from Water

Various methods of treating cyanide wastes are in current use. The process more frequently employed is cyanide destruction by chlorination, according to J.W. Patterson and R.A. Minear (1). Cyanide may be either partially oxidized to cyanate (CNO$^-$), or completely oxidized to carbon dioxide ($CO_2$) and nitrogen ($N_2$). Another cyanide destruction process, called electrolytic decomposition, is applicable where waste cyanide concentrations are very high. Ozone oxidation of cyanide has also been employed with some success, and is claimed to be equivalent to destruction by chlorination, on the basis of effectiveness and costs.

*Chlorination:* Destruction of cyanide by chlorination may be accomplished by direct addition of sodium hypochlorite, or by addition of chlorine gas plus sodium hydroxide to the waste. Sodium hydroxide reacts with the chlorine to form sodium hypochlorite. One decides on the basis of economics which of the two methods to use. Chlorine gas treatment is about half as expensive as the direct hypochlorite treatment, but handling is more dangerous and equipment costs are higher. The hypochlorite added or produced oxidizes cyanide to cyanate. Oxidation of cyanide to cyanate is accomplished most completely and rapidly under alkaline conditions at pH 10 or higher, and this cyanide treatment process is often termed alkaline chlorination.

The resulting cyanate is much less toxic than cyanide (2), but can if necessary be further oxidized to carbon dioxide and nitrogen, two biologically nontoxic compounds. Oxidation of cyanate to carbon dioxide and nitrogen can be accomplished by additional chlorination, or by acid hydrolysis. Acid hydrolysis must take place at pH 2 to 3 however, usually obtained by addition of sulfuric acid. Thus, acid hydrolysis requires addition of acid to achieve the required pH, and subsequent neutralization of the acidic wastewater before the effluent can be discharged. These processes of pH adjustment increase the total dissolved solids of the wastewater. Cyanate will be destroyed by acid hydrolysis within about five minutes contact time.

Complete cyanide oxidation to $CO_2$ and $N_2$ can be accomplished solely through chlorination, providing close pH control is maintained. After initial oxidation to cyanate, further oxidation to yield $CO_2$ and $N_2$ will occur slowly over several hours at pH 10+. However, if the waste pH is reduced to 8 to 8.5, cyanate oxidation by chlorine can be completed within one hour. At the lower pH, sufficient chlorine must be added to insure an excess beyond that needed to oxidize the cyanide to cyanate, in order to avoid liberation of highly toxic cyanogen chloride. Cyanogen chloride is the intermediate product of the oxidation of cyanide to cyanate. It breaks down very rapidly and poses no problem at pH 10+. However, at the lower pH, excess chlorine is needed to speed the breakdown (2).

Several manufacturers have developed small "package" cyanide treatment plants which employ the chlorination process. Zievers, et al (3) have described basic types of package plants. A tank type plant provides cyanide treatment to the level of cyanate only. If complete destruction is required, a second tank is employed for secondary pH control and second-stage chlorination. A reaction tower type plant provides complete cyanide treatment to the level of $CO_2$ and $N_2$. Package plants such as these range from treatment capacities of 800 gallons of waste per hour containing a maximum of 450 mg./l. of cyanide, to units of 1,800 gallons per hour and a maximum cyanide content of 350 mg./l. A unit of 800 gallons per hour capacity costs about $15,000 complete (4).

Larger volumes of cyanide waste require especially designed and constructed treatment plants, tailored to the individual industrial process and flow configuration producing the waste. Fisco (5) has described such a treatment plant, which oxidizes the cyanide component of a waste stream totaling 1.5 million gallons per day. Treatment involves total cyanide destruction by a two step process of chlorination at high and low pH as discussed above.

Schink (6) has reported treating a cyanide waste flow of 2,500 to 3,000 gallons per hour at the Tektronix (Oregon) electronics plant. Treatment was by alkaline chlorination to the level of cyanate. Raw waste cyanide concentration was not reported. An effluent cyanide content of 0.1 mg./l. was achieved by the process. Complete oxidation of cyanide to $CO_2$ and $N_2$ was achieved in treating the IBM Plant waste by alkaline chlorination plus acid hydrolysis. Cyanide content of the rinse water was reduced from 700 mg./l. to 0 mg./l. cyanide plus cyanate (7).

The cyanide contents of electroplating rinse waters are quite variable, ranging from the 700 mg./l. concentration discussed above to 10 to 25 mg./l. as reported for a small plating shop (8). Watson (9) has also reported the treatment of a low dilute cyanide waste, for a General Electric metal plating plant (Erie, Pennsylvania). The waste contained an average cyanide content of 32.5 mg./l. and was treated to 0.0 mg./l. cyanide, on a batch basis at the rate of 200,000 gallons per month. Effluent cyanate concentration was not reported. Cyanide treatment was to the level of cyanate by alkaline chlorination, 3 pounds of chlorine being fed per pound of cyanide treated. It has been reported elsewhere (2) that oxidation of cyanide to cyanate requires approximately 1.75 pounds of chlorine per pound of cyanide, and complete oxidation to $CO_2$ and $N_2$ requires about 4.3 pounds of chlorine per pound of cyanide. The zero cyanide residual reported for the GE waste likely resulted from addition of chlorine in excess (insofar as oxidation to cyanate).

Complete destruction of cyanide to $CO_2$ would increase the treatment costs above those for simple oxidation to cyanate. Zievers et al (3) have published cost curves for single stage cyanide treatment (to cyanate), and for complete cyanide oxidation.

Shockcor (10) has pointed out that unless metal-bearing wastes are segregated from the cyanide wastes, an extra chlorine demand will be exerted as a result of metal oxidation by the chlorine (e.g., $Ni^{+2}$ to $Ni^{+3}$, $Fe^{+2}$ to $Fe^{+3}$, $Cr^{+3}$ to $Cr^{+6}$, etc.). As the cost of chlorine is approximately 13¢ per pound in the Midwest (3), any extra chlorine demand exerted by a mixed metal-cyanide waste would increase cyanide treatment costs significantly.

A process for the treatment of acid and cyanide industrial wastes has been described by C.W. Rice (11). Acid and cyanide wastes are formed adjacently in various industrial operations. For example, in the steel industry pickling operations and cyaniding operations are sequentially performed on bodies of metal for plating with nickel, chrome or the application of other protective or ornamental coatings. Wastes containing sulfuric acid or one of the other mineral acids are highly detrimental if present in streams in substantial concentration, and wastes containing cyanides are poisonous to stream life and dangerous to human life if carried into streams in any appreciable quantity. Certainty in operations for disposing of those wastes is therefore a matter of necessity.

Certainty in neutralizing the acid wastes and in combining or decomposing the cyanide wastes having been assured, efficiency and economy in the methods of disposal are greatly to be desired. Assuming that the reagents as for neutralization and decomposition must be present in quantity in excess of that theoretically required for those reactions, it is desirable to utilize the reagents in as slight excess as gives certainty in disposal. Also, it is desirable, still within the bounds of safety, to minimize the number of tanks required for treating the wastes and for settling the products of the treatment.

One object of the process described by Rice is to insure complete neutralization of the acid waste and to insure complete decomposition of the cyanide waste. Another object is to provide a system so safeguarded that it is made certain a safe excess of the reagents for rendering the wastes harmless will at all times be supplied, and to minimize the quantity of reagents required for positive safety.

Another object is to provide simplifed plant structure of such sort that it reduces the initial cost of plant installation, while also attaining the other objects relating to the maintenance of assuredly safe conditions with a minimum expenditure of reagents. Another object is to provide a system for disposing of industrial wastes of the type indicated above which is under automatic and correlated control of both the intake of wastes to the system and the supply of requisite reagents thereto to maintain safe conditions, and automatically to safeguard the supply of reagents.

It may be stated generally that in attaining the above-cited objects, one course is to maintain a large body of sludge in the conditioning tank for treating the acid waste and in the conditioning tank for treating the cyanide waste. Also, a large body of sludge may be maintained in a settling tank which is common to both industrial wastes. The conditioning tanks are so arranged that effluent cannot pass from a tank without passing through the body of sludge. In accordance with one preferred procedure, one greatly overtreats the cyanide waste in the cyanide conditioning tank to give an effluent which contains a relatively high alkalinity. This effluent is passed from the cyanide conditioning tank into the acid conditioning tank and then from the acid conditioning tank into the common settling tank. In passage through each of these conditioning tanks, the liquid wastes pass through a large body of sludge and carry some of the sludge to the settling tank.

A waste treatment apparatus for the handling of metal finishing wastes containing cyanides has been described by J.F. Zievers, C.W. Riley and R.W. Crain (12). Such a system includes at least one motor driven impeller mounted in a vertical mixing chamber and a pump for continuously passing the waste liquid to be treated through the mixing chamber. The impeller recirculates a portion of the flowing fluid to form a vortex, and a treating chemical is introduced into the vortex via a feed conduit for mixing with the rinses to be treated.

Another method of treating metal finishing wastes containing cyanides has been described by J.F. Zievers, C.W. Riley and R.W. Crain (13). It includes a plurality of vertically-spaced, motor-driven impellers in a vertical mixing chamber and a sump pump for discharging continuously a portion of the waste liquid to be treated into one inlet of the mixing chamber, and for discharging continuously the remaining portion of the waste into another inlet of the chamber near one of the impellers via an eductor which continuously premixes a treating chemical, which may be toxic or corrosive, with the waste liquid by diffusing the treating chemical into the waste liquid flowing through the eductor so that the toxic or corrosive chemical does not enter the chamber in the absence of waste liquid.

An integrated wastewater treatment process for cyanide plating wastes has been described by L.E. Lancy (14). In this process used or waste wash water treatment solution from individual cyanide type plating lines that apply coatings of different metals of workpieces, having a chelating compound and a dissolved content of plating metal that have been picked up from drag-out on the workpieces, is treated with a salt of metals of the class consisting of calcium, barium, lithium, aluminum and magnesium. The salt acts as an exchange element in the chelating compound to replace a more noble dissolved metal content in the wast water treatment solution before it is reused to wash off drag-out from workpieces being processed in a different plating line.

*Electrolytic Decomposition:* Wastes containing high concentrations of cyanide are most successfully treated by electrolytic decomposition. Electrolytic decomposition is primarily used by industry for the destruction of cyanide in concentrated spent metal plating solutions. It is not practical to treat rinse waters by this method, due to the initial low cyanide concentrations (8).

The concentrated cyanide waste is subjected to electrolysis at high temperature (approximately 200°F.) for several days. Initially, cyanide is completely broken down to carbon dioxide and ammonia, with cyanate as an intermediate (15). As the process continues, the waste-electrolyte becomes less capable of conducting electricity, and the reaction may not go to final completion. Some residual cyanate may be formed, which requires chlorination to complete the waste treatment.

Data published by Easton (16) indicates that low levels of residual cyanide can be achieved by electrolytic decomposition, providing sufficient treatment time is allowed. Wastes containing several tens of thousands of mg./l. of cyanide were treated, over 7 to 18 day periods, down to cyanide levels of less than 0.5 mg./l. Easton also reported electrolytic decomposition as being less effective on cyanide wastes containing sulfate. Cyanide reductions to only 695 to 750 mg./l. were achieved, before heavy scaling at the anode prevented further electrolysis. Sulfate concentrations in the cyanide wastes were not reported by Easton.

Operating costs were reported by Easton for electrolytic decomposition (16). Total treatment cost was 8.2¢ per pound of cyanide destroyed. As a cyanide waste concentration of 75,000 mg./l. is equivalent to 62.55 pounds/1,000 gallons, these costs convert to $53.29 per 1,000 gallons at that cyanide concentration. Higher or lower cyanide concentrations would alter the cost per 1,000 gallons proportionately. For Easton's process copper recovery, by virtue of being plated out in the electrolysis process, and the value of copper recovered, defrayed in part the total treatment cost to yield a net cost of 4.3¢ per pound of cyanide destroyed, or $26.90 per 1,000 gallons at a cyanide concentration of 75,000 mg./l.

Stauffer Chemical Company markets a package electrolytic decomposition treatment plant (17). The system was developed specifically to treat rinse waters containing low cyanide content, and is claimed to purify a waste stream with 20 mg./l. to less than 0.5 mg./l. of cyanide. Unit capacity is 5 to 10 gallons per minute and capital cost is less than $20,000 for the typical plating shop (17).

A process has been developed by H.A.H. Ericson and G.B. Norstedt (18) for electrolytic oxidation of cyanide wastes. This process, which includes an electrolytic alkaline cleaning bath and a cyanide-containing surface treating bath, comprises passing the waste cyanide-containing water from the cyanide treatment bath to an alkaline electrolytic cleaning bath, whereby the cyanide is electrolytically oxidized. Figure 43 shows the process as applied to electrolytic zinc plating in cyanide-containing electrolytic baths.

As shown in the drawing, the installation comprises a sequence of baths 1 through 12 for treating metal articles for zinc plating. These baths have the following functions:

| | |
|---|---|
| 1 | Alkaline degreasing |
| 2 | Cathodic degreasing |

| 3 | Anodic degreasing |
|---|---|
| 4 | Cold water rinse |
| 5 | Cold water rinse and recovery rinse |
| 6 | Zinc plating |
| 7 | Cold water rinse |
| 8 | Cold water rinse |
| 9 | Acid pickling |
| 10 | Chromating |
| 11 | Cold water rinse |
| 12 | Hot water rinse |

The metal articles to be treated pass through the baths mainly in the order mentioned. The thin arrow lines to the right in the figure illustrate the flow of liquid to and from and between the baths as explained in detail on the following page.

FIGURE 43:  SEQUENTIAL APPARATUS FOR TREATING CYANIDE-CONTAINING WASTES

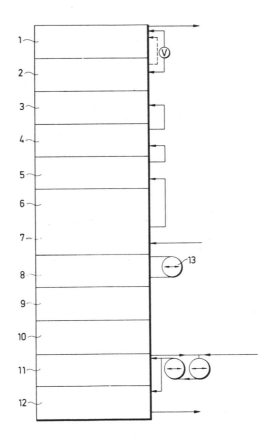

Source:  H.A.H. Ericson and G.B. Norstedt; U.S. Patent 3,645,867; February 29, 1972

More particularly, the baths may operate in the following manner. Bath 1 is a degreasing bath containing sodium hydroxide, sodium carbonate and metasilicate, operating at a temperature of about 80°C. Bath 2 is an electrolytic degreasing bath of the same composition as bath 1. The article is connected as a cathode. The anode consists of graphite. The bath operates at a temperature of about 80°C. and at a current density (anodic and cathodic) of 5 to 10 amperes per square decimeter. Bath 3 is also an electrolytic degreasing bath of the same composition as bath 2 and operated under similar conditions, except that the article is coupled as the anode and the cathode consists of steel. Suitably, a complexing agent of gluconate type is added to this bath.

The evaporation of water from baths 1, 2 and 3 (indicated by the outwardly directed arrow at bath 1) is about 8 to 10 liters per square meter of bath surface an hour. To compensate this liquid loss, and to achieve the desired destruction of cyanides, liquid is passed from bath 4 to baths 1, 2 and 3 as indicated by the arrows. Bath 4 is a cold water bath for rinsing the articles after they have been treated in bath 3. Bath 5 is also a cold water bath for further rinsing the articles before they are treated in the plating bath. The articles may also be rinsed in this bath after they have been treated in the plating bath.

Bath 6 is a cyanide-containing electrolytic zinc plating bath, containing about 35 g./l. Zn, 30 g./l. NaOH and 110 g./l. NaCN. The bath operates at a temperature of not above 30°C. and at a current density of 3 to 6 amperes per square decimeter. Bath 7 is a cold water bath for rinsing the electroplated articles, which may previously have been rinsed in bath 5. Fresh water is introduced into bath 7 at a rate corresponding to the evaporation in baths 1, 2, 3, the water flowing through baths 7, 5, 4 and thence to baths 1, 2, 3, as indicated by the arrows in the diagram on the drawing.

Thus, zinc and cyanide accompany the water to the degreasing baths 3, 2, 1. The zinc is precipitated in bath 3 on the cathodes. The cyanide is decomposed by electrolytic oxidation in baths 3 and 2. To avoid introducing cyanide into bath 1 it may be suitable in some cases to introduce water from bath 4 only into baths 3 and 2 and take water for bath 1 from bath 2 as indicated by the valve and the broken line arrow.

Bath 8 is a cold water bath for further rinsing the article. From this bath, water is circulated through a basic anion exchange unit 13, where any residual zinc cyanide complexes are taken up. The ion exchange composition is regenerated, when necessary, with NaOH and the regenerate may be passed to the zinc bath 6.

Bath 9 is an acid pickling bath containing about 1% $HNO_3$. Bath 10 is a so-called one-step chromating bath containing about 2 g./l. alkali bichromate and fluoride and sulfate ions. Total salt content is about 4 g./l. Bath 11 is a cold water rinsing bath, wherein the water is circulated through a two bed (anionic and cationic) ion exchange filter unit 14, 15 so that the water will be completely deionized. Fresh water is introduced into this bath (through the filter unit) to compensate evaporation in bath 12. Bath 12 is a final hot water rinsing bath operating at about 80°C. Evaporation is about 8 liters per square meter an hour, which is replaced by water taken from bath 11.

*Ozonation:* Ozonation shows some promise as a substitute for chlorination of cyanide waste. Besselievre (4) has pointed out that ozonation offers perhaps the cheapest method of destroying cyanide, all things considered because the cost of producing ozone, without fixed costs, is low, and the amount of ozone required per pound of cyanide is small. Kandzas and Mokina (19) have reported complete ozone oxidation of cyanide (via cyanate intermediate) to end products. Complex zinc, nickel and copper cyanides were easily destroyed, but complex cobalt cyanide was resistant to treatment by ozonation.

Ozonation has been employed to treat a cyanide waste at the Boeing Airplane (Wichita, Kansas) metal working plant (20). Waste flow through the plant is 500 gallons/minute, and cyanide concentrations to 25 mg./l. were encountered. Complete oxidation of cyanide to cyanate is reported, with partial oxidation of cyanate to final end products. Ozone production costs were 14 to 15¢ per pound (20).

*Oxidation:* A process developed by H.A. Gollmar (21) is one in which raw blast furnace gas is washed by suitable gas washers well known in the art for the removal of solid particles. HCN is also removed from this gas if it contains HCN, and the resulting effluent is further treated so that the solid particles entrained therein are settled out. The dissolved HCN is removed from the clarified effluent into a mixture of air and $CO_2$ gas by circulating the effluent by suitable means through a plurality of aeration towers in series which utilize air and $CO_2$ for the aeration so that the HCN is disposed into the atmosphere.

The HCN removal may be accomplished by one aeration tower but it has been found desirable to use more than one aeration tower. In order to eliminate scale formation which may accumulate due to the aeration in a plurality of aeration towers in the pumps, pipes and nozzles which conduct the effluent containing HCN, and for a more effective HCN removal, $CO_2$ gas is introduced from a suitable source at the bottom of the aeration towers along with air for accomplishing the aeration and for preventing scale formation in the aerated effluent.

A process has been described by J.-P. Zumbrunn (22) for the treatment of effluents in which cyanides are directly converted to nontoxic and hydrolyzable cyanates, through the action of a peroxide of the inorganic or organic peroxoacid type, as the free acid or as a salt.

A process described by O.B. Mathre (23) provides for the destruction of cyanide anions in aqueous solutions using hydrogen peroxide and a soluble metal compound catalyst such as a soluble copper compound to increase the reaction rate. An aqueous composition containing hydrogen peroxide and the catalyst is also provided which can be added directly to the cyanide solution. The pH of the cyanide solution to be treated is adjusted with acid or base to a pH of about 8.3 to 11. The soluble metal compound catalyst is added at a level to give about 5 to 1,000 ppm of catalyst, the hydrogen peroxide is added to give a molar ratio of hydrogen peroxide to cyanide anions of at least about 0.8, and the temperature of the solution is maintained within the range of $20°$ to $75°C$.

A process developed by D.G. Hager and J.L. Rizzo (24) is one in which copper salts and oxygen are added to the water upstream of a bed of activated carbon. The copper and cyanide form a complex of uncertain composition. The copper complexed cyanide thus formed is adsorbed on the activated carbon, which catalyzes the oxidation of the cyanide to carbon dioxide and nitrogen.

*Aldehyde Treatment:* A process developed by J.L. Morico (25) is a method for the destruction of free and/or chemically combined cyanide present in waste solutions so as to render the solution suitable for safe disposal. The process involves mixing together the cyanide-containing waste solution and an aldehyde and/or a water-soluble bisulfite addition reaction product thereof, with the aldehyde or addition reaction product thereof supplied in the stoichiometric amount and preferably in excess over the stoichiometric amount required for reaction with all the cyanide present. The selected material and free and/or chemically combined cyanide of the waste solution are then reacted together at room temperature or elevated temperature until the cyanide has been converted into nontoxic materials to that degree required for safe disposal of the solution.

A process developed by B.C. Lawes and O.B. Mathre (26) for detoxifying cyanide wastewaters such as cyanide electroplating rinse waters and chemical process wastewaters is provided which allows the treated wastewaters to be discharged to a sewer or to a solids separation step. The process uses a combination of hydrogen peroxide and formaldehyde in certain ratios and preferably a magnesium salt, such as Epsom salt. Hydrogen peroxide is used in amounts so as to provide an initial molar ratio of $H_2O_2/CN^-$ within the range of 0.6 to 3, preferably 0.75 to 1.5, while the formaldehyde is used in amounts so as to provide an initial molar ratio of $CH_2O/CN^-$ within the range of 0.5 to 3, preferably 0.6 to 2. When the magnesium salt is used, it is used in an amount to provide at least about 0.3 ppm of $Mg^{++}$, preferably about 0.3 to 100 ppm of $Mg^{++}$.

It is preferred that the temperature for carrying out the process be within the range of 50° to 180°F. and the pH of the wastewaters treated be within the range of about 9 to 12.5. When zinc cyanide electroplating rinse waters are treated, the settled or filtered basic zinc salts can be recycled to the zinc electroplating step. For the first time, this process has brought economy and efficiency to the destruction of cyanide using peroxygen compounds without the need for a catalyst.

This patent is apparently the basis for a new proprietary process offered by Du Pont to the trade and known as the "Kastone" process (27).

*Starch Treatment:* A process developed by A.C. Lauria and J.L. Owens (28) is one in which undesirable cyanide compounds in waste effluents can be converted to nontoxic biodegradable materials by treating such waste effluents with a starch conversion syrup. Preferably the waste effluent is also treated with a metal chelating composition when such effluent contains heavy metal cyanides.

*Ion Exchange:* A process has been developed by W.J. Sloan (29) for removing cyanide values from aqueous solutions containing sodium or potassium cyanide, or the sodium or potassium salts of the complex cyanides of the metals copper, nickel, zinc, silver, gold, and cadmium, utilizing a mixed bed of ion exchange resins.

*Radiation Treatment:* A method developed by R.F. Byron, J. Danaczko, Jr., A.L. Dixon and L.M. Welker (30) involves the decomposition of cyanide ions contained in aqueous solutions into harmless constituents by exposure to penetrative ionizing radiation, preferably gamma radiation, until the cyanide ions have been decomposed into nontoxic constituents to that degree required for safe disposal, after which the treated solution may be discharged into native streams and the like. In general, best results are obtained by diluting relatively concentrated solutions down to a cyanide ion ($CN^-$) concentration at least as low as about 1 to 5 grams per liter prior to exposure.

The percentage of decomposition increases more or less directly with the total radiation dosage up to conversions of approximately 90 to 95%, after which markedly increased dosages are required to produce further increased decomposition yields. While substantially complete conversion can be obtained for all concentrations with sufficient exposure, a determination of the optimum dosage for each individual application depends on the initial concentration, the degree of predilution desirable, the percentage or absolute conversion required, and such other factors as the presence of extraneous constituents in the solution. However, total dosages between about $10^6$ to $10^9$ rads appear to be effective over a wide range of solutions and concentrations with the lower dosages corresponding to the more dilute solutions.

The intensity and time period of irradiation are not particularly critical, independently, but correlation between the two must be maintained to obtain the total dosage necessary to decompose the cyanide constituents to the required extent; for example, 95 to 98%. Thus, the radiation source may be selected in accordance with availability and expense, and the time varied to suit the specific situation.

*HCN Liberation and Recovery:* A method developed by L.F. Scott (31) for converting cyanide wastes into sodium cyanide is one in which the wastes are heated and pumped into a first tank containing an acid solution to produce hydrogen cyanide. The hydrogen cyanide is then passed through a first tower of sodium hydroxide and a second tank containing a first solution of sodium hydroxide and sodium cyanide. The materials in the second tank are recycled therethrough and through the first tower to produce the sodium cyanide reaction product, which is then withdrawn from the second tank.

During recirculation, vaporized hydrogen cyanide is withdrawn from the second tank and passed and recycled through a second tower of sodium hydroxide and a third tank to produce a second solution of sodium hydroxide and sodium cyanide, which is pumped into the second tank at the start of the next conversion cycle to form the first solution therein.

*Summary:* Treatment of strong cyanide wastes appears to be most feasible by electrolytic decomposition according to Patterson and Minear (1). Low to intermediate effluent cyanide content is reported by this process, and either partial treatment to cyanate or complete destruction can be achieved as required. Chemical oxidation by chlorination or ozonation is most appropriate for low cyanide wastes, whether carried over from the electrolytic decomposition process or as dilute rinse water wastes. Operating costs of the two chemical oxidation treatment processes are equivalent, although capital costs for ozone treatment appear high in comparison to chlorination equipment costs. Chlorination adds chloride ions to the waste and thus increases both chloride and total dissolved solids content, while ozonation has no residual effect of itself.

New treatment methods for cyanide wastes have, however, been described by the editors of *Chemical Week* (32). These include the Du Pont "Kastone" process referred to earlier in this section of this book. They also include an improved electrolytic process developed by Resource Control, Inc. of West Haven, Connecticut, which is claimed to remove dissolved cyanide completely at a cost of 15¢/pound of removed cyanide.

Techniques for removal of cyanides from electroplating wastes have been reviewed by Battelle Memorial Institute staff (33).

### References

(1) Patterson, J.W. and Minear, R.A., "Wastewater Treatment Technology," *Report PB 204,521;* Springfield, Va., National Technical Information Service (August 1971).

(2) American Electroplates Society, "A Report on the Control of Cyanides in Plating Shop Effluents," *Plating,* 56, 1107-1112 (1969).

(3) Zievers, J.F., Crain, R.W. and Barclay, F.G., "Waste Treatment in Metal Finishing: U.S. and European Practice," *Plating,* 55, 1171-1179 (1968).

(4) Besselievre, E.B., *The Treatment of Industrial Wastes,* New York, McGraw Hill Book Company, (1969).

(5) Fisco, B., "Plating and Industrial Waste Treatment at the Expanded Plant of Fisher Body, Elyria, Ohio," *Proc. 25th Purdue Industrial Waste Conf.* (1970).

(6) Schink, C.A., "Plating Wastes: A Simplified Approach to Treatment," *Plating,* 55, 1302-1305 (1968).

(7) Hansen, N.H., "Design and Operation Problems of a Continuous Automatic Plating Waste Treatment Plant at the Data Processing Division, IBM, Rochester, Minnesota," *Proc. 14th Purdue Industrial Waste Conf.,* 227-249 (1959).

(8) Palla, L.T. and Spicher, R.G., "Cyanide Treatment in Profit and Cure," *Proc. 26th Purdue Industrial Waste Conf.* (1971).

(9) Watson, K.S., "Treatment of Complex Metal-Finishing Wastes," *Sew. Indust. Wastes,* 26, 182-194 (1954).

(10) Shockcor, J.H., "Waste Treatment Cost Reduction," *Wat. Sew. Works,* 112, R220-R229 (1965).

(11) Rice, C.W.; U.S. Patent 2,640,807; June 2, 1953.

(12) Zievers, J.F., Riley, C.W. and Crain, R.W.; U.S. Patent 3,391,789; July 9, 1968; assigned to Industrial Filter & Pump Mfg. Co.

(13) Zievers, J.F., Riley, C.W. and Crain, R.W.; U.S. Patent 3,531,405; September 29, 1970; assigned to Industrial Filter & Pump Mfg. Co.

(14) Lancy, L.E.; U.S. Patent 3,682,701; August 8, 1972; assigned to Lancy Laboratories Inc.

(15) "New Process Detoxifies Cyanide Wastes," *Envir. Sci. Tech.,* 5, 496-497 (1971).

(16) Easton, J.K., "Electrolytic Decomposition of Concentrated Cyanide Plating Wastes," *Jour. Wat. Poll. Control Fed.,* 39, 1621-1625 (1967).

(17) "Electrolysis Speeds up Waste Treatment," *Envir. Sci. Tech.,* 4, 201 (1970).

(18) Ericson, H.A.H. and Norstedt, G.B.; U.S. Patent 3,645,867; February 29, 1972; assigned to Nordnero AG, Switzerland.

(19) Kandzas, P.F. and Mokina, A.A., "Use of Ozone for Purifying Industrial Waste Waters," *Chem. Abst.,* 71, 237 (1969).

(20) "Ozone Counters Waste Cyanide's Lethal Punch," *Chem. Engr.,* 63-64 (March 24, 1958).

(21) Gollmar, H.A.; U.S. Patent 2,989,147; June 20, 1961; assigned to Koppers Company, Inc.

(22) Zumbrunn, J.-P.; U.S. Patent 3,510,424; May 5, 1970; assigned to L'Air Liquide Societe Anonyme pour l'Etude et l'Exploitation des Procedes George Claude.

(23) Mathre, O.B.; U.S. Patent 3,617,567; November 2, 1971; assigned to E.I. du Pont de Nemours and Company.

(24) Hager, D.G. and Rizzo, J.L.; U.S. Patent 3,650,949; March 21, 1972; assigned to Calgon Corporation.

(25)  Morico, J.L.; U.S. Patent 3,505,217; April 7, 1970; assigned to Enthone, Incorporated.
(26)  Lawes, B.C. and Mathre, O.B.; U.S. Patent 3,617,582; November 2, 1971; assigned to E.I. du Pont
       de Nemours and Company.
(27)  *Chemical Week*, November 18, 1970; p. 113.
(28)  Lauria, A.C. and Owens, J.L.; U.S. Patent 3,697,421; October 10, 1972; assigned to Miles Labora-
       tories, Inc.
(29)  Sloan, W.J.; U.S. Patent 3,656,893; April 18, 1972; assigned to E.I. du Pont de Nemours and Co.
(30)  Byron, R.F., Danaczko, J., Jr., Dixon, A.L. and Welker, L.M.; U.S. Patent 3,147,213; September 1,
       1964; assigned to Western Electric Company, Inc.
(31)  Scott, L.F.; U.S. Patent 3,592,586; July 13, 1971; assigned to Franke Plating Works, Inc.
(32)  *Chemical Week*, December 16, 1970; pp. 54-55.
(33)  Battelle Columbus Labs., "An Investigation of Techniques for Removal of Cyanide from Electro-
       plating Wastes," Columbus, Ohio, Battelle Memorial Institute (November 1971).

# CYCLOHEXANE OXIDATION WASTES

In the cyclohexane oxidation to cyclohexanone with oxygen-containing gases, secondary oxidation products form which are present as acid materials in the waste liquor during the separating and refining steps following oxidation. Working off of these by-products and their removal has presented grave problems to the industry.

It has been attempted in the past to make use of these waste liquors by separating them into their components, esterification, hydrogenation, solvent extraction, conversion of lactones into lactams, etc. However, the results have been most unsatisfactory. Even burning of the wastes is impractical because it is very expensive. The unconverted products cannot be dumped into rivers or lakes because of the ensuing pollution, so that some disposal is necessary.

## Removal from Water

A process developed by W. Griehl and C. Suter-Homuth (1) is a method for the production of cellular protein from the waste liquors obtained in the cyclohexane oxidation. Microorganisms of the family of Pseudomonadaceae convert these waste products to proteins in alkaline medium at ambient temperatures at 24° to 30°C. and pH levels of substantially 6.7 to 9.2 within 50 to 90 hours. The process has the twofold advantage of manufacturing a useful product from wastes and preventing water pollution.

### References

(1)  Griehl, W. and Suter-Homuth, C.; U.S. Patent 3,523,064; August 4, 1970; assigned to Inventa AG
      für Forschung und Patentverwertung, Switzerland.

# DETERGENTS

The problem of detergent pollution is a complex one involving both the development of substitute nonpolluting materials as well as the removal of detergents from aqueous streams. Both aspects of the problem have been reviewed by H.R. Jones (1).

## Removal from Water

A process developed by W.A. Jordan, H.N. Dunning and L.T. Ditsch (2) involves the treatment of detergent-laden wastes to remove and/or recover various materials or contaminants therefrom by ion exchange. The removal of surface active materials is of particular interest in the treatment of sewage and industrial wastes. The separation of surfactants from sewage became more acute with the widespread use of anionic detergents, particularly of the alkyl aryl sulfonate type. Unlike the previously used soaps, these detergents are not readily consumed by bacteria and other organisms. Accordingly, when sewage containing an alkyl aryl sulfonate is discharged into a river or lake or disposed of underground, the alkyl aryl sulfonate becomes a more or less permanent addition to the body of water. In rivers, the amount of detergent builds as the river flows past communities discharging detergent-containing sewage. In many parts of the country, lakes and ground water are already contaminated with detergents.

The levels of detergent in these bodies of water could continue to grow, and in the future present a hazard not only to drinking water for human consumption, but also to fish and other life dwelling in our lakes and rivers. Further, it may render the water unusable in many present commercial operations. At this time in some localities of the country the level of detergent in the available water supply is sufficiently high to affect adversely the foaming properties of beer made therefrom. In anticipation of the continuing buildup of detergents, some breweries are considering processes for removing detergent from their water supply. This is just one example of many industries whose products or processes require water of high purity.

In the United States, launderettes have become quite popular. If the detergents were removed from the sewage of laundries and laundromats, the amount of contamination of water supplies would be considerably reduced. In addition to this immediate problem, is the treatment of the entire sewage and industrial wastes of factories and communities.

There has been very little success in adopting known separative processes to the removal of detergents and other surfactants from sewage and industrial wastes. This is primarily because of the large volumes of material to be handled and the low concentration of detergents. In order for any method to be practical, there must be a very low cost per unit of material treated. The difficulties in adapting known separative processes to sewage treatment are illustrated by considering a process such as solvent extraction. This process has been mentioned in the prior art as a method of purifying sewage by removal of greases and other organic-soluble materials.

In a solvent extraction process large amounts of solvents are generally required. In many cases these solvents present hazards of fire and explosion. Low volatility solvents often do not provide the necessary extraction efficiency. There are also the problems of the contamination of the treated material with the extracting solvent and the expense involved in the attrition of the solvent. Perhaps the greatest difficulty with solvent extraction is the cost of separating the solvent and the extracted materials. Generally, distillation or some other expensive process is required.

A method which has also been suggested for the treatment of wastewater is the foam fractionation method. Using this technique, liquids containing surface active compounds are foamed by passing a gas through the liquid. The foam is then recovered and condensed. It is found that the concentration of the surface active agents is higher in the condensate than in the residual liquid or the original feed solution. The difficulty with the foam fractionation process is that considerable equipment is required to handle and condense the large amounts of foam which are produced. In addition, the treatment generally requires several treating stages in order to remove a substantial portion of the surface active material.

While this method is particularly suited to quite dilute solutions, when the solution becomes so dilute that it does not foam readily, additional removal of surface active agents becomes difficult and expensive. Even under the best conditions the removed surfactant is diluted by large amounts of water.

Froth flotation has also been employed to remove suspended particles from aqueous solutions. Froth flotation has sometimes been confused with foam fractionation, but the two are very different in principle. In froth flotation, the particles are treated to make them hydrophobic and then are removed from the water by being attached to rising air bubbles.

Emulsion fractionation has also been considered for sewage treatment, although the process is notoriously expensive. This technique involves mixing an aqueous phase and an immiscible organic phase to form an emulsion. When the emulsion is partially broken, the concentration of surface active materials in the emulsion phase is greater than in either the aqueous or the organic liquid phase. By separating the emulsion phase and breaking the emulsion, a concentration of the interfacially active compounds will be found.

It has been discovered that organic anionic materials can be removed from their aqueous solutions by complexing them with an anionic ion exchanger in an immiscible phase and thereafter breaking the complex to allow reuse of the ion exchanger and, where desired, recovery of the anionic materials. The process comprises admixing a solution of organic anionic material with an anionic ion exchanger in an immiscible phase to form a complex in the immiscible phase between the organic anionic material and anionic ion exchanger and separating the immiscible phase containing the complex. The latter phase can then be treated to break the complex and the anionic ion exchanger can be reused. The organic anionic material can, where desirable, be concentrated and recovered.

Figure 44 is a block flow diagram showing the essential features of the process.

**FIGURE 44: BLOCK FLOW SHEET OF PROCESS FOR REMOVAL OF ANIONIC DETERGENTS FROM WATER BY SOLVENT EXTRACTION**

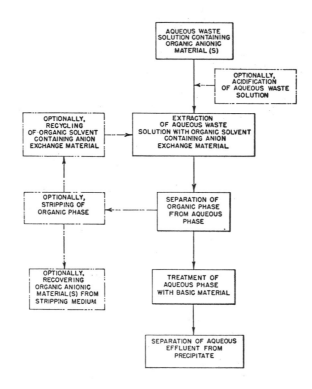

Source: W.A. Jordan, H.N. Dunning and L.T. Ditsch; U.S. Patent 3,259,568; July 5, 1966

The process is effective for the removal of anionic detergents from sewage. The removal drastically reduces the amount of detergents being pumped from sewage disposal plants into rivers which are the source of potable water supplies for downstream communities. Such process is adaptable to continuous operation and, once established, the amount of treating chemicals used is very small. The cost of the treatment to remove anionic material may be largely defrayed by the value of the recovered chemical materials. The process is effective for the extraction of soaps as well as synthetic detergents.

Further in the process, the aqueous phase is specifically treated with a basic material selected from the oxides, hydroxides and basic salts of the alkaline earth, alkali and/or rare earth metals. The treatment with the basic material produces a precipitate which can be separated by conventional means to yield an aqueous phase having a reduced solids content, phosphate content and chemical oxygen demand. The aqueous phase also has increased clarity. Thus the process provides substantially complete treatment of detergent-laden waste waters. Treatment of the original aqueous waste solution with the basic material prior to the extraction thereof with the anion exchange material is not effective.

A process developed by I.M. Abrams (3) involves the removal of the anionic surfactant by contacting the surfactant-containing water, which may contain both cations and anions, with a weak base anion exchange resin in the acid, i.e., acid-salt, form. The surfactant may then be removed from the resin simply by contacting it, as by eluting it, with an aqueous alkali solution, without the necessity of employing a polar solvent. The early effluent fractions from the resin in the acid-salt form may have a pH as low as about 2.5 due to hydrolysis of the resin's amine-salt groups. If an acid effluent is objectionable, the effluent may be neutralized by passing it through a second column containing a weak base resin in the free base form. The process is shown schematically in the block flow diagram in Figure 45.

**FIGURE 45: PROCESS FOR REMOVAL OF ANIONIC DETERGENTS FROM WATER BY ION EXCHANGE**

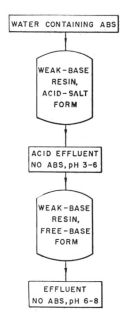

Source: I.M. Abrams; U.S. Patent 3,232,867; February 1, 1966

The second column serves a double purpose, in that it not only increases the pH of the solution, thus serving to facilitate removal of any acid escaping the first column, but also acts as a scavenger for traces of surfactant and other material which may remain in the effluent from the first column. In the treatment of detergent-containing sewage effluent, the second column also serves to scavenge organics and thus reduce COD (chemical oxygen demand). When the first column is exhausted or saturated with surfactant, most of the acid will have been transferred to the second column, which may then be placed in the first position without further acid treatment. The first column may be regenerated with alkali to remove the surfactant, then rinsed with water and placed in the second position.

A method developed by J.C. Lowe (4) involves removal of detergents from liquid by foam fractionation. More specifically, the process comprises passing wastewater containing detergent, or other forms of colored or colorless impurities, into contact with a stream of inert gas sufficient to foam the detergent out of the water while simultaneously applying a negative pressure (that is, less than atmospheric) above the water to encourage foaming, to prevent the weight of top layers of foam from crushing lower layers, and to carry off the foam. Preferably, after contacting the gas stream the detergent-containing water may be held in one or more zones of relative quiescence to give enough time for the slight vacuum applied above the body of water to remove all the foam formed by the gas contact from the water.

The foam is then passed through a comminuting device to break the foam bubbles and produce a concentrate of detergent in water. Preferably, the inert gas employed is air and, also, preferably, this air is injected into the waste at a multiplicity of points along the flow path to insure complete and intimate contact between the detergent-containing water and the inert gas. Also, it has been found that remarkably effective results are obtained when the gas-water contact takes place by gas injection into a plurality of points along an upflow zone of restricted cross section, the upflow zone terminating at an orifice within the zone of relative quiescence. A sufficient number of such upflow and quiescent zones are employed to insure the desired amount of detergent removal without allowing a weight of foam to accumulate at the top of any zone which would cause undue breakage of foam bubbles within those zones. The air or other inert gas may be introduced at a plurality of vertically and horizontally spaced points along the vertical flow path through each zone.

In removing certain types of impurities from used industrial water and especially those which have discolored the water, color-absorbing materials, such as unslaked lime, may be introduced into the water and suitably mixed therewith to absorb the coloring impurities, following which the absorbing material which has absorbed most, if not substantially all, of the discoloring material, is separated from the water by the above described procedures.

A process developed by F.F. Sako and J.A. Abbott (5) is one in which a large number of interfaces, providing a large total contact area, are formed between discrete droplets of oil flowing countercurrently to the water under treatment in a tower or the like. These oil droplets are removed from one end of the treatment tower, are coalesced, and the concentrated solution is separated out from the resulting body of oil. There is no foaming or emulsification involved.

A process developed by J.-Y. Shang (6) is one in which ABS detergent is separated from water containing same by a process which comprises mildly contact the water with a water-immiscible organic liquid and settling the resulting mixture so as to form a lower layer of water and an upper layer of water-immiscible organic liquid. Upon such contacting and settling, the ABS is concentrated, i.e., collected, adjacent to the interface between the two immiscible liquids. The organic liquid and the ABS concentrated adjacent to the interface are then separated from the water phase in any convenient manner, whereby water of reduced ABS content is obtained.

A process developed by H.N. Dunning and J.S. Gulenchyn (7) is based on the discovery that a sedimentation or flotation operation may be combined with a flocculation or precipitation operation and thus simplify the equipment which is otherwise necessary in order to effectively treat a detergent-laden waste solution. It has been found that the oils and solids which previously had been removed in the sedimentation operation may now be retained with the lime-sludge and removed with this sludge. This technique avoids the difficult separation step which was normally encountered in the sedimentation operation; thus the most difficult step in the operation is effectively eliminated. The apparatus necessary for carrying out the technique is highly effective, more easily operated, less expensive, and requires less space for installation and maintenance.

A process developed by W.W. Eckenfelder, Jr. and E.L. Barnhart (8) provides a method of removing alkyl aryl sulfonates and phosphates from wastewater by complexing the sulfonates with a quaternary ammonium compound, forming a disperse precipitate; complexing the sulfates with a calcium salt, forming an insoluble phosphate, which then flocs the disperse precipitate from the wastewater.

### References

(1) Jones, H.R., *Detergents and Pollution: Problems and Technological Solutions*, Park Ridge, New Jersey, Noyes Data Corp. (1972)
(2) Jordan, W.A., Dunning, H.N. and Ditsch, L.T.; U.S. Patent 3,259,568; July 5, 1966; assigned to General Mills, Inc.
(3) Abrams, I.M.; U.S. Patent 3,232,867; February 1, 1966; assigned to Diamond Alkali Company.
(4) Lowe, J.C.; U.S. Patent 3,434,968; March 25, 1969; assigned to Broadway Research and Development Corporation.
(5) Sako, F.F and Abbott, J.A.; U.S. Patent 3,247,104; April 19, 1966; assigned to FMC Corporation.
(6) Shang, J-Y.; U.S. Patent 3,247,103; April 19, 1966; assigned to Sun Oil Company.
(7) Dunning, H.N. and Gulenchyn, J.S.; U.S. Patent 3,259,567; July 5, 1966; assigned to General Mills, Inc.
(8) Eckenfelder, W.W., Jr. and Barnhart, E.L.; U.S. Patent 3,389,081; June 18, 1968; assigned to McGraw-Edison Company.

# DYESTUFFS

**Removal from Wastewater**

A process developed by J.B. Story (1) relates to the treatment of waste liquors containing alkali sulfides, and more particularly to the treatment of sulfur dye wastes.

The manufacture and use of sulfur dyes results in the formation of a toxic and highly alkaline waste liquor which cannot be disposed of in a conventional manner, since it unduly pollutes and stagnates fresh water rivers and streams. These waste liquors are highly colored and odoriferous. The sodium sulfide or other alkali sulfides contained therein hydrolyzes with water and forms hydrogen sulfide which is toxic to marine life and has a foul, disagreeable odor. No wholly suitable means for the treatment or disposal for these liquors is known. Neutralization treatments of these alkaline liquors, such as with sulfuric acid, result in the formation of a highly turbid suspension and the evolution of large quantities of hydrogen sulfide gas. Such acid treatments do not suitably improve the color of these waste liquors.

It has now been found that sulfur dye waste liquors can be deodorized, decolorized, neutralized and otherwise purified by treatment with an aqueous sulfurous acid solution or other sulfur dioxide-containing compounds. However, it is necessary that the liquor be mixed with the sulfurous acid within a limited time, and that the ratio of liquor to acid or the initial pH be controlled within relatively close limits. Surprisingly, under such conditions, the sulfurous acid reacts with the toxic, color and odor producing components of the waste liquor, coagulating the undesired components thereof and forming a stabilized clear solution phase. The coagulated fraction can then be removed from the clear solution by filtration, decantation, centrifugation or any other separation technique and the solution disposed of in a river or other waterway without danger of pollution or stagnation of the water.

The following is a specific example of the conduct of the process. A volume of sulfur black dye waste liquor (T-1636 Sulfogene carbon HCF grains) containing sodium sulfide (1,000 parts by weight of liquor) having a foul odor and a pH greater than 11 was treated at room temperature with sulfurous acid to coagulate the dye constituents thereof and to eliminate the stagnating and polluting contents of the liquor. The liquor contained about 2 parts of sodium sulfide. In the treatment, the acid was poured into the waste liquor, and mixed thoroughly within five seconds. A sufficient amount of sulfurous acid (65 parts by weight of 0.5 molar solution) was required to give a 6.8 pH for the treated liquor, immediately following the mixing. Particles immediately appeared in the mixture.

To so-treated liquor was then allowed to settle, leaving a clear, colorless, odorless and essentially neutral supernatant liquid and a black, curdy precipitate. Upon separation in a filter, the supernatant liquid was disposed of in a conventional sewerage system. The precipitate is thereafter redissolved with sodium sulfide solution and reused for further dyeing operations.

A process developed by G. Hertz (2) is one in which waste dye liquor is decolorized by electrolytic treatment of an aqueous solution containing chloride ions. The organic dyestuff is oxidized by hypochlorite formed thereby permitting the treated material to be discharged into sewers and sewage systems.

**References**

(1) Story, J.B.; U.S. Patent 2,877,177; March 10, 1959.
(2) Hertz, G.; U.S. Patent 3,485,729; December 23, 1969; assigned to Crompton & Knowles Corporation.

## FATS

### Animal Fat Removal from Water

A process developed by H.T. Anderson (1) involves treating wastewater containing appreciable amounts of fats, grease, oily materials and polar materials such as proteins. The process is especially useful in connection with packinghouse and edible oil operations, but it can also be advantageously used for cleaning other types of fat-contaminated water, such as from industrial plants using mineral oils. The process is applicable to the economic recovery of valuable oils from emulsified systems as well as the mere removal of spent oil or greases from wastewater so as to prevent contamination of rivers, lakes and seas.

Generally speaking, free fat, oil, i.e., nonemulsified fat and oil, present no serious problems in regard to separation from water as they will float to the surface and can be skimmed off. Emulsified fats, on the other hand, stay in solution causing severe separation problems.

It has been the usual practice in the past to run the wastewater from the packinghouse to a settling tank or basin having baffles wherein the water would set for about one-half hour or more and the free fat would rise to the top and be skimmed off. The emulsified fat would of course remain in the water and would accompany it to the sewers. Various means such as aeration and complex apparatus have been employed in attempts to de-emulsify the wastewaters. Usually, however, unless the emulsified oil was very valuable, no effort was made to separate it from the water that was passed to the sewers and hence to the rivers and seas.

In a few industries, solvent extraction and other unit processes are used to recover the valuable oil. Also, in processes where the water is reused, the oil can be removed from the system by coagulation with aluminum sulfate and alkali, followed by filtration through a nonsiliceous filter medium. The oil is caught in the floc formed, and the floc, with the entrained oil, strained out of the condensate by the filter. Periodical backwashings of the filter with hot caustic soda are required. It should be noted that those processes used to completely remove the oil from the water are clearly uneconomical for use in cleaning up wastewater from packinghouse and edible oil operations. Yet on the other hand, it is desirable to remove these low-grade oils from sewage as they may be used as industrial cutting oils, etc. Further, it has now become necessary to clean up the water prior to its discharge into the rivers and seas.

Thus, this process has as its goals the removal of emulsified fats, oils and greasy materials and polar materials such as protein from wastewater systems in an economic manner without using solvent extraction or other costly means.

The process then comprises impressing direct electrical current in such a manner that a carefully defined, relatively deep, anolyte stream (i.e., a path of liquid in the immediate neighborhood of the anodes during electrolysis) is formed near the bottom of a receptacle at the entry end and becomes narrower in depth as it progresses upwardly to near the water surface at the opposite end or exit end of the receptacle. The anolyte stream, having a very low pH due to multiple anodes, breaks the emulsified fat enabling it to rise to the surface and be skimmed off. The system comprises multiple, submerged anodes, usually approximately equal distance apart, parallel to the longitudinal walls of a rectangular tank or other receptacle but inclined upwardly. In a circular tank, the anodes are parallel to the walls but substantially vertically mounted.

The following is one specific example of the operation of the process. A steel tank about 60 ft. long by 10 ft. wide and 6 ft. deep was approximately three-quarters full with packinghouse wastewater containing principally fat, proteins and cellulose. Ten high silica-iron anodes, each 1½ in. in diameter and 5 ft. long were connected in parallel and positioned below the water surface in two rows parallel the longitudinal walls of the tank. The anodes were about 2½ ft. from the longitudinal walls and were inclined progressively upward from inlet end to outlet end with the two anodes nearest the inlet end positioned so as to be slightly below the point of entry of the incoming wastewater. An electrical wire was connected to a high point of the tank wall, which acted as the cathode, and grounding this to the rectifier completed the circuit. Using a rectified DC current to energize the anodes, various voltages were applied. Voltages between about 6 and 9 volts and a current of between 20 and 35 amperes resulted in an anolyte stream with fat particles and other solids floating to the surface where they were removed by mechanical sweeps.

### References

(1) Anderson, H.T.; U.S. Patent 3,673,065; June 27, 1972; assigned to Swift & Company.

## FLOUR

### Removal from Air

A process developed by R.E. Meade (1) provides an improved means for the removal of dusts such as flour from air. Prior dust filters exhibit an undesirable back pressure across the filter screen after a short period of operation. In this process, droplets of liquid are introduced to a moving stream of gas that contains dust. The droplets make the dust particles cohere in three ways: (a) by the wetting of their surfaces; (b) dissolving potentially tacky solubles contained in the dust; and (c) by the deposition of an adhesive contained in the droplets of liquid but insufficient liquid is introduced to cause the particles to coalesce. A porous collecting element such as a woven metal screen extends entirely across the air stream. The air stream is forced through the collecting element while the particles

are in a tacky coherent condition.  The mat of dust particles which forms on the surface
of the collecting member is maintained in a highly porous condition because the bonds
that form between the dust particles hold the particles apart.  The mat of dust itself acts
as the filter medium.  It is self-supporting, often friable and offers much less resistance
to the flow of gas than the same dust particles in an unbonded condition.

Figure 46 shows a suitable form of apparatus for the conduct of the process.  The apparatus comprises a chamber or vessel such as a vertically disposed housing **120** communicating through an inlet port **123** with a supply duct **122** and a heater **124** which in turn
communicates through a duct **126** with a damper **128** to which air containing the dust
that is to be collected is supplied by blower **130**.

## FIGURE 46:  FILTER FOR REMOVAL OF FLOUR FROM AIR

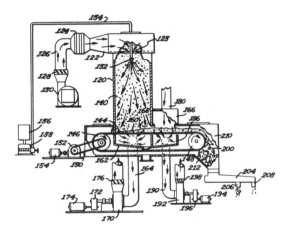

Source:   R.E. Meade; U.S. Patent 3,675,393; July 11, 9972

Mounted concentrically within the upper end of housing **120** is a nozzle **132**.  The activating liquid is supplied to the nozzle through a pipe **134**.  The liquid is supplied from a
storage tank **136** by means of a pump **138**.  The liquid sprayed from the nozzle **132** is
initially present in the form of wet droplets **140**.  These droplets are suspended in the
stream of dust particles and air passing downwardly through housing **120**.  As the dust
particles fall, they become moistened and strike a foraminous collecting element or screen
**144** wrapped over a pair of horizontally disposed spaced rollers **146** and **148**, the former
being driven by means of a belt **150** connected to a speed reducer **152**.  The speed reducer
is in turn operated by means of a motor **154**.  The screen has the form of an endless belt
and can comprise a variety of materials such as a woven wire screen or perforated metal
sheet.  A variety of other screen materials will be apparent to those skilled in the art.

As clearly shown, a mat of dust **160** forms on the upward surface of the screen **144**.  The
air flowing into the chamber **120** from the duct **122** passes through the mat **160** as indicated by arrows into a compartment **162** below the screen and is exhausted through a
duct **164**.  This duct is connected to an exhaust blower **170** driven by the provision of a
speed reducer **172** and a motor **174**.  From the blower the air passes through a suitable
manually controllable damper **176** which like damper **128** can be opened or closed as
conditions of the operation require.

The foraminous screen **144** is moved so that its upper surface travels toward the right as seen in the figure, thereby carrying the mat into a second chamber **166** through an opening **168.** The chamber communicates with an inlet duct **180** connected to an air cooler **182** of any suitable known construction. On the opposite side of the screen **144** from the chamber **166** is a chamber **186** which communicates with an outlet duct **190** to a blower **192** driven by means of motor **194** coupled through a speed reducer **196.** Air passing through the blower **192** is exhausted through a flow control damper **198.** The foraminous member **144** and the rollers are mounted within a chamber **210.** At the lower end of the chamber **210** a doctor knife **212** is provided for scraping the dust from the screen **144** before it returns to the inlet end of the chamber **120.**

During operation, as the screen **144** travels over the rollers **146** and **148,** the dust mat breaks into chunks **200** which pass downwardly into a collecting trough **204.** When the mat is to be used for some purpose, a suitable sifter can be provided for separating and grading chunks of different sizes and exhausting them through a pair of ducts **206** and **208.** The following are specific examples of the operation of the process.

*Example 1:* Flour entrained in an air stream is removed in the following manner. A stream of air containing suspended flour particles is passed through a heater until its temperature has reached about 325°F. About 114 parts of water by weight is sprayed continuously onto the stream for each 100 parts of flour. The flour particles are collected when in a tacky condition on a woven wire screen having openings of about one-sixteenth inch in width. The moisture content of the deposited material may be about 6%. It will be porous, friable and will consist of a multiplicity of microscopic particles bonded together at their points of contact and having a multiplicity of interconnecting passages extending from the influent to effluent surfaces. It will be dry to the touch and can be crushed to pieces of any desired size.

*Example 2:* A dusty material produced in the process of making a dry cake mix is collected and removed from an air stream in the same manner as described in Example 1 with the following changes. The air stream is heated to a temperature of from 315° to 325°F. and is introduced at the rate of 2,500 cfm. Water is heated to 125°F. and sprayed into the stream at a distance of about 10 feet from the collecting screen in the same ratio as described above. The moisture content of the porous, friable mat formed on the screen may be about 2.3% by weight.

**References**

(1) Meade, R.E.; U.S. Patent 3,675,393; July 11, 1972; assigned to The Pillsbury Company.

## FLUORINE COMPOUNDS

Industries which discharge significant quantities of fluorides in process waste streams are: glass manufacturers, electroplating operations, steel and aluminum producers, and pesticide and fertilizer manufacturers.

Glass and plating wastes typically contain fluoride in the form of hydrogen fluoride (HF) or fluoride ion ($F^-$) depending upon the pH of the waste. The greatest amounts of fluoride discharged from fertilizer manufacturing processes are in the form of silicon tetrafluoride ($SiF_4$), as a result of processing of phosphate rock (1). The aluminum processing industry utilizes the fluoride compound cryolite ($Na_3AlF_6$) as a catalyst in bauxite ore reduction. Previously, the gaseous fluorides resulting from this process charged directly into the atmosphere. Wet scrubbing of the process fumes, an air pollution abatement procedure, now results in transfer of the fluorides to aqueous waste streams according to Patterson and Minear (2).

Removal of Fluorine Compounds from Air

The reader is also referred to the section of this handbook on Aluminum Cell Exit Gases, Removal from Air for additional discussion of removal of fluorine compounds from air.

A process developed by W.J. Sackett, Sr. (3) involves a dust-fume control system having a series of dry cyclones to remove a major portion of the dust, a wet impingement type separator wherein the gas impinges on the surface of a liquid to entrap any particles of dust remaining and to remove at least some of the fumes, and a tangential wet wall type scrubber to remove any residual dust and fume from the gas issuing from the wet impingement type separator. Such a system is shown in Figure 47.

FIGURE 47: APPARATUS FOR DUST AND FLUORIDE FUME REMOVAL IN
FERTILIZER MANUFACTURE

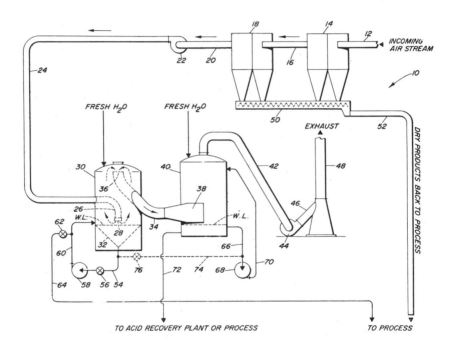

Source: W.J. Sackett, Sr.; U.S. Patent 3,494,107; February 10, 1970

Reference numeral 12 indicates generally a duct conveying incoming, heavily polluted air to a dust-fume system 10. This air is typically that which arises from the vats, dens, drums, mixers, reactors and dryers of industrial plants of the chemical fertilizer type, for example. Such air is usually laden with dust and chemical fumes. A typical example is that from a superphosphate granulation process plant where crushed phosphate rock is treated with acids, and the duct 12 carries 25 pounds of dust, and as fumes 1.25 pounds of fluorine, 1.92 pounds of silicon tetrafluoride per minute. The duct 12 leads into a first set of two paralleled dry cyclones 14. A duct 16 connects the first set of cyclones 14 to a similar second set of paralleled dry cyclones 18.

The source of propulsion for this air is a centrifugal fan 22 which has its intake connection attached to the outlet duct 20 of the second set of cyclones 18. A duct 24 leads from the blower **22** into an impingement scrubber **30** where its exit end **26** terminates in an adjustable sleeve **28**. The scrubber **30** contains, in its lower portion, a funnel **32** which retains water whose surface level **W.L.** is just beneath the adjustable sleeve **28**. The upwardly open receiving end **36** of a duct **34** leads from near the top of the scrubber through a sidewall to terminate in a tangential discharge end **38** in the sidewall of a packed wet cyclonic scrubber **40**.

The exhaust duct **42** of this scrubber **40** is given a negative pressure by a centrifugal fan **44** which discharges through a duct **46** into a stack **48**, exhausting to the atmosphere. The dust collected to the sets of cyclones **14** and **18** is removed by a screw conveyor **50** as a dry product for return through a duct **52** to process. In the example given, 23 of the 25 pounds per minute of dust is thus recovered.

In the impingement scrubber **30**, the remaining dust particles are forcefully driven into the water with the result that a slurry is gradually established. This slurry is recirculated through a circuit consisting of a pipe **54** connected at the bottom of funnel **32**, a valve **56**, a recirculating pump **58**, and a return pipe **60** back to the mouth of the funnel **32**. When the slurry builds up to a sufficient density in this recirculating system, it is tapped off from the pipe **60** by a valve **62** and returned over a pipe **64** to process. Fresh water as indicated is released into the impingement scrubber **30** to replenish the funnel **32**. Practically all the dust is now removed, a typical figure in this example is 1.8 pounds per minute here withdrawn. For this reason, the following packed wet cyclonic scrubber **40** operates at high efficiency to react the vapors with relatively small requirement for fresh water replenishment except to maintain a sump with a water line **W.L.** as shown.

A circulating pump **68** withdraws liquid from this sump through a pipe **66** and reintroduces it through a pipe **70** as spray within the scrubber **40**. A pipe **72** leads to an acid recovery plant (not shown) or if desired to process. Typically here the saving is 0.14 pound of dust (in suspension), 0.37 pound of silica, and 1.15 pounds of fluorine as hydrosilicofluoric acid in 8.25 pounds per minute of water. An optional cross connection of a pipe **74** and valve **76** between the liquid circulating systems of the two scrubbers **30** and **40** is installed when it is unnecessary to keep separate the two effluents.

While this process has been described in connection with a phosphate fertilizer plant, other uses will suggest themselves wherever an air laden with a large volume of dry dust and water-soluble or water-reactant vapors are to be cleansed before release to the atmosphere and where it is important not to release an effluent to a stream or drain.

A process developed by A.J. Teller and E.S. Wyatt (4) treats the products of the acidulation of phosphate rock by acids. A gas is generated which might contain silicone tetrafluoride, hydrogen fluoride, water vapor, dust and mists, as well as the carrier gas, air. The recovery of these fluorides is an important aspect of the control of air pollution. To avoid contaminating air it is desirable that this recovery be complete in excess of 99%.

The system includes a cylindrical recovery zone into the top of which the gaseous mixture is introduced tangentially. The gaseous mixture is contacted by a radially directed spray of a washing liquid. The particular washing liquid utilized will depend upon the foreign values to be removed. Where fluorides are to be removed, the washing liquid might be water, recycled pond water or any solution of basic material such as sodium hydroxide or calcium hydroxide. In other services, the washing liquid might be an acid.

The radially directed spray washes the gaseous mixture with a finer spray than can be utilized in apparatus having a wall spray. Thus, a greater degree of gas absorption is achieved. The majority of the foreign material is recovered in this zone in which no plugging areas exist. The high velocity spiral motion of the gaseous mixture resulting from its tangential inlet causes centrifugation of the liquid spray and any resulting solid. Thus, these materials are spun outward toward the walls of the vessel from which they drain off behind a barrier. The remaining gas is relatively clean of foreign materials, but still contains some entrained liquid. This gas proceeds downward within the vessel, through a cocurrent spray of washing liquid such as water. The gas and washing liquid pass through a bed of packing material which removes substantially all of the remaining foreign material. The liquid then leaves the bottom of the vessel, and the gas leaves through a side outlet including a second packed section which further removes entrained liquid from the gas.

The process performs in a single piece of equipment the several functions which other systems have performed in separate pieces of equipment. Thus, in the apparatus the foreign gases and particulates are washed from the carrier gas, and entrained liquids are removed from the carrier gas, all within one piece of equipment. This single shell system substantially reduces the complexity and cost of the apparatus. Figure 48 shows such an apparatus.

The gas containing foreign materials, for example air carrier gas with fluoride contaminants, enters recovery zone 8 within elongated, vertical, cylindrical vessel 10 tangentially through inlet 12 near the top of the vessel. Pipe 14 extends vertically downward through the top 16 of vessel 10, a short distance into recovery zone 8. Exteriorly of the vessel, pipe 14 is connected to a source of washing liquid such as water (not shown). The pipe is closed at its lower end within zone 8 and includes a plurality of openings 18 along its length. As a result, the washing liquid within pipe 14 is directed radially outward in a spray.

The gas enters recovery zone 8 from inlet 12 at a velocity high enough to result in a cyclonic action within the recovery zone, e.g., in a ten foot diameter recovery zone, a velocity in the range of about 2,000 to 5,000 feet per minute and preferably a velocity of about 3,000 feet per minute. While velocities above 5,000 feet per minute could be utilized, the increased pressure drop resulting from high velocities makes them uneconomical. The nozzle pressure of the liquid spraying from pipe 14 is great enough to cause the spray to reach cylindrical sidewall 20 of vessel 10, e.g., a pressure in the range of from about 10 to 500 psi, preferably in excess of 40 psi. The gas velocity and nozzle pressure are controlled for optimum operation of the apparatus and might vary with the carrier gas and foreign material encountered and the washing liquid utilized, as well as with the size of vessel 10 and the size of spray droplets emitted by pipe 14.

The gaseous mixture entering recovery zone 8 through inlet 12 is contacted by the radially directed spray from pipe 14. As a result a large portion of the foreign material within the carrier gas is removed. This removal, by way of examples, might be by absorption of foreign gas from the carrier gas or by capturing of foreign particulates by the washing liquid. Thus, a great proportion of the fluorides and most foreign particulates of a size above about 3 microns are removed by this spray. The radially directed spray and the cyclone action of the gas, created by its high velocity tangential inlet, cause centrifugation of the liquid spray droplets, the absorbed gases, the trapped foreign particles, and the solids formed. Consequently, these materials are spun outward toward cylindrical sidewall 20. To break the vortex flow of liquid within the recovery zone 8, which otherwise might result in liquid flowing directly downward in the center of the vessel, an obstacle such as core buster 22 can be placed within the center of the recovery zone, a distance below pipe 14.

A liquid slurry containing fluoride contaminants accumulates at sidewall 20 and is drained off behind barrier 23, passing out of vessel 10 through outlet pipes 24. Advantageously, this liquid can be treated for removal of fluorides, for example by passage to a settlement pond in which calcium ions cause precipitation of fluorides which then settle to the bottom of the pond, thereby substantially reducing the concentration of fluorides in the liquid. The liquid is then either discharged or recycled into the system. If the foreign material within the carrier gas is a desired end product, the liquid from vessel 10 can be recycled with little or no treatment to increase the concentration of that material within the resulting liquor. The remaining gas, which is relatively clean of foreign material but which still contains some entrained liquid and smaller foreign particles, proceeds downward through the central aperture of barrier 23.

Inlet pipe 26 extends horizontally into vessel 10 below barrier 23 and exteriorly is connected to a source of washing liquid such as water (not shown). The end of pipe 26 within vessel 10 is sealed, and a plurality of openings are provided on the lower side of pipe 26

## FIGURE 48: GAS SCRUBBER FOR FERTILIZER PROCESS AIR POLLUTION CONTROL

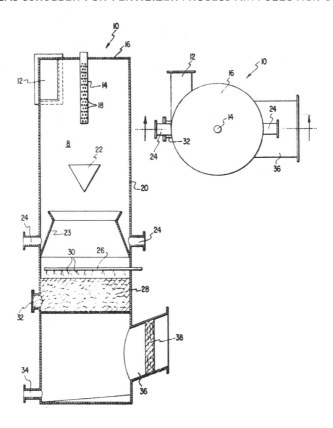

Source: A.J. Teller and E.S. Wyatt; U.S. Patent 3,505,788; April 14, 1970

to cause even distribution of the washing liquid over packed bed **28**. Thus, for example, the openings in pipe **26** might cause the liquid to pass from the pipe in a spray, preferably in sprays such as full-cone sprays **30**. The gas passing through barrier **23** and containing entrained liquid likewise encounters the packed bed and passes therethrough cocurrently with the liquid from sprays **30**. Bed **28** is packed with a material having a substantially nonadhering surface and a high free volume. The surface renewal provided by such a packing results in removal of the fluorides or other foreign materials to levels in excess of 99.9%. Covered opening **32** provides access to bed **28** for the insertion and removal of packing.

Over 80% of the fluorides have been removed from the liquid leaving the vessel at outlet **24**. Consequently, even though the packing within bed **28** has a complicated geometry to achieve the surface renewal, the small quantity of solids deposition, the nonadhering surface of the packing, the high free volume within bed **28**, and the cocurrent flow of the gas and liquid inhibit the buildup of solids within the system. Liquid passing out of bed **28** falls to the bottom of the vessel and leaves through lower outlet **34**. As with the liquid from outlet pipes **24**, this liquid can be recycled, treated, or discharged. The gas passing from bed **28** leaves the vessel through gas outlet **36** which preferably includes a vertical packed bed **38** to further remove any liquid entrained in the gas. Consequently, the gas leaves vessel **10** in an essentially fluoride-free condition.

A process developed by M. English (5) is a method for recovering fluorine compounds from vapors containing water vapor and fluorine compounds comprising the steps of scrubbing the vapors with an aqueous liquid at such a temperature as not to cause any substantial condensation of the water vapor, whereby impurities are scrubbed out of the vapors and fluorine compounds remain therein; subsequently scrubbing the vapors with an aqueous fluorine compound absorbing liquid at such a temperature as not to cause any substantial condensation of the water vapor, thereby removing fluorine compounds in the fluorine compound absorbing liquid; and recovering fluorine compounds from the fluorine compound absorbing liquid.

Figure 49 shows a sutiable form of apparatus for the conduct of this process. The apparatus indicated in the drawing comprises a vacuum evaporating vessel **1** with an inlet (not shown) for dilute phosphoric acid (about 26 to 32% by weight $P_2O_5$), means **2** for heating the acid, an outlet (not shown) at the bottom controlled by a valve for removal of concentrated phosphoric acid and an outlet **3** at the top for vapors from the dilute acid. The concentrated phosphoric acid produced commercially in the apparatus will have a concentration of about 54% by weight $P_2O_5$ and contains a smaller weight of fluorine compounds than the dilute acid. The vapors leaving the evaporating vessel via the outlet **3** at the top contain water vapor and fluorine compounds, mainly HF and $SiF_4$.

**FIGURE 49:  APPARATUS FOR FLUORINE COMPOUND REMOVAL FROM PHOSPHORIC ACID CONCENTRATOR EXIT GASES**

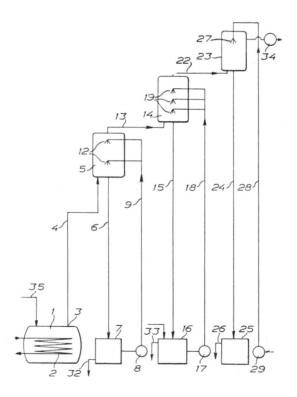

Source:  M. English; U.S. Patent 3,512,341; May 19, 1970

A form of entrainment separator (not shown) is included in the top of the evaporating vessel to remove large droplets of phosphoric acid. The separator can be located in a separate vessel between the evaporator and the first scrubber.

A conduit 4 is connected to the outlet 3 at the top of the evaporating vessel 1 and leads to the bottom of a first scrubber tower 5. Also connected to the bottom of the first scrubber tower is a barometric leg 6 ending in a seal box 7. A pump 8 and conduit 9 lead from the seal box to spray nozzles 12 in the scrubber tower 5.

A conduit 13 leads from the top of the scrubber tower 5 to the bottom of a second scrubber tower 14 also having a barometric leg 15 ending in a second seal box 16. From the second seal box a further pump 17 and a conduit 18 lead to a spray nozzle 19 in the second scrubber tower 14. A conduit 22 leads from the top of the second scrubber tower 14 to the bottom of a third scrubber tower 23 also having a barometric leg 24 ending in a third seal box 25 and overflow 26 from which leads to waste. A spray nozzle 27 in the third scrubber tower is connected to a source of condensing liquid (not shown), in this case cold water, which is supplied to the nozzle along a conduit 28 by a pump 29.

A second outlet 32 from the first seal box 7 leads to a storage tank (not shown) for dilute phosphoric acid and an overflow 33 from the second seal box 16 leads to a storage tank (not shown) for the intermediate product, a solution of fluosilicic acid containing fluorine compounds. A reduced pressure is maintained in the apparatus by means of a vacuum pump 34 connected to the third scrubber tower 23.

Operation of this apparatus under test conditions will now be described. In the test, dilute phosphoric acid was heated in the evaporation vessel 1 and fluosilicic acid was introduced into the vessel through a pipe 35 at a rate to give approximately 3 to 5% by weight of fluosilicic acid in the water vapor emerging from the vessel along the pipe 4. A vacuum of between 27.5 and 28 inches of mercury is maintained in the apparatus by the pump 34 and the acid in the evaporating vessel is heated to drive off water vapor. The acid was of a concentration boiling at 78° to 83°C. at the pressure maintained in the apparatus.

The vapor enters the first scrubber 5 at about 73° to 79°C. and is thoroughly washed by the first scrubber liquor which is pumped through the nozzles 12 by the pump 8. The first scrubber liquor comprises a solution of fluosilicic acid containing about 54% by weight of fluosilicic acid containing 7% by weight of phosphoric acid. The temperature of the scrubber liquor is slightly below the temperature of the vapor but is such that no substantial condensation of water vapor is brought about. The scrubber liquor is too strong a solution of fluosilicic acid to absorb any substantial proportion of fluorine compounds but washes impurities such as phosphorus peroxide out of the vapors. The vapors passing to the second scrubber tower 14 comprise water vapor, fluorine compounds and some non-condensables.

Two spray levels are provided in the scrubber tower 5, as shown. From time to time, liquor is removed from the seal box 7 along the pipe 32 and make-up water is added to maintain the phosphoric acid concentration in the first scrubber liquor at an acceptable level, below 10 to 15% by weight.

In the second scrubber 14, the vapor is scrubbed with the second scrubber liquor pumped through the nozzles 19 by pump 17 and which comprises, in this case, a solution of fluosilicic acid containing about 18% by weight of fluosilicic acid. The second scrubber liquor is at a temperature slightly below that of the vapors entering the second scrubber tower. The liquor causes no substantial degree of condensation of water vapor in the second scrubber but takes up 90 to 95% of the fluorine compounds in the vapor. The vapors passing from the second scrubber tower 14 to the third scrubber tower 23 comprise water vapor, some fluorine compounds and some noncondensables. In the third scrubber tower 23, enough water at 25° to 30°C. is brought into contact with the vapors to condense the water vapor and the remaining fluorine compounds and the noncondensables are removed from the apparatus through the vacuum system.

In commercial operation, a proportion of the first scrubber liquor would be removed continuously from the first seal box **7** and returned to the dilute phosphoric acid tank (not shown) and make-up water would be mixed with the liquor taken from the first seal box to the spray nozzles **12** in the first scrubber tower **5**. A portion of the second scrubber liquor is removed continuously from the second seal box **16** over the overflow **33** to the intermediate product storage tank (not shown) and make-up water is mixed with the liquor taken from the second seal box to the spray nozzles **19** in the second scrubber tower **14**. The fluorine compounds are recovered from the intermediate product by conventional methods.

A process developed by R.G. Hartig (6) involves the separation of fluorine compounds from waste gases by absorption-desorption with NaF, BaF$_2$ or KF, and absorption of SiF$_4$ in water. HF may also be recovered.

Figure 50 is a flow diagram showing the essential features of the process. A gaseous effluent stream from a phosphate complex or other phosphate process is delivered through conduit or line **10** to the upper portion of a dust scrubber **11**. The effluent gas contains SiF$_4$, some HF, air, and has entrained therein phosphates in both solid and liquid forms. The liquid droplets contained in the gaseous effluent may consist of droplets of solutions of phosphoric acid (H$_3$PO$_4$) and solutions of phosphate salts. The dusts may contain phosphate rock dust, mono- and dicalcium phosphate, and many other materials depending on the type of plant and manner of its operation.

Dust scrubber **11** is a water scrubber, water slurry flowing downwardly through the scrubber and the gas flowing downwardly concurrent with the scrubber liquor. The scrubber contains a liquid compartment **12** at its lower end from which the scrubber liquor drains through conduit **13** to pump tank **14**, from which it is pumped by pump **15** through line **16** for recycle to the top of the scrubber. Makeup water is added at **17** as required to maintain the proper dilution of the scrubber liquor. The scrubber liquor is usually maintained to have a solid content of from 20 to 30% solids, by weight, the phosphate (expressed as P$_2$O$_5$) concentration being from about 40 to 50% by weight. These concentrations will vary depending on the dust and liquid phosphate contents of the gas stream. In the scrubber liquor, SiF$_4$ (silicon tetrafluoride) is dissolved or reacted with water according to the following chemical reaction:

$$3SiF_4 + 2H_2O \longrightarrow 2H_2SiF_6 + SiO_2$$

In the scrubber liquor, the fluosilicic acid concentration reaches a strength of about 25% H$_2$SiF$_6$ by weight. A small portion of the scrubber liquor is recycled to the processing units through line **19**. All P$_2$O$_5$ lost as liquid or dust carryover with the effluent gases is recovered on return to the processing plant, the recycled H$_2$SiF$_6$ being decomposed during processing and coming off at a later time in the effluent gases in the form of HF and SiF$_4$.

In view of the fact that the scrubber liquor in scrubber **11** is maintained at saturation, with a H$_2$SiF$_6$ concentration of about 25% by weight, the bulk of the SiF$_4$ passes through dust scrubber **11** in the vapor stage and is recovered as Na$_2$SiF$_6$, K$_2$SiF$_6$, BaSiF$_6$, or other similar salt, in fluorine scrubber **22**. This scrubber has a liquid compartment **23** at its lower end, from which scrubber liquor is delivered through conduit **24** to pump tank **25**. The scrubber liquor in this case is delivered through line **26** by a pump **27**, to a settling tank **29**. A slight excess of one or more of NaF, BaF$_2$ and KF is maintained in scrubber **22**. The SiF$_4$ reacts therewith to form Na$_2$SiF$_6$, BaSiF$_6$, or K$_2$SiF$_6$ according to one or more of the following reactions:

$$2NaF + SiF_4 \longrightarrow Na_2SiF_6$$

$$BaF_2 + SiF_4 \longrightarrow BaSiF_6$$

$$2KF + SiF_4 \longrightarrow K_2SiF_6$$

## FIGURE 50: ABSORPTION-DESORPTION PROCESS FOR HF RECOVERY

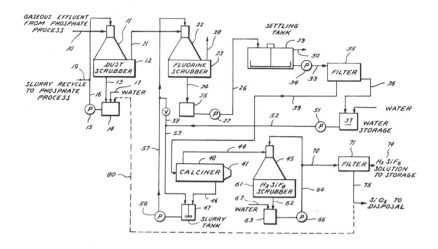

Source:  R.G. Hartig; U.S. Patent 3,642,438; February 15, 1972

Because of the excess of NaF, BaF$_2$ and/or KF maintained in the scrubber, the vapor pressure of SiF$_4$ in scrubber **22** is nil, and therefore complete recovery of the fluorine content of the effluent gases from the plant is continuously maintained.  Scrubbed gas is discharged through line **28**.

If the gases delivered to the process contain appreciable amounts of HF (hydrofluoric acid), additional NaF (or KF) can be used to form NaHF$_2$ (or KHF$_2$) in either of the scrubbers **11, 22**.  Since BaHF$_3$ does not form, BaF$_2$ cannot be used for HF precipitation.

In settling tank **29**, the mother liquor from scrubber **22** is settled, the Na$_2$SiF$_6$, BaSiF$_6$, and/or K$_2$SiF$_6$ settling out, while the effluent liquor is removed by overflowing through conduit or line **30**.  The thickened slurry from the bottom of settling tank **29** is delivered through line **33** by pump **34** to a filter **35**, where the water phase is removed and passed through line **36** to a water storage tank **37**.  The solids are delivered via conveyor **39** to a calciner **40**, heated by burner **41**, wherein the solids are decomposed to form solid NaF, BaF$_2$, or KF, and gaseous SiF$_4$.  The SiF$_4$ is delivered through line **44** to the upper end of the scrubber **45**.  The recovered solids are delivered via **46** to a slurry tank **47** wherein water delivered from tank **37** by pump **51** and lines **52,53** is mixed therewith to form the scrubber liquor for scrubber **22**.  The scrubber liquor is delivered by pump **56** and lines **57** to the upper part of scrubber **22**.  Water may be added to the scrubber liquor at valved line **58** if required.

The SiF$_4$ exiting from calciner **40** through line **44** is scrubbed with water in scrubbing tower **45**. Tower **45** has a lower liquor compartment **61** from which scrubber liquor is passed through line **62** to pump tank **63**.  The scrubber liquor is circulated to the tower through line **64** by a pump **66**.  Makeup water may be added as required at **67**.  A portion of the scrubber liquor is continuously drawn off through line **70** to a filter **71**, where H$_2$SiF$_6$ in water solution is separated and delivered to storage as indicated at **74**.  SiO$_2$ is delivered from the filter by way of **75** and disposed of or used as convenient.

In scrubber **45**, $H_2SiF_6$ and $SiO_2$ are formed according to the same reaction as occurs in tower **11**, namely,

$$3SiF_4 + 2H_2O \longrightarrow 2H_2SiF_6 + SiO_2$$

The $H_2SiF_6$ water solution is maintained at a concentration of about 15 to 25% $H_2SiF_6$, by weight, by adjustment of water addition at **67**.

The solid decompositions in calciner **40** occur according to the following chemical reactions:

$$Na_2SiF_6 \xrightarrow{\text{(heat)}} SiF_4 + 2NaF$$

$$BaSiF_6 \xrightarrow{\text{(heat)}} SiF_4 + BaF_2$$

$$K_2SiF_6 \xrightarrow{\text{(heat)}} SiF_4 + 2KF$$

An alternative method useful in connection with the flow diagram shown is indicated by dashed line **80**. If the incoming gas at line **10** contains an appreciable amount of HF, then the $SiO_2$ issuing from filter **71** at **75** may be delivered via **80** to the scrubber liquor of the scrubber **11**, so that the HF is converted to $SiF_4$ according to the following chemical reaction:

$$4HF + SiO_2 \longrightarrow SiF_4 + 2H_2O$$

The following reaction also takes place:

$$SiF_4 + 2HF \longrightarrow H_2SiF_6$$

The $SiF_4$ passes through the dust scrubber (except for that dissolved to maintain the $H_2SiF_6$ concentration in the scrubber liquor) to be converted in scrubber **22** as has been described. This procedure is an alternative to the use of additional NaF and/or KF in the scrubber to remove HF from the gases. The alternative whereby $SiO_2$ is delivered via **80** to scrubber **11** is especially valuable in case the incoming gases in conduit **10** contain calcium. The formation of $CaF_2$ is prevented, in which form the fluorine would be lost to recovery. Without the presence of $SiO_2$ in scrubber **11**, the calcium would react as follows:

$$Ca^{++} + 2HF \longrightarrow CaF_2 + 2H^+$$

The presence of $SiO_2$ in the scrubber liquor results in the presence of the $SiF_6$ ion solution in the scrubber liquor, to prevent the foregoing chemical reaction from taking place.

In other types of phosphate processing, such as in plants producing defluorinated calcium phosphate for use as animal feeds (as mineral feeds supplements), sintering or nodulizing processes, production of calcium and other metaphosphates, pyrophosphates, and the like, effluent gases are produced which contain the fluorine essentially in the form of HF, with only small amounts of $SiF_4$. In most of these processes, the volume of HF in the gas stream is very small, and the concentration of $SiF_4$ very much smaller. Since HF has a high vapor pressure over aqueous solutions, concentrated solutions of HF cannot be obtained by scrubbing the gases with water. Consequently, copious amounts of water are required with the addition of lime or limestone to remove the fluorine in the form HF, from the gas stream or from liquid streams, in the processing plant.

A modified process is provided for treatment of such gases, high in HF as compared with the $SiF_4$ content, to recover the fluorine in the form of saleable products, including anhydrous HF and aqueous fluosilicic acid ($H_2SiF_6$), and thereby eliminating pollution of the atmosphere and of streams and other bodies of water in the phosphate processing plant area. In addition, the process eliminates phosphate and other dust losses from the gas streams.

A process developed by H.H. Predikant, H. Betz and J. Schäffer (7) is one in which fluorine values are recovered from industrial waste gases containing fluorine and/or hydrogen fluoride, notably from gases evolved in the manufacture of aluminum through electrolysis of alumina in the presence of cryolite.

This process involves scrubbing with a highly acidic aqueous solution containing hydrogen fluoride in a first washing zone and then with an aqueous sodium hydroxide solution with formation of sodium fluoride in solution without precipitation in a second washing zone, splitting the stream of solution from each washing zone into a recycle stream for return to the zone wherein it was formed, and into a precipitation stream. The two precipitation streams are combined to precipitate sodium fluoride, which is recovered with recirculation of resultant mother liquor to the first washing zone. Figure 51 shows the form of apparatus which may be used to carry out such an operation.

## FIGURE 51: CAUSTIC SCRUBBER FOR HF REMOVAL FROM ALUMINUM CELL EXIT GASES

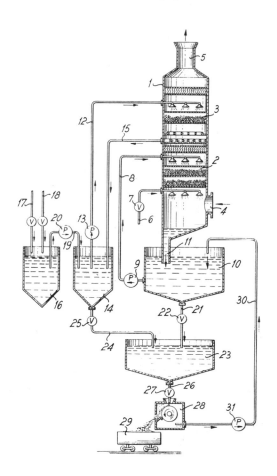

Source: H.H. Predikant, H. Betz and J. Schäffer; U.S. Patent 3,660,019; May 2, 1972

Referring to the drawing, the scrubber 1 consists of first washing zone 2 and a washing zone 3 provided above it. Below the washing zone 2 the waste gas to be purified enters the scrubber 1 through waste gas inlet 4 and flows through the scrubber upwardly, passing first through the washing zone 2 and then through the washing zone 3 and finally leaving the scrubber through waste gas outlet 5. Water is introduced into the lower section of the scrubber through conduit 6 equipped with valve 7, which water adiabatically cools the waste gas entering the scrubber through gas inlet 4 to a temperature of, for example, 70°C. and balances the liquid losses caused during operation.

Aqueous hydrogen fluoride is passed through the conduit 8 equipped with pump 9 from the collecting container 10 to the washing zone 2 and is there contacted countercurrently with the waste gases flowing upwardly through the scrubber. After contact with the waste gases to be purified, the aqueous hydrogen fluoride enters the collecting container 10 via the outlet 11 at the bottom of the scrubber 1 and is passed from container 10 again to the washing zone 2 through conduit 8 by means of the pump 9.

Aqueous sodium hydroxide is pumped through conduit 12 equipped with pump 13 from the collecting container 14 into the second washing zone 3 and leaves the latter after countercurrent contact with the waste gases to be purified through conduit 15 which discharges into the collecting container 14. Aqueous sodium hydroxide is delivered, if desired, from the storage container 16 equipped with supply conduits 17 and 18 to the collecting container 14 through conduit 19 equipped with pump 20.

Continuously or intermittently aqueous hydrogen fluoride, for example, concentrated by recycling to a content of 5 to 6% by weight, is passed through conduit 21 equipped with valve 22 to the neutralization container 23, and aqueous sodium hydroxide also enriched with sodium fluoride by recycling is passed from the collecting container 14 to the neutralization container 23 through conduit 24 equipped with valve 25. The valves 22 and 25 are controlled in such a manner that a small molar excess of sodium hydroxide relative to the amount of hydrogen fluoride flows to the neutralization container.

Within the neutralization container 23 the hydrogen fluoride entering from the collecting container 10 and additionally containing dust and possibly sulfur dioxide is neutralized completely by means of the sodium hydroxide supplied from the collecting container 14, whereby the solubility product of the sodium fluoride is exceeded and sodium fluoride is precipitated. The neutralization mixture is discharged from the bottom of neutralization container 23 through conduit 26 equipped with valve 27 and separated within the centrifuge 28. The wet filter cake is discharged from the centrifuge and disposed at 29. Through the conduit 30 equipped with pump 31, the mother liquor of the neutralization mixture is recycled to the collecting container 10 and thus introduced again into the cycle of the washing liquid for the first washing zone.

Suitably, at the collecting container 10 a device not shown can be provided which senses the level of the liquid within the collecting container 10 and opens or closes the valve 7 of the conduit 6 in response to variations of the level of the liquid so as to maintain the liquid level substantially constant or at least between predetermined upper and lower level limits.

The following is a specific example of the operation of the process. Waste gas to be purified from the production of aluminum by smelting flux electrolysis was supplied in an amount of 195,000 Nm³/hr. to the apparatus described above. The waste gas had a temperature of 70°C., moisture content of 0.04 kg. of $H_2O$/kg. of dry air, a dust content of 140 mg./Nm.³, a content of fluorine and hydrogen fluoride, calculated as fluorine, of 60 mg./Nm.³ as well as a content of sulfur dioxide of 60 mg./Nm.³. Thus, by introducing the waste gas into the scrubber 1, 27.3 kg. of dust, 11.7 kg. of fluorine and 11.7 kg. of sulfur dioxide have been introduced per hour.

First within the first washing zone 2, the waste gas flowing upwardly is washed with water from the collecting container 10 which was recycled until a 6% solution of hydrogen fluoride had been obtained. The amount of the cycled washing liquid was 236.7 kg./hr. and after the mentioned enrichment with hydrogen fluoride contained 23.2 kg. of dust, 10.5 kg. of hydrogen fluoride, 10.0 kg. of sulfur dioxide, 7.0 kg. of sodium fluoride and 21.5 kg. of sodium sulfite. Thus, the efficiency of the first step was about an 85% removal of the dust originally contained, a 95% removal of the fluorine originally present and an 85% removal of the sulfur dioxide originally present.

Within the second washing zone 3, the waste gas flowing upwardly was washed with a 30% aqueous sodium hydroxide solution, whereby additionally 2.73 kg./hr. of dust, 0.351 kg./hr. of fluorine and 1.17 kg./hr. of sulfur dioxide have been removed from the waste gas. Thus, within the second washing zone additionally about 10% of dust, 3% of fluorine and 10% of sulfur dioxide have been removed. Now, the waste gas had such a purity that it could be delivered to the atmosphere without hesitation. Thus, the recycled aqueous hydroxide absorbed 0.795 kg. of sodium fluoride, 4.6 kg. of sodium sulfite per hour and additionally dust.

After a starting period in which the hydrogen fluoride in the solution in the collecting container 10 had been concentrated to 6%, 236.7 kg./hr. of hydrogen fluoride solution were passed through the conduits 21 and 25 continuously from the collecting container 10 and 215.8 kg./hr. of aqueous sodium hydroxide lye from collecting container 14 were passed into the neutralization container 23. The resultant neutralization mixture contained 23.2 kg. of dust, 30.4 kg. of sodium fluoride, 45.8 kg. of sodium sulfite and 353.1 kg. of water. On separation within a filter centrifuge there were obtained 100.7 kg. of filter cake comprising 23.2 kg. of dust, 20.0 kg. of sodium fluoride, 7.5 kg. of sodium sulfite and 50.0 kg. of water. The filter cake was discharged and converted to cryolite. The mother liquor from the filter centrifuge comprised 351.8 kg./hr. and consisted of 303.1 kg. of water, 10.4 kg. of sodium fluoride and 38.3 kg. of sodium sulfite. The mother liquor was recycled to the collecting container 10 through conduit 30.

A process developed by A.E. Henderson, Jr. (8) involves fluorine recovery from the process of making superphosphate from phosphate rock and mineral acid. The chief aim is to convert phosphate rock (generally apatite in composition and principally fluoapatite) to monocalcium phosphate, which is the predominant chemical form of available phosphoric acid in various superphosphates.

Most phosphate rocks contain considerable amounts of fluorine. When these rocks are reacted with various mineral acids to produce superphosphates, generally between 10% and 30% of the fluorides present in the raw materials are released and must be subsequently removed, usually scrubbed in water, to avoid air pollution.

While single and intermediate superphosphates continue to evolve fluorides during curing and storage, the degree is less than in triple superphosphate. In the latter case, serious air pollution problems result and producers are currently considering the difficult task of scrubbing the gases vented from very large storage and curing sheds common in the phosphate industry.

The triple superphosphate now produced by the industry with wet process phosphoric acid and Florida phosphate rock analyses about 2.4 to 2.6% fluorine (dry basis) and evolves approximately 0.6 pound of fluorine per ton of product per day. This evolution rate is the average for the first 24 hours after manufacture and continues at a decreasing rate for at least several weeks. If the residual fluorine in the product can be reduced to below about 2.0% (dry basis), the evolution of fluorine is drastically reduced immediately and reaches essentially zero at about 48 hours. The use of furnace grade phosphoric acid in place of wet process acid will, of course, lower the starting fluorine level; however, the cost is considerably higher and as a result furnace acid is not generally used in superphosphate manufacture.

The need for a process by which superphosphates can be produced with the maximum elimination of fluorides and the minimum of storage has long been recognized in order to eliminate or reduce the problem of fluoride evolution during storage. No process has, however, been available which successfully accomplishes this purpose. Producers of superphosphate have long recognized that the problems of fluorine control and of acid mixing were closely related and that the mixing of phosphate rock and acid must be carried out under well defined conditions to avoid producing products which cannot be adequately handled or which contain large percentages of unavailable forms of phosphate which will not cure to available form.

The process described below markedly reduces the fluorine level of the superphosphate product and eliminates the need for scrubbing the off-gases from curing and storage buildings in order to comply with normal state emission regulations. Figure 52 is a flow diagram showing the arrangement of essential equipment in the process.

## FIGURE 52: APPARATUS FOR MANUFACTURE OF LOW FLUORINE SUPERPHOS-PHATE AND HF SCRUBBING FROM VENT GASES

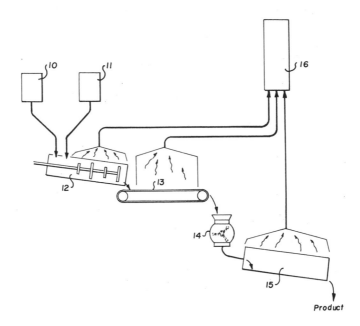

Source: A.E. Henderson, Jr.; U.S. Patent 3,672,828; June 27, 1972

Referring to the drawing, there are illustrated a hot acid supply **10** and a hot rock supply **11**, both feeding into a high speed pug mill mixer **12**. The mixer discharges onto a curing or settling belt **13** on which the product is denned. The belt **13** discharges to a pulverizer **14** which in turn discharges to a cooler **15**. The product from the cooler is ready for storage, shipment or further processing. All gases from the pug mill **12**, the belt **13**, and the cooler **14** are carried to scrubber **16** for removal of fluorides.

This process depends upon the reaction of hot acid and hot phosphate rock to produce superphosphate of substantially lower fluorine content than that presently available. The resulting product is substantially free of fluorine evolution from product storage.

A process developed by W.E. Rushton and G. Kleinman (9) is one in which fluorine compounds evolved in equilibrium concentrations from an evaporator can be recovered in increased amounts by scrubbing the vapor product from such evaporator in a scrubbing tower operated at a reduced pressure with respect to the pressure in the evaporator. For example, in an evaporator for concentrating a dilute phosphoric acid feed wherein the evaporator is operated at an absolute pressure of about 8½ inches Hg, the scrubbing tower could be operated at an absolute pressure of 2 inches Hg. Under such circumstances, the fluorine compounds evolved from the evaporator (e.g., $SiF_4$ and HF) are recovered with appreciably higher efficiencies at any given concentration by an aqueous fluosilicic scrubbing solution. These increased efficiencies become particularly significant in those situations wherein the scrubbing solution has a fluosilicic acid concentration of 20% or more.

The pressure in the scrubber should be at least ½ inch Hg lower than in the evaporator, however, improved results are obtained where greater differential pressures are employed. As such, the process advantageously provides improved fluorine recovery together with substantially reduced stream and air contamination.

## Removal of Fluorine Compounds from Water

A variety of treatment methods are available for fluoride-bearing waste streams. A summary of the more important of these treatment processes and their removal efficiencies is presented in Table 9. Current treatment methods can be divided into two categories: additive methods and adsorptive methods.

## TABLE 9: SUMMARY OF FLUORIDE TREATMENT PROCESSES AND LEVELS OF TREATMENT ACHIEVED

| Process | Initial Fluoride Conc., mg./l. | Final Fluoride Conc., mg./l. | Current Application | Reference |
|---|---|---|---|---|
| Lime addition | – | 10 | Industrial | 10 |
| Lime addition | 1,000 – 3,000 | 20 | Industrial | 1 |
| Lime addition | 1,000 – 3,000 | 7 – 8 (after 24 hr. settling) | Industrial | 1 |
| Lime addition | 500 – 1,000 | 20 – 40 | Industrial | 11 |
| Alum coagulation | 3.6 | 0.6 – 1.5 | Municipal | 12 |
| Hydroxylapatite beds: | | | | |
| Synthetic | 12 – 13 | 0.5 – 0.7 | Municipal | 13 |
| Synthetic | 10 | 1.6 | Municipal | 14 |
| Bone char | 6.5 | 1.5 | Municipal | 15 |
| Bone char | 9 – 12 | 0.6 | Municipal | 16 |
| Alumina contact beds | 8 | 1 | Municipal | 15 |
| Alumina contact beds | 9 | 1.3 | Industrial (lab scale) | 1 |
| Alumina contact beds | 20 – 40 | 2 – 3 | Industrial | 11 |

Source: Report PB 204,521

*Additive Methods:* These involve addition of treatment chemicals, and formation of fluoride precipitates or adsorption of fluoride on a resulting precipitate. Removal is accomplished by solids separation of the precipitate. Treatment efficiency is therefore in part dependent upon the effectiveness of liquid–solid separation. Chemicals employed include lime (CaO), magnesium compounds (e.g., dolomite), and aluminum sulfate (alum).

Lime — Direct addition of lime is the standard technique for removal of high concentrations of fluoride ion. The lime reacts with fluoride in the wastewater to produce calcium fluoride. Calcium fluoride has a theoretical maximum solubility of approximately 8 mg./l. of fluoride at pH 11. Thus, concentrations of calcium fluoride above this solubility limit will form a precipitate.

Reported fluoride removals down to residuals of only 10 to 20 mg./l. reflect the slow rates of precipitate formation (1)(10). Zabban and Jewett (1) report, however, that 24 hour contact with lime reduced fluoride to near the theoretical limit of 8 mg./l. Addition of excess lime further reduces fluoride concentration, at the expense of additional treatment chemical costs and increased calcium concentration in the waste effluent. Recarbonation of the effluent provides a means of removal of excess calcium and pH adjustment.

The lime treatment process for fluoride waste is composed of standard unit operations of the same sort employed in lime softening of municipal water. These units include a rapid mixing basin for chemical addition, a flocculating unit, and a settling basin for solids separation. Costs have been reported for the Tektronix, Inc. waste treatment facility of Beaverton, Oregon (10) where fluoride wastes are collected as a separate stream and treated independent of other wastes originating in the plant (10). Fluoride was reduced to 10 mg./l. by lime treatment. Treatment costs for the Tektronix, Inc. plant were not itemized for the individual types of waste. The individual stream flows were:

| | |
|---|---|
| Chromium wastes | 2,000 gal./hr. |
| Cyanide wastes | 3,000 gal./hr. |
| Fluoride wastes | 200 gal./hr. |
| Other wastes | 15,800 gal./hr. |

Costs associated with overall treatment of the four waste streams (1964-1968 prices) were $80,000 for capital investment and $3,000/month for operating cost. Based on a 20 year amortization and 7 day per week operation, this represents an average cost of approximately 20¢/1,000 gallons of waste. The Tektronix plant utilized a 44,000 square foot lagoon to handle sludge disposal from the waste treatment process. The final lagoon effluent contained 2.0 mg./l. of fluoride, accomplished primarily through dilution, upon combining the treated waste streams.

General sludge handling techniques should apply to the precipitated calcium fluoride sludge. Dewatering may be accomplished by one of the standard processes of sand bed drying, lagooning, vacuum filtration or centrifugation, depending upon availability of land and the overall nature of the final sludge (i.e., other sludges combined). General costs given by Burd (17) for sludge dewatering are:

| | |
|---|---|
| Lagooning | $1 to $5/dry ton |
| Sand bed drying | $3 to $20/dry ton |
| Vacuum filtration | $8 to $50/dry ton |
| Centrifugation | $5 to $35/dry ton |

Land disposal is estimated to cost an additional $1 to $5/dry ton.

A process developed by A.N. Baumann and R.E. Bird (18) is one in which fluorine values are removed from an aqueous fluosilicic acid solution by adding limestone to precipitate calcium fluoride, separating the calcium fluoride by filtration or otherwise, adding lime to a portion of the filtrate to form a slurry, combining the slurry with the remaining portion of the filtrate, allowing solids to form and settle, and removing the settled solids. In a preferred embodiment, the amount of fluorine removed from aqueous phosphorus contaminated fluorine-containing acid is improved by adding lime after the limestone addition but prior to the calcium fluoride separation step.

Figure 53 is a block flow diagram of this process, and the process will now be described in further detail with reference to that figure. Wastewater, such as the pond water from a superphosphate and wet process phosphoric acid manufacturing plant, is introduced into a first reactor 12 where it is mixed with a charge of limestone. The limestone is added in an amount sufficient to raise the pH of the wastewater from below 3.0 to 3.2 The limestone treated wastewater is maintained in the reactor for a period of about 30 minutes to allow the reaction of the limestone with the fluosilicic acid content of the wastewater to

## FIGURE 53:  FLOW SCHEME OF PROCESS FOR FLUORINE REMOVAL FROM WASTEWATER AS CALCIUM FLUORIDE

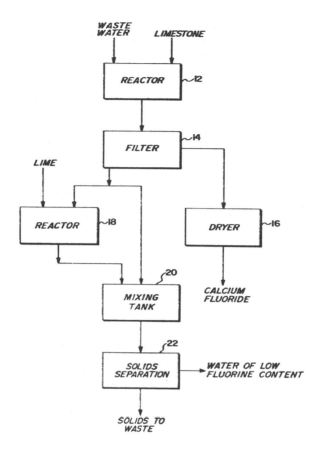

Source:  A.N. Baumann and R.E. Bird; U.S. Patent 3,551,332; December 29, 1970

go to completion and precipitate calcium fluoride.  The limestone treated wastewater is fed to filter **14** where a calcium fluoride filter cake is separated.  The calcium fluoride filter cake is conveyed by suitable means from filter **14** to dryer **16**.  The dried calcium fluoride filter cake obtained from dryer **16** may then be treated by any suitable method so as to remove some of the contaminants therefrom and improve the purity of the calcium fluoride.

A portion of the filtrate, i.e., about 20%, from filter **14** is passed to a second reactor **18** where it is mixed with sufficient lime to raise the pH thereof to 11.00.  About 30 minutes are allowed for the lime to react with the portion of the filtrate and form a slurry.  The slurry is then introduced into the mixing tank **20**, where it is combined and thoroughly mixed with the remaining 80% of the filtrate from filter **14**.  The combined liquid is maintained in tank **20** for 30 minutes.  The combined liquid is finally introduced into solids separation section **22**, where water having a substantially reduced fluorine content is separated from the solids content thereof such as by filtration.  The water is either discharged, or recycled to the plant and reused. The solids obtained from separation section 22, which are primarily a complex mixture of calcium fluoride salts, are discharged as waste.

A process developed by D.R. Randolph (19) involves purifying dilute aqueous acidic phosphate solutions typically containing calcium, sulfate ion, silica, iron, and aluminum and less than about 3% $P_2O_5$ and 1.5% fluorine, and recovering the $P_2O_5$ and fluorine values therefrom in a useable form.

The manufacture of wet process phosphoric acid typically involves treating phosphate rock with a strong mineral acid to solubilize the phosphate constituents thereof as phosphoric acid. The ground rock is introduced into a reactor as a fine powder, at least 80% 100 mesh, and digested with sulfuric acid to yield a mixture of phosphoric acid and calcium sulfate. However, since the phosphate rock is a complex material and contains in addition to fluorapatite, its principal constituent, contaminants such as silica, iron oxide, aluminum oxide and organic matter, the reaction of the mineral acid with the contaminants occurs simultaneously with the fluorapatite-acid reaction and results in the dissolution of the contaminants along with the dissolution of phosphate constituents. Thus, in order to recover the phosphoric acid from the reaction mixture, it is necessary to subject the mixture to several processing steps designed to separate the undesirable constituents from the phosphoric acid.

Gypsum (i.e., calcium sulfate dihydrate), the major contaminant in the reaction mixture is formed during the reaction and is removed by the relatively simple expedient of filtration. The filter cake is generally washed with water to recover as much $P_2O_5$ as possible. The cake is then slurried in water and pumped to settling ponds for disposal. The recovered phosphoric acid is further concentrated in evaporators. During the concentration process, a large amount of fluoride-bearing gases are evolved; the fluorides are dissolved in water and the water is then pumped to the settling ponds along with the gypsum. After the gypsum has settled, the water in the pond may be reused in the processing plant or, if abundant, as for example during the rainy season, it is discharged to streams or lakes. It is the process water from these ponds, or process water of similar chemical composition, which is meant by the term "dilute acidic aqueous phosphatic solutions" as used herein.

Before such water can be discharged, it must be treated to eliminate or reduce the phosphorus and fluorine content to acceptable levels and adjust the pH of the water to about neutral. The pH adjustment is generally achieved by "liming" to a pH of about 6 to 7. The limed water is then subjected to an additional settling period before discharge. While this practice is effective, it nevertheless leaves much to be desired, especially from the standpoint of economics. When it is recognized that generally about 2.0% to 5.0% of the total $P_2O_5$ and from 50% to 55% of the fluorine fed to the system are lost and that typically a single processing plant utilizes several hundred billion gallons of processing water per year containing some 15,000 to 25,000 tons of waste including 3,000 to 4,000 tons of $P_2O_5$ and 13,000 to 14,500 tons of fluorine, the need and importance of providing a satisfactory process for separation and recovery of these values from processing water becomes exceedingly clear.

Figure 54 is a flow diagram showing the essential features of the process. As shown there, pebble lime from storage vessel 10 is introduced into slaker 11 where it is made up as a 10% aqueous milk of lime solution. This solution is then introduced into agitated first liming vessel 12 where it is admixed with process water, containing from 0.3 to 3.0% $P_2O_5$ and 0.15 to 1.5% fluorine, in sufficient amount to adjust the pH thereof to a predetermined value between about 2.6 and 5.0.

The pH to which the dilute aqueous acidic phosphatic solution is initially adjusted is critical. If the pH is below about 2.6 substantially no precipitation will occur. On the other hand, if the pH exceeds about 5.0, precipitation of fluorine will occur; however, the precipitate will be enriched with $SiO_2$ to the extent that the $SiO_2$ content of the precipitate will exceed 1.5% by weight.

## FIGURE 54: PROCESS FOR RECOVERY OF FLUORINE AND P₂O₅ FROM DILUTE AQUEOUS ACIDIC PHOSPHATIC SOLUTIONS

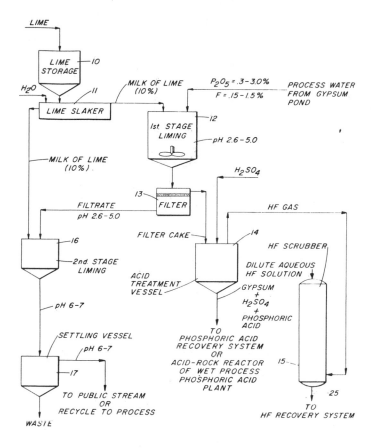

Source: D.R. Randolph; U.S. Patent 3,625,648; December 7, 1971

When it is desirable to produce a very high quality hydrofluoric acid, liming is terminated when the pH of the process water approaches 3.5 and preferably between about 3.2 and 3.5. Where quality is less important, the pH may be adjusted to as high as 5.0. At a pH of 5.0, silica is still at an acceptable level (i.e., below 1.5%) and more than 99% of the fluorine and 80% of the $P_2O_5$ are precipitated and can be recovered. The primary reactions occurring in the first stage liming under the above specified conditions of this process may be shown as follows:

$$2HF + Ca(OH)_2 \longrightarrow CaF_2\downarrow + 2H_2O$$

$$H_2SiF_6 + 3Ca(OH)_2 \longrightarrow 3CaF_2\downarrow + SiO_2 + 4H_2O$$

$$2H_3PO_4 + Ca(OH)_2 \longrightarrow Ca(H_2PO_4)_2 \cdot H_2O + H_2O$$

$$Ca(H_2PO_4)_2 \cdot H_2O + Ca(OH)_2 + H_2O \longrightarrow 2CaHPO_4 \cdot 2H_2O$$

The suspension obtained from first stage liming vessel **12** is then subjected to a separation treatment to remove the solid materials from the liquid. In practice, the suspension is generally filtered as shown at **13**. Other methods of separation may be employed as, for example, centrifugation, settling, or a combination of such treatments, the important step of course being the separation and recovery of both the filter cake and the filtrate.

Filter cake from filter **13** is introduced into acid reactor **14** where it is admixed with an acid, preferably anhydrous, such as sulfuric acid, glacial acetic acid or perchloric acid, and preferably sulfuric acid. The reaction may be conducted at a temperature between about 20° and 100°C. at atmospheric pressure. A slight excess of acid over the stoichiometric requirement is usually employed; however, a substantial excess of acid may be used if desired (i.e., 2 or 3 mol equivalents). This reaction liberates gaseous HF and is represented by the following equation:

$$CaF_2 + CaHPO_4 + 2H_2SO_4 \longrightarrow 2CaSO_4 + H_3PO_4 + 2HF\uparrow$$

If silica is present in the filter cake in significant amounts (more than 1.5%) it will react with the HF thus produced to form silicon tetrafluoride which hydrolyzes to fluosilicic acid, an undesirable occurrence where HG recovery is desired. Thus, the importance of proper pH control in the first stage liming is again evident.

The HF produced by the above reaction is conducted from acid reactor **14** to HF scrubber **15** where it is absorbed in dilute aqueous HF solution, water, sodium hydroxide solution, or a similar solution from which HF is readily recovered; dilute aqueous hydrofluoric acid is preferred. The reaction mixture from which HF has been evolved is primarily a mixture of sulfuric acid, phosphoric acid, and gypsum; as such it can be introduced directly into the recovery system of the phosphoric acid plant or it may be introduced into the acid-rock reactor. Where excess sulfuric acid is used in the filter cake treatment, the latter course is frequently most desirable. In either case the $P_2O_5$ content is thus recovered.

The filtrate from filter **13** has a pH between 2.6 and 5.0 and as such is not suitable for discharge to public streams or lakes. It is therefore introduced into a second stage liming vessel **16** where it is treated with a 10% milk of lime solution in sufficient amount to adjust the pH thereof to between 6 and 7. Following adjustment the aqueous gypsum suspension which is formed is introduced into settling vessel **17** and permitted to stand until the solids in this second stage liming have settled. These solids are waste materials and may be treated as such. The water remaining after solids separation is effected is suitable for discharge to public waters or may be recycled to the phosphoric acid plant for use.

**Magnesium** — Early work in the treatment of drinking waters containing excessive fluoride concentrations demonstrated that fluoride ion was removed during the standard lime-soda water softening process. Removal was in direct proportion to the amount of magnesium hardness removed, and was attributed to adsorption of fluoride ion onto the magnesium hydroxide floc formed in the softening process.

Fluoride concentrations of 3 to 4 mg./l. were reported reduced to 0.8 to 0.12 mg./l. (12). However, in the absence of appreciable magnesium naturally occurring in the water or wastewater to be treated, the requirement for supplementary addition of magnesium may make this process economically unfeasible. No fluoride removal plant has been specifically designed to use this process. Fluoride reduction is, however, accomplished in conjunction with magnesium hardness removal at several municipal water treatment plants in the United States where fluoride is present in the raw water at concentrations of 2 to 3 mg./l. (20).

**Aluminum Sulfate (Alum)** — For a water of low hardness containing 3.6 mg./l. fluoride ion, simple alum addition plus small quantities of lime for pH control resulted in fluoride reduction to 1 mg./l. (12). For hard waters, calcium and magnesium ions interfered, and other methods of removal were more effective. Culp and Stoltenberg (12) report alum treatment to reduce a soft water fluoride concentration of 3.6 mg./l. to 1.0 mg./l. No cost data are given for plant construction but a standard alum treatment sequence consisting of 1 to 2 minutes rapid mix, 30 minutes flocculation, lime feed in last section of the floc-culation basin, and 2 to 4 hours detention in a settling basin equipped with mechanical sludge removal was used. Polishing by rapid sand filters was included. The authors rec-ommended lagooning for disposal of the resultant alum sludge.

Other Methods — The direct addition of bentonite, fuller's earth, diatomaceous earth, silica gel, bauxite, sodium silicate, ferric salts, and sodium aluminate has also been exam-ined (21). These processes all require low pH (about 3), which makes them less attractive in terms of additional chemical costs for pH adjustment, unless the wastewater pH is in the required range.

*Adsorption Methods:* These involve the passage of the wastewater through a contact bed, with fluoride being removed by general or specific ion exchange or chemical reaction with a solid bed matrix. Although solids removal is not required for these operations, bed re-generation and subsequent treatment of the concentrated regenerant are an integral part of the overall treatment process.

Since the basic mechanism of a contact process is one of ion exchange or surface reaction, these methods are usually appropriate only for low level fluoride wastes, or polishing proc-esses after fluoride reduction by previous treatment to the 10 to 20 mg./l. level. Other-wise, the requirement for frequent bed regeneration makes the process economically un-feasible. Bed media that have been used are hydroxylapatite (synthetic, processed bone and bone char), ion exchange resins, and activated alumina.

Hydroxylapatite — Reduction of fluoride from 12 to 13 mg./l. down to 0.5 to 0.7 mg./l. was reported by Wamsley and Jones (13), upon passage through a full scale contact bed of synthetic hydroxylapatite. Although the authors reported the operation to be highly effective, later analysis (21) indicated that this unit and others employing the same proc-ess were ultimately abandoned due to high chemical costs and bed attrition (42% loss per year at Britton, S.D.). Cillie, et al (14) reported for a similar pilot plant scale operation in South Africa that 10 mg./l. fluoride could be reduced to 1.5 mg./l. The presence of chlorides in the water reduced filter bed capacity and increased regeneration chemical re-quirements.

Natural hydroxylapatite in the form of bone char was reported capable of reducing fluo-ride from 10 mg./l. to 0.6 mg./l., but regeneration with phosphoric acid and sodium hy-droxide resulted in excessive chemical costs at the Britton, S.D. unit (which had converted from synthetic hydroxylapatite) and at Camp Irwin, California (16).

Ion Exchange — Natural and synthetic zeolite ion exchangers have been examined and found to be effective for removal of fluoride. The exchangers were both pretreated and regenerated with aluminum salts. Further investigation indicated that fluoride removal was attributable to aluminum oxide which precipitated in the column bed (15)(22). A strong cation exchange resin selective for fluoride was developed by Rohm and Haas but overall treatment and chemical costs have been reported as excessive (15).

Alumina — Contact beds of activated alumina have been used for many years in municipal water treatment plants for removal of fluoride ion. One unit in Bartlett, Texas, has oper-ated successfully since 1952 using a 500 cubic foot bed regenerated by sodium hydroxide and neutralized with sulfuric acid. Fluoride is reduced from 8 mg./l. to 1 mg./l. (15). Zabban and Jewett (1) investigated use of an alumina bed as a polishing unit to follow lime precipitation of high fluoride (1,000 to 3,000 mg./l.) wastes and found that 30 mg./l. residual fluoride carryover from the precipitation process could be reduced to approximately 2 mg./l. upon passage through the contact bed. At a pH of 11.0 to 11.5, they were able to reduce a concentration of 9 mg./l. of fluoride to 1.3 mg./l. In regenerating, they found that 4% of the alumina was lost for every 100 regeneration cycles.

*Summary:* Industrial wastes containing high fluoride levels require two-stage treatment. Lime precipitation removes fluorides down to 10 to 20 mg./l. Further reductions can be accomplished down to the 1 mg./l. level with other existing techniques. The preferred method seems to be activated alumina, with alum regeneration. The units involved are essentially the same as those used for ion exchange water softening, and the regenerant can be treated in the high lime removal unit by recycling. Although treatment levels obtainable by the various fluoride removal techniques are well defined in the literature, there is much less information available on costs. Published data however indicate that costs are generally in the range of 10 to 50¢/1,000 gallons.

### References

(1)  Zabban, W. and Jewett, H.W., "The Treatment of Fluoride Wastes," *Proc. 22nd Purdue Industrial Waste Conference*, pp. 706-16 (1967).
(2)  Patterson, J.W. and Minear, R.A., "Wastewater Treatment Technology," *Report PB 204,521*, Springfield, Va., National Tech. Information Services (August 1971).
(3)  Sackett, W.J., Sr.; U.S. Patent 3,494,107; February 10, 1970.
(4)  Teller, A.J. and Wyatt, E.S.; U.S. Patent 3,505,788; April 14, 1970; assigned to Wellman-Lord, Inc.
(5)  English, M.; U.S. Patent 3,512,341; May 19, 1970; assigned to L.A. Mitchell Limited, England.
(6)  Hartig, R.G.; U.S. Patent 3,642,438; February 15, 1972.
(7)  Predikant, H.H., Betz, H. and Schäffer, J.; U.S. Patent 3,660,019; May 2, 1972; assigned to Universal Oil Products Company.
(8)  Henderson, A.E., Jr.; U.S. Patent 3,672,828; June 27, 1972.
(9)  Rushton, W.E. and Kleinman, G.; U.S. Patent 3,415,039; December 10, 1968; assigned to Whiting Corporation.
(10) Schink, C.A., "Plating Wastes: A Simplified Approach to Treatment," *Plating*, 55, 1302-1305 (1968).
(11) Petrova, T.N. and Bakhurov, U.G., "Purification of Waste Solutions from Fluorine," *Atomn. Eng.*, 26, 552-553 (1969); *Wat. Pollut. Abstr.*, 43, No. 609 (1970).
(12) Culp, R.L. and Stoltenburg, H.A., "Fluoride Reduction at La Crosse, Kansas," *Jour. Amer. Water Works Assoc.*, 50, 423-431 (1958).
(13) Wamsley, R. and Jones, W.F., "Fluoride Removal," *Wat. and Sew. Works*, 94, 372-376 (1947).
(14) Cillie, G.G., Hart, O.O., and Standy, G.J., "Defluoridation of Water Supplies," *Jour. Inst. Water Eng.*, 12, 203-210 (1958).
(15) Maier, F.J., "Defluoridation of Municipal Water Supplies," *Jour. Amer. Water Works Assoc.* 45, 879-888 (1953).
(16) "Treating a High Fluoride Water," *Public Works*, 86, 67 (1955).
(17) Burd, R.S., "A Study of Sludge Handling and Disposal," *Publication No. WP-20-4*, U.S. Dept. of Interior, Washington, D.C. (1968).
(18) Baumann, A.N. and Bird, R.E.; U.S. Patent 3,551,332; December 29, 1970; assigned to International Minerals & Chemical Corporation.
(19) Randolph, D.R.; U.S. Patent 3,625,648; December 7, 1971; assigned to American Cyanamid Co.
(20) Maier, F.J., "Fluorides in Water," *Water Quality and Treatment*, New York, McGraw-Hill Book Company (1971).
(21) Savinelli, E.A. and Black, A.P., "Defluoridation of Water with Activated Alumina," *Jour. Amer. Water Works Assoc.*, 50, 33-44 (1958).
(22) Maier, F.J., "Methods of Removing Fluoride from Water," *Amer. Jour. Public Health*, 37, 1559-1568 (1947).

# FLY ASH

## Removal from Air

A process developed by G.H. Flowers, Jr. (1) may be applied to an incinerator for burning in large volume all types of rubbish or waste material, the exhaust gases from the main combustion zone being further subjected to another burning zone insuring complete combustion by combustion of any combustible products remaining in the exhaust gases and to a zone removing fly ash from the exhaust gases, thereby reducing air pollution to a minimum. The incinerator utilizes an apparatus for combustion of combustible products in the exhaust gases, the apparatus also removing the fly ash by discriminately directing the exhaust gases so that fly ash is separated therefrom.

A process developed by J.P. Tomany and W.A. Pollock (2) involves an $SO_2$ and fly ash removal system for coal burning power plant stack gases which provides for limestone-dolomite addition to the coal carrying through the power plant to form stable sulfate-sulfite compounds and unstable calcium and magnesium oxides which will carry along with the fly ash particles to a scrubbing zone. The fly ash and the sulfate-sulfite compounds are countercurrently contacted with a descending alkaline scrubbing stream in the presence of self-cleaning mobile contact elements in the scrubbing zone to effect the absorption of $SO_2$ and physical removal of fly ash and the stable sulfate and sulfite materials. The continuously circulating alkaline stream used in the scrubbing zone is obtained in part from the reaction of portions of the clacium and magnesium carbonates and oxides from the limestone-dolomite addition, with recirculating water.

Figure 55 shows a suitable form of apparatus for the conduct of the process. There is shown a boiler or furnace zone 1 having a fuel inlet means 2 suitable for introducing powdered coal and a heated air stream, which in turn is shown as being carried to such inlet by way of duct 3. In a usual arrangement, the powdered coal and forced draft air stream carries upwardly through a vertically oriented boiler unit and thence over into an economizer unit, such as 4. That portion of fly ash material together with some calcium and magnesium sulfates which are relatively heavy and fail to carry over from the upper boiler unit 1 will settle into a lower hopper zone 5 to be removed through outlet means 6 into an ash pit or other removal means not shown.

The hot flue gas stream, with entrained material, is indicated in this embodiment as carrying through the economizer unit 4 by way of duct 7 into air preheater 8 and thence outwardly through duct means 9. At the air preheater 8, a suitable flow of atmospheric air is passed thereto by way of inlet 10 and thence outwardly from outlet 11 to duct means 3 such that a heated air stream is made available for introduction into a high temperature boiler zone 1.

In accordance with this process, crushed or powdered limestone and/or dolomite additive is made available from hopper means 12 and valve means 13 to a separate inlet duct 2', such that approximately 10%, or more, by weight of the coal, of alkaline additive is available to carry through the high temperature boiler zone. Injecting air or other dispersion gas may be supplied to duct 2' from a suitable supply source not shown. Thus, in this system, the resulting compounds from the alkaline material added into the high temperature burning zone will carry through from the latter into the economizer 4, the air preheater 8, the transfer duct means 9 and then into the subsequent scrubbing zone or zones, such as provided by tower means 14.

Where limestone and dolomite are both used as alkaline materials, there will be an initial conversion to calcium and magnesium oxides and then a mixture of these active oxides combine with $SO_2$ to provide resulting calcium and magnesium sulfate and sulfite compounds which are in a relatively stable form so as to carry entirely through the duct system to the scrubber zone 14. Generally, about 50% of the $SO_2$ has been shown by tests to be converted to the fairly stable sulfate and sulfite compounds. Some of the remaining calcium and magnesium oxide materials which are formed in the high temperature boiler zone will tend to revert to calcium and magnesium carbonates, as previously noted, as the temperature decreases through the rest of the heat exchange and duct portion of the system. As a result, relatively insoluble calcium and magnesium carbonate materials are entrained with the fly ash and remaining $SO_2$ in the flue gas stream such that they are continuously available to add to the alkalinity of the recirculating aqueous stream descending through scrubber tower 14 by way of an inlet line 15 and liquor distributor means 16.

For purposes of simplification, there is illustrated a single scrubber tower 14 having two superposed contact stages 17 and 18, with each stage having a multiplicity of mobile lightweight contact elements 19. However, for a power plant unit having a large volume of flue gas, there may be two or more similar type scrubber towers provided to adequately effect the countercurrent scrubbing of the total gaseous stream. In each instance, the gas stream enters the lower portion of a scrubber tower 14 from inlet conduit means 9 such that there may be a vertically rising gas flow through the entire height of the tower. Also,

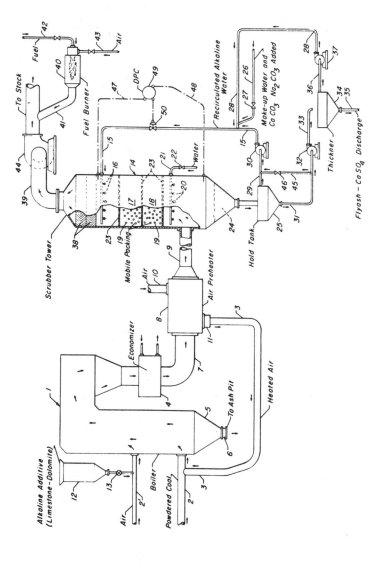

FIGURE 55: SYSTEM FOR REMOVAL OF SO₂ AND FLY ASH FROM POWER PLANT FLUE GASES

Source: J.P. Tomany and W.A. Pollock; U.S. Patent 3,520,649; July 14, 1970

preferably, in the lower zone, there is effected a presaturation or humidification of the gas stream by means of water flow from spray means **20** which in turn is supplied by line **21** having control valve **22**. Such water spray means will provide a precooling of the gas stream and an initial separation of entrained particles, as well as the desired humidification thereof prior to its flowing upwardly through the superposed contact stages **17** and **18**.

In order that there is obtained the highest efficiency from the special form of scrubber tower, the opposing gas and liquid flows are adjusted so that there is random turbulent motion of all of the contact elements **19** in each vertical stage, whereby there is, in turn, a maximum efficiency obtained from the constant rotational effect and large surface area being presented by each individual element to the rising gas stream and to the descending liquid absorption stream from spray nozzles **16**. Generally, for each contact stage, the contact elements **19** will fill the space or volume between the perforated retainer plates **23** to less than 30% of such volume when the elements are at rest, with neither liquid nor gas flow through the unit.

In view of the substantial quantity of alkaline additive (by way of oxides and carbonates of calcium and magnesium), carrying through to the scrubber zone **14** there will be a resulting continuous addition of alkalinity to the descending scrubbing stream so as to effect absorption of the sulfur dioxide into the liquor that carries into lower zone **24** and subsequently into a hold tank **25**. However, for purposes of closely controlling pH of the recirculating stream through line **15** there may be provision for adding further calcium carbonate, or other alkaline material, such as by way of line **26** with control valve **27** into line **28** which in turn is connective with the aforesaid line **15**. In this instance, recirculation is indicated as being provided through line **29** and pump means **30** which takes suction from the hold tank **25**. In the hold tank there is an initial separation of the alkaline slurry stream from particulates, including fly ash and precipitated sulfate-sulfite compounds, whereby there may be the desired continuous circulation of the alkaline water stream through scrubber tower **14**.

Inasmuch as the preferred system minimizes the addition of limestone or other alkaline material by conserving a greater portion of the alkaline slurry for recirculation purposes, there may be additional particle thickener or removal means so as to effect further concentration of the solids prior to their discharge from the system. In other words, from hold tank **25** there may be withdrawal of settled slurry by way of line **31**, pump **32** and line **33** into a suitable second hold tank or thickener unit **34**. The latter provides for still further concentration of solids materials which may be withdrawn from the lower portion thereof by way of line **35** while alkaline liquid or slurry is withdrawn by way of line **36** and pump **37** for discharge into line **28** in turn carrying to the recirculation line **15**. On the other hand, where it is desired to retain the maximum use of certain of the calcium and magnesium values, some recirculation of slurry may be made from line **31** through line **45** and control valve **46**.

The scrubbed, substantially $SO_2$ free, flue gas stream leaving the upper portion of the scrubber tower **14** is shown as passing through suitable mist extractor means **38** and carrying into an outlet duct **39** and fan **44**. The duct **39** will normally connect directly to suitable stack means, not shown in the drawing.

Where desired, and in order to preclude a white plume from entrained moisture carrying into the stock outlet, there may be the addition of heat to the cleaned flue gas stream ahead of the outlet. For example, this embodiment indicates a fuel burner means **40** as discharging into duct **41** which in turn connects with duct means **39** so as to supply high temperature burner gas into admixture with the flue gas stream leaving tower **14**. Fuel and air to the burner means **40** are indicated as being supplied respectively through lines **42** and **43**. In all cases, the heated air stream added to the stack gas from fuel burner means **40** should be kept to a minimum in order to conserve fuel but, at the same time, shall add sufficient heat to the discharge gas stream so as to preclude having a large quantity of white plume from moisture condensation at the stack discharge.

It should be pointed out that this embodiment is only diagrammatic with respect to the means for thickening and discharging the precipitates and fly ash recovered from the scrubbing zone as well as diagrammatic with respect to the means for effecting the recirculation of the alkaline water scrubbing stream that is carried down through the scrubber tower **14** in a continuous manner.

For example, a preferred system may provide for connecting a differential pressure controller **49**, by lines **47** and **48** across the height of the scrubber unit **14** to maintain collection and absorption efficiencies. In other words, a constant pressure drop can be maintained across the tower **14** by means of the differential pressure controller **49**, to control valve **50** located in the recirculation slurry line **15**. By this means, any variation in gas flow can be compensated by either increasing or decreasing this recirculated liquor flow so as to maintain constant pressure drop across the scrubber.

Obviously, modifications may also be made with respect to the particular means or with regard to the types of thickener apparatus for effecting the handling of the slurry leaving the scrubber zone **14** at the lower reservoir end **24** at the zone of the hold tank **25**, or at thickener **34**. Also, it should be noted that the drawing is diagrammatic with respect to the power plant unit, including the economizer **4** and air preheater **8**, inasmuch as various types or forms of heat recovery units and duct means may be utilized to advantage to provide the transfer of the laden hot flue gas stream leaving the boiler unit **1** and carrying to the integrated scrubber section as provided by this system.

The description previously set forth has indicated that a combination of limestone and dolomite may be injected into the boiler zone; however, where desired, either limestone or dolomite alone may be used as the alkaline additive material and good results provided in this integrated system. On the other hand, where the cost is not prohibitive, still other powdered alkaline materials as have been noted, may be added so as to be carried through the high temperature boiler zone; subsequent heat recovery zones and duct work to the scrubbing unit, where the particular alkaline additive material will combine with the recirculated liquor stream to supply at least a portion of the alkalinity required for the completion of $SO_2$ removal in the overall system. Generally, better than 90% $SO_2$ removal is accomplished along with a 98.5%+ removal of fly ash.

A process developed by L.J. Kinney (3) involves removing ash and other particles suspended in a moving stream of hot gases, such as smoke, as well as cleansing these gases of any nauseous, toxic or otherwise undesirable pollutants by means of treating the gases with a series of cleansing and cooling steps, including water sprays, screens and baffles, and additionally includes as one of the final steps a spray of a chemical reagent such as ammonium hydroxide for combining with any of the remaining pollutants such as sulfur dioxide to form water-soluble components which are then removed by means of a final water spray.

A process developed by H.L. Richardson (4) is one in which improved electrostatic precipitation of entrained solids from flue gas, which is generated by burning sulfur-containing solid carbonaceous fuel, is attained by burning a vanadium-containing liquid hydrocarbon such as high vanadium content crude oil or refinery residual oil together with the solid fuel. The vanadium which is thus added to the combustion process is converted to vanadium pentoxide which causes at least partial catalytic oxidation of sulfur dioxide to sulfur trioxide, which provides improved results and greater efficiency during subsequent treatment of the flue gas by electrostatic precipitation, in terms of greater solids removal from the flue gas.

A process developed by L.M. Roberts (5) is one in which suspended matter is removed from combustion gases containing sulfur dioxide by removing a major portion of the particulate matter from a minor portion of the gases, catalytically converting a major portion of the sulfur dioxide content of the minor portion of the gases to sulfur trioxide, mixing the minor portion of the gases with the main gas stream, and subjecting the mixed gases to electrical precipitation.

Difficulty has been encountered in the electrical precipitation of suspended particles or fly ash from combustion gases arising from the phenomenon of "back discharge" due to high resistivity of the particulate material, and it has been proposed to overcome this difficulty by catalytically converting sulfur dioxide in the gases to sulfur trioxide to increase the conductivity of the particulate material in the gases. However, the catalyst rapidly loses its efficiency and the catalyst bed rapidly increases its resistance to the flow of gases therethrough due to the accumulation of particulate matter thereon.

Since combustion gases typically contain a much larger amount of sulfur dioxide than is required to produce the amount of sulfur trioxide desired for effective conditioning of the particulate material for efficient electrical precipitation, it has been found that these difficulties may be avoided by subjecting only a minor portion of the gases to catalytic sulfur dioxide conversion after removing from such minor portion of the gases a major portion of their suspended particulate matter, preferably by electrical precipitation at a high temperature, for example, from about 800° to 1000°F. This minor portion of the gas is then mixed with the main gas stream either before or after it has been cooled, for example, in a combustion air preheater, and the mixed gases are subjected to a conventional electrical precipitation operation.

In a typical powdered coal boiler installation, the sulfur dioxide content of the combustion gases coming from the boiler would be about 0.1% by volume whereas conversion of from 0.001 to 0.005% of sulfur dioxide by volume to sulfur trioxide would be effective to adequately condition the gases for efficient electrical precipitation of the fly ash. Thus the auxiliary gas cleaner and catalyst chamber needed for this method would only have to be large enough, in this case, to handle from 1 to 5% by volume of the combustion gases.

Figure 56 shows a suitable form of apparatus for the conduct of this process. In the drawing, 10 is the main flue leading from the combustion chamber of a coal fired boiler (not shown) through air preheater 11, electrical precipitator 12 and fan 13 to a chimney stack (not shown). An auxiliary fan 14 draws a small portion of the gases from the main flue a and passes them successively through high temperature electrical precipitator 15 and catalyst chamber 16 back to the main flue upstream of the main precipitator 12.

**FIGURE 56: ARRANGEMENT OF APPARATUS FOR ELECTROSTATIC PRECIPITATOR TREATMENT OF COMBUSTION GASES CONTAINING FLY ASH**

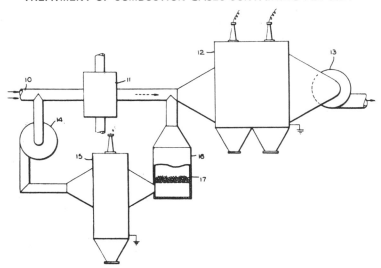

Source: L.M. Roberts; U.S. Patent 3,581,463; June 1, 1971

In a typical operation, 2% of the combustion gases coming from the boiler at about 800°F. and containing about 0.1% by volume of sulfur dioxide are passed by fan **14** through electrostatic cleaner **15** through catalyst chamber **16**, where substantially all of the sulfur dioxide is converted to sulfur trioxide by catalyst **17** which may be, for example, a vanadium pentoxide catalyst on silica gel granules or any of the commercially available sulfur dioxide oxidation catalysts. The sulfur trioxide containing gas is then returned to the main gas stream which has been cooled to about 300°F. in heat exchanger **11** and the mixed gases are passed through electrical precipitator **12**.

A process developed by L.C. Hardison (6) for reducing the fly ash and sulfur dioxide content of a flue gas stream involves passing the flue gas through a first mobile-packing scrubber stage to effect collection of the bulk of the fly ash. The flue gas is then passed through subsequent mobile-packing scrubber stages. In these subsequent stages the flue gas is washed with a liquid in which an oxygen-containing compound of magnesium or calcium is present. The sulfur dioxide in the flue gas stream reacts with these compounds and the resulting liquids and solids are carried to a kiln, where the reaction products are decomposed. Sulfur dioxide is driven off and collected, and the magnesium or calcium compound is regenerated for reuse in part.

A process developed by J. Misarek and W.W. Jaxheimer (7) for sequential removal of various air pollutants or contaminants such as fly ash, chemicals and hydrocarbons from exhaust systems consists of a series of afterburning, steam, fog, spray and washing chambers which remove toxics, hydrocarbons, fly ash and other undesirables before the fumes are emitted to the atmosphere. The sulfides or other chemicals removed during this process are placed in water solution which can be further refined, reclaimed, neutralized or disposed of.

### References

(1) Flowers, G.H., Jr.; U.S. Patent 3,489,109; January 13, 1970; assigned to Waste Combustion Corp.
(2) Tomany, J.P. and Pollock, W.A.; U.S. Patent 3,520,649; July 14, 1970; assigned to Universal Oil Products Company and Wisconsin Electric Power Company.
(3) Kinney, L.J.; U.S. Patent 3,522,000; July 28, 1970; assigned to Chillum Sheet Metal, Inc.
(4) Richardson, H.L.; U.S. Patent 3,568,403; March 9, 1971; assigned to Chemical Construction Corp.
(5) Roberts, L.M.; U.S. Patent 3,581,463; June 1, 1971; assigned to Research-Cottrell, Inc.
(6) Hardison, L.C.; U.S. Patent 3,632,305; January 4, 1972; assigned to Universal Oil Products Company.
(7) Misarek, J. and Jaxheimer, W.W.; U.S. Patent 3,668,839; June 13, 1972; assigned to Combustion Control Devices.

# FORMALDEHYDE

### Recovery from Wastewater

A process developed by W. Riemenschneider and O. Probst (1) permits obtaining a sewage that is practically free from formaldehyde when concentrating aqueous formaldehyde solutions by pressure distillation. The process comprises introducing boiling water into the sump of the distilling zone, continuously feeding in the aqueous formaldehyde solution to be concentrated at a height of one-quarter to four-quarters of the total height of the distilling zone and discharging a product that is practically free from formaldehyde from the sump of the distilling zone. The discharged sump product can be drained off without additional intermediate treatment as sewage which is unobjectionable from a biological point of view. Figure 57 shows a suitable form of apparatus for the conduct of the process.

The following is a specific example of the operation of the process. A pressure distilling apparatus consisting of a still of 300 l. volume, a column having a diameter of 150 mm. and a height of 10 m. which was packed with 15 mm. Raschig rings, a condenser operated with water having a temperature of 40° to 45°C., a reflux divider adjustable by rotameter and a corresponding pressure recipient was operated under the following conditions.

In the sump, 200 l. of water were brought to the boil under a pressure of 4 atmospheres gauge by means of steam pressure, and distillation commenced. The sump temperature was 152°C. Then 40 kg. of an aqueous 9% formaldehyde solution was introduced per hour by pumping into the column at three-quarters height of the total height of the column. The reflux ratio at the head of the column was adjusted at 1:5. There were obtained per hour 12 kg. of a head distillate containing 30% of formaldehyde and 28 kg. of a sump product containing 30 ppm of formaldehyde and 0.04% of formic acid. The loss owing to the Cannizzaro reaction amounted to 0.3% calculated on the formaldehyde applied.

FIGURE 57:  PROCESS FOR THE QUANTITATIVE SEPARATION OF FORMALDE-
HYDE FROM AQUEOUS SOLUTIONS BY PRESSURE DISTILLATION

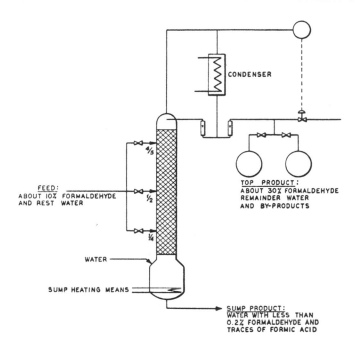

Source:  W. Riemenschneider and O. Probst; U.S. Patent 3,328,265; June 27, 1967

References

(1)  Riemenschneider, W. and Probst, O.; U.S. Patent 3,328,265; June 27, 1967; assigned to Farbwerke Hoechst AG, Germany.

## FOUNDRY EFFLUENTS

Air pollution aspects of the iron foundry industry have been reviewed by the staff of A.T. Kearney & Co. in a series of reports (1)(2)(3)(4). The economic impact of air pollution controls on the gray iron foundry industry has been reviewed by the staff of the U.S. Public Health Service (5). A more recent study on this topic has been prepared by A.T. Kearney & Co. (6)(7)(8).

### Removal of Particulates from Air

A process developed by J.S. Schumacher (9) provides for the reduction of dust pollution in foundries that use sand molds for casting and recycle the used sand for subsequent molding operations. A quantity of sand is added to the used sand, preferably at the shakeout, so that the total amount of sand weighs at least twelve times the weight of the metal previously cast.

One of the principal sources of pollution in a foundry is from sand which has been used in the molding operation. After the molding, the sand molds and the castings are separated and the sand is reclaimed for reuse. During at least a part of this reclamation process the sand is hot. Consequently, a great deal of dust is expelled into the atmosphere. A large part of this dust occurs at the shakeout since this is the point at which the sand is at its hottest temperature before the next molding operation. As the sand is conveyed from the shakeout additional pollution occurs.

In order to combat this type of pollution, foundries have utilized various types of devices. An idea of how sizeable the market for such devices is can be gleaned from an examination of the foundry periodicals. Such periodicals also present a reliable indicator for how serious the problem is, some even dedicating entire sections of each issue to the problems and supposed solutions.

Many of the various types of pollution equipment proposed for foundry use are expensive and require constant maintenance. By reason of the initial costs and the subsequent maintenance costs, many smaller foundries are not able to afford the required equipment. Unfortunately too, is the fact that some of the equipment is unable to satisfactorily perform its function.

In the casting of metals into sand molds, the amount of sand used to form a mold is usually expressed as a function of the amount of metal to be poured into it. This relationship is called the sand to metal ratio. For example, sufficient sand can be used for the mold to give a sand to metal ratio of from 3 to 1 to 20 to 1. While these ratios are frequently used, others are often times used too. A common ratio used is a ratio of about 6 to 1. That is, a mold is formed which consists of 6 pounds of sand for each pound of metal to be cast.

Foundry sand is, of course, not sand alone but contains additional constituents such as clays and/or carbons and/or other additives and/or temper water. Carbons in common use are powdered coal, coal tar, pitch, asphalt, graphite and coke. Other additives can be added such as celluloses, cereal binders, etc. After one or more molding operations new sand and/or clays and/or carbons can be added to the used sand to replace the portion of the original sand which has been made unuseable.

Conventional air pollution due to pollutants expelled from hot foundry sand can be very substantially reduced if enough relatively cool sand is added to the hot sand so that a sand to metal ratio of 12 to 1 or greater is obtained. Preferably the relatively cool sand is added at or near the shakeout since this is the point at which this type of air pollution begins. Of course, the relatively cool sand could be added later but then all of the benefits of this process would not be obtained. It also may be desirable to add at the shakeout whatever additives are to be added to reconstitute the new sand. Additions at such point also serve to reduce air pollution.

At a sand to metal ratio of 12 to 1 or above, the mixture of hot and cool sand at the shakeout is relatively cool and relatively moist. Thus, a major source of pollution is contained. While sand to metal ratios of less than 12 to 1 may be produced at the shakeout, the lower the ratio the more pollution will occur.

A process developed by F.O. Ekman (10) involves controlling dust and fume emissions from a cupola furnace of the type including a stack having a charging opening therein spaced upwardly of a blast gas tuyère system adjacent the lower end of the stack. The method includes the steps of removing a portion of the gases flowing in the stack at a takeoff level intermediate the tuyere system and the charging opening. The portion of the gases removed are passed through a dust collector for treatment to remove dust, fume and contaminants therefrom. Inert, noncombustible gas is introduced into the stack forming a blanket between the charging opening and the takeoff level so that outside air infiltrating into the stack through the charging opening is greatly reduced, thereby preventing the possibility of explosion in the stack by rapid oxidation of the gases generated in the cupola.

Figure 58 shows the arrangement of apparatus which may be employed in the conduct of this process. Referring now to the drawing, there is illustrated a typical cupola furnace **10** including a circular base **12** supported from the floor or other surface on a plurality of support legs **14**. The furnace includes an outer steel shell **16** which is lined with an acid or basic refractory lining material **18**, and the shell includes a cylindrical lower portion and an upper portion tapered to slope inwardly and progressively smaller in diameter toward the top or upper end. The refractory lining material **18** is formed to fit the tapering contour of the shell, and together the shell and lining form an elongated vertically extending tapering cupola stack **20**.

The base structure **12** supports a sand bottom **22** for the furnace which is sloped downwardly towards one side to a radial tap hole **24** for removing molten metal **34** from the cupola. A removable plug **26** is inserted to close the tap opening when molten metal **34** is not being withdrawn, and the sloping sand bottom **22** extends outwardly beyond the cylindrical shell **16** and is provided with a pouring lip **28** at the outer end to guide the molten metal into a hopper or bucket when the furnace is tapped.

Spaced upwardly above the bottom **22** is provided an annular wind chamber **30** for supplying heated blast air to the furnace for blowing the charge or burden therein. A plurality of radially spaced tuyère openings **32** are formed in the furnace wall above the base and these tuyère openings **32** are connected to the wind chamber by a plurality of elbow-like windpipes **36**.

In order to charge the cupola stack with alternate layers of pig iron and/or scrap iron **40** on one hand, and coke and limestone **42** on the other, the stack **20** is provided with an enlarged charging opening **44** spaced some distance above the lower end of the furnace. The bottom of the opening is approximately level with a charging floor structure **46** and charging of the cupola is accomplished by the use of a movable charging bucket **48** having a discharge opening **50** in the bottom controlled by a movable gate valve **52**. The charging hopper is supported from a traveling rail **54** carried on a pair of roller brackets **56** which are in turn supported on a suitable track member **58** mounted above the charging floor. As the heating and refining process in the cupola furnace is carried on, additional layers of pig and/or scrap iron and limestone and coke are introduced through the charging opening **44** by means of the charging bucket **48**, and a pool of molten metal **34** collects at the bottom of the cupola.

In order to supply heated and pressurized air to the wind chamber **30** for blowing the charge in the cupola through the tuyère openings **32**, the system includes an air blast heater generally indicated at **60**, which is capable of supplying a large volume of blast air at elevated temperatures in the range of 400° to 1500°F. and at pressures in the range of 1 to 3 psi. The air blast heater includes a housing **62** supported on a plurality of legs **64** and a suitable, tubular, heat exchanger **66** is mounted in the housing. The lower end of the heat exchanger extends downwardly through the bottom wall of the housing for connection with a blower compressor which is capable of delivering the required volume of airflow at the desired pressure. The upper or outlet end of the heat exchanger extends outwardly through a sidewall of the housing and is connected to the wind chamber **30** by a supply duct **70**.

As the blast air passes through the heat exchanger **66**, it is heated by hot gases flowing

### FIGURE 58: APPARATUS FOR CONTROLLING FUME AND DUST EMISSIONS FROM CUPOLA FURNACES

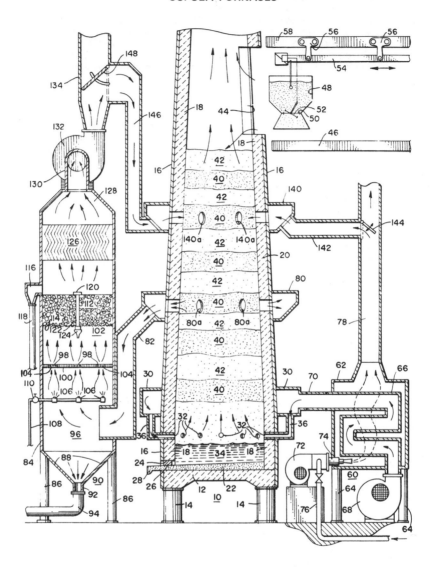

Source: F.O. Ekman; U.S. Patent 3,499,264; March 10, 1970

around the outside of the heat exchanger, and these heated gases are produced in a burner assembly including a blower **72** and a nozzle **74** which has an outlet extending in to the lower end of heater housing **62**. Natural gas, oil, or other fuel is supplied to the burner assembly through a supply line **76**, and suitable thermostatic controls and regulator valves are provided for regulating and controlling the temperature of the blast air as it leaves the heat exchanger **66** into the supply duct **70**. The hot products of combustion generated by the burner assembly pass through the heater housing **62** and around the heat exchanger **66** thereby heating the blast air flowing internally through the heat exchanger and the products of combustion leave the heater housing through a suitable exhaust or outlet stack **78**. The heated gas passing upwardly through the exhaust stack **78** of the air blast heater contains a large proportion of carbon dioxide, water, small amounts of carbon monoxide, and other constituents and the gas can be generally classified as inert because complete combustion is obtained by the use of modern burners available.

The cupola furnace **10** is provided with an annular, bustle offtake chamber **80** mounted on the stack **20** and spaced upwardly of the lower end between the wind chamber **30** and the charging opening **44**. The bustle offtake is in communication with the interior of the stack **20** through a plurality of radially spaced openings **80a** and is connected to a separate dust collector **90** closely adjacent to the cupola furnace through a downtake duct **82** having its lower end in communication with an inlet chamber **96** in the lower end of an upright housing **84** of the dust collector. The dust collector **90** may be of a known construction, such as shown in U.S. Patent 2,645,304 and preferably is of the wet scrubber, high efficiency type, such as that shown in U.S. Patent 3,348,825.

Briefly, the dust collector **90** includes a rectangular or cylindrical upstanding housing **84** supported by a plurality of legs **86** at the base. The bottom of the housing is closed by a plurality of inwardly and downwardly sloping sidewalls **88** forming a liquid sump for collecting contaminated scrubbing liquid used in the dust collecting process. The lower end of the sump is connected by a suitable fitting **92** to a drainpipe **94** for carrying away the contaminated scrubbing liquid to a treatment station or for discharge to a sewer or settling pond.

Immediately above the sump in the housing **84** is the inlet chamber **96** in communication with the lower end of downtake duct **82** from which the contaminated gases from the bustle offtake **80** are supplied. The incoming gases flow upwardly in the housing **84** and pass through one or more openings **98** of relatively small cross-sectional area which are formed in a horizontal dividing wall **100** separating the inlet chamber **96** from a filter bed containing chamber **102**. As the gases flow through the openings **98** or nozzles, the gas velocity is substantially increased, as indicated by the converging and diverging arrows, and each of the openings **98** is provided with a plurality of spherical filter elements **104** therein, which further reduce the effective flow cross section and further increase the velocity of the gas to provide more efficient scrubbing as more fully described in U.S. Patent 3,348,825.

The incoming gases flowing upwardly toward the nozzle openings **98** and filter elements **104** are prewetted by means of a plurality of spray nozzles **106** which introduce finely divided droplets or sprays of contaminant collecting scrubbing liquid into the gas flow to wet and collect the fume, dust, and other contaminants in the gas. Scrubbing liquid for the nozzles **106** is supplied by a piping manifold **108**, and the amount of liquid supplied is controlled by a suitable valve **110**.

The wetted gases flowing upwardly through the nozzles or openings **98** in the wall **100** pass upwardly in the filter chamber **102** which contains a large filter bed **112** containing a plurality of spherical filter elements which are supported on a screen or mesh **114**. An intense foaming action takes place within the filter bed **112** and dust, fume, and other contaminant particles in the gases are wetted and collected in the scrubbing liquid. Scrubbing liquid and foam reaching the upper level of the filter bed is removed from the housing **84** by a weir system **116** which is connected to a suitable drain or water treatment system by a drain line **118**, and thus the height of the foam bed formed above the filter bed **112** is limited by draining off excess liquid and foam. Also, a suitable liquid drain weir **120** may be provided to remove excess liquid from the center of the filter bed and discharge the liquid downwardly through a pipe **122** and rubber, sock-type valve **124** into the lower portion of the filter chamber **102** beneath the filter bed **112**.

By the time the gases pass through the upper level of the filter bed, almost all of the fume, dust, and other contaminant particles in the gas have been removed therefrom and are collected in the scrubbing liquid which is drained off. The cleansed gases continue flowing upwardly and pass through a demisting vane assembly **126** comprising a plurality of closely spaced zigzag vanes. The zigzag vanes cause abrupt reversal of flow direction of the gas and drops of liquid remaining in the gas collect on the vanes and eventually gravitate back into the filter bed. The demisting vane assembly **126** provides the final step in cleansing the gas which then flows upwardly into an outlet chamber **128** at the upper end of the housing **84**. The cleansed gases pass from the housing into a duct **130** connected to an exhaust blower **132** which discharges the gas upwardly through an outlet or exhaust stack **134**. While the cleansed gas passing through the stack **134** may contain considerable amounts of carbon monoxide, for most purposes the gas can be considered as inert because of the relatively high moisture content and relatively low temperature thereof.

The cupola furnace **10** is also provided with an annular, bustle intake chamber **140** that is mounted on the stack between the offtake bustle **80** and the charging opening **44**. Inert or noncombustible gas is supplied to the cupola stack **20** from the intake bustle **140** through a plurality of radially spaced openings **140a**, and this gas serves to blanket or block any appreciable amount of outside air from infiltrating into the stack through the charging opening **44** and flowing downwardly to mix with the combustible gases generated at the lower end of the stack and flowing upwardly toward the charging opening. The inert or noncombustible gas introduced into the stack through the intake bustle **140** helps to divert the gases flowing upwardly in the stack from the lower end of the furnace into the bustle offtake **80** connected to the dust collector **90**.

Some of the gas introduced into the stack through the air blast from the tuyère **32** and the gaseous products generated by the refining process in the lower portion of the cupola may flow upwardly through the blanket of inert gas and past the intake bustle **140** for eventual discharge out the top of the cupola stack **20**. However, most of the contaminants and harmful dust and fume emissions in the gas have been taken off through the offtake bustle **80** and have been treated in the dust collector **90** before discharge from the exhaust stack **134** so that only a relatively small volume of pollutants are discharged from the system into the atmosphere.

Two sources of noncombustible or inert gas are available to the intake bustle **140** for introduction into the stack through the openings **140a**. One source of gas is provided by the heated products of combustion from the air blast heater **60**, and a supply duct **142** is provided to interconnect the exhaust stack **78** of the air blast heater with the intake bustle **140**. A suitable control damper **144** is provided in the stack **78** to regulate the amount of hot gas diverted from the stack for flow into the intake bustle **140**.

Because the gaseous products from the stack **78** are for the most part completely oxidized and contain relatively low percentages of carbon monoxide and oxygen, these gases, when diverted into the cupola stack through the intake bustle, are very effective in forming a blanket of inert gases in the stack for preventing explosive conditions from occurring. Such explosive conditions could be present if excessive outside air enters into the stack through the charging opening **44** and flows downwardly in the stack to mix with the unoxidized gases produced in the cupola process.

The relatively inert, heated products of combustion from the air blast heater exhaust stack **78** flow from the intake bustle openings **140a** in a slight downdraft in the stack and for the most part are taken off in the offtake bustle **80** along with the major portion of gases generated in the refining process taking place in the cupola. This mixture is treated in the dust collector **90** and, accordingly, pollutant emission problems from the discharge of gases from the system are reduced. The damper **144** is adjusted as desired to regulate the amount of gas diverted from the stack **78** into the cupola stack **20** for providing the necessary blanket or barrier therein to prevent excessive outside air from infiltrating through the charging opening **44** and flowing downwardly in the stack.

An alternate source of inert or noncombustible gas available for introduction through the intake bustle **140** is provided by using the treated exhaust gases from the outlet stack **134** of the dust collector **90**. These gases are extremely well suited for use in the blanketing process because during their treatment in the dust collector the gas temperature is reduced significantly and the moisture content is increased significantly by the liquid introduced in the cleaning process. The treated gas in the exhaust stack **134** of the dust collector **90** is almost free of contaminating impurities, and if not used as a source of inert gas instead of the exhaust gas from the air blast heater, can be discharged into the atmosphere with relatively little or no pollution problems. Moreover, because the exhaust gas from the dust collector contains cooled carbon monoxide, any excess gas introduced into the stack by way of the intake bustle **140** will tend to migrate up through the burden or charge in the stack and burn in a controlled manner without danger of explosion.

A duct **146** is interconnected between the bustle intake **140** and the exhaust stack **134** of the dust collector, and a suitable control damper **148** is provided to regulate the amount of gases diverted from the stack into the intake bustle **140**. If desired, gas from the dust collector stack **134** and from the air blast heater exhaust stack **78** can be supplied to the intake bustle **140**, and the desired proportions can be regulated by adjusting the dampers **144** and **148**.

In theory, a small volume of inert gas introduced through the intake bustle **140**, as compared to the volume of air blast input through the tuyères **32** may cause a slight downdraft in the stack between the bustle intake **140** and the bustle offtake **80** so that the inert gas forms an effective buffer zone in the stack **20**, resulting in the fact that almost none of the gas in the stack containing large amounts of impurities can escape upwardly and be discharged at the upper end of the cupola stack without having been treated to remove the objectionable contaminants therefrom in the dust collector **90**.

A process developed by O.M. Arnold, V.W. Hanson, R.M. Jamison, N.J. Panzica, and E. Umbricht (11) provides an air pollution control system for cleaning contaminated gases (including the hot gases such as issue from cupola furnaces) which preconditions the contaminated gases by saturating with a selected liquid (usually water), cooling, and condensing the resulting vapor causing it to adsorb on the solid contaminants and thereby to form aggregations of droplet-borne contaminants which having a larger size either fall out of the gases or are more easily removed by the conventional gas washer to which they are then presented.

The saturating is accomplished by a high velocity, intense spray which is generated to uniformly treat the contaminated gases with a cleaning, as well as a saturating, spray pattern. Ambient air may be introduced beyond the saturating spray, or the saturated gases may be cooled, to aid the condensation. Afterburners may be introduced before the spray to burn out many of the contaminants and simultaneously activate the remaining particles to cooperate with the foregoing improvement to result in even greater aggregation for facilitating contaminant removal.

### References

(1) Kearney, A.T. & Co., "Air Pollution Aspects of the Iron Foundry Industry," *Report PB 204,712,* Springfield, Va., National Technical Information Service (February 1971).
(2) Kearney, A.T. & Co., "Systems Analyses of Emissions and Emissions Control in the Iron Foundry Industry," *Report PB 198,348* (February 1971).
(3) *Ibid. Report PB 198,349.*
(4) *Ibid. Report PB 198,350.*
(5) Public Health Service, U.S. Department of Health, Education and Welfare, "Economic Impact of Air Pollution Controls on the Gray Iron Foundry Industry," *Report No. AP-74*, Raleigh, N.C., National Air Pollution Control Administration (November 1970).
(6) Kearney, A.T. & Co., "Study of Economic Impacts of Pollution Control on the Iron Foundry Industry," *Report PB 207,147* (November 30, 1971).
(7) *Ibid. Report PB 207,148.*
(8) *Ibid. Report PB 207,149.*
(9) Schumacher, J.S.; U.S. Patent 3,646,987; March 7, 1972; assigned to International Minerals and Chemical Corporation.
(10) Ekman, F.O.; U.S. Patent 3,499,264; March 10, 1970; assigned to National Dust Collector Corp.

(11) Arnold, O.M., Hanson, V.W., Jamison, R.M., Panzica, N.J., and Umbricht, E.; U.S. Patent 3,475,881; November 4, 1969; assigned to Ajem Laboratories Inc.

## FRUIT PROCESSING INDUSTRY EFFLUENTS

The reader of this handbook is referred to the volume on waste disposal control in the fruit and vegetable processing industry by H.R. Jones (1).

### References

(1) Jones, H.R., *Waste Disposal Control in the Fruit and Vegetable Industry*, Park Ridge, New Jersey, Noyes Data Corporation (1973).

## GLYCOLS

### Removal from Wastewaters

The removal of glycols from wastewater has been described by M.A. Zeitoun and W.F. McIlhenny (1) of the Texas Division of The Dow Chemical Co., Freeport, Texas.

A number of extremely useful and widely produced compounds are produced by the alkaline hydrolysis of chlorohydrins. Ethylene and propylene glycols and glycerin may be produced in this manner. The brine wastewater resulting from the production of the glycols is characterized by a high salt content (8 to 10% NaCl), excess alkalinity, and the presence of several organic compounds.

Several processes were examined in the laboratory to determine the usefulness for the treatment of the brine wastewater. Solvent extraction of the glycol wastewater with secondary or tertiary amines produces a raffinate that is salt-saturated and low in glycol and a product enriched in the glycol and nearly free of salt. The required large solvent-to-feed ratio, the requirement for near freezing temperatures, and the high reflux to produce a pure product, make solvent extraction, as a treatment method, uneconomical.

Adsorption of glycols on activated carbon was found to be unfeasible because of the low capacity of carbon for the glycols. Commercially available cellulose acetate membranes with low salt rejection were found to exhibit a high pore flow of propylene glycol and were therefore unable to significantly separate the salt and the glycol.

Biological oxidation of propylene glycol wastewater in batch and continuous laboratory units gave a total oxygen demand (TOD) removal efficiency of 86 to 88% at a residence time of 12 hours.

An activated sludge pilot plant with a feed rate of 0.5 gpm was constructed and successfully operated as a completely mixed aerator on an equalized propylene glycol wastewater. Removal efficiencies of over 90% at a retention time of 8.0 hours at loadings of 2.0 to 3.0 pounds total oxygen demand (TOD) per pound of mixed liquor volatile suspended solids (MLVSS) per day were obtained. The effluent quality was improved by operation at recycle ratios of 24 to 40%.

The operational and design parameters determined from the pilot plant operation were used to design an activated sludge plant to treat 6 MGD of wastewater resulting from the production of 1.2 million pounds per day of propylene glycol at an estimated operating cost of 3.3¢ per pound of TOD removed or less than 0.2¢ per pound of propylene glycol produced. The total fixed capital requirement for a 6 MGD plant was estimated to be 1.4 million dollars.

References

(1) Zeitoun, M.A. and McIlhenny, W.F., "Treatment of Wastewater From the Production of Polyhydric Organics," *Report 12020 EEQ-10/71*, Washington, D.C., U.S. Environmental Protection Agency (October 1971).

# GREASE

## Removal from Air

A process developed by N.F. Costarella and A.A. Giuffre (1) provides a ventilating hood and duct structure for stoves wherein a baffle and water spray arrangement within the hood forces the smoke to pass through successive water curtains. Grease and other impurities are drained from the hood structure through a rectangular trough which overlies the fire area.

References

(1) Costarella, N.F. and Giuffre, A.A.; U.S. Patent 3,628,311; December 21, 1971; assigned to Nino's Inc.

# HYDRAZINE

## Removal from Air

A process developed by W.M. Gardner and W.H. Revoir (1) provides a special filter for removing contaminants of hydrazine and its organic derivatives from fluids. The filter operates by having at least two distinctly different porous materials which are known for their sorbing properties arranged in layers or in a mixed relation. In the layer method the upstream layer is impregnated with a strong oxidizing agent while the downstream layer remains unimpregnated. The vapors of hydrazine or organic derivatives of hydrazine on first entering the filter react with the oxidizing agent to form volatile and nonvolatile substances. The nonvolatile substances are sorbed by the impregnated material and the volatile substances that pass through the impregnated material are then sorbed by the nonimpregnated material.

References

(1) Gardner, W.M. and Revoir, W.H.; U.S. Patent 3,489,507; January 13, 1970; assigned to American Optical Corporation.

# HYDROCARBONS

## Hydrocarbon Removal from Air

Hydrocarbon removal from the air may involve recovery of the hydrocarbon by absorption or adsorption. Alternatively, it may involve destruction by combustion in a flare stack or an incinerator.

A process developed by E.E. Frost, H.C. Hottel and J.P. Longwell (1) provides a method and apparatus for flaring combustible gaseous materials wherein combustion and burning of such materials is accomplished substantially without air pollution by smoke. The process further involves flaring such materials substantially without producing a flame visible in any direction except from substantially directly above the zone of combustion.

In the refining of petroleum oils in many industrial operations, and particularly in the refining of petroleum oils, large volumes of combustible gaseous materials are produced. Some of these materials have no real economic value as fuel or otherwise and must be disposed of. Additional gaseous materials result from upset conditions in the normal operation of a refinery wherein gases which ordinarily might be subjected to further processing in order to obtain valuable products must be vented in order to avoid the occurrence of dangerously abnormal pressures in operating equipment.

The total volume of all such accumulations of gases in normal refinery operation, and the hazardous nature thereof make it impossible for such gases to be discharged directly into the atmosphere. In practice, therefore, it is customary to burn such gases as they are discharged from an exhaust or flare stack at a considerable distance above the ground, although on occasion for the sake of economy, burning may be accomplished in relatively low, large diameter enclosures or flare structures.

In the burning of such gaseous materials in the conventional manner, difficulty is experienced in providing access of sufficient quantities of air to accomplish complete combustion. As a result, whether burned in elevated flare stack structures or otherwise, the burning is usually accomplished by the formation of large quantities of heavy, sooty smoke. Whether this smoke is released at higher or lower levels, an undesirable pollution of the atmosphere is created in the vicinity of the burning operation. An additional objection to conventional burning methods and the means provided therefor is to be found in the luminosity of the flame produced. Both luminosity and smoke primarily are due to the presence of carbon particles which in the area of the flame are incandescent, and upon cooling at some distance from the point of ignition, form dense, sooty clouds of smoke.

In general, the prior art has suggested two methods of flaring and burning such gaseous materials. According to the one method, gas is discharged from the open upper end of a flare stack, without substantial premixing of air therewith, and ignited and burned at the point of discharge. Combustion of the discharged gases thus takes place in the ambient atmospheric air rather than as a combustible mixture of air and gas. Although a portion of the gas may burn immediately, a deficiency of oxygen induces carbon formation and especially where olefinic constituents are present. In addition, it has been found that the heat generated by combustion of a portion of the gas stream may cause cracking of the unburned portions to form additional olefinic and paraffinic materials as well as carbon and hydrogen. At the same time, some molecules may polymerize to form long chain hydrocarbons. The combustion or partial combustion of such compounds increases carbon formation particularly in the presence of hydrogen and a reducing atmosphere.

The other most conventional method of burning combustible gaseous materials involves the adaptation of the principles of Bunsen or other similar gas burners. The use of flare stacks operated according to such principles has been largely unsuccessful for the reason that such type of operation requires that the gas be supplied to the burners at a substantially constant rate and high flow velocity in order to prevent reverse flow of the combustible gas through the air induction inlets provided. Furthermore, as the gases to be burned are usually exhaust gases from process equipment which must be released without substantial back pressure, they are usually discharged at pressures close to atmospheric pressure, and it would be impractical and uneconomic to repressurize such gases once they had been released from the process pressures.

Where air dilution or premixing of the gas stream with a forced air supply has been attempted, it has been found that the volume of air required plus the volume of combustible gas to be burned would necessitate employment of flare stacks of excessively large diameter. For example, whereas a conduit six inches in diameter would be adequate as a flare stack according to conventional procedures, to obtain an optimum flame condition by dilution of the gas with air prior to discharge of the gas from the stack, in the burning of each 1,000 cubic feet of gas, provision would have to be made for the stack to handle about 6,000 cubic feet of air per minute in addition. With such volumes of air and gas, the stack would have to be about 16 inches in diameter. Inasmuch as the volumes of gas contemplated for disposal according to this process may be in excess of 2,000 cubic feet per minute, combustion air at the rate of at least 12,000 cubic feet per minute would have to be handled by the stack in addition. Further, any system wherein the disposable gases were to be diluted with combustion air prior to flaring of the mixture would require an elaborate and expensive control system, and accordingly be most uneconomical.

The process involves dividing an initial, confied stream of hydrocarbon-bearing gaseous materials into a plurality of confined, substantially coplanar, laterally spaced, parallel streams of lesser cross section than the initial stream, discharging the streams of lesser cross section into an ambient atmosphere of air as a plurality of jetted streams of further reduced cross section, and in a spaced pattern in a substantially common plane. The gaseous materials thus discharged are dispersed to form an unconfined turbulent mixture of materials in the ambient air in the immediate vicinity of the jet streams. The mixture is ignited in the vicinity of the jet streams, whereby the gaseous mixture is at least partially consumed in an initial, unconfined combustion zone in the immediate vicinity of the jet streams, inducing upward convection gas flow currents by the heat of combustion from the initial combustion zone.

The combustion products, unconsumed gaseous materials, and an excess of combustion air from the ambient atmosphere of the initial combustion zone are passed upwardly into a second, confined, vertically extending combustion zone in open communication at each end with ambient atmospheric air. The discharge of the gaseous materials is regulated to a flow rate in the range between about 20,000 and 2,000 cubic feet per day for each square foot of the cross-sectional area of the secondary combustion zone. The combustion of the discharged gaseous materials is substantially completed within the vertical limits of the second zone.

A process developed by W.W. Ford and A.M.L. Kube (2) relates to a system for the recovery of valuable constituents in gases normally sent to the flare line. It has been proposed to return such gases to a low pressure point in the system and to provide a liquid trap in the flare line to prevent venting of gases to the flare line under normal operation conditions.

The gases in a flare line system usually represent a heterogeneous mixture of gases as a result of their having originated from a variety of sources, and it is not always desirable to introduce such mixture to a low pressure point in the system. Furthermore, a liquid trap has not always proven successful because a water trap has a tendency to freeze during unexpected cold weather and an oil trap has a tendency to accumulate condensable hydrocarbon materials within itself and to pass lugs of liquid to the flare as a result of such increase in volume through condensation of liquefied material.

The process provides for separating the condensable materials present in a flare line from the noncondensable materials, sending the noncondensable materials to be used as fuel, and processing the condensable materials for further use. Figure 59 shows the essential pieces of equipment which may be used in the conduct of this process.

Referring now to the drawing, gases enter the flare line **10** from a gathering system (not shown), via conduit **11** and pass through an induction line **12** to beneath the surface of the liquid **13** in tank **14** so as to produce a pressure of 2 or 3 pounds in flare line **10**. Gases which escape through the liquid trap in tank **14** pass through line **15** to the flare via conduit **10a**. Liquid trap **16** is positioned in flare line **10** and is connected in parallel with the liquid trap of tank **14** and contains liquid **17**. The liquid head in trap **16** will normally be maintained at a level to produce a pressure 2 or 3 pounds higher than that produced by the liquid head in tank **14**. A scrubber **7** can be placed in conduit **10** downstream from liquid trap **16** to recover liquid blown from trap **16** by a surge of high pressure in flare conduit **10**. The liquid is returned to trap **16** via conduit **8** containing check valve **9** if scrubber **7** is elevated sufficiently for gravity flow. Check valve **9** can be replaced by a pump (not shown) if scrubber **7** is not elevated sufficiently for gravity flow.

FIGURE 59:  SYSTEM FOR RECOVERY OF HYDROCARBONS FROM FLARE GASES

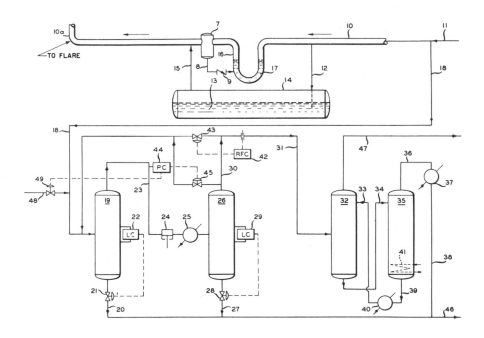

Source:  W.W. Ford and A.M.L. Kube; U.S. Patent 2,978,063; April 4, 1961

A branch line **18** conducts gas from flare line **10** to a first liquid-gas separator **19** where any liquid which is present in the gas accumulates in vessel **19** and is removed by means of conduit **20** through motor valve **21**, which is actuated by liquid level control **22**.  Gases removed from vessel **19** via conduit **23** are compressed by compressor **24**, cooled by a cooler **25** and passed into a second liquid-gas separator **26**.  Liquid is removed from vessel **26** via conduit **27** through motor valve **28** which is controlled by liquid level control **29**.  Gas from vessel **26** passes through conduits **30**, **31** to absorber **32** where it is contacted with a liquid absorbent introduced via conduit **33**.

The rich absorbent passes from absorber **32** through conduit **34** to stripper **35** from whence desorbed materials are removed via conduit **36**, cooled in cooler **37** and removed via conduit **38**.  The lean solvent or absorbent is removed from stripper **35** via conduit **39**, is cooled in cooler **40** and returned to absorber **32** via conduit **33**.  Heat is added to stripping vessel **35** by means of a heater **41**, which can be any conventional heating means, such as steam, electricity and the like.

A recording flow controller **42** operatively connected to conduit **31** actuates a normally open motor valve **43** so as to limit the flow to absorber **32** to a predetermined maximum value.  A pressure controller **44** operatively connected to conduit **23** actuates a normally closed motor valve **45** so as to open the valve when a pressure in conduit **23** falls below a predetermined value so as to recycle compressed gas back to the compressor inlet at times when the pressure in the flare line system is abnormally low.  The condensed materials recovered from vessels **19**, **26** and **35** are removed via a common conduit **46**.  Non-condensed gases are passed to a fuel supply system via conduit **47**.

Positive control of compressor suction pressure can be obtained by supplying makeup fuel gas from an outside source via conduit **48**. If fuel gas is supplied via conduit **48**, pressure controller **44** will actuate normally closed motor valve **49** instead of, or in addition to, motor valve **45**.

The water level in vessel **14** can be maintained by conventional means such as a water inlet valve operated by a liquid level control means. Heating means can be supplied to heat the water in vessel **14** during cold weather if such is desired. The head of water in vessel **14** is selected so that a back pressure of from about 1 to 3 psig is maintained in the flare system **10** and the liquid head maintained in liquid trap **16** is usually about 1 to 2 pounds higher than that of the liquid trap of vessel **14**. The gas in the conduit **31** is compressed to some preselected value, such as about 68 psig, and the recording flow controller **42** is set at a preselected value, for example, about 3 mm. standard cubic feet per day. The pressure controller **44** is set to open valve **45** at some predetermined minimum pressure in conduit **23**, for example, about ½ psig.

From the above it can be seen that all of the noncondensable gases which enter the flare line system are used for fuel under normal operation and that the condensable materials are recovered for further processing. When gases pass into the flare line system at an abnormally high rate, the excess over that which can be used as fuel is passed to the flare. By this practice, the flare line is always capable of passing large quantities of gas at relatively low pressure and at the same time reverse flow of gas through the flare line is substantially completely prevented.

If desired, the pressure controller **44** can be operatively connected to a means for shutting down the compressor **24** if the pressure in conduit **23** falls to a predetermined low value which might result in harm to the compressor. Thus, the pressure controller can be caused to actuate a switch which opens the operating electrical circuit to the motor operating compressor **24** if an electric motor is used, or can be caused to close a valve in a steam line if steam power is used to operate the compressor.

A process developed by R. Bloxham (3) relates to the depollution of air by oxidation of the contaminants in the air.

Numerous attempts have already been made to alleviate the atmospheric pollution problem created when exhaust gases from industrial processes are discharged into the atmosphere. The main attempts have been along the line of what is considered a conventional afterburner, through which the exhaust gases of the industrial processes are passed before being discharged into the atmosphere. Generally, a conventional afterburner consists of a heating chamber which is usually heated by direct fire. The entire mass is usually raised to about 1500° to 1600°F. and the polluted exhaust effluent of the industrial process is passed through the heating chamber.

While this is a very effective method for oxidation of contaminants, it is exceedingly wasteful of fuel. For example, in a process which requires a working temperature of 200°F. in order to produce the desired final product, it would require five to seven times the fuel costs of the process to operate the afterburner and depollute the exhaust gases coming from the process. This precludes the use of conventional afterburners for most industrial and commercial applications.

The other principal method developed for control of atmospheric pollution is a type of equipment employing catalytic combustion. This type of equipment is very expensive to manufacture and has the serious disadvantage of its catalytic elements being susceptible to contamination, especially from silicones, making them ineffective through this so-called "poisoning." This requires periodic and all too frequent rejuvenation or replacement. Moreover, the operation of catalytic combustion equipment requires that the effluent gases be raised to at least 600°F. for the catalytic reaction to start. Therefore, to make the equipment reasonably efficient, an expensive heat exchanger is required.

Both of these methods, especially the conventional afterburner method, create nitrogen oxides which are considered a serious air pollutant in themselves. This is due to the fact that the creation of nitrogen oxide compounds is in direct relation to the gas temperature and both methods employ relatively high gas temperatures. The depollution oxidation process described below is highly efficient because it has the feature of permitting heating of only the contaminants and does not require heating of the ambient atmosphere.

The selective oxidation by heating is accomplished by bombarding the contaminated gas in the presence of oxygen with, in the preferred embodiment, infrared radiation. The contaminating smoke particles, hydrocarbons, carbon monoxide, and the like, absorb sufficient radiant heat energy to raise them to their ignition temperature whereby they are oxidized in the presence of excess oxygen, which is of course present in air. The usual predominant gases of air, that is, oxygen and nitrogen, do not absorb significant amounts of radiant energy. Therefore, the average exit temperature of gases subjected to this process is relatively low and the heat quantity required to initiate the oxidation reaction of the contaminants is extremely low.

The specific apparatus used is an open ended, annular, elongated furnace having input and output ends permitting a flow of air therethrough, the furnace having on its inner surface an infrared emitting material, means for heating the material, and a plurality of transverse spaced apart baffles disposed proximate to the material to substantially reduce the longitudinal velocity of air flowing through the furnace member in the vicinity of the emitting material, thereby to minimize the cooling effect of longitudinally flowing air on the material.

A device developed by R.D. Reed and J.S. Zink (4) provides a burner assembly for the combustion of flared gases at the upper end of a stack wherein steam and air are mixed with the gas in the combustion zone and turbulence is developed promoting mixing of the steam and air with the gases to provide substantially smokeless burning.

A process developed by E.W. Merrill (5) is one in which organic pollutants, such as hydrocarbons, can be removed from air by mixing the air with an aerosol of particles of water containing a surfactant which presents an oleophilic surface on the particles. The air is preferably humidified to 100% RH prior to mixing it with the aerosol, so that evaporation of the aerosol particles will not occur. The aerosol particles are effective by absorbing organic pollutants into their oleophilic surfaces, and removal of the particles by filtration or electrostatic precipitation leaves the air substantially purified of such pollutants.

A process developed by L.C. Hardison (6) is one for effecting the continuous catalytic oxidation of a waste gas stream by providing for the cocurrent flow of subdivided catalyst particles therewith upwardly in the lower part of a reactor-stack unit, separating the catalyst particles from the contacted gas stream at the upper end of the reactor section, and returning them to a collecting and flow regulating means for reintroduction into a waste gas inlet zone to the unit. Burner means, mounted in combination with the inlet zone, is utilized to provide a hot gas stream to contact the recirculated catalyst particles and thus heat and regenerate them for use in the continuous catalytic oxidation system.

A process developed by V. Jasinsky and A.T. Upfold (7) is one in which flare stack smoke is reduced or eliminated by controlling the temperature of the flare. The temperature is sensed by a sensing element, e.g., one or more thermocouples connected in parallel and located in the vicinity of the flare; the sensing element produces a signal which directly or indirectly activates a steam control valve to adjust the flow of steam to the tip of the flare stack so as to raise or lower the flare temperature to the temperature of optimum combustion of flare stack effluent.

## Hydrocarbon Separation from Plant Wastewaters

Just as with hydrocarbons in air, the removal of hydrocarbons from water may involve separation techniques or destruction techniques.

A process developed by J.B. Davis and R.L. Raymond (8) relates to the removal of oil from industrial wastes by microbial action.

It is well known that microbes are used in treatment of waste products, as, for example, in the disposal of sewage. Use of activated sludge, aeration, diffusers and other accelerating means are well known. Further, oily products can be treated to remove objectionable sulfur compounds using microbial materials such as hydrogenase. Despite the knowledge on hand about treatment of waste products, the treatment of oily sludge or of industrial wastes containing hydrocarbons requires attention. One of the reasons for this is that many microbes will feed first upon the readily digestible materials such as carbohydrates and proteins and crowd out hydrocarbon oxidizers. Another reason is that the microbes are acting in an aqueous medium in which the oily material is immiscible.

Even though microbial action may produce detergents and even though aeration is usually used, the desired and necessary contact of the microbes with the hydrocarbons to be removed previously has not been accomplished effectively. Also, while there may be present a myriad of microbes of various kinds that are active on nonoily wastes, the waste may be devoid of microbes that utilize hydrocarbons, and hence there results no activity on the oily matter in the wastes. Therefore, the desire of industry to remove oily materials from its waste has been hampered. Even though the amounts of oil, for example, in refinery or in oil field wastewaters are as small as 100 ppm, it is desired to remove that portion.

The process described here is carried out by intimately mixing the oily waste to be treated with a microbial culture or sludge enriched with microbes active on hydrocarbons and maintaining the resultant mixture in continual contact with air or oxygen. The oily substrate, an aqueous medium containing water-insoluble oily or hydrocarbon materials, is agitated to decrease the oily globules to the smallest practical size without addition of emulsifiers. Since the amount of oil is very small, this reduction into globules occurs quite readily for there is a great deal of water around to keep an oil droplet once formed from getting together with another. Further, the absorption of the oil by the microbial cells assists in this size reduction.

Thus, moderate agitation obtained using air spargers, aeration through a perforated pipe under several pounds of pressure, or by a mechanical impeller is generally adequate. Prior to or during the agitation an acclimated microbial flora grown on oily wastes or containing hydrocarbon oxidizers is added to the aqueous system. Thus, the mass is rendered as homogeneous as possible, small oil droplets are formed, the surface of oil exposed to microbial action is vastly increased, oxygen is available, and the microbial action at the oil-water interface is vastly increased with resultant oxidation and adsorption of the wanted materials.

The oils and hydrocarbons are caused to volatilize in part, to be oxidized in part to carbon dioxide and other harmless materials, to be converted to microbial cell materials such as protein and lipids, and to be adsorbed unchanged on the cells. While some extracellular material may be produced, most of the hydrocarbonaceous material is converted to intracellular products. These are contained in the sediment as part of the cells while any solid extracellular products are precipitated by their own weight but being a part of the sediment all of which is easily removed. Periodically, portions fo the sludge are removed for charging or recharging vats that are to receive untreated wastes for processing.

Since the continuous method affords a more ready maintenance of the concentrate of oxidizers than does the batch process, it is preferred. These acclimated sludges are rich in microbes effective in acting on oily materials. While there may be present microbes which do not oxidize hydrocarbons, there are generally little or no other foods in the hydrocarbon wastes treated by this process. Thus, the hydrocarbon oxidizers used in the process have very little competition for minerals, food or oxygen. Fast disposal results.

Figure 60 is a block flow diagram showing the essentials of the process. In the figure it can be seen that the influent is sent to an ordinary API separator, which removes the large majority of oil and is, as known, a normal step in the recovery of oils or hydrocarbon matter at refineries or in oil fields. These separators, while very effective, do not recover the entire amount of hydrocarbons in the influent, a very small amount being discharged which is in the order of 100 parts per million. It is desired to prevent effluents containing even that small amount from being discharged into rivers, and this microbial process is designed to remove these small amounts quantitatively.

From the separator the influent will be charged into a vat containing an activated sludge which is rich or contains microbial hydrocarbon oxidizers to a much greater extent than sludges previously have contained. As the influent is being charged into the vat or prior thereto it is treated, if necessary, with neutralizers. For example, if the influent is acidic, ammonia may be bubbled into the aqueous matter, or if basic, nitric acid may be added. In this way, an inorganic form of nitrogen is made available along with neutralization. In certain instances, it will be necessary to add other inorganics. Usually, a standard mixture is kept on hand and is charged to the influent.

### FIGURE 60:  BLOCK FLOW DIAGRAM OF PROCESS FOR MICROBIAL DISPOSAL OF OILY WASTES

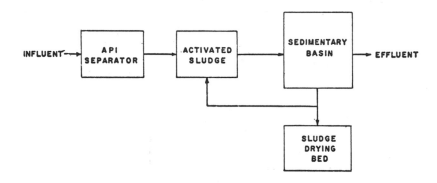

Source:  J.B. Davis and R.L. Raymond; U.S. Patent 3,152,983; October 13, 1964

Such a standard normally contains the following salts, the amount in parenthesis indicates the number of parts used per 1,000 parts of water: ammonium sulfate (1.0), disodium hydrogen phosphate (0.3), potassium dihydrogen phosphate (0.2), sodium carbonate (0.1), calcium chloride (0.01), ferrous sulfate septahydrate (0.005), and manganous sulfate (0.002). These and other inorganic salts may be simply added in the solid form to the aqueous influent or they may be dissolved first and then added. The amounts mentioned are merely suggestive for one will determine the mineral content of the given influent prior to starting the process and adjust the salt content in accordance not only with that determination but with the requirements of the given microbe or microbes being used. Very frequently, it is found that the main mineral salts needed to be added to the wastewater are nitrogen and phosphate salts. For this purpose diammonium hydrogen phosphate is used. Its addition in amounts of about 0.1% generally satisfies the need.

With the influent now under optimum conditions as to growth material promoters and temperature, being normal room temperatures of about 20° to 35°C., the waste is acted upon by an activated sludge. As stated previously, this sludge contains predominantly microbes that attack hydrocarbons. At this state the influent and the activated sludge are quite vigorously agitated and air is uniformly passed into the stirred mass, assisting in the agitation. The vat is open to the atmosphere, and volatile materials are readily discharged into the air. After several hours the resultant mass is passed to a sedimentary basin. From this basin will be taken aliquots of enriched cultures which are added to an activated sludge in use or which are used to form the culture or sludge to be used for the incoming new influent, as shown in Figure 60.

It will be appreciated that while only one activated sludge treatment is shown, the influent may be passed from one oxidator to another, in series, as desired before it is passed to the sedimentation basins. In such step-wise processes, the influent may first be subject to the action of microbes which remove harmful sulfur compounds or convert them to noninjurious forms. If the waste contains more than about 100 parts per million of oily matter, it is sent to an API separator or similar device which is used to skim off or similarly remove most of the oil. Thus, the modified wastes, whether they come from an oil operation, a city or whatever, are reduced in oil content to about 100 parts per million or less.

The modified influent is then passed to another oxidator, or from an API separator directly to a first oxidator as the case may be, which oxidator contains a concentrate of oil microbial oxidizers. In any event, the effluent that comes from the sedimentation basis is oil-free or substantially so, and the effluent can be discharged into streams and rivers without any harmful effects upon nature. In the sludge drying bed are the solids and occluded liquids, leaving the effluent clean.

Thus, oily wastes of various kinds and sources can be effectively treated by this process. These wastes contain about 0.01% by weight or less of the oily matter or about 10 mg. per 100 grams. By this process about 8 mg. is converted to 6 or 7 mg. of microbial cells and about 2 mg. of the oily waste is absorbed by 6 to 7 mg. of cells. The effluent becomes substantially oil-free, and this is accomplished in less than 24 hours.

A process developed by W. Teske and E. Ringel (9) is one in which hydrocarbon-containing wastewaters are treated with oxides or hydroxides of metal occurring in at least two different valency stages, the higher valency stage of which has a sufficient oxidation potential with respect to the impurities to be removed, and the lower valency stage of which is capable of being reoxidized to the higher valency stage by oxidants with simultaneous precipitation in the form of a difficultly soluble deposit, at temperatures within the range of about 50°C. to the boiling point of the wastewaters, preferably at temperatures exceeding 80°C. The oxides or hydroxides that have passed to the lower oxidation stage with reoxidation to the higher valency stage by means of a suitable oxidant are precipitated from the wastewater treated, separated therefrom in known manner and reused for the treatment of further wastewaters.

Figure 61 is a flow sheet of this process. According to this flow sheet, the pH of the crude wastewater to be treated is adjusted in stage I (if necessary after a mechanic preliminary purification) to a value within the range of about 1 to 3 by the addition of an acid (for which purpose residuary acids may advantageously be used), and the wastewater is maintained in the presence of the metal oxide or metal hydroxide of the higher oxidation stage at a temperature within the range of about 50° to 100°C., preferably with agitation and steam heating.

In stage II, the pH of the mixture is adjusted to a value within the range of about 8 to 10 by means of an alkali, for example sodium hydroxide solution or burnt or slaked lime while simultaneously blowing in air or oxygen, whereby the metal oxides present in the lower valency stage, which may also have been dissolved in the valency stage, are simultaneously oxidized and precipitated.

The precipitate, if necessary after a partial preliminary neutralization of the mixture, is separated in stage III from the treated wastewater in known manner, for example by sedimentation, filtration or centrifuging, and reconducted to stage I for the treatment of further amounts of wastewater. The treated wastewater from stage III is, if necessary, completely neutralized in stage IV and may then be drained off into the draining ditch or, if necessary, into a central biological purification installation, either undiluted or diluted with other suitable wastewaters or cooling waters, which depends on the absolute amount

FIGURE 61:  OXIDATION PROCESS SCHEME FOR PURIFICATION OF OILY
WASTEWATERS USING METAL OXIDE CARRIER

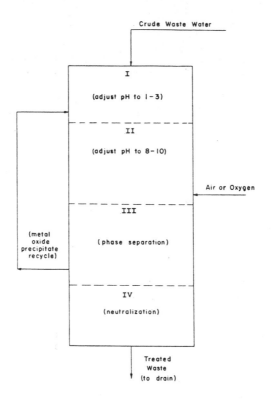

Source:  W. Teske and E. Ringel; U.S. Patent 3,337,452; August 22, 1967

of the wastewater and the purification degree attained.  If lead dioxide is used as oxidant, the reoxidation of the lead dioxide that has been reduced to the lower valency stage must be effected by means of chlorine or hypochlorite, advantageously in a separate operation after the separation in stage III.  It is, of course, possible to carry out the process in a fully continuous manner.

The following is a specific example of the conduct of this process.  Two liters of wastewater discharged by a large petrochemical installation and containing as main impurities benzene and the homologs thereof, was boiled in a flask provided with reflux condenser with 50 grams of hydrated dioxide of manganese for 20 minutes, at a pH that had been adjusted to 1 to 2 by means of hydrochloric acid.  The manganese that had partially passed into solution as bivalent ion was then precipitated with sodium hydroxide solution as $Mn(OH)_2$ in the alkaline range of 8 to 10, and reoxidized into hydrated dioxide of manganese by passing in air for half an hour.  The sediment of hydrated dioxide of manganese that had deposited was reused for oxidizing the impurities contained in a further part of the wastewater.  The hydrated dioxide of manganese applied was used altogether eighteen times for the purification of wastewater.

The determinations which were carried out in compliance with German standard methods gave the average permanganate values, as follows: untreated wastewater — 1,573 mg. $KMnO_4$ per liter of wastewater; wastewater treated according to this process — 288 mg. $KMnO_4$ per liter of wastewater. The average permanganate value thus decreased by 82%. The wastewater treated in the above manner did not contain detectable amounts of manganese.

A process developed by J.R. Strausser and R.S. Kurland (10) is one in which the dispersed phase of stable emulsions comprising aromatic hydrocarbon-containing oils in aqueous media is demulsified by intimately contacting the aqueous media with an aromatic hydrocarbon solvent. This results in an oil-rich solvent extract phase and an emulsified oil-depleted aqueous phase. The oil-depleted aqueous phase is passed through a finely divided crystalline silica coalescing medium to demulsify the dispersed phase of the remaining emulsified oil.

Figure 62 shows the application of this process to the removal of hydrocarbons from ethylene plant quench water. Referring to the drawing, the gases to be cracked, such as ethane, propane, and/or butane, for example, are introduced into the system by a line 11, preferably at a temperature of about 60° to 300°F. Steam is introduced, if desired, by means of a line 12 along with the charge stock for the purpose of controlling the space velocity of the gases through the cracking heater 13 and to inhibit coking. Suitable amounts of steam include from about 10 to 100% by weight based upon the gases introduced by means of the line 11. The gases are passed through the cracking heater at a space velocity of between about 4,000 and 14,000 pounds per square foot per hour for about 0.005 to 0.025 minutes. A temperature of about 1400° to 1550°F. is maintained in the cracking heater by means not shown, and the pressure is in the range of between atmospheric and about 50 pounds per square inch gauge.

A cracked gas mixture comprising $C_1$ to $C_6$ unsaturated hydrocarbons, such as ethylene, propylene, 1,3-butadiene, hydrogen, carbon dioxide and aromatic hydrocarbons comprising benzene, toluene, xylene and tars, is removed from the heater 13 by means of a line 14 and introduced into a prequench pot 15, and/or a quench tower 15a wherein it is cooled by cooling water in the form of a spray which is introduced therein by one or more lines 16. The cooling water is suitably at a temperature of between 40° and 450°F. Ordinarily about 3 to 30 parts of water per part of gas is sufficient to effect the desired cooling. So as to inhibit secondary reactions from arising within the cracked gas mixture, it is necessary that the cooling be effected in less than about 1 second.

Gaseous products of the cracking reaction including olefinic, saturated and aromatic hydrocarbons, hydrogen and carbon monoxide are cooled to a temperature of between about 50° and 300°F. and are removed by means of a line 17 for subsequent processing. The cooling water contains the remainder of the hydrocarbons which are dissolved and entrained therein. The quench water stream is discharged from the quench tower 15a by means of a line 18 and is introduced into a quench water drum 19. The hydrocarbons are suspended in the quench water in the form of an emulsion. The quench water issuing from the quench tower is at a temperature in the range of between 100° and 200°F. If desired, the temperature of the stream may be reduced in a cooler 20 so as to favor a physical separation of the tars from the quench water by gravity settling in the drum 19.

In the quench water drum 19 a heavy tar-water emulsion 21 collects at the bottom and is discharged from the drum by means of a line 22. Some of the aromatic hydrocarbons in the quench water float to the top of the drum and form a layer 23. Between the layers 21 and 23 forms an oily mixture of water, olefins, aromatics and lighter polymers having a density near that of water. Contained in the layer 24 is the stable emulsion which is to be treated by this process. The hydrocarbon oil content of the layer may be, for example, in the range of between about 2,000 and 7,000 parts per million.

The quench water of the layer 24 is discharged from the quench drum 19 by means of a line 25 through twin basket strainers 26a and 26b for the purpose of removing any

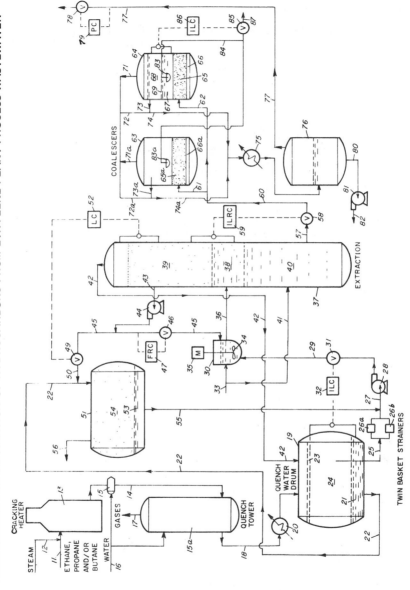

FIGURE 62: FLOW SCHEME FOR SEPARATION OF HYDROCARBONS FROM ETHYLENE PLANT PROCESS WASTEWATER

Source: J.R. Strausser and R.S. Kurland; U.S. Patent 3,507,782; April 21, 1970

suspended solid materials, e.g., carbon particles, from the waste quench water. Any suitable strainers or other filtering devices may be employed for this purpose. The waste quench water is passed from the strainers by means of a line 27 and a pump 28 into a line 29 by means of which the stream is introduced into an inline mixer 30. The flow rate of quench water to the mixer is regulated by means of a control valve 31 which is activated by an interface level controller 32 connected to the quench drum 19. The quench water stream in the line 29 may be introduced into the mixing vessel 30 at a temperature in the range of between 140° and 180°F.

Meanwhile, an aromatic hydrocarbon solvent, such as aromatic distillate having a boiling range between about 100° and 350°F. and containing 40 to 70% by weight benzene is introduced by means of a line 33 into the mixer 30. Suitable aromatic hydrocarbon solvents include benzene, toluene, xylene, etc. or as previously mentioned, it is especially preferred to employ an aromatic distillate fraction that was produced in the cracking process as the solvent for the waste quench water. Thus, the solvent stream 33 may suitably be the benzene-rich fraction that is recovered from the cracked gases issuing from the quench tower 15a by means of the line 17. The olefinic and aromatic hydrocarbon recovery facilities (not shown) may include compression, absorption, stabilization, etc. operations. In this manner, the process is wholly self-contained, and thereby no outside solvent is introduced which could dilute or contaminate the recovered hydrocarbons or add other pollutants to the water.

The aromatic hydrocarbon solvent and the waste quench water are thoroughly blended in the mixer 30 by means of an agitator 34 that is driven by a motor 35. The degree of mixing should be sufficient to enable the highly aromatic solvent to unify most of the aromatic, olefinic and polymeric materials in the quench water into a distillate fraction, but should not be so extensive as to homogenize or create a new and more stable emulsion. For example, the agitator may be suitably employed at a speed of about 1,100 to 1,200 revolutions per minute, while a contact time for liquid mixture may be in the range of between about 0.036 and 0.042 minutes.

Although an agitated vessel is illustrated as the mixing device in the drawing, any suitable mechanical mixing device including an orifice-type mixer, etc., may be employed provided the mixing obtained is not so extensive as to cause the formation of a stable emulsion. The mixture is discharged from the mixing device 30 by means of a line 36 whereby it is introduced into an extraction tower 37. In the extraction or deoiling water tower 37, the mixture separates into a distillate layer and a water layer which are separated by a water-oil interface 38. Disposed above the interface is the distillate layer 39 comprising the aromatic solvent and most of the aromatic hydrocarbons and low molecular weight polymers that were unified or agglomerated by the solvent. Below the water-oil interface layer is the water layer 40 which is substantially depleted of emulsified hydrocarbons.

If desired, the extraction tower 37 may be provided with suitable means such as packing or trays which will aid in the resolution of the above described layer. Suitable processing conditions which may be used in the extraction tower include a top pressure in the range of between 5 and 100 psig, preferably between 15 and 20 psig, while employing a temperature in the range of between about 45° and 200°F., preferably between 140° and 160°F. The residence time for the quench water may suitably be in the range of between about 6 and 60 minutes, preferably between 32 and 45 minutes, while a water superficial velocity of between about 0.002 and 0.03 feet per second, preferably between 0.011 and 0.13 feet per second may be suitably employed.

Alternately, or in addition to the aromatic hydrocarbon solvent introduced into the mixer by means of line 33, fresh aromatic solvent may be sparged into a lower portion of the tower 37 by means of a line 41 which is positioned below the process line 36 which introduces the wastewater.

Small amounts of vapor which are disengaged from the solvent and quench water in the top of the tower 37 are vented from the tower by means of a vent line 42. The vapor is then passed into the vapor space of the quench water drum **19**. Instead of employing a mixer-extraction tower combination as shown in the drawing, the aromatic hydrocarbon solvent and waste quench water may be intimately contacted in a continuous, counter-current manner.

Referring again to the drawing, a solvent extract or distillate stream is removed from the extraction tower **37** by means of a line **43** and a pump **44** from which it is introduced into a line **45**. A portion of the solvent extract is recycled to the mixer **30** at a rate controlled by a valve **46** and a flow rate control assembly **47**. A suitable ratio of quench water to the recycle solvent extract stream is in the range of between about 3 and 40 volumes of quench water per volume of recycled solvent extract, preferably between 15 and 25 volumes per volume.

The makeup solvent which is introduced by means of the line **33** may be introduced into the mixer **30** at a rate corresponding to between about 1 and 50 volumes of quench water per volume of makeup solvent, preferably between 20 and 30 volumes of quench water per volume of makeup solvent. The aromatic hydrocarbon solvent may be introduced into the mixture at a temperature in the range of between about 45° and 120°F., preferably between 70° and 90°F. while at a specific gravity in the range of between about 0.600 and 0.900, preferably between 0.750 and 0.850 measured at 60°F./60°F.

The portion of the solvent extract in the line **45** which is not introduced into the mixer **30** is transferred by means of a valve **49**, a line **50** and the line **22** into a tar settling drum **51**. The discharge of the excess solvent and extract stream into the drum is controlled by the level control system **52**. Two individual layers are formed in the drum **51**, which layers comprise a lower water layer **53** and an excess solvent extract layer **54**. This layer also contains heavy tar that was removed from the quench water drum **19**.

The heavy tar which is removed from the bottom of the drum **19** by means of the line **22** is absorbed from the tar-in-water emulsion with the water combined therewith by the solvent introduced by the line **50**. The surplus solvent, extract, and extracted tar are sent to a storage tank, not shown, by means of the line **56**. The specific gravity of the heavy tar is in the range of between about 1.00 and 1.05.

The tar settling drum **51** may be operated at a temperature in the range of between about 100° and 200°F., preferably between 140° and 160°F. The liquid residence time therein may be varied in the range of between 15 and 120 minutes, preferably between 60 and 100 minutes. A suitable pressure is in the range of between about 10 and 15 psig, preferably 12 psig.

The water layer **53** is drawn off the bottom of the drum **51** and is then introduced by means of a line **55** into the process line **27** so as to become a part of the main waste quench water stream that is introduced into the mixer **30** for blending with the aromatic solvent. Other miscellaneous aromatic-laden wastewaters from the cracking process may also be introduced into line **27** at this point.

An extracted water stream is removed from extraction tower **37** by means of a line **57** and a valve **58** at a rate controlled by the interface level rate controller **59**. Thus, the liquid inventory in the tower **37** is maintained by the two controllers **52** and **59** which regulate the extract-solvent inventory and water discharge flow, respectively. The extracted water from the tower **37** may be passed by means of the lines **60**, **61** and **62** into the degasser-coalescers **63** and **64**. As shown in the drawing, the coalescer **64** is in operation, while the coalescer **63** is a spare which may be employed in a standby position.

The extracted water which is removed from the lower portion of tower **37** has a greatly reduced hydrocarbon content as compared with the composition of the waste quench water that was withdrawn from the quench water drum **19**. However, the extracted water in the line **60** is still not of a sufficient purity that it may be discharged into local waters or reused in various plant processing operations. Thus, the extracted water is contacted with a coalescing medium comprising crystalline silica having a particle size in the range of between 4 and 100 mesh. This simple contacting results in a coalescing and agglomeration of the remaining dissolved and emulsified hydrocarbons and permits their removal from the quench water.

Referring again to the figure, the extracted water in the line 60 is introduced upflow by means of the line 62 into the coalescer 64 where it contacts a bed of fairly high purity sand particles 65 having a particle size of between 4 and 100 mesh, preferably between 8 and 20 mesh. The silica particles may be provided in a bed having a height in the range of about 3 and 4 feet, for example. A suitable support means such as a grid 66 is provided to hold the coalescing medium in place. The coalescer 64 may be suitably operated at a temperature of between 45° and 180°F., preferably between 130° and 150°F. while employing a top pressure in the range of about 2 and 25 psig, preferably between 15 and 18 psig.

The water to be purified may be introduced into the coalescer 64 at a superficial velocity in the range of 0.001 and 0.015 feet per second, preferably between 0.005 and 0.013 feet per second. The coalescing medium may be employed in amounts sufficient to provide between 0.9 and about 7.0 volumes of quench water per volume of coalescing medium per hour, preferably between 4 and 6 volumes per hour per volume. A residence time in the range of 6 and about 60 minutes, preferably between 8 and 16 minutes may be suitably employed in the coalescer.

The coalescing medium is crystalline silica of a fairly high purity and is preferably quartz. However, sand of a lower purity than quartz but without excessive organic materials, or other inert particles may be successfully employed. The action of the coalescing medium causes the water and the oil to separate into layers 67 and 68, respectively. A water-oil interface 69 is therefore established above the sand bed 65.

Entrained gases are released by the action of the coalescing medium and are vented from the vessel 64 by means of a line 71. This gaseous stream is passed from the line 71 into a line 72 and is then combined with the separated oil from the layer 68 by means of a line 73. The combined stream is passed by means of the line 74 to a cooler 75 where the temperature may be reduced from between 60° and 100°F., preferably between 60° and 80°F. so as to cause the separation of noncondensible gases from the oil in the vessel 76 by means of a line 77 to a flare system (not shown) at a rate determined by the valve 78 which is activated by pressure control system 79.

The oil that accumulates in the vessel 76 is discharged therefrom by means of the line 80 into the pump 81 and may be passed by means of the line 82 into the line 50 (by means not shown) for introduction into the tar settling drum 51. In this manner, all of the hydrocarbons that are recovered from the process, including the solvent extracted hydrocarbons and those coalesced from the quench water, are combined and transferred to a by-product storage tank (not shown).

The water in the layer 67 rises around a large dish shaped baffle 83 which causes the water to change its direction of flow. The water is thus drawn off by means of a line 84 and a valve 85 which is activated by an interface level controller 86. The water discharged by means of a line 87 has a hydrocarbon content in the range of between 17 and 45 ppm, so that over 99% by weight of oil which was originally present in the waste quench water is removed.

Accordingly, the purified water may be discharged to local water streams or may be used in selected process operations of the olefin producing unit. This process permits, for example, the purification of quench water, which has an emulsified oil content of 3,125 and 6,192 ppm to high purity water streams containing 29 and 45 ppm oil, respectively. Thus, a removal of over 99% by weight oil is obtainable.

A process developed by W. Weiler (11) relates to the removal of hydrocarbons such as benzene, gasoline and oil from polluted water. It is known that conducting a liquid containing hydrocarbons through a filter containing an oleophilic compound will effect separation of the hydrocarbons from the water because of its selective absorption characteristics for oil, benzene and other similar hydrocarbons.

Other methods of separation operate by utilizing the fact that hydrocarbons are lighter than water and will rise to the surface of a relatively quiescent body so separation can be effected by skimming. However, this method removes only about 95% of the hydrocarbons present and to meet some purification standards, further treatment is necessary.

With previous filters incorporating an oleophilic compound, efficiency, capacity and, ultimately, the life of a filter is diminished because of a tendency to clog, fill up and create pockets of untreated liquid. Besides the obvious operational inefficiencies of such equipment, additional expense is incurred initially because the filter must be overdesigned to allow for the decreased steady state operation capacity.

This process eliminates clogging in the filter so that it is utilized to a maximum degree on a continuous basis. A filter packing formed of particles containing an oleophilic compound such as charcoal, bituminous coal, inflated mica material or vulcanic limestone is placed in a supporting bed with an outlet beneath the packing. A plurality of high pressure liquid nozzles are movably mounted above and proximate to the filter packing surface.

These nozzles project a jet of liquid, which may have previously been run through a skimming separator of the type previously mentioned, onto the filter packing with force sufficient to joggle the filter particles to cause a slow, steady shift and circulation of the filter as a whole which prevents pockets of stagnant liquid from forming, promotes the exposure of the surface areas of the entire filter to incoming liquid and provides a pressurized contact between the pollutants and filter media.

To facilitate contact between liquid and filter, the outlet can be dammed to maintain a desired level and degree of saturation. The dammed liquid also makes it easier to mix and circulate the filter material under the power of the nozzle jets. In addition, it assures uniform filtering conditions, even when the incoming liquid supply is unsteady. The filter packing material may be of a type that floats or sinks in water. In the latter case, a coarse grained, porous material such as gravel is used to support the filter and allow easy drainage.

Figure 63 shows a suitable form of apparatus for the conduct of this process. In the figure, a separator 2 is connected to a circular purification tank 1 via pipeline 3. The separator functions by providing a quiescent basin into which water polluted with hydrocarbons such as oil, benzene and gasoline are introduced at level 9. The hydrocarbons, being lighter than water, tend to be displaced upwardly in the separator so that the liquid discharged into the pipeline is about 95% water and 5% pollutants and other impurities. This aids the purification process in the tank by eliminating large glut portions of hydrocarbons, thus decreasing the filter size required.

Within tank 1, a layer of filter packing 10 is placed on bottom 19. The filter packing preferably consists of granules or small particle formed pieces of material having oleophilic and, preferably, also hydrophobic properties. Oleophilic substances such as charcoal, inflated mica containing minerals and vulcanic limestone are used because of their high affinity for hydrocarbons which are absorbed and separated from the water. Hydrophobic properties aid in the filtering process by not allowing the filter packing compounds to become saturated with water.

Pipeline 3 enters tank 1 at the bottom and turns vertically upwardly near the center and is capped with a rotary coupling 7. A sprinkler 4 having a plurality of radially extending sprinkler arms 6, each supported by cables 5, is mounted on the coupling. Several nozzles 8 are mounted on the underside of each sprinkler arm and are canted at an angle with a vertical plane to provide vertical and horizontal reactive force components. Water pressure is supplied by either connecting a pump (not shown) to line 3 or making the outlet of

FIGURE 63:  OLEOPHILIC FILTER FOR HYDROCARBON REMOVAL FROM
WASTEWATER

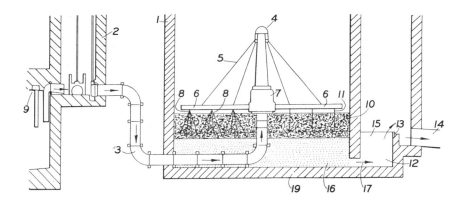

Source:    W. Weiler; U.S. Patent 3,527,701; September 8, 1970

pipeline 3 to separator 2 much higher than nozzles 8, or both.  Regardless of the method,
sufficient water pressure is developed to turn the sprinkler with the reactive forces from
the liquid jets emerging from the nozzles.  The force of the jets impinging on the filter
packing particles is sufficient to cause agitation and movement of the particles as units and
circulation of the filter as a whole.

Thus, under the influence of the constantly moving and variably positioned water jets, the
entire filter packing and even the portion below the surface, is exposed to fresh incoming
untreated water.  The movement of the filter particles under the impact of the water jets
acts in combination with the fine droplets so produced to provide maximum contact between the hydrocarbons and oleophilic compounds to optimize the filtering process.  It is
therefore possible with a smaller amount of filter packing to achieve results equal or superior
to prior filters.

In the event that the design parameters are such that the depth of filter packing 10 is too
thick for even high pressure jets to completely mix, a plurality of agitating rods 11 are
mounted on sprinkler arms 6 and extend downwardly into the filter particles.  As the
sprinkler rotates, possibly with the help of a motor (not shown), rods 11 agitate and mix
the filter materials.  Resistance to the passage of the rods is slight because the filter packing components are generally light.

To further enhance the contact and filtering process, a weir-like dam 13 is placed in outlet
17 of tank 1 so that the liquid level 15 within is maintained at a desired level above the
bottom level of filter packing 10.  To further facilitate uniform contact between the liquid
jets and the filter, its components can be made of materials capable of floating in water.
In this case, a coarse grained permeable material 16, like gravel, is interposed between the
tank bottom 19 and the filter packing to allow drainage to the outlet but prevent the
floating particles from washing away under the impact of the water jets.

Dam 13 allows the steady discharge from outlet basin 12 into discharge pipe 14 to the sewerage system.  Tank 1, including the outlet basin and the bottom plate, can be designed
as a compact unit capable of being transported and erected as a unit and is preferably constructed of steel.  In experiments conducted with wastewaters polluted with a benzene-
gasoline mixture comprising 40 g./l. of water before introduction into skimming separator
2, and about 0.2 g./l. as introduced into purification tank 1, this filtering apparatus reduced

the benzene-gasoline content to 10 mg./l. at a liquid flow rate of 10 l./sec./2 m.² of filter surface in a filter composed of particles sized from about 0.5 to 5 mm. and having a packing height of about 15 to 20 cm.

### References

(1)  Frost, E.E., Hottel, H.C. and Longwell, J.P.; U.S. Patent 2,971,605; February 14, 1961; assigned to Esso Research and Engineering Company.

(2)  Ford, W.W. and Kube, A.M.L.; U.S. Patent 2,978,063; April 4, 1961; assigned to Phillips Petroleum Company.

(3)  Bloxham, R.; U.S. Patent 3,190,823; June 22, 1965; assigned to Coast Manufacturing and Supply Company.

(4)  Reed, R.D. and Zink, J.S.; U.S. Patent 3,512,911; May 19, 1970.

(5)  Merrill, E.W.; U.S. Patent 3,593,496; July 20, 1971; assigned to Charles River Foundation.

(6)  Hardison, L.C.; U.S. Patent 3,632,304; January 4, 1972; assigned to Universal Oil Products Co.

(7)  Jasinsky, V. and Upfold, A.T.; U.S. Patent 3,667,408; June 6, 1972; assigned to Polymer Corporation, Limited.

(8)  Davis, J.B. and Raymond, R.L.; U.S. Patent 3,152,983; October 13, 1964; assigned to Socony Mobil Oil Company, Inc.

(9)  Teske, W. and Ringel, E.; U.S. Patent 3,337,452; August 22, 1967; assigned to Farbwerke Hoechst AG, Germany.

(10)  Strasser, J.R. and Kurland, R.S.; U.S. Patent 3,507,782; April 21, 1970; assigned to Gulf Oil Corp.

(11)  Weiler, W.; U.S. Patent 3,527,701; September 8, 1970; assigned to Passavant-Werke, Germany.

## HYDROGEN CHLORIDE

Hydrogen chloride reacts rapidly with the moisture in the air and is generally found in the ambient atmosphere as hydrochloric acid. The acid at low concentrations, 15,000 to 75,000 $\mu$g./m.³ (10 to 50 ppm), irritates primarily the mucous membranes of the eyes and upper respiratory tract in both humans and animals. Prolonged exposures to low concentrations of hydrochloric acid can also erode the teeth.

Work is intolerable after more than 60 minutes in atmospheres containing approximately 75,000 to 150,000 $\mu$g./m.³ (50 to 100 ppm) of hydrochloric acid. Higher concentrations (approximately 1,500,000 $\mu$g./m.³ or 1,000 ppm) can attack the mucous membranes, causing inflammation of the upper respiratory system and resulting in pulmonary edema or spasm of the larynx, which can be fatal according to Stahl (1).

A wide variety of plant life is susceptible to the toxic effects of hydrogen chloride or hydrochloric acid. Several examples of plant damage due to hydrochloric acid emissions have been reported in the literature. The primary effect on plants is a discoloration or bleaching of the leaves. The threshold for visible damage was originally reported as 75,000 to 150,000 $\mu$g./m.³ (50 to 100 ppm) of hydrogen chloride. However, recent studies indicate that the threshold for many plants is less than 10 ppm for 4 hour exposures. The strong acidic properties of hydrochloric acid make it extremely corrosive to most metals.

Some countries have established ambient air quality standards for hydrochloric acid, including West Germany (approximately 750 $\mu$g./m.³ or 0.5 ppm as a mean 30 minute average) and Russia (approximately 15 $\mu$g./m.³ or 0.009 ppm as a 24 hour average).

Hydrogen chloride is produced by three main processes in the United States: acid-salt; direct synthesis; and the by-product process from chlorination of organic compounds. The latter process, the major production source, has shown a steady increase. Many organic chlorinating processes and other organic processes involving chlorinated compounds may produce hydrogen chloride as a by-product, but it may not be economically feasible to recover the product for other uses. Hydrochloric acid is produced by absorbing the hydrogen chloride gas into water. Hydrochloric acid (or hydrogen chloride) is primarily used to manufacture inorganic and organic chemicals, the latter accounting for almost 50% of the production in 1963. Other major uses include metal production, oil well acidizing, metal and industrial cleaning and food processing. Another potential source of emissions of hydrochloric acid is the heating or burning of chlorinated materials.

## Removal of HCl from Air

The high solubility of hydrogen chloride in water and the low vapor pressures of even 20% hydrochloric acid solutions make collection of hydrogen chloride in water an effective and inexpensive method of control.

Thus, in the manufacture of hydrochloric acid (or hydrogen chloride), the emission control system, which is also part of the system for obtaining hydrochloric acid from the hydrogen chloride, consists mainly of water absorption facilities. Different types of absorbers are used, but the systems generally consist of a packed tower or a cooling absorption tower followed by a packed tail tower (2)(3)(4). The packed tower systems (2)(3) normally include a connected set of S-bend tubes, followed by one or more towers in series.

Cold water is added at the last or tail tower and flows over into the previous tower. The concentration of hydrochloric acid in this water is thus increased from tower to tower until it reaches the S-bend tubes where the acid solution attains its final strength. However, the packed tower systems are rapidly being replaced by cooled absorption systems because the latter are more efficient, economical and compact. The cooled absorption tower designs may consist of either a countercurrent or cocurrent flow of gas and water. One type uses water jacketed packed tantalum towers with countercurrent flow (3). The most common system is the cocurrent falling film absorbers (3)(5)(6). In this system, gas and weak acid solution from the packed tail tower flow downward over vertical, water cooled, wetted wall columns. Where anhydrous chloride is the desired product, absorption systems are not used (2).

It has been reported that the use of water scrubbing systems can reduce the emissions to less than 0.1 pound of hydrogen chloride per ton of acid produced, although emissions can be 30 times that amount with less effective equipment (7). According to W.L. Faith, D.B. Keyes and R.L. Clark (8), hydrogen chloride emissions can be reduced to a range of 0.1 to 0.3% by volume by the use of two or more tail towers in series. The emissions of hydrochloric acid may be high if an upset occurs in the absorption system due to either improper temperature control or insufficient feed water (4).

Other systems that appear to be effective for the control of hydrochloric acid or hydrogen chloride emissions are the rotary brush scrubber (6) and the ejector venturi scrubber (9). With the former system, collection efficiencies of 99.995% have been reported with an initial hydrogen chloride content of 610 g./m.$^3$. In the latter system, scrubbing efficiency as high as 99% has been given for a single-stage unit for fumes containing up to 20% hydrogen chloride. The high solubility of hydrogen chloride in water accounts for the high degree of efficiency of these systems. Dry solid adsorbents have also been studied for removal of hydrogen chloride vapors. Adsorbents investigated include chromium oxinate and basic salts (10).

A process developed by J.E. Seebold (11) relates to the manufacture of hydrochloric acid by absorption from waste gases containing hydrogen chloride. In this process, gases rich in hydrogen chloride may be passed through an absorption tower without external cooling means, the tower being of such size, velocity and temperature that 20° Baumé hydrochloric acid is produced.

A process developed by H.C. Wohlers, F.N. Grover, C.W. Lentz and L.E. Pauling (12) relates to the separation and recovery of hydrogen chloride, hydrogen bromide and other gaseous substances from waste gases discharged such as from brominators or from chlorinators engaged in the preparation of chloral and related products by alcohol chlorination, paraldehyde or acetaldehyde chlorination. Recovery of hydrogen chloride from the vent gases of a chlorinator, as in the manufacture of chloral, poses many of the problems characteristic of the recovery of hydrogen chloride from other gas systems but, in addition, a number of particularly difficult problems are imposed because of the presence also of large amounts of free chlorine and relatively large amounts of chloral hydrate, chloral and other impurities such as monochloroacetaldehyde, dichloroacetaldehyde resulting from the chlorination of acetaldehyde or paraldehyde or ethyl chloride resulting from chlorination of ethyl alcohol.

The removal of organic matter, such as chloral hydrate and chloral, which are soluble in hydrochloric acid and would contaminate product acid resulting from the recovery has offered the greatest difficulty. Attempts have been made to remove the organic materials by adsorptive charcoal but such techniques have proven unsuccessful. Other systems have made use of a series of coolers and separators ranging in temperature from +20° to –30°C. but the initial cost for coolers, refrigerators and separators and the operation thereof has rendered this technique impractical. Many other systems have been advanced by the art but for various reasons were incapable of use in an economical manner to effect the desired results.

In order to achieve efficient and effective removal of the organic impurities soluble in hydrochloric acid at an initial stage of the process, it is desirable to carry out the scrubbing action in a scrubber such as a packed column having hydrochloric acid passing downwardly therethrough by gravitational force from an acid inlet at the top to a discharge opening at the bottom while the vent gases discharged from the reactor or chlorinator are fed through a gas inlet near the bottom of the column and rise upwardly through the column to a gas outlet at the top. The efficiency of the stripping action of hydrochloric acid depends greatly upon the degree of contact between the acid and the gas and for this purpose, packings such as Berl saddles, Raschig rings and the like are provided in the scrubber to increase the area of contact.

It is desirable to maintain flow of sufficient acid down through the column substantially completely to wet out the surfaces of the packing. Suitable wetting out may be achieved with as little as 5% by weight hydrochloric acid calculated on the amount of hydrochloric acid which is formed as a product by the recovery of the hydrogen chloride gas.

While the removal of chloral, chloral hydrate and other accompanying hydrochloric acid soluble impurities increases in proportion to the amount of acid used, it is economically inadvisable to make use of hydrochloric acid in amounts greater than 25% by weight of the product acid unless subsequent separation of the scrubber acid from the impurity can be carried out in a simple and easy manner.

As in the scrubber, it is desirable to pack the absorber with Berl saddles, Raschig rings or the like to increase the surface area and provide means for more substantial contact between the gases passing upwardly from the gas inlet at the bottom of the column to the gas outlet at the top and the aqueous medium introduced at the inlet at the top which passes by gravitational flow to the outlet at the bottom.

Since the other gaseous materials soluble in hydrochloric acid were previously removed in the scrubber, only hydrogen chloride is dissolved in the aqueous medium although small amounts of chlorine may become dissolved therein but is not considered harmful in the amounts present in the final product. While the amount of chlorine may range to about 2,000 to 3,000 parts per million in the product acid, dechlorination may be effected in a simple and easy manner by any of a number of well-known systems to reduce the chlorine content to an insignificant 10 to 20 parts per million.

It is desirable to balance the rate of flow of gases and aqueous medium fed to the absorber substantially completely to remove the hydrogen chloride so that practically none is to be found in the gaseous material discharged and to dissolve sufficient hydrogen chloride to form concentrated hydrochloric acid by the time that the aqueous medium is discharged at the bottom of the absorber. Solution of hydrogen chloride in the aqueous medium to form hydrochloric acid takes place with the generation of heat.

For most efficient operation it is desirable to maintain temperatures in the absorber to as low a level as possible and for this purpose cooling may be provided by means of a cooling jacket or cooling coils extending through the absorber. The concentrated hydrochloric acid containing impurities dissolved therein may be processed for recovery of the acid and impurities in a number of ways depending upon the character of the impurities.

If the impurities boil above 110°C., water may be added to the stripping acid to reduce the acid concentration to that of constant boiling acid. Then distillation at 110°C. will remove all of the acid, leaving impurity as a residue. If the impurities volatilize below 110°C., water may be added in amounts to reduce the acid to constant boiling and then distillation at 110°C. will volatilize off the impurity, leaving constant boiling hydrochloric acid as residue. The corresponding breaking point for the hydrobromic acid system is at its constant boiling temperature of about 126°C. which contains about 47% hydrogen bromide.

*Example:* The vent gases from a chlorinator were analyzed to contain the following in percent by weight: 74 to 75% hydrogen chloride; 15 to 16% chlorine; 9 to 10% ethyl chloride; and 0.1 to 2.5% chloral hydrate. In the experimental work two scrubbers were used in series each measuring 1 inch in diameter and packed to a depth of 40 inches with ¼ inch Berl saddles. The hydrogen chloride absorber comprised a column 1 inch in diameter packed to a length of about 44 inches with Berl saddles and jacketed for cooling.

In operation, vent gases from the chlorinator were first passed through a trap to remove solids and then introduced into the second scrubber at a flow rate of about 10 cubic feet per hour. The scrubbing acid, such as product acid or reagent hydrochloric acid was introduced into the first scrubber at a rate of about 10% of the product acid. From the first scrubber the acid was introduced to the top of the second scrubber while the gas discharged from the second scrubber was drawn into the bottom of the first scrubber.

The concentrated hydrochloric acid discharged from the base of the second scrubber contained from 20 to 30 mg. chloral hydrate per milliliter. The gas discharged from the first scrubber was metered into the base of the absorber while water was fed into the top at a rate to extract substantially all of the hydrogen chloride gas and produce concentrated hydrochloric acid as product acid having the following analysis.

| | |
|---|---|
| HCl assay | 33.16 percent |
| Organics | 0.08 percent |
| Solids (dissolved) | 0.009 percent |
| Color | Yellow |
| Fe | 2.9 ppm |
| $Cl_2$ | nil |
| $SO_4$ | 0.001 percent |

Without treatment in the manner described to remove impurity in advance, the product acid would contain 0.5 to 0.6% organics calculated on the amount of chloral. Other organics will probably increase in proportion as impurity in the final product.

A process developed by W. Kunzer and F. Leutert (13) relates to an apparatus for working up waste gases containing hydrogen chloride and impurities and originating from organic chlorination reactions. It is known to recover hydrochloric acid from waste gases which contain hydrogen chloride and which have been obtained from organic chlorinations, by allowing the waste gas to flow upwardly through a tower filled with a suitable packing and to which an aqueous absorption liquid is supplied at the top.

When working in this way there is the disadvantage that the waste gas passing to the further working up is in general saturated with the vapors of impurities contained therein and originating from the organic chlorination reaction and that there is a risk of these vapors being condensed upon the entry of the waste gas into the absorption zone of the tower. Even when only small amounts of liquid or solid condensate are thereby formed, it is not always certain that at the speed at which the hydrochloric acid flows away from the tower the condensate will completely vaporize again before the hydrochloric acid enters the cooler through which it passes before leaving the tower. The hydrochloric acid obtained may therefore be substantially contaminated especially by substances which evaporate only slowly or of which the boiling point is so high that they are only volatile azeotropically with steam.

The disadvantage can be avoided by heating the contaminated gas containing hydrogen chloride, prior to the absorption, to a temperature which is above the dewpoint of the impurities. This heating above the dewpoint of the impurities may advantageously be effected by leading the gas coming from the chlorination apparatus in heat exchange relation to the hot acid flowing from the absorption zone of the tower. For this purpose it is thus only necessary to install in the absorption zone a suitable device for heating up the gases to be worked up, through which the acid produced flows downwardly.

In normal operation there is sufficient heat set free by the absorption for the gas supplied to the absorption zone to be heated above the dewpoint of the impurities without any appreciable change of temperature taking place in the absorption zone. The separation of the liquid droplets or solid particles entrained with the gases containing hydrogen chloride, from these gases prior to the supply of the gases to the absorption zone is preferably also effected in the device serving for the heating up of the gases. Figure 64 shows a suitable form of apparatus for the conduct of this process.

### FIGURE 64: APPARATUS FOR THE PRODUCTION OF HYDROCHLORIC ACID FROM WASTE GASES CONTAINING HYDROGEN CHLORIDE

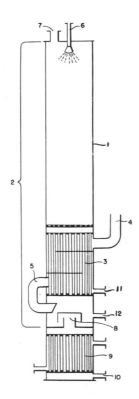

Source: W. Kunzer and F. Leutert; U.S. Patent 3,036,418; May 29, 1962

A tower 1 has an absorption zone 2 charged with packings in which there is also installed a heat exchanger 3 to which is supplied through a pipe 4 the contaminated gas containing hydrogen chloride and in which this gas is heated up, in heat exchange with the hot hydrochloric acid formed in the absorption zone, to a temperature above the dewpoint of the impurities contained therein. This gas then passes through a pipe 5 into the absorption zone of the tower, first completely saturates the hot acid flowing down in the heat exchanger and is then absorbed in the packed portion of the absorption zone above the heat exchanger by the water sprayed in at 6 with the formation of hydrochloric acid.

The residual gas, which contains steam produced by the heat of absorption and organic substances, leaves the tower through a pipe 7 and is supplied to plant for the recovery of the organic substances. The hydrochloric acid flowing away from the heat exchanger 3 passes a dip seal 8 and is then cooled in a heat exchanger 9 and led away from the tower through a pipe 10 as concentrated hydrochloric acid which is practically free from organic impurities and chlorine. Condensate entrained by the gases containing hydrogen chloride can be withdrawn through a pipe 11.

In order to heat up the gas supplied to the tower when the latter is set in operation, steam is led in at 12 until the heat exchanger has reached a temperature of about 80°C. When the absorption is proceeding, the temperature is maintained by the absorption heat formed so that it is not necessary to supply steam for a prolonged period.

The heat exchanger for the heat exchange between the gases containing hydrogen chloride to be supplied to the absorption zone and the hydrochloric acid formed in the absorption zone can also be arranged in the space between the heat exchanger 9 and the dip seal 8. The heat exchanger for the heating up of the gases containing hydrogen chloride may also be installed outside the tower in the pipe 4 and brought to the desired temperature by means of a heating medium such as steam, water, fuel, oil, diphenyl or the like.

Whereas according to the methods hitherto known there is obtained from the waste gas containing hydrogen chloride contaminated with benzene from a chlorination plant, a 30% hydrochloric acid which contains 0.2 g. of benzene per liter, the working up of the same gas according to this process makes possible the recovery of a 30% hydrochloric acid of which the content of benzene amounts to only 0.0015%.

In the working up of waste gases which contain carbon tetrachloride in addition to hydrogen chloride, it is possible according to this process to obtain a 30% hydrochloric acid of which the content of carbon tetrachloride amounts to only 0.006% as compared with a content of 0.01% in 30% acid recovered by the known methods.

A process developed by W.R. Fuller, B.H. Bieler and D.C. Morgan (14) involves reducing the amount of halogen emitted during the incineration of halogen-containing plastics. The method involves applying an alkali to the plastic before it is burned. The method can reduce the emission of halogen by greater than 75% when properly employed.

A process developed by S. Ezaki (15) for recovering hydrogen chloride from a spent organo-chlorine compound involves carrying out a combustion treatment of spent liquor or waste gas containing the organo-chlorine compound, the discharge of which in the untreated form is undesirable from the viewpoint of minimizing public nuisance.

Almost all the chlorine contained in the organic spent liquor or waste gas is converted to hydrogen chloride, recovering the thus converted hydrogen chloride as hydrogen chloride gas or hydrochloric acid having a concentration of higher than the azeotropic composition of $HCl-H_2O$ system, i.e., approximately 20% by weight, and at the same time, discharging other components into the atmosphere in the form of completely harmless and odorless gases. Figure 65 is a flow diagram showing the essentials of this process. The spent organic compound to be subjected to combustion treatment and the combustion air are both led to burner 3, where the spent organic compound is burned, via lines 1 and 2, respectively.

**FIGURE 65: PROCESS FOR RECOVERING HYDROGEN CHLORIDE FROM THE COMBUSTION OF AN ORGANO-CHLORINE COMPOUND**

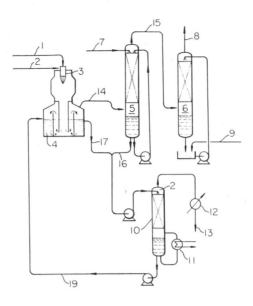

Source:  S. Ezaki; U.S. Patent 3,589,864; June 29, 1971

The hot combustion gas containing hydrogen chloride is injected into an extracting agent liquid bath (extracting agent solution concentrator) **4**, where the gas is cooled by the direct heat exchange between the gas and the liquid, and simultaneously, the water in the extracting agent liquid bath is caused to evaporate by the combustion gas, and thus the concentration of the extracting agent solution is carried out.  The method by which a liquid is heated and caused to evaporate by contacting with a hot combustion gas injected directly into the liquid stored in a tank is generally called the "submerged combustion method."  The concentrated extracting agent solution is transported to the extractive distillation step.

The gas cooled by heat exchange is introduced to the lower part of absorber **5** via line **14**, and is contacted countercurrently with water provided via line **7** that joins the upper part of the column.  Some of the column types usable therefor include plate tower such as bubble cap plate and perforated plate, and packed tower packed with Raschig ring, Lessing ring, cross-partition ring, Pall ring, Berl saddle, Intalox saddle, Tellerette, spiral ring, etc. Corrosive materials are involved in the operation, so a packed tower is preferable, since it is easier to maintain.

The hydrogen chloride vapor in the gas is absorbed by contacting, directly and countercurrently with a water spray, the water preferably being sprayed from the upper part of the column.  As a result, a 16 to 18% hydrogen chloride solution, and in some cases, a solution approximating the azeotropic composition can be obtained from the bottom of the column.

In case the concentration obtained is low, it is possible to recirculate a part of the hydrogen chloride solution obtained from the bottom of absorber **5** to the top.  In case a packed tower is used as an absorber, the recirculation is required in some cases in order to secure the liquid load necessary for providing appropriate gas-liquid contact.

The gas released from the upper part of the absorber 5 consisting mainly of nitrogen, carbon dioxide and steam can be transported to the lower part of scrubber 6 via line 15 in order to eliminate any remaining harmful materials.  In the scrubber, the gas is countercurrently contacted with a dilute alkaline solution which is introduced via line 9, and circulated through the column.  The dilute alkaline solution contains, for example, sodium carbonate, sodium bicarbonate, sodium hydroxide and calcium hydroxide or the like.

The small amount of unrecovered hydrogen chloride can be removed in column 6 through neutralization.  Thus, the gas is made completely harmless and odorless and is released into the atmosphere from the top via line 8.  Thereafter, in order to concentrate the dilute hydrogen chloride solution withdrawn from the bottom of absorber 5 via line 16 to a concentration of higher than the azeotropic composition, the dilute hydrogen chloride solution is subjected to a continuous mixing with the concentrated extracting agent solution passed from the liquid bath 4 via line 17, and is provided to the top of the extractive distillation column 10.

Reboiler 11 provided at the bottom of the extractive distillation column supplies the heat necessary for effecting the distillation and thus the extractive distillation is carried out. The vapor withdrawn from the top of the distillation column is condensed as it is in condenser 12, and is recovered in the form of hydrogen chloride solution having a concentration higher than the azeotropic composition and high purity via line 13.  As can easily be understood to those in the art, when it is required, a part of the aforementioned hydrogen chloride solution can also be returned to the extractive distillation column via line 18 as a reflux stream.

### Removal of HCl from Water

A process developed by L.J. Lefevre (16) is an ion exchange process for the recovery of hydrochloric acid from spent hydrochloric acid pickle liquor.  The process includes regeneration of the ion exchange material with a complex sulfonic acid, and treatment of the regenerant effluent to recover the complex sulfonic acid regenerant and to produce iron oxide suitable for use in steel making.

Figure 66 shows a suitable form of apparatus for the conduct of this process.  A countercurrent continuous ion exchange reactor of the type described in U.S. Patent 2,815,322 is particularly suitable for use in the process and is schematically shown in the drawing. The continuous reactor, generally indicated at 10, includes an ion exchange zone 12, a regeneration zone 14, water rinse zones 16 and 18 immediately above the ion exchange and regeneration zones, respectively, and a recycling path 20.

A shown in the figure, spent hydrochloric acid pickle liquor stream 22 is initially contacted with the ion exchange material in zone 12 of continuous reactor 10.  The liquor contains an aqueous solution of ferrous chloride formed in the pickling treatment of steel and some hydrochloric acid.  Broadly stated, the purpose of the ion exchange in zone 12 is to remove the ferrous ions in the liquor and replace them with hydrogen ions which form hydrochloric acid with the chloride ions present in the liquor.  The hydrochloric acid is, of course, then available for further use as a pickling agent in the steel pickling process.

The ion exchange material is a cation exchange resin in the hydrogen form, capable of providing hydrogen ions for exchange with the ferrous cations in the spent pickle liquor.  Exemplary of cation exchange resins which may be used in the process are strong acid resins such as the sulfonated styrene-divinylbenzene copolymers Dowex 50W–X8, Amberlite IR-120, Ionac C-240, and the like.  A preferred cation exchange resin for use in the process is Dowex 50W–X8 resin.

The spent pickle liquor to be treated is passed through the resin in continuous reactor 10 to remove its ferrous ions.  The reaction that occurs in ion exchange zone 12 can be exemplified by the following equation in which $RSO_3^-$ represents the cation exchanger resin:

$$FeCl_2 + 2RSO_3H \rightarrow Fe(RSO_3)_2 + 2HCl$$

## FIGURE 66: APPARATUS FOR TREATMENT OF SPENT HYDROCHLORIC ACID PICKLE LIQUOR FOR RECOVERY OF HYDROCHLORIC ACID

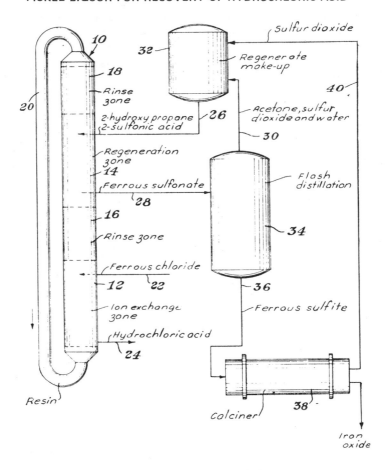

Source: L.J. Lefevre; U.S. Patent 3,522,002; July 28, 1970

The ferrous chloride in spent pickle liquor **22** is thus converted to its acid component producing an effluent **24** containing hydrochloric acid. Spent cation exchange resin from zone **12**, containing ferrous ions, is then passed through rinse zone **16** where the resin is washed with water to prevent cross-contamination between the hydrochloric acid formed in ion exchange zone **12** and the acid used in regeneration zone **14**.

While it is not essential to rinse the spent resin following ion exchange, such a rinse step is desirable because the presence of chloride ion in the resin would present a serious corrosion problem in the subsequent regenerant recovery step of this process.

The spent, and desirably rinsed, cation exchange resin containing ionically bound ferrous cations is regenerated in regeneration zone **14** by contacting the resin with a complex sulfonic acid solution such as described in U.S. Patent 3,248,278. This sulfonic acid solution is prepared by adsorbing sulfur dioxide in an aqueous solution of a suitable water-soluble aldehyde or ketone to produce a strong, complex sulfonic acid by the reaction:

$$H_2O \ + \ R\overset{\overset{\displaystyle O}{\|}}{C}R' \ + \ SO_2 \ \rightleftharpoons \ \underset{R'}{\overset{R}{>}}C\underset{SO_3H}{\overset{OH}{<}}$$

This complex sulfonic acid ($ZSO_3H$) is a highly effective resin regenerant in this system. Exemplary of carbonyl compounds suitable for use in the preparation of the complex sulfonic acid are water-soluble $C_1-C_8$ aldehydes and ketones, such as acetone, methyl ethyl ketone, acetaldehyde, formaldehyde, furfuraldehyde, isobutyraldehyde, cyclohexanone, benzaldehyde, methyl isobutyl ketone, mesityl oxide, and salicylaldehyde. Acetone is preferred and reacts with sulfur dioxide and water to produce 2-hydroxypropane-2-sulfonic acid.

At room temperature maximum ionization of the complex sulfonic acid can be most economically achieved using a weight ratio of about 7 parts water to about 1 part of acetone. An optimum sulfonic acid composition should utilize a minimum amount of acetone or other carbonyl compound, both because the carbonyl compound is the most expensive reagent in the system and because the solution should be as aqueous as possible to maximize ionization of the acid. At a weight ratio of 7 parts water to 1 part acetone, the complex sulfonic acid solution contains about 11% acetone, 12% sulfur dioxide and 77% water, and has a normality roughly equivalent to a 2N acid.

Complex sulfonic acid solution 26 containing $H^+$ cations and $ZSO_3^-$ anions is brought into contact with the spent cation exchange resin in regeneration zone 14 of continuous reactor 10. In this regeneration reaction, the hydrogen ions of the sulfonic acid solution are exchanged for the ferrous ions in the cation exchange resin to produce an effluent 28 containing ferrous sulfonate. The cation regeneration reaction can be exemplified by the following equation:

$$Fe(RSO_3)_2 \ + \ 2H^+ \ + \ 2ZSO_3^- \ \longrightarrow \ 2RSO_3H \ + \ Fe(ZSO_3)_2$$

The ferrous cations are thus removed from the resin as sulfonates and the resin is returned to the hydrogen form for further use in the initial treatment of spent pickle liquor. Ferrous sulfonate is relatively soluble in water and thus regenerant effluent 28 is an aqueous solution containing dissolved ferrous sulfonate.

The regenerated ion exchange resin from regeneration zone 14 of continuous reactor 10 is then passed through rinse zone 18 where it is washed with water to avoid loss of residual sulfonic acid in the resin following regeneration. While it is not essential to rinse the regenerated resin, the rinsing step is preferred because the carryover of sulfonic acid in the regenerated resin represents a loss of sulfur dioxide and the carbonyl compound. The regenerated and desirably rinsed, cation exchange resin is recycled from rinse zone 18 through passage 20 for reuse in ion exchange zone 12 of reactor 10.

Regenerant effluent 28 is treated to recover its carbonyl and sulfur dioxide components for further use in the regeneration of cation exchange resin. This recovery procedure is preferably carried out by thermal distillation of the regenerant effluent at about 100°C. in a suitable vessel 34.

Such distillation effectively strips the carbonyl compound, a portion of the sulfur dioxide, and some water from effluent 28 as distillate 30. Since all of distillate 30 is to be recombined to form the complex sulfonic acid regenerant, the distillate can be immediately passed to regenerant makeup tank 32 without the need for separating it into its various components. Distillation of regenerant effluent 28 also produces an insoluble precipitate of ferrous sulfite as can be exemplified by the equation shown on the following page.

$$Fe(ZSO_3)_2 \xrightarrow{\triangle} FeSO_3\downarrow + 2Z'\uparrow + SO_2\uparrow + H_2O$$

In the equation above Z' is the aldehyde or ketone originally used to form the sulfonic acid regenerant. While ferrous sulfonate can be decomposed in this equation at a temperature as low as about 100°C., the decomposition reaction is preferably carried out at about 250°C. to substantially remove all water from the product as water vapor, leaving an essentially dry ferrous sulfite product 36.

The dry ferrous sulfite product is calcined in rotary kiln 38 at a temperature of about 450°C. to decompose it and form sulfur dioxide and ferrous oxide. Decomposition of ferrous sulfite can be exemplified by the following equation.

$$FeSO_3 \xrightarrow{\triangle} FeO + SO_2$$

The sulfur dioxide given off by decomposition of ferrous sulfite is recycled at 40 to regenerate makeup vessel 32, thus providing for the substantial recovery of all of the sulfur dioxide used in preparing the cation regenerant solution. The iron oxide produced in calciner 38 may contain trace amounts of elemental sulfur, but it is not contaminated with chloride and therefore can be used in the steel making process.

It will be apparent from the foregoing description that this process provides a simple, economical and effective method for the recovery of hydrochloric acid from spent hydrochloric acid pickle liquor having several important advantages not heretofore realized in prior art processes. Such advantages include elimination of the problem of disposing of pickle liquor wastes, since the hydrochloric acid and iron oxide recovered by the process are both useful in the steel making process. Further, the chemical agent used to regenerate the ion exchange resin in the process is recovered substantially in its original form, and thus little or no cost is encountered in the replenishment of the regenerant once the process is on stream.

A process developed by A.B. Pashkov, N.M. Vdovin, O.N. Voronkova, R.R. Dranovskaya, A.M. Egorov, A.F. Kljushnev, P.I. Shatrin, M.Y. Zeigman, V. Patrushev and Y.V. Epshtein (17) involves processing acid wastes, which consist of monochlorodimethyl ether, hydrochloric acid and salts thereof and which appear as a result of the process of chloromethylating copolymers of styrene and vinylaromatic compounds with monochlorodimethyl ether in the presence of a Friedel-Crafts catalyst.

The wastes are treated with an excess of methanol. Next the treated wastes are rectified first at a temperature of 41° to 44°C. to recover methylal and then at a temperature of 60° to 70°C. to recover methanol. The still bottoms remaining after rectification may be heated to carbonize organic impurities contained therein, after which the precipitated carbonized impurities are removed, and the remaining still bottoms are evaporated to recover the Friedel-Crafts catalyst.

A process that has been developed by Y. Morimoto (18) for recovering hydrochloric acid from a spent hydrochloric acid pickle liquor is characterized by adding sulfuric acid to hydrochloric acid waste to convert $FeCl_2$ in the waste to HCl and $FeSO_4$ and to obtain a mixture containing at least 38% by weight of free sulfuric acid, distilling the resultant mixture to vaporize substantially all of the HCl therefrom together with water and to precipitate ferrous sulfate, condensing the HCl and water thus vaporized to recover hydrochloric acid, separating the precipitated ferrous sulfate from the residual liquid and circulating the resultant liquid free of ferrous sulfate as a sulfuric acid source.

### References

(1) Stahl, Q.R., "Air Pollution Aspects of Hydrochloric Acid," *Report PB 188,067*, Springfield, Va., National Technical Information Service (September 1969).

(2) *Hydrochloric Acid Manufacture*, Report No. 3, Air Pollution Control Association, TI-2 Chemical Committee (1962).

(3)  Kleckner, W.R., "Hydrochloric Acid," in *Kirk-Othmer Encyclopedia of Chemical Technology*, vol. 11 (New York: Interscience, p. 307, 1966).

(4)  Stern, A.C. (ed.), *Air Pollution, III*, 2nd ed. (New York: Academic Press, pp. 192-197, 1968).

(5)  "Hydrochloric Acid, Aqueous and Hydrogen Chloride, Anhydrous," *Chemical Safety Data Sheet SD-39*, Manufacturing Chemists Association, Washington, D.C. (1951).

(6)  Knott, K.H. and Turkolmez, S., "Krupp Rotary Brush Scrubber for the Control of Gas, Vapour, Mist and Dust Emissions," *Krupp Technical Review* (Essen) 24(1):25 (1966).

(7)  "Atmospheric Emissions from Hydrochloric Acid Manufacturing Processes," Joint National Air Pollution Control Administration and Manufacturing Chemists' Association, Inc. (To be published).

(8)  Faith, W.L., Keyes, D.B. and Clark, R.L., *Industrial Chemicals* (New York: Wiley, 1957).

(9)  Harris, L.S., "Fume Scrubbing with the Ejector Venturi System," *Chem. Eng. Prog. 62*(4):55 (1966).

(10) Gadomski, S.T., *Dry Packed Beds for the Removal of Strong Acid Gases from Recycled Atmospheres*, Chemistry Division, Naval Research Laboratory, Washington, D.C., NRL Rept. 6399 (1966).

(11) Seebold, J.E.; U.S. Patent 2,545,314; March 13, 1951; assigned to Hercules Powder Company.

(12) Wohlers, H.C., Grover, F.N., Lentz, C.W. and Pauling, L.E.; U.S. Patent 2,730,194; January 10, 1956; assigned to Michigan Chemical Corporation.

(13) Kunzer, W. and Leutert, F.; U.S. Patent 3,036,418; May 29, 1962; assigned to Badische Anilin- & Soda-Fabrik AG, Germany.

(14) Fuller, W.R., Bieler, B.H. and Morgan, D.C.; U.S. Patent 3,556,024; January 19, 1971; assigned to The Dow Chemical Company.

(15) Ezaki, S.; U.S. Patent 3,589,864; June 29, 1971; assigned to Yawata Chemical Engineering Co., Ltd., Japan.

(16) Lefevre, L.J.; U.S. Patent 3,522,002; July 28, 1970; assigned to The Dow Chemical Company.

(17) Pashkov, A.B., Vdovin, N.M., Voronkova, O.N., Dranovskaya, R.R., Egorov, A.M., Kljushnev, A.F., Shatrin, P.I., Zeigman, M.Y., Patrushev, V., and Epshtein, Y.V.; U.S. Patent 3,592,587; July 13, 1971.

(18) Morimoto, Y.; U.S. Patent 3,635,664; January 18, 1972; assigned to Daido Chemical Engineering Corporation, Japan.

# HYDROGEN CYANIDE

## Removal from Air

A process developed by P. Kaunert and H. Doering (1) involves the separation and recovery of HCN and $H_2S$ from a waste product gas which also contains ammonia.

It is known that for example in waters of coke oven plants the following toxic substances are present: ammonia, hydrogen sulfide, hydrocyanic acid and phenols. While the complete removal of ammonia and phenols does not offer much difficulty, the problem of removing hydrogen sulfide and hydrocyanic acid has not previously been satisfactorily solved. Both components pass into the wastewater in the course of the usual gas scrubbing processes.

For example, more than half of the hydrocyanic acid present in coke oven gas is stripped in the wet removal of hydrogen sulfide, in addition to the hydrogen sulfide, and also in the recovery of ammonia and benzene. Appreciable amounts of hydrocyanic acid are contained above all in the deacidification and stripping vapors from the circulatory-type ammonia and hydrogen sulfide wash usually employed for the scrubbing of coal distillation gases. Minor amounts also occur in the water discharged from ammonia expelling columns, in the wastewater from end stage washers serving for the scrubbing of ammonia remainders and in aqueous condensates from crude benzene stripping columns.

The deacidification and stripping vapors which contain considerable proportions of hydrocyanic acid are led, when using the so-called indirect process for the ammonium sulfate manufacture, through saturators in which the ammonia is combined with sulfuric acid. The effluent vapors, which contain hydrogen sulfide and hydrocyanic acid, are cooled and supplied, for example, to the combustion furnace of a sulfuric acid plant. At the same time, in the direct cooling, the vapors containing hydrogen sulfide and hydrocyanic acid are scrubbed, the whole of the hydrocyanic acid with a small amount of hydrogen sulfide passing into the washing water. The removal of the hydrocyanic acid is necessary to protect the sulfuric acid plant from corrosion. The wastewaters thus obtained containing hydrogen sulfide and hydrocyanic acid have previously in general been discarded without further treatment, a course which can no longer be permitted by reason of the pollution of streams and rivers which is increasing to a threatening degree.

It is true that distillative methods for the removal of hydrogen sulfide and hydrocyanic acid from industrial wastewaters have in principle been known, but all of these methods have such disadvantages that they could not be accepted in practice. The reasons for this are as follows.

The distillative stripping of hydrogen sulfide and hydrocyanic acid, in particular by reason of the water occurring in large amounts with only small concentration, requires a considerable consumption of energy which as a rule is covered by feeding steam to the stripping column. In some descriptions of the process there is found the intimation that the steam consumption can be considerably reduced by transferring the heat content of hot media from other sources to the wastewater to be stripped by heat exchangers. This of course overlooks the fact that in modern plants all available possibilities of heat exchange are already utilized and consequently the proposal to diminish the stream consumption of distillative stripping of wastewater by heat exchange only results in a shifting of the cost to other points of the plant.

The mere distillative stripping of wastewaters without the recovery of hydrogen sulfide and hydrocyanic acid, i.e., the discharge of the impurities from the aqueous phase into the ambient air, is after all to be rejected. Such methods merely shift the water and air pollution problem from the water side to the air side. For the same reason, all those methods of wastewater purification are to be rejected in which water containing hydrogen sulfide and hydrocyanic acid is delivered to natural draught or fan cooler towers. Apart from the fact that only part of the substances contained in the water can be removed in this way, the pollution of the air with the injurious substances discharged is intolerable.

Conventional distillative methods, which provide for the stripped-off hydrocyanic acid being recovered, have the disadvantage, in addition to requiring high cost of energy, that hydrogen sulfide remains in the wastewater so that for the complete purification further expensive measures are necessary.

Within the compass of quite different problems in which the recovery of hydrocyanic acid and not the purification of coke oven or other industrial wastewaters has been aimed at, attempts have been made to strip hydrocyanic acid from solutions with air or other gases. The methods based on this technique have not influenced the problem of purifying industrial wastewaters and exhibit at least for this problem considerable defects. With such measures the selectivity as regards hydrogen sulfide and hydrocyanic acid is not satisfactory so that the recovery of pure hydrocyanic acid is rendered difficult.

In order to be able to work selectively to some extent, it would be necessary that relatively large amounts of hydrogen sulfide should be retained in the water, which apart from the waste of hydrogen sulfide and hydrocyanic acid, is not permissible with regard to the desired purity of the wastewater and for the removal of the remainders of hydrogen sulfide from the water containing hydrocyanic acid the use of chemicals by which the hydrogen sulfide is bound would be necessary. This would result in a further contamination of the wastewaters. In a similar manner, the addition of acids usually made to acidify the wastewater prior to its aeration in one of the conventional aeration processes has the undesirable effect that apart from the consumption of acid with only a trivially increased selectivity, neutralizing agents, as for example lime, must be added before the evacuation of the wastewaters into public waters.

The process described here is shown schematically in the flow diagram in Figure 67.

FIGURE 67:  MULTISTAGE PROCESS FOR THE SEPARATION AND RECOVERY OF
HCN AND H₂S FROM WASTE PRODUCT GASES
CONTAINING AMMONIA

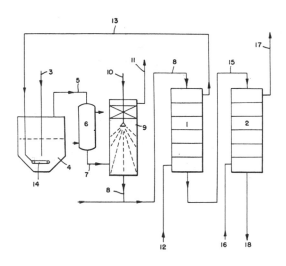

Source:  P. Kaunert and H. Doering; U.S. Patent 3,096,156; July 2, 1963

Vapors originating from the deacidification and stripping columns of a wet desulfurization
and ammonia production plant are fed in known manner, at a temperature of 94°C. and
a pressure of 875 mm. Hg, through an inlet pipe 3 into a saturator 4, which contains a
sulfuric acid bath of about 2 to 3% strength. The quantity of vapors fed in amounts to
820 cubic meters (NTP) per hour. The concentrations in hydrogen sulfide and hydrocyanic acid are 400 grams of H₂S per cubic meter (NTP) corresponding to 328 kilograms of
H₂S per hour, and 16.2 grams of HCN per cubic meter (NTP) corresponding to 13.3 kilograms of HCN per hour.

In the saturator 4 the ammonia contained in the vapors is combined with sulfuric acid to
form ammonium sulfate. Since the reaction proceeds exothermically, the temperature in
the saturator bath rises to 96°C. For the thorough agitation of the bath there are supplied
to the saturator through line 13 and an annular rose 14, instead of the usual agitation air,
286 cubic meters (NTP) per hour of a stripping air containing 20.6 grams of H₂S per cubic
meter (NTP) corresponding to 5.9 kilograms of H₂S per hour and 11.2 grams of HCN per
cubic meter (NTP) corresponding to 3.2 kilograms of HCN per hour, from a bubble-tray
column 1 of the first process stage.

In the saturator the ammonia contained in the vapors is completely combined with sulfuric
acid. 700 cubic meters (NTP) per hour of vapors leave the saturator through line 5 at a
temperature of 96°C. and a pressure of 760 mm. Hg. The vapors leaving the saturator contain 477 grams of H₂S per cubic meter (NTP) corresponding to 334 kilograms of H₂S per
hour and 23.6 grams of HCN per cubic meter (NTP) corresponding to 16.5 kilograms of
HCN per hour.

Since the water vapor partial pressure in the vapors above the sulfuric acid bath liquid
saturated with ammonium sulfate is about 90% of the vapor pressure above pure water,
the vapors leaving the saturator entrain about 2,000 kilograms of water vapor per hour.
The vapors are cooled consecutively in a heat exchanger 6 and a direct cooler 9. The

vapors which leave the heat exchanger **6** at a temperature of 72°C. are led through line **7** with the water vapor condensed in the heat exchanger into the direct cooler **9** which is fed through a feed pipe **10** with 12 cubic meters of water per hour having a temperature of 22°C. The water supplied through line **10** to the direct cooler **9** strips the hydrocyanic acid completely and the hydrogen sulfide to a small extent with reference to the total concentration of hydrogen sulfide in the vapor.

The vapors which are free of hydrocyanic acid and enriched with hydrogen sulfide are fed through line **11** at a temperature of 32°C. to a sulfuric acid plant (not shown) and processed to sulfuric acid. These vapors contain 473 grams of $H_2S$ per cubic meter (NTP) corresponding to 328 kilograms of $H_2S$ per hour, given a quantity of vapor of 694 cubic meters (NTP) per hour. The 14 cubic meters of water per hour flowing from the direct cooler **9** and composed of 12 cubic meters per hour of water fed in at **10** and 2 cubic meters per hour of water vapor condensate from line **5**, have a mixing temperature of 41°C.

The water discharged, which contains 422 milligrams of $H_2S$ per liter corresponding to 5.9 kilograms of $H_2S$ per hour and 1,180 milligrams of HCN per liter corresponding to 16.5 kilograms of HCN per hour, is led through line **8**, which also serves to feed in washing waters or condensates containing hydrocyanic acid, and minor amounts of hydrogen sulfide, coming from other operations, into the bubble-tray column **1**. 280 cubic meters (NTP) per hour of air are fed through line **12** and passed in countercurrent through column **1**. In this way, practically the whole of the hydrogen sulfide, down to 0.8 milligram of $H_2S$ per liter (corresponding to 11 grams of $H_2S$ per hour), passes from the water into the stripping air, i.e., 99.8% of the hydrogen sulfide contained in the water fed into the column is stripped. At the same time, the water only gives off 19.4% of the hydrocyanic acid content to the stripping air.

The air containing hydrogen sulfide and hydrocyanic acid which leaves at the top of column **1** is fed, as described above, through line **13** into the saturator **4**. The water drained from the column at a temperature of 40°C. still contains, besides a trivial residual amount of 0.8 milligram of $H_2S$ per liter, 80.6% of the hydrocyanic acid present in the feed water. The hydrocyanic acid concentration in the water drained from column **1** amounts to 950 milligrams of HCN per liter which corresponds to 13.3 kilograms of HCN per hour given an unchanged drainage of 14 cubic meters of water per hour.

The amount of air required in the first process stage depends, inter alia, on the amount of hydrogen sulfide contained in the water. In general, 20 to 50 cubic meters (NTP) of air per cubic meter of water supplied are sufficient for the complete removal of the hydrogen sulfide therefrom. In the above example, a ratio of air to water of 20:1 is assumed.

In the second process stage, the water leaving the column **1** practically free of hydrogen sulfide is fed through line **15** to the bubble-tray column **2**. The stripping air enters the lower part of column **2**, becomes enriched with hydrocyanic acid and leaves overhead through line **17**. The amount of air blown in is 2,400 cubic meters (NTP) per hour. On flowing through column **2**, about 11 cubic meters (NTP) of HCN per hour pass into the stripping air.

From the sump of column **2** the water is drained at 38°C. with a negligible hydrocyanic acid concentration of 5.4 milligrams of HCN per liter corresponding to 75 grams of HCN per hour, i.e., 99.4% of the hydrocyanic acid contained in the water fed into column **2** is stripped in the second process stage. The effluent water also contains 0.7 milligram of $H_2S$ per liter; this means that the extremely small amount of hydrogen sulfide left in the water discharged from the first process stage is not stripped in the second process stage either. Consequently, the stripping air enriched with hydrocyanic acid which leaves by way of line **17** is entirely free from hydrogen sulfide.

The water leaving column **2** is fed through line **18** to a main canal or to the sewage system, but can also be used partly or wholly as cooling water in the direct cooler **9**. The air required in the second process stage ranges from 100 to 200 cubic meters (NTP) per cubic meter of water depending on the hydrocyanic acid concentrations in the feed water and on the manner of operation and the design of the column 2. In the above example, the ratio of air to water is about 170:1.

The hydrocyanic acid contained in the stripping air, free of hydrogen sulfide, of the second process stage which leaves through line 17 can be processed by conventional methods into pure liquid hydrocyanic acid or, for example, sodium or potassium cyanide. The hydrocyanic acid having thus been removed again from the stripping air of the second process stage, the completely clean stripping air is either released into the atmosphere or recycled and reused as stripping air in the second process stage.

**References**

(1) Kaunert, P. and Doering, H.; U.S. Patent 3,096,156; July 2, 1963; assigned to Gewerkschaft Auguste Victoria, Germany.

# HYDROGEN FLUORIDE

**Removal of HF from Air**

See Fluorine Compounds, Removal from Air.

# HYDROGEN SULFIDE

Hydrogen sulfide ($H_2S$) is a colorless gas that has an obnoxious odor at low concentrations. The odor threshold is in the $\mu g./m.^3$ range. In higher concentrations, the gas is toxic to humans and animals and corrosive to many metals. It will tarnish silver and react with heavy metals in paints to discolor the paint. In humans, it will cause headache, conjunctivitis, sleeplessness, pain in the eyes, and similar symptoms at low air concentrations and death at high air concentrations. However, the majority of the complaints arising from hydrogen sulfide air pollution are due to its obnoxious odor in extremely low air concentrations.

Air pollution by hydrogen sulfide is not a widespread urban problem but is generally localized in the vicinity of an emitter such as kraft paper mills, industrial waste disposal ponds, sewage plants, refineries, and coke oven plants according to Miner (1).

Hydrogen sulfide is highly toxic to humans, and at concentrations of over 1,000,000 $\mu g./m.^3$ quickly causes death by paralysis of the respiratory system. At lower concentrations, hydrogen sulfide may cause conjunctivitis with reddening and lachrymal secretion, respiratory tract irritation, psychic changes, pulmonary edema, damaged heart muscle, disturbed equilibrium, nerve paralysis, spasms, unconsciousness, and circulatory collapse.

The odor threshold for hydrogen sulfide lies between 1 and 45 $\mu g./m.^3$. Above this threshold value, the gas gives off an obnoxious odor of rotten eggs, which acts as a sensitive indicator of its presence. At these concentrations, no serious health effects are known to occur. At 500 $\mu g./m.^3$, the odor is distinct; at 30,000 to 50,000 $\mu g./m.^3$, the odor is strong, but not intolerable; at 320,000 $\mu g./m.^3$, the odor loses some of its pungency, probably due to paralysis of the olfactory nerves. At concentrations over 1,120,000 $\mu g./m.^3$, there is little sensation of odor and death can occur rapidly. Therefore, this dulling of the sense of smell constitutes a major danger to persons exposed to high concentrations of hydrogen sulfide. Hydrogen sulfide produces the same health effects on domestic animals as on man, and at approximately the same concentrations.

An episode occurred at Poza Rica, Mexico, where the accidental release of hydrogen sulfide from a natural gas plant killed 22 persons, hospitalized 320 people and killed 50% of the commercial and domestic animals and all the canaries in the area. No measurements were made of the environmental hydrogen sulfide concentrations at the time of the episode. In Terre Haute, Indiana, hydrogen sulfide emanations from an industrial waste lagoon caused foul odor, public complaints, and discomfort. However, very few people sought medical attention. Hydrogen sulfide concentration in the atmosphere during the episode ranged between 34 and 450 $\mu$g./m.$^3$.

There is little evidence that significant injury to field crops occurs at hydrogen sulfide concentrations below 60,000 $\mu$g./m.$^3$. At higher concentrations, the hydrogen sulfide injures the younger plant leaves first, then middle-aged or older ones.

Hydrogen sulfide combines with heavy metals in paints to form a precipitate which darkens or discolors the paint surface. Air concentrations as low as 75 $\mu$g./m.$^3$ have darkened paint after a few hours' exposure. White lead paints often fade in the absence of hydrogen sulfide due to oxidation of the sulfite to sulfate. Paint darkening has occurred in Jacksonville, Florida, New York City, South Brunswick, N.J., Terre Haute, Indiana, and in the areas near Lewiston, Idaho, and Clarkston, Washington. Hydrogen sulfide will also tarnish silver and copper. The sulfide coating formed on copper and silver electrical contacts can increase contact resistance and even weld the contacts shut.

The states of California, Missouri, Montana, New York, Pennsylvania, and Texas have ambient air quality standards. The standards vary from a level of 150 $\mu$g./m.$^3$ averaged over one hour in California to a 24 hour average of 7.5 $\mu$g./m.$^3$ for Pennsylvania.

Hydrogen sulfide is produced naturally by biological decay of protein material, mainly in stagnant or insufficiently aerated water such as swamps and polluted water. The background air concentrations due to this source is estimated to be from 0.15 to 0.46 $\mu$g./m.$^3$. The average concentrations of hydrogen sulfide found in the urban atmosphere will vary from undetectable amounts to 92 $\mu$g./m.$^3$ (based on limited data). However, measurements as high as 1,400 $\mu$g./m.$^3$ have been recorded, but these have generally been in the vicinity of high hydrogen sulfide emissions.

The general area of hydrogen sulfide removal processes has been reviewed by P.G. Stecher (2).

### Removal of H$_2$S from Air

A number of systems and types of equipment have been developed for removal of hydrogen sulfide from gas streams. Many of these systems are designed to recover the hydrogen sulfide for subsequent conversion to valuable by-products, such as sulfur and sulfuric acid. Many of the removal systems are based on scrubbing the gas streams with a suitable absorbent and then removing the absorbed gas from the absorbent for disposal by burning or conversion to a valuable by-product. Some of the absorbents convert the hydrogen sulfide to an innocuous compound which may be useful in some cases as a fertilizer. Such chemicals as aqueous solutions of diethanolamine and monoethanolamine, sodium hydroxide, tripotassium phosphate, and aqueous solutions of chlorine and sodium carbonate have been used as absorbents. Different types of contacting devices (wet scrubbers) have been used, including conventional and novel design spray towers, plate towers, and venturi scrubbers.

*Kraft Paper Mills:* In kraft paper mills, the greatest reduction of hydrogen sulfide emissions was achieved by the black liquor oxidation process. This process consists of oxidizing the sulfides in the weak black liquor (before the multiple effect evaporation) or strong black liquor (after the multiple effect evaporation) by contacting it with air in a packed tower, thin film, or porous plate black liquor oxidizing unit. The oxidation converts the sulfides to less volatile compounds which are also less odorous and have less tendency to escape (3). This conversion has the effect of reducing the hydrogen sulfide emissions from the direct contact evaporator and the recovery furnace stack by 80 to 95% (3)(4)(5).

Since these evaporators and furance stacks are the principal emitters of hydrogen sulfide in kraft mills (6), the net result is a substantial reduction of hydrogen sulfide emission. The weak black liquor oxidizing process also reduces emission from the multiple effect evaporators.

The majority of the black liquor oxidizing systems installed in the United States are based on oxidation of weak liquor, and are located in the western part of the country. In the southern part of the country, the woods used in kraft processes cause excessive foaming problems in the weak black liquor oxidizing process (5)(7). To alleviate this, a few Southern mills have installed a process based on oxidizing the strong black liquor (7)(8).

The key to minimizing hydrogen sulfide emissions from the recovery furnace, even in those systems employing black liquor oxidizing systems, is proper furnace-operating conditions (4). At furnace loading greater than design capacity, hydrogen sulfide emissions rise substantially. For minimum emissions from the recovery furnace, the furnace should not be operated above design conditions. There should be 2 to 4% excess oxygen leaving the secondary burning zone (i.e., leaving the furnace), and there should be adequate mixing (turbulence) in the secondary combustion zone.

In the direct contact evaporator, where the flue gases from the recovery furnace are used to concentrate the black liquor, the carbon dioxide in the flue gases reacts with the sulfite in the black liquor to release hydrogen sulfide (7). Even where the black liquor oxidizing process is employed, some sulfite remains after oxidation. This releases some hydrogen sulfide when contacted with flue gases. Therefore, removal of the direct contact evaporator from the stream further reduces sulfide emissions.

In the Scandinavian countries the recovery furnaces are designed and operated in such a manner that black liquor can be fed directly to them from the multiple effect evaporators. These furnaces efficiently burn all malodorous compounds; the necessity for oxidizing black liquor prior to burning is therefore eliminated. Approximately 45 mills in Sweden use additional multieffect evaporators to replace the direct contact evaporator and burn unoxidized black liquor (9). Three such installations are being installed in North America (9). The exact location of these units was not specified.

Another approach to burning unoxidized black liquor while still minimizing the emission of malodorous compounds was described by Hochmutch (10). In this system, the combustion gases from the recovery furnace are used to preheat air in a recuperative air preheater. The hot air is then used to concentrate the black liquor in the direct contact evaporator. The air from the direct contact evaporator is then used as primary and secondary air in the recovery furnace, where malodorous compounds in the air are incinerated in the high temperature combustion zones.

To reduce recovery furnace particulate emissions, some mills have installed a secondary wet scrubber to follow the primary scrubber (direct contact evaporator). Secondary scrubbing does not remove hydrogen sulfide unless a basic solution such as weak caustic is used. Limited pilot plant studies and some plant experience have shown that weak wash (i.e., weak caustic solution) has removed hydrogen sulfide from the stack gases. In other instances, no hydrogen sulfide removal has been obtained under such a system. In general, the removal of hydrogen sulfide from flue gases containing 11 to 14% carbon dioxide with a caustic solution has not been developed (5)(11)(12)(13).

Other sources of hydrogen sulfide emissions from kraft mills are the noncondensible gases released from digesters and multiple effect evaporators. Various systems developed and installed for minimizing these emissions are generally based on collecting the noncondensible gases in a gas holder, then oxidizing or burning them at a constant flow rate. The various methods used are the following.

[1]  Burning the gases in the recovery furnace or lime kiln (16).

[2]  Oxidizing the gases in a separate catalytic oxidizing furnace or
a direct flame incinerator (5)(14).

[3]  Oxidizing the gases in an absorption tower with aqueous chlorine
solutions, such as chlorine bleach water from the bleach plant,
waste chlorine, hypochlorite, etc.  Sometimes this is followed by
processing in another absorption tower, where the absorbent is
either a weak chlorine solution or a caustic solution (15)(16)(17).

[4]  Absorbing the gases with a caustic solution in a scrubber (15).

In the lime kiln, the use of wet scrubbers with an alkaline-absorbent, efficient control of
combustion, and proper washing of lime mud will substantially reduce hydrogen sulfide
emissions.  Scrubbing smelt tank gaseous emissions with weak wash or green liquor in an
absorption tower will reduce hydrogen sulfide emissions from this source (16).

Around 1951, masking of odors by adding aromatic compounds to the digester, the black
liquor, and the stack gases was tried in the United States.  This strictly makeshift approach
did not solve the basic pollution problem and is not used at the present time (17).

*Petroleum Industry and Petrochemical Plants:*  In refineries and petrochemical plants, small
quantities of hydrogen sulfide associated with gas streams can be burned in the plant full
system or in a flare (18).  In refineries and natural gas plants where larger quantities of
hydrogen sulfide are associated with gas streams (sour gas), the hydrogen sulfide is generally extracted in an absorption tower using a number of different absorbents, such as aqueous amines (Girbotol process); sulfalone (Shell process); alkaline arsenites and arsenates
(Gianmarco-Vetrocoke process); organic solvents such as propylene carbonate, glyceral triacetate, butoxy-diethyleneglycol acetate, methoxy-triethylene glycol acetate (Fluor solvent
process); and many others.  These processes are regenerative — that is, the absorbent is regenerated by removing the hydrogen sulfide and reused.

In the case of the Gianmarco-Vetrocoke process, sulfur is recovered directly as part of
the absorbent regeneration process.  In the other processes the hydrogen sulfide from the
regeneration process is converted to sulfur by the Claus process (19)(20)(21).  The sulfur
can be further processed into sulfuric acid if this is the end product desired.

*Coke Oven Plants and Chemical Plants:*  In coke oven plants, the coke oven gases are often
purified of hydrogen sulfide by passing the gases through iron oxide impregnated wood
shavings (19)(22)(23).  This process is generally nonregenerative, although methods for regenerating the iron oxide have been developed (24).  Regenerative liquid absorption systems using such absorbents as ammonium carbonate, sodium thioarsenate, and sodium ar-
senate solutions have also been used (19)(25).

Similarly, various liquid absorbents have been used in the chemical industry for removal of
hydrogen sulfide in gas streams.  For example, hydrogen sulfide liberated in the production
of sulfur dyes in aniline plants is effectively absorbed by alkali in scrubbers (26).  Another
commonly used method for preventing release of hydrogen sulfide to the atmosphere in
the chemical industry is to collect the various gaseous vents and destroy them by incineration (27).

*Coal Piles:*  The pollution of the atmosphere in the vicinity of burning coal refuse piles
can be minimized by constructing the refuse piles in such a manner that ignition is minimized and it is possible to easily extinguish a fire (28).

*Tanneries:*  In the tanning industry, practices adopted in modern tanneries in Russia have
essentially eliminated the hydrogen sulfide problem.  These consist of more rapid processing
of raw material, use of lime solutions to destroy hide proteins and alkalize sodium sulfide,
and neutralization of the semimanufactured products to eliminate residual sodium sulfide
which previously contaminated acid-tanning solutions (29).

In sewage plants, the most comprehensive elimination of hydrogen sulfide is accomplished by enclosing the process and venting the gases to an incinerator (30). Other methods of removing hydrogen sulfide are absorbing or chemically oxidizing the gas. The oxidation process is utilized in New York City and Sarasota, Florida (30). Other methods consist of odor-masking with scented mint, catalytic combustion, and odor counteraction.

In sewers, the production and release to the atmosphere of hydrogen sulfide can be minimized by maintaining sufficient velocities of sewage to avoid sulfide buildup and minimizing lines of pressure and points of high turbulence. Atmospheric pollutions may also be controlled by adequate ventilation, injection of air to maintain aeration conditions, cleaning of sewers to remove slime and silt, use of chemicals such as chlorine and ozone for suppressing biological activity (31), and addition of specific biological life to suppress the development of organisms producing the hydrogen sulfide (14). A method of preventing release of hydrogen sulfide to the atmosphere which has had some degree of success is trapping the gas in laterals, branches and mains by use of specially designed junctions (14). A method utilized by the County Sanitation District of Los Angeles to control hydrogen sulfide is to add lime slurry to sewage periodically in relatively large quantities.

*General Abatement Systems:*  Kalyuzhnyi et al (32) reported that reduction of hydrogen sulfide and other pollutants in air was achieved by placing a green vegetation belt between the industrial emitter and the residential areas. They observed that hydrogen sulfide concentrations outside the green belt then decreased from 70 to 30 $\mu$g./m.$^3$ at 500 meters, while inside the green vegetation belt the hydrogen sulfide concentrations decreased from 70 to 25 $\mu$g./m.$^3$; this difference was considered significant by the author.

A process developed by F.X. Pollio, K.A. Kun and R. Kunin (33) is a process for removing hydrogen sulfide and/or ammonia from gaseous streams. The $H_2S$ gas is removed by treatment with substantially anhydrous macroreticular or macroporous ion exchange resins having a prescribed minimum surface area and ion exchange functionality. The $NH_3$ is removed by treatment with carboxylic type cation exchange resins.

Figure 68 shows the essentials of this process. In a typical mixture which was treated in this manner the composition of the gases was as follows (in volume percent): 55.50 $CH_4$; 42.80 $N_2$; 0.85 $H_2S$; and 0.85 $NH_3$. Column 1 was a premixer containing no resin, and serving merely to mix the gases thoroughly before proceeding to the next column. Column 2 contained a carboxylic macroreticular structured cation exchange resin in the hydrogen form.

A particularly preferred carboxylic cation exchanger is one prepared by suspension copolymerizing a mixture of methacrylic acid and about 3 to 10% divinylbenzene. Column 2 optionally can precede or follow Column 3. A further option is to eliminate Column 2 altogether, in the event that it is not desired or required to remove $NH_3$, and only $H_2S$ is to be removed. The next column through which the gases are passed is Column 3, containing Resin B.

A resin of the type which works effectively to remove $H_2S$, hereinafter identified as Resin B, is the chloride form of a cross-linked quaternary ammonium synthetic polymer, preferably of styrene and divinylbenzene, which is prepared in the presence of a diluent that is substantially removed from the polymer thus formed. The resin may be made of other polymers and has a macroreticular structure.

A description of such polymers and structure, along with detailed explanations for making them, will be found in U.S. Patent 3,247,242. A critical requirement for the resin is that it must have a minimum surface area of approximately 5 m.$^2$/g., and it is also essential that it be employed in an anhydrous system. A further critical requirement is that it have a minimum of 1.0 meq./g. anion exchange capacity. Resins failing to meet either of these minima are incapable of performing satisfactorily. After the mixture of gases had been processed to a point where some $H_2S$ leakage from Column 3 was detected, the system was shut down and Resin B in Column 3 was thermally regenerated with nitrogen at 100°C.

FIGURE 68: APPARATUS FOR REMOVAL OF $H_2S$ AND $NH_3$ FROM GAS STREAMS

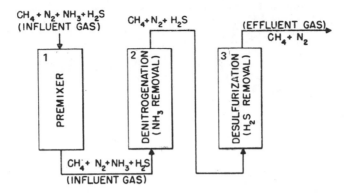

Source: F.X. Pollio, K.A. Kun and R. Kunin; U.S. Patent 3,556,716; January 19, 1971

until substantially all the $H_2S$ which had been picked up on that resin was removed, as established by a standard moist lead acetate paper test which can detect as little as a few parts per million $H_2S$.

A process developed by R. Hohne (34) involves removal of $H_2S$ by partial oxidation of the $H_2S$ to $SO_2$ and recovery of resulting elemental sulfur by adsorption and activated carbon. The following is an example of the operation of this process.

*Example:* Exhaust gas composed of 99% $CO_2$ and 1% $H_2S$ by volume was passed through the plant at a rate of 300 standard cubic meters per hour, which means a rate of 150 standard cubic meters of exhaust gas flowing per hour through each of the two activated carbon bins included in the crude gas stream. Before entering the bins, the exhaust gas was heated to 130°C. and was mixed with such a quantity of air that oxygen slightly in excess of the stoichiometric quantity with respect to $H_2S$ entered the crude gas. The hydrogen sulfide was thus oxidized in the activated carbon beds to form sulfur and small amounts of free sulfur dioxide.

The connection of the bins was changed after intervals of 17.5 hours so that each bin was included in the exhaust gas stream for 35 hours and then in the regeneration cycle for 17.5 hours. The exhaust gas was thus purified to 2 to 10 ppm $H_2S$ and 300 to 500 ppm $SO_2$. The regeneration was effected over a 17.5 hour period by passing nitrogen, at a temperature of 500°C., through the activated carbon for 13 hours.

A process developed by S. Suzuki, G.H. Tjoa and K.H. Kilgren (35) involves selectively removing $H_2S$ and like sulfides from fluids containing them by contact with a cyanopyridine (e.g., a mixture of ortho and meta-cyanopyridines) and an alkali hydrosulfide, preferably in a substantially hydroxyl-free solvent such as N-methyl pyrrolidone. Preferably, in the process an admixture of $H_2S$ and $CO_2$ in natural gas is contacted with the cyanopyridine-containing contacting solution to react the $H_2S$ with the cyanopyridine, the $CO_2$ and/or like hydrocarbons are rejected from the contacting solution by mild heating and/or pressure reduction and thereafter $H_2S$ is regenerated by heating the remaining solution.

A process developed by M.L. Roberts, C.E. Johnson and R.C. Miller (36) involves removing hydrogen sulfide from gaseous streams by contacting the gaseous stream with an absorbing solution of from 0.005 to 20% of a ferric ion complex, from 25.0 to 99.945% water, from 0.05 to 10.0% of a buffering agent, and from 0 to 74.945% of a water-soluble organic solvent selected from the group consisting of dimethylformamide and dimethyl-sulfoxide.

## Removal of H₂S from Water

The removal of $H_2S$ from water has been reviewed by P.G. Stecher (2).

A process developed by L.L. Wilkerson (37) involves treating wastewater associated with hydrocarbon production by steam or hot water secondary recovery methods wherein residual hydrogen sulfide gas concentrations in the water are substantially removed therefrom by spraying the water vertically into the air and the temperature of the water is lowered to a point whereby it will not upset the ecological balance of surrounding bodies of water.

Such hydrogen sulfide concentrations may be formed by bacterial action or sulfite or sulfate ions found in the formation water and/or reactions with respect to sulfur compounds found in the formation caused by the heat of the steam. For example, sulfur may be naturally present in the oil itself or in materials, e.g., iron pyrites, found in the formation.

A suitable form of apparatus for the conduct of this process is shown in Figure 69. In the figure, an oil production well **10** is shown schematically within the earth **11** with production fluid including wastewater flowing upwardly through line **11a** out of the well and into a separating station **12**. The separating station serves to substantially separate the hydrocarbon components of the production fluid from the wastewater associated therewith in a well-known manner. For example, a flotation cell system or any other desired type of separator system may be used for this purpose.

While most of the water separates freely from the crude, some of the water often occurs as microscopic droplets in a water-in-oil emulsion and it may be necessary to treat the production fluid in the separating station by means of one or more of the known methods to ensure substantial separation of the water and hydrocarbon constituents. For example, the emulsion may be heated and chemicals added or it may be passed between electrodes maintained at a high alternating voltage potential, to mention just a few of the commonly used techniques.

In any event, the separated oil is removed from the separating station through drain line **13** and removed to a suitable storage facility (not shown) for subsequent treatment. The separated wastewater is sucked from the separating station through a suction line **14** by means of a pump **15**, which is preferably a constant pressure pump, or a pump provided with means for dampening pressure variations.

The pump forces the wastewater effluent from separating station **12** into a conduit **16** under a substantially constant pressure. From the conduit the pressurized water flows into a suitable spray manifold **17** which may comprise a hollow substantially tubular-shaped member or housing means **18** closed at the one end thereof remote from conduit **16** and having a plurality of spray nozzles **19** through **23** communicating with the interior thereof and directed in an upward direction.

Spray manifold **17** is disposed over a collection basin **24** of any suitable design. Since the water is maintained under pressure, it emerges from tubular-shaped member **18**, through spray nozzles **19** through **23**, in the form of water sprays **19'** through **23'** which spray into the surrounding atmosphere in the manner illustrated.

After reaching a maximum predetermined spray height, the water falls downwardly into collection basin **24** and collects therein. The basin may be in the form of a natural or manmade depression on the earth's surface. The maximum height of the sprays **19'** through **23'** will, of course, depend upon the pressure of the water and the dimensions of the nozzles **19** through **23**. The spray height must be sufficient to aerate the water to such an extent that the undesirable quantities of $H_2S$ are removed from the water and dissipated to the surrounding atmosphere, and the water is cooled to a point whereby disposal thereof

FIGURE 69:  APPARATUS FOR TREATING WASTEWATER ASSOCIATED WITH
HYDROCARBON PRODUCTION TO REMOVE H$_2$S

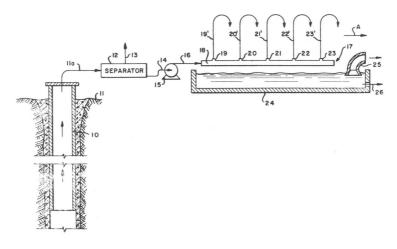

Source:   L.L. Wilkerson; U.S. Patent 3,547,190; December 15, 1970

will not upset the ecological balance of surrounding bodies of water in which it will
eventually be disposed.  The arrow A represents in schematic fashion the removal of the
H$_2$S from the sprayed water.  The spray nozzles may be of any suitable construction.  It
has been found that when five nozzles having 2 inch openings are employed in the instant
arrangement, wastewater will be sprayed approximately 75 feet into the air when pump
pressure is maintained at about 350 pounds.  This spray height has been found sufficient
to reduce measured concentrations of H$_2$S in the order of 95 ppm to trace levels in the
wastewater processed through the spray system.

In addition, wastewater may be reduced from 100°-125° down to 75°F. so as not to disturb
the ecological balance of surrounding bodies of water.  Spray height may of course be
varied as desired to compensate for changing conditions with respect to H$_2$S concentrations,
water treatments, air movement, etc.  This may be done simply by varying the size of the
nozzle openings and/or the pump pressure.  In addition one or more of the plurality of
spray nozzles (which may be of any desired total number) may be shut down if so desired.

The water collecting in basin 24 after the aeration thereof may contain traces of oil or
other hydrocarbon residue not completely removed by separating station 12.  It is desirable to remove this oil from the water before releasing it from the basin to flow into the
surrounding area.  Since the oil would tend to float on top of the water in the collection
basin, removal of the oil may be accomplished readily by skimming it off the surface by
means of vacuum line 25 which may be connected to a vacuum storage truck (not shown).
The oil thus collected may then be disposed of subsequently at any desired location.
Wastewater may then be drained from the collection basin through outlet or drain pipe 26
into the surrounding countryside.

It should thus be readily apparent that the above described method and apparatus provide
an inexpensive approach for the treatment of wastewater associated with oil production.
Residual H$_2$S concentrations in the water may be drastically reduced and the water may be
cooled without the necessity of constructing relatively expensive cooling tower systems,
which may cost in the order of $45,000 or more, for this purpose.  Although the system
has been described with particular reference to a well wherein steam or hot water secondary

recovery techniques have been employed, it should be understood that it has equal application to those situations wherein water actually produced with the oil from underground formations is to be treated.

### References

(1) Miner, S., "Air Pollution Aspects of Hydrogen Sulfide," *Report PB 188,068,* Springfield, Va., National Technical Information Service (September 1969).

(2) Stecher, P.G., *Hydrogen Sulfide Removal Processes,* Park Ridge, N.J., Noyes Data Corp. (1972).

(3) Kencine, P.A. and Haces, J.M., "Air Pollution and the Kraft Pulping Industry," *Public Health Service Publ.* 999-AP-4, Washington, D.C. (1963).

(4) Hendrickson, E.R. and Harding, C.I., "Black Liquor Oxidation as a Method of Reducing Air Pollution," *J. Air Pollution Control Association* 14 (12) (1964).

(5) Landry, J.E. and Longwell, D.H., "Advances in Air Pollution Control in the Pulp and Paper Industry," *Tappi* 4 (6) (1965).

(6) "Permissible Emission Concentrations of Hydrogen Sulfide," Subcommittee on Effects of Hydrogen Sulfide of the Committee on Effects of Dust and Gas of the Verein Deutscher Ingenieure Committee on Air Purification, *VDI* 2107 (1960).

(7) Owens, V.P., "Considerations for Future Recovery Units in Mexican and Latin American Alkaline Pulping Mills," *Combustion* (November 1966).

(8) Owens, V.P., "Trends in Odor Abatement from Kraft Mills Recovery Units," *Paper Trade J.* (August 1968).

(9) Clement, J.L. and Elliott, J.S., "Kraft Recovery Odor Control," presented to the 4th. Paper Industry Air and Stream Improvement Conference, Halifax, Nova Scotia (September 17, 1968).

(10) Hochmutch, F.W., "Odor Control Systems for Chemical Recovery Units," *Paper Trade J.* (September 1968).

(11) Blosser, R.O. and Cooper, H.B.H., Jr., "Secondary Scrubbing of Kraft Recovery Stack Gas," presented at the 61st. Annual Meeting, Air Pollution Control Association, St. Paul, Minnesota, *Paper No. 68–129* (June 23 to 27, 1968)

(12) Blosser, R.O. and Cooper, H.B.H., Jr., "Trends in Atmospheric Particulate Matter Reduction in the Kraft Industry," *Tappi* 51, 5 (1968).

(13) Broaddus, T.E., et al, "Air Pollution Abatement at S.D. Warrens Kraft Mill in Westbrook, Maine," *Tappi,* 50, 8 (1967).

(14) Santry, I.W., Jr., "Hydrogen Sulfide Odor Control Measures," *J. Water Pollution Control Federation* 38, 3, 459 (1966).

(15) Jensen, G.A., et al, "Absorption of Hydrogen Sulfide and Methyl Mercaptan from Dilute Gas Mixtures," *J. Air Pollution Control Association,* 16, 5, 248 (1966).

(16) Shan, I.S., "Pulp Plant Pollution Control," *Chem. Eng. Prog.* 64, 9 (1968).

(17) Wenzl, H.F. and Ingruber, O.V., "Controlling Problems of Air and Water Contamination," *Paper Trade J.* 151, 3:42 (1967).

(18) Mencher, S.K., "Change Your Process to Alleviate Your Pollution Problem," *Petro/Chem Eng.* (1967).

(19) Falgout, D.A. and Harding, C.I., "Determination of H$_2$S Exposure by Dynamic Sampling with Metallic Silver Filters," *J. Air Pollution Control Association* 18:15 (1968).

(20) Maddox, R.N. and Burns, M.D., "Physical Solution is the Key to these Treating Processes," *Oil Gas J.* (January 1968).

(21) Munro, A.J.E. and Masdin, E.G., "A Study of a Method for Desulfurizing Fuel Gases," *Brit. Chem. Eng.* 12, 3 (1967).

(22) Altybaev, M. and Streltsov, V.V., "Removal of Sulfur Compounds from Gaseous Fuels," *Coke Chem. (USSR)* 8:43 (1966).

(23) Strimbeck, D.C., "Clean Gas from Coal May Be Economical Fuel for Gas Turbines," *Power Engineering* (July 1966).

(24) Maddox, R.N. and Burns, M.D., "Solids Processing for Gas Sweetening," *Oil Gas J.* 66, 25:167 (1968).

(25) Ganz, S.N. and Likshin, M.A., "Coke Gas Purification from Hydrogen Sulfide in High Speed Rotary Absorbers," *USSR Literature on Air Pollution and Related Occupational Diseases* 4:85 (1960).

(26) Matsak, V.G., "The Purification of Air Pollution by Vapors and Gases from the Central Sanitary and Hygienic Laboratory in Moscow," *Gigiena i Sanit.* 8 (1950).

(27) Stutz, C.N., "Treating Parathion Wastes," *Chem. Eng. Progr.* 62, 101 (1966).

(28) Sussman, V.H. and Mulhern, J.J., "Air Pollution from Coal Refuse Disposal Areas," *J. Air Pollution Control Association* 14, 7 (1964).

(29) Sinitsyna, E.L., "Investigations into Certain Aspects of the Health of People Working in Main Shops of Tanneries," *Hyg. Sanitation* 30, 6:336 (1965).

(30) Ledbetter, J.O., "Air Pollution from Wastewater Treatment," *Water Sewage Works* 113, 2:43 (1966).

(31) Mohanrao, G.J., Sastry, C.A. and Garber, W.F., "Hydrogen Sulfide in Concrete Sewers and Digesters," *J. Inst. Eng.* (India) 46, 6:90 (1966).
(32) Kalyuzhnyi, Y., et al, "Effectiveness of Sanitary Clearance Zones Between Industrial Enterprises and Residential Quarters," *Gigiena i Sanit.* 4:179 (1952).
(33) Pollio, F.X., Kun, K.A. and Kunin, R.; U.S. Patent 3,566,716; January 19, 1971; assigned to Rohm and Haas Company.
(34) Hohne, R.; U.S. Patent 3,634,028; January 11, 1972; assigned to Metallgesellschaft AG, Germany
(35) Suzuki, S., Tjoa, G.H. and Kilgren, K.H.; U.S. Patent 3,656,887; April 18, 1972; assigned to Chevron Research Company.
(36) Roberts, M.L., Johnson, C.E. and Miller, R.C.; U.S. Patent 3,676,356; July 11, 1972; assigned to Nalco Chemical Company.
(37) Wilkerson, L.L.; U.S. Patent 3,547,190; December 15, 1970; assigned to Shell Oil Company.

# IODINE

Iodine-131 is a gaseous nuclear fission by-product found in effluent gases associated with the atmosphere in a reactor containment system, particularly from fuel rupture or fuel melt down incidents. It is also found in the effluent gases associated with the processing of spent nuclear fuels.

The iodine generated under such conditions takes on two principal forms, as molecular iodine, hydrogen iodide, and a small but radiologically significant fraction consisting of low molecular weight alkyl iodides, principally methyl iodide. It is known that radioactive elemental or molecular iodine, when taken into the body, concentrates in the thyroid gland. Subsequent studies have also shown that low molecular weight iodides, particularly methyl iodide, behave in the same way as elemental iodine in the human body. It is therefore imperative from a radiological health standpoint to control the concentration of iodine, in any form, present in gaseous effluents of the type described.

### Removal of Iodine from Air

A process developed by F.N. Case (1) is one whereby iodine removal from an iodine-containing off-gas stream is provided by contacting the stream with a metallic fatty acid solution wherein the metal is selected from silver, copper or mercury. Also provided is an iodine removal filter medium which comprises a polyurethane backing impregnated with copper oleate.

A process developed by B.A. Soldano and W.T. Ward (2) utilizes a solution which can effectively remove molecular iodine as well as organic iodine compounds from gaseous atmospheres. Such a solution is (a) an aqueous solution adjusted to a pH in the range 9 to 10 with an alkali metal hydroxide such as potassium hydroxide or sodium hydroxide, the solution containing (b) up to 0.3 weight percent boron as borate; (c) a reducing agent for molecular or atomic iodine selected from the group consisting of sodium thiosulfate and formaldehyde and (d) from 0 to an effective amount of a free radical getter, or a material which reacts with hydrated electrons.

A process developed by W.J. Maeck (3) involves absorbing and retaining air-borne inorganic iodine and organic iodine species by passing a gaseous stream containing these iodines through a filter bed of synthetic zeolite in a metal ion exchanged form, which metal is reactive with iodine.

### References

(1) Case, F.N.; U.S. Patent 3,429,655; February 25, 1969; assigned to the U.S. Atomic Energy Comm.
(2) Soldano, B.A. and Ward, W.T.; U.S. Patent 3,630,942; December 28, 1971; assigned to U.S. Atomic Energy Commission.
(3) Maeck, W.J.; U.S. Patent 3,658,467; April 25, 1972; assigned to U.S. Atomic Energy Commission.

# IRON

Potential industrial sources of dissolved iron species are reported by Patterson et al (1) to be mining operations, ore milling, chemical industries, dye industries, metal processing industries, textile mills, food canneries, tanneries, and titanium dioxide production.

Iron exists in the ferric or ferrous form, depending upon conditions of pH and dissolved oxygen concentration. At neutral pH and in the presence of oxygen, soluble ferrous iron ($Fe^{+2}$) is oxidized to ferric iron ($Fe^{+3}$), which readily hydrolyzes to form the insoluble precipitate, ferric hydroxide, $Fe(OH)_3$. Therefore, acidic and/or anaerobic conditions are necessary in order for appreciable concentrations of soluble iron to exist. In addition, at high pH values ferric hydroxide will solubilize due to the formation of the $Fe(OH)_4^-$ anion. Ferrous and ferric iron may also be solubilized in the presence of cyanide, due to the formation of ferro- and ferricyanide anions.

Perhaps the most significant industrial source of soluble iron waste is spent pickling solutions. Pickling baths are employed to remove oxides from iron and steel during processing or prior to plating of other metals on the surface. These baths contain strong acid solutions, usually sulfuric acid of 5 to 20% concentration by weight, although more recently hydrochloric acid has come into use.

The baths accumulate ferrous and ferric ions until the iron concentration interferes with product quality. At that time the bath must be replaced. According to Bramer (2), few industrial wastes have had the quantity of research and process development effort directed toward their treatment as has spent pickling solution. Iron concentrations reported for some industrial wastewaters are given in Table 10.

## TABLE 10: IRON CONCENTRATIONS REPORTED FOR INDUSTRIAL WASTEWATERS

| Process | Iron Concentration, (mg./l.) | Reference |
|---|---|---|
| Mine Drainage | | |
| Ferrous iron | 36 | 3 |
| Total iron | 360 | |
| Motor Vehicle Assembly | | |
| Body assembly | 4 | 4 |
| Vehicle assembly | 3 | |
| Steel Processing | | |
| Waste pickle liquor | 70,000 | 5 |
| Pickle bath rinse | 200 – 5,000 | 6 |
| Metal Processing | | |
| Appliance manufacturer | 0.09 – 1.9 | 7 |
| Automobile heating | | |
| controls | 1.5 – 31 | 8 |
| Appliances | | |
| Mixed wastes | 0.2 – 20 | 9 |
| Spent acids | 25 – 60 | |
| Alkaline wastes | Trace | |
| Chrome Plating | 40 | 5 |

Source: Report PB 204,521

The second type of iron as a pollutant, aside from dissolved iron, is suspended particulate iron of one chemical form or another. McKee and Wolf (10) have pointed out that ferric oxides are used as pigments in certain paints. In addition, the authros indicate iron oxide is used as a polishing powder for glass, metal and ceramic materials. Such operations very likely represent a source of iron pollution, due to the colloidal nature of polishing compounds and the fact that water is usually involved as a coolant-lubricant and rinsing agent.

In the processing of phosphoric acid, iron phosphates are formed and collected as a sludge, representing a potential waste source in the absence of efficient solids removal (11).

Iron and steel production represents a major source of small iron particles from the furnaces (2)(12) by virtue of air pollution control measures involving flue gas scrubbing. Steel mill wash waters may contain from 1,000 to 2,000 mg./l. suspended solids, with their metal composition roughly reflecting the furnace charge. Fifty percent of blast furnace suspended solids may be less than 50 microns in diameter, while open hearth steel furnace solids may be even smaller (12). Ferromanganese steel processing produces the highest percentage of semicolloidal particles with 50% being less than 10 microns in diameter (2).

Milling of iron or steel products yields solids in the quench tank in the form of rust scale. The total iron loss during milling is roughly 2½% of the finished product weight (2). Much of this scale is of large, readily settleable size, but as metal product thickness decreases so does the scale size, with finishing mill scale at the micron level. Quenching water (12) and casting equipment cleaning water (13) in iron foundries represent another source of finely divided iron and iron oxide particles.

### Removal of Iron from Air

The air pollution aspects of iron and its compounds have been reviewed by Sullivan (14). The major iron pollutant is iron oxide.

### Removal of Soluble Iron from Water

The primary removal process for iron is conversion of ferrous iron to the ferric state, and precipitation of ferric hydroxide near pH 7, where its solubility is at a minimum. Conversion of ferrous to ferric iron occurs readily upon aeration at pH 7.5. Spontaneous formation of ferric hydroxide then results in iron removal by precipitation.

Iron removal from municipal water supplies has been successfully accomplished by the water treatment industry for many years. Characteristics of high iron waters are that they usually have high $CO_2$ concentrations and little or no oxygen. Removal of $CO_2$ and aeration to oxidize ferrous to ferric iron comprises the standard treatment. Lime is frequently employed for pH adjustment. The iron treatment levels and costs discussed below are for municipal water treatment systems; industrial wastes at similar iron levels are generally equivalent in performance, as described later in this section.

Haney (15) reports 1.5 mg./l. of iron was reduced to less than 0.05 mg./l. by alum-lime-sodium silicate precipitation, preceded by chlorination. Chlorination oxidized the soluble ferrous iron to the insoluble ferric form. Settling for 4 to 10 hours was followed by sand filtration to remove finely divided ferric hydroxide particles.

Owens (16) reports "complete" iron removal with lime addition and settling, followed by sand filtration. Original iron concentration was 2.4 mg./l. Capital costs were not given, but chemical costs were approximately 10¢/1,000 gallons for a 24 MGD capacity operation. Dalton (17) reports a 2.25 MGD (imperial gallons) iron removal unit to have cost $50,000 and to reduce iron from a range of 2.25 to 4.70 mg./l. to less than 0.3 mg./l. The process is patented under the name Walker Process and does not employ lime, although pH adjustment is required for acid waters. The raw water is distributed over a graded anthracite coal bed by a system designed to entrain air into the solution. The bed surface contains residual iron hydroxide which catalyzes oxidation and precipitation of the dissolved iron. The unit must be backwashed periodically, with subsequent solids removal and disposal.

Gardner and Stika (18) describe the use of a sand filter which can handle 2.5 gal./min./ft.$^2$, and produce a final iron concentration of 0.13 mg./l. from an original, preaerated, solution concentrations of 2.5 mg./l.

A 1.5 MGD municipal plant, in operation as early as 1927, reduced 5.7 mg./l. of iron to

0.13 mg./l. (19). The treatment sequence consisted of aeration, filtration through a coke bed (probably for color removal), plain sedimentation, and sand filtration. McCracken (19) reports results of pilot plant operation where 9.0 mg./l. iron was reduced to 0.5 mg./l. with chlorination, lime addition and sand filtration.

Kollin (20) reports that diatomate filtration methods developed by Johns-Manville Co. can remove 10 mg./l. of iron to produce finished water containing less than 0.1 mg./l. iron. The process is basically the same as others described above, and aerates the influent water to remove $CO_2$ and initiate oxidation in an aeration chamber. Upon passing into the detention tank, pH adjustment is accomplished by either soda ash or lime addition to pH 8.0 to 8.5, or by adding calcined magnesite to pH 7.1. Diatomaceous earth is added as body feed in the detention tank and the outflow passes through a diatomaceous earth filter. The detention time is only ten minutes, substantially reducing the usual capital costs.

Iron treatment processes for other industries differ from that of the water industry in that for iron removal it is often necessary to neutralize fairly high concentrations of acid contained in the iron-bearing wastewaters. Neutralization increases treatment costs considerably, and since lime is the preferred base and sulfuric acid is generally present in the waste, large quantities of insoluble calcium sulfate are produced which precipitate, and which complicate the sludge disposal problem.

One of the most costly processes involving iron removal is the treatment of waste pickle liquor. In addition to high levels, the waste is strongly acidic and must be neutralized. Capital and operating costs for a typical acid neutralization plant, at various flows and waste acid concentrations, are given by the U.S. Department of the Interior (11). Solids disposal has not been considered in these data. Burd (21) indicated that land fill disposal costs for sludge range from $1 to $5 per dry ton, with increased costs for long hauling distances.

Neutralization and subsequent precipitation with lime is most commonly used by steel mills to treat iron-bearing spent pickling liquors. Wing (22) reports the capital costs for a plant capable of treating 15,000 gallons per day of pickle liquor. His data indicate submilligram per liter concentrations in the sludge vacuum filter filtrate. Heise and Johnson (23) report capital costs for a plant treating 60,000 gallons per 8-hr. day, and give operating costs, including sludge disposal. A recent steel industry study reported that costs for neutralization of pickling wastes totaled $20.00/1,000 gallons. This included an indirect cost of $1.44/1,000 gallons (24). Limestone treatment of the more dilute rinse waters from a hydrochloric acid pickling process achieved 100% neutralization of acid and 98% removal of iron at a cost of 24¢/1,000 gallons (25). Influent soluble iron averaged 210 mg./l., while effluent total iron ranged from 1.0 to 10.0 mg./l., and averaged 4.0 mg./l. Waste flows of 100 to 1,450 gallons /minute were reported.

A recently reported acid neutralization cost (26) is 1¢ to 4¢ per gallon for neutralization but this figure includes lagooning of the difficult to dewater ferrous hydroxide sludge, rather than more expensive sludge handling operations.

Heynike and von Reiche (27) provide cost figures for an integrated steel mill system in which waste treatment facility costs are given as 1.5 to 2% of total plant costs. These costs are for South Africa. Coulter (28), however, indicates that in general, metal processing and plating waste treatment capital costs should be expected to be 12 to 15% of new plant costs, with operating costs an additional 12 to 15% of the plant process operating costs. These two estimates probably represent the extremes, except in cases of older plants which require extensive modifications and installation of expensive waste collection systems.

Mihok et al (2) report the use of a 400 mesh limestone slurry instead of lime to neutralize acid mine drainage and remove iron contained in the waste stream. The pH 2.8 waste contained 360 mg./l. total dissolved iron and 36 mg./l. ferrous iron, with a total acidity of 1,700 mg./l. The slurry-waste mixture was aerated, initially settled in a primary system, and then settled overnight. The primary effluent contained 7 mg./l. of iron while after overnight settling, no iron was detectable. Major advantage claimed for the process is a one-half to two-thirds reduction in operating costs. A water high in calcium would presumably result from this form of treatment, however.

Frequently two waste streams can be utilized to provide treatment for one or the other, or both. The most obvious case is combination of acidic and basic solutions. Not infrequently, chromate wastes may be present in the same plant with waste pickle liquor or rinse water containing appreciable ferrous iron. Instead of separate treatment, the pickle liquor can be employed as a reductant for chromate, followed by coprecipitation of chromium and iron hydroxide by lime addition. Cupps (29) reports such a process reduces iron and chrome in the final effluent to 0.5 and 0.05 mg./l., respectively.

A process developed by T.K. Roy and L.N. Allen, Jr. (30) for the treatment of waste ferrous sulfate liquors containing ferrous sulfate and sulfuric acid comprises neutralizing the liquors with lime, heating an aqueous slurry of the resulting precipitate with carbon dioxide and ammonia and thereby converting its calcium sulfate content to calcium carbonate, and simultaneously forming ammonium sulfate in solution, and separating and drying the resulting mixture of calcium carbonate and iron hydroxide.

A process developed by R.B. Fackler and R.R. Davis (31) involves soluble iron (and manganese) removal from water by treatment of the water with permanganate ion and filtration through a manganese oxide zeolite bed.

A process developed by A.H. Rice and R.L. Culp (32) is based on the discovery that iron and manganese values may be removed effectively from water by adding to such waters an oxidizing agent and a polyelectrolyte or conditioning agent, and passing such treated water immediately and without prior settling through filter media constructed of granular material sized and arranged so that the flow is always from an area of larger to smaller grain size.

A process developed by G.R. Bell and G.J. Coogan (33) is a method for removing soluble iron from water comprising continuously preconditioning the water by adding thereto prior to filtration body feed consisting essentially of between 10 and 100 parts per million of water to be filtered of pulverulent filter aid and between 2.5 and 60 parts per million of water to be filtered of a water-soluble alkali to produce a pH of the water between 7 and 9. The alkali may be selected from the group consisting of alkali metal carbonates, alkaline earth metal carbonates, alkaline earth metal hydroxides, alkaline earth metal oxides, alkali metal aluminates, alkali metal ferrates, and mixtures thereof. The water containing the body feed is mixed and retained in a detention area to insure uniform distribution of the body feed and to affix the contaminant on the filter aid. Then the mixture containing water is filtered through a filter septum.

A process developed by H.G. Glover, J.W. Hunt and W.G. Kenyon (34) is an activated sludge process for the bacteriological oxidation of ferrous salts in acidic solution, comprising the steps of subjecting the solution to the action of an active sludge comprising autotrophic iron oxidizing bacteria carried on particles of solid material in a liquid suspension, by passing the solution through at least two containers in series, in which containers the active sludge is maintained; recovering by sedimentation sludge appearing in the outflow of the last container of the series; and returning the recovered sludge to the first container of the series.

A process developed by A. Cywin (35) is one in which acid wastewaters containing ferrous iron are neutralized using limestone in a finely divided state. Substantial amounts of a mixed valence, hydrous iron oxide sludge are recycled back to the neutralization and aeration steps of the process to produce a dense, easily dewatered sludge having improved handling characteristics.

Another process developed by A. Cywin (36) is one in which ferrous iron-containing acid wastewaters are neutralized to form a dense, compact, and easily settleable sludge. Ferrous to ferric iron ratios are adjusted prior to neutralization by catalytic oxidation to conform approximately to that of magnetite; $1Fe^{++}$ to $2Fe^{+++}$. Neutralization of the acid waste and precipitation of a mixed valence iron oxide is accomplished using a finely divided limestone slurry as the preferred neutralizing agent.

*Deep Well Disposal:* According to a recent survey article (26), deep well injection has become increasingly popular with steel companies as a disposal technique. Costs are widely variable depending upon depth of well and pressure requirements. Wells may cost from $50,000 to $75,000 to drill and case, and operating costs may be of the order of $2.50 to $10.00/1,000 gallons according to one source (26). Another report (37) indicates that capital costs may vary from $30,000 for an 1,800 foot deep well where no surface equipment is required to $1,400,000 for another 12,000 feet depth plus a clarifier, dual filters and four displacement type pumps. The filter requirement applies to any waste containing suspended solids, as they will clog the well. A cost of $1.00/1,000 gallons has been reported for deep well disposal of steel mill waste (24). Deep well disposal is not feasible for rinse water disposal, due to the large waste volumes involved.

*Recovery:* The recovery of by-products from waste pickle liquor is normally not economical for steel mills. Direct treatment of the pickle liquor can be avoided in the event that dragout is sufficiently high that the solution is continuously regenerated by makeup acid addition and contaminants remain essentially at a steady-state concentration (12). The problem of treating rinse water remains, however.

Delaine (38) has reported that neutralization of waste pickle liquor is uneconomical and impractical, due to the quantities of sludge produced. He has described recovery processes for sulfuric acid pickle liquor and for hydrochloric acid pickle liquor. Since other bath and rinse tank wastes are still in need of treatment, he prescribes a closed loop system. No operating costs data are presented however.

Various pickle liquor recovery processes are in operation, and their nature and some cost data have been summarized by Besselievre (39). For each process discussed, however, the primary emphasis is on recovery of acid rather than removal of iron. In brief, the processes evaporate the waste liquor and remove iron as ferrous oxide, ferrous sulfate or ferric oxide, and acid as either concentrated hydrochloric or sulfuric acid.

*Summary:* Existing technology is capable of soluble iron removal to levels well below 0.3 mg./l. Failure to achieve these levels appears to be the result of improper pH control resulting in increased solubility of ferric hydroxide during coagulation and/or in incomplete oxidation of the more soluble ferrous iron to the ferric state. Apparent soluble iron may actually be present in effluent waters in the finely divided solid state, due to inefficient settling of the ferric hydroxide. Precipitation treatment systems which consistently produce low iron waters also employed some polishing treatment, such as rapid sand filters.

Oxidation and precipitation, as described in this section, is capable of providing effective iron removal from most wastes, providing that prior treatment has removed iron complexing agents (e.g., cyanides) prior to precipitation, and that the waste pH is suitable. However, treatment costs may be extremely high for chemical requirements and sludge dewatering and disposal, when wastes contain high concentrations of iron and acid, particularly sulfuric acid, due to precipitation of calcium sulfate sludge. Although dollar values are high, estimates for waste treatment facilities range from 5 to 15% of the installation and operation costs of the process.

Recovery processes appear to be of economical value currently only for large industries such as steel mills, although certain process modifications may make recovery more feasible and enhance treatment efficiencies at reduced costs. Other than precipitation and recovery, only deep well disposal seems economically competitive to industries with high acid, high metal waste solutions. The cost, if geological conditions are favorable, are likely to be more attractive to smaller industries than other treatment alternatives.

### Removal from Water of Suspended Particulate Iron and Iron Compounds

The techniques for removal of soluble iron are well established, as described before. However, insoluble iron oxide frequently exists in suspension as a colloidal dispersion, which resists sedimentation due to the small particle size. In addition, the particles tend not to agglomerate into larger particles due to their characteristic surface charge, which causes them to repel each other upon approach. Ferric hydroxide, another frequently encountered insoluble iron compound, ultimately becomes ferric oxide in solution through chemical reaction. However, freshly precipitated ferric hydroxide has the characteristic of having a very low specific gravity, which makes settling difficult without long detention times or additional treatment.

Removal of total iron, which consists of the solid particulate forms, consists essentially of converting soluble to insoluble iron, and removing the suspended solids. The suspended solids may be iron oxide or hydroxide in a precipitation unit effluent, or colloidal iron particles from a particular industrial process.

In the case of ferric hydroxide overflow from a sedimentation clarifier or settling basin which does not provide adequate settling, chemical conditioning of the floc to obtain better settling characteristics may be implemented at increased operating costs, particularly for polyelectrolytes. Such chemical treatment to improve settling will normally increase effluent dissolved solids. Data presented in the preceding section on soluble iron indicate that proper clarifier design is capable of producing an effluent containing total iron content of 1 to 2 mg./l., and with iron generally in the submilligram/liter range if some type of filtration (generally sand) follows the iron precipitation and sedimentation system.

For larger particles, sedimentation is adequate with proper design. Bramer (2) reports that older installations treating iron and steel mill wastewaters typically produce total effluent suspended solids up to 250 mg./l. Effluent solids of 80 mg./l. have been reported by the same author for modern installations. No data on the relative composition of the residual suspended material was given although it should be primarily iron, having resulted from a ferrous metal processing industry. Open hearth steel furnaces are reported to produce extremely small iron particles, which when scrubbed from flue gas require chemical coagulation to achieve satisfactory settling. This increases treatment costs considerably (12). The alternative is a dry collection system such as electrostatic precipitation or bag filters.

Milling operations utilize a rough settling basin referred to as a scale pit, which collects the iron rust scale for recycle (12). Older practice was to discharge the scale pit overflow, containing approximately 200 mg./l. of suspended solids, without further treatment. Modern practice is to follow the scale pit with a clarifier to remove residual rust particles, and to recycle the clarifier effluent water (2).

Nemerow (6) reports that suspended solids in foundry waste can be reduced from about 3,760 mg./l. to 10 mg./l. by a sequence of primary sedimentation, addition of lime and alum coagulant, secondary sedimentation, further addition of lime and alum, and flotation. Standard operations for particulate iron removal include sedimentation with or without coagulation, and filtration through sand, gravel, crushed anthracite, etc.

In summary then, reduction of the solids fraction of total iron can be readily accomplished by existing technology and is essentially a matter of suspended solids removal. Two distinct situations exist: [1] the iron occurs in a definite particle form, and [2] the particulate iron has been formed in the treatment process either by soluble iron removal per se or through the use of iron as a flocculating agent in treating other waste materials. Particulate iron removal is then through sedimentation with or without coagulant aids, or by filtration. The use of treatment chemicals, if necessary, will have a marked effect upon treatment costs.

**References**

(1) Patterson, J.W. and Minear, R.A., "Wastewater Treatment Technology," *Report PB 204,521,* Springfield, Va., National Technical Information Service (August 1971).

(2) Bramer, H.C., "Iron and Steel" in *Chemical Technology, Vol. 2, Industrial Wastewater Control,* C. Fred Gurnham, ed., New York, Academic Press (1965).

(3) Mihok, E.A., Duel, M., Chamberlain, C.E., and Selmeczi, J.G., "Mine Water Research. Limestone Neutralization Process." *U.S. Bur. Mines, Rep. Invest. No. 7191* (1969).

(4) *The Cost of Clean Water, Vol. III, Industrial Waste Profiles, No. 2, Motor Vehicles and Parts,* Washington, D.C., U.S. Dept. Interior (1967).

(5) Parsons, W.A., *Chemical Treatment of Sewage and Industrial Wastes,* Washington, D.C., National Lime Association (1965).

(6) Nemerow, N.L., *Theories and Practice of Industrial Waste Treatment,* Reading, Mass., Addison Wesley Publishing Co. (1963).

(7) Watson, K.S., "Treatment of Metal Finishing Wastes," *Sew. Ind. Wastes,* 26, 182-194 (1954).

(8) Gard, C.M., Snavley, C.A., and Lemon, D.J., "Design and Operation of a Metal Works Waste Treatment Plant," *Sew. Ind. Wastes,* 23, 1429-1438 (1951).

(9) Anderson, J.S. and Iobst, E.H., Jr., "Case History of Wastewater Treatment in a General Electric Appliance Plant," *Jour. Wat. Poll. Control Fed.,* 40, 1786-1795 (1968).

(10) McKee, J.E. and Wolf, H.W., "Water Quality Criteria," California State Water Quality Control Board Publ. No. 3-A (1963).

(11) "The Cost of Clean Water," Vol. III, *Inorganic Chemical Industry Profile,* Washington, D.C., U.S. Dept. of the Interior (1970).

(12) Kemmer, F.N., "Pollution Control in the Steel Industry" in *Industrial Pollution Control Handbook,* H.F. Lund, ed., New York, McGraw-Hill Book Co. (1971).

(13) Barzler, R.P. and Giffels, D.J., "Pollution Control in Foundry Operations," in *Industrial Pollution Control Handbook,* H.F. Lund, ed., New York, McGraw-Hill Book Co. (1971).

(14) Sullivan, R.J., "Air Pollution Aspects of Iron and Its Compounds," *Report PB 188,088,* Springfield, Va., National Technical Information Service (September 1969).

(15) Haney, C.G., "Potassium Permanganate Solves a Problem at Charlottesville, Va.," *Wat. Sew. Works,* III R353-R354, 1964.

(16) Owens, L.V., "Iron and Manganese Removal by Split Flow Treatment," *Wat. Sew. Works,* 110:R76-R86, 1963.

(17) Dutton, C.S., "Iron Removal from Richmond Hill's Water Supply," *Can. Munic. Util,* 99:19, 1961.

(18) Gardner, J.R. and Stika, A.J., "Valveless Filter Installed for Iron Removal," *Pub. Wks.,* 95:109-110, 1964.

(19) McCracken, R.A., "Study of Color and Iron Removal by Means of Pilot Plant at Amesberry," *New Eng. Wat. Wks. Assoc.,* 75:102-114, 1961.

(20) Kollin, W., "Iron Removal Methods by Diatomite Filtration," *Wat. Sew. Wks.,* 109:R182, 1962.

(21) Burd, R.S., "A Study of Sludge Handling and Disposal," *Pub. WP-20-4,* Washington, D.C., U.S. Dept. of the Interior, 1968.

(22) Wing, W.E., "A Pilot Plant for Lime Neutralization Studies," *Proc. 5th Purdue Industrial Waste Conf.,* pp. 252-260, 1949.

(23) Heise, L.W. and Johnson, M., "Practical Development Aspects of Waste Pickle Liquor Disposal," *Proc. 13th Purdue Industrial Waste Conf.,* pp. 40-150, 1958.

(24) *The Cost of Clean Water, Vol. III, Industrial Waste Profiles, No. 1., Blast Furnace and Steel Mills,* Washington, D.C., FWPCA Publication U.S. Dept. of the Interior, 1967.

(25) Armco Steel Corp., *Limestone Treatment of Rinse Waters From Hydrochloric Acid Pickling of Steel,* Water Poll. Control Research Series No. 12010 DUL 02/71, Washington, D.C., Environmental Protection Agency, Water Quality Office, 1971.

(26) "Water Pollutant or Reusable Resource?" *Env. Sci. Technol.,* 4:380-382, 1970.

(27) Heynike, J.J.C. and von Reiche, F.V.F., "Water Pollution Control in the Iron and Steel Industry, with Special Reference to the South African Iron and Steel Corporation," *Wat. Poll. Control,* 68:569-573, 1969.

(28) Coulter, K.R., "Pollution Control and the Plating Industry," *Plating,* 57:1197-1202, 1970.

(29) Cupps, C.C., "Treatment of Wastes for Automobile Bumper Finishing," *Indus. Wat. Wastes,* 6:111-114, 1961.

(30) Roy, T.K. and Allen, L.N., Jr.; U.S. Patent 2,798,802; July 9, 1957; assigned to Chemical Construction Corporation.

(31) Fackler, R.B. and Davis, R.R.; U.S. Patent 3,167,506; January 26, 1965; assigned to Hungerford & Terry, Inc.

(32) Rice, A.H. and Culp, R.L.; U.S. Patent 3,171,800; March 2, 1965; assigned to General Services Co.

(33) Bell, G.R. and Coogan, G.J.; U.S. Patent 3,235,489; February 15, 1966; assigned to Johns-Manville Corporation.

(34) Glover, H.G., Hunt, J.W. and Kenyon, W.G.; U.S. Patent 3,218,252; November 16, 1965; assigned to Coal Industry (Patents) Limited, England.

(35) Cywin, A.; U.S. Patent 3,617,559; November 2, 1971; assigned to the U.S. Secretary of the Interior.

(36) Cywin, A. and Mihok, E.A.; U.S. Patent 3,617,562; November 2, 1971; assigned to the U.S. Secretary of the Interior.

(37) Spencer, E.F., Jr., "Pollution Control in the Chemical Industry," in *Industrial Pollution Control Handbook*, H.F. Lund, ed., New York, McGraw-Hill Book Co. (1971).

(38) Delaine, J., "Management to Achieve Water Economy in the Iron and Steel Industry – Conclusion," *Eff. Wat. Treat. J.,* 6:543-551, 1966.

(39) Besselievre, E.B., *The Treatment of Industrial Wastes*, New York, McGraw-Hill Book Co., 1969.

# IRON OXIDES

Inhalation of iron and iron oxides is known to produce a benign siderosis (or pneumoconiosis). However, in addition to the benign condition, there may be very serious synergistic effects as well as other undesirable effects, such as chronic bronchitis. In the laboratory, iron oxide has been shown to act as a vehicle to transport the carcinogens in high local concentrations to the target tissue. Similarly, sulfur dioxide is transported in high local concentrations deep into the lung by iron oxide particles. The relationships between dose and time and these conditions have not been determined according to Sullivan (1).

## Removal of Iron Oxides from Air

The reader is also referred to the section of this handbook on Steel Mill Converter Emissions.

A process developed by H. Werner (2) is one in which the flue dust contained in the waste gases of metallurgical furnaces, preferably converters for the manufacture of steel, is precipitated in an electric filter, and the precipitated dust, in accordance with its hydrated component parts, is pelletized and the dust in the form of fresh pellets immediately recycled into the furnace.

Figure 70 illustrates the essential equipment involved in the process. The waste gases from the furnace or converter 1 flow into chimney 2. After flowing through measuring nozzles, the gases pass through pipe 3 into the dry electric filter 4 at a temperature of about 150°C. and are exhausted from the filter through pipe 5. The precipitated dust collected in hoppers 6 drops to conveyor belt 7 which takes the dust to an inclined rotating pelletizing disc 8 whereon the dust is mixed with water and formed into pellets. The freshly formed pellets having a diameter of up to about 20 mm. fall from the pelletizer 8 into a trough 9 from which they can be dropped through one or more openings 10 by means of a flap valve 11.

Beneath each opening is a drying hopper 12. Each drying hopper is provided with a plurality of vertically separated individual grates 13, each of which is movably mounted in its hopper so that the grates can be displaced, moved or folded. This permits the pellets to be formed in spaced layers at varied heights with a necessary intermediate space between each layer. The upper grates are moved out of the way so that the lowermost grates can be first covered.

Blower 14 is movable beneath the bottom opening 15 of each hopper so that air or waste gases can be blown into the lower portion of a hopper through a nozzle 16. If necessary, waste gases can be used containing carbonic acid, but preferably converter gas is used. The drying gas is blown into the hopper as long as it is necessary, as, for example, 8 hours, until the pellets in that hopper are hardened. As shown in the drawing, the middle hopper 12 is being blown. After the drying and hardening of the pellets is completed, blower 14 is removed from the opening of that particular hopper 12, the intermediate grates 13 are removed so that the pellets now dried to a prescribed limit drop onto a conveyor belt 17

FIGURE 70:  APPARATUS FOR ELECTROFILTER RECOVERY, PELLETING AND
RECYCLE OF FLY DUST FROM A STEEL CONVERTER

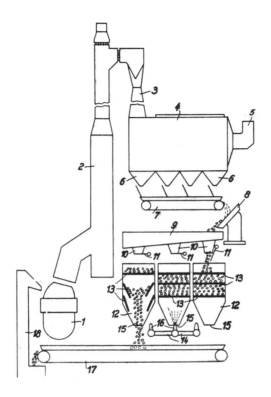

Source:  H. Werner; U.S. Patent 3,169,054; February 9, 1965

and are carried either directly to a converter or to an elevator **18** for introduction into
the converter **1**, or may be temporarily stored in a bin.

A process developed by N.F. Tisdale (3) is a method for the economical recovery of iron
oxide particles suspended in fumes produced in the melting, refining and production of
steel, sintering of ores, or blast furnace operations. It also relates to the recovery of iron
from iron oxide scale, produced in the heating operation of iron and steel production.
The oxygen type furnace, which is being widely adopted, is a producer of copious iron
oxide fumes. The open hearth furnace, with or without the oxygen lance, also produces
these iron oxide laden fumes. Several other melting and refining processes all produce
similar fumes.

Communities which have grown up around such melting plants have suddenly enacted
laws to force the reduction of these fumes and their solids from going into the air because
of their adverse effect on health. Since there is approximately 1½% of the charge volatized
into the air as iron oxide during the melting operation, this becomes a substantial economic loss unless recovered and a large amount of iron oxide is put into the atmosphere.
Methods to recapture the particles from these fumes require high initial expense, substantial operating cost and an elaborate cooling system to cool the fumes. There are several
methods which are used singly or in combination, in an attempt to do an efficient job.

All of the basic oxygen furnaces come equipped with dust precipitators which are essentially water scrubbers, electrostatic precipitators and bag houses. This kind of equipment costs from 18 to 25% of the total cost of installing the oxygen furnace installation, where the system described here costs from 6 to 12% of the total cost.

The system essentially is to take the fumes and have them impinge in a heated liquid bath of silica slag which has been altered just a small amount by materials to make it more fluid. The furnace is so constructed that there may be one or more baffles along its length with the result of forcing impingement of the gas on the surface of the slag. This can be augmented by pressure or by a pump to force this material to hit the slag. Figure 71 shows the arrangement of equipment which may be used.

### FIGURE 71: METHOD OF REMOVING IRON OXIDE PARTICLES FROM STEEL CONVERTER FUMES BY IMPINGEMENT ON MOLTEN SLAG

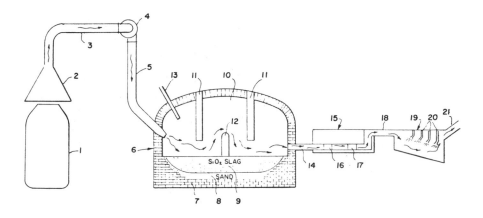

Source: N.F. Tisdale; U.S. Patent 3,365,340; January 23, 1968

Referring to the drawing, numeral **1** denotes a furnace, such as, for example, a basic oxygen furnace which emits fumes containing iron oxide, which fumes are directed into a hood **2** and through a duct **3**. A pump **4** may be provided between ducts **3** and **5** to increase the velocity of flow of the fumes as they impinge the $SiO_2$ slag denoted by numeral **9** contained within an acid air furnace **6**, which slag rests upon a bed of sand **8** contained within the basic brick work **7** of the furnace. Fuel is introduced through tube **13** into the furnace. Preferably, one or more baffles such as **11** are provided to more effectively deflect the fumes onto the slag and thus prevent the fumes from bypassing the slag and going directly through the furnace. A sand separator **12** may also be used, if desired, which acts similarly to baffles **11**.

As shown by the arrows, after the fumes impinge on slag **9** they flow through ducts **14**, **16** and **17** in a waste heat chamber **15** and thence through duct **18** into a spray chamber **19** where a plurality of sprays **20** will cool the fumes and clean them of the last traces of dust contained therein before passing through exit **21** into the air.

### Removal of Iron Oxides from Water

The reader is also referred to the earlier section of this handbook under Iron dealing with Removal from Water of Suspended Particulate Iron and Iron Compounds.

A process developed by J.E. Cooper (4) involves joint treatment of the extremely fine dusts produced by the blast furnace and the pickle liquor produced by the sulfuric acid pickling of steel.

The dust blown out of a blast furnace during operation has substantially the same chemical composition as the charge, and varies in size from sizable chunks down to particles of submicron size. By weight, by far the greater portion of the blast furnace dust is recovered in the initial dry dust catcher. However, to clean the blast furnace gas sufficiently to permit its efficient utilization as a furnace fuel, resort is had to three further cleaning apparatuses. Arranged in series, these are first, an ordinary wet gas scrubber; second, a so-called disintegrator; and third, an electrostatic Cottrell precipitator. In each of these pieces of apparatus, the dust is finally removed as a suspension in water. In a typical blast furnace installation producing 3,000 tons per day of pig iron, the water flow through these three final gas cleaners amounts to about 8,250 gallons per minute and is burdened with approximately 229 grains per gallon of solid matter. Over a 24 hour period these wash waters remove from the three final cleaners about 194 tons per day of dry solids.

In conventional blast furnace practice, these wash waters are combined and processed through a settling chamber and the effluent from the settling chamber is discharged into the sewer or other adjacent water course. The average dry solids content of this effluent which is discharged into the sewer is about eight grains per gallon or, in the course of 24 hours, about 7 tons of dry solids are so lost.

The iron value of these dry solids is not particularly serious economically even though its recovery would, of course, be desirable. The unfortunate feature of this process is that the material escaping in the effluent from the settling chamber represents the very finest dust produced in the blast furnace, and hence the dust which is the most objectionable from a standpoint of disposal in natural water courses, since these dusts are the most objectionable from a chromogenic standpoint. These dusts impart a persistent iron red color to the water course which is highly objectionable to the general public, despite the fact that quantitatively their effect upon the water course may be insignificant.

As shown in Figure 72, the combined flows of dust laden water from the blast furnace wet washer, disintegrator and Cottrell precipitator are emptied into mixer **10**. While being intensively agitated in the mixer, pickle liquor containing ferrous sulfate and sulfuric acid is slowly added. The agitating apparatus employed in mixer **10** is preferably one which is capable of inspiring air into the liquid being mixed. This type of agitator is well known in the art and any of several types may be employed. It is essential for this mixing operation that the pickle liquor be added slowly, so that at no time does the concentration of calcium sulfate produced by the interaction of the calcium carbonate in the blast furnace dust and the free acid in the pickle liquor exceed the solubility of calcium sulfate in the existing liquid.

In this mixer the natural alkalinity in the incoming water as well as the alkalinity available from the iron, lime and magnesium compounds from the blast furnace serve to at least partially neutralize the acidic constituents of the pickle liquor. It is to be understood that the free acid in the pickle liquor will be first neutralized with the resultant production of carbon dioxide. The air added to this mixer serves the purpose of scrubbing the carbon dioxide from the mix as well as partially oxidizing any iron hydrates which may be precipitated at this stage. The removal of carbon dioxide, of course, permits the neutralization of more of the acidic constituents in the pickle liquor by the calcium carbonate and bicarbonate available in mixer **10**.

It is difficult in the mixer to raise the pH of the liquid to a value between 7 and 8, which is necessary for the quantitative precipitation of the iron compounds. Accordingly, the liquor from mixer **10** is flowed to a precipitator **11** which may be structurally similar to the mixer. Here calcium hydroxide, preferably in the form of slurry is slowly added. The agitation in the precipitator is accompanied by intensive aeration again to remove carbon dioxide from the slurry and to oxidize precipitated ferrous hydrate to ferric hydrate.

FIGURE 72: PROCESS OF RECOVERING IRON VALUES FROM BLAST FURNACE DUST BY INTERACTION WITH PICKLE LIQUOR

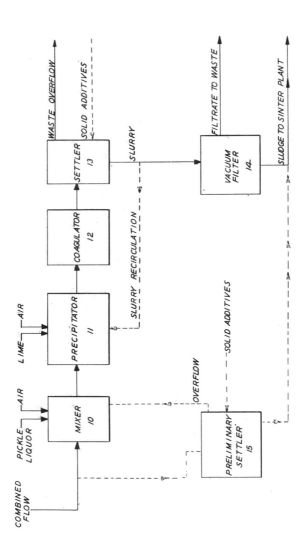

Source:   J.E. Cooper; U.S. Patent 2,810,633; October 22, 1957

Contrary to most literature teachings, the oxidation of ferrous to ferric iron in this very dilute solution proceeds quite well and yields a precipitate much more amenable to the subsequent steps of coagulating, settling and filtering. A typical time in mixer 10 and precipitator 11 would be two to five minutes each.

The effluent from precipitator 11 is quite free of soluble iron and is passed to coagulator 12. In this vessel the material is very gentle agitated for a period of time of about twenty minutes. During the residence in the coagulator, the precipitated iron hydrates and the iron oxides, lime compounds and coke carried over from the blast furnace co-act to produce a mass which will settle in a practicable length of time and which will produce a sludge which can be handled on a vacuum filter.

The effluent from the coagulator 12 is conducted to any conventional quiescent settling apparatus of which the Dorr Thickener is a typical example. The size of the thickener is selected so that the residence of the material therein will be sufficiently protracted to produce an effluent having a solids content below the desired minimum. In most cases, it is possible to produce an effluent from settler 13 having a solids content below ten parts per million. An effluent of this caliber may be added directly to streams with no adverse effects.

The underflow or slurry produced by settler 13 is conducted to vacuum filter 14. This vacuum filter may be any of the well known continuous rotary vacuum filters, such as the Oliver or Emico, or resort may be mad to continuous centrifuges for this separation. A particularly satisfactory filter for this separation is known as the string filter. Any of these devices will produce a substantially clear filtrate and a sludge having a water content between 50 and 80%, which is sufficiently dense to enable its transportation by conveyor to a sintering plant. Here it may be sintered along with the dry pulverulent solids recovered from the blast furnace gas by the dust catcher.

Modification of the above described flow sheet may be necessary to accommodate the conditions obtaining at individual plants. For example, a preliminary settler 15 may be interposed ahead of mixer 10 to give a preliminary separation of the more coarse solids and to relieve vacuum filter 14 of a portion of its load. The underflow from preliminary settler 15 may be added directly to the sludge from vacuum filter 14. Similarly, a portion of the slurry obtained from settler 13 may be recirculated back to precipitator 11. This results in a further utilization of the alkalinity available in the slurry which would otherwise be wasted in the vacuum filter filtrate. This recirculation also provides solids for the nucleation of the ferrous and ferric hydrates.

To either settler 13 or preliminary settler 15 may be added any carbonaceous or ferruginous materials which are either unobjectionable or desirable in the blast furnace after passage through the sintering plant. Typical examples of material which may be added at these points, and hence disposed of to advantage, are coke breeze or fines, mill scale, fine iron ore, fine coal, or wood flour. These materials will aid in the settling or filtration or add valuable iron to the blast furnace or provide fuel for the sintering process.

This process has been described particularly with reference to treatment of whole pickle liquor. However, it is equally applicable to the disposal of copperas solutions which are comparatively acid-free. The necessity of disposing of copperas has hindered the utilization of many of the processes which have been suggested for the recovery of sulfuric acid from spent pickle liquors. These processes would be more feasible economically if the copperas produced could be considered as an asset if handled by the above described process and not as an industrial liability.

In any event, it is necessary that the concentration of the reacting ingredients be kept sufficiently low and the agitation sufficiently intense that at no time is calcium sulfate precipitated. This is in accordance with well known chemical engineering techniques. In this way, all of the sulfate ions originating in the pickle liquor are disposed of harmlessly as soluble sulfates and no sulfur is carried into the sintering plant or blast furnace.

A process developed by F.J. Sines (5) relates to the removal of small scale particles from the flushing water of a steel mill after the recovery of larger scale particles for use in the blast furnace or open hearth.

In the purification of scale flushing water, the water is first treated by aeration during which the insoluble oils held in suspension are floated to the top. Then, the larger scale particles are removed by passing the water through a bar screen, and the oils on the surface of the water are skimmed off and carried to an oil storage tank. With prior methods and means of purifying scale flushing water, the water is then pumped into large sedimentation basins, where it sits for long periods of time to allow the fine mill scale particles to settle to the bottoms of the basins. After removal of these scale particles, the purified water is either directed into a storage lagoon from where it is pumped into the plant service water system for reuse, or it is discharged into a river.

The main problem with the conventional type of purification system lies in the long time that is required for the water to stand in the sedimentation basins before it is pure enough to be reused in the plant service water system or to be discharged into a river. Due to this long time, numerous large sedimentation basins are needed to purify the amount of scale flushing water that is discharged from a steel mill. These basins are costly to maintain and occupy much valuable land area.

In the process described below, such particles are removed from water by a combination of filtering and sedimentation processes that are carried out in a relatively compact tank, as compared to the large sedimentation basins used previously. The filter elements are continuously cleaned by backwashing, and this cleaning is performed on one part of the screens while other areas are in use gathering more particles from contaminated water.

Thus, the process need not be interrupted for the purpose of changing filter elements or cleaning them. In addition, the efficiency of the system is increased by reintroducing the filter backwash water into the tank. This allows the particles which settle to the bottom on their first entrance into the tank to gather additional particles on their way down and thus increase the overall rate of sedimentation within the tank.

A process developed by M. Steinberg and J. Pruzansky (6) is a method for removing dissolved iron oxides from acidic aqueous solutions, such as mine wastewaters, which comprises exposing the aqueous solution to gamma irradiation while aerating and contacting the solution with calcium carbonate to induce precipitation of the contained iron oxides from the solution.

The process is especially useful in treating acidic mine wastewater which has soluble ferrous iron compounds contained therein. It will remove substantially all ferrous iron compounds from the aqueous solution. Exemplary of ferrous iron compounds which can readily be removed from aqueous solutions are ferrous sulfate, ferrous bicarbonate, ferrous chloride, etc. When the process is employed, the ferrous iron compound concentration can efficiently and economically be lowered from the saturation point to under six parts per million. Conventional nuclear and chemical engineering techniques and equipment can be employed to carry out the practice of this process.

### References

(1)  Sullivan, R.J., "Air Pollution Aspects of Iron and Its Compounds," *Report PB 188,088,* Springfield, Va., National Technical Information Service (September 1969).
(2)  Werner, H.; U.S. Patent 3,169,054; February 9, 1965; assigned to Metallgesellschaft AG, Germany.
(3)  Tisdale, N.F.; U.S. Patent 3,365,340; January 23, 1968.
(4)  Cooper, J.E.; U.S. Patent 2,810,633; October 22, 1957.
(5)  Sines, F.J.; U.S. Patent 3,351,551; November 7, 1967; assigned to United States Steel Corporation.
(6)  Steinberg, M. and Pruzansky, J.; U.S. Patent 3,537,966; November 3, 1970; assigned to the United States Atomic Energy Commission.

## LAUNDRY WASTES

**Removal from Water**

The reader of this handbook is also referred to the section on Detergents.

A process developed by W.H. Hoffman (1) is concerned with an effective method of breaking emulsions during the treatment of waste laundry water without the use of either expensive additives or expensive equipment.

This process for the purification of wastewater containing fat, oil, greases, or the like in emulsion includes the step of treating the water in a flotation vessel in the presence of lime and air bubbles and the step of recycling partially purified water to the flotation vessel. One improvement comprises the step of diluting the wastewater entering the flotation vessel with a sufficient volume of the recycled water to break the emulsion.

More particularly, the improvement comprises the steps of diluting each volume of wastewater before entering the flotation vessel with a first dilution stream which consists essentially of about an equal volume of water containing a substantial proportion of the soil-lime agglomerates which form in the flotation vessel; diluting each volume of wastewater entering the flotation vessel with a second dilution stream which consists essentially of at least one volume of recycled water for each one thousand parts per million impurities in the wastewater, the second dilution stream being substantially purer water than the first dilution stream; and adding the air and lime to the second dilution stream.

**References**

(1)  Hoffman, W.H.; U.S. Patent 3,350,301; October 31, 1967.

## LEAD

Lead is used as an industrial raw material for battery manufacture, printing, paint and dyeing processes, photographic materials and matches and explosives manufacturing. Despite its wide use by industry, and the industry associated lead-bearing wastes which must result from its use, there is extremely little information in the technical literature on lead concentrations in wastewaters, or on methods of treating lead-bearing wastes. No waste treatment costs directly bearing on lead wastewaters have been uncovered, despite an extensive search of the industrial and pollution control literature, according to Patterson, et al (1).

The use of lead (as sodium plumbite) in petroleum refining produces a lead sludge, from which lead is normally recovered. This lead recovery process has been briefly discussed by Hill (2). Lead is also a common constituent of plating wastes, although not so frequently encountered as copper, zinc, cadmium and chromium. Fales (3) has reported the lead content of wastewater from one engine parts plating plant as ranging from 2.0 to 140.0 mg./l. The higher concentrations resulted from dumping of spent plating bath solution. The author reported that the quantity of lead discharged per day might amount to 112 lbs. from a single lead-plating bath. Pinkerton (4) has reported lead content of plating bath rinse waters which, as expected, contain lower lead concentrations than presented by Fales for spent plating bath solutions. Rinse waters contained 0 to 30 mg./l. of lead (4).

Pollution from lead mines, mining, and smelting have also been reported in the literature (5)(6)(7). Wixson, et al, (5) discussed the possibility of stream pollution from a lead mining region in southeastern Missouri, but provided only general information. In a related and more recent publication, Jennett and Wixson (6) report that a stream receiving effluent from a lead mining operation contained 0.00 to 0.04 mg./l. of lead over a 22 month survey period.  A nearby control stream, tested over the same period, contained 0.00 to 0.01 mg./l. of lead.  No trend of increasing or decreasing lead content with time was found.  Ettinger (8) has mentioned the occurrence of lead in streams at concentrations similar to those described previously, particularly in waters receiving metal processing industrial wastes.  Municipal water pumped from an abandoned mine shaft in Australia which was used as a municipal water supply infiltration gallery was reported to contain 0.31 mg./l. of lead (7).

## Removal from Water

Treatment of dissolved lead involves for the most part reaction to form a lead precipitate, and sedimentation.  Liebig, et al, (9) have reported however the treatment of a lead contaminated water by ion exchange.  Lead content was reduced from 0.055 to 0.0015 mg./l.  However, the process was bench-scale, and very few details were provided.  Undoubtedly, lead-contaminated water is susceptible to ion exchange treatment, on the same basis and at similar costs as reported in, for example, the hexavalent chromium and the copper treatment feasibility reports.  As discussed in those reports, the net cost of ion exchange treatment is reduced if recovery of the removed metal is economically feasible.

In the precipitation processes, lead is normally precipitated as the carbonate, $PbCO_3$ or the hydroxide, $Pb(OH)_2$.  The solubilities of both compounds are very low at alkaline pH, and formation of either product is effective in reducing dissolved lead concentrations (10).  Lead is commonly reported as being recovered from the sludge formed in the precipitation processes (11).

In forming insoluble lead hydroxide, lime is the treatment chemical of choice.  Fales (3) reports the use of lime in precipitating a plating bath waste containing 2.0 to 140 mg./l. of lead, plus other metals.  The author states that after lime treatment and settling for 1 hour the supernatant was clear and contained only a trace of heavy metals.  Lead in a municipal drinking water supply has also been reported reduced by lime treatment and settling from an initial value of 0.31 mg./l. down to 0.1 or less mg./l. of lead (7).

Treatment of a wastewater from a tetraethyl lead plant has also been described (11).  The effluent was alkaline, and contained inorganic lead salts in solution and suspension, plus organic lead salts.  No values of lead concentration, either prior to or after treatment, were included in that report (11).  The waste pH was adjusted to 8 to 9, and lead precipitated by addition of lime.

The precipitation was assisted by the addition of ferrous sulfate, which acted as a flocculating agent.  The treated effluent flowed to a Dorr-Oliver Clariflocculator, in which the lead sludge was collected at the bottom, dewatered in a filter, and reclaimed by a lead slag refinery.  No costs were reported for the process.

Nozaki and Hatotani (12) also have described treatment of wastewaters from a tetraethyl lead manufacturing process.  The two major categories of waste were inorganic lead wastewaters and organic lead wastewaters.  After sedimentation in a holding basin to recover solid lead and lead oxide, the inorganic lead waste fraction was treated by coagulation with ferric and ferrous sulfate.

Influent inorganic lead of 66.1 mg./l. was reduced to "trace" levels in the effluent.  Treatment of combined organic and inorganic lead waste streams containing respective concentrations of 13.3 mg./l. and 84.9 mg./l., was not as effective, however.  Effluent contained trace to 1.4 mg./l. of inorganic lead, and most of the influent organic lead.  A separate bench scale treatment process was therefore developed for the organic lead waste stream.

The organic lead compounds were removed effectively on a strongly acid cation exchange resin.  Reductions of organic lead from 126.7 to 144.8 mg./l. down to 0.020 to 0.53 mg./l. (as lead) were reported.  The organic lead compounds absorbed on the ion exchange resin were eluted by caustic soda.  This eluate was collected in a reaction basin, into which chlorine gas was injected at 95°C. for 45 minutes. The organic lead compounds were thereby almost completely converted to inorganic lead compounds, which could be treated as described above. Overall lead removals to less than 1 mg./l. residual were expected for the final plant effluent, after installation of the organic lead wastewater treatment sequence (12).

A third report on treatment of lead bearing waste from a tetraethyl plant reported a similar treatment sequence for inorganic lead (13). Lead occurring in the manufacturing process effluent at a concentration of 45 mg./l. was reduced to 1.7 mg./l. by precipitation at pH 10.4 to 10.8, aided by addition of ferrous sulfate. The lead sludge was recovered, and lead was extracted and reclaimed.

The author reports that one-half million dollars (1960 prices) was spent on waste treatment facilities, out of a total plant investment of twelve million dollars. Waste treatment capital costs thus represented 4.2% of the total plant costs. Waste treatment included settleable solids and chloride control, in addition to lead removal.

Similar treatment has been reported for lead removal from a municipal water supply (14). Lime and ferrous sulfate are added to the water to precipitate and coagulate lead leached from a nearby abandoned mine. As is common for most water treatment plants, the water is filtered, which undoubtedly acts as polishing treatment by removing nonsettleable lead hydroxide particles carried through the sedimentation basins.

The treatment costs associated with lead removal as discussed above are due to the chemical and operating costs of the lime precipitation and sludge disposal processes. Anderson and Iobst (15) have reported costs associated with a mixed metal plating waste treatment by lime precipitation and sludge disposal. The treatment plant had an initial installation cost of $1,108/1,000 gallons treated per day.

Operating costs were 80¢/1,000 gallons. The waste, while not containing lead, was a typical heavy metal waste which required treatment equivalent to the lead precipitation process. Watson (16) has reported lime treatment costs for a mixed metal plating rinse water as 62¢/1,000 gallons, and as $1.62/1,000 gallons for operating plus amortization costs.

Precipitation of dissolved lead by dolomite ($CaCO_3 \cdot MgCO_3$) to yield lead carbonate has also been reported (17). The process involved filtration of lead wastewater through a calcined dolomite bed, which reacted with and trapped the dissolved lead. The dolomite filters were regenerated by backwashing with water, which removed the lead carbonate formed in the process. Neither treatment efficiencies nor costs are reported for the process. Based upon a comparison of chemical strength and commercial cost, however, dolomite is approximately equivalent to lime on the basis of net operating cost.

The reaction of lead with dolomite, yielding insoluble lead carbonate, probably explains why very little lead enters receiving streams from the southeastern Missouri lead mines. The lead vein is located in a dolomite deposit, and the mine tailings thus contain crushed dolomite. As the mine tailings are channeled through sedimentation basins prior to release of the waste effluent, dolomite plus reacted lead carbonate are readily removed. As the authors point out, under existing conditions any heavy metals in the wastewater are precipitated rapidly in the settling lagoons at the slightly basic pH (7.5 to 8.2) of the wastewater.

In addition to the standard lead removal processes discussed above, there are several patented commercial processes available such as a Du Pont process (18). All are based upon the recovery and recycle of lead, rather than directed specifically toward waste treatment.

In summary, sparse information is available in the technical literature on the treatment of lead bearing wastewaters. Precipitation by lime or dolomite, and sedimentation with sludge disposal or lead recovery have been reported. Little data is available on effluent lead values after treatment, and no operating treatment costs have been found. However, the extreme insolubilities of both lead hydroxide and lead carbonate, the two most common precipitation products, would indicate that good conversion of dissolved lead to insoluble lead should be achieved. These precipitated lead compounds can then be removed by settling, with or without a coagulant aid or removed by filtration. Capital and operating costs should be equivalent to those reported for lime precipitation of other heavy metals (e.g., copper, zinc, nickel).

References

(1) Patterson, J.W. and Minear, R.A., "Wastewater Treatment Technology," *Report PB 204,521,* Springfield, Va., National Technical Information Service (August 1971).
(2) Hill, J.B., "Waste Problems in the Petroleum Industry," *Indust. Engng. Chem.,* 31 (11), 1361-1363 (1939).
(3) Fales, A.L., "A Plating Waste Disposal Problem," *Sew. Wks. Jour.,* 20, 857–860 (1948).
(4) Pinkerton, H.L., "Waste Disposal," in *Electroplating Engineering Handbook,* 2nd Ed., A. Kenneth Graham, editor-in-chief, New York, Reinhold Publishing Co. (1962).
(5) Wixson, B.G., Bolter, E.A., Tibbs, N.H. and Handler, A.R., "Pollution from Mines in the New Lead Belt of South Eastern Missouri," *Proc. 24th Purdue Industrial Waste Conf.,* pp. 632–643 (1969).
(6) Jennett, J.C. and Wixson, B.G., "Treatment and Control of Lead Mining Wastes in S.E. Missouri," *Proc. 26th Purdue Industrial Waste Conf.,* (1971).
(7) Anonymous, *Annual Report Western Australian Government Chemical Laboratories,* Dept. of Mines, Australia (1962).
(8) Ettinger, M.B., "Lead in Drinking Water," *Water Waste Engng.,* 4 (3), 82–84 (1967).
(9) Liebig, C.F., Jr., Vanselow, A.P. and Chapman, H.D., "The Suitability of Water Purified by Synthetic Ion-Exchange Resins for Growing of Plants in Controlled Nutrient Cultures," *Soil Sci.,* 55, 371–376 (1943).
(10) Sillen, L.G. and Martel, A.E., "Stability Constants of Metal-Ion Complexes," *2nd Ed., Sp. Pub. No. 17, Chemical Soc.,* London (1964).
(11) Anonymous, "An Integrated Plant for Tetraethyl Lead," *Indust. Chemist,* 30, 429-436 (1954).
(12) Nozaki, M. and Hatotani, H., "Treatment of Tetraethyl Lead Manufacturing Wastes," *Water Res.,* 1 (2), 167–177 (1967).
(13) Robb, L.A., "Waste Water Treating Facilities at Ethyl Corporation of Canada, Ltd.," *Proc. 8th Ontario Industrial Waste Conf.,* pp. 90-96 plus 11 figures (1961).
(14) Anonymous, "East Midland Districting Meeting," *J. Instit. Municip. Engrs.,* 84 (November): XXV–XXVCC (1957).
(15) Anderson, J.S. and Iobst, E.H., Jr., "Case History of Wastewater Treatment in a General Electric Appliance Plant," *Jour. Wat. Poll. Control Fed.,* 10, 1786-1795 (1968).
(16) Watson, K.S., "Treatment of Complex Metal-Finishing Wastes," *Sew. Ind. Wastes,* 26, 182-194 (1954).
(17) Voznesenskic, S.A., Evalanova, A.V., and Suvorova, R.V., "Physico-Chemical Methods of Purification of Industrial Waste Waters," *Wat. Poll. Abst.,* 13, 135 (1940).
(18) E.I. du Pont de Nemours & Co.; French Patent 1,361,089; May 15, 1964.

# LEAD TETRAALKYLS

## Removal of Tetraalkyllead Compounds from Air

A process developed by W.C. Jaasma (1) involves the recovery of tetraalkyllead compounds from an inert gas stream by contacting the stream at least once with an inert organic scrubbing liquid.

Figure 73 shows the configuration of apparatus for the process as applied to the production of tetramethyllead in the presence of toluene. Referring to the drawing, inert gas vapor from a steam still or other product recovery or production means containing quantities of tetramethyllead and toluene or other volatile condensables such as aluminum compounds enters the process by line 10 and is fed to fume scrubber 11 where it is contacted with aqueous caustic solution entering fume scrubber 11 by line 12 and line 5 to produce a scrubbed gas phase indicated in first sump vessel **13** at **14** and an aqueous liquid phase **15**. Should the quantity or quality of volatile aluminum compounds, or other material, in the inert gas stream of line **10** be such that condensation thereof may cause plugging of fume scrubber **11**, additional quantities of aqueous caustic solution may be added to or blended with the entering inert gas stream by line **9** to prevent such plugging. It is clear also that when the process is used for the recovery of tetraalkyllead product from an inert gas stream containing no or substantially no volatile aluminum compound or like material, fume scrubber **11** may be bypassed or eliminated, and the product containing inert gas fed directly to vapor space **14** without any pretreatment.

A first sump in vessel **13** containing the aqueous liquid phase **15** and a first organic liquid **16** from this first sump is removed by line **17** through pump **18** and line **19** to cooler **20** where the first organic liquid phase may be cooled to a temperature of from -10° to 50°F. by brine entering cooler **20** by line **21** and leaving by line **22**. Cooled first organic liquid phase from cooler enters tower **30** by line **31** where it contacts scrubbed gas phase from vessel countercurrently.

Suitable packing section **32** or other contacting means such as distillation plates may be provided in the tower to permit good contact of the liquid and gas phases. The first organic liquid phase after contact returns to the vessel and settles to the organic liquid phase with extracted tetramethyllead and toluene contained therein, and tetramethyllead product and toluene recovered from the scrubbed gas phase, along with first organic liquid are removed from the system as product by line **33** to pump **34** and then to process product recovery by line **35**.

By the use of weir **40** located within the vessel, first organic liquid phase is retained within the vessel, except for that removed to product recovery, and aqueous liquid phase is permitted to overflow into aqueous sump **41** where it may be removed by line **42** through pump **43** and recycled to fume scrubber **11** by line **44** joining line **12**. As necessary, portions of contaminated aqueous caustic solution may be removed from the system to waste by line **45** and fresh aqueous caustic solution added to the system by line **46**.

### FIGURE 73: APPARATUS FOR RECOVERY OF TETRAALKYLLEAD FROM AIR BY SOLVENT SCRUBBING

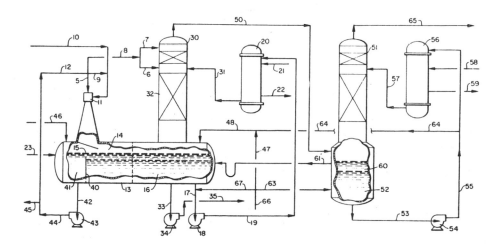

Source:  W.C. Jaasma; U.S. Patent 3,403,495; October 1, 1968

In operation of the process over long periods, it may be necessary or desirable to provide a continuing or intermittent aqueous caustic wash for the tower to prevent buildup of deposits in the contacting section 32 or the demisting section, if any, above the contacting section. When this is necessary or desirable, aqueous caustic solution from line 12 may be supplied in a desired quantity by line 8 through line 6 or 7 or both. As necessary, and generally in a volume approximating that volume removed to product recovery, substantially pure organic scrubbing liquid may be added to the system from a supply not shown by line 66 to line 47 and then to line 48.

In most process applications, the concentration of desirable products in the second scrubbed gaseous phase exiting the tower by line 50 is reduced to a minimum and the venting gas phase either may be recovered for reuse as inert gas or vented to the atmosphere without pollution. In operation in this manner as a single tower system, it is desirable to supply substantially pure organic scrubbing liquid in a necessary volume to that portion of the scrubbing liquid being pumped to cooler 20 and this may be accomplished as shown by line 67 from supply line 66.

Under certain conditions of production and high volumes of vent gas, it is sometimes desirable to remove only a major portion of product from the inert gas system in the tower and to remove the final minor portion of the product in tower 51 which may be substantially identical to tower 30. When operating the process in this manner, second scrubbed gaseous phase exiting tower 30 by line 50 is fed to the bottom portion of tower 51 where it is contacted countercurrently with organic scrubbing liquid from sump 52 in the bottom portion of tower 51.

The organic liquid is removed by line 53 and pumped by pump 54 through line 55 to cooler 56 and from cooler 56 to the top portion of tower 51 by line 57. As in the case with cooler 20, chilled brine enters cooler 56 by line 58 and exits by line 59 to cool the organic liquid to a temperature of from -10° to 50°F. The second contacting organic liquid phase flows down tower 51 to sump 52 which is connected by line 61 to first sump 16 so that product containing organic liquid does not build up within sump 52 beyond the level of connecting line 61 and any excess product containing organic liquid flows to sump 16 within vessel 13 for use in the first contacting in tower 30 and subsequent removal to product recovery. Any aqueous phase which may be present in second sump 52 as indicated at 60 also may flow to the first sump in vessel 13 for joining with the aqueous phase therein as indicated at 15.

When the system is operated using two contacting towers, substantially pure organic scrubbing liquid may be charged to the system of both sumps; however, it is preferred to charge it by line 63 only in a quantity substantially equal to that being removed to product recovery. In this manner, substantially pure organic contacting liquid is used to remove the remaining trace quantities of product from gaseous phase entering tower 51 to obtain a more efficient operation. Connecting line 64 joining line 48 may be provided to pump quantities of organic liquid to first sump 16 to maintain the desired level, if needed, and the substantially product free inert gas exiting tower 51 by line 65 may be vented to the atmosphere without air pollution problems or recovery for reuse in an inert gas system, if desired.

In a typical example of the operation of the process for the recovery of tetramethyllead and toluene from a nitrogen gas stream, a vapor feed composition having a concentration of 57% nitrogen, 40.2% tetramethyllead-toluene mixture and 2.8% water, all percentages by weight, and having a temperature of 100° to 110°F., was fed to a fume scrubber where it was scrubbed with a 4% by weight aqueous solution of sodium hydroxide. Scrubbed vapors continued to a first tower where they were contacted countercurrently with ethylene dichloride at a temperature of approximately 35°F. in a quantity sufficient to cause the gaseous phase leaving the first tower to have a temperature of approximately 50°F.

The contacted vapors from the overhead of the first tower were fed to the bottom of a second tower where they were contacted countercurrently with ethylene dichloride entering the tower at a temperature of approximately 32°F. Contacted vapors entered the second tower at a temperature of approximately 50°F. and exited at a temperature of approximately 35°F. Both towers were substantially identical and contained 2 inch steel Pall rings in a depth in the tower sufficient to provide approximately three theoretical distillation plates in each tower and each tower was operated at substantially atmospheric pressure.

The concentration of the sodium hydroxide in the aqueous caustic solution was maintained at approximately 4% by weight by the addition of a 25% by weight aqueous sodium hydroxide solution as necessary and fresh ethylene dichloride was added to the sump of the second tower as necessary to compensate for the volume of ethylene dichloride-tetramethyllead-toluene mixture removed from the sump below the first tower.

Approximately 90% of the tetramethyllead-toluene mixture in the vapor feed stream was recovered in the first tower and a total of approximately 99% of the tetramethyllead-toluene mixture in the original vapor feed stream was recovered in the two tower system. Based upon total quantity of tetramethyllead produced from a reaction process not utilizing the process, sufficient tetramethyllead-toluene mixture was recovered using the process to increase the overall yield of the tetramethyllead produced approximately 4 to 5%.

### Removal of Tetraalkyllead Compounds from Water

A process developed by H.E. Collier, Jr. (2) involves the removal of dissolved alkyllead compounds from the aqueous effluent used in the manufacture of alkyllead compounds. Tetraalkyllead compounds are highly useful antiknock compounds and are manufactured commercially by alkylating a lead-sodium alloy with an alkyl chloride, such as ethyl chloride, methyl chloride or a mixture of methyl and ethyl chlorides.

In these processes, after the excess unreacted alkyl chloride has been distilled off, the rest of the reaction mass is drowned in a large volume of water and the tetraalkyllead steam distilled therefrom. The still residue is settled, the solids separated from most of the water and further washed with water to remove various lead-containing salts. The water from the settling, separating and washing steps constitutes waste aqueous effluent from the tetraalkyllead manufacturing process.

In order to remove certain sludge-forming impurities from the steam-distilled tetralkyllead, the impure tetralkyllead product is further subjected to an aeration or oxygen purification treatment in the presence of water as disclosed in U.S. Patent 2,400,383. The purified tetralkyllead is then separated from the aqueous phase. This aqueous phase, containing some soluble alkyllead materials formed in the purification, also constitutes part of the aqueous effluent from the tetraalkyllead process.

The aqueous effluent from the tetraalkyllead process may be neutral but generally is strongly alkaline, usually having a pH between 11 and 12, due to the reaction of the water with unreacted sodium-lead alloy to produce sodium hydroxide. It contains large amounts of sodium chloride and small amounts of water-soluble organic lead compounds, such as those formed in the purification process described above. In addition to the above, a very small amount of water-soluble inorganic lead compounds are sometimes present in the aqueous effluent. The disposal of the aqueous effluent from the tetraalkyllead process is a serious problem, since, in some locations, the maximum amount of soluble lead compounds allowed in aqueous effluent discharges is 5 parts per million parts of water.

The sodium chloride and sodium hydroxide compounds in the aqueous effluent naturally do not present a serious disposal problem. Moreover, the water-soluble inorganic lead compounds in the aqueous effluent do not pose a serious disposal problem since they can be easily removed from the effluent before discharge by simple adjustment of the pH to a range of from 8 to 9.5 in the presence of water-soluble carbonates as described in Canadian Patent 572,192. However, the soluble alkyllead compounds in the aqueous effluent, such as trialkyllead chlorides, trialkyllead hydroxides, dialkyllead dichlorides and dialkyllead

dihydroxides, cannot be discharged into lakes and streams without causing a potential contamination problem. Moreover, the disposal of such soluble alkyllead compounds constitutes a significant loss of lead.

While the soluble inorganic lead compounds may be removed from the effluent by adjustment of the pH to 8 to 9.5 in the presence of water-soluble carbonate as described above,
or by precipitation with hydrogen sulfide or other soluble sulfides, the organic lead compounds are not satisfactorily removed by such treatment, and furthermore, any excess
soluble sulfide in the effluent would be highly objectionable and would have to be removed
from the aqueous effluent before its disposal.

The process constitutes a process for treating an aqueous effluent from the manufacture of
alkyllead compounds and containing about 5 to 5,000 ppm of lead as dissolved organic
lead not precipitatable by pH adjustment to 8 to 9.5, which process comprises the steps of:

   [1]   Adjusting the pH of the effluent to between 8.0 to 9.5.
   [2]   Intimately contacting the aqueous effluent with an ozone-containing gas.
   [3]   Precipitating the converted lead compounds.
   [4]   Separating the precipitated lead-containing compounds from the aqueous
         effluent.

This process is based on the surprising discovery that ozone is highly effective in removing
water-soluble organic lead compounds from wastewater by producing water-insoluble products. Indeed, the discovery that ozone successfully converted the soluble products to
water-insoluble materials was most surprising since air and other oxidizing agents have
been found unsatisfactory to effect this removal.

By this process the dissolved alkyllead can be removed from the aqueous effluent to such
an extent that the aqueous effluent from tetraalkyllead manufacture can be safely discharged into lakes and streams. The process is easy to operate, and the quantity of materials, the time and intimacy of contact are easily coordinated to reduce the dissolved organic lead content to an acceptable level, below 5 ppm, or substantially nil if desired.

A process developed by E.R. Taylor, Jr. (3) involves the removal of dissolved organic lead
from an aqueous effluent produced in the manufacture of alkyllead compounds by contacting a metal more electropositive than lead but essentially nonreactive with water with
the effluent until the dissolved organic lead is converted to an insoluble lead-containing
product, leaving a reduced dissolved organic lead content in the effluent. The following
is one specific example of the conduct of this process.

250 ml. of effluent water from the tetraethyllead process with a pH of 11.2 and containing 9 ppm lead [90% as organic lead, tri(mixed $C_1$ and $C_2$ alkyl)lead, and 10% inorganic
lead as $Pb^{++}$] were passed at 10 ml. per minute through a column of No. 40 mesh iron
filings 3 cm. in diameter and 13 cm. deep. The effluent water had a contact time of 9.2
minutes with the column of iron. The effluent water after the iron contact contained no
organic phase and a maximum lead content of 0.2 ppm, thereby showing a greater than
97% reduction of lead content.

### References

(1)  Jaasma, W.C.; U.S. Patent 3,403,495; October 1, 1968; assigned to Ethyl Corp.
(2)  Collier, H.E., Jr.; U.S. Patent 3,308,061; March 7, 1967; assigned to E.I. du Pont de Nemours & Co.
(3)  Taylor, E.R., Jr.; U.S. Patent 3,697,567; October 10, 1972; assigned to E.I. du Pont de Nemours
       & Co.

## MAGNESIUM CHLORIDE

**Removal from Air**

A process developed by W.L. Young et al (1) involves a method of efficiently concentrating a magnesium chloride solution without producing an exhaust gas plume. The process includes the steps of burning an air and gas mixture to produce hot combustion gases, controllably intermixing secondary air with the hot combustion gases to provide hot drying gases having a maximum temperature less than that at which, under the given environmental pressure the magnesium chloride has a vapor pressure of less than $1/10,000$ of an atmosphere, and injecting the hot drying gases directly into the magnesium chloride solution.

**References**

(1) Young, W.L. and Douglas, C.J.; U.S. Patent 3,526,264; September 1, 1970; assigned to American Magnesium Co.

## MANGANESE

Inhalation of manganese oxides may cause chronic manganese poisoning or manganic pneumonia. Chronic manganese poisoning is a disease affecting the central nervous system, resulting in total or partial disability if corrective action is not taken, according to Sullivan (1). Some people are more susceptible to manganese poisoning than others. Manganic pneumonia is a croupous pneumonia often resulting in death. The effect of long exposure to low concentrations of manganese compounds has not been determined.

Manganese compounds are known to catalyze the oxidation of other pollutants, such as sulfur dioxide, to more undesirable pollutants — sulfur trioxide, for example. Manganese compounds may also soil materials.

The most likely sources of manganese air pollution are the iron and steel industries producing ferromanganese. Two studies, one in Norway and one in Italy, have shown that the emissions from ferromanganese plants can significantly affect the health of the population of a community.

Other possible sources of manganese air pollution are manganese fuel additives, emissions from welding rods, and incineration of manganese-containing products, particularly dry-cell batteries. Manganese may be controlled along with the particulates from these sources. Air quality data in the United States showed that the manganese concentration averaged $0.10$ $\mu g/m.^3$ and ranged as high as 10 $\mu g/m.^3$ in 1964.

McKee and Wolf (2) have reported that manganese and its salts are used in the following industries: steel alloy, dry cell battery, glass and ceramics, paint and varnish, ink and dye and match and fireworks.

Among the many forms and compounds of manganese, only the manganous salts and the highly oxidized permanganate anion are appreciably soluble. The latter is a strong oxidant which is reduced under normal circumstances to insoluble manganese dioxide. The highly soluble manganous chloride is used in dyeing operations, linseed oil driers and electric batteries, while the equally soluble manganous sulfate is used in porcelain glazing, varnishes and production of specialized fertilizers (2).

Since the chemistry of manganese is similar to that of iron (3) it would be expected that any pickling operation involving manganese steel alloy results in dissolution of manganous ion and the presence of this ion in pickling and rinse solutions. In contrast to iron, the divalent (manganous) form is not readily oxidized to the insoluble manganic form, other than at elevated pH.

## Removal of Manganese from Air

The emission of manganese particulates will be controlled at the same time as the particulates from the steel furnaces and incinerators. No special equipment is required to remove manganese. However, the ferromanganese furnace presents special problems (4)(5).

Control of manganese from steel furnaces is accomplished by various types of collectors, including electrostatic precipitators, high-efficiency wet scrubbers, and fabric filters (6)(7). Four physical factors make the dust collection economically difficult: the small particle size (as low as $0.03\mu$), the large volume of gas, the high gas temperature, and the low value of the recovered material.

Control of emissions from a ferromanganese blast furnace is more difficult than from other furnaces because the waste gas temperature is hotter and the dust is finer. Electrostatic precipitators have been successful in removing approximately 80 to 90% of the dust. No economical way of using the collected dust has been developed (5). Control of emissions from burning fuels containing antiknock and smoke-inhibiting additives may require special systems. The hazards of these organic manganese additives are to be studied as required by the Clean Air Act.

## Removal of Manganese from Water

Removal of soluble manganese has been practiced for many years in the treatment of municipal water supplies and industrial process waters. Low manganese waters are essential for many industries, including the food and beverage industry. Treatment technology basically involves conversion of the soluble manganous ion to an insoluble precipitant. In water treatment, this generally means simultaneous removal of iron, since conditions under which high iron levels occur in water supplies are essentially the same as those for soluble manganese. Removal is effected by oxidation of the manganous ion, and separation of the resulting insoluble oxides and hydroxides of manganese. Welch (8) in a survey of manganese removal processes has listed those shown in Table 11.

## TABLE 11: MANGANESE REMOVAL PROCESSES

| | |
|---|---|
| Aeration | Slow and ineffective below pH 9. |
| Chlorine and hypochlorite oxidation | Not particularly effective for organically bound manganese. |
| Adjustment of pH | Lime soda type treatment gives removal at pH 9.5. |
| Catalysis | Copper ion enhances air oxidation. |
| Ion exchange | Effective for small quantities of iron and manganese. Resins quickly fouled by iron and manganese oxides. |
| Chlorine dioxide | Rapidly oxides manganese to the insoluble form, but expensive. |
| Manganese dioxide or potassium permanganate | Regeneration, green sand filter process. Economic disadvantage of requiring excess permanganate. |
| Direct potassium permanganate addition | Requires sand filtration. |

Source: Report PB 204,521

Because of the low reactivity of manganous ion to oxygen, simple aeration is not an effective oxidation technique below pH 9, and even above that pH the oxidation reaction is slow (8). Borgren (9) has reported that even at high pH, organic matter in solution can combine with manganese and prevent its oxidation by simple aeration. Holland (10) reported that at lime and sodium alginate (a coagulant aid) dosage producing optimum iron coagulation, no manganese removal was obtained. The effect of pH on manganese removal by lime has been reported by Owens (11). A reaction pH above 9.4 is required to achieve significant manganese reduction. The use of chemical oxidants to convert manganous to insoluble manganese, in conjunction with coagulation and filtration, has been recommended, although it may cause filtration problems. Borgen (9) states that the character of manganese flocs produced when strong oxidants are used tends to rapidly clog slow sand filters. Although Welch (8) indicates that manganese oxidation by chlorination may be effective, successful manganese removals have been reported.

Owens (11) reports removal of manganese at concentrations as high as 5.7 mg./l. with chlorination, followed by lime coagulation and rapid sand filtration. Split flow treatment was employed, to achieve intermediate manganese levels in the final effluent. Chemical costs, adjusted to the treated stream only, are about 2.1 ¢/1,000 gallons. Sodium silicate was used as a coagulant aid. Harris (12) states that manganese oxidation requires a free chlorine residual of 0.5 mg./l. and requires 1.29 mg. of chlorine per milligram of manganese oxidized.

The successful use of potassium permanganate as a chemical oxidant has been reported. Haney (13) reports that addition of 1.5 to 2.0 mg./l. of potassium permanganate reduces 1.0 mg./l. of manganese to less than 0.05 mg./l. Addition is 30 minutes upstream from the prechlorination point, which is followed by lime-alum-sodium silicate coagulation (4 to 10 hours detention), and rapid sand filtration. Costs attributed to permanganate addition are less than 0.3 ¢/1,000 gallons.

Since iron is generally present, the quantity of permanganate used must include that required for oxidation of the iron content. The relationship between permanganate required and the quantity of iron and manganese present has been reported by Willey et al (14). The ability to remove manganese at lower pH (an advantage when simultaneous iron removal is desired) by permanganate oxidation is demonstrated by the data of Holland (10). The addition of 0.04 mg./l. of sodium alginate reduced 0.35 mg./l. of manganese to less than 0.02 mg./l. in the filtered supernatant. Borgren (9) cites the absence of residual taste or odor as an additional advantage of permanganate over chlorination.

Permanganate may be used indirectly for manganese removal in the manganese greensand filter method. Willey and Jennings (14) describe the filter as a combination of New Jersey Glauconite (greensand), manganese sulfate and potassium sulfate, which results in a filter bed of sand grains containing high oxides of manganese. Upon contact with manganous ion an oxidation-reduction reaction occurs, resulting in the formation of insoluble manganese dioxide which remains in the filter bed. The bed is periodically reoxidized with permanganate, and solids are back washed from the bed.

Alling (15) reports manganese was reduced from 1.7 to 0.0 mg./l. at pH 7.2 using a pressure greensand system. The plant of 500 to 700 gpm capacity cost $26,000 completely installed and operating, compared with an estimated $75,000 for a conventional aeration-lime coagulation-clarifier-filter-recarbonation system. Chemical costs were 3.5 ¢/1,000 gal. for a permanganate dosage of 13 mg./l. The water had a manganese concentration of 13 milligrams per liter, which required excess permanganate. Treatment costs were roughly equivalent to lime chemical costs required for a similar treatment level. Regeneration for this plant was on a continuous basis with periodic backwashing of the greensand filter. Generalized removal capacities for greensand filters are given as 0.09 lb. of manganese per cubic foot of filter volume with regeneration requirements of 0.18 lb./ft.$^3$ of potassium permanganate. Flow rates of 3 gal./min./ft.$^2$ are normal (16).

Alsentzer (16) reports on the use of ion exchange for iron and manganese removal. Many advantages are cited, including smaller capital investment than coagulation-filtration, nearly four times greater flow rates, smaller plant requirements and simple operation. The obvious drawback to ion exchange is the nonselective removal of other ions, which rapidly increases operation costs. Regeneration brine disposal poses an additional treatment problem since the manganese remains in soluble form in the brine. An anaerobic waste stream is an absolute necessity for the ion exchange system, to avoid air oxidation of the manganese with resultant precipitation and clogging of the ion exchange bed.

The application of the above processes to manganese removal from wastewaters should be direct within the constraint that the presence of acids, other oxidizable metals, or organics in the waste stream may substantially increase costs related to manganese removal. The unit operation costs for coagulation-sedimentation-filtration should be similar to those for iron treatment systems.

Additional costs would be attributable to chlorine (or chlorine dioxide) and chlorination equipment, plus lime requirements to meet the higher pH requirements. Capital and operation costs for chlorination given by Smith (17) are for treatment of secondary sewage effluents. Chemical costs could be somewhat lower or higher, depending upon the nature of the industrial waste to be treated.

A process developed by G.R. Bell (18) relates to the removal of manganese as a contaminant from water sources, and particularly to filter and filtration techniques for effectively achieving such removal. More specifically, this process relates to preconditioning of both ground and surface water to remove soluble manganese by means of a simple filter aid filtration system. Removal of soluble iron compounds will also be achieved by the same process.

This technique involves the incorporation in the liquid to be filtered of a small amount of finely divided particulate material, the filter aid, which functions to form continuously a porous cake upon the filtering surface and to entrap impurities by various mechanisms. The materials most generally used as filter aids are siliceous materials such as diatomaceous silica and perlite, carbon, and fibrous matter such as asbestos and cellulose, and the properties of these materials, e.g., porosity, diversity of shape, incompressibility, etc., make them unique for this purpose. A particularly important feature of filter aid filtration is the fine porosity of the filter aid cake which enables the removal of substantial portions of the suspended particles.

In order to increase the efficiency of the filtration process, a precoat of filter aid particles may be provided on the filter septum in addition to incorporating the particles within the filter feed. This keeps the main filter cake containing the impurities from coming into direct contact with the filter medium and consequently prevents the gummy particles from clogging the medium and lessening the filtration efficiency.

It has been shown that when a strong oxidant such as $KMnO_4$ is used in a filtration process to assist in manganese removal, iron, if present in quantity, is oxidized to ferric hydroxide which deleteriously affects the filtration of both iron and manganese. It has been found that with iron removed or reduced to a low concentration, a filter aid filtration process with suitable preconditioning utilizing filter aid, alkali and a strong oxidant, can effectively reduce manganese to low levels with low capital and operating expenses. The iron concentration may be effectively removed by utilizing concepts of preconditioning with subsequent filter aid filtration.

Briefly, the iron and manganese containing water is first treated with an appropriate alkali or mixture of alkalis to impart to the water a pH of between 7 and 10. The alkali is so selected as to be a slow reacting material, e.g., $Na_2CO_3$, CaO, $Ca(OH)_2$, and the addition controlled so as to attain a stable pH with controlled agitation. Filter aid is added along with the alkali and the resulting preconditioned water filtered to remove the iron without prior settling. Once having removed the iron, or at least reducing it to a workable level, the manganese concentration may be reduced to about 0.05 ppm, or lower, by a similar preconditioning technique.

For instance, with a detention time of less than 10 minutes in the preconditioning tank at pH values of about 8.6 obtained by adding $Na_2CO_3$, filtration preconditioned water containing 1 ppm of $KMnO_4$ and 20 ppm filter aid, reduced an initial manganese content of 1.5 ppm to less than 0.05 ppm with a filter cycle of more than 12 hours and nominal head loss rate.

A process developed by G.B. Hatch et al (19) involves oxidizing the manganous ion to higher oxides, hydrous oxides, and hydroxides, to permit its subsequent removal from water. Dissolved manganese is generally present in natural waters in the form of the manganous ion. Oxidation of the ion frequently occurs to some extent on exposure to air. Such oxidation leads to precipitation of dark brown or black hydrous oxides or hydroxides of the higher oxidation states of manganese which are very insoluble.

When these precipitates remain suspended in the water, they cause objectionable discoloration known as "black water"; when they settle out, black deposits form which can block lines, or act as catalysts causing further manganese deposition. These deposits are very deleterious in textile and laundry operations as they interfere with dyeing processes and leave spots which are difficult to remove. They may increase the corrosion of copper, although this has not been completely proven. They are also troublesome in municipal water distribution systems where their presence makes it extremely difficult to maintain a chlorine residual.

The method usually employed to remove manganese is to oxidize the manganous ion to insoluble higher oxides, hydrous oxides, or hydroxides, which precipitate and may be removed by coagulation and settling, filtration, or both. The oxidiation may be effected by raising the pH of the water to 8 or higher where naturally occurring dissolved oxygen or mechanical aeration brings about oxidation, or by the use of chlorine or permanganate.

However, all of these methods have difficulties which limit their usefulness and effectiveness. The use of a high pH to facilitate oxidation by dissolved oxygen is expensive and tends to cause scale deposition. Chlorine is only slightly more active than dissolved oxygen for oxidation of manganese; chlorine also requires pH elevation and to only a slightly lower level than that required for oxygen. Permanganate is expensive and one may not wish to have the permanganate ion in the water he is treating because of its intense color, and because excesses may themselves cause manganese deposits.

In this process, there is added a salt of iron, copper, or cobalt and any compound yielding bisulfite ions in solution to the manganese-containing water. This results in the formation and subsequent precipitation of manganese dioxide or other higher oxidation state oxides, hydrous oxides, or hydroxides which may be removed by filtration. The iron, copper, or cobalt salt already added frequently also serves as a coagulant. Should additional coagulant be desired, it is preferable to add alum or more iron.

*Summary:* Manganese removal to levels of 0.05 mg./l. or less have been accomplished for many years in the treatment of municipal drinking and industrial process waters. Several methods of removal are available, each of which may have distinct advantages depending upon the overall nature of the waste, other required waste treatment operations, and the total treatment goals for a particular waste. The most common general approach seems to involve oxidation of the soluble manganous form to insoluble manganese hydroxide or oxide, with subsequent precipitation. Ion exchange treatment has proven effective however, and has several advantages. No information regarding treatment of wastewaters, industrial or municipal, specifically for manganese removal was found.

References

(1) Sullivan, R.J., "Air Pollution Aspects of Manganese & Its Compounds," *Report PB 188,079,* Springfield, Va., National Technical Information Service (September 1969).

(2) McKee, J.E. and Wolf, H.W., "Water Quality Criteria," *California State Water Quality Control Board Publication No. 3-A* (1963).

(3) Cotton, F.A. and Wilkinsen, G., *Advanced Inorganic Chemistry,* New York, Interscience Publishers (1962).

(4) Nabi, S.E.D.A. and Kayed, K.S., "EMG and Conduction Velocity Studies in Chronic Manganese Poisoning," *Acta Neurol, Scandinav.* 41, 159 (1965).

(5) Schueneman, J., High, M.D. and Bye, W.E., "Air Pollution Aspects of Iron and Steel Industry," *U.S. Dept. of Health, Education and Welfare, Public Health Service, Publication No. 999-AP-1* (1963).

(6)  Benz, D.L. and Bird, R., "Control of Pollution from Electric Arc Furnaces at Oregon Steel Mills
     Co.," *Preprint. Oregon Steel Mills Co.,* Portland, Oreg. (1965).
(7)  Danielson, J.A., "Air Pollution Engineering Manual," *U.S. Dept. of Health, Education,and Welfare,
     Public Health Service Publication No. 999-AP-40,* U.S. Government Printing Office, Washington,
     D.C. (1967).
(8)  Welch, A., "Potassium Permanganate in Water Treatment," *Jour. Am. Wat. Wks. Assoc.,* 55, 735–
     741 (1963).
(9)  Borgren, G.R., "Removal of Iron and Manganese from Ground Waters," *J. New Eng. Wat. Wks.
     Assoc.,* 76, 70–76 (1962).
(10) Holland, G.J., "Removal of Dissolved Aluminum, Iron and Manganese," *Chem. Ind.,* 2016, London
     (1964).
(11) Owens, L.V., "Iron and Manganese Removal by Split Flow Treatment," *Wat. Sew. Wks.,* 110,
     R76–R87 (1963).
(12) Harris, W.C., "Iron and Manganese Removal," *Eng. Ex. Sta. Bull. No. 68,* Louisiana State Univer-
     sity (1962).
(13) Haney, C.G., "Potassium Permanganate Solves a Problem of Charlottesville, Va.", *Wat. Sew. Works*
     111, R353–R354 (1964).
(14) Willey, B.F. and Jennings, H., "Iron and Manganese Removal with Potassium Permanganate,"
     *Jour. Am. Wat. Wks. Assoc.,* 55, 729–734 (1963).
(15) Alling, S.F., "Continuously Regenerated Greensand for Iron and Manganese Removal," *Jour. Am.
     Wat. Wks. Assoc.,* 55, 749–752 (1963).
(16) Alzentzer, H.A., "Ion Exchange in Water Treatment," *Jour. Am. Wat. Wks. Assoc.,* 55, 742–748
     (1963).
(17) Smith, R., "Cost of Conventional and Advanced Treatment of Wastewater," *Jour. Wat. Poll. Con-
     trol Fed.,* 40, 1564–1574 (1968).
(18) Bell, G.R.; U.S. Patent 3,340,187; September 5, 1967; assigned to Johns-Manville Corp.
(19) Hatch, G.B. and Guthrie, E.A.; U.S. Patent 3,349,031; October 24, 1967; assigned to Calgon Corp.

## MEAT PACKING FUMES

In the processing of various materials including waste in meat packing plants, a very strong
stench permeates the atmosphere in the neighborhood of these plants because of the odoriferous materials which escape into the atmosphere as a result of rendering or cooking operations. While attempts to decrease or avoid this stench have been made, they have been
generally unsuccessful in view of the fact that a considerable amount of noncondensible
gas emanates from these cookers.

Since the removal of such odors does not involve recovery of any materials from the atmosphere which would have any economic advantage, the purpose of such removal is
merely for improving the working conditions of plant employees and the living conditions
for persons living in or passing through the neighborhood. Therefore, the expense of such
odor-removal is not offset in any way by recovery of chemicals, and in order to be attractive for such plants, must be inexpensive as well as efficient.

### Removal from Air

A process developed by A.J. Teller (1) relates to deodorizing gases from rendering vessels
or cookers. In this process, the odoriferous gas is passed into a spray chamber and thereafter is passed through a demisting zone before passage through a bed of adsorbent material such as activated carbon, the noncondensible gases being pulled through the bed by
a vacuum applied on the opposite side of the adsorbent bed, preferably by a steam ejector.

The gas from the cookers, etc. enters into the spray chamber in a lower region and is
drawn upward toward the water sprays by a vacuum which is applied at the gas outlet
end of the apparatus which is at a point remote from the gas inlet. The spray section is
located between the gas inlet and the gas outlet. Steam and other easily condensible vapors are condensed by the cooling effect of this spray and the condensate falls to the bottom of the chamber with the spray water and passes through an outlet having a hydrostatic

leg to insure the reduced pressure conditions. The noncondensed gas is then passed through a demister which can comprise a finely knit wire mesh preferably having an open space therein of 80 to 90%. This demister removes entrained liquid and solid particles so that they are not deposited in the bed of adsorbing material through which the gas is next passed.

This adsorbent material is preferably activated carbon, although any other adsorbent material can be used that will serve as an adsorbent or otherwise effect removal of the odoriferous element. By condensing the condensible vapors from the steam as well as removing entrained liquid from the steam before the noncondensible gases are admitted to the bed of adsorbing material, it has been found that the adsorbent is allowed to act more efficiently and in a more concentrated fashion on the offensive or odoriferous material and thereby to effect more efficient removal of odor from the gas stream.

The separation of the noncondensible gases from the condensible material is effectively controlled and directed by the application of the vacuum on the outlet side of the adsorbing bed. This vacuum or decreased pressure is inexpensively and advantageously effected by a steam ejector or aspirator which has the low pressure side connected to the outlet of the deodorizing equipment. The exhaust from the ejector or aspirator can be connected to the outside atmosphere and the exhaust steam from the ejector together with the deodorized gas can be passed directly into the outside atmosphere.

Figure 74 shows the form of an apparatus which is suitable for the conduct of the process. The odoriferous gas is fed through inlet **1** into the lower section of chamber **2** in which sprays **3** are located at an intermediate point so that the odoriferous gas must pass through the spray region. In this case, the sprays are supplied with spray water from manifold **4**.

## FIGURE 74: SCRUBBER FOR DEODORIZATION OF MEAT PACKING PLANT FUMES

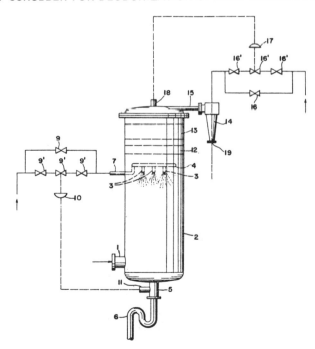

Source: A.J. Teller; U.S. Patent 3,183,645; May 18, 1965

The spray water and condensate are drained from the bottom of the chamber through outlet 5 which has hydrostatic leg 6 to prevent air from being drawn into the chamber by the vacuum which is maintained therein. Water is supplied to the manifold and sprays through water inlet 7. Valves 9 and 9' are appropriately adjusted and the water flow is controlled by control 10 which is automatically adjusted in accordance with the temperature of the water in the outlet from the spray chamber. This control is actuated through actuating means 11 which is responsive to the temperature of the outlet water.

After the spray zone, the noncondensible gas is passed through demister 12, which can comprise a series of finely knit wire mesh screens, for the removal of entrained liquid and solid particles. Following this, the gas is passed into an adsorbing bed 13 which can comprise finely divided particles of activated carbon. After passing through the bed of adsorbing material, the gas is completely deodorized and pulled through the top of a chamber by the vacuum effected on the upper region by steam ejector 14 which is connected to outlet means 15 of the chamber.

The rate of steam feed into the ejector is controlled by valves 16 and 16' which are adjusted to give the appropriate conditions and are controlled by steam flow control 17 which reacts to the pressure in the top region of chamber 2 by virtue of pressure sensing means 18 connected at the top of chamber 2. Exhaust steam and deodorized gas pass through outlet 19 of steam ejector 14.

As an illustration of the effective operation of this process, a fully automated unit having a chamber three feet in diameter and eight feet high of the design shown, utilizing 125 gallons per minute of water and 50 lbs. of steam per hour, was able to completely deodorize the noncondensible gases and to effectively condense the steam from 4 rendering units and one blood cooker which provided an average of 4,000 lbs./hr. of steam together with the noncondensible and odoriferous gases.

In addition to the solution of the air-pollution problem, the efficiency of the cookers was increased 5 to 10% because of the improved conditions resulting. In addition, the water consumption was decreased 25% compared with that used in previous operation whereby the gas was merely sprayed with water. Moreover, the resultant dried blood product was increased in value because of the improved drying conditions that were effected. Furthermore, because of the lower boiling temperatures resulting from the use of this process, the tallow product recovered from the rendering units had considerably less color than had previously been possible.

References

(1)  Teller, A.J.; U.S. Patent 3,183,645; May 18, 1965; assigned to Mass Transfer, Inc.

## MERCAPTANS

Due to the foul odor of mercaptans, it is commonly a requirement in industry that such compounds be removed from exhaust gases. For example, in the production of gasoline, it is common to treat the gasoline to remove mercaptans therefrom. This results in a gaseous or liquid waste material containing the mercaptans and a problem arises with respect to disposal of the waste material.

The intensity of the odor of mercaptans is extraordinarily great. Concentration data given in the literature as to the amount of mercaptans which can just barely be sensed are not entirely in agreement due to the subjective nature of the sense of smell. The reported values indicate the order of significant concentration. Ethyl mercaptan is said to be clearly detectable by odor in a dilution of $1:10^{11}$ (*Chemisches Zentralblatt*, 1936, p. 4178). A dilution of $1:10^{12}$ is considered as clearly perceptible [*Gas- und Wasserfach*, 74 p. 248 (1931)]. In any case, the high degree of dilution in which mercaptans can be smelled is surprising.

## Mercaptan Removal from Air

A process developed by H. Friess (1) involves removing mercaptans from gases containing oxygen comprises contacting the gases with an adsorbent which has been impregnated with an alkaline liquid containing a thickening agent such as starch to convert the mercaptans to disulfides and thereafter removing the disulfides from the gases by adsorption. The following is one specific example of the conduct of the process.

250 cc or 100 grams of pumice were impregnated with a 20% solution of NaOH of viscosity 1.29° Engler (20°C.). The pumice thus treated was used as a catalyst bed. A column of dry active charcoal with a 1 liter charge was connected to its output. This apparatus was charged with 200 l./hr. of air which had been bubbled through a waste liquor containing mercaptans. Every 24 hours the decrease of mercaptans in the air was compensated by adding fresh waste liquor.

The liquor was kept at a temperature of 70°C. The air contained 10 to 20 grams of mercaptan sulfur per cubic meter. In all cases, the mercaptans were reacted to more than 98%. The disulfides that had formed were driven out of the active charcoal column by steam distillation. After 30 days the transformation rate dropped under 95%.

A process has been developed by Trobeck et al (2) for mercaptan removal from gases. This process is applicable to both pulp making and to petroleum refining processes venting mercaptans but will be discussed here solely with respect to pulp manufacture. In the form illustrated for pulp mills, but also usable for deodorization in the oil industry, the gases are shown, Figure 75, coming from a digester 10 to gas relief and blowoff tank 11, and then to a heat recovery chamber of any conventional type 12, or in some other conventional manner.

FIGURE 75: APPARATUS FOR REMOVING OBNOXIOUS MERCAPTAN ODORS FROM GASES

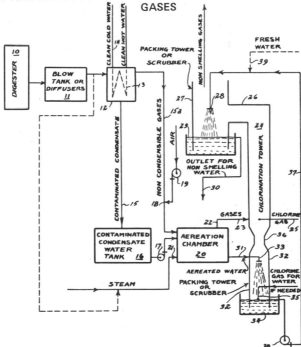

Source: K.G. Trobeck, A. Tirado and W. Lenz; U.S. Patent 3,028,295; April 3, 1962

The contaminated condensate water obtained here passes through a drawoff pipe **15** to a storage tank **16**. The noncondensable gases coming from the heat recovery chamber **12** pass off through a pipe **15a** into which air is admitted at **18** by an air blower or pump **19**, and the air and treated gases enter the aeration chamber **20** to which steam can be added at **21** to obtain the desired reaction temperature. The steam can preferably be taken from the steam line between the blow tank **11** and the heat recovery chamber **12**. By means of pump **17** the contaminated water in tank **16** is introduced into the aeration chamber **20** where the sulfur compounds carried in the noncondensable gases and in the contaminated water are converted into less odorous substances in a single operation. Thus, no fresh water is needed to treat the obnoxious gases coming from the digester.

From the aeration chamber **20** gases carrying some residual odor pass off at the top at **22** and are conducted by a pipe **23** into a chlorination tower **24** near an inlet **25** admitting chlorine gas to the tower. In the embodiment illustrated, the tower is shown as empty but, if desired, it may be packed as with broken brick to ensure contact of flowing gases. The tower delivers the gases through a horizontal outlet **26** to a final scrubbing tower **27** open at both ends, so that gases escape at the open top, flowing through a water spray coming from an overhead sprayer **28**. The condensed material carried down by the spray falls into a partly filled tank **29** from which odorless water is carried off by an overflow **30**.

The water for the sprayer **28** may be satisfactorily drawn from the aerated condensed water of the chamber **20** by connecting the drawing-off pipe **31** directly to the suction of the pump **38**. If that water in pipe **31** needs deodorizing, it may fall through a scrubber or loosely packed tower **32** being delivered by a sprayer **33** so as to fall into a bottom tank **34**. While falling, it may be chlorinated as by chlorine gas delivered into the tower **32** by an inlet **35**. The tower **32** is preferably connected below the bottom of the tower **24** through a constricted section **36**.

The bottom tank **34** has been found a satisfactory source of water for the spray **28** and may deliver the water through a pipe **37** being driven through the pipe **37** by a pump **38**. Fresh water may be added at **39** to the water flowing in the pipe **37**. The temperature of the aerator chamber **20** is preferably 30° to 95°C., and is more efficient at the higher temperature. In chlorination tower **24**, chlorine was added continuously at about 0.5 to 2 kg. per ton of pulp, with an additional 0.1 to 0.3 kg./ton to eliminate the odor of the blow-off from the digesters, which is added in batches if the blow-off is in batches. In chlorination tower **24** the temperature should usually be between 30° and 70°C. to avoid attacking the tower material. The heavy stream of white fumes produced at the chlorination tower carries no odor, or at times a faint sweet odor, disappearing within a distance of 60 to 100 feet.

Figure 76 shows in detail an aerating device that has been found highly efficient for treating the gases and contaminated water in a single operation. The modified aerating device shown in Figure 76 with modified connections may include an inlet **58** for material to be treated, an air fan **59**, an inlet **60** for oxygen, a casing **61**, a shaft **62** shown as operated by a motor above the casing and carrying discs **63** rotating with it preferably at a peripheral speed of at least twenty meters a second. A condensate supply **64** delivers condensate on the rotating discs so that the condensate is scattered through the chamber formed by the casing **61**.

The condensate thrown by the top disc **63** against the inner face of the casing **61** flows down into an annular trough **66a** lying against that face and delivers the condensate to the next disc **63** through radial pipes **67**. That disc throws the condensate again against the face where a second trough delivers it to a third disc. It should be understood that the number of discs may be increased for greater efficiency. The effluent water in this form of device passes off through a level maintaining tank **68** and may be carried through to a final spray **65** for washing the escaping gases. These gases come from the scrubber **61** through a flue **69** into which chlorine is introduced at **69a** and are then led to a chlorinating tower **70** and then to the tower **72** containing packing **71** where ammonia may be added and any surplus water is drained from the bottom thereof to the sewer.

# FIGURE 76: DETAIL OF AERATING DEVICE FOR PLANT FOR REMOVAL OF OBNOXIOUS MERCAPTAN ODORS FROM GASES

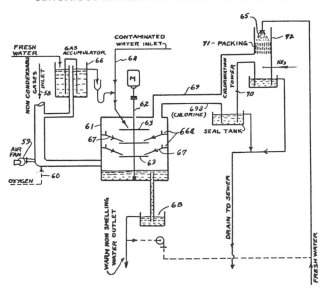

Source:  K.G. Trobeck, A. Tirado and W. Lenz; U.S. Patent 3,028,295; April 3, 1962

Some details of the treatment of gases from the recovery boiler **40** are diagrammatically shown in Figure 77. There the gases are shown as conducted from the boiler **40** by a pipe **41** to a direct contact evaporator **42** and then by a pipe **43** to a dust collector system **44**. The gases, dust free, then pass by a pipe **45** to a scrubber **46** shown as supplied with spent or dilute alkaline solution by an overhead spray **47**, from an alkali treatment described below. If properly constructed, the direct contact evaporator **42** should wash out the $SO_2$ and the scrubber **46** may be eliminated. Final odorless discarded water flows from the scrubber **46** through a bottom pipe **48** while gases free from $SO_2$ escape through a top pipe **49** to a chlorination chamber **50**.

A small amount of air is needed to promote the reaction between the mercaptans and the water before the gases are introduced into the chlorination chamber **50**. Normally, this air enters through small leaks in the apparatus and/or ducts numbered **42** to **49** inclusive, or if needed, additional air may be introduced separately. Chlorine gas enters the chamber **50** through a pipe **51** and chlorinated gas is led from the top by a pipe **52** into a final wash chamber **53** where it is sprayed by fresh alkali solution through a spray pipe **54**. If desired, another oxidizing agent, such as sodium hypochlorite may be added instead of the chlorine. Absolutely odorless gases escape from the wash chamber **53** through a pipe **55** which delivers the gases to a chimney **56** discharging to the atmosphere.

Ordinarily the gases coming from the recovery boiler, if the black liquor is first properly oxidized, are odorless and the scrubber **46** and chlorination at **50** may not be needed. If chlorination seems needed here, it was found that 0.25 kg. of gaseous chlorine per ton of pulp was adequate. A total alkali requirement of 1 to 3 kg. NaOH per ton of pulp was all that was needed, or the equivalent $Na_2CO_3$. A temperature of 50° to 80°C., within the chlorination chamber **50** was adequate and the pH of the final washing water between 6 and 9. Since hypochlorous acid is a very efficient agent to destroy odorous sulfur compounds, the lower the pH the better the result.

## FIGURE 77: DETAIL SHOWING TREATMENT OF GASES FROM THE RECOVERY BOILER FOR MERCAPTAN REMOVAL

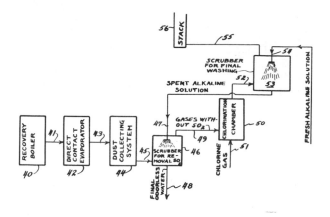

Source: K.G. Trobeck, A. Tirado and W. Lenz; U.S. Patent 3,028,295; April 3, 1962

The following table shows the concentration in percent per millions of volume of sulfur compounds in gases from different sources during the pulp production, as well as in some other gases for the purpose of comparisons, as determined by means of a titrilog reading. The titrilog is a chemical-electronic instrument which continuously records concentration of oxidizable sulfur compounds. The measurement is accomplished by a titration with bromine. The bromine is electrolytically generated in a solution in which the sulfur compounds are absorbed, a feedback amplifying system controls the bromine generating current so that the net rate of bromine generation is at all times equivalent to the rate of absorption of the sulfur compounds. A recording milliammeter records the generating current. The net current is proportional to the sulfur compound concentration in the sample (see *Industrial Engineering Chem.*, vol. 46, page 1422, July 1954).

| Sources | Titrilog Net Reading | Sulfur Compounds Given as $H_2S$, ppm |
|---|---|---|
| Liquor tank (all measurements obtained by taking samples at a point 6" inside vent pipe): | | |
| Vent of weak black liquor tank from washers | 40 - 50 | 2.44 - 3.55 |
| Same before oxidation | 18 - 22 | 1.09 - 1.34 |
| Same after oxidation | 11 | 0.67 |
| Vent from cyclone of oxidizing tower | 17 - 23 | 1.04 - 1.40 |
| Sealed tank of barometric column multiple effect evaporation | 3 | 0.18 |
| Condensate water from multiple effect evaporator | 2 | 0.12 |
| Chimney of smelt-dissolving tank | 7 | 0.43 |
| Vent of white liquor tank | 11 - 15 | 0.67 - 0.92 |
| Concentrated black liquor tank | 22 - 28 | 1.34 - 1.70 |
| Water and gases from digesters untreated: | | |
| Relief gases after dilution with 900 ft.³ of air/min. | Variable from 30 - 300 | from 1.8 - 18.3 |
| Gases from chlorination outlet within fume steam | 3 - 6 | 0.18 - 0.36 |
| Gases from recovery boiler (the black liquor was previously oxidized, with an oxidation of 100%, measured as conversion of sulfides into thiosulfates): | | |
| Before scrubbing and chlorination | 10 - 26 | 0.61 - 1.58 |

(continued)

| Sources | Titrilog Net Reading | Sulfur Compounds Given as $H_2S$, ppm |
|---|---|---|
| After scrubbing and chlorination, final wash at pH of 6 | 8 - 20 | 0.49 - 1.22 |
| Same: Final wash at pH of 9 | 3 - 4 | 0.18 - 0.24 |
| For Comparison: | | |
| Gases from power boilers, burning fuel oil | 2 - 3 | 0.12 - 0.18 |
| Cigarette smoke. Cigarette 2" from intake of sampling apparatus | 90 | 5 - 49 |

## Mercaptan Removal from Water

A process developed by W.C. Meuly et al (3) involves abating malodors and reducing health hazards, vegetation and wild life destruction and other pollution nuisances caused by the hydrogen sulfide and/or mercaptans present in fluid streams passing through or leaving various industrial processes such as Kraft papermaking and petroleum refining and the like.

Briefly stated, the process is based on the discovery that the pollution of fluid streams polluted with at least one of the components of hydrogen sulfide and mercaptans may be abated by adding to the polluted stream a chelate of a polyvalent metal with a substance selected from the class consisting of acetyl acetone, cyclopentadiene, ethylene diamine tetraacetic acid, N-hydroxyethyl ethylene diamine triacetic acid, gluconic acid, tartaric acid and citric acid in an amount that is stoichiometrically substantially less than the amount of the pollutant, and contacting with gaseous oxygen the polluted stream containing the chelate. In a preferred form, the gaseous oxygen is present as a component of atmospheric air. Preferably, the oxygen is present in an amount at least about twice the stoichiometric quantity of the pollutant, more preferably at least about four times the quantity.

The polyvalent metals include all polyvalent metals. Among these, the most commercially attractive are chromium, cobalt, copper, iron, lead, manganese, mercury, molybdenum, nickel, palladium, platinum, tin, titanium, tungsten and vanadium. Iron, cobalt and nickel are preferred, and iron particularly is preferred.

Among the organic components of the chelate, N-hydroxyethyl ethylene diamine triacetate is particularly preferred, and its chelate with iron is the most preferred chelate. Pressure is not critical. The reaction proceeds at atmospheric pressure as well as at other pressures normally encountered in the industrial processes in question.

Similarly, temperature is not critical. The reaction proceeds at least within the range -50° to 400°F., and evidently also outside this temperature range. Accordingly, room temperature or the outside temperatures encountered at any season of the year are quite suitable for the practice of the process.

Similarly, concentration of the pollutant is not critical. The process of the method is capable of reducing or substantially destroying concentrations of hydrogen sulfide and/or mercaptans in excess of about 100,000 parts per million (ppm) and at the other extreme, in concentrations even below 1 ppm.

Within normal ranges, pH is not critical. Operation within the pH range of 4 to 11 is feasible. During the reaction, the pollutant is oxidized by the oxygen present, and the chelate catalyzes the reaction. In view of its nature as a catalyst, therefore, the concentration of the chelate is also not critical.

The catalyst of this process is preferably added to the polluted stream in liquid phase, in solution in a suitable carrier which preferably is either water or an organic liquid miscible with water, such as ethanol, methanol or the like. To prepare a catalyst suitable for use in this process, 28 parts of cobaltic chloride may be dissolved in 50 parts of water and 135 parts of a 45% solution of tetrasodium ethylene diamine tetraacetate are added with stirring. The resultant product is filtered and filtrate is reserved for use as the catalyst. The following is one typical example of the application of such a catalyst to mercaptan removal.

Crude sulfate turpentine containing high concentrations of lower alkyl mercaptans is scrubbed with a 25% aqueous solution of the catalyst with a stream of air bubbling through the mixture. After 10 minutes the catalyst is washed out, the turpentine is dried over sodium sulfate and the treated material is evaluated against the untreated crude turpentine for odor and is found to be much superior in odor.

### References

(1)  Friess, H.; U.S. Patent 3,391,988; July 9, 1968; assigned to Gelsenberg Benzin AG, Germany.
(2)  Trobeck, K.G., Tirado, A. and Lenz, W.; U.S. Patent 3,028,295; April 3, 1962; assigned to
       AB B.T. Metoder, Sweden and Fabricas de Papel Loreto y Pena Pobre, SA, Mexico.
(3)  Meuly, W.C. and Seldner, A.; U.S. Patent 3,226,320; December 28, 1965; assigned to Rhodia, Inc.

## MERCURY

The general aspects of mercury emission, detection and control have been reviewed by H.R. Jones (1).

### Mercury Removal from Air

Mercury vapor is introduced into various gases such as air, hydrogen and carbon dioxide in various ways. A common source of mercury vapor is the mercury electrolytic cell used in electrolysis reactions, e.g., in the formation of hydrogen gas. The hydrogen gas formed is frequently used in the hydrogenation of fats and oils in the production of foodstuffs. Since even small amounts of mercury are poisonous it is important to remove the mercury from the hydrogen gas.

The removal of mercury vapor from air or other gases is desirable for various reasons. It is considered that health hazards exist if the concentration of mercury vapor in air rises above about $3 \times 10^{-4}$ grams per cubic meter and such hazards may exist in chemical laboratories where mercury is used or in plants, such as those utilizing mercury boilers, where leakage of mercury vapor may possibly occur accidentally.

It may be readily appreciated that the dangerous concentration may readily exist in enclosed places when there is considered the fact that air saturated with mercury vapor at ordinary temperatures may contain around $1.5 \times 10^{-2}$ grams of mercury per cubic meter. Exposed or spilled liquid mercury may thus raise the concentration of mercury vapor substantially above the danger point. The removal of mercury vapor from air or other gases may also be desirable for technological reasons as well as for health reasons.

A process developed by M. Manes et al (2) achieves mercury removal from gases by impregnating materials which dissolve mercury on activated carbon. It has been found that these products will rapidly and in many cases quantitatively remove mercury vapor from gases including hydrogen, air, carbon dioxide, nitrogen and oxygen. The high surface area of the activated carbon which is impregnated with the mercury reactant appears to be in part responsible for the greatly improved adsorption of the mercury vapor and the carbon appears to activate the metal or other material which dissolves the mercury.

The action of activated carbon appears to be specific. Thus when silica gel is employed in place of activated carbon as the support for the mercury solvent it is not possible to get a satisfactory coating. Additionally, the silica gel readily picks up water vapor which is undesirable.

As the material employed to impregnate the activated carbon there can be used metals which will amalgamate with mercury. Examples of such metals are gold, silver, cadmium, indium, thallium, aluminum, lead, gallium and copper. Metals which are not easily oxidized such as the noble metals are preferred and silver is especially preferred metal and copper is one of the poorer materials.

The metal can be formed on the activated carbon by impregnating the carbon with a solution of a reducible salt of the metal, e.g., aqueous silver nitrate, auric chloride or cupric chloride, and then reducing the salt to the free metal in any conventional manner. After the mercury has been adsorbed on the impregnated carbon it can be regenerated by heating. Thus 100% regeneration of activated carbon impregnated with silver, for example, can be obtained by heating to around 300°C. The mercury is recovered in pure form by this regeneration, e.g., by distillation.

Silver impregnated activated carbon picked up mercury from contaminated air in an amount of over 3% by weight of the carbon. This was 100 times as much mercury as was picked up by the unimpregnated activated carbon. Usually 5 to 50% of the impregnant is used with the activated carbon.

The granular activated carbon employed generally has a particle size between 4 and 60 mesh although this can be varied. In the specific examples there was employed Pittsburgh type carbon BPL which can be made as described in U.S. Patent 2,763,580. There can also be used other activated carbons such as Pittsburgh type CAL, Columbia activated carbon Grade SXAC, Darco activated carbons, etc.

A process developed by S.H. Williston et al (3) utilizes a mercury removal device which comprises a chamber containing a porous absorbent which presents an extended surface of a mercury-wettable metal. The metals desirably used are silver, gold or combinations of these metals, preferably carried in the form of very thin films on an extended support which may take the form of glass or other wool, granular material, or the like.

Figure 78 shows the essentials of such an apparatus. A chamber 2 is provided with inlet and outlet openings 4 and 6 which may respectively communicate with the source of gas containing mercury vapor, for example, air, and with the device or space within which mercury-free gas is desired. A chamber having this construction, for example, is used in the apparatus described in the prior application.

The chamber 8 contains the absorbent material and, as a specific example, this absorbent material may be glass wool (providing high porosity for the flow of the gas) the fibers of which are coated with pure gold. The gold may be deposited in the form of a quite thin film on the glass wool by any of numerous processes which are known for the deposition of gold, an ordinary and convenient method merely involving the wetting of the glass wool with a gold salt such as the chloride which is then decomposed for deposition of the gold by heat.

For use in a measuring apparatus the cost of using gold is immaterial since only small quantities of gold are involved. In place of glass wool as a carrier nickel wool may be used on which the gold may be deposited in the same fashion as described or by precipitation by the nickel from a solution of the gold salt. The carrier may be of many types so long as it is of a physical form to present a maximum absorbing surface, reasonably low resistance to flow, and adhesion to its absorbent coating.

Granular alumina may be used, for example, completely coated to prevent absorption of water. Porous ceramic pellets may also be used. If it is desired to regenerate the absorbent after it absorbs sufficient mercury to lower its effectiveness, and since regeneration is most readily effected merely by heating the absorbent material to drive off the mercury, the carrier should be resistant to heat, a property shared by the materials referred to such as glass wool, nickel wool, alumina or ceramic material.

FIGURE 78:  GENERAL SCHEMATIC VIEW OF MERCURY VAPOR ABSORPTION
DEVICE

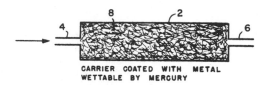

CARRIER COATED WITH METAL
WETTABLE BY MERCURY

Source:  S.H. Williston and M.H. Morris; U.S. Patent 3,232,033; February 1, 1966

In larger installations, for example for the removal of mercury vapor from rooms or other
enclosures, the deposited absorbent metal may well be silver which has the advantages of
lower cost and very simple deposition procedure, carriers being used such as those indicated for gold.  Silver has a slight disadvantage in that it may be rendered ineffective by
hydrogen sulfide or chlorine in the atmosphere; but it is found that its life even under the
adverse conditions of such atmosphere is so long that it is actually preferred for the provision of absorbent cartridges which may be disposable without attempts at regeneration.

The silver may be deposited in numerous fashions utilizing baths such as are commonly
known for the deposition of silver for mirrors, or silver salts may be precipitated or used
to wet the carriers which are readily decomposable by heat.  Among such salts are the
nitrate and a large groups of silver salts of organic acids as well as silver oxide.  The length
of active life of a silver coating in atmospheres of hydrogen sulfide or chlorine may be
lenghthened by coating a deposit of silver with an extremely thin film of gold which, in
itself, might not have a sufficiently long life for practical use, the mercury being then, in
effect, absorbed by the gold and passing into the silver so that the life is essentially that
of the silver from the standpoint of absorbability.

The same end may be achieved by precipitating mixtures of gold and silver which at elevated temperatures will form alloys resistant to corrosion, such alloys being as effective
as the separate metals.  Alloys of gold and silver with other metals which are high in
gold or silver content may also be used.

Figure 79 shows an installation which may be used for removing mercury vapor from relatively large flows of air as, for example, for reducing the concentration of mercury vapor
in an enclosed atmosphere.

A housing **10**, open at both ends, may be mounted in a wall or partition **12** and is provided with a flanged structure **14** of any suitable type for holding a large cartridge **16**
containing the absorbent material **8** such as already described.  The chamber provided at
**16** has inlet and outlet passages in the form of holes **18** through which the air may enter
and leave.  Flow may be provided by a motor driven fan indicated at **20**.

The cartridge **16** in its simplest form may be merely a cardboard box provided with openings and containing the absorbent material.  When the effectiveness of its absorption is reduced to a degree making replacement desirable, the entire cartridge may be thrown away
and replaced by a new one.

It is in such a case that the relatively inexpensive absorbent material using silver as the
coating for the carrier may be used.  Even a very thin coating of silver will ordinarily
take care of thoroughly effective removal of mercury from a very large volume of air or
other gas.

# FIGURE 79: INSTALLATION FOR MERCURY VAPOR REMOVAL FROM LARGE FLOWS OF AIR

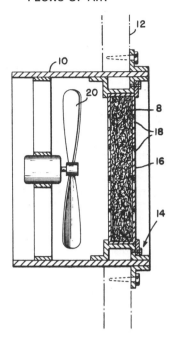

Source: S.H. Williston and M.H. Morris; U.S. Patent 3,232,033; February 1, 1966

A process developed by J.C. Park et al (4) is an improved process for reducing the concentration of mercury vapor in hydrogen. Hydrogen produced by electrolysis of brine at a mercury cathode contains substantial quantities of mercury vapor, e.g., more than 20,000 micrograms of mercury per cubic meter, and is usually saturated with water vapor. It is desirable to remove substantially such mercury concentrations from hydrogen since the presence of mercury in hydrogen in amounts greater than about 25 micrograms, and in many cases about 10 micrograms per cubic meter, interferes with the use of the hydrogen in catalytic reductions of organic compounds.

The prior art methods of mercury removal are not entirely satisfactory. For example, it is known to remove mercury from hydrogen gas by chemical means, e.g., by converting mercury to mercurous chloride by treatment with chlorine gas in brine, but such methods require delicate control and may introduce additional objectionable impurities into the hydrogen. Further, it is known to absorb mercury from hydrogen by passing the gas through a bed of activated charcoal. Charcoal, however, is a relatively inefficient absorbent for mercury, and when saturated with mercury can only be partially reactivated.

It is also known to adsorb mercury from hydrogen by passing the hydrogen through a bed of particulate matter comprising metallic silver supported on particles of siliceous earth. This method is unsatisfactory, in that it is only effective with hydrogen gas which is substantially completely free from water vapor, even very small proportions of water vapor resulting in wetting of the siliceous earth support, causing it to crumble and disintegrate and rendering the material nonadsorbent. Moreover, even with moisture-free hydrogen gas, the siliceous earth supported silver particles become disintegrated and pulverized in short periods at the high gas velocities required for efficient commercial operation.

In this process hydrogen gas contaminated with mercury vapor is passed at a temperature below the boiling point of mercury through a bed of particulate adsorbent comprising silver metal on a fused aluminum oxide support at a rate sufficient to substantially reduce the mercury content of the gas.

According to a preferred embodiment of the process, hydrogen contaminated with mercury in amounts in excess of about 25 micrograms per cubic meter is passed at a temperature of 0° to 50°C., through an adsorbent bed of fused aluminum oxide containing at least about 1%, but not more than about 15% of metallic silver by weight, preferably between 5 and 12% by weight metallic silver.

Under the preferred conditions of the process concentrations of mercury vapor in hydrogen from as low as about 25 micrograms to as high as 20,000 to 35,000 or more micrograms per cubic meter can be reduced to less than about 25 micrograms, or even less than about 10 micrograms per cubic meter.

In carrying out the process, particles of fused aluminum oxide containing at least about 1% by weight of silver metal supported thereon, preferably about 10%, are packed into a suitable column, either vertical, horizontal or inclined. Hydrogen gas contaminated with mercury, is then passed through the adsorbent at a space velocity effective to cause adsorption of substantially all, or the desired portion, of the mercury in the hydrogen.

Adsorption is extremely rapid so that space velocities are not unduly critical and may range between 0.0016 and 16 in terms of volume of hydrogen (STP) per second per unit volume of adsorbent. Such velocities are usually sufficient for substantially complete removal of mercury from the gas, i.e., to values of not more than about 25 micrograms per cubic meter. The hydrogen gas contaminated with mercury should be passed through the adsorbent bed at temperatures below the boiling point of mercury, i.e., 357°C. At greater temperatures the high partial pressure of the mercury makes adsorption of mercury negligible.

Preferably the temperature of the hydrogen gas to be purified is about 0° to 50°C. Conveniently hydrogen gas containing mercury vapor which is at a temperature above the preferred temperature range may be cooled by suitable means before contact with the adsorbent. On cooling, a portion of the mercury contained in the hydrogen usually condenses and thus can be removed. The process can be carried out with hydrogen gas at atmospheric, subatmospheric, or superatmospheric pressures.

The adsorbent employed in this process is of sufficient density and hardness to resist abrasion under flows of gas at high velocity. Furthermore, the adsorbent is also inert towards moisture which may be present in the hydrogen being purified. An adsorbent support useful in this process is fused aluminum oxide, Alundum, and when containing 10% by weight silver metal is available as Davison Catalyst SMR-55.

However, any silver-fused aluminum oxide composition can be employed. For example, a suitable adsorbent may be prepared by breaking into fragments an aluminum oxide (Alundum) which contains 10% by weight clay as bonding additive. These fragments of suitable dimensions, e.g., roughly 3/8 by 1/2 inch, are then soaked in an aqueous solution of silver nitrate of sufficient strength to afford about 10% of silver on the finished adsorbent.

The solution containing the carrier is evaporated to dryness. Size is not unduly critical and particles with diameters as large as an inch or more are satisfactory. The carrier with the silver nitrate thereon is then subjected to reduction with hydrogen at a temperature of approximately 330°C. The carrier is thus impregnated with metallic silver, and may be used in this form or crushed and screened to obtain particles of convenient size for packing.

The adsorbent, when saturated with mercury vapor, is readily reactivated by heating at a temperature of at least about 200°C. or higher for at least about 1 hour, preferably from 1 to 6 hours, under a stream of dry, inert gas, e.g., nitrogen. Heating the saturated adsorbent at temperatures lower than 200°C. for periods shorter than about 1 hour results in only partial reactivation of the adsorbent. Heating the saturated adsorbent at higher temperatures, e.g., about 600°C. or higher, for longer periods, e.g., about 6 hours or more, while effective, is unnecessary and may be destructive to the adsorbent. An excellent result is generally obtained by heating the saturated adsorbent at a temperature of 230° to 250°C. for about 1 hour while passing a stream of dry nitrogen with a linear velocity of about 0.5 to 1 foot per second through the adsorbent.

Conveniently the nitrogen used is preheated to the temperature employed in the reactivation. Advantageously, the adsorbent can be nitrogen treated before use in the process in order to drive off gases adsorbed from the atmosphere. The adsorbent of this procedure retains its adsorptive capacity for mercury after being reactivated as many as seven times.

The reactivation of adsorbent in this process is carried out when the mercury concentration of hydrogen gas leaving the adsorption bed exceeds the maximum level desired, e.g., 10 micrograms per cubic meter. For continuous operation two adsorption beds can be connected in parallel, so that mercury removal from hydrogen can continue in one bed while the other bed is shut down for reactivation, conveniently an activated charcoal adsorption bed can be placed at the end of adsorption train to insure against carry-over of mercury in case of an accidental breakdown in the operation of the silver-aluminum oxide adsorption beds.

A process developed by M. Manes (5) involves removing mercury from a gas stream contaminated with mercury by passing the gas stream containing mercury over an activated carbon impregnated with a metallic silver having a crystallite size of not over 250 A. The metallic silver is formed by reducing a silver thiocyanate complex impregnated on the activated carbon.

A process developed by D. Bell (6) involves the removal of mercury from a gas stream by washing the gas with alkaline hypochlorite containing added alkali metal or calcium chloride. The alkaline hypochlorite solution is suitably sodium hypochlorite. Sodium hypochlorite solutions are well-known in commerce and normally contain sodium hypochlorite and sodium chloride of approximately equimolar proportions. When mercury vapor is reacted with such solutions, or with solutions prepared by diluting the commercial solutions with water, a precipitate of insoluble mercury compounds is formed. This is inconvenient because it tends to settle out in vessels and pipelines and the like and makes the recovery of the mercury more difficult.

It has been discovered that if additional alkali metal chloride or calcium chloride is added to the alkaline hypochlorite solution, the mercury remains in solution, possibly in the form of a complex anion. The amount of additional sodium or calcium chloride to prevent mercury compounds precipitating depends upon the composition of the solution particularly with respect to pH.

The mercury vapor can be absorbed effectively from gas streams by dilute solutions of such hypochlorite containing additional alkali metal chloride. The solution should have an available chlorine content of at least about 0.02 g./l. About 25 g./l. available chlorine is a usually convenient upper limit, although it is possible to use higher available chlorine contents. The pH should be controlled between approximately 8 and 12, and preferably between pH 9 and 10.5. The minimum sodium chloride content of a solution containing 5 g./l. of available chlorine in the preferred pH range is 40 g./l. at pH 9 and 140 g./l. at pH 10.5 for operation at ambient temperature.

Washing the mercury-containing gas streams may be carried out in any gas-liquid contacting device, for example, a column packed with Raschig rings or on diffuser plates. It may be carried out at ambient temperature or at any other convenient temperature. Mercury may be recovered from the solution either chemically or electrolytically. A suitable electrolytic cell for recovering the mercury contains a graphite or platinized titanium anode and a mercury cathode. The mercury in solution is reduced to mercury at the cathode. A preferred method of recovering mercury from the absorbing solution is to blend it slowly into the feed brine stream supplying one or more commercial mercury cells. The mercury in solution is then recovered electrolytically at the cathode.

A process developed by I. Ferrara et al (7) is one in which gases contaminated with mercury are purified by bringing the same into intimate contact with a catalytic composition comprised of mercuric sulfide and sulfur.

It has been established that such composition has a very high mercury absorption power. It has also been ascertained that this power is due to the catalytic action developed by the mercuric sulfide on the reaction of absorption of the mercury and carried out by the sulfur.

The catalytic composition is fixed on a suitable carrier such as, for example, active carbon or alumina. The quantity of sulfur and mercuric sulfide fixed per unit weight of carrier may vary over a wide range; as a matter of fact, very good results are obtained with catalysts containing only about 10 mg. of HgS and about 1 mg. of S per gram of carrier, as well as with catalysts containing far greater quantities thereof, for example, on the order of several hectograms HgS and fewer hectograms S per gram of carrier, as well as with catalysts that contain intermediate quantities.

For example, very good results are obtained with different qualities of active carbon or alumina by fixing from 60 to 80 mg. of mercuric sulfide and from 5 to 10 mg. of sulfur per gram of carrier. The sulfur itself may serve as a carrier. The catalyst in this instance will consist of crystalline sulfur, whose granules have been partially coated with mercuric sulfide.

The catalyst can be used over wide ranges of temperatures and pressures. Good results have been achieved at atmospheric pressure, as well as at pressures of up to about 30 atm. In regards temperature, one can operate at room temperature, as well as at lower temperatures (from $0°$ to $10°C.$, for example) or at higher temperatures (for example at $75°C.$).

When operating at temperatures higher than room temperature, with certain catalyst formulations, one may encounter the drawback of the gas to be purified contaminated with small quantities of sulfur. Such a drawback can be eliminated by allowing the sulfur to be absorbed by active carbon in a successive absorption tower kept at room temperature.

The concentration of the mercury in the gas to be purified does not, in practice, have any influence on the quality of the results obtained. Good results are obtained regardless of whether it has been operated with a gas having a low content of mercury (i.e., $250\gamma/m.^3$ – standard conditons) or with gases saturated or substantially saturated with mercury. The symbol gamma is utilized to indicate microgram.

The space velocity of the gas to be purified may also vary over a wide range. Good results have been obtained with space velocity of about 10,000 per hour. In one of the preferred embodiments of this process, use is made of space velocity of from 3,000 to 4,000 per hour.

The catalyst may be prepared in the following manner. The carrier, for example, active carbon, alumina or sulfur, is soaked with mercuric chloride and then subjected to the action of a gas containing hydrogen sulfide; thereby the complete conversion of the mercuric chloride into mercuric sulfide is effected. For this purpose gases may be used containing for example, from 0.5 to 500 mg. of hydrogen sulfide per liter. The slight charge of sulfur necesary for a working catalyst with active carbon or alumina may be obtained simultaneously with the formation of the mercury sulfide when a gas containing hydrogen sulfide and small quantities of oxygen is used.

For this purpose, gases containing from 0.5 to 500 mg. of hydrogen sulfide and from 0.02 to 2 mg. of oxygen per liter can be used. Similar compositions are readily available, e.g., stripping gases obtained in various processes.

The catalysts thus obtained have an excellent mechanical resistance and a long life. The humidity of the gas to be purified does not alter operation. The active carbon or alumina catalyst will work as long as it contains sulfur. Once it has been exhausted, the catalyst can be reactivated by charging it with traces of sulfur.

A process developed by W.H. Revoir et al (8) provides adsorbents for removal of mercury vapor from air or gas including activated carbon granules impregnated with interhalogen compounds, particularly iodine monochloride and iodine trichloride.

A process developed by P. Bryk et al (9) is one in which the concentration of circulating sulfuric acid is adjusted to 80 to 98% by weight and used to wash hot gases containing mercury. The temperature of the acid is maintained between 70° and 250°C. and the solid material separating from the circulating wash solution is recovered.

Figure 80 is a block flow diagram showing the essentials of the process. Mercury-containing gases, which have gone through the normal dry dust separation, are conducted to a special wash tower for gases where they are washed with a circulating sulfuric acid solution, the concentration of which is at least 80% by weight. The temperature of the circulating wash solution saturated with metal sulfates and other metal salts is regulated during the washing phase with the circulation rate and outside heat exchange so that the mercury contained in the incoming gas can be made to sulfatize and form mercuric compounds during the washing phase as completely as possible.

**FIGURE 80: SULFURIC ACID SCRUBBING PROCESS FOR MERCURY VAPOR REMOVAL FROM GAS STREAM**

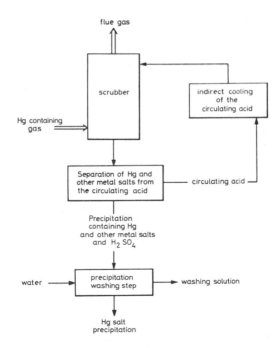

Source:  P. Bryk, M. Haani, J.B. Honkasalo, J. Kinnunen, O. Lindsjo, E. Nyholm,
J. Poijarvi, J. Rastas and J. Kangas; U.S. Patent 3,677,696; July 18, 1972

At this time it is important that the wash solution is not diluted by the humidity contained in the gases. The concentration of the wash solution should be held constant, and the concentration and the washing temperature adjusted so that the mercury will be sulfatized or form mercuric compounds as completely as possible.

The dilution of the wash solution can be prevented and the concentration of the wash solution held constant if the temperature during the washing of the gases is adjusted by the circulation of the wash solution and by outside heat exchange such that the water vapor pressure of the wash solution during the washing phase is the same as the partial pressure of the water vapor contained in the gas to be washed, the temperature generally being in the range of from 70° to 170°C.

If the circulating wash solution is saturated by mercuric and other metal salts (metals contained in the dust), the salts being particular to each system, these salts separate continuously from the circulating wash solution. The solid matter containing these salts is then separated from the wash solution with some suitable method.

The separated precipitate is washed, in which case the sulfuric acid or other circulating solution and most of the other metal salts are dissolved in the washing solution and the mercury is left in the final precipitate. The precipitate may contain, also, other valuable elements, such as selenium. When the gases to be washed contain selenium, it, also can be bound into the solid matter separating from the circulating wash solution and made to remain in the final precipitate after the washing of the solid material. Mercury and selenium can then be separated into a pure form from the other elements contained in the final precipitate with some known production processes for mercury and selenium.

**Mercury Removal from Water**

A process developed by G.L. Bergeron et al (10) involves recovering mercury from brine effluent from mercury cathode electrolytic cells. The mercury cathode electrolytic cells are constructed with a relatively small gap between a fixed anode and a steel plate or other current conducting material. In the operation of these cells, saturated sodium chloride or potassium chloride brine and mercury are passed through this gap during the electrolysis. The mercury upon entering the cell spreads over the steel plate or other conducting material and acts as a cathode for the cell.

In the process saturated brine solutions are used. After passing the brine once through the cell, the brine discharged from the cell is dechlorinated by air stripping or other means, resaturated, and recycled again through the cell. When a sodium chloride brine is used, the brine generally will contain around 300 grams of sodium chloride per liter, while for potassium chloride solution, the chloride concentration may be up to around 350 grams of potassium chloride per liter. In passing through the cell the chloride concentration of the brine is seldom reduced over 20%. Thus, the brine discharge from the cell is still relatively saturated.

While the mercury cathode cells have many advantages over other conventional cells, a small but significant amount of mercury is lost in the process. A major portion of the mercury loss results from the chlorination of the mercury to a soluble salt which dissolves in the brine as it passes through the cell. This mercury which reacts with the chlorinated brine is often lost in the resaturation step of the process. The brine leaving the cell may contain as much as 50 parts of mercury per million parts of brine and in some cases a considerably greater amount. This loss of mercury while small per pass becomes considerable when a battery of cells is operated over an extended period of time.

It is therefore, a principal object of this process to provide for the recovery of mercury from the brine discharged from mercury cathode cells. A further object is to recover the mercury by an economical method which will not contaminate the brine and make it unsuitable for further electrolysis. The above objects may be accomplished by adding a soluble sulfide such as sodium sulfide, potassium sulfide or lithium sulfide to the brine to convert the reacted mercury in the brine to a mercury sulfide dispersion and then separating the dispersion from the brine through flocculation. A ferric compound and starch or gum arabic are added to the brine and the mercury sulfide flocculated by adjusting the pH of the mixture. The combined precipitate is then recovered from the brine by mechanical means and processed by usual methods to recover the mercury.

Figure 81 is a block flow diagram showing the essentials of this process. Dechlorinated brine discharged from the electrolytic cell is passed into agitated precipitator 1. A sodium sulfide solution is added to the precipitator in an amount such that a stoichiometric excess in the range of 50 to 80% is maintained to convert the dissolved mercury present in the brine to mercury sulfide. The brine containing the fine colloidal particles of mercury sulfide is passed to mixer 2 where ferric chloride and starch are added.

The amount of ferric chloride added is in the range of 5 to 20 parts by weight per million parts of brine and the starch in the range of 0.5 and 5 parts by weight per million parts of brine. Also, sodium hydroxide solution is added to the mixture until the pH in mixer 2 is in the range of 8 to 10. The retention time in the mixer is in the range of 10 to 30 minutes. From the mixer the brine containing the ferric hydroxide starch, and mercury sulfide is passed to settler 3 where the ferric hydroxide and mercury sulfide settle out as a sludge and are withdrawn from the settler. Ther overflow from the settler is a substantially mercury-free brine which can then be further saturated and returned to the electrolytic cells. The sludge containing the mercury sulfide is then processed by known methods to recover the mercury.

FIGURE 81: PROCESS FOR THE RECOVERY OF DISSOLVED MERCURY SALTS FROM BRINE EFFLUENT FROM MERCURY CATHODE ELECTROLYTIC CELLS

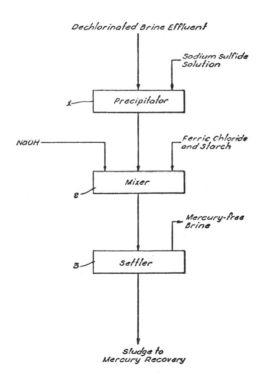

Source:  G.L. Bergeron and C.K. Bon; U.S. Patent 2,860,952; November 18, 1958

A process developed by M.P. Neipert et al (11) involves intermixing the brine effluent containing the mercury in a reacted form as a mercury compound with at least a stoichiometric amount of an aldehyde elected from the group consisting of formaldehyde and acetaldehyde at a pH of at least 7. The soluble mercury salts in the brine are reduced by the aldehyde to metallic mercury which is then recovered by settling.

It is surprising and unexpected to find that the small amount of mercury salt present in the relative saturated brine can be substantially completely reduced to metallic mercury by the addition of a small amount of formaldehyde or acetaldehyde and thus recovered. By this process the mercury may be recovered from natural or artificial sodium chloride brines or natural or artificial potassium chloride brines. Thus, the term brine, as used herein, means natural or artificial aqueous solutions of sodium chloride or potassium chloride.

A batch or continuous process may be used. A continuous process is preferred. By using a continuous process the mercury recovery step may be very conveniently integrated into the electrolysis process by placing the recovery step after the dechlorination of the brine effluent and prior to the resaturation. The dechlorinated brine discharge from the dechlorinated unit instead of going to the resaturation step is introduced into an agitated reactor to which formaldehyde or acetaldehyde is also continuously charged.

A retention time in the agitated reactor of from ½ to 4 hours is provided. Within this time the reaction of the aldehyde in reducing the mercury salt is completed. The mixture is then passed to a settler or other classification equipment where the mercury settles out. The overflow from the settler is a substantially mercury-free brine which may then be resaturated in the normal method used in the process.

After the addition of the aldehyde to the brine, the pH of the brine is adjusted to at least a pH of 7, preferably in a range of 10 to 12. The brine effluent leaving the cell and after dechlorination is slightly acidic and thus a basic compound must be added. Any soluble alkaline compound, such as alkali metal hydroxides, alkaline earth metal hydroxides, carbonates, etc. may be used. For potassium chloride brines potassium hydroxide is preferred and for sodium chloride brines sodium hydroxide is preferred. By the use of these compounds, the brine is not contaminated by other metals. The reaction temperature and pressure are not critical. Generally, atmospheric pressure is used and the brine is treated at its discharge temperature from the dechlorination unit which may be at a temperature in the range of 40° to 80°C.

While both formaldehyde and acetaldehyde may be used to reduce the mercury salt in the brine, formaldehyde is preferred. Formaldehyde is more active and complete recovery of the mercury can be obtained with a less amount of this aldehyde. Theoretically, only a stoichiometric amount of the aldehyde to react with the mercury salt is necessary. However, generally, even after the brine is dechlorinated, it will contain small amounts of chlorine which will react with the aldehyde.

Consequently, the amount of aldehyde necessary will vary with the amount of chlorine present in the brine as well as the amount of mercury. Since the amount of aldehyde added is small, generally a sufficient excess is used to take care of any variation of chlorine or mercury the brine may contain. Normally, after the conventional methods of dechlorination, from 0.01 to 1.0 pound mol of formaldehyde or acetaldehyde per thousand gallons of brine will give substantially complete reduction of the mercury. For formaldehyde the amount preferred is from 0.05 to 0.1 pound mol per thousand gallons of brine while for acetaldehyde the amount is a little greater being from 0.15 to 0.3 pound mol per thousand gallons of brine. If the dechlorination of the brine is not effective, larger amounts may be necessary.

A process developed by R.S. Karpiuk et al (12) involves recovering mercury from aqueous solutions containing compounds thereof by passing the solution and a liquid alkali metal amalgam concurrently through a bed, preferably a vertical column, of steel turnings or pieces, preferably first amalgamated, whereby the mercury component of the mercury compounds therein is caused to form additional metallic mercury and/or diluted amalgam, some of which adheres to the surface of the amalgamated steel pieces or turnings and some of which drips from the surface of the turnings to the lower part of the bed or column and collects there from which it can be subsequently removed.

Another process developed by R.S. Karpiuk et al (13) for recovering mercury from aqueous solution comprises passing the solution and concurrently therewith an alkali metal amalgam through a bed, preferably longitudinally, through a vertical column of glass, ceramic, or plastic beads or fragments or Berl saddles or other nonmetallic material which is substantially inactive to the brine, to the alkali metal amalgam, and to compounds normally found in the brine or amalgam, whereby a substantial proportion of the mercury in the brine forms additional dilute amalgam with the alkali metal amalgam and to some extent subsequently forms metallic mercury, mercury thereby being separated from the brine, either as diluted amalgam or as mercury metal.

Both the diluted amalgam and the metallic mercury may be returned to the denuder or decomposition chamber of a mercury chlorine cell (where the amalgam is converted by hydrolysis to mercury and NaOH) or the mercury may be conveniently trapped below the bed of beads and thereafter tapped off as desired.

A process developed by J.F. Gilbert et al (14) is based on the discovery that by bringing an aqueous solution having a pH between 2 and 11 and containing from 1 to 500 ppm of dissolved mercury into intimate contact in a reaction zone with a substantially water-stable solid metallic reducing agent having a greater solution potential than mercury, elemental metallic mercury is liberated. The liberated mercury amalgamates the surfaces of the reducing agent, and also coalesces into droplets on the surfaces.

Depending on the manner of carrying out the process, particles of amalgam and mercury droplets are either allowed to fall from the reducing agent and collected from time to time, or the amalgam and mercury droplets are flushed from the surfaces of the reducing agent along with inert solid settlings formed thereon and recovered from the flushing liquid as by settling or filtration. Impure mercury so recovered is purified by standard methods, such as acid washing or retorting or by a combination of methods. If desired, mercury may also be recovered by removing the reducing agent from the reaction zone periodically along with accumulated reaction products and retorting the entire mass.

As reductor metals those readily reducing or liberating mercury from solution as metallic mercury include iron, zinc, bismuth, tin, nickel, magnesium, manganese and copper. Of these iron and zinc are to be preferred because of the lower cost of the metals as well as generally lower solution losses and higher reaction rates when used in this process. Iron may be used for those solutions advantageously treated at a moderate pH, for example to avoid or minimize precipitation of solids, such as oxides and hydroxides at higher pH values.

Optimum mercury removal per pound of iron consumed is obtained upon adjusting the pH of the solution to a value between 6 and 9. At pH values below 6 solution losses of iron become increasingly larger and below a pH of 5 hydrogen evolution reduces the effective surface area of the metal and the competition of hydrogen ions with mercury ions becomes significant.

At a pH of 7 to 8 the consumption of iron in the form of steel turnings may be expected to be of the order of 0.1 pound per thousand imperial gallons (1,200 U.S. gallons) of solution treated by this method. Zinc metal may be used for those solutions that are advantageously treated at a higher pH. Zinc is best used with solutions brought to a pH between 9 and 11. Although zinc readily liberates mercury in less alkaline or in acidic solutions zinc solution losses become increasingly larger at lower pH values.

Zinc amalgams such as are produced by the liberation of mercury at the surface of zinc metal are physically more stable than iron amalgams similarly produced and do not suffer the disadvantage of being rather readily carried away in a fine state of subdivision during a flushing cycle as are iron amalgams. On the other hand, iron has the advantages of having a slightly higher reaction rate and of being a less expensive metal than zinc.

A process developed by R.C. Calkins et al (15) is based on the discovery that mercuric ions can be selectively removed from brines containing halide ions and metal ions, particularly the alkali and alkaline earth metal ions, e.g., sodium, calcium, and the like, by contacting the brine at a pH of 9.5 or above, with an insoluble homopolymer or copolymer prepared from at least one monomer of the group consisting of vinylphenyl aliphatic primary and secondary aminomonocarboxylic acids and mixtures containing a predominant amount by weight of at least one such aminomonocarboxylic acid and a minor proportion, e.g., up to 20% of divinylbenzene.

It has been further found that the mercuric ions may be readily recovered from the resin chelate by elution with chloride solutions having a pH value below 9.5. It is necessary that the pH of the brine be maintained at 9.5 or above for efficient mercury removal, and is desirably maintained at 10 to 11, although higher pH values may be used. It is desirable to have resin properties which will permit a minimum contact time.

A desirable physical form which these resins could assume would be as small, 0.15 mm. diameter or less, resilient beads, which would swell and shrink in use by as much as 50%. In this form the resins pick up mercury most rapidly, and are most easily eluted. The resilient beads are also more stable to mechanical breakdown. The contact time is dependent upon such variables as the physical condition of the resin, the temperature of operation, the concentration of ions in the feed solution, and the allowable concentration in the effluent and must be determined by the economic considerations for each particular case.

Among the homopolymers and copolymers suitable for use in this process are those containing monocarboxylic amino acid residues such as, for example, vinylbenzylisovaline, vinylbenzylalanine, vinylbenzyl-2-aminobutyric acid, vinylbenzylvaline, vinylbenzylleucine, vinylbenzylglycine, N,N-bis(vinylbenzyl)glycine and vinylphenylglycine. A method for making suitable resinous polymers is described in U.S. Patent 2,840,603.

Mercury is recovered from the resin by elution with solutions having pH values below 9.5. Elution operation is advantageously performed at pH values less than 2 in concentrated brine, e.g., 5N sodium chloride. Removal is facilitated by the presence of a high concentration of halide ions, e.g., chloride ions, to tie up mercury as the complex chloro anion, $HgCl_4^=$, as fast as it is released. Thus, a solution of 1N HCl in 5N NaCl is more efficient as an eluting agent than a solution of 1N HCl alone.

A process developed by H.G. Scholten et al (16) involves passing the brine effluent, containing dissolved mercury salts, through a strongly basic anion exchange resin; washing the basic anion exchange resin with an excess of an aqueous sulfide solution; and separating the mercury from the aqueous sulfide solution.

A process developed by G.E. Edwards et al (17) permits recovery of chlorine by at least partial dechlorination of a brine which contains free and available chlorine and which also permits in addition simultaneously recovery of mercury by at least partial removal of mercury from a brine which contains both free and available chlorine and mercuric chloride.

The chlorine and mercury can be removed simultaneously or concurrently from the effluent brine of mercury cells by flowing the effluent brine through the cathode compartment of for instance a diaphragm cell having a graphite anode and a steel wool cathode where an EMF is applied between the anode and the cathode at least equal to the sum of the overvoltage of the chlorine at the anode and of the electromotive force required to overcome the ohmic resistance of the brine.

Figure 82 shows one form of apparatus which is suitable for the conduct of the process. 1 is a glass vessel which is provided with an overflow 2, 3 is a porous ceramic pot which is closed by a plug 4 and which separates the graphite anode 5 within it from the outer annular space within the glass vessel **1**. The annular space is packed with steel wool **6** which acts as cathode. **7** is brine from a laboratory mercury cell (not shown). There is catholyte-anolyte flow as the brine **7** enters the glass vessel **1** as chlorinated brine through tube **8**, passes through the walls of the porous ceramic pot **3** to fill this pot up to the overflow level of the overflow **2** and is bled off from within porous ceramic pot **3** as chlorinated brine by anolyte bleed pipe **9**.

On the application of a suitable electromotive force the brine **7** leaves the glass vessel **1** as dechlorinated brine at the overflow **2**. **10** is a pipe for chlorine gas which is formed within the porous ceramic pot **3** on the application of the electromotive force. Mercury is deposited on the steel wool cathode **6**.

FIGURE 82: ELECTROLYTIC DEVICE FOR REMOVAL OF CHLORINE AND
MERCURY FROM BRINE

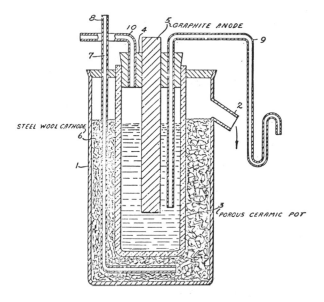

Source: G.E. Edwards and N.J. LePage; U.S. Patent 3,102,085; August 27, 1963

A process developed by M.G. Deriaz (18) is one in which waste brine can be rendered substantially free from mercury by the precipitation of mercury as mercuric sulfide by treatment in two stages with a material which will introduce sulfide ions into the brine such as a sodium sulfide solution.

The waste brine coming from the mercury cell is acidic and hot. If sodium sulfide solution is added to the brine any chlorine remaining is converted to hydrochloric acid. However, if the addition of sulfide is continued so that a quantity sufficient to precipitate the mercury present in brine is added, no clear end point is obtained with the first small excess of sulfide and then a large excess of sulfide tends to be added to the brine. This is undesirable as the mercuric sulfide separates in a form difficult to settle and separate.

On the other hand, if the brine is made alkaline on leaving the cell and the addition of sulfide carried out, a clear end point is obtained, but sulfur separates due to the reaction between hypochlorite and sulfide. This reaction does not occur in two-stage addition as little or no chlorine is present having already been converted to hydrochloric acid before the brine is made alkaline.

According to the process, the substantial removal of mercury from waste brine comprises the following steps. The addition to brine containing chlorine of sulfide ions until the redox potential relative to the saturated KCl-calomel electrode is in the range +0.85 volt to +0.93 volt, raising the pH by the addition of alkali to a pH in the range 7 to 9, preferably 8, then continuing the introduction of sulfide ions into the brine until the redox potential is in the range 0 to –0.2 volt, flocculating the precipitate obtained, settling the precipitate and separating the substantially mercury-free brine.

A process developed by G.E. Crain et al (19) involves the recovery of mercury from effluent brine by adsorptive contact with a strong anion exchange organic resin of the quaternary ammonium, cross-linked type.

Figure 83 is a flow sheet of this process. Referring to the drawing, sodium chloride, or potassium chloride, is combined with water to form a substantially saturated brine which is introduced into an electrolytic cell having a flowing mercury cathode. Chlorine gas is evolved from the electrolytic cell and is passed to apparatus for the recovery of chlorine in a pure state. The flowing mercury cathode, having picked up sodium or potassium from the brine forming an amalgam, the amalgam is sent to a denuder for the purpose of recovering the sodium or potassium hydroxides, or other suitable compounds which may be obtained from the amalgam.

FIGURE 83:  BLOCK FLOW DIAGRAM OF PROCESS FOR MERCURY RECOVERY
FROM BRINE BY ION EXCHANGE

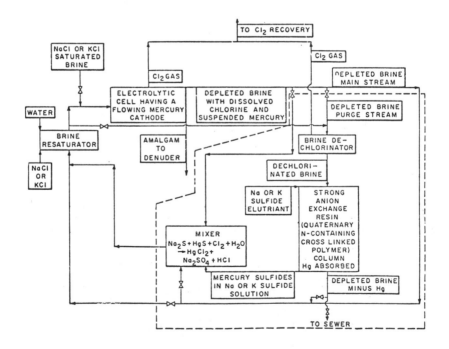

Source:  G.E. Crain and R.H. Judice; U.S. Patent 3,213,006; October 19, 1965

In the brine circuit, the main portion of the depleted brine from the flowing mercury cathode electrolytic cell with chlorine dissolved therein, as well as mercury in the form of the double salt or complex, noted hereinabove, is resaturated with solid salt, e.g., NaCl, and recycled to the mercury cell. A portion of the main stream of the depleted brine effluent from the cell is taken off as a side stream. This side stream may be the purge stream from which undesirable impurities are to be removed before the stream is recombined with the main depleted brine stream, prior to resaturation.

This side or purge stream of depleted brine, like the main depleted brine stream, has chloride dissolved therein, as well as mercury in the form of the double salt or complex of mercury, and is passed to apparatus where the dissolved chlorine in the brine is evolved, which evolved chlorine is sent to apparatus for the recovery of chlorine in its purest state. In the operation of this process, it has been found to be necessary to remove any free chlorine from the brine stream, prior to passing it in contact with the strong anionic exchange quaternary ammonium base resin, so as not to destroy the activity of this resin.

Following the dechlorination step, the brine is passed to the strong anion exchange quaternary base absorption column, where the mercury is adsorbed. The dechlorinated brine containing the mercury in dissolved or suspended form is introduced into the anion exchange resin column until such time as the mercury in the effluent from the column suddenly rises, at which time the effluent stream is shut off, and the elutriant is then introduced into the column.

As noted hereinabove, such elutriant is preferably sodium or potassium sulfide, corresponding to the brine initially fed to the electrolytic mercury cathode cell. The elutriant or regenerant sodium sulfide solution is preferably relatively concentrated, e.g., 2N, and it will be found that, during the first part of the elutriation, or regeneration cycle, the effluent contains in suspended form a black precipitate of mercury sulfide, and as the mercury concentration in the solution is diminished by further additons of $Na_2S$ solutions, the precipitate dissolves and the solution assumes an overall light yellow color. In a sodium sulfide solution of this concentration the mercuric polysulfide is believed to be formed, and it is found that the polysulfide is soluble to the extent of about 16,800 ppm as Hg at a pH of about 13 and at room temperature.

The flow rate of the elutriating, or regenerating medium, is preferably within the range of 4 to 8 gallons per minute per square foot of column cross section area, and the flow rate for the introduction of the dechlorinated brine to the anion exchange resin column is generally within this same range during the adsorption portion of the cycle.

In general, it will be found that the capacity of the anion quaternary ammonium base exchange resin, which may be of the type obtained by copolymerizing styrene with about 8% of divinylbenzene, and then forming the quaternary ammonium nuclei will have a capacity for the removal of mercury from the sodium chloride or potassium chloride brine of the order of about 4 lbs. of mercury per cubic foot of resin up to the point where the concentration of mercury in the effluent from the absorption column suddenly rises. It will be understood by those skilled in the art that this reference to capacity is intended to mean that the figure given for pounds per cubic foot is in terms of mercury (Hg), regardless of the chemical compound or complex in which the mercury is adsorbed in the anion exchange resin.

The elutriate from the anion exchange column, containing mercury sulfides in sodium sulfide or potassium sulfide solution, is then passed to a reactor which contains depleted brine from the electrolytic mercury cathode cell, which depleted brine may be the main stream of depleted brine or, alternatively, may be taken off as a second side stream from the main depleted brine stream. This depleted brine contains dissolved halogen, e.g., chlorine, in order to decompose the sulfides, including mercury sulfide, in the elutriate from the ion exchange resin, and thereby form mercuric chloride and/or the double salt or complex of mercuric chloride with either the potassium chloride or sodium chloride contained in the brine, as the case may be.

From the reactor, the brine solution is returned to the brine resaturator where additional sodium chloride or potassium chloride is dissolved to the point of substantial saturation with respect to these salts, and from the brine resaturator the brine passes back again to the electrolytic cell for the recovery of mercury by plating out on the flowing mercury cathode.

As a further alternative, the entire stream of depleted brine from the cell may be passed to the resaturator, before a side stream is treated for the removal of mercury. After the depleted brine has been resaturated, a side or purge stream is taken off and passed through the dechlorinator and ion exchange resin for treatment in accordance with the method. The sulfide elutriate from the ion exchange resin is then admixed with the chlorinated brine, e.g., the main brine stream, prior to the time it is resaturated. It is, thus, seen that the method is applicable both to saturated brine, i.e., brine which has been resaturated, and to depleted brine.

During the adsorption cycle where the mercury containing brine is passed through the anion exchange quaternary ammonium resin adsorption column for the adsorption of mercury, the effluent is substantially free from Hg compounds, i.e., less than 0.01 ppm, and this effluent may be subjected to purification steps to remove undesirable impurities, such as Ca, Mg, Fe and $SO_4^=$, before being recombined with the main depleted brine stream to be returned to the brine resaturator for combining with additional quantities of sodium chloride or potassium chloride to bring the concentration of the brine substantially to saturation, prior to passing the reconstituted saturated brine back to the electrolytic mercury cathode cell.

Additionally, it is to be noted that, where the depleted brine is not recycled to the mercury cell but is discarded after one pass through the cell, the effluent from the ion exchange resin will simply be purged from the system to a sewer or drain. In this instance, of course, it is obvious that substantially all of the depleted brine effluent from the cell will be passed through the dechlorinator and the ion exchange resin bed, rather than only a small purge stream.

A process developed by J.W. Town (20) involves the selective precipitation of mercury from a solution containing mercury and antimony. Sulfide ores or concentrates containing both mercury sulfide (HgS) and antimony sulfide ($Sb_2S_3$) are conventionally leached with sodium sulfide solution or with a sodium sulfide-sodium hydroxide solution to extract both the mercury and antimony. Such solutions contain mercury and antimony in concentrations of from 5 g./l. to saturation, depending on the sodium sulfide concentration of the leach solution.

Typically, the sodium sulfide concentration in the leach solution ranges from 25 to 200 grams per liter, with sodium hydroxide used to replace up to about 50% of the sodium sulfide. A solution concentration of 150 g./l. sodium sulfide and 50 g./l. sodium hydroxide is generally optimum for extracting the mercury sulfide and antimony sulfide from ore or concentrate.

Aluminum is then conventionally added to the mercury and antimony-containing extracts to precipitate the two metals. The resulting precipitate, containing mercury, antimony, sodium aluminate and aluminum, requires considerable washing before the mercury will coalesce. Some mercury is thus lost along with the antimony and unreacted aluminum.

It has been found that this disadvantage may be overcome by treating the mercury and antimony-containing solution with antimony metal to precipitate the mercury in a clean metallic form, without precipitation of antimony. The reaction is illustrated by the following equation:

$$3Na_2HgS_2 \ + \ 2Sb \ \longrightarrow \ 3Hg \ + \ 2Na_3SbS_3$$

The mercury coalesces readily for recovery and is tapped from the bottom of the reaction vessel.

When antimony is used as the precipitant no additional elements (such as aluminum) are added to the solution and the antimony can be electro-deposited from solution for reuse, thus reducing reagent consumption. The amount of antimony required to precipitate the mercury depends on the concentration of mercury in solution, temperature of solution, time allowed for reaction, particle size of antimony precipitant and concentration of antimony in solution and is best determined experimentally. An excess of antimony is usually required to obtain high mercury precipitation and the unused antimony is recovered for reuse.

A process developed by D.W. Rhodes et al (21) involves recovering mercury metal, present in acidic aqueous solutions as the mercuric ion, by adding hydrazine hydrate to the solution and refluxing the solution until the mercuric ion is reduced to mercury metal.

More specifically, this process involves recovering mercury metal from acidic nuclear fuel reprocessing waste solutions where it is present as the mercuric ion. Important to processing of irradiated nuclear reactor fuels is the disposition of the radioactive reprocessing waste solutions. Many of these fuel reprocessing waste solutions are calcined or heated to a high temperature to drive off the moisture and concentrate the remaining material into a solid suitable for storage for long periods of time until the radioactivity reaches more tolerable levels.

Many of these reprocessing waste solutions contain mercuric ion, generally present as mercuric ion, generally present as mercuric nitrate. The mercury is a catalyst added to the nitric acid solution to promote dissolution of aluminum fuel cladding. During calcination, in which temperatures may exceed 500°C., the mercury present in the waste solution is volatilized and subsequently contaminates the off-gas cleaning equipment. This necessitates periodic cleaning and decontamination of the equipment at considerable expense.

Additionally, this results in the loss of valuable mercury metal which could otherwise be reclaimed and reused at a substantial saving in cost. Efforts by others to develop a method to recover the mercury present in these solutions resulted in a method whereby the solution was allowed to seep through a metal or glass column packed with either copper shot or aluminum turnings.

Although the method was successful in removing the mercury from the solution by formation of an amalgam, the high concentrations of acid present in most waste reprocessing solutions attacked the aluminum and copper and the resulting gas made the columns practically inoperable. The cost of processing the amalgam to recover the mercury also makes the method economically unattractive.

The process for recovering mercury from acidic solutions eliminates the difficulties described above, and yet permits recovery of more than 99% of the mercury present in the solution. The process involves adding a small amount of hydrazine hydrate to the acidic solution containing the mercuric ion. The solution is then heated to refluxing temperature and the mercuric ion digested for a period of time until the mercury can be recovered as the pure metal.

A process developed by J.B. MacMillan (22) is a process for removing mercury from a mercury cell caustic soda liquor whereby the crude liquor is allowed to pass in a generally upward direction through a composite bed of particulate material selected from polyethylene shreds, polytetrafluoroethylene shreds, graphite, charcoal and activated carbon and pieces of a metal selected from nickel, stainless steel and tantalum. Caustic soda solutions are obtained containing as low a concentration of mercury as 0.1 ppm and 75% of the mercury suspended in the crude caustic liquor can be recovered.

Figure 84 shows a suitable form of apparatus for the conduct of this process. In the drawing, 1 is a squat column closed at the top by a removable top 2 provided with an opening communicating with a line 3 the end of which constitutes a vent or water connection. Branched on line 3 is a discharge line 4 for evacuation of the treated caustic soda liquor.

FIGURE 84:   FILTER FOR MERCURY REMOVAL FROM LIQUIDS

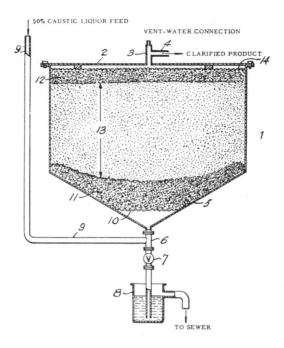

Source:   J.B. MacMillan; U.S. Patent 3,502,434; March 24, 1970

The bottom of the column **1** is in the form of a downwardly tapering cone **5** truncated at the apex to provide an opening.  To this opening is connected a line **6** provided with a drain valve **7** and ending into a drain receiver **8**.  A caustic soda feed line **9** is branched to line **6** above valve **7**.

Column **1** contains a filtering bed as hereinafter defined, such bed being supported on a nickel screen **10**, preferably a 4 mesh nickel screen.  The bed generally consists of two layers **11** and **12** made of nickel turnings or nickel wire mesh and of a compacted layer **13** of particulate activated carbon interposed between the nickel layers **11** and **12**.  The bed may advantageously be covered with a nickel screen **14**, preferably a 20 mesh nickel screen.

There is no limitation to the shape or form of the nickel from which the layers **11** and **12** are made except that such shape or form must lend itself to the formation of a pad or layer which is permeable to liquids.  There is no limit either to the particle size of the activated carbon except that it must be such as to permit the formation under pressure of a liquid permeable activated carbon bed.

It should be understood that column **1** may contain several alternate layers of nickel mesh and carbon instead of the two layers of nickel and single layer of carbon shown in the drawing.  It has been found, however, that the provision of a multiplicity of alternate layers does not give appreciable improvement in performance.  In actual operation, a 50% aqueous caustic soda solution from mercury cells (not shown in the drawing) is gravity fed or pumped through line **9**, valve **7** being so positioned as to close off line **6**.  The aqueous solution flows vertically upwards through line **6** and then into column **1** in contact with layers **11, 13** and **12.** The clarified aqueous caustic soda solution flows out of column **1** through line **3** and discharge line **4.** Discrete hydrogen gas bubbles separate out from the effluent liquor, the nickel mesh becomes shiny with mercury amalgamation, and globules of mercury separate on the activated carbon. During operation, the flow rate decreases gradually but can be easily restored by backwashing with water. Recovery of mercury from the backwash liquor as finely divided globules is possible by simple gravity separation.

A process developed by E.J. Botwick et al (23) involves the recovery of mercury chloride from weak brines effluent from mercury cathode electrolytic cells by contacting the effluent with activated carbon, washing the carbon with water, then heating in a stream of inert gas and condensing mercury chloride by cooling the gas.

A process developed by K.O.H. Fuxelius (24) involves the purification of polluted water from heavy metal ions such as lead and mercury, present therein. The process is characterized by treating the water with an ion exchange composition on the basis of epoxidized black liquor obtained from the sulfate pulp process. The product as used for this purification may be a reaction product of (a) black liquor, (b) epichlorohydrin or 1,2-dichlorohydrin and if desired (c) a thiol compound.

A process developed by W.E. Dean et al (25) for removing mercury from an alkaline solution having a pH of at least 7 is provided comprising intermixing with the solution a sulfur compound in an amount sufficient to provide sulfide ion in an amount greater than the stoichiometric equivalent of mercury, oxidizing the soluble mercury polysulfide and recovering the insoluble mercury sulfide.

A process developed by W. Knepper et al (26) is one in which mercury recovery from wastewaters is accomplished by converting the mercury to low-water-solubility compounds and/or to metallic mercury by the use of precipitants and subsequent separation thereof from the wastewater.

The mercury is preferably converted to mercuric sulfide by precipitation with the preferred compound hydrogen sulfide or other compounds which yield sulfide ions. The mercury can also be converted to metallic mercury by reduction with reducing agents such as nonnoble metals and low-molecular aldehydes. The precipitant is used in chemical equivalent amounts directly proportional to the dissolved mercury content of the wastewater, and it is advisable to add an excess amount of precipitant up to $10^4$-molar and preferably up to 100-molar.

After addition of the precipitant, a holding period of at least 5 seconds, preferably 15 to 500 seconds, and especially 30 to 120 seconds is required to permit the precipitation to take place before separation of the formed low-water-solubility mercury compounds and/or metallic mercury. Conventional means such as settling tanks, filters, and the like can be used for separating the precipitated mercury compounds and/or metallic mercury from the wastewater.

### Mercury Removal from Sediments and Sludges

A process developed by K. Yamori et al (27) involves recovering mercury from sludge from a purification tank for the purification of saturated alkali chloride solution obtained in the production of caustic alkali and chlorine by the electrolysis of alkali chloride solution in the so-called "mercury process."

In the electrolysis of alkali chloride solution by the mercury process, alkali chloride is usually dissolved in water to a concentration of about 300 g./l. and this saturated alkali chloride solution is introduced into an electrolytic cell fitted with a mercury cathode. The electrolysis is then carried out and thereby sodium amalgam is produced at the mercury cathode and chlorine gas is generated at the anode and collected therefrom. According to the above electrolytic step, about 10% of the alkali chloride in the influent alkali chloride solution is electrolyzed, after which it is exhausted from the electrolytic cell. (Hereinafter this alkali chloride solution will be called "depleted brine.")

Additional alkali chloride is dissolved in this depleted brine to produce again the saturated alkali chloride solution and the saturated alkali chloride solution, from which impurities such as $Ca^{++}$, $Mg^{++}$ and $SO_4^=$ mixed together with the additional alkali chloride are removed in a purification step, is again circulated into the electrolytic cell.

According to the above system, the depleted brine flowing out of the electrolytic cell contains mercury in the range of several milligrams per liter to two score or more milligrams per liter. This is believed due to mercury used as cathode being oxidized by chlorine generated at the anode, with production of an aqueous soluble compound. Previously, the mercury contained in above brine was removed from the abovementioned purification step and discarded as mud. The loss of mercury is a serious problem from the viewpoint of economics and public hazard.

The process comprises dissolving the sludge or a cake filtered from the sludge into an acidic liquid, for the sake of converting the mercury in the sludge into soluble salt form or the like, after which the mercury solution obtained is contacted with anion exchanger, with or without prior neutralization. The mercury which is adsorbed on the anion exchanger is then eluted, preferably with hydrochloric acid. When the eluate is used for adjusting the pH value of the purified, saturated alkali chloride solution, the recovered mercury is collected as metallic mercury by means of electrolytic reduction in the electrolytic cell.

Figure 85 is a schematic representation of the overall process. Reference numeral 1 designates an electrolytic cell, where purified saturated brine is electrolyzed and run out as a depleted brine via conduit 2. The brine 2 is dechlorinated in a dechlorinating apparatus 3 and then runs into a salt-dissolving tank 4. Sodium chloride is added at 5 to the depleted brine in the salt-dissolving tank 4 to produce a saturated salt solution and the resultant solution is fed into a brine purification tank 6, where the necessary amount of sodium carbonate and caustic soda are added, via inlets 7 and 8 respectively, for removing impurities such as $Ca^{++}$, $Mg^{++}$ and other heavy metal ions therefrom, these impurities being precipitated. In this case, mercury dissolved in the depleted brine is precipitated together with other impurities.

The sludge from tank 6 passes via conduit 9 to dissolving tank 13 while filter cake 11 filtered from the sludge by means of a filter 10 also goes to tank 13, where dissolution is effected with the aid of an acidic liquid, supplied through a conduit 14. The filtrate, namely saturated salt solution from the filter 10 is recovered in the brine purification tank 6, to which it is led through conduit 12.

The solution produced in the dissolving tank 13 in which the sludge from 9 is dissolved is subjected to removal by mercury adsorption by bringing it in contact via conduit 15 with anion exchanger 18. At this point, the solution, provided via 15, may be neutralized with alkali, supplied at 17, in a neutralizing tank 16 and then brought into contact with the anion exchanger 18. The effluent, leaving at 19, is then discarded. Subsequently, the mercury adsorbed in the anion exchanger 18 is eluted with hydrochloric acid supplied at 20; and the eluate, containing mercury, is supplied via conduit 21 to a storage tank 22.

On the other hand, the saturated salt solution purified in the brine purification tank 6 has a pH value in the range of 10 to 11, so that it has to be neutralized; this is accomplished with the aid of acid in a neutralizing tank 23.

The eluate, containing mercury, from the anion exchanger 18 is added, as an acidic liquid for neutralization, into the neutralizing tank 23 which contains purified salt solution, supplied via conduit 24, the purified salt solution being simultaneously neutralized and the mercury in the eluate transferred to the purified saturated salt solution. Thus, the mercury adsorbed on the anion exchanger 18 is dissolved in the purified, saturated salt solution, coming via conduit 24, and the solution, retaining its purified and saturated state is circulated back into the electrolytic cell 1.

The sodium chloride is electrolyzed in the electrolytic cell **1**, where the mercury in the purified saturated salt solution is simultaneously electrolyzed and is reduced to metallic mercury, which is then recovered.

**FIGURE 85: BLOCK FLOW DIAGRAM OF PROCESS FOR RECOVERING MERCURY FROM A MERCURY-CONTAINING SLUDGE**

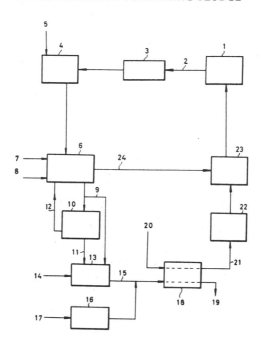

Source: K. Yamori, M. Takatoku, A. Miyahara, T. Omagari and M. Kitamura; U.S. Patent 3,536,597; October 27, 1970

A process developed by V.A. Stenger (28) involves recovering mercury from sediment at the bottom of a body of water by overlaying the sediment with a layer of metal containing at least one of aluminum, magnesium and zinc. Preferably, the metal is in sheet form for more efficient interception of mercury vapor rising from the sediment. After removal of the metal from the water body, the mercury, which forms an amalgam on the surface of the metal, may be recovered by distillation from the metal surface.

A study of mercury pollution control in stream and lake sediments has been made by the Advanced Technology Center (29). The study was directed at a means for removing mercury from sediments without destroying the environment or otherwise upsetting the ecological balance. The problem was addressed by evaluating a number of recoverable sulfur-based getter systems. Examples of the systems tested were elemental sulfur employed as a coating on a recoverable meshwork and thiourea-type organic compounds dispersed in a recoverable matrix.

**References**

(1) Jones, H.R.; *Mercury Pollution Control,* Park Ridge, N.J., Noyes Data Corp. (1971).
(2) Manes, M. and Grant, R.J.; U.S. Patent 3,193,987; July 13, 1965; assigned to Pittsburgh Activated Carbon Co.

(3) Williston, S.H. and Morris, M.H.; U.S. Patent 3,232,033; February 1, 1966; assigned to Cordero Mining Co.

(4) Park, J.C. and Winstrom, L.O.; U.S. Patent 3,257,776; June 28, 1966; assigned to Allied Chemical Corp.

(5) Manes, M.; U.S. Patent 3,374,608; March 26, 1968; assigned to Pittsburgh Activated Carbon Co.

(6) Bell, D.; U.S. Patent 3,647,359; March 7, 1972; assigned to BP Chemicals (UK) Limited, England.

(7) Ferrara, I. and Castellani, I.; U.S. Patent 3,661,509; May 9, 1972; assigned to Simcat-Societa Industriale Catanese SpA, Italy.

(8) Revoir, W.H. and Jones, J.A.; U.S. Patent 3,662,523; May 16, 1972; assigned to American Optical Corp.

(9) Bryk, P., Haani, M., Honkasalo, J.B., Kinnunen, J., Lindsjo, O., Nyholm, E., Poijarvi, J., Rastas, J. and Kangas, J.; U.S. Patent 3,677,696; July 18, 1972; assigned to Outokumpu Oy, Finland.

(10) Bergeron, G.L. and Bon, C.K.; U.S. Patent 2,860,952; November 18, 1958; assigned to The Dow Chemical Co.

(11) Neipert, M.P. and Bon, C.K.; U.S. Patent 2,885,282; May 5, 1959; assigned to The Dow Chemical Co.

(12) Karpiuk, R.S. and Hoekstra, J.J.; U.S. Patent 3,029,143; April 10, 1962; assigned to The Dow Chemical Co.

(13) Karpiuk, R.S. and Hoekstra, J.J.; U.S. Patent 3,029,144; April 10, 1962; assigned to The Dow Chemical Co.

(14) Gilbert, J.F. and Rallis, C.N.; U.S. Patent 3,039,865; June 19, 1962; assigned to The Dow Chemical Company.

(15) Calkins, R.C., Mock, R.A. and Morris, L.R.; U.S. Patent 3,083,079; March 26, 1963; assigned to The Dow Chemical Co.

(16) Scholten, H.G. and Prielipp, G.E.; U.S. Patent 3,085,859; April 16, 1963; assigned to The Dow Chemical Co.

(17) Edwards, G.E. and LePage, N.J.; U.S. Patent 3,102,085; August 27, 1963; assigned to Imperial Chemical Industries Ltd., England

(18) Deriaz, M.G.; U.S. Patent 3,115,389; December 24, 1963; assigned to Imperial Chemical Industries Limited, England.

(19) Crain, G.E. and Judice, R.H.; U.S. Patent 3,213,006; October 19, 1965; assigned to Diamond Alkali Co.

(20) Town, J.W.; U.S. Patent 3,361,559; January 2, 1968; assigned to U.S. Secretary of Interior.

(21) Rhodes, D.W. and Wilding, M.W.; U.S. Patent 3,463,635; August 26, 1969; assigned to U.S. Atomic Energy Commission.

(22) MacMillan, J.B.; U.S. Patent 3,502,434; March 24, 1970; assigned to Canadian Industries Limited, Canada.

(23) Botwick, E.J. and Smith, D.B.; U.S. Patent 3,600,285; August 17, 1971; assigned to Olin Corp.

(24) Fuxelius, K.O.H.; U.S. Patent 3,617,563; November 2, 1971; assigned to Research Ab Sundbyberg, Sweden.

(25) Dean, W.E. and Dorsett, C.M.; U.S. Patent 3,674,428; July 4, 1972; assigned to PPG Industries, Inc.

(26) Knepper, W. and Austin, S.; U.S. Patent 3,695,838; October 3, 1972; assigned to Chemische Werke Huels, AG, Germany.

(27) Yamori, K., Takatoku, M., Miyahara, A., Omagari, T. and Kitamura, M.; U.S. Patent 3,536,597; October 27, 1970; assigned to Toyo Soda Manufacturing Co., Ltd. and Japan Organo Co. Ltd., Japan.

(28) Stenger, V.A.; U.S. Patent 3,679,396; July 25, 1972; assigned to The Dow Chemical Co.

(29) Advanced Technology Center, Inc., *Mercury Pollution Control in Stream and Lake Sediments*, Washington, D.C., U.S. Government Printing Office (1972).

# METAL CARBONYLS

## Removal from Air

A process developed by E.R. Breining et al (1) relates to the recovery and purification of metal carbonyl compounds and particularly to the recovery of such compounds from the exhaust gases of metallizing operations.

Nickel carbonyl, which at room temperature is a liquid, is commonly employed in metallizing by thermal decomposition of the carbonyl in contact with the heated substrate to be metallized. Frequently such decomposition is incomplete, and undecomposed carbonyl, together with products of decomposition such as CO, and carrier gases such as $CO_2$ and $N_2$

form the exhaust gases of the operation. Carbonyl recovery is important as many of the carbonyls are expensive and even in the case of nickel carbonyl efficient recovery is an important factor in commercial aspects of the operation. The exhaust gases of such procedure are passed in the process into a liquid, such as water at relatively low temperature to effect condensation of the carbonyl; being insoluble and heavier than water the condenser carbonyl is readily separable from the water, dried and passed to storage for further use in metallizing.

Figure 86 shows a suitable form of apparatus for the conduct of the process. The numeral 1 designates a conduit having a valve 2 which is communicable with a metallizing chamber, a fragment of which is indicated by the numeral 3. Accordingly gases emanating from the chamber 3, as indicated by the arrow, will pass through the conduit 1 and the depending portion thereof to the interior of a tank 4, which is sealed by a suitable closure designated at 5. Surrounding the tank 4 is a water jacket 6 which may be utilized for cooling the interior of the contents of the tank 4 if necessary. Usually the contents of the tank will tend to be somewhat warm since the gases entering are generally hot as they leave the metallizing chamber at about 80° to 100°F.

**FIGURE 86: AQUEOUS SEPARATION CHAMBER FOR NICKEL CARBONYL RECOVERY FROM GAS STREAM**

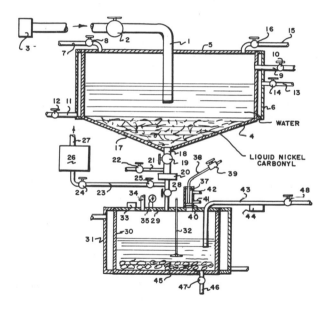

Source: E.R. Breining and W.M. Bolton; U.S. Patent 2,985,509; May 23, 1961

The numeral 7 designates an inlet pipe having a valve 8 through which water may be introduced to the tank 4. An outlet pipe 9 having a valve 10, and in the nature of an overflow, is provided in the upper portions of the tank 4. The inlet to the water jacket is indicated at 11 and is provided with a suitable valve 12. The outlet for the jacket is designated at 13 and has a valve indicated at 14.

In the upper reaches of the tank and preferably extending through the closure 5 is a conduit 15 having a valve 16 and which conduit leads to the exhaust of the recovery system. Such exhaust as indicated may be a vent to an off-gas burner.

With the equipment thus far described it will be apparent that exhaust gases flowing from the metallizing chamber **3** through the conduit **1** will pass into the water contained in the tank **4**. Insoluble compounds condensible at the temperature of the water to the liquid state will tend to separate out in the tank **4**. In the case of nickel carbonyl the temperature of the water is preferably maintained slightly above the freezing point of water (32°F.) and the nickel carbonyl, as indicated by the liquid, will separate to the bottom of the tank **4**, carrying with it some slight amount of water.

The exhaust bases from the plating chamber will generally contain carbon monoxide and other products of decomposition, depending upon the specific metallizing operation; also, in such operations a carrier gas, such as nitrogen, argon or carbon dioxide, is customarily employed, and such constituents will, of course, be present in the exhaust gas going to the water in the tank **4**.

While these gases will dissolve to some extent, the water will become saturated therewith and a vapor pressure of such gas is built up above the water and the gases flow through conduit **15** and valve **16** to the off-gas burner and into discharge. Gases such as carbon monoxide will be oxidized and rendered noninjurious by the off-gas burner. The liquid nickel carbonyl being heavier than water separates therefrom readily as illustrated.

The bottom of tank **4** is suitably tapered as indicated at **17** and directs the liquid carbonyl to a depending conduit **18** having a valve **19** and a sight glass **20**. A branch conduit **21** having a valve **22** provides for draining of the water from the tank **4** at the completion of an operation, for example. Another branch conduit below the sight glass **20** and designated at **23** has valves **24** and **25** that communicate with an interphase tank **26** provided with an outlet **27**. This interphase tank **26** may be utilized to assist in the mechanical separation of the carbonyl from the water, and accordingly the interphase tank aids the efficiency of the recovery procedure.

The conduit **18** is provided with a valve **28** through which flow of the nickel carbonyl may be directed to a tank **30** having a cover **29** through which the conduit passes. The tank **30** is suitably surrounded with a heating jacket **31**. In addition the tank has a stirrer **32** and an access hole **33** through which a chemically inert drying agent, such as calcium chloride, calcium sulfate, phosphorous pentoxide, silica gel, activated alumina, or other drying agents, may be passed to within the tank. The numeral **34** designates an inlet conduit for a gas such as carbon dioxide while at **35** there is designated a pressure gauge.

Supported by the cover **29** is a reflux condenser **37** which communicates with a conduit **38** having a valve **39** and through which gases may flow to an off-gas burner. The numeral **40** designates a jacket surrounding the reflux condenser and which jacket has an outlet **41** and an inlet **42** for the flow of cooling water through the jacket.

The outlet **43** through which the dried carbonyl is to pass to storage is provided with a filter **44**, which is effective to screen out small particles which might become entrained in the liquid. The numeral **45** designates particles of the drying agent, such as calcium chloride, and provision is made with a pipe **46** having a valve **47** for the removal of constituents through the base of the tank **30** when so desired.

Referring to the liquid nickel carbonyl within the tank **4**, such is separable readily by gravity from the water and may be passed through the conduit **18** to the tank **30** where it is mixed with the drying agent, for example, calcium chloride. Stirrer **32** provides for good contact of the liquid carbonyls with the calcium chloride particles and is effective to achieve pick up of any water retained by the carbonyl.

The reflux condenser inhibits the loss of carbonyl during the period when the combination is in contact with the drying agent. Normally heating is not necessary, but if gently used may speed drying time. The reflux condenser will, of course, pass any gases, such as carbon monoxide or carbon dioxide, which may have been entrained in the liquid carbonyl. Suitably the reflux condenser operates at about 0°C.

Where calcium chloride is employed as the drying agent it is preferable, before permitting the passage of carbonyl through the valve **48** to storage, to allow the particles of the drying agent to settle out. Such may be done by shutting off the stirrer **32** and permitting the calcium chloride particles to settle out.

Removal of the liquid carbonyl from the tank **30**, is achieved by pressuring carbon dioxide or other inert gas through the conduit **34** to the surface of the liquid carbonyl within the tank **30**, the carbonyl then flowing under the applied pressure through the upstanding arm of the conduit **43** and the valve **48** to storage.

A process developed by A. Schmeckenbecher (2) is one in which nickel carbonyl vapor can be removed from a gas stream, or a stream containing a mixture of gases, very effectively by passing the gas through any conventional air or gas scrubber containing a solvent which is selective for nickel carbonyl. Such specific and selective solvents are benzene, toluene, tetrahydrofuran, carbon tetrachloride and glacial acetic acid.

If exposure to light is avoided a substantial part of the nickel carbonyl vapor is recovered from the selective solvent as such. Since these selective solvents do not react with the nickel carbonyl, the latter is not decomposed during the time required to pass the gas through the solvent in any conventional air or gas scrubber washer. In other words, the solution of the nickel carbonyl in the solvent is stable provided undue exposure to light is avoided for a sufficient period of time to permit recovery of the nickel carbonyl by the conventional methods, preferably by fractional distillation.

Figure 87 shows a suitable form of apparatus for the conduct of this process. The scrubber shown may be connected to two or more of the same type scrubbers in series. In operating the apparatus shown, the selective solvent is introduced into the solvent inlet and the flow adjusted by means of any conventional flow indicator. The scrubbing tower may range anywhere from several feet to as high as 30 feet maximum with a width of 1 to 3 feet.

FIGURE 87: SOLVENT SCRUBBER FOR NICKEL CARBONYL REMOVAL FROM GAS STREAM

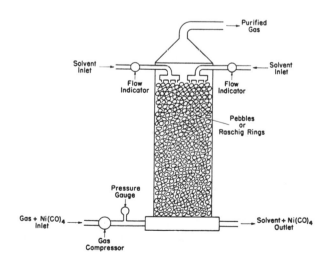

Source: A. Schmeckenbecher; U.S. Patent 3,086,340; April 23, 1963

The scrubbing tower is filled with either pebbles or raschig rings or glass beads having a diameter of ½ to 2 inches. The beads or pebbles may be either round or irregular surfaces. As for the raschig rings, the preferred diameter is from 1 to 3 inches. After the scrubber has been filled with pebbles or raschig rings, the selective solvent is introduced until all of the pebbles or rings are completely submerged in the solvent.

The gas or gas mixture containing nickel carbonyl vapor is introduced at room temperature at its normal pressure. If its normal pressure is too low, i.e., slightly above atmospheric, a gas compressor is put into operation and the pressure adjusted by means of the pressure gauge so that it is sufficient to push the gas through the solvent against the hydrostatic pressure. If the tower is 30 feet high it will be necessary to increase the normal pressure by about 1 atmosphere; if the tower is shorter or if other types of scrubbing towers are used, such as the spray scrubbers mentioned above, the pressure needed is correspondingly smaller.

After the gas has surged through the scrubbing bed, the solvent outlet is opened and the solvent containing the nickel carbonyl vapor is directed to a fractional distillation column and the recovered solvent passed to the solvent inlet. The solvent inlet flow indicator is adjusted to give a continuous flow of solvent countercurrent to the gases to be purified. In lieu of the two solvent inlets shown, four to six separate inlets may be provided and the flow control through each one adjusted.

The only limitation, in fact a prerequisite, being that the interior of the inlet and outlet lines as well as the interior of the scrubber apparatus be constructed of material resistant to nickel carbonyl, such as for example, glass, ceramic, stoneware, stainless steel, etc.

### References

(1) Breining, E.R. and Bolton, W.M.; U.S. Patent 2,985,509; May 23, 1961; assigned to Union Carbide Corp.
(2) Schmeckenbecher, A.; U.S. Patent 3,086,340; April 23, 1963; assigned to General Aniline & Film Corp.

## MINE DRAINAGE WATERS

The reader is referred to the section of this handbook on Sulfuric Acid, for a discussion of the treatment of acid mine drainage waters.

## NAPHTHOQUINONE

### Removal from Air

A process developed by G. Hoffmann et al (1) involves the purification of waste gases resulting from the manufacture of phthalic anhydride by the gas phase oxidation of naphthalene and mixtures of naphthalene and o-xylene.

In the piror art processes for the production of phthalic anhydride from the gas phase oxidation of naphthalene, o-xylene, and mixture thereof, as disclosed in Kirk & Othmer's *Encyclopedia of Chemical Technology* (1953), vol. 10, pp. 584–595, 606–607, and Shreve's *Chemical Process Industries* (1945), pp. 859–861, the corrosive waste gases resulting have presented a number of problems. It is a general object of the process to provide an improved and economical process for the recovery of phthalic anhydride from the gas phase oxidation of naphthalene and mixtures of naphthalene and o-xylene.

Another object is to provide for the recovery of 1,4-naphthoquinone from the process for the production of phthalic anhydride. Another object is to provide a water wash solution for the improved recovery of 1,4-naphthoquinone by the addition of maleic acid. A still further object is to recover 1,4-naphthoquinone from a water solution of maleic acid and 1,4-naphthoquinone by liquid-liquid extraction using a water-immiscible solvent. Still another object is to provide for the recovery of fumaric acid from the liquid-liquid extracted solution of maleic acid and water.

In the processes for the preparation of phthalic anhydride by the gas phase oxidation of naphthalene and mixtures of naphthalene and o-xylene, it is necessary for reasons of safety, to carry out the oxidation with a large excess of air. After the phthalic acid anhydride has been separated from the reaction gas, a considerable amount of waste gas remains. The waste gas contains in addition to about 81 to 82 volume percent of nitrogen, 16 to 17 volume percent of oxygen, 1 to 2 volume percent of carbon dioxide, 0 to 0.5 volume percent of carbon monoxide, as well as small amounts of sulfur dioxide (about 100 to 150 mg. per cubic meter of gas at normal conditions), and sulfur trioxide (about 15 to 30 mg. per cubic meter of gas at normal conditions), along with varying amounts of organic substances and entrained amounts of finely divided reaction products.

The organic and entrained substances present in the waste gas are mainly small quantities of phthalic anhydride, maleic anhydride, benzoic acid, and 1,4-naphthoquinone. Because of the strong corrosive properties of these substances, it is not possible to release into the atmosphere the waste gases of the naphthalene oxidation without previously carefully purifying the same.

According to the prior art, various processes have been recommended for purifying the waste gases. For example, the waste gases have been burned catalytically or thermally directly after they leave the apparatus used for condensing the phthalic anhydride. During such as oxidation process, all organic substances present in the waste gases are eliminated. Such methods are uneconomical and require a high energy expenditure for the necessary heating up of the waste gases.

Depending upon the type of heating means employed in the direct oxidation process, a considerable additional charging of the atmosphere with sulfur dioxide often cannot be avoided. A further disadvantage of burning up the waste gases is that all the potentially valuable reaction products still present in the waste gas are destroyed. This makes it impossible to recover part of the considerable expenditure for this purification process by recovering these reaction products.

A water washing step is often employed for the waste gas purification. Suitable processes and apparatus are disclosed in Kirk & Othmer's *Encyclopedia of Chemical Technology*, vol. 7 (1951), pp. 97–101. Power washers, for example ventilator washers, (suction washers) are preferred. With respect to the purity of the waste gases such washing systems are quite satisfactory. However, it is obvious that the water washing method only prevents the contamination of the atmosphere, while the drain water of the washing step is loaded with impurities. As a result, an indispensable second step of this method is the purification of the drain water.

A particular and inherent disadvantage of purifying the waste gases of the naphthalene oxidation by water washing resides in the poor water-solubility of the 1,4-naphthoquinone present in the waste gases in varying amounts. When the water is sprayed into the waste gas, the 1,4-naphthoquinone is eliminated therefrom in the form of a solid, and not in dissolved form. As a result, not only extensive clogging of pipelines and apparatus occurs, but an imbalance is created in the apparatus and there is danger of destroying or damaging washing devices having rapidly moving parts incorporated therein. Substances which under normal circumstances are quite soluble in water, such as for example, the dicarboxylic acids, are partially prevented from being dissolved by the presence of solid 1,4-naphthoquinone. Because of this phenomenon which is probably the incrustation of the organic acids by the 1,4-naphthoquinone, the quantity of solids obtained is always considerably larger than the theoretical amount of 1,4-naphthoquinone contained in the waste gas. The elimination of such precipitates is necessary for avoiding pressure increases in the overall apparatus and pipelines. The capacity of the process is decreased by the necessity to interrupt at regular and relatively long intervals the otherwise continuously conducted naphthalene oxidation.

The cleaning operations conducted during the down period of the plant represent an annoying and also very expensive working step since only manual operations are involved. The toxicity of the 1,4-naphthoquinone makes it additionally necessary to maintain special protective measures for the personnel since it is a strong skin irritant.

A need exists for a simple method permitting the purification of the waste gases obtained during the abovementioned oxidation process, preferably by means of a water washing step even though 1,4-naphthoquinone has poor water-solubility.

According to this process, it has been discovered that the waste gases of the gas phase oxidation of naphthalene and mixtures of naphthalene with o-xylene can be treated with concomitant purification by washing the waste gases with an aqueous maleic acid solution, extracting the used washing solution with an organic solvent which is water-immiscible, and subsequently concentrating the residual washing solution by evaporation.

The process uses the surprising discovery that an aqueous maleic acid solution dissolves 1,4-naphthoquinone. The prior art water solutions without maleic acid precipitated the entrained 1,4-naphthoquinone or removed the 1,4-naphthoquinone in the form of dust from the waste gas. According to the process the problems resulting from precipitating solids in the washing step are avoided. The waste gas purifying effect of prior art water washing methods is maintained and a clear solution of 1,4-naphthoquinone in acidic water is obtained where the 1,4-naphthoquinone content increases with an increasing amount of maleic acid.

The aqueous solution of maleic acid which is to be used as a washing liquid can be prepared in the usual manner by dissolving maleic acid in water. It is, however, expedient and economical to begin the washing of the exhaust gases with water, and to keep the latter in closed circulation until the washing liquid has acquired the desired maleic acid concentration by washing out from the exhaust gases the maleic anhydride which is present therein only in small amounts. After this concentration has been reached, a portion of the circulating solution is then continually drawn off and replaced by fresh water.

During this process it is often necessary to filter off the solid phthalic acid which has precipitated from the maleic acid solution before returning the latter to the closed washing circuit. The fact that a washing solution produced in the manner described is also saturated with phthalic acid in addition to the maleic acid content desired for these purposes, has been found, contrary to expectation, not to have an unfavorable influence on the ability of the washing liquid to dissolve the 1,4-naphthoquinone.

For the washing step, conventional apparatus is suitable, for example spray towers, washing cascades, and preferably high power washing devices, for example ventilator washers or brush washers or scrubbers. The aqueous solution obtained also contains a concentration of naphthoquinone, maleic acid, benzoic acid, and phthalic acid, permitting an economical further processing for obtaining the abovementioned substances.

The product resulting from the washing step requires further processing. The maleic acid solution branched off from the cycle of the washing step and replaced by fresh water contains 1,4-naphthoquinone and phthalic acid and must be removed before it is optionally discharged into the wastewater. The organic substances contained therein, particularly the 1,4-naphthoquinone and the maleic acid, otherwise would lead to extensive contamination of the wastewater. In the treatment of the wastewater of the waste gas washing process of a gas phase oxidation of naphthalene and of naphthalene o-xylene mixtures, for example by evaporation, the strong reactivity of 1,4-naphthoquinone creates a number of problems.

When such solutions are warmed, part of the 1,4-naphthoquinone is separated in the form of resins and condensation products which contaminate the heating surfaces and interfere with the operation. Since the 1,4-naphthoquinone is steam volatile, it is impossible to obtain a wastewater which is noncorrosive by simply distilling the solution.

According to a further aspect of the process, 1,4-naphthoquinone is washed out of the wastewater of the waste gas purification processes by extraction with organic solvents which are water-immiscible. Suitable water-immiscible organic solvents are, for example, chlorinated hydrocarbons, such as methylene chloride, chloroform, carbon tetrachloride, trichloroethylene, and hexachlorobutadiene, aromatics, such as benzene, toluene, xylene, ethyl benzene, cumene, or higher alkyl aromatics. Water-immiscible organic solvents are defined as having less than 5% water-soluble therein at standard conditions. A preferred water-immiscible solvent would have less than 0.5% water-soluble therein at standard conditions.

The extraction can be conducted by conventional methods available for this purpose, such as disclosed in Kirk & Othmer's *Encyclopedia of Chemical Technology,* (1951), vol. 6 pp. 122–139, for example by countercurrent extraction in towers filled with the liquid to be washed and the extraction agent. Also possible is the treatment of the wastewater with the extraction agent in vessels with high speed stirrers, and furthermore, for example, by intense mixing of both components in pump cycles.

In the latter operations the separation of both phases in a separating flask or with the aid of a separator is subsequently required. As a result of the 1,4-naphthoquinone content, it is, however, necessary in most cases to conduct the extraction in several stages in countercurrent flow of wastewater and extraction agent. An extraction with a very small space requirement is possible by using a multistage extraction centrifuge where wastewater freed from 1,4-naphthoquinone and extraction agent loaded with 1,4-naphthoquinone are obtained simultaneously. By means of all the above described operations, the 1,4-naphthoquinone is easily removed from the wastewater to a degree below the analytical determination limit of about 5 mg./kg. wastewater.

From the wastewater freed from the naphthoquinone by the extraction step, all of the acids are recovered, for example by evaporation. It is also possible to convert the maleic acid solutions into fumaric acid and recover the latter because of its poor solubility. Fumaric acid may be obtained by heating the maleic acid solutions in an inert atmosphere at 145° to 260°C. or by boiling solutions of maleic acid in the presence of various catalysts such as thiazole, thiourea, thiazolines, thiourane, disulfide, dithiocarbamates, mercaptans, alkyl disulfides, strong neutral acids, heavy metal salts, ammonia, pyridine and primary and secondary amines. The water produced during this evaporation step as the distillate is free from interfering by-products and can be fed to the drains or sewers. Advantageously, it can be recycled as fresh water to the washing cycle of the waste gas washing process.

References

(1) Hoffmann, G. and Striebeck, A.; U.S. Patent 3,370,400; February 27, 1968; assigned to Chemische Werke Huls AG, Germany.

# NICKEL

Nickel and its compounds are of concern as air pollutants because harmful effects of exposure to these materials have been observed among industrial workers as described by Sullivan (1). Exposure to airborne nickel dust and vapors may have produced cancers of the lung and sinus, other disorders of the respiratory system, and dermatitis. There is a substantially higher mortality rate among nickel workers due to sinus cancer, up to 200 times the expected number of deaths. However, since other metal dusts have also been present in industrial exposures to nickel, it has not been possible to determine whether nickel is the carcinogen. Yet experiments have shown that nickel carbonyl and nickel dusts can induce cancer in animals. Nickel contact dermatitis was found in 77% of the females and 10% of the males suspected of having allergenic reactions to metals. No information on the effects of nickel air pollution on commercial and domestic animals, plants, or materials was found in the literature.

The most likely sources of nickel in the air appear to be emissions from metallurgical plants using nickel, engines burning fuels containing nickel additives, and plating plants, as well as from the burning of coal and oil, and the incineration of nickel products. In 1964, urban air concentrations of nickel averaged 0.032 $\mu$g./m.$^3$ and ranged up to a maximum of 0.690 $\mu$g./m.$^3$ in East Chicago, Ind.

Wastewaters containing nickel originate primarily from metal industries; particularly plating operations. High levels of nickel have also been reported in wastes from silver refineries (2). Nickel concentrations typically found in plating and metal processing wastes are summarized in Table 12. Industries concerned primarily with copper or brass plating or processing may also contain nickel in plant wastewaters, but generally at low levels of 1 mg./l. or less, and are therefore of little concern.

### TABLE 12: SUMMARY OF NICKEL CONCENTRATIONS IN METAL PROCESSING AND PLATING WASTEWATERS

| Industry | Nickel Concentration, mg./l. Range | Average | Reference |
|---|---|---|---|
| Tableware Plating | | | |
|   Silver bearing waste | 0 – 30 | 5 | 3 |
|   Acid waste | 10 – 130 | 33 | 3 |
|   Alkaline waste | 0.4 – 3.2 | 1.9 | 3 |
| Metal Finishing | | | |
|   Mixed wastes | 17 – 51 | | 4 |
|   Acid wastes | 12 – 48 | | 4 |
|   Alkaline wastes | 2 – 21 | | 4 |
| Business Machine Manufacture | | | |
|   Plating wastes | 5.0 – 35 | 11 | 3 |
|   Pickling wastes | 6.0 – 32 | 17 | 3 |
| Plating Plants | | | |
|   (4 different plants) | 2 – 205 | | 5 |
|   Rinse waters | 2 – 900 | | 6 |
|   Large plants | up to 200 | 25 | 7 |

Source: Report PB 204,521

Nickel exists in waste streams as the soluble ion. In the presence of complexing agents such as cyanide, nickel may exist in a soluble complexed form. The presence of nickel cyanide complexes interferes with cyanide and nickel treatment, and the presence of this species may be responsible for increased levels of both cyanide and nickel in treated wastewater effluents.

### Removal of Nickel from Air

Nickel air pollution usually occurs as particulate emissions and may be controlled with the particulate by the usual dust handling equipment such as bag filters, precipitators, and scrubbers, as described by Boldt (8). Nickel poses no unique control problems except with gaseous nickel carbonyl. Control of nickel carbonyl may be accomplished by taking advantage of the fact that nickel carbonyl decomposes at temperatures above 60°C.

Gases containing nickel carbonyl may be passed through a furnace, where the nickel carbonyl decomposes into its two components, nickel and carbon monoxide. The resulting nickel metal can then be removed as particulate, as described by Kirk et al (9). No control methods were found for nickel emissions from vehicle engine exhausts.

**Removal of Nickel from Water**

Regardless of the type of treatment employed, the level of treatment required and associated costs can often be reduced by good housekeeping practices to reduce accidental loss, spillage, or leaks of plating or pickling solutions containing nickel. Process modifications to reduce the major contributor, dragout loss of plating bath solution into rinse waters, have been recommended by Domey and Stiefel (10).

Lancy (11) indicates that waste treatment facilities can be expected to cost 10% or less of the installed cost of the associated industrial process equipment. Depending upon the industrial process, wastewater treatment method, and size of operation, capital costs totaling 3 to 5% appear to be mentioned most frequently. Treatment processes can be classed as destructive, based upon nickel precipitation and disposal, and nickel recovery, which can be accomplished by several means. Recovered nickel is normally recycled through the metal processing operations.

*Destructive Treatment:* The formation and precipitation of nickel hydroxide is the basis for destructive treatment of nickel bearing wastes. Nickel forms insoluble nickel hydroxide upon addition of lime. The nickel hydroxide has a minimum theoretical solubility of 0.01 mg./l. at pH 10 (12). The precipitation is most effective at a high pH, although little efficiency is gained above pH 10. Therefore, close pH control is not essential other than from considerations of the chemical costs associated with pH adjustment.

Kantawala and Tomlinson (13) have reported the reduction of 100 mg./l. of nickel to 1.5 mg./l. at pH 9.9, by addition of 250 mg./l. lime and resultant formation of nickel hydroxide. The use of lime, followed by nickel hydroxide sludge conditioning with ferric chloride (to improve dewatering), and sand filtration reduced a combined waste effluent averaging 21 mg./l. of nickel down to 0.09 to 1.9 mg./l. Total waste flow was 200,000 gallons per day.

Anderson and Iobst (4) reported 35 mg./l. nickel could be reduced to 0.2 mg./l. in the final effluent by use of lime. Since the final effluent included an equal volume of rinse water which acted to dilute the treated waste stream, treatment actually accomplished reduction to 0.4 mg./l. for the nickel waste. Total flow was 0.45 MGD. Operating costs for the metal bearing waste stream were estimated at 80¢/1,000 gallons (1966 prices), and installation costs as $1,108/1,000 gallons per day capacity.

Hanson and Zabban (14) report that lime treatment for a waste which employed the coagulant aid, Separan NP-10 at a concentration of 1 to 2 mg./l. was able to reduce nickel from 39 mg./l. to 0.17 mg./l. No costs data were given. Stone (15) reported that nickel in the lime treated effluent of a copper, brass, cupro-nickel metal processing plant was 0.1 mg./l., prior to sand filtration of the treated wastewater. Filtration acts as a polishing process to remove small nonsettleable metal hydroxide particles.

Tupps (16) reported a final concentration of 15 mg./l. nickel in a treated plating waste in which chromium had been reduced to 0.05 mg./l. and iron to 0.5 mg./l. The effluent pH was 8.1, which resulted in poor nickel hydroxide formation. Weiner (17) indicates that a pH greater than 8.5 is required for effective nickel hydroxide precipitation.

The Lancy process (11) is designed to treat plating bath dragout solution directly, instead of diluting it into a rinse water and treating the larger volume. The plated metal, with dragout adhering to it, is submerged in a vat containing a concentrated solution of the chemical (e.g., lime), employed to precipitate the nickel. This solution is continuously circulated from the vat to a reservoir, with sufficient detention time to allow settling of the metal hydroxide precipitate.

Sludge removal is provided in the reservoir. Makeup chemicals are added to the reservoir in proportion to the amount of precipitate formed. Subsequent rinses contain salts derived from the precipitation vat. Lancy (11) claims this process is the only means of guaranteeing nickel levels of less than 1 mg./l., although these levels have been reported by other plants using direct lime precipitation (4)(14)(15).

A process developed by G. Hirs et al (18) involves removing nickel from plating rinse water or the like by upwardly adjusting the pH of an aqueous nickel solution to precipitate the nickel as a colloidal suspension, passing the precipitate suspension through a deepbed filter having a granular, synthetic organic medium therein to deposit the nickel precipitate on the filter medium, backwashing the filter medium to remove the deposited nickel precipitate therefrom and then filtering the backwash liquid through a conventional filtration mechanism.

*Recovery Processes:*  Recovery of nickel from process waste streams is more attractive than for some other metals, due to the high value of nickel.  A recent report (6) estimates rinse water recovery values to be $0.80 to 7.00/1,000 gallons for nickel concentrations ranging from 150 to 900 mg./l.  Recovery processes currently in use are ion exchange, evaporative recovery, electrolytic recovery, and direct reuse by countercurrent rinse techniques in systems where evaporative losses are high.  In addition, reverse osmosis has been evaluated for recovery.  Not all systems provide complete metal recovery, and final polishing treatment may be required to remove residual nickel.

Usually, total flow reduction is an integral part of recovery systems, in order to reduce capital costs.  Barnes (19) used countercurrent flow multiple rinse tanks to allow direct recovery of nickel plating solution dragout.  The first (and therefore most concentrated) rinse solution served as plating bath solution makeup to compensate for evaporation from the hot plating bath.  Use of four sequential rinse tanks resulted in a nickel concentration of only 0.9 mg./l. in the final rinse.

Ion Exchange – Recovery of nickel by ion exchange has been employed for some time. Reents and Stromquist (20) report on the use of cation-anion exchange to treat nickel sulfate plating bath rinse water containing 870 mg./l. nickel.  Complete nickel removal was reported, and the reclaimed water was acceptable for reuse in the rinse tank.  The unit processed 1,300 gal./hr.

Mattock (21) has reported that the extra cost of ion exchange equipment for recovery of nickel is rapidly offset by the value of the recovered material.  He emphasizes the necessity for waste stream segregation, particularly in the presence of cyanide, in order that the nickel, once concentrated in the ion exchange regenerant, can be recovered by precipitation as the carbonate.

Heidorn and Keller (22) also employed a scheme of cation exchange for nickel removal, followed by weak base anion exchange.  Reclaimed concentrations of 60,000 mg./l. nickel and 24,000 mg./l. sulfuric acid were achieved upon regenerating the exchange resins.  Neutralization with nickel carbonate allowed reuse of the solution.  Chemical costs were given as approximately 52¢/lb. of nickel recovered.

With mixed metal wastes where no chemical recovery can be practiced, Ross (23) has demonstrated that for small waste volumes (35,000 gal./day), ion exchange is more than twice as costly as chemical precipitation (70¢ versus 34¢/1,000 gal.), in terms of operating plus capital costs.  Detailed comparative costs for a waste containing nickel, zinc and tin are presented in his discussion.

Evaporative Recovery – To be economical, evaporative recovery requires reduction in rinse water volume, thereby increasing nickel concentrations in the rinse waters to be processed by the recovery system.  Countercurrent rinse serves this need.

Culotta and Swanton (24)(25) report costs for nickel plating solutions to range from $0.60 to $1.23/gallon, and provide formulas for comparison of destructive treatment costs versus evaporative recovery costs (24).  Graphical analysis based upon their data was provided by the authors for zinc cyanide with a lower unit cost of zinc plating bath solution

($0.45/gal.). The nickel recovery system would be expected to exhibit more favorable economics than zinc recovery, due to nickel's higher recovery value.

Besselievre (26) reported that a commercially available evaporative recovery system that was capable of treating 400 gal./hr., would cost $45,000, and could recover $100,000/yr. in chemicals.

Other Processes — Banerjee and Banerjee (2) have reported on nickel recovery by electrolysis from silver refinery waste. However, even at high nickel concentrations, recovery was only 65 to 70% effective. Thus, additional treatment was required to reach low nickel levels. No costs were given, but electrolytic recovery would not appear to be practical for plating applications.

Golomb (27) investigated the use of reverse osmosis as a bench scale process, for the recovery of nickel from plating bath rinse solutions. Treatment costs were quite high on both a volume and daily operating expense basis. For each level of operation, the evaporation rate was assumed to be ten times the dragout rate (D). With the reverse osmosis process, not all the permeate (i.e., filtered) solution returned to the process. The excess contained 22 mg./l. of nickel, therefore requiring additional treatment.

Pinner (28) has reported on the use of the Lancy process for recovery of nickel by precipitation directly as the carbonate, and reuse after washing and redissolution of the precipitate. Both batch and continuous recovery systems have been developed. If recovered nickel carbonate amounts to less than 1 ton per year, recovery is accomplished once a week on a batch basis. Capital costs for simple batch systems vary from 800 to 1,200 English pounds ($1,920 to $2,880) and for continuous flow systems from 2,000 to 8,000 pounds ($4,800 to $19,200).

*Summary:* The primary sources of nickel wastes are plating and metal processing industries. Thus nickel treatment is frequently combined with treatment for other contaminants, and specific costs of nickel removal are difficult to isolate. Destructive treatment for nickel by lime precipitation is capable of reducing residual nickel concentrations to less than 1 mg./l. Costs and efficiency of treatment can be substantially reduced by process control and modification.

Implementation of process modifications also makes nickel recovery an attractive proposition when waste stream separation is practiced, as nickel is a relatively expensive metal. Several recovery systems have been demonstrated to be economically feasible, including precipitate recovery after ion exchange, and evaporative recovery. For hot nickel plating baths, process modifications have led to direct nickel recovery, by return of rinse water to the plating bath.

### References

(1) Sullivan, R.J., "Air Pollution Aspects of Nickel & Its Compounds," *Report PB 188,070;* Springfield, Va., National Technical Information Service (September 1969).

(2) Banerjee, N.G. and Banerjee, T., "Recovery of Nickel and Zinc from Refinery Waste Liquors: Part I - Recovery of Nickel by Electro-deposition," *J. Sci. Industr. Res.* 11B: 77-78, 1952.

(3) McGarvey, F.X., Tenhoor, R.E. and Nevers, R.P., "Brass and Copper Industry: Cation Exchangers for Metals Concentration from Pickel Rinse Waters," *Ind. Eng. Chem.*, 44: 534-541, 1952.

(4) Anderson, J.S. and Iobst, E.H., Jr., "Case History of Wastewater Treatment in a General Electric Appliance Plant," *Jour. Water Poll. Control Fed.*, 40: 1786-1795, 1968.

(5) Wise, W.S., "The Industrial Waste Problem: IV, Brass and Copper Electroplating and Textile Wastes," *Sew. Industr. Wastes*, 20: 96-102, 1948.

(6) *State of the Art: Review on Product Recovery,* Water Pollution Control Research Services, U.S. Department of Interior, Washington, D.C., November 1969.

(7) Pinkerton, H.L., "Waste Disposal," in *Electroplating Engineering Handbook,* 2nd ed., A. Kenneth Graham, ed.-in-chief, Reinhold Pub. Co., New York, 1962.

(8) Boldt, J.R., Jr., *The Winning of Nickel,* Toronto, Longmans Canada, Ltd. (1967).

(9) Kirk, R.E. and Othmer, D.F., "Nickel and Its Alloys" in *Encyclopedia of Chemical Technology,* 2nd. ed., 13, 735-753, New York, Interscience (1967).

(10)  Domey, W.R. and Stiefel, R.C., *Wastewater Reduction in Metal Finishing Operations,* presented at
      43rd Ann. Conf. Wat. Poll. Control Fed., Boston, Massachusetts (October 5, 1970).
(11)  Lancy, L.E., "An Economic Study of Metal Finishing Waste Treatment," *Plating,* 54, 157–161 (1967).
(12)  Jenkins, S.H., Keight, D.G. and Humphreys, R.E., "The Solubility of Heavy Metal Hydroxides in
      Water, Sewage and Sewage Sludge I. the Solubility of Some Metal Hydroxides," *Internat. Jour.
      Air Wat. Poll.* 8, 53–56 (1964).
(13)  Kantawala, D. and Tomlinson, H.D., "Comparative Study of Recovery of Zinc and Nickel by Ion
      Exchange Media and Chemical Precipitation," *Wat. Sew. Works,* 111, R281–R286 (1964).
(14)  Hansen, N.H. and Zabban, W., "Design and Operation Problems of a Continuous Automatic Plating
      Waste Treatment Plant at Data Processing Division, IBM, Rochester, Minnesota," *Proc. 14th
      Purdue Industrial Waste Conf.,* pp. 227–249 (1959).
(15)  Stone, E.F.F., "Treatment of Non-ferrous Metal Process Waste at Kynoch Works, Birmingham,
      England," *Proc. 22nd Purdue Industrial Waste Conf.,* pp. 848–865 (1967).
(16)  Tupps, C.C., "Treatment of Wastes for Automobile Bumper Finishing," *Industr. Wat. and Wastes,*
      6, 111–114 (1961).
(17)  Weiner, R.F., "Acute Problems in Effluent Treatment," *Plating,* 54, 1354–1356 (1967).
(18)  Hirs, G. and Kozar, R.S.; U.S. Patent 3,630,892; December 28, 1971; assigned to Hydromation
      Filter Co.
(19)  Barnes, G.E., "Disposal and Recovery of Electroplating Wastes," *Jour. Wat. Poll. Control Fed.,* 40,
      1459–1470 (1968).
(20)  Reents, A.C. and Stromquist, D.M., "Recovery of Chromate and Nickel Ions from Rinse Waters by
      Ion Exchange," *Proc. 7th Purdue Industrial Waste Conf.,* pp. 462–473 (1952).
(21)  Mattock, G., "Modern Trends in Effluent Control," *Metal Finishing Jour.,* 14, 168–175 (1968).
(22)  Heidorn, R.F. and Keller, H.W., "Methods for Disposal and Treatment of Plating Room Solutions,"
      *Proc. 13th Purdue Industrial Waste Conf.,* pp. 418–426 (1958).
(23)  Ross, R.D., *Industrial Waste Disposal,* New York, N.Y., Reinhold Book Corp. (1968).
(24)  Culotta, J.W. and Swanton, W.F., "The Role of Evaporation in the Economics of Waste Treatment
      for Plating Operations," *Plating,* 55, 957–961 (1968).
(25)  Culotta, J.W. and Swanton, W.F., "Case Histories of Plating Waste Recovery Systems," *Plating,* 57,
      251–255 (1970).
(26)  Besselievre, E.B., *The Treatment of Industrial Wastes,* New York, McGraw-Hill Book Co. (1969).
(27)  Golomb, A., "Application of Reverse Osmosis to Electroplating Waste Treatment: Part I, Recovery
      of Nickel," *Plating,* 57, 1001–1005 (1970).
(28)  Pinner, R., "Quantitive Recovery of Nickel Dragout Losses by the Integrated Method," *Metal
      Finishing Jour.,* 14, 272–273 (1968).

# NITRATES

Nitrogen is present in domestic wastewaters in the range of 15 to 25 mg./l. (1). The nitrogen may be divided between organic nitrogen (40 to 45%), as ammonia nitrogen to about 55 to 60% and in the oxidized form ($NO_2$ nitrogen and $NO_3$ nitrogen) to about 0 to 5% (2). Organic nitrogen compounds are usually transformed in a typical biological wastewater treatment plant into ammonia through hydrolysis. Ammonia is then utilized by the bacteria for cell synthesis and, depending on the operational characteristics of the aeration tank, the excess ammonia may leak out in the final effluent or be oxidized to nitrites or nitrates.

## Removal of Nitrates from Water

The removal of nitrates from wastewaters has been reviewed by A. Shindala (3) as well as in a publication of the Federal Water Quality Administration (4).

*Biological Denitrification:* Biological removal of nitrogen through nitrification-denitrification has been extensively studied and reported in the literature (3). Nitrification results from the oxidation of ammonia by Nitrosomonas to nitrite and the subsequent oxidation of the nitrite to nitrate by Nitrobactor. The rate of conversion to nitrite was reported to control the rate of the overall reaction.

$$NH_4^+ + 1.5\ O_2 \xrightarrow{\text{Nitrosomonas}} NO_2^- + H_2O + 2H^+ \qquad NO_2^- + 0.5\ O_2 \xrightarrow{\text{Nitrobacter}} NO_3^-$$

The nitrifiers are chemo-autotrophic organisms which utilize $CO_2$ as a source of carbon for cell synthesis and obtain their energy from the oxidation of inorganic substrate. These organisms are much slower growing organisms than the heterotrophic bacteria responsible for the oxidation of carbon in a biological waste treatment system.

Therefore, in order to achieve both a complete oxidation of the organic matter as well as a maximum degree of nitrification in an activated sludge process, cell residence time greater than the growth rate of the nitrifying bacteria must be allowed. Shorter cell residence times will result in washout of the nitrifying organisms.

Cell residence time, which has also been referred to as sludge age or solids retention time, is defined as the ratio of the solids in the system to the solids leaving the system per day. Wuhrmann (2) from a pilot plant operation concluded that the sludge age would be the pertinent design factor to consider in the nitrification process. Wuhrmann reported that a sludge age of 2 to 3 days at temperature greater than 14°C. and 4 to 5 days at temperatures 8° to 10°C. produced a stable nitrification condition.

Downing (5) presented a relationship whereby the minimum sludge retention time necessary for a consistent nitrification in activated sludge is related to temperature, mixed liquor suspended solids, and waste strength. According to Downing (5) an activated sludge plant treating a waste with a $BOD_5$ concentration of 250 mg./l. and carrying a mixed liquor suspended solids of 2,000 mg./l. will require a minimum residence time of about 12 hours and over 24 hours at 17°C. and 7°C., respectively.

In addition to sludge retention time, several other factors have been reported to influence nitrification in activated sludge plants. Johnson and Schroepfer (6) reported that a highly nitrified effluent would be obtained providing that the organic loading (lb. BOD/lb. MLVSS-day) of the aeration unit would be less than 0.30. Downing and Hopwood (7) limited the dissolved oxygen in the aeration tank to a minimum of 1 mg./l. for effective nitrification. Downing (5) listed the pH and the temperature as important environmental factors affecting the growth of the Nitrosomonas.

Biological denitrification is the process where nitrate is reduced to nitrogen gas. The denitrifiers are heterotrophic organisms utilizing organic sources of carbon for energy and growth. These include species of *Pseudomonas, Achromobacter, Bacillus and Micrococcus* (8).

Since the denitrifiers require organic carbon for their metabolic activities and because in a typical activated sludge plant effluent most of the available carbon has already been oxidized, an external source of carbon has usually been added to the denitrification process. Methanol has received the widest application as a supplementary source of carbon.

Methanol requirements for effective denitrification have been reported by many investigators. McCarty et al (9) and St. Amant and McCarty (10) reported that total methanol requirements would consist of (a) that needed for the reduction of nitrate to nitrogen gas calculated as 1.9 mg./l. methanol per mg./l. $NO_3$ nitrogen reduced according to the following reaction:

$$NO_3^- + \tfrac{5}{6} CH_3OH \longrightarrow \tfrac{1}{2} N_2 + \tfrac{5}{6} CO_2 + \tfrac{7}{6} H_2O + OH^-$$

(b) that consumed by the presence of oxygen, calculated at 0.67 mg./l. of methanol per mg./l. dissolved oxygen present according to the following reaction:

$$O_2 + \tfrac{2}{3} CH_3OH \longrightarrow \tfrac{2}{3} CO_2 + \tfrac{4}{3} H_2O$$

(c) that needed to satisfy the requirements for bacterial growth estimated at 30% in addition to the stoichiometrical requirements of (a) and (b). The authors (9)(10) presented the following relationship for estimating the total methanol requirements:

$$C_m = 2.47 N_o + 1.53 N_1 + 0.87 D_o$$

where:

$C_m$ =  total methanol requirement, mg./l.
$N_o$ =  initial nitrate-nitrogen, mg./l.
$N_1$ =  initial nitrite-nitrogen, mg./l.
$D_o$ =  initial dissolved oxygen concentration, mg./l.

English (11) is a pilot plant investigation in a downflow columnar reactor suggested the following equation to predict total methanol requirements:

$$Q_t \ = \ 2.0 \ (NO_3 \ nitrogen) \ + \ 1.5 \ (D.O)$$

where:

$Q_t$ =  total methanol requirement, mg./l.
$NO_3$ nitrogen = nitrate nitrogen removal, mg./l.
D.O = influent dissolved oxygen concentration, mg./l.

In a recent study, conducted by Gulf South Research Institute, New Iberia, Louisiana, the effects of temperature and dissolved oxygen on methanol requirements for effective denitrification were evaluated in two types of continuous flow reactors, packed column and suspended growth (12). The results indicated that the optimum methanol: $NO_3$ nitrogen ratio for both reactors lies between 2:1 and 3:1 for 20°C. and 30°C. and slightly greater than 3:1 to 5°C. At these ratios, denitrification was little affected by temperature, overall efficiency in excess of 90% was obtained at all temperatures.

Dissolved oxygen, at the levels tested, did not appear to affect biological denitrification for methanol: $NO_3$ nitrogen ratios between 2:1 and 3:1. Overall efficiencies were greater than 90% for the columnar reactor at all dissolved oxygen levels (0.3 to 0.5 mg./l., 1.3 to 2.0 mg./l. and >4.0 mg./l.) while the suspended growth reactor efficiency only fell below 90% (84 to 87%) at dissolved oxygen level >4.0 mg./l. The authors concluded that, based on detention time only, the packed column reactor (detention time 15 minutes) is a more efficient denitrifying unit than the suspended growth reactor (detention time 210 minutes).

Another recent detailed study (8) of the process of nitrification-denitrification was conducted by Envirogenics, a division of Aerojet-General Corporation, El Monte, California. The investigation was carried out in a laboratory scale unit consisting of an activated bed where the waste is continuously recirculated over a dense mass of microorganisms.

Results indicated that nitrification was complete at an organic loading up to 1.1 grams COD/day/g. MLVS which is higher than the loading of 0.30 lb. BOD/lb. MLVS-day reported by Johnson and Schroepfer (6). Dissolved oxygen utilization in the nitrification unit was typical of the endogenous respiration of bacteria. Both the rate of nitrification and the form of nitrogen present were influenced by the residence time.

A residence time of 71 minutes resulted in a 95% nitrification. The rate of denitrification was reported to double for a 10°C. rise in temperature. Downing (5) reported a double in the growth rate of the Nitrosomonas for a 7°C. rise in temperature while studies at the Gulf South Research Institute (12) reported that temperature between 5° and 30°C. had no appreciable effect on the denitrification process as long as adequate amounts of methanol are added.

Dissolved oxygen level in the nitrified effluent appeared to have an influence on the denitrification process. The authors reported a residence time of 400 minutes above which dissolved oxygen was reported to have no effect on the denitrification process and other studies (12) showed that dissolved oxygen did not affect biological denitrification. The authors concluded that the energy derived from the oxidation of the residual COD entering the denitrification unit is sufficient to account for only 12% of the $NO_3$ nitrogen reduction and, therefore, the bacteria must be utilizing stored reserves as the principal energy source. Denitrification efficiency was reported to be 90% at a residence time of 8 hours while the overall efficiency of the nitrification-denitrification process was reported to range between 70 and 90% of the total Kjeldahl nitrogen in the feed.

Biological denitrification for nitrogen removal has been tested by a number of investigators using different process designs and different operating conditions and the process proved to provide consistently high removals. Smith, et al (13) from pilot plant studies obtained 90% denitrification using packed columns at surface loading of 7.0 gpm/ft.$^2$ and average temperature of 27°C. The actual contact time for the 90% denitrification varied from 5 minutes for coarse sand to 15 minutes for ¾ inch stones. Methanol:nitrate-nitrogen ratio of 2.5:1 was required for a feed having a dissolved oxygen of 2.5 mg./l. The cost of chemicals to remove 20 mg./l. of nitrate-nitrogen in the presence of 6 mg./l. of dissolved oxygen was reported at 1.9¢/1,000 gallons.

At Firebaugh, California (14) larger media, 1 to 2 inch aggregates were successfully employed to denitrify agricultural surface drainage in an upflow operation. Nitrate reduction exceeding 90% was achieved in contact time of 1 hour for 1 inch aggregate and 2 hours for the 2 inch aggregates at a temperature of 12°C. Methanol requirement was reported at 3:1 methanol to nitrate-nitrogen. In another pilot plant study (15), the anaerobic filter (1½ inches gravel) was reported to provide over 90% denitrification of secondary effluent in a detention time of 1.5 hours. The upflow operation gave the best performance and the least short circuiting.

Mulbarger (16), utilizing the three-stage system as described by Barth et al (17) reported a total Kjeldahl nitrogen reduction from 25.2 mg./l. to 5.4 mg./l. over two seventeen-day studies. From the preceding discussion, one may conclude:

[1] Nitrification-denitrification is a reliable process for nitrogen removal. It has been tested by many workers under a variety of operating conditions and proved to achieve a high and consistent removal at an acceptable cost.

[2] The columnar reactor appears to be more efficient than the suspended growth reactor and holds a promising potential for a wider application.

[3] Methanol appears to be the most economical and the most practical source of carbon for the denitrification process. Total methanol requirements may be predicted by using the relationship developed by McCarty and his co-workers (9)(10). Optimum methanol: $NO_3$ nitrogen ratio appears to range between 2:1 to 3:1.

[4] Sludge age seems to be the pertinent factor controlling the degree of nitrification. Sludge age of 2 to 3 days appears adequate.

[5] Biological-denitrification will be best suited at locations where cold-freezing weather may prevail for a long period of time.

A process developed by G.V. Levin et al (18) is an activated sludge sewage treatment process in which the nitrogen content of raw sewage is removed. In the process, raw sewage is mixed with activated sludge to form a mixed liquor and the mixed liquor is aerated at a rate sufficient to convert ammonia present in the sewage to nitrate. The mixed liquor is then passed to a zone where it is maintained under conditions in which there is insufficient oxygen present to satisfy the needs of the microorganisms in the mixed liquor. This causes the microorganisms to break down the nitrate and to fulfill their oxygen needs by obtaining oxygen from the nitrate. Nitrogen gas is formed in the process and is evolved from the system.

There is also involved a process whereby the phosphate content of sewage is also reduced. In this embodiment, conditions are controlled so that the sludge which is withdrawn from the mixed liquor contains a substantial portion of the phosphate content. The final effluent which is passed out of the system is substantially free of phosphate and nitrate.

Figure 88 illustrates the essentials of the process. A raw sewage influent stream 1 is passed through conventional screening and grit removing units and is optionally subjected to primary settling in a tank **2** from which primary sludge is removed in line **3**. The primary settled sewage is mixed with recycled, activated sludge hereinafter described to form a mixed liquor and is passed by line **4** to the aeration tank **5**.

## FIGURE 88: NITRATE REMOVAL FROM SEWAGE

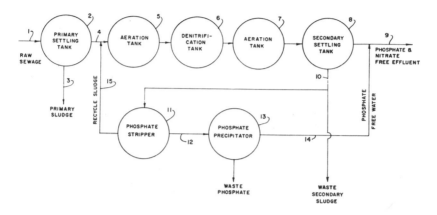

Source:  G.V. Levin and G.J. Topol; U.S. Patent 3,654,147; April 4, 1972

In the aeration tank **5**, the mixed liquor is aerated at a rate sufficient to convert ammonia present in the sewage to nitrate.  During aeration, the bacteria present take up phosphate and consume organic matter present in the sewage.  A high degree of BOD removal is obtained during aeration.

After aeration, the mixed liquor is fed into a tank **6** where it is maintained under conditions in which there is insufficient oxygen present to satisfy the needs of the microorganisms in the mixed liquor.  This induces the microorganisms to consume the nitrate content of the sewage.

After depletion of the nitrates and release of the nitrogen as nitrogen gas, the mixed liquor is passed to the aeration tank **7** in which it is again aerated.  In this tank, the microorganisms in the sludge take up any phosphate which has leaked out during the period the mixed liquor was in the tank **6**.  This step, and the phosphate stripping operation hereinafter described, may be omitted if it is only desired to remove nitrogen in the sewage, i.e., where phosphate removal is not required.

After areation in the tank **7**, the mixed liquor is fed into a secondary settling tank **8**.  In the secondary settling tank **8**, phosphate-enriched sludge settles and thereby separates from the mixed liquor.  The sludge contains a substantial portion of the phosphate present in the sewage.  The substantially phosphate-free and nitrate-free effluent is discharged for disposal in a conventional manner by line **9**.

The phosphate-enriched sludge is removed from the settling tank by line **10**.  A portion of the sludge may be delivered to waste and the remainder is passed to the phosphate stripper **11**.  In the phosphate stripper, the phosphate-enriched sludge is treated to cause the microorganisms in the sludge to release phosphate.  This treatment may be accomplished by holding the mixture under anaerobic conditions as described in U.S. Patent 3,236,766 or by aerating the mixture, or by appropriate pH adjustment, i.e., adjusting the pH to less than 6.5 and maintaining it at this pH for at least 10 minutes.  This treatmetn causes the organisms in the sludge to release the phosphate which they have taken up in the aeration tank **5**. The phosphate leaks out of the sludge into the liquid phase. A phosphate-enriched supernatant liquor is produced upon settling of the sludge. After settling, the sludge is passed by line **15** for mixing with the raw sewage which is being fed to the aeration tank **5**.

A phosphate-enriched supernatant liquor is produced by the phosphate stripper **11** and is passed by line **12** to the phosphate precipitator **13**. A phosphate precipitant, such as aluminum or iron salts or lime, is mixed with the phosphate-enriched supernatant liquor in the phosphate precipitator **13** to precipitate phosphate.

The phosphate precipitate may be combined with any waste phosphate-enriched sludge removed from the secondary settling tank **8** and converted into a fertilizer or otherwise disposed of by conventional methods. A phosphate-free supernatant liquor is withdrawn from the phosphate precipitator **13** and passed by line **14** to line **9** where it is combined with the phosphate-free effluent from the secondary settling tank **8**.

A process developed by E.S. Savage (19) provides for denitrification of the effluent from activated sludge sewage treatment. Effluent from the settling zone of an activated sludge sewage treatment process containing nitrogen compounds such as nitrates and nitrites, is passed through a deep bed filter, the filter media of which has been innoculated with bacteria that converts the nitrogen compounds to nitrogen gas. The filter, in addition to removing the nitrogen compounds, removes any suspended solids from the settling zone effluent, so that the final effluent from the filter is concurrently clarified and denitrified. By controlling the backwash of the filter, bacteria is retained thereon so as to enable continuous use of the filter for denitrification.

*Ammonia Stripping:* Ammonia stripping operates on the principle of converting the ammonium ions to ammonia gas and then stripping off the ammonia gas by contacting the waste with ammonia-free air. Ammonium ions in wastewater exist in equilibrium with ammonia and hydrogen ions as follows:

$$NH_4^+ \rightleftharpoons NH_3 + H^+$$

As shown in the above equation raising the wastewater pH above 7 will shift the equilibrium to the right and ammonia gas is formed. Raising the pH to a high level, greater than 11, will, therefore, convert most of the ammonium ions to ammonia gas which can be liberated by agitation of the wastewater in a stripping tower equipped with an air blower. The most work on ammonia stripping has been conducted at the South Tahoe Public Utility District Tertiary Wastewater Treatment Plant.

Pilot plant studies (20) indicated that at pH of 11, a 24 foot high tower with 1½ x 2 inch packing and water loading of 2 gpm/ft.$^2$ will give an ammonia removal of about 95% with an air supply of 400 ft.$^3$/gallon. Cost for South Lake Tahoe facility has been estimated at 2.9¢/1,000 gallons of wastewater treated (21). This does not include the cost of the lime and facilities to raise the pH to about 11. If 90% ammonia removal is required even in cold weather, these costs should be increased by about 50% to provide for a higher air-to-water ratio (22).

Limitations of the ammonia stripping towers were first revealed during the initial operations of the Lake Tahoe project which was in the winter season. These limitations include (21):

[1] At air temperatures below 0°C., freezing of water occurs at the air inlet making the towers inoperable.

[2] Since ammonia solubility increases with decreasing temperatures, more air is required to remove it, 800 ft.$^3$/gallon at 0°C., which results in increasing the cost of treatment.

[3] Formation of scale in the tower which is mostly calcium carbonate.

This scale formation is due to the lime treated waste being supersaturated with $CaCO_3$. This scale has been found to vary in nature and composition. For example, at Tahoe project, the scale could be flushed from the tower, while at the FWQA's Blue Plain Washington Pilot Plant, the scale was found hard and adheres to the tower fill (21). The best approach to minimize scale and its effects appears to be the use of a pH of 10.8, countercurrent rather than cross-flow operation, and an open fill to allow for easy flushing of accumulated solids.

From the preceding discussion, one may conclude that although ammonia stripping provides very efficient removals, its use will be limited to warm climates. In addition, the scale problem has not yet been solved.

*Ion Exchange:* Information relative to the use of ion exchange for the removal of ammonium and nitrate ions from wastewater is scarce. Pilot plant studies (20) at Lake Tahoe making use of a commercial resin (Duolite C25, product of Diamond Alkali) showed that 400 bed volumes could be treated prior to breakthrough to 1 mg./l. ammonia-nitrogen. Influent ammonia-nitrogen was reported at 12 to 28 mg./l. Regeneration was with salt solution which was estimated at 0.5% of the flow treated.

Recently, Pacific Northwest Laboratory reported the development of a selective ion exchange process utilizing a natural zeolite, clinoptilolite, which was reported to be selective for ammonium ions in the presence of sodium, magnesium, and calcium ions (22)(23). Clinoptilolite occurs naturally in several extensive deposits in the western United States.

The process bascially concentrates the ammonia in a small volume of regenerant which can be recovered by air stripping or electrolyte treatment to remove the ammonia-nitrogen. Pilot plant application to clarified secondary effluent containing 16 mg./l. ammonia nitrogen resulted in an average of 97% ammonia removal at a flow rate of 6 gal./ft.$^2$/min. for a total of 70,000 gallons per day.

Two 500 gallon clinoptilolite columns were used (22). An average of 93% ammonia removal was also obtained in 330 gallon clinoptilolite columns at a flow rate of 8.4 gallons per square foot per minute with a filtered secondary effluent having an initial ammonia-nitrogen content of 15 mg./l. Laboratory studies showed that organic fouling reduced the available exchange capacity by about 25% during two upflow runs with no pretreatment of secondary effluent. Cost estimates for 95% ammonia removal were reported to be 10 to 14¢/1,000 gallons in the 10 to 300 MGD plants (23).

Nitrate specific ion exchange resin was developed by TYCO Laboratories, Inc., which incorporated selected primary amines in polystyrene (24). The best selectivity of nitrate ion over chloride ion ranged from 7.5 to 14 with 1-naphthyl methylaminomethyl(1-NMA). Nitrate was reported to be adsorbed quantitively from feed solutions containing five times as much chloride ions as nitrate ions. Regeneration was reported by 1N HCl. From the preceding discussion one may conclude:

[1]   The selective ion exchange for ammonia removal (22)(23) using clinoptilolite holds a promising potential for full scale application.

[2]   Ion exchange for nitrate removal will only be feasible when a resin is developed that has a high selectivity for nitrate over other anions present in the waste.

[3]   Ion exchange for ammonia-nitrogen removal may find wide application at locations where freezing weather prevails and where low ammonia residuals are required.

*Chlorination of Ammonia:* Ammonia can be oxidized to nitrogen gas by chlorinating to the breaking point with either chlorine gas or sodium hypochlorite. The theoretical reactions are as follows:

$$3Cl_2 + 2NH_3 \longrightarrow N_2 + 6HCl \qquad 3NaOCl + 2NH_3 \longrightarrow N_2 + 3NaCl + 3H_2O$$

Four mols of chlorine or hypochlorite per mol of nitrogen gas liberated or 10 mg. chlorine per mg. ammonia nitrogen were reported (25) necessary to complete the reaction. Hypochlorite is more expensive than chlorine but is much safer to transport and handle. Break point chlorination will also disinfect the wastewater along with removing the ammonia. However, chlorination will add chloride ions which may be considered as pollutants. Chlorine cost estimate based on 20 ppm of ammonia nitrogen in secondary effluent was reported at 6¢/1,000 gallons not including the cost of handling the corrosive hydrochloric acid produced. Sodium hypochlorite may cost twice as much as chlorine but associated costs are greatly reduced.

*Summary:* The most successful nitrogen removal processes include biological denitrification, ammonia stripping and ion exchange. Biological denitrification, both columnar and suspended growth, proved to be a reliable and economic process in achieving a high degree of nitrogen removal. Control of the nitrification step is a must in providing an effective reduction. Any reduced form of nitrogen reaching the denitrification step would leak through and reduce the efficiency of the process. Methanol appears to be the most practicable and economical source of carbon. Optimum methanol to nitrate-nitrogen ratio seems to range between 2:1 and 3:1. The three-stage system appears to be the best approach to control nitrification.

Ammonia stripping, although it produces very low ammonia residuals, is limited to locations which do not experience prolonged periods of freezing weather. The problem of calcium carbonate scale is still unsolved and needs consideration.

Selective ion exchange utilizing clinoptilolite appears to be very attractive, not only in providing high removals of ammonia but because the process produces practically no liquid waste from the regeneration cycle. This process appears to be best suited to locations where freezing weather prevails over a long period and where consistently high removals need to be maintained.

### References

(1) Barth, E.F. and Dean, R.B., "Nitrogen Removal from Wastewater: Statement of the Problem," *Symposium on Advanced Waste Treatment,* Dallas, Texas (January 12 to 14, 1971).

(2) Wuhrmann, K., "Objectives, Technology and Results of Nitrogen and Phosphous Removal Processes," in *Advances in Water Quality Improvements,* Austin, Texas, University of Texas Press (1968).

(3) Shindala, A., "Evaluation of Current Techniques for Nutrients Removal from Wastewater," *Report PB 201,751,* Springfield, Va., National Tech. Information Service (June 1971).

(4) Federal Water Quality Administration, "Nitrogen Removal from Wastewaters," *Report PB 212,280,* Springfield, Va., National Tech. Information Service (October 1970).

(5) Downing, A.L., "Factors to be Considered in the Design of Activated Sludge Plants," *Advances in Water Quality Improvements,* University of Texas Press (1968).

(6) Johnson, W.K. and Schroepfer, G.T., "Nitrogen Removal by Nitrification and Denitrification," *Journal Water Pollution Control Federation,* 36, (8), 1015 (August 1964).

(7) Downing, A.L. and Hopwood, A.P., "Some Observation on the Kinetics of Nitrifying Activated Sludge Plant," *Schweizerische Zeitschrift für Hydrologie,* XXVI, Fase. 2, 271 (1964).

(8) Mechalas, B.J., Allen, P.M. III and Matyskiela, W.W., "A Study of Nitrification and Denitrification," *Envirogenics,* For the Federal Water Quality Administration, Report No. 17010 DRD 07/70, Cincinnati, Ohio (July 1970).

(9) McCarty, P.L., Beck, L.A. and St. Amant, P.P., "Biological Denitrification of Waste Waters by Addition of Organic Materials," *24th Purdue Industrial Waste Conference,* Purdue University (May 1969).

(10) St. Amant, P.P. and McCarty, P.L., "Treatment of High Nitrate Waters," *Journal American Water Works Association,* 61, 12 (December 1969).

(11) English, J.N., "Nitrate Removal in Granular Activated Carbon and Sand Columns," *Internal Report of the Advanced Waste Treatment Research Laboratory,* Federal Water Quality Administration, Pomona Water Reclamation Plant, Pomona, California (February 1969).

(12) English, J.N., "Methanol Requirements and Temperature Effects in Waste Water Denitrification," *Gulf South Research Institute for the Environmental Protection Agency,* Report No. 17010 DHT 09/70, Cincinnati, Ohio (September 1970).

(13) Smith, J.M., Masse, A.N., Feige, W.A. and Kamphake, L.J., "Nitrogen Removal from Municipal Waste Water by Columnar Denitrification," *160th National ACS Meeting,* Chicago, Ill. (Sept. 13-18, 1970).

(14) Tamblyn, T.A. and Sword, B.R., "The Anaerobic Filter for the Denitrification of Agricultural
     Subsurface Drainage," *24th Purdue Industrial Waste Conference,* Purdue University (May 1969).
(15) Seidel, D.F. and Crites, R.W., "Evaluation of Anaerobic Denitrification Process," *Proc. American
     Society of Civil Engineers,* SA 2, 267 (April 1970).
(16) Mulbarger, M.C., "Modification of the Activated Sludge Process for Nitrification and Denitrification,"
     *43rd Conference Water Pollution Control Federation,* Dallas (October 1969).
(17) Barth, E.F., Brenner, R.C. and Lewis, R.F., "Chemical-Biological Control of Nitrogen and Phosphorus
     in Waste Water Effluent," *Journal Water Pollution Control Federation,* 40 (12) 2040 (Dec. 1968).
(18) Levin, G.V. and Topol, G.J.; U.S. Patent 3,654,147; April 4, 1972; assigned to Biospherics, Inc.
(19) Savage, E.S.; U.S. Patent 3,709,364; January 9, 1973; assigned to Dravo Corp.
(20) Slechta, A.F. and Culp, G.L., "Water Reclamation Studies at the South Tahoe Public Utility
     District," *Journal Water Pollution Control Federation,* 39 (5) 787 (May 1967).
(21) Farrell, J.B., "Ammonia Nitrogen Removal by Stripping with Air," *Symposium on Advanced
     Waste Treatment,* Dallas, Texas (January 12 to 14, 1971).
(22) Farrell, J.B., "Ammonia Removal from Agricultural Runoff and Secondary Effluents by Selected
     Ion Exchange," *Pacific Northwest Laboratories,* Report No. TWRC – 5, Richland, Washington,
     (March 1969).
(23) Mercer, B.W. and Ames, L.L., "Mobile Pilot Plants for the Removal of Ammonia and Phosphates
     from Waste Waters," *International Water Conferences of the Engineers Society of Western
     Pennsylvania,* Pittsburg, Pennsylvania (October 27 to 29, 1970).
(24) Wallitt, A.L. and Jones, H.L., "Basic Salinogen Ion-Exchange Resins for Selective Nitrate Removal
     from Potable and Effluent Waters," *TYCO Laboratories for the Federal Water Quality
     Administration,* Report No. 17010 FKF 12/69, Cincinnati, Ohio (December 1969).
(25) Dean, R.B., "Other Methods for Removing Nitrogen," *Symposium on Advanced Waste Treatment,*
     Dallas, Texas (January 12 to 14, 1971).

## NITRITES

**Removal from Water**

A process developed by J. Müller et al (1) for the detoxification of cyanide and nitrite containing aqueous solutions is accomplished by mixing the cyanide and nitrite solutions and
reacting the cyanide with the nitrite at a pH of not over 5 in the presence of a contact
catalyst.

Cyanides, especially those which are not bound in complexes, and nitrites are exceptionally poisonous substances which are allowed to occur in wastewaters and main drainage
channels only in exceedingly small concentrations (cyanide 0.1 mg./l., nitrite about 10 to
20 mg./l.). Since both substances are used, particularly in the hardening industry, for different processes, the removal of the waste in many cases creates a serious problem. Of
course, completely automatic plants for detoxification are known today.

They are, however, expensive and the use of chemicals for oxidation of nitrites and cyanides, for acidification and finally for neutralization is considerable. Especially as oxidizing
agents, there is employed mainly chlorine gas, sodium hypochlorite or, in certain cases,
hydrogen peroxide which make for a considerable cost factor in the detoxification of cyanide and nitrite containing wastewater.

The detoxification of cyanides is successful chiefly through oxidiation with the above
named agents to cyanates or other methods whereby nonpoisonous cyanates are obtained.
The detoxification of nitrites can also be obtained by oxidation to nitrates. However,
there is also considered a reduction to nitrogen, as is accomplished, for example, in using
urea.

The process is directed to the idea of allowing cyanides and nitrites to react with each
other with the result that the nitrites oxidize the cyanides to cyanates or other materials
and at the same time they themselves are reduced to nonpoisonous substances, expecially
nitrogen. This reaction is not achieved, however, by bringing the reactants together in

neutral and alkaline conditions, so that utilization of this possibility has not been practical. It has been surprisingly proven, that the reaction occurs with interesting great rapidity if the reacting solutions are acidified to a hydrogen ion concentration of about 5 or less and if at the same time a contact catalyst is introduced to hasten the reaction.

The pH can be as low as 0. Of course, in most cases such a reaction must occur in a completely closed apparatus which, however, presents no problem. As experiments have shown, the cyanide is converted partially to cyanate and partially to carbonate while the nitrite is compeltely reduced so that finally only pure nitrogen remains.

As catalysts, one uses suitable highly activated carbon or highly activated silica. Other high surface area catalysts can be used, such as activated alumina. There can also be employed other catalysts, for example, metallic catalysts, in which case it is only necessary that the chosen metal does not go into solution (as does, for example, copper or nickel) and thereby create a new problem of detoxification.

Examples of suitable metals include activated metals from the Pt group such as Pt or Pd. The preferred acid for reducing the pH is sulfuric acid but there can be employed other acids such as muriatic acid, hydronitric acid and others. The original cyanides and nitrites can be, for example, alkali compounds such as sodium cyanide, potassium cyanide, sodium nitrite and potassium nitrite.

References

(1) Müller, J. and Kuhn, R.; U.S. Patent 3,502,576; March 24, 1970; assigned to Deutsche Gold- und Silber- Scheideanstalt, Germany.

## NITROANILINES

A number of highly colored nitrophenols and nitroanilines are becoming increasingly important commercial products. Of special interest are those products which are commercially available as herbicides. In the manufacture of these nitrophenols and nitroanilines waste streams containing significant quantities of these materials are generated. Because of their toxicity to plant and aquatic life, it is essential that these nitrophenol and nitroaniline compounds be degraded before the waste streams are released to nature. The intense color of these compounds emphasizes the desirability of their degradation and also acts as a built-in indicator for their presence.

These nitrophenol and nitroaniline compounds are not satisfactorily degraded by conventional disposal systems. Thus, if waste streams containing such compounds are treated in the conventional manner, the effluent from such treatment contains unacceptably high concentrations of the nitro compound.

### Removal from Wastewaters

A process developed by R.H.L. Howe (1) is based on the discovery that a waste stream containing color producing nitrophenols or nitroanilines can be disposed of in conventional disposal systems if such stream is first subjected to a pretreatment comprising acidifying to a pH of less than about three, adding an adsorbent material to take up colored components, adding a metallic oxide or hydroxide to adjust the pH to more than about five and also to form a precipitate, and separating the mixture into an effluent and a sludge or foam (scum).

The effluent and sludge (or foam) can then be separately treated by conventional disposal means without interference from the original nitrophenol and nitroaniline compounds. Such a process results in treated wastes free of harmful amounts of nitrophenols and nitroanilines.

References

(1) Howe, R.H.L.; U.S. Patent 3,458,435; July 29, 1969; assigned to Eli Lilly & Co.

## NITROGEN OXIDES

The general topic of nitrogen oxides emission control has been reviewed by A.A. Lawrence (1). Air quality criteria for nitrogen oxide have been reviewed by the Environment Protection Agency (2). Combustion gases from automobiles are one major source of nitrogen oxide pollutants.

Estimates show that automobile exhaust provides 65% (900 tons/day) of the hydrocarbons and 60% [430 tons/day as nitrogen dioxide ($NO_2$) or 280 tons/day as nitric oxide (NO)] of the oxides of nitrogen found in the atmosphere in Los Angeles County.

The hydrocarbons in exhaust gases arise from the incomplete combustion of the fuel. The nitrogen oxides, which result from the fixation of nitrogen during combustion, are formed during all phases of operation in the internal combustion engine. Their concentration in the exhaust gases varies from 20 ppm at idle motor conditions to 1,000 ppm during acceleration.

Most of the total oxides of nitrogen emitted are present as nitric oxide (NO) which is readily converted to nitrogen dioxide ($NO_2$) in the atmosphere. The dioxide absorbs ultraviolet radiation in sunlight and is thereby dissociated into a nitric oxide molecule and an oxygen atom. The oxygen atom enters into a host of chemical reactions, with the organic constitutents of exhaust gases, particularly the hydrocarbons. Furthermore, the photolysis of nitrogen dioxide ($NO_2$) appears to be important in ozone formation.

A substantial reduction of the oxides of nitrogen from automobile exhaust gases should minimize the undesirable properties of smog. Among the methods which have been suggested for their removal are: (a) the decomposition of nitric oxide on activated carbon, (b) the use of a maximum performance carburetor and (c) water injection. None of these methods, however, have been found practical and none have afforded a solution to the problem of removing nitrogen oxides from combustion gases.

Stationary sources such as power plants are a second major source of nitrogen oxide pollutants. The combustion of fuels produces a quantity of nitrogen oxides in addition to the usual better known combustion products such as carbon dioxide, sulfur oxides and water vapor. Apparently, the quantity of nitrogen oxides formed during the combustion process is primarily a function of the temperature developed during the combustion of the fuel, as such temperatures are influenced by the burner characteristics and the configuration of the combustion space.

While it has been found that the quantity of nitrogen oxides formed during combustion of the common fuels, such as coal, fuel oil and natural gas, will vary to some extent with different furnace and burner arrangements, all fuels will produce some nitrogen oxides during combustion. The amount of nitrogen oxides will usually lie in the range of 150 to 1,500 parts per million.

In recent years, it has been determined that irritating atmospheric contaminants observed during smog conditions, are primarily due to a photochemical reaction with organic materials, mostly hydrocarbons, in the presence of oxides of nitrogen. Thus, one desirable means for reducing atmospheric contamination, such as smog, is to reduce the amount of nitrogen oxides formed during the combustion of fuels and discharged into the atmosphere. While central station power plants are operated efficiently and usually the production of nitrogen oxides is small per unit of fuel burned, the large quantities of fuel consumed is such as to make even a small reduction in the parts per million of nitrogen oxides in the effluent gases highly desirable.

## Removal of NO$_x$ from Air

Control techniques for nitrogen oxide emissions from stationary sources (3) and mobile sources (4) have been reviewed by the National Air Pollution Control Administration.

A process developed by J.A. Brennan (5) involves the removal of nitrogen oxides from exhaust gases from internal combustion engines. The process accomplishes nitrogen oxide removal by bringing automotive exhaust gases into contact at a temperature between 400° and 600°C. with a crystalline alkali metal or alkaline earth metal aluminosilicate having rigid three dimensional networks made up of unit cells characterized by the substantial absence of change in unit cell dimensions upon dehydration and rehydration and pores of uniform dimensions distributed throughout the crystalline structure which material has been subjected to a thermal pretreatment in an inert atmosphere prior to the contact at a temperature within the approximate range of 400° to 550°C. for a period of between 1 and 10 hours. The following is one specific example of the conduct of the process.

Three pounds of 13X molecular sieve in the form of $^1/_{10}$ inch pellets were thermally treated at 400°C. for 2 hours in an atmosphere of nitrogen. The thermally treated material was then utilized for removing nitrogen oxide from the exhaust gas from a CLR test engine of the type described in report by Coordinating Research Council-Project No. CM-20-58 "Development of Research Technique for Study of the Oxidation Characteristics of Crankcase Oils in the CLR Oil Test Engine," March 1959. Such exhaust gas initially contained 0.55 milligram NO/liter. The reaction temperature was 420°C. The exhaust gas charge was passed over the thermally treated solid at a space velocity of 520 liters of gas at reactor conditions per liter of solid per minute.

A sample of effluent gas was collected after the thermally treated crystalline aluminosilicate had been in contact with the exhaust gases for three hours. The collected sample was found to have a reduced nitrogen oxide content of 0.13 mg. NO per liter.

A process developed by R.M. Hardgrove (6) involves burning fuel with a consequent reduction in the quantity of nitrogen oxides formed during the combustion of the fuel, and more particularly to the regulation of the combustion air and its admission to and mixing with the fuel in a combustion space, so as to reduce the formation of nitrogen oxides in the effluent gases.

It has been found that the production of nitrogen oxide in central station power plants can be reduced as much as 50% or more by operation of the combustion process in accordance with the process. Moreover, this reduction can be accomplished without a reduction in the efficiency of combustion or substantial adverse effects on the heat transfer within the vapor generating unit.

In this process, the usual fuel burner is supplied with air in an amount less than that theoretically required for complete combustion. The remainder of the combustion air is supplied to the furnace at a position spaced from the burner so that the air mixes with the burning fuel after the initial high temperature associated with the combustion process has been attained. In effect, complete combustion of the fuel is retarded to reduce the maximum temperature of the flame, while at the same time completing the combustion within the confines of the furnace.

More specifically, the process contemplates that from 80 to 95% of the theoretical combustion air will be introduced with or in close proximity to the fuel for burning in the furnace, while the remainder of the combustion air, including the desired excess air, will be introduced into the furnace from a position spaced from the burner or burners.

A process developed by E. Childers et al (7) involves subjecting stack gases from chemical processing operations such as nitrations or nitric acid oxidations, or other gases containing nitrogen oxides, to contact, at elevated temperatures and in the presence of a reducing gas such as natural gas, with or without small amounts of hydrogen, with a highly effective catalyst for the reduction.

Figure 89 shows a suitable form of apparatus for the conduct of the process. The reaction convertor **1** is adapted and arranged to hold a mass of catalyst in a bed. Above the catalyst bed **2**, a gas sparger **4** is located in the gas inlet space **5**. A major gas inlet **6**, auxiliary gas inlet manifold **7**, and treated gas exit **8** provide feed inlets and innocuous gas outlets from the convertor **1**. A spark plug **9** is placed in the inlet gas space **5** for igniting the heating gases for use in the light-off of reactions using little or no hydrogen in the natural gas. Heat exchanger **10** aids in maintaining a heat balance in the process but is not essential to the process. A thermocouple well **11** is placed in the catalyst bed **2** for temperature measurements.

Gas containing nitrogen oxides is treated in this manner. The convertor **1** is charged with a suitable catalyst which is disposed in bed **2**. A natural gas-air mixture is fed through manifold **7**, into the gas space **5**, and ignited therein by a spark from plug **9**. The resulting gas flame is passed through the catalyst bed **2** until the catalyst has attained the desired temperature for the reaction. When that temperature has been reached, natural gas through sparger **4** and valve **b**, valve **a** may be closed, with or without hydrogen through valve **c**, nocuous gas through heat exchanger **10**, and feed inlet **6**, and a controlled amount of steam or nitrogen or other inert gases through manifold **7** are introduced into the convertor **1**.

**FIGURE 89: CATALYTIC CONVERTOR FOR REDUCTION OF NITROGEN OXIDES WITH REDUCING GASES**

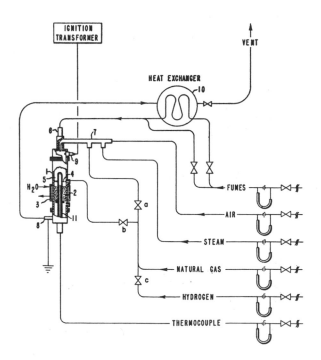

Source: E. Childers, C.W. Ellis and D.J. Ryan; U.S. Patent 3,125,408; March 17, 1964

The air is turned off prior to the introduction of fumes and the temperature of the natural gas (with or without hydrogen) and the reaction temperature are controlled by preheating the feed, or by regulating the amount of steam or nitrogen or other inert material used as a cooling medium. The nitrogen oxide gases are reduced by the natural gas in the catalyst bed 2 and the reaction products are discharged through heat exchanger 10 to a stack not shown. The stoichiometric reducing gas requirements are based on the following equations:

$$CH_4 + 2NO_2 \rightleftharpoons CO_2 + N_2 + 2H_2O$$
$$CH_4 + 2O_2 \rightleftharpoons CO_2 + 2H_2O$$
$$CH_4 + 4N_2O \rightleftharpoons CO_2 + 4N_2 + 2H_2O$$
$$CH_4 + 4NO \rightleftharpoons CO_2 + 2H_2O + 2N_2$$
$$H_2 + NO \rightleftharpoons H_2O + \tfrac{1}{2}N_2$$
$$2H_2 + O_2 \rightleftharpoons 2H_2O$$
$$2H_2 + NO_2 \rightleftharpoons 2H_2O + \tfrac{1}{2}N_2$$
$$H_2 + N_2O \rightleftharpoons H_2O + N_2$$

The reactions are conducted at a temperature of the catalyst bed 2 between 100° and 1000°C. For noble catalysts, temperatures may range from 100° to 1000°C., and for metal hydrogenation catalysts, or reforming catalysts, from 450° to 1000°C., depending on the composition of the reducing gas. Pressure is generally used to the extent of 0.5 psi or higher in the inlet gas space to provide an adequate flow of gases through the catalyst bed.

Higher or lower pressures may be used, however. For example, higher pressures would be desirable when power recovery is used. The amount of low temperature steam or other inert gas used is regulated to maintain the temperature within the range that is most effective for the particular catalyst.

Natural gas is preferably used in excess of that necessary to convert stoichiometrically the nitrogen oxides of the nucuous gas and/or other reducible compounds to nitrogen and water or other relatively nontoxic concentrations or materials by, inter alia, the processes of the above equations. Natural gas, methane alone, or a gas containing an alkane, may be used, such as ethane, propane, butane, etc., or mixtures thereof, or, if desired, any hydrocarbon gas of the oil or coke oven industry.

While natural gas or other alkane-containing gas may be used solely as the reducing gas, one feature of the process in which excellent heat control is attained and light off of the reaction improved, is realized by the use of a relatively small amount of hydrogen with the methane. When hydrogen is used, it should be present in amounts up to 10 to 40% of the stoichiometric amount necessary to reduce the nitrogen oxides with the hydrocarbon constituent of the reducing gas present in sufficient amounts to complete the reaction. To insure complete reduction, the reducing gas may be used in amounts up to about 25% in excess. The gas flow, after the preheating operation, should range between 25,000 and 150,000 reciprocal hours or more, i.e., 25,000 to 150,000 SV the number of cubic feet (STP) of nocuous gas flowing per hour over one cubic foot of catalyst.

Catalyst requirements for the reaction are quite critical. It is necessary to have a catalyst that has a comparatively low light-off temperature for the reaction, and it has been found that noble metal catalysts, supported or unsupported, favor light-off of the reaction at ambient temperatures or slightly higher. These catalysts may be used in the form of gauze, mats, or the like or may be supported on alumina or any other type of suitable inert catalyst support such as infusorial earth, kieselguhr, etc. Platinum, rhodium and palladium catalysts or mixtures of two or more noble metal catalysts are especially well adapted for the decomposition reaction and light-off at low temperatures.

Other types of catalysts may be employed providing they are sufficiently active for the decomposition reaction. Metal oxide containing catalysts may be employed although usually they are not as effective in complete cleanup of the nitrogen oxide fumes as are the noble metal catalysts. For those reactions in which maximum cleanup is not required, such metal catalysts as nickel supported on activated alumina, copper chromite, fused metal oxide catalysts, such as are described in U.S. Patent 2,061,470 may be used. An especially suitable catalyst for this reaction with a light-off temperature, however, somewhat above that of the noble metal catalysts is one prepared from basic metal carbonate and ammonium carbonate disposed on finely divided alumina hydrate as described in U.S. Patent 2,570,882. Reforming catalysts may likewise be used, examples of which are the nickel promoted catalysts and the other metal of the ion group metal oxide promoted catalysts of U.S. Patent 2,119,565 and the catalysts of U.S. Patent 2,064,867.

A process developed by T.D. Felder, Jr. (8) relates to the disposal of industrial waste gases and more particularly relates to converting gases containing undesirable quantities of nitrogen oxides (NO and $NO_2$), which should not be discharged directly into the atmosphere, to innocuous gases that can be so discharged.

The device which may be used, as shown in Figure 90 is composed of inlet conduits 1 and 2, a stack or base cylinder 3, and a flared conical discharge cap 4. Off-gas, or any gas containing nitrogen oxides in concentrations that are noxious, is fed into conduit 2; a reducing gas such as natural gas is fed into conduit 1; these gases enter the cylindrical portion 3 of the stack via ports 5 and 6 and as the resulting mixture flows into the conical (inverted) discharge cap 4 the mixture is ignited and the nitrogen oxides present converted to nitrogen and water.

As shown in the drawing, the mixing chamber is a hollow cylinder, and the buring chamber is the inverted frustum of a cone. The internal diameter of the mixing chamber and the internal diameter of the small end of the frustoconical burning chamber are the same.

FIGURE 90:  MIXING APPARATUS FOR CONVERTING OXIDES OF NITROGEN TO
INNOCUOUS GASES

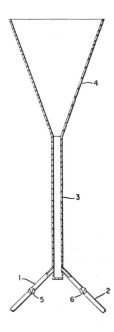

Source:  T.D. Felder, Jr.; U.S. Patent 3,232,713; February 1, 1966

The waste gas must contain or be supplemented with sufficient oxygen to form a combustible mixture with the combustible gas. In the usual case, oxygen will need to be added to the mixture to insure combustion, as the oxygen which is supplied from the NO, $NO_2$ and $N_2O_3$ is insufficient, but in some instances, the amount of nitrogen oxides in the waste gas may be high enough that no additional oxygen need be added. The amount of oxygen, calculated as $O_2$, necessary to maintain combustion is 15 to 30 mols per 100 mols of waste gas reacted.

A process developed by R.F. Bartholomew et al (9) involves removing sulfur dioxide and/or nitrogen dioxide and/or carbon dioxide from gaseous mixtures comprising contacting the gaseous mixture first with a molten nitrate selected from the group consisting of sodium nitrate, silver nitrate and potassium nitrate and then with a molten hydroxide selected from the group consisting of sodium hydroxide and potassium hydroxide or with a molten mixture of at least one of the nitrates and hydroxides.

A process developed by R.E. Stephens (10) involves decomposing nitrogen oxide by contacting an oxide of nitrogen-containing gas with a suitable catalyst, the catalyst consisting of 0.001 to 25 weight percent of neodymium in an oxide form, usually with a catalyst support.

In order to use the catalyst in an internal combustion engine exhaust system, the catalyst is incorporated in a suitable manner into the exhaust system of the engine. One method commonly used is to place the catalyst in a so-called "catalytic muffler." Examples of these are disclosed in U.S. Patents 3,154,389, 3,149,925, 3,149,926 and 3,146,073, among others.

Essentially these are containers having an opening to receive and discharge the exhaust gas. To firmly retain the catalyst material, the receiving and discharge openings are covered with wire screen. The container may have internal baffling to allow greater contact between catalyst and exhaust gas or to use the hot reaction gases to heat the incoming exhaust gases. The container may actually replace the vehicle muffler or may be incorporated into the conventional exhaust system of current vehicles. The catalyst bed may also be located in the exhaust manifold of the engine.

The neodymium oxide catalyst may be used by itself or it may be used in conjunction with a second catalyst whose function is to oxidize the hydrocarbon or carbon monoxide constituents of the exhaust gas. A catalyst eminently suited for this purpose is a supported copper-palladium catalyst as described in U.S. Patent 3,224,981. The neodymium oxide catalyst may be intimately mixed with the oxidation catalyst or the different catalysts may be stratified.

When used to decompose oxides of nitrogen in streams other than the exhaust stream of internal combustion engines the catalyst is merely incorporated in the oxide of nitrogen-containing stream so that intimate contact is obtained between the catalyst and the oxides of nitrogen. For example, in the discharge stream of a nitric acid plant employing the ammonia process for synthesizing nitric acid the spent gas containing nitric oxide is passed through the neodymium oxide catalyst bed and the temperature of the bed maintained at a temperature of from 400° to 1000°C.

A process developed by T.C. Wooton et al (11) involves treating gases containing noxious oxides, particularly oxides of nitrogen, without subjecting the gases to catalytic action, where the gases are intimately contacted with a porous body composed of fibers plated with gold, the porous body being negatively charged.

A process developed by A. Warshaw (12) is one in which a gas stream containing nitrogen oxides, such as the tail gas from a nitric acid plant, is scrubbed with an aqueous urea solution. The nitrogen oxides dissolve in the solution to form nitrous acid, which reacts with the urea to form nitrogen, carbon dioxide and water. The resulting scrubbed gas stream is of reduced nitrogen oxides content and, in the case of nitric acid plant tail gas, may be safely discharged to the atmosphere without causing air pollution.

Figure 91 is a flow diagram of the process. The process source **1**, which typically consists of a nitric acid facility, discharges a gas stream **2** containing nitrogen oxides, generally as nitric oxide and nitrogen dioxide. When source **1** is a nitric acid plant which produces nitric acid by the catalytic oxidation of ammonia, stream **2** will consist of the tail gas from the nitrogen oxides absorber.

The stream is passed into gas-liquid contact tower **3** below section **4**, which usually consists of a bed of suitable packing such as rings, spheres or saddles. In other instances, section **4** may consist of other gas-liquid contact means such as bubble cap plates or sieve trays. In any case, the gas stream rises through the section countercurrent to the liquid stream **5** which is sprayed or otherwise dispersed into the tower above section **4**.

**FIGURE 91: AQUEOUS UREA SCRUBBER FOR NITROGEN OXIDE REMOVAL FROM GAS STREAM**

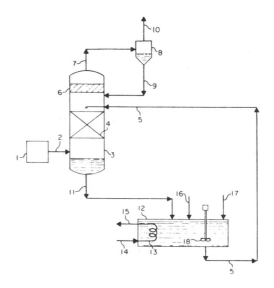

Source: A. Warshaw; U.S. Patent 3,565,575; February 23, 1971

Stream **5** consists of an aqueous acidic urea solution, and stream **5** typically contains in the range of about 1 to 30 grams of dissolved urea per 100 ml. of solution, together with dissolved free acid in a proportion up to about 10% by volume. Any suitable free acid which ionizes in aqueous solution may be present in stream **5**, and the free acid may be an inorganic or organic acid.

Conventional acidic agents such as nitric acid, sulfuric acid, hydrochloric acid or acetic acid are preferred. In most instances, stream **5** will be at an initial temperature in the range of 30° to 90°C., and it has been determined that the reaction rate of the reaction which takes place in section **4** increases at higher temperatures. Consequently, stream **5** will usually be a hot aqueous acidic urea solution, which may even be heated to a temperature above 90°C. and in some instances stream **5** will be at a temperature above 50°C.

The reaction between the gaseous and liquid phases which takes place in section **4** results in the dissolving of nitrogen oxides in the liquid phase, with the resultant formation of nitrous acid in solution. The nitrous acid and dissolved nitrogen oxides react with the dissolved urea, and in the acid solution the reaction products are nitrogen, carbon dioxide and water, which are innocuous components. The resulting gaseous phase rising within the tower above section **4** is substantially free of nitrogen oxides, however, the gaseous phase contains entrained liquid droplets which must be removed prior to discharge or further utilization of the gas stream. The gas stream flows through mist separator **6**, which consists of one or a plurality of wire mesh filter pads or other suitable mist removal devices.

The separated liquid phase flows downwards from section **6** into section **4**, while the resulting mist-free gaseous phase is discharged from the top of the tower via stream **7**. In some instances, stream **7** is suitable for discharge to atmosphere, however, stream **7** is usually passed through cyclonic gas-liquid separator **8** for the removal of residual liquid droplets. The separated liquid phase is recycled to unit **3** below section **6** via stream **9**, while the liquid-free gaseous phase stream **10**, which is of a substantially innocuous composition, is discharged to atmosphere or further utilized as desired.

Returning to tower **3**, the spent scrubbing solution is removed from the bottom of tower **3**, via stream **11**, which flows into mixing tank **12** in which the solution is prepared for further gas scrubbing. The body of liquid solution in tank **12** is heated by heating coil **13**, through which steam or other suitable heating fluid is circulated via stream **14**, with condensate water or cooled fluid being removed via stream **15**.

Makeup solid urea or aqueous urea solution is passed into tank **12** via stream **16**, and a water stream **17** may also be passed into tank **12** to compensate for water losses from the circulating solution in tower **3**. The body of aqueous solution in tank **12** is stirred and mixed in tank **12** by agitator **18**, in order to dissolve and blend the streams **16** and **17** into the aqueous solution. Regenerated solution is withdrawn from tank **12** via stream **5**, which is passed to tower **3** as described above.

A process developed by P. Kandell et al (13) is a process for the catalytic reduction of nitrogen oxides in tail gas from nitric acid synthesis, where water is added to the tail gas prior to and after final reduction so as to give increased reduction, longer catalyst life, and increased energy available for recovery in the treated effluent.

Figure 92 is a block flow diagram showing the essential steps in the process. The apparatus used in the following example consisted of two conventional combustors each 2½ feet deep, modified to accept water spray nozzles at the outlet. Eight water nozzles spaced around the internal circumference of the shell are used for good dispersion or alternatively sprays can be installed in the outlet piping. The nozzle heads are adjusted to give a fairly fine spray. The two combustors are operated in series. The tail gas from a nitric acid absorption tower varies somewhat, but the composition given below is typical:

|  | Volume Percent |
|---|---|
| $O_2$ | 2.62 |
| $N_2$ + Ar | 96.47 |
| NO | 0.32 |
| $NO_2$ and $N_2O_4$ | 0.04 |
| $H_2O$ | 0.55 |

This tail gas is fed directly as incoming feed to the first combustor chamber, at a rate of about 8,500 scfm at about 482°C. and at about 104 psig. As already noted, the effect of the first catalyst is to remove residual oxygen and to substantially convert $NO_2$ (and $N_2O_4$) to NO. The hot gases leaving the first catalyst bed are therefore substantially free of $NO_2$ and $N_2O_4$.

Water at about 40°C. is sprayed into this effluent, at the approximate rate of 6 gallons per minute, or about 0.0007 gallon/cubic foot of effluent at standard conditions, and passes to the second catalyst bed (second combustor). Conversion of NO to $N_2$ is rapid and substantial, the NO content dropping to about 400 ppm or less. As the thus treated gas emerges from the second catalyst bed, it is quenched a second time with 0.0006 to 0.0007 gallon of water per scf of gas.

FIGURE 92:  TWO-STAGE COMBUSTOR FOR NITROGEN OXIDE REDUCTION IN
TAIL GASES

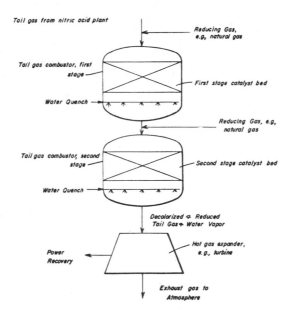

Source:  P. Kandell and G. Nemes; U.S. Patent 3,567,367; March 2, 1971

The temperature of the vapor at this point, and as it now exits the second combustor to enter the power recovery section, is about 532° to 538°C. This effluent is particularly adapted as feed to a turbine to operate compressors for ammonia and/or air, or feed to a waste heat boiler for generation of steam, which steam can be used in the compressors aforesaid, or for other conventional uses.

The effluent from the turbine or waste heat boiler as exhausted finally to the atmosphere generally contains no more than 400 ppm nitrogen oxides and may be substantially less (e.g., 50 to 100 ppm). It meets nitric acid plant pollution regulations for most areas.

A process developed by S.G. Hindin et al (14) is one in which improved catalysts for use in the purification of waste gases containing oxides of nitrogen comprise high surface area thoria or zirconia as a support for a catalytic deposit. The catalysts exhibit exceptionally high stability in the reaction environment.

A process developed by W.D. Stevens (15) is one in which fluid which is heated in a furnace having several rows of burners is directed through tubes aligned in rows between the burner rows to minimize the concentration of nitrogen oxides in the combustion products.

A process developed by P.P. Gertsen et al (16) is a process for the purification of exhaust gases from nitric oxides where the exhaust gases are subjected to sorption with an alkaline reagent for removing higher oxides, then the gas is contacted with pyrite $FeS_2$ sprayed with water, in which case the basic portion of the nitric oxide is reduced to elementary nitrogen, after which the purified gas with residual nitric oxide is contacted with carbon monoxide so that the residual nitric oxide is reduced to elementary nitrogen.

**References**

(1) Lawrence, A.A., *Nitrogen Oxides Emission Control,* Park Ridge, N.J., Noyes Data Corp. (1972).

(2) Environmental Protection Agency, "Air Quality Criteria for Nitrogen Oxides," *Publication No. AP-84,* Wash. D.C., U.S. Govt. Printing Office (January 1971).

(3) U.S. Dept. of Health, Education & Welfare, Public Health Service, "Control Techniques for Nitrogen Oxide Emissions from Stationary Sources," *Publication AP-67,* Wash. D.C., National Air Pollution Control Admin. (March 1970).

(4) U.S. Dept. of Health, Education & Welfare, Public Health Service, "Control Techniques for Carbon Monoxide, Nitrogen Oxide & Hydrocarbon Emissions from Mobile Sources," *Publication No. AP-66,* Wash. D.C., National Air Pollution Control Admin. (March 1970).

(5) Brennan, J.A.; U.S. Patent 3,015,369; January 2, 1962; assigned to Socony Mobil Oil Co., Inc.

(6) Hardgrove, R.M.; U.S. Patent 3,048,131; August 7, 1962; assigned to The Babcock & Wilcox Co.

(7) Childers, E., Ellis, C.W. and Ryan, D.J.; U.S. Patent 3,125,408; March 17, 1964; assigned to E.I. du Pont de Nemours & Co.

(8) Felder, T.D., Jr.; U.S. Patent 3,232,713; February 1, 1966; assigned to E.I. du Pont de Nemours & Co.

(9) Bartholomew, R.F. and Garfinkel, H.M.; U.S. Patent 3,552,912; January 5, 1971; assigned to Corning Glass Works.

(10) Stephens, R.E.; U.S. Patent 3,552,913; January 5, 1971; assigned to Ethyl Corp.

(11) Wooton, T.C. and Mangold, W.F.; U.S. Patent 3,562,127; February 9, 1971; assigned to Scientific Industries of California.

(12) Warshaw, A.; U.S. Patent 3,565,575; February 23, 1971; assigned to Chemical Construction Corp.

(13) Kandell, P. and Nemes, G.; U.S. Patent 3,567,367; March 2, 1971; assigned to W.R. Grace & Co.

(14) Hindin, S.G. and Dettling, J.C.; U.S. Patent 3,615,166; October 26, 1971; assigned to Engelhard Minerals & Chemicals Corp.

(15) Stevens, W.D.; U.S. Patent 3,675,629; July 11, 1972; assigned to Foster Wheeler Corp.

(16) Gertsen, P.P. and Suverneva, T.G.; U.S. Patent 3,695,828; October 3, 1972; assigned to Permsky Politekhnichesky Institut, USSR.

# OIL

Oily waste materials may be measured in terms of their hexane solubility for purposes of pollution evaluation. Hexane is an organic solvent employed to separate oily organic compounds from wastewaters. Oily wastes include greases, as well as many types of oils. Grease is not a specific chemical compound, but a rather general group of semiliquid materials which may include fatty acids, soaps, fats, waxes and other similar materials extractable into hexane.

Unlike some industrial oils which represent precise chemical compositions, greases are, in effect, defined by the analytical method employed to separate them from the water phase of the waste (1). A waste treatment manual recently published by the American Petroleum Institute (2) suggests the following classification for types of oily wastes:

[1] Light hydrocarbons - including light fuels such as gasoline, kerosine and jet fuel, and miscellaneous solvents used for industrial processing, degreasing or cleaning purposes. The presence of waste light hydrocarbons may make removal of other, heavier oily wastes more difficult.

[2] Heavy hydrocarbons, fuels and tars - include the crude oils, diesel oils, No. 6 fuel oil, residual oils, slop oil and even in some cases, asphalt and road tar.

[3] Lubricants and cutting fluids - oil lubricants generally fall into two classes; nonemulsifiable oils such as lubricating oils and greases, and emulsifiable oils such as "water-soluble" oils, rolling oils, cutting oils and drawing compounds. Emulsifiable oils may contain fat, soap, or various other additives.

[4] Fats and fatty oils - these materials originate primarily from processing of foods and natural products. Fats result from processing of animal

flesh. Fatty oils for the most part originate from the plant kingdom. Quantities of these oils result from processing soybeans, cottonseed, linseed and corn.

There are many industrial sources of oily wastes. Table 13 presents the major types of industry producing oil and grease-laden waste streams, and lists characteristic types and sources of oily wastes associated with each industry. By far the three major industrial producers of oily waste are petroleum refineries, metals manufacture and food processors.

## TABLE 13: INDUSTRIAL SOURCES OF OILY WASTES

| Industry | Waste Character |
|---|---|
| Petroleum | Light and heavy oils resulting from producing, refining, storage, transporting and retailing of petroleum and petroleum products. |
| Metals | Grinding, lubricating and cutting oils employed in metal-working operations, and rinsed from metal parts in clean-up processes. |
| Food processing | Natural fats and oils resulting from animal and plant processing, including slaughtering, cleaning and by-product processing. |
| Textiles | Oils and grease resulting from scouring of natural fibers (e.g., wool, cotton). |
| Cooling and heating | Dilute oil-containing cooling water, oil having leaked from pumps, condensers, heat exchangers, etc. |

Source:   Report PB 204,521

Petroleum refineries produce large quantities of oil and oily emulsion wastes. Because of the long history of pollution problems associated with petroleum refining, the American Petroleum Institute (API) has exerted a good deal of effort in developing and publishing methods of oily waste control [e.g., (2)(3)]. Nevertheless, a recently published comprehensive federal industrial waste profile on the petroleum refining industry reported no values for oil content of refinery wastewater, and stated that "data concerning the amounts of oil (in wastewater) are not complete enough to justify inclusion" in that report (4). Indicative of the oil content of refinery waste, however, is a reported value of 154 mg./l. of oil, after preliminary skimming to partially remove floating oils (5).

In the metals industry the two major sources of oily wastes are steel manufacture and metal working. Oily wastes include both emulsified and nonemulsified or floating oils. In steel manufacture, steel ingots are rolled into desired shapes in either hot or cold rolling mills. Oily wastes from hot rolling mills contain primarily lubricating and hydraulic pressure fluids. In cold strip rolling, however, the steel ingot is usually oiled prior to rolling, to lubricate and to reduce rusting.

Additional oil-water emulsions are sprayed during rolling to act as coolants. After shaping, the steel is rinsed to remove the adhering oil. Rinse and coolant waters from the cold rolling mills may contain up to 700 mg./l. of oil, of which over 200 mg./l. may be emulsified and thus difficult to separate from the wastewater (6). Emulsified oil in hot rolling mill effluents rarely exceeds 20 mg./l. (2). More concentrated oily wastes, such as from batch dumps of spent coolant or lubricating fluid must also be treated.

Metal working produces shaped metal pieces such as pistons and other machine parts. Oily wastewaters from metal working processes contain grinding oils, cutting oils, and lubrication fluids. Coolant oil-water emulsions are also employed in many metal working processes. Soluble and emulsified oil content of wastewaters may vary from 100 to 5,000 mg./l. (7).

The third major source of oily wastes, and particularly greases, is the food processing industry. In the processing of meat, fish and poultry, oily and fatty materials are produced primarily during slaughtering, cleaning and by-product processing. The major grease sources are the rendering areas, in particular from the wet (or steam) rendering process which gives the highest levels, pound per pound of scrap processed, of hexane extractables in food processing waste streams (2). Grease content in meat packinghouse waste streams may run several thousand mg./l. (8), and it has been reported that waste from a fish processing plant contained 520 to 13,700 mg./l. of fish oil (9).

### Removal of Oil from Air

A process developed by R.H. Illingworth (10) is one in which oil-laden air from machine shops and kitchens and the like is purified by scrubbing with an aqueous solution of a nonionic wetting agent as the air is drawn through a spray chamber. In such a process, entrained oily matter often constituting a health and fire hazard is removed from the air.

In high speed machine factories and in rooms where comestibles are prepared, such as kitchens in restaurants, homes, galleys in seagoing vessels, industrial bakeries and the like, large amounts of oily foreign particles are entrained in the air. These oily particles comprise an art recognized class of water-immiscible, generally combustible, ether-soluble, usually liquid or at least easily liquefiable on warming, unctuous substances, which leave a greasy stain on paper or cloth.

Merely by way of illustration can be mentioned mineral oils, fuel oils, lubricants derived from petroleum, fatty substances of vegetable and animal organisms, glycerol esters of fatty acids and the like. Unless removed, the oily foreign particle content rapidly builds up to a level which constitutes a serious health hazard to operating personnel and other occupants. It is a matter of common knowledge and experience also that a high oil content in the air also is a very serious fire hazard.

Previously, in factories for example, it has been customary to use a "fast air change" to prevent buildup of oily particles, such as cutting oils, in the air, i.e., fresh air is continuously sucked into the room and foul air continuously exhausted to the atmosphere. This procedure is disadvantageous for several reasons. It requires large blowers and extensive ducting systems. Because of the high content of oil, exhaust of the foul air to the atmosphere creates serious air pollution problems.

It has also been proposed heretofore to remove oily particles, such as cooking fats, from the air in rooms by filtration. A variety of filtration means have been suggested whose purpose is to eliminate the foreign matter from the air. Such filtration means employ filtering elements such as screens, liquid baths, sprays and the like. Conventional filtering means of the type previously employed however require periodic replacements of the filter elements, which become inactive due to clogging by the oily materials.

The improved process overcomes these disadvantages and uses an apparatus such as that shown in Figure 93. There is illustrated an air purification unit 2 comprising a housing 4 formed of sheet metal or other suitable material. Housing 4 may be arranged in a suitable air-ducting system and includes an inlet 6 and an outlet 8. While unit 2 is not limited to any particular size, it has been found that better air circulation is achieved by mounting several smaller units throughout the room to be purified rather than using one large unit.

A fluid supply conduit leads from a separation tank, not shown, and has at least one branch 23 which extends into air purification unit 2 through housing 4. Nozzles 24 are mounted on branch pipe 23 near the terminus thereof in a manner whereby a spray of the air-purifying solution may be directed substantially in a pattern to form a spray zone highly saturated with the solution evenly distributed over the chamber from wall to wall. The air passes through the spray zone.

FIGURE 93: SCRUBBER FOR THE REMOVAL OF OILY MISTS FROM AIR

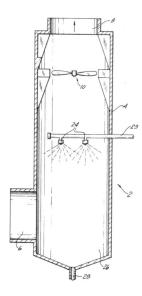

Source:   R.H. Illingworth; U.S. Patent 3,633,340; January 11, 1972

Exhaust fan **10** is mounted within housing **4** on the outlet side of the spray zone, drawing the contaminated air through inlet **6**, past the spray, and exhausting the purified air
out of housing **4** through outlet **8**.  There can be optionally provided a heater unit re-
movably mounted within housing **4** for heating the purified air.

Conduit **28** is provided in the lower casing **26** of housing **4** below the spray zone.  As
herein preferably embodied, there is no spillover of the purifying solution from inlet **6**
into the room being purified.  The air-purifying solution drains down after being sprayed
through the spray zone where it contacts and entrains foreign particles from the air stream
drawn through inlet **6** by fan **10**.  Upon draining down into lower casing **26**, the solution
and entrained particles therein pass through conduit **28** to a sewer or to a recovery system.

Especially good air purification is achieved when the nonionic surfactant is either an ethylene oxide adduct or dodecyl phenol having the general formula $R(CHR'CHR'O)_nH$ where R is
the residue of an alkyl phenol, R' is hydrogen or lower alkyl and n is 8 to 15, or a condensation product of nonylphenol and ethylene oxide having the structural formula:

$$C_9H_{10} \text{—} \bigcirc \text{—} O\text{—}(CH_2CH_2\text{—}O)_n\text{—}H$$

where n is an integer between 8 and 15.  Such nonionic surfactants constitute highly preferred species for use herein.  A small, effective amount of nonionic wetting agent will be
employed in the aqueous solution used to scrub the air.  In general, the aqueous solution

will contain less than 1% and generally less than 0.5% by weight of the nonionic wetting agent. Especially good results are obtained with aqueous solutions containing between 0.005 and 0.25% by weight of nonionic wetting agent, and such solutions are preferred.

If desired, a small effective amount of an antifoaming agent may be utilized in the aqueous solution of the nonionic wetting agent. Typical antifoaming agents include the water-soluble type, such as glyceride oils and fatty acids; and the water-insoluble type, such as octyl alcohols, cyclohexanol, lauryl and cetyl alcohol, 1,2- and 1,3-glycols, water-insoluble esters of phosphoric acid, and vegetable oils, such as castor oil, ethyl oleate and the like.

In operation, a suitable air-purifying solution is made up in a tank or, for example, in the barren solution chamber of a separation tank, whereupon exhaust fan 10 is set into operation and the solution is pumped from the makeup tank through a supply line into feed line 23 to nozzles 24 where it is sprayed into the spray zone continuously contacting and wetting the air stream. Exhaust fan 10 draws oil-laden contaminated air from the high speed machine room, kitchen or the like, through inlet 6 and into the spray zone where it contacts the purifying solution.

As the air passes through the spray zone, the solution scrubs the air, removing the oily foreign particles therefrom. The purified air is then either exhausted to the atmosphere or recycled to the room. The solution carrying the oily particles scrubbed from the air drains into lower chamber 26 and moves through conduit 28 either to a sewer or to a separation tank.

In the tank the oily particles will rise to the top and collect and can then be discharged into a sewer system. The decontaminated solution, which settles, may then be recycled to the filter area. Additional amounts of water and chemical may be added as needed.

### Removal of Oil from Water

Treatment of oily waste is similar in concept to treatment of domestic sewage. In domestic sewage treatment a primary level of treatment is employed to separate the easily settleable solids from the liquid, and in treatment of oily waste primary treatment separates the floatable oils from the water and emulsified oily material. A secondary treatment phase is then required to break the oil-water emulsion and separate the remaining oil and water (11).

*Primary Treatment:* Primary treatment takes advantage of the difference in specific gravity of oils and greases versus water. The treatment process normally involves retaining the oily waste in a bholding tank and allowing gravity separation of the oily material, which is then skimmed from the wastewater surface. Gravity type separators are the most common devices employed in oily waste treatment. The effectiveness of a gravity separator depends upon proper hydraulic design, and design period of wastewater retention.

Longer retention times allow better separation of the floatable oils from the water. Wallace, et al (12) have reported the influence of retention time on separator efficiency based on data for an oily wastewater from the SOHIO Toledo, Ohio refinery. Short detention times of less than 20 minutes resulted in less than 50% oil-water separation, while more extended holding periods improved oil separation from the waste stream.

Gravity separators are equally effective in removing both greases and nonemulsified oils. The standard unit in refinery waste treatment is the API separator, based upon design standards published by the American Petroleum Institute (3). Separators used for metal and food processing oil wastes work, of course, upon the same principle of floating the oil, and many are designed in a similar fashion to the API process insofar as skimming, retention time, etc. Separators may be operated as batch vats, or as continuous flow-through basins, depending upon the volume of waste to be treated.

Table 14 presents the percent efficiencies of several oil separation processes, including the API separator. The values given in this table are based upon treatment of petroleum refinery waste (4).

## TABLE 14: EFFICIENCIES OF OIL SEPARATIONS PROCESS*

| | | Percent Removal | |
| Treatment | Source of Influent | Floating Oil, % | Emulsified Oil, % |
|---|---|---|---|
| API separator | Raw waste | 60 – 99 | Not applicable |
| Air flotation, without | API | | |
| chemicals | Effluent | 70 – 95 | 10 – 40 |
| Air flotation, with | API | | |
| chemicals | Effluent | 75 – 95 | 50 – 90 |
| Chemical coagulation and | API | | |
| sedimentation | Effluent | 60 – 95 | 50 – 90 |

*This table includes only commonly used processes.

Source:  Report PB 204,521

However, efficiencies presented are also generally applicable to oily wastes from other sources undergoing the indicated treatment.  For example, Symons (6) has reported 90% separation of floating oil in a holding tank receiving an influent oil concentration of 700 mg./l. (470 mg./l. floating and 230 mg./l. emulsified). The waste stream originated in a cold rolling mill of the Bethlehem Steel Corporation (Lackawanna, New York). The flow stream volume was 3,000 gal./min.  Additional treatment, described subsequently in this report, was required for the emulsified oil, in order to meet an effluent requirement of 15 mg./l. of oil.

Garrison and Geppert (8) have reported on primary treatment of grease-laden wastewater from a Rath Packing Company (Waterloo, Iowa) plant.  Grease content of the waste stream was 2,850 mg./l. for a flow volume ranging from 1 to 7 MGD.  A 90 minute detention period in the gravity separator, and skimming of the floating greases reduced the waste stream grease content by about 75% to approximately 750 mg./l.  Grease was recovered for further processing and sale.

Recovery of skimmed oil or grease for all major types of oily waste producers is increasingly common as the value of the recoverable oil is realized.  Frequently a substantial savings is possible through recovery or recycle of oily material (13).  If skimmings cannot be reused, they are typically disposed of by burial, lagooning, or perhaps incineration.  Odor and nuisance-free grease incineration has been reported by Dreier and Walker (14).

Capital costs of $1,200,000 have been reported (1966 prices) for an API separator with a capacity of approximately 2,000 gal./min. (11).  Refinery oily waste treatment costs have also been reported for both capital and maintenance and operation (4).  A federal report (15) on treatment costs for the meat processing industry has presented capital, and maintenance and operation costs for grease and oil treatment of food industries.

*Secondary Treatment:* Unlike primary treatment, which consists only of gravity separation plus skimming, any of several different processes may be employed for secondary treatment of oily wastes.  All processes are directed toward breaking the oil-water emulsion which has passed through the primary separator, and separating the demulsified oil from the water phase.

Emulsions may be broken by chemical, electrical, or physical methods.  Chemical methods are in widest use for treatment of oily wastewaters.  The electrical process is directed toward emulsions containing mainly oil, with small quantities of water.  Most oily wastes in contrast are primarily water with lesser amounts of oil.

Physical emulsion-breaking methods include heating, centrifugation, and precoat filtration, with the latter two most common.  Centrifugation breaks oil emulsion by separating the oil and water phases under the influence of centrifugal force.  Centrifugation is best applied to oily sludges, and is generally not used in the treatment of the typical dilute oily wastewater stream unless the volume is small (2).

Filtration has been employed to treat a steel mill waste containing 230 mg./l. of emulsified oil. The waste volume was 3,000 gal./min. The waste stream first passed, through an oil-water separator to remove the floating oil (470 mg./l.) by skimming (6). The overall goal of treatment was stated as to produce an effluent containing less than 15 mg./l. of oil. Three alternative levels of secondary treatment were evaluated with respect to achieving the required level of oil removal. These were:

(a) a high rate sand and gravel filter (HRF)
(b) addition of a polyelectrolyte coagulant prior to high rate filtration
(c) passage of the HRF effluent through a diatomaceous earth filter (DE).

Chemical treatment of an emulsion is usually directed toward destabilizing the dispersed oil droplets, or destroying any emulsifying agents present. The process may consist of rapidly mixing coagulant chemicals with the wastewater, followed by flocculation and flotation or settling. Acidification may also be effective in breaking an oil-water emulsion. Mentens (16) has listed the alternative chemical demulsifying processes as:

[1] adding of coagulating salts
[2] adding of acids
[3] adding of salts and heating the emulsion
[4] adding of salts and treatment by electricity
[5] adding of acids plus organic cleaving agents
   (demulgaters).

In a discussion of these processes, it has been emphasized that coagulation with aluminum or iron salts is generally effective for demulsifying oily wastes (17). However, the aluminum or iron may form hydroxide sludges, which are difficult to dewater. Acids generally cleave emulsions more effectively than coagulant salts, but are more expensive and the resultant acid wastewater must be neutralized after oil-water separation.

The "salting out" of the emulsifier by adding large quantities of an inorganic salt may create additional pollution problems by significantly increasing the dissolved solids in the effluent. Organic demulgaters are extremely effective demulsifying agents, but due to their high cost are considered impractical at high rinse water flow rates and low oil concentrations (17). Thus acid or coagulant salts find the widest acceptance as demulsifying agents in industry. Frequently, coagulant aids are added to assist in flocculating the coagulant particles of oil sludge.

Following destabilization of the oily emulsion, a process called air flotation is commonly employed to separate the oil and water. Air flotation consists of saturating a portion of the wastewater with dissolved air under high pressure. The pressure is then suddenly released, resulting in the formation of thousands of microscopic air bubbles which attach themselves to oil droplets and float them to the surface. The oil-air bubble mixture forms a froth layer at the surface, which is skimmed away. Chemical flocculating agents such as salts of iron and aluminum, with or without organic polyelectrolytes, are particularly helpful in improving the effectiveness of the air flotation process.

Expected efficiencies of air flotation, with and without use of chemical aids, are presented in Table 14. Coagulant aids can double the effectiveness of air flotation in removing emulsified oils, as indicated in that table.

Quigley et al (11) have quoted Rohlich as reporting 62% oil removal by direct air flotation, and 94% removal upon addition of 25 mg./l. aluminum sulfate coagulant. These authors have quoted Simonsen as reporting 70% oil removal from a refinery wastewater by air flotation, while addition of polyelectrolyte and bentonite clay increased oil removal to 95%.

Oil removal in the latter case was presumably by sedimentation, with the bentonite clay providing settling weight to the floc particles. Simonsen also reported, for another refinery, 79% removal of oil by flotation, increased to 87% removal at an aluminum sulfate coagulant dosage of 25 mg./l. Quigley and Hoffman have reported for their own refinery operation an air flotation efficiency of 70 to 80%, and up to 90% oil removal upon addition of aluminum sulfate at concentrations of 30 to 70 mg./l. Lime, added at 75 to 100 mg./l. provided even better removal to 95% (11). Capital costs for the air flotation unit was reported at $360,000 (1966 price); waste volume treated was approximately 2,000 gal./min. Maintenance and operating costs were 11.5¢/1,000 gallons of waste. Effluent oil concentrations below 15 mg./l. were required (11).

Wigren and Burton (5) achieved oil reduction, by air flotation of petroleum refinery waste, from 154 mg./l. down to 40 mg./l. This represents a reduction of approximately 70%, without chemical aids. No capital or operating costs were reported. Garrison and Geppert (8) reported for an industrial pilot process 93% grease reduction of a meat packing waste by air flotation. Grease content of the waste was reduced from 1,944 mg./l. to 142 mg./l., with 300 mg./l. of aluminum sulfate added. The process resulted in decreased volumes and qualities of recoverable grease however. Cost of chemicals was 13.4¢/1,000 gallons of waste treated, and capital costs were estimated at $30,000 for a one MGD unit.

Chemical destabilization of oily waste is sometimes followed by flocculation and sedimentation, rather than air flotation and skimming. Table 14 has presented expected treatment efficiencies for chemical coagulation and sedimentation. Quigley and Hoffman (11) have reported that Simonsen achieved 95% oil removal by coagulation and addition of bentonite clay. Link-Belt treats oily waste from a ball and roller bearing plant (Indianapolis, Indiana) by coagulation with sodium carbonate, lime and a polyelectrolyte flocculating agent to achieve grease and oil reduction from 302 down to 28 mg./l. This exceeds 90% removal, in a process treating 30,000 to 60,000 gal./day.

Oil and grease laden waters from food processing are, on occasion, biologically treated in lagoons and oxidation ponds. Saucier (18) has reported the use of an anaerobic lagoon to achieve 88% grease removal. Influent grease content during four 4-day sampling periods within one month ranged from 425 to 1,270 mg./l., while effluent grease concentrations of 69 to 147 mg./l. were recorded. The lagoon treated 528,000 gal./day, at a capital cost (excluding land) of $44,000.

Wigren and Burton (5) reported 88% reduction of oil from a refinery waste in an oxidation pond. Oil content was reduced from 154 to 18 mg./l., in a waste stream of approximately 8 MGD. No capital or operating costs were presented by these authors.

A process developed by A. Gasser et al (19) involves purifying wastewater contaminated by emulsified oil droplets, to wit, wastewater which may be considered as a stable oil-in-water emulsion.

The process is a two-stage process. In the first stage, the wastewater is mixed at an alkaline pH value with extraneous oil and a solution of a ferric salt capable of forming iron hydroxide in an alkaline medium. The ferric salt may be, for example, ferric sulfate, ferric chloride or the like. Due to the alkalinity of the wastewater, the ferric salt precipitates in the form of a flocculent iron hydroxide and the iron hydroxide therefore acts in nascent state.

The system is then allowed to stand, whereby the iron hydroxide, together with the contaminating oil droplets and the added oil, rise to the surface to form a top layer while the purified water forms a subjacent water column. This water column, however, is not entirely devoid of oil, and therefore in the second stage the water column is mixed with a purifying material of iron hydroxide, oil and a stabilizing agent. This mixture is allowed to stand whereby the remaining oil and purifying material rise to the surface to form a surface layer, the pure water which may then be discharged being in the form of a subjacent column below this layer.

A process developed by G.P. Canevari (20) is one in which fine droplets of oil are separated from an aqueous phase using a mixture comprising a sodium montmorillonite clay and an agent selected from the class consisting of organic cationic agents and glycols. The organic cationic agent is preferably an amine.

*Summary:* Treatment of floating oils is achieved economically and efficiently by gravity separation and skimming. Oil removals as high as 99% have been reported (4). Treatment of emulsified oil-water mixtures is more complex and costly, and represents a secondary phase of treatment after the primary gravity separation process. Chemical demulsifying is commonly employed, with air flotation to separate the oil and water phases. Removals equivalent to that achieved by air flotation have, however, been reported for coagulation and precipitation, and for biological lagoons and oxidation ponds.

### References

(1) Public Health Assoc., *Standard Methods for the Examination of Water and Wastewater,* 13th ed., APHA, New York, 1971.

(2) Amer. Petroleum Instit., *Industrial Oily Waste Control,* API, New York, (undated).

(3) Amer. Petroleum Instit., *Manual on Disposal of Refinery Wastes,* 7th ed., API, New York, 1963.

(4) *The Cost of Clean Water: Vol. III, Industrial Waste Profiles, No. 5 - Petroleum Refining,* U.S. Dept. Interior, Washington, D.C., 1967.

(5) Wigren, A.A. and Burton, F.L., *Modernizing Refinery Wastewater Control,* presented at 43rd Ann. Conf. Wat. Poll. Control Fed., Boston, October 1970.

(6) Symons, C.R., *Treatment of Cold Mill Wastewaters by Ultrahigh-Rate Filtration,* presented at 43rd Ann. Conf. Wat. Poll. Control Fed., Boston, October 1970.

(7) Brink, R.J., "Operating Costs of Waste Treatment in General Motors," *Proc. 19th Purdue Indust. Waste Conf.,* pp. 12-16 1964.

(8) Garrison, V.M. and Geppert, R.J., "Packinghouse Waste Processing Applied Improvement of Conventional Methods," *Proc. 15th Purdue Indust. Waste Conf.,* pp. 207-217, 1960.

(9) Chun, M.J., Young, R.H.F. and Burbank, N.C., Jr., "A Characterization of Tuna Packing Waste," *Proc. 23rd Purdue Indust. Waste Conf.,* pp. 786-805, 1968.

(10) Illingworth, R.H.; U.S. Patent 3,633,340; January 11, 1972; assigned to The Evening News Publishing Co.

(11) Quigley, R.E. and Hoffman, E.L., "Flotation of Oily Wastes," *Proc. 21st Purdue Indust. Waste Conf.,* pp. 527-533 (1966).

(12) Wallace, A.T., Rohlich, G.A. and Villemonte, J.R., "The Effect of Inlet Conditions on Oil-Water Separators at SOHIO's Toledo Refinery," *Proc. 20th Purdue Indust. Waste Conf.,* pp. 618-625, (1965).

(13) Schutt, G.J., Keil, C.C. and Hallasz, S.J., "Recovery and Reuse of Oil Extracted from Industrial Wastewater," *Proc. 23rd Purdue Indust. Waste Conf.,* pp. 493-496 (1968).

(14) Dreier, D.E. and Walker, J.D., "Grease Incineration," *Proc. 19th Purdue Indust. Waste Conf.,* p. 161 (1964).

(15) *The Cost of Clean Water: Vol. III, Industrial Waste Profiles No. 8 - Meat Products,* U.S. Dept. Interior, Washington, D.C. (1967).

(16) Mentens, A., "Treatment of Wastes Originating from Metal Industries," *Proc. 22nd Purdue Indust. Waste Conf.,* pp. 908-925 (1967).

(17) Barker, J.E., Foltz, V.W. and Thompson, R.J., *Treatment of Waste Oil-Waste Water Mixtures,* presented at Ann. Conf., AIChE, Chicago, November 1970.

(18) Saucier, J.W., "Anaerobic Lagoons Versus Aerated Lagoons in the Treatment of Packing-House Wastes," *Proc. 24th Purdue Indust. Waste Conf.,* pp. 534-541 (1969).

(19) Gasser, A. and Ronge, G.; U.S. Patent 3,446,732; May 27, 1969; assigned to Entwicklungs- und Forschung- Anstalt, Liechtenstein.

(20) Canevari, G.P.; U.S. Patent 3,487,928; January 6, 1970; assigned to Esso Research & Eng. Co.

# OIL (INDUSTRIAL WASTE)

## Removal from Water

A process developed by M.E. McMahon (1) is one in which industrial wastes from various sources containing, for example, water and oil impurities, and solid impurities, are crudely separated to give an oil phase, a water phase, and a solids phase. The water phase is passed to a surge pond which contains large quantities of water from various sources and the water phase is blended with the water in the surge pond to produce a substantially homogeneous water phase. A portion of the water phase is taken from the pond and admixed with the solids phase and the mixture is passed to a treater where substantially pure water is produced. In a more specific aspect, the water phase is passed into an eductor tube which is located in the surge pond to aid in homogenizing the water in the pond. Further, the mixture of the solids phase and the homogeneous water phase is treated with a coagulant and chemical treatment agents to produce the substantially pure water.

Figure 94 shows the essential features of the process. Referring to the drawing, industrial wastes pass into phase separator 2 through line 1 and therein are separated into a liquid phase; an oil phase, which separates as a supernatant liquid on the liquid phase and is removed through line 3; and a solid phase. The solid phase is removed from the phase separator 2 through line 4. The liquid phase, consisting substantially of water, fine solids, and dissolved minerals, is passed from phase separator 2 through line 5 and into surge pond 6.

The surge pond is a large vessel or pond which can hold large quantities of water. For example, a surge pond could have a capacity of 120,000 bbl. of water. The water phase passes into surge pond 6 through eductor tube 7 which is larger than line 5 and is open to liquid in the surge pond at the point where line 5 enters tube 7.

The water phase passing through line 5 into eductor tube 7 will draw water from the surge pond and admix the water with the water phase from line 5. The water phase passes through tube 7 into conduit 8 where further mixing and homogenization takes place. Conduit 8 can be provided with a series of holes for allowing the water phase from line 5 and tube 7 to admix with the water in the surge pond.

FIGURE 94:  APPARATUS FOR PURIFYING INDUSTRIAL WASTEWATERS
            CONTAINING OIL AND SOLIDS

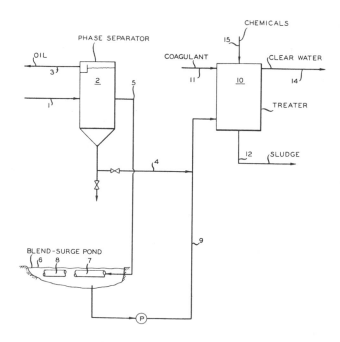

Source:  M.E. McMahon; U.S. Patent 3,297,568; January 10, 1967

The surge pond can be placed on a geographically lower level than phase separator 2 so that the water phase from separator 2 will flow by gravity into the surge pond under pressure to aid in the mixing and blending of the liquids. Alternatively, a pump can be provided in line 5 to supply the requisite pressure to the water phase passing through the line 5. The surge pond serves as an area where water phase from different industrial wastes can be blended to provide a water phase with a substantially homogeneous mineral content.

Substantially homogeneous water phase is passed from the surge pond through line 9 and admixed with the solids phase passing through line 4. The solids aid in precipitating finely divided entrained solids. The mixture is passed into treater 10 where the substantially complete separation takes place. The solids from line 4 act as a weighing agent and help precipitate solids in line 9.

Coagulant to help in precipitation of the solids can be added through line 11. Any standard coagulant used to precipitate industrial solid impurities in water can be used. Further, chemical treating agents can be added through line 15 to remove mineral impurities from the water. When the mineral content of the water entering the treater remains substantially constant, a constant amount of chemical treating agent can be added through line 15. Treater 10 can consist of a single treater or a series of treaters which cooperate to produce substantially clear water which is removed through line 14 and a sludge phase which is removed through line 12.

A process developed by E. Baer et al (2) involves the purification of wastewaters by the addition of metal salts or metal oxides and the flocculation of same as their hydroxide by setting a suitable corresponding pH in the wastewater, with the distinction that: (1) the metal hydroxide after the first flocculation is redissolved; (2) the pH value, of the wastewater, after the removal of the separated impurities, is again set to the flocculation point, and (3) the alternation of precipitation, redissolving and separation of the impurities is repeated as many times as necessary.

The especially hard-to-purify wastewaters of the auto industry, which consist, for example, of the washing water and bore emulsions, present particular difficulties in clarification because of their high content of mineral oils and detergents or emulsifiers. These wastewaters are usually alkaline. In such and other alkaline wastewaters, it is recommended that, for the first precipitation, a metal salt be used which precipitates a hydroxide at about the pH range of the wastewater, and then, for the last precipitation, that a metal salt be used of which the hydroxide precipitates at a lower pH value.

For example, in the case of a wastewater with a pH of about 11, magnesium salts may be used for the precipitation in the first and following steps, and only in the last step has the pH fallen low enough so that the hydroxide of the second metal salt precipitates. This salt, for example, an aluminum or iron salt, may be added just before the last precipitation step, or it may be added even before the first precipitation step with the magnesium salt. This procedure has the advantage of only a very slight consumption of the relatively expensive aluminum salts but, for the last precipitation step, advantage is still taken of the especially good adsorption power of the aluminum hydroxide.

A further saving in cost is permitted by the fact that the metal hydroxide slime from the last precipitation step, that is, the aluminum or iron hydroxide, can be used for the treatment of newly arriving wastewater. Moreover, in this variant of the process, the total clarification of the alkaline wastewater takes place in an alkaline to neutral range. In this way there is rendered superfluous a final neutralization before discharging the clarified water to the drainage, and the costs of a plant for carrying out this variation of the process are much lower because the tanks need not be of acid-resistant material.

Figure 95 shows a suitable form of equipment for the conduct of the process. The incoming wastewater (line 1) is treated through a dosing line 2 with metal salt and mixed in chamber 3. The pH value necessary for the precipitation of the metal salt in question is adjusted through feed 5 according to the pH measurement 6 by means of a regulating device.

## FIGURE 95: APPARATUS FOR CLARIFICATION OF OILY INDUSTRIAL WASTEWATER USING METAL HYDROXIDE FLOCCULATING AGENT

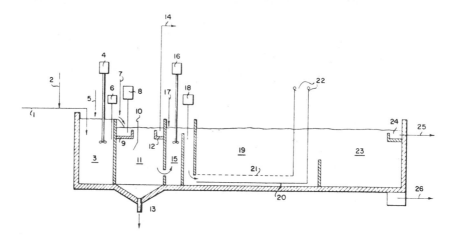

Source:  E. Baer and K. Xylander; U.S. Patent 3,347,786; October 17, 1967

The flocculated wastewater arrives by an overflow dam into the channel **9** where, by means of a dosing device **7**, controlled by the pH measuring or regulating device **8**, an acid or an alkali is fed, in order to redissolve the hydroxide precipitated in the chamber **3**. The water can stand in the chamber **11** while the heavy components sink into the funnel **13** and the floating components are collected in the channel **12**.

The deposited or floating slime is drawn off through line **13** or **14**, as the case may be. The partition wall **10** serves as a conduction surface for the incoming water. From chamber **11** the prepurified water flows into the chamber **15**, whereby dosing with an acid or alkali through line **17** and corresponding pH value regulation **18**, the flocculation pH value is set again.

The agitator **16** serves to mix the acid or alkali. The flocculate and water mixture then proceeds into the flotation cell **19**, where the metal hydroxide is floated by the very fine gas bubbles resulting in the electrolysis. The electric current is fed through connection **22**. The electrode **21**, located above the electrode **20** is perforated, so that the gas bubbles leaving the electrode **20** can pass through the electrode **21**.

The wastewater then reaches a standing zone **23**. The slime floated in the electrolysis is drawn off through channel **25**, while the clarified wastewater can be removed from line **26** and conducted into the drainage.

In cases where two different metal salts are used, for example, magnesium sulfate and aluminum sulfate, in the case of alkaline wastewaters, the magnesium sulfate is added through dosing line **2**. Since magnesium precipitates in a very wide range (pH 10 to 13), the pH regulation **6** is superfluous when the wastewater is very alkaline. After adding an acid through dosing device **7** and pH regulation, **8** the magnesium sulfate is dissolved at about pH 8. The broken, but no longer flocculated emulsion is then subjected to a preliminary purification in chamber **11**, in which the heavy components sink (into funnel **13**) and the light components, such as floating oil, for example, quickly rise and can be drawn off through the line **14**. In the chamber **15**, after adding aluminum sulfate and setting the pH value for flocculation (about 5.5 to 7.5) there is another flocculation by addition of an acid. In the flotation cell **19**, then, as already described, the corresponding aluminum hydroxide flakes are separated from the water. If the hydroxide slime drawn from the line 25 is dissolved with 2% by volume concentrated sulfuric acid, the dirt particles themselves are separated. After their removal, the remaining dissolved aluminum sulfate can be added, instead of fresh aluminum sulfate, through the dosing line 2 or through dosing line 17 to the wastewater again.

The separation of the dirt particles separated in the tank 11 is greatly accelerated when, as in the flotation tank 19, the electrodes are arranged one above the other, the upper being perforated. In this case, even the solid particles settling downward are floated upward by the gas bubbles, where they can be deposited together with the rest of the floated slime.

A process developed by J.T. Jockel (3) involves disposing of waste oil-in-water emulsions by acidifying the emulsion, centrifuging the resulting emulsion to separate an oil phase which is recovered and a water phase, passing the water phase through an adsorbent and treating it with a base and disposing of the treated water phase.

The pollution of public waterways is a constant concern to manufacturers with regard to the disposal of industrial waste, particularly of waste oils. The quantity of foreign matter which may be introduced into rivers and streams is being continually decreased by governmental restrictions. The concentration of oil in effluent plant streams entering public waterways, in some cases, may be no greater than 20 parts per million. The metal rolling industry, for example, uses oil and water emulsions containing from 50,000 to 100,000 ppm of oil. Emulsions at these concentrations of oil cannot, therefore, be passed directly into a public waterway without prior treatment.

It has previously been the practice to transfer the waste oil emulsions to one or more large settling tanks capable of holding from 50,000 to 100,000 gallons. There they are treated with an acid or acid salt, such as alum, and allowed to settle into separate oil and water phases. This settling may take from a few days to almost two weeks. Frequently, the large tanks have to be heated to as high as 180°F. to hasten the settling. At the end of this extensive period, the water phase, still containing non-oil impurities, is passed into the nearest sewer or directly into a waterway.

The oil phase, usually containing up to 40% by weight of water and often still emulsified, is transferred to nearby lagoons or open pits. This means of disposing of waste oil not only requires a large capital expenditure for extra settling tanks and for heating facilities, it also takes up valuable land which could be used for more productive purposes. If these pits are near rivers or streams, there is a danger that seepage may cause pollution of these public waterways. Moreover, if the oil is to be burned off as a means of final disposal, the burning may also create an air pollution problem. Another difficulty with this procedure is that oil containing 5% or more water cannot be burned and then must undergo further separation therefrom before final disposal.

In the process, a waste oil emulsion containing by weight of total emulsion, from 90 to 95% of water and from 10 to 5% of oil and typical oil additives may be disposed of by the steps of: [1] adding a mineral acid to the emulsion; [2] subjecting the acid-treated emulsion to centrifuging, and removing the oil phase; [3] adding to the remaining water phase an inorganic alkaline substance, while keeping the pH below neutral; [4] passing the water phase through a bed of activated carbon; [5] adding an inorganic alkaline substance, increasing the pH of the material to above neutral; and [6] subjecting the water phase to centrifuge.

If desired, the first alkalizing treatment, step [3], may be performed prior to centrifuging, step [2], to avoid contacting the centrifuge surfaces with highly acidic materials. Regardless of which alternative procedure is used, the first centrifuge step should be performed as soon as possible, after the acidification or acidification-alkalization treatment.

The water phase taken from the centrifuge is then passed through a bed of activated carbon, after which the pH is raised once more by a further addition of the inorganic alkaline substance and finally centrifuged. The effluent water phase from the process contains less than 20 ppm of oil and usually is as low as about 5 to 10 ppm and it is almost entirely free of other impurities. This water phase may be safely passed into public waterways. Surprisingly, the oil phase contains less than 5% by weight of water and may be burned off easily in controlled in-plant equipment, thereby avoiding the fire hazard and air pollution of open-pit burning.

A process developed by W.C. Harsh (4) involves the treatment of effluents containing cutting oils used in machining. Such oils effectively suspend fine powdered materials such as dust, dirt, fine particles of metals and oxides, and the like in any aqueous medium in which they are placed.

It is well-known, of course, that these cutting oils wear out and eventually find their way into the sewer water. Also, coolants for grinding machines contain emulsifiers for the oil content and these readily suspend the ground off metal particles, metal oxides, particles of grinding stone materials such as silicon carbide, aluminum oxide and the like. These will also effectively suspend rouge particles used for polishing operations in glass factories and the like.

The result is that these small particles become so effectively suspended by the oily phase of the wastewater that ordinary coagulants and treating procedures for the suspended particles are rendered entirely ineffective.

The problem is further complicated by the fact that the oily content is mixed with the water phase so thoroughly that settling processes are ineffective to bring the oil to the top for skimming and removal. Thus, attempts at skimming are ineffective. Therefore, a substantial step forward in the art would be provided by a coagulant for finely divided materials in aqueous media containing residues of immiscible liquids, and by a process for its use.

It has been found that fly ash in combination with alum and/or ferric chloride is an unexpectedly active coagulant for particulate materials suspended in an aqueous medium, which medium also includes immiscible liquid, such as oil residues. The unexpected aspect of the process is that fly ash per se has some coagulating effect, but the effect is weak, and relatively slow to be of any reasonable benefit in a commercial treating operation. Fly ash contains iron and aluminum hydroxides which would have some coagulating tendencies but these tendencies are so small that by themselves they are considered inconsequential.

However, when fly ash is used in combination with alum and/or ferric chloride, the amount of the latter can be reduced, but their same coagulating effect is retained. Thus, fly ash per se does not perform any substantial coagulating function, but when admixed with alum and/or ferric chloride, the coagulating effect is substantially intensified, much more than would be expected from the addition of the actually used amount of alum or ferric chloride. There is an unexpected increase in coagulating power. In other words, there is a definite synergistic effect.

The process will now be described in stepwise fashion with reference to the flow diagram in Figure 96. Step 1 — Providing a waste effluent. By reference to the drawing, note that shop sewer effluent, made up largely of water from a shop **10** is designated. This may represent a railroad repair machine shop, a metal working factory, a tool room grinding establishment or other. In such a shop **10**, substantial quantities of water are usually used in various processing.

Particularly, water is used as a coolant in grinding operations, and in cutting fluids. It is also used for soap showers, hand washing, cleaning and the like. The water thus picks up any oily residue phase, termed herein a normally immiscible liquid phase. However, agitation causes the normally immiscible contaminants such as saponified grease, emulsifiers, cutting oils and the like to become at least partially emulsified in the aqueous phase. This produces a suspension of particulate materials such as dust, metal cuttings, and the like rendering separation extremely difficult. Normally shop effluents contain both organic and inorganic particulate materials.

Step 2 — According to this step, the effluent in toto is conveyed via line **12** to a mixing chamber **14**. A stirrer **16**, actuated by a motor **18**, is provided in the mixing chamber **14**. At the mixing chamber **14** fly ash is added by means of a dry chemical feeder **20** at a prescribed ratio, as indicated above. After thorough admixture of the fly ash, alum and/or ferric chloride is added to provide a desired pH. An effluent pump **24** and line **26** convey the stirred effluent to the next step.

### FIGURE 96: INDUSTRIAL WASTEWATER TREATMENT SCHEME USING FLY ASH AND ALUM AS TREATING AGENTS

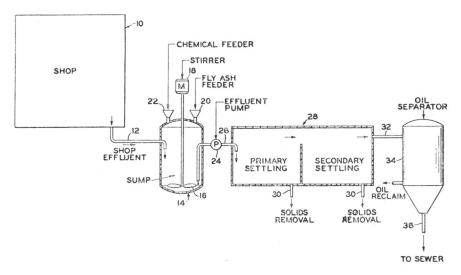

Source: W.C. Harsh; U.S. Patent 3,446,731; May 27, 1969

Step 3 — After thorough mixing, the treated effluent is allowed to settle in a settling apparatus **28**. This may comprise one or more stages as required by the volume per unit of time being treated. In such settling stage, a solids removal conduit **30** is provided so that the heavier sludge which accumulates on the bottom of the settling chamber can be removed for appropriate disposition.

Step 4 — During the settling operation, the oily phase is broken away from the aqueous phase and rises to the top so that it can be skimmed off. Therefore, from the settling zone **28**, the effluent, still containing the oily phase, but having been converted to an immiscible state, is conveyed via a line **32** to an oil separator **34**. Here, the immiscible phase, including oils, detergent residues and other materials that float are removed by skimming, by centrifuging or other.

Step 5 — The clarified and oil-free effluent is then ready for discharge via a sewer line **36** to public waters. Depending upon the bacteria content, acid content or other antibacterial treatment and/or pH adjustment may be affected at this time, just prior to the final discharge.

A process developed by J.I. Wallover (5) is one in which waste industrial oil is filtered through a bed of activated earth to remove chemicals and extremely small foreign particles. The filtered oil is conducted away from the bed and to distilling apparatus if the oil contains distillates. Spent earth in the filter bed is removed periodically and passed through an incinerator to clean and reactivate it. Its place in the bed is taken by activated earth delivered from a substantially air-tight reservoir, to which reactivated earth from the incinerator is returned. The smoke produced by the incinerator is burned and the resulting gases are scrubbed to remove solids from them in order to provide clean air which can be delivered to the atmosphere. The solids are collected for suitable disposal.

There are vaious kinds of industrial oils, such as gear oil, hydraulic pressure oil, quenching oil, transfer oil, mineral spirits and kerosene. Some of these may contain chemical additions. These oils oxidize or otherwise deteriorate in use and must be replaced with fresh oil. Attempts in the past to reclaim the waste oil by reconditioning it have been expensive and have produced problems of disposal of waste products removed from the oil. On the other hand, a much greater disposal problem is presented if reclamation is not attempted and the waste oil is merely dumped. Finding suitable places to safely dispose of waste oil is very difficult. Of course, it cannot be allowed to pollute rivers or lakes.

Figure 97 shows a suitable arrangement of apparatus for the conduct of the process. Referring to the drawing, it is assumed that the waste industrial oil that is to be reconditioned or reclaimed contains distillates, chemicals and water, and solids in the form of foreign particles. The waste oil may be pumped from a storage tank through a line **1** to the top of a settling tank **2**. After the tank has been filled the pump **3** is turned off and the oil in the tank is allowed to remain quiet for at least several hours. During this time water and solids settle to the bottom of the tank.

### FIGURE 97: APPARATUS FOR RECLAIMING WASTE INDUSTRIAL OILS

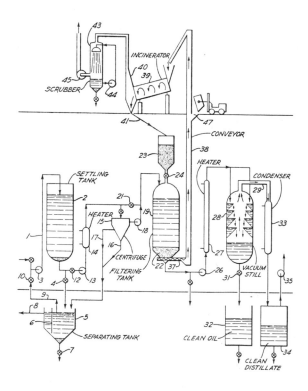

Source:  J.I. Wallover; U.S. Patent 3,527,696; September 8, 1970

At the desired time a valve **4** is opened below the tank to draw off most of the settled solids and water and deliver them to a separating tank **5**. In this tank there is a vertical baffle **6** that divides the tank into a receiving chamber and a discharge chamber that are connected only through a passage beneath the baffle. Solids settle to the bottom of the tank and can be drawn off periodically through a valve **7** and then used for land fill or other purposes. Water will settle also and can be drawn off at **8** from the other side of the baffle and drained into a sewer, as it will be clean enough for that purpose. The fairly clean oil left in the separating tank can be conducted back through a line **9** and a valve **10** to the pump **3**, which will recirculate it to the settling tank **2**.

After solids and water have been drawn out of the settling tank through drain valve **4**, that valve is closed and a valve **12** is opened that leads to a pump **13** which pumps oil from the settling tank through a heater **14** that may be a heat exchanger heated by steam. The oil is heated to a temperature of about 150°F. and is delivered to a centrifuge **15**, where most of the remaining solids and water are separated from the oil.

The settling solids and some of the water leave the bottom of the centrifuge through a pipe **16**, and the rest of the water leaves through a lateral pipe **17**. These two pipes may join and they empty into separating tank **5**, where the solids and water again separate and are removed as explained above. Oil, now free of water and most solids, is removed from the upper part of the centrifuge by a pump **18** that delivers it to the top of a filtering tank **19**.

When the waste oil does not contain water and solids or when most of those two elements are removed in the settling tank, the centrifuge can be bypassed through a valve **21** connecting the heater **14** with the outlet of pump **18**. In either case, the oil, which still contains distillates and chemicals and perhaps fine solids, is delivered to the upper part of the filtering tank. In the bottom of the tank there is a bed **22** of granular filtering material that will remove chemicals and extremely small foreign particles from the oil. This material is an activated earth, such as fuller's earth, bauxite or alumina, for example.

The filtering material for the bed is supplied from a substantially air-tight reservoir **23** mounted on top of the filtering tank and connected with it through a normally closed valve **24**. For best results the activated earth should be of a size to pass through a 60 to 90 mesh screen. Finer material may plug the filter, while coarser material has a smaller total surface area and therefore requires more time for filtering.

The filtered oil is withdrawn from the bottom of the filtering tank by a pump **26** that pumps it through a heater **27** and into the top of a vacuum still **28**, where the distillates in the oil separate and pass out of an outlet pipe **29** at the top of the still. The clean, reclaimed oil in the bottom of the still is drawn off through a valve **31** that leads to a suitable storage vessel **32**. The vapors that leave the top of the still pass through pipe **29** through a condenser **33** and the resulting liquid distillate falls into a tank **34**, the top of which is connected with a vacuum pump **35** that forms the vacuum in the still.

The main feature of this is the provision for reconditioning or reactivating the spent activated earth in the filter bed so that it can be used over and over, thereby greatly reducing requirements for new filtering material and making it unnecessary to dispose of oil saturated material. Accordingly, suitable means, such as a screw conveyor **37** or the like, is provided for removing spent earth from the bottom of the filtering bed. This material is delivered to a conveyor **38** of any suitable form that will elevate it to a point above the filtering tank and discharge it into the upper end of an inclined rotary incinerator **39**.

In this incinerator carbonaceous material in the spent filtering material is burned out, thereby producing reconditioned or reactivated earth that is discharged from the lower end of the incinerator. This material and the products of combustion enter an afterburner **40**, in which the products of combustion are burned as completely as possible. The hot reactivated earth, now as good as new, leaves the bottom of the afterburner through a conduit **41** and is delivered to the top of the storage reservoir **23** on the top of the filtering tank. Since the activated earth in the reservoir is quite warm or hot, there is no danger of condensation on it, which would deleteriously affect the material. The temperature of the incinerator should be at least as high as about 1000°F. to burn the carbonaceous material out of the spent earth, but for fuller's earth it should not go above about 1100°F. or it will destroy the filtering material. On the other hand, bauxite can be heated to 1600°F. without harming it.

The smoke from the afterburner enters the top of a scrubbing tower 43, where it is scrubbed with water pumped from the lower part of the tower by a pump 44 to remove solids and any chemicals from the smoke. The solids settle to the bottom of the tower and can be drawn off periodically and used for land fill or the like. The smoke, which is now clean air, is drawn out of the tower by an exhaust fan 45 that delivers it to the atmosphere.

Even though the activated earth is reused, some of it is lost in the operation and must be replaced occasionally. This can be done conveniently by providing the side of the vertical conveyor 38 with an opening 47, through which new activated earth can be dumped into the conveyor when required. This material will be heated in the incinerator and will therefore be hot and dry when it enters the storage reservoir.

It will be understood that with the process and apparatus disclosed, waste oil reclamation becomes feasible. The cost of activated earth, which has made earlier attempts at reclamation uneconomical, is greatly reduced because the earth can be reactivated repeatedly and therefore used over and over again. This reuse has the additional great advantage that it no longer is necessary to dispose of spent earth, which was a great problem because of the water pollution factor.

The solids that are removed during the processing are clean and entirely suitable for land fill. Likewise, the removed water is clean enough to be acceptable in sanitary sewers, while the gases that are produced will not contaminate the atmosphere. Although a batch operation has been described, it can be made continuous by using several settling tanks and shifting to another as soon as one is emptied.

A process developed by H.N. Skoglund et al (6) involves the treatment of dyed penetrant wastes used in metal flow detection. The penetrant waste liquor, containing oily penetrant materials, an emulsifier and wash water are introduced into an agitation zone where the waste liquor is combined with an electrolyte emulsion breaker and a clay, the mixture is thoroughly agitated, and then stratified to produce a plurality of layers one of which is essentially an oil free layer of water which can be reused in the washing zone or disposed of without causing pollution.

The penetrant method of flaw detection has been in use since the early 1940's. Basically, the method involves applying an oily penetrant containing either a visible or a fluorescent dye onto the surface of a test piece, removing excess penetrant while leaving the penetrant lodged in any surface flaws which may exist, applying a developer which is typically a particulated material or a suspension of finely divided absorptive particles in a volatile liquid, and then inspecting the part for the presence of flaws.

The penetrant is drawn into the developer by capillary action whereupon the indication is enlarged and thereby rendered more readily visible to the inspector, using white light if a visible dye was used, or ultraviolet light where a fluorescent penetrant was used. Processes of this type are described in U.S. Patent 2,259,400.

Improvements in rendering the penetrants removable by water were effected during the commercialization of this type of process. U.S. Patent 2,405,078 describes penetrant compositions which contain water emulsifiable materials therein so that water alone could be used to effect their removal.

Further improvements were made by De Forest and Parker (U.S. Patent 2,806,959) where they described a more sensitive system where an oily penetrant devoid of emulsifier was employed in combination with a wash water which contained an emulsifier. Using this system, the emulsifier diffused into a superficial portion of the penetrant coating and was thereupon washed away leaving an emulsifier free penetrant trapped in the discontinuities which could bleed out to the surface and be detected by the developer.

Whether the emulsifying agent is included in the penetrant, per se, or added as part of the wash water, a penetrant waste liquor is produced which contains an oil-in-water emulsion of the oily constituents of the penetrant, in combination with emulsifier, water-soluble components of the penetrant and the wash water. Previously it has been customary to discharge this penetrant waste liquor from the washing zone in which the emulsion is formed to a sewage system without making any attempt to recover the water content of the waste liquor or to eliminate its oily content.

This type of disposal is objectionable for two reasons. For one, the presence of the oily materials in the waste liquor is objectionable from the standpoint of pollution of water courses into which the waste liquor is dumped. In addition, large amounts of water are used for washing purposes in large plants and with rising water costs, the water discharged into the sewer may represent a considerable sum.

It has been proposed to treat the effluent from the washing station to break the oil and water emulsion and thereafter separate the oil from the water by sedimentation or froth flotation techniques. However, such treatments are predicated on the assumption that penetrant wastes can be completely broken down into two nonmiscible liquid phases, one water and the other oil. This, however, is not a valid assumption, since most penetrant waste liquors contain organic emulsifiers, such as petroleum sulfonates which are water-soluble, or organic solvents that are water-miscible. Consequently, penetrant waste liquors as previously treated may result in a wastewater component that still contains oil and organic materials as contaminants in sufficient quantity to strain local sewage treatment facilities.

A suitable form of treating apparatus for the conduct of the process is shown in Figure 98. The oil-in-water emulsion, of whatever source is introduced into a treating vessel 20, together with an electrolyte, such as sodium nitrite, which is added by means of a line 21. An adsorbent clay material, such as bentonite, is added to the vessel 20 through a line 22. The resulting contents of the vessel 20 are then thoroughly agitated by means of compressed air introduced near the bottom through a header 23 and through upwardly directed discharge nozzles 24 spaced along the length of the header.

The vessel 20 is also provided with a bottom discharge outlet 25 from which a pump 26 pumps settled material for discharge through piping 27. The vessel 20 is further provided near its top with a swivel drawoff pipe 28, swiveled at its upper end, as at 29, to permit the lower end 36 thereof to be raised or lowered. The drawoff pipe 28 is connected through a pipe 30 to a pump 31 which serves to draw off supernatant liquid from the vessel by means of a pump 31. The oil-free water which is thus withdrawn can be pumped back into the washing station or it can be discharged directly into the sewer since it is quite innocuous.

After the air agitation and consequent mixing of the contents of the vessel, the air is shut off and the contents of the vessel are allowed to remain quiescent for a number of hours, usually overnight or for an equivalent length of time. While in its quiescent state, the mass within the sedimentation vessel stratifies into a lower layer 33 which is a relatively heavy pumpable slurry of adsorbent material and of the material adsorbed thereon, including oily materials and coloring matter, and a relatively thin supernatant layer 34 which may be in the nature of a scum of foreign floatable solids and contaminants brought to the surface with air bubbles from the header 23.

This layer of scum may be easily skimmed off. A main body layer 35 which is of considerable depth, consists largely of water containing water-soluble organic materials from the penetrant referred to herein as water-miscible organic solvents. The water in the layer 35 is of sufficient purity for reuse at the washing station.

### FIGURE 98:  APPARATUS FOR PURIFYING PENETRANT WASTE LIQUORS

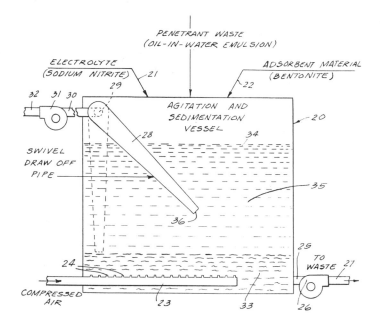

Source:  H.N. Skoglund and A. Mlot-Fijalkowski; U.S. Patent 3,528,284; September 15, 1970

After skimming off the layer **34**, the water containing water-miscible organic solvents is drawn off through the swivel drawoff pipe **28** until a level of the water reaches about that of the level of the slurry, or the free flowing slurry may be pumped out in advance of drawing off the water layer **35**.  The pump **31** serves to draw the water out through the swivel drawoff pipe **28** from the various depths to which the open end **36** of the drawoff pipe is lowered (as indicated by the pipe **28** in dotted lines) into the pipe **32** for further disposition.

The purpose of the electrolyte is to aid in causing the adsorbent material to settle out as completely as possible and leave a main layer **35** of clear liquid.  Sodium nitrite is the preferred electrolyte because it is also a corrosion inhibitor, but any highly ionizable inorganic salt can be used that will aid in causing an effective settling out of the adsorbent material.

As the adsorbent material, it is preferable to use bentonite, but any equivalent oil-adsorbent material having adsorbent properties toward water-immiscible contaminants, such as oil, can be used.  The adsorbent material during air agitation is throughly dispersed and becomes coated with the water-immiscible components and with the coloring matter of the penetrant waste liquor.  Upon settling, there is left a clear intermediate or body layer **35** of water containing water-miscible organic solvents.  The layer **35** has little or no fluorescence in cases where fluorescent dye is used in the penetrant.

As to the concentrations that are employed, sufficient quantities of sodium nitrite and bentonite are added to give a solution of about 2.5 grams of sodium nitrite per liter, and 10 grams of bentonite per liter of the liquid contents of the vessel **20**.  With a penetrant waste liquor containing about 0.1% by weight of total penetrant components, these respective concentrations of sodium nitrite and bentonite result in a separate layer **35** of water and water-miscible organic solvents of sufficient purity so that the resulting water solution can be reused in the washing station, or discharged into water courses.

In a typical operation, after overnight sedimentation, the free-flowing slurry of bentonite contaminated with the oily and color constituents of the penetrant waste liquor is drained or pumped off at the beginning of the day. Layer 35 may be returned to the washing station, together with fresh water equal to the combined volumes of layers 33 and 34, to make up the original volume of the system. The cycle is then repeated.

A process developed by M. Sheikh (7) involves treating soluble oil waste emulsions in an industrial plant effluent with treatment chemicals selected from the group consisting of inorganic salts and bases to split the soluble oil waste emulsions into an oil phase containing the oil and a water phase containing the other pollutants. An emulsifier is added in the process to eliminate the tendency of the oil phase to become sticky from the splitting by the treatment chemicals.

The oil and water phases are separated from each other and thereafter separately treated without discharging oil or the other pollutants into streams, treating the oil phase to render it reusable as soluble oil, concentrating the water phase to render it reusable as the treatment chemicals for treating further amounts of the soluble oil waste emulsions, and repeating the above treatment in a closed cycle in respect to the industrial plant effluent by reusing the concentrated water phase as the treatment chemicals to treat further amounts of soluble oil waste emulsions in the industrial plant effluent.

A process developed by R.A. Willihnganz (8) involves passing an immiscible mixture of oil in water through a knitted polyethylene polypropylene or polyvinyl chloride packing which has an affinity for oil whereby the oil coalesces on the surfaces of the knitted polymer packing. The oil having a lower specific gravity than the water phase and the size of the knit being such that the movement of the oil is not retarded rises along the surfaces of the knitted polymer and is released on the surface of the water phase as a layer which is continuously drawn off.

A specific example in which three or more phases are involved to which this process has particular applicability is the removal of machine oil from a water-based emulsion coolant. An emulsion is generally a two-phase system consisting of two incompletely miscible liquids, one being dispersed and held in stable suspension in the other by the aid of an emulsifying agent whose function is to reduce the interfacial tension between the two phases to increase the ease of formation and to promote the stability of the emulsion.

In a typical water-based emulsion coolant one part of a mixture of about 85% mineral oil and 15% emulsifier is dispersed in about 32 parts of water to form a water-based emulsion having about 3.1 to 3.2% mineral oil stably suspended therein. During a machining operation the coolant can pick up as much as 3.7% machine oil creating a three-phase mixture, two oil phases and a water phase, to about 6.8 to 6.9%. Since the emulsion can only hold 3.1 to 3.2% oil in stable suspension the oil that is picked up is excess and must be separated and removed if the emulsion coolant is to be reused. In passing the immiscible mixture of excess oil in the stable emulsion coolant through the apparatus described, the excess oil coalesces on the surfaces of the knitted polymer packing while the water and that amount of oil which is held in stable suspension passes through the packing substantially unaffected.

In one series of tests the mixture containing about 6.9% oil was passed through the apparatus previously described having a container volume of 7.22 cubic feet at a rate of 1 gallon per minute. The initial rate of oil removal was about 0.25 gallons/cubic foot/hour which rose to about 0.41 after an hour and then slowly decreased reaching a steady-state rate of removal of 0.305 to 0.31 gallons/cubic foot/hour after about 3 hours.

Therefore, in a 24-hour period about 1,440 gallons of a mixture of excess oil in stable emulsion of oil and water can be passed through the apparatus whereby about 54 gallons of excess oil can be separated and collected in that time. Analysis of the water-based emulsion following separation showed only a 0.1% drop in the mineral oil concentration. It is apparent then that this offers an extremely efficient method of selectively removing excess oil from a stable emulsion of oil and water.

A process developed by J.R. Keogh, Jr. (9) is one in which oil-laden metal waste such as turnings, chips and the like are cleaned to remove oil and other impurities from the metal waste enabling recovery of the metal and reclaiming of the oil. The cleaning is achieved through counterflow of a detergent solution in a wash bath, removing the cleaned metal waste from the bath and separating a substantial quantity of the remaining moisture therefrom, and removing the detergent solution from the wash bath and running it through a rehabilitation circuit where it is subjected to centrifugal force to separate sludge and oil from the solution. Then concentrated makeup detergent solution is added, the makeup and restored detergent solution heated and returned from the rehabilitation circuit to the wash bath.

### References

(1) McMahon, M.E.; U.S. Patent 3,297,568; January 10, 1967; assigned to Phillips Petroleum Co.
(2) Baer, E. and Xylander, K.; U.S. Patent 3,347,786; October 17, 1967.
(3) Jockel, J.T.; U.S. Patent 3,414,523; December 3, 1968; assigned to Mobil Oil Corp.
(4) Harsh, W.C.; U.S. Patent 3,446,731; May 27, 1969; assigned to Cleveland Technical Center, Inc.
(5) Wallover, J.I.; U.S. Patent 3,527,696; September 8, 1970; assigned to Wallover Oil Co.
(6) Skoglund, H.N. and Mlot-Fijalkowski, A.; U.S. Patent 3,528,284; September 15, 1970; assigned to Magnaflux Corp.
(7) Sheikn, M.; U.S. Patent 3,595,787; July 27, 1971.
(8) Willihnganz, R.A.; U.S. Patent 3,617,548; November 2, 1971; assigned to General Motors Corp.
(9) Keogh, J.R., Jr.; U.S. Patent 3,639,172; February 1, 1972; assigned to FMC Corp.

## OIL (PETROCHEMICAL WASTE)

The reader is also referred to the sections of this handbook entitled:

> Adipic Acid, Removal from Wastewaters
> Cyclohexane Oxidation Wastes, Removal from Water
> Hydrocarbons, Separation from Plant Wastewaters
> Oxydehydrogenation Process Effluent, Removal of Carbonyl Compounds
>     from Water

for related discussions of petrochemical wastes.

### Removal from Water

A process developed by J.M. Collins (1) involves neutralizing and cleaning an oil contaminated acid solution, such as results from the hydrolysis of a heavy ends stream from a process for the manufacture of ethyl chloride by reaction of ethylene and hydrogen chloride in the presence of aluminum chloride catalyst, by passing the solution downward through a flooded bed containing a carbonate, such as in limestone or clam shells, reacting the acid with the carbonate to form carbon dioxide, and allowing the carbon dioxide to pass upward through the bed.

The oil is stripped from the water and held at the surface of the water as froth for removal. If particulate matter is present in the acid solution, it will be removed with the oil.

### References

(1) Collins, J.M.; U.S. Patent 3,536,617; October 27, 1970; assigned to Ethyl Corp. of Canada Limited, Canada.

## OIL (PRODUCTION WASTE)

**Removal from Water**

Large volumes of water are treated prior to disposal or reuse in the oil industry in crude oil production. Primary oil production produces crude oil with as much as 20 to 50% water. In water flooding operations, commonly used as a secondary recovery method for oil production, oil is produced by injecting seawater into an oil bearing sand to push the oil and water in the sand formation towards adjacent wells where it is pumped to the surface. The oil thus produced is an oil-water mixture containing 60 to 70% water. The oil in such production mixtures is normally separated from the water by physical separation means, such as settling tanks.

The wastewater from which the oil in the mixture is separated, still contains on the order of 50 to 500 ppm oil and must be subjected to further cleanup operations to remove such residual oil therefrom prior to discharge into adjacent bodies of water. The discharge of wastewater into the ocean, for example, is regulated by governmental water quality control agencies which establish quality requirements for the wastewater discharged into the ocean in order to protect the beneficial uses of the ocean.

The wastewater cleanup operation employed conventionally is a flotation process by which air is injected into the water and the mixture pressurized by pumping the oil-water mixture from a surge tank into a flotation cell where the pressure is released, so that as the air bubbles through the water attaches to the suspended oil globules or particles which reduces the effective specific gravity of the particles and causes the oil to move upwards where it can be skimmed off in the flotation cell.

Another requirement for wastewater discharge into the ocean, is a limitation of the amount of dissolved sulfide, which is usually present as hydrogen sulfide either un-ionized or ionized depending upon the pH of the water.

Hydrogen sulfide may be removed from water by several methods, e.g., air or gas stripping, chlorination, precipitation and air oxidation. Gas or air stripping physically removes the hydrogen sulfide from the water and transfers it to the offgas and ultimately to the atmosphere, which is objectionable from an air pollution standpoint. The stripping process also tends to raise the pH of the wastewater which increases the scaling tendency of the water.

The removal of hydrogen sulfide by chlorination is applicable from a practical standpoint, only to water of low sulfide content. Where the amount of hydrogen sulfide to be removed is substantial, the cost of the chlorine gas becomes excessive. Precipitation methods, with zinc chloride, for example, are also limited to wastewaters having relatively small sulfide contents because of cost considerations. Air oxidation processes for removing hydrogen sulfide from wastewaters on an economically feasible basis, require an alkaline solution, since air oxidation of hydrogen sulfide in neutral or acidic wastewaters is normally too slow to be economically feasible.

A process developed by S.M. Verdin (1) involves extinguishing gas and oil well fires, particularly in multiple-well, offshore installations, and preventing loss of oil to the surrounding area after fire extinction. The apparatus used a hood adapted to be placed over the well site, the hood having a curved interior upper wall which deflects the gushing oil into a catch basin from which it can be pumped away.

A process developed by J. Duffy (2) involves removing residual oil and dissolved sulfides from oil production wastewaters prior to disposal of the wastewater into the ocean, by the injection of controlled amounts of air and soluble nickel catalyst into the wastewater. The mixture is pumped under pressure into a flotation cell or tank where the pressure is released and the air effects oil separation and oxidation of the dissolved hydrogen sulfide.

A process developed by L.P. Teague (3) is one in which, in a well drilling operation, drilling cuttings are sequentially carried, together with drilling mud, through a first shaker. This device comprises in essence a screen separator of such a mesh that a major component of the fluid mud will pass through the screen. The more solid drilling cuttings remain on, and are carried to the screen discharge. The substantially mud-free cuttings are then passed through a cleaning cycle where they are washed by contact with a liquid detergent.

The cuttings, together with used detergent, are again separated at a second shaker. The detergent passes through the shaker screen and is recycled for subsequent reuse. The substantially oil-free cuttings are discharged from the shaker to a collector where they are subjected to a final bath comprising relatively clean water, which in this instance can be seawater. The clean, and detergent-free cuttings are then deposited beneath the water's surface where they gravitate to the ocean floor.

In a normal well drilling operation at an offshore site, a pressurized stream of liquefied drilling mud is introduced down the drill string as the latter rotates. The drilling mud functions both as a lubricant and as a vehicle whereby to facilitate the cutting and removal of materials comprising the substratum. This rather heavy effluent stream carried from the well bore usually includes drilling mud, drilling cuttings, seawater and possibly oily constituents picked up from the substratum. As the drilling mud passes upwardly through the annulus defined by the rotating drill string and the bore hole, it acts as a vehicle for sand, clay, stone and other loosened solids which constitute the substratum.

These latter mentioned materials after being separated from the mud, as a matter of practicality, are normally returned to the water where they sink to the ocean floor. However, the cuttings are often coated with oily materials such as crude oil from the well bore, or other nonwater-soluble constituents which make up the drilling mud mixture.

The discharge into the surrounding water of such nonwater-soluble materials, can lead to a water polluting condition. Even the discharge of minor cutting amounts will tend to cause a visible discoloration at the water's surface.

Usually the drilling mud comprises essentially a water based, flowable composition of adequate weight and chemical quality to facilitate operation under a particular set of circumstances. However, the mud is frequently compounded with a lubricant material such as diesel, crude oil, or other nonwater-soluble petroleum base constituent, whereby to facilitate the mud's lubricating characteristics. In either event, the process is applicable toward the removal of petroleum base, water contaminating or polluting matter that would ordinarily adhere to the drilling cuttings and tend to float to the water's surface when the cuttings are discharged overboard.

Figure 99 shows a suitable arrangement of apparatus for the conduct of the process. The well bore cuttings are carried from a well bore **4** during the drilling operation so long as the drilling mud flows. Thus, as drill string **6** is rotatably driven to urge the drill bit **7** downward, liquefied mud is forced under pressure through the drill string **6** to exit at the lower drill bit **7**. The mud thereby lubricates the downhole operation, and in passing upwardly through the annulus **5** between drill string **6** and the well bore wall, carries with it various forms of drilling cuttings as previously mentioned.

Further in regard to the drilling mud, as is generally known, the composition of the mud is usually compounded to the particular drilling situation and condition. More specifically, the weight and the chemical makeup of the mud are initially determined and subsequently altered as needed and as the drilling progresses.

While not shown in great detail, the mud flow is urged under pressure from well bore **5**, upwardly to the drilling deck of the offshore platform, and discharged as an effluent stream by way of line **9**, into a tank **8** that is ancillary to shaker **10**. From tank **8** the mud mixture overflows onto the perforated face **11** of shaker **10**. Shaker **10** as shown, comprises a vibratory or stationary type separator having a tilted screen working surface **11** upon which the mud mixture overflows from tank **8**.

## FIGURE 99: APPARATUS FOR AVOIDING WATER POLLUTION AT AN OFFSHORE DRILLING SITE

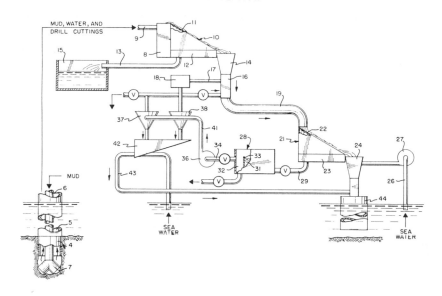

Source: L.P. Teague; U.S. Patent 3,693,733; September 26, 1972

The mesh of the screen utilized on shaker face **11** is variable, being contingent on the characteristics of the substratum being drilled and the type of drilling cuttings being carried by the mud flow.

In shaker **10**, the liquefied mud vehicle will by and large traverse the screen openings and is received in an underpositioned pan **12**. The latter is communicated by a conduit **13**, with a mud storage tank **15**. The remainder of the mixture deposited on shaker face **11**, and which does not pass through the screen, will comprise essentially an aggregate of solids being of sufficient size to remain at the screen surface. Solid matter, through the screen's vibratory action or through gravity flow, advances along the screen face to be discharged at the lower side thereof.

A collector **14** communicated with shaker face **11** receives the stream of drilling cuttings which in essence comprises a conglomerate of solid matter as well as some liquid. This flowable mass further embodies the previously mentioned nonwater-soluble, oily base constituents which normally cling to the cuttings. A wash chamber **16** is communicated with the collector discharge outlet to receive a stream of unprocessed drilling cuttings. Wash or spray chamber **16** includes a compartment adapted to receive the downwardly passing drilling cuttings, with means in the compartment to retain the cuttings sufficiently long to be brought into contact with the liquid detergent.

The wash chamber, in the arrangement under discussion, a spray chamber, further includes a spray nozzle system **17** disposed thereabout and appropriately arranged to deliver detergent streams against the cuttings. Spray nozzle system **17** is communicated with a pressurized liquid detergent source represented by reservoir **18**.

Toward cleaning or scouring the cuttings of oily matter, the cuttings as an alternative can be immersed in a bath rather than being sprayed. The apparatus used in this latter step will be adapted in accordance with the consistency and the volume flow of cuttings as well as with other features of the drilling process. Within spray chamber **16**, detergent is brought into contact with the cuttings under sufficient pressure and/or turbulence to remove substantially all of the extraneous matter clinging thereto. An elongated conduit **19** directs a stream of liquid detergent and drilling cuttings from spray chamber **16** whereby to physically expose the cuttings to the cleaning and separating action. For this use, and toward achieving the necessary scouring and cleaning function, the detergent liquid can include any of a number of commercial solutions as for example, a biodegradable phosphate-free detergent.

Conduit **19** is connected at the discharge end thereof to a second separator **21**. The latter, as in the instance of shaker **10**, is a vibratory unit having a screen-type face **22** across which the detergent and cuttings flow is directed. The mesh size or openings of the screen face **22** are usually smaller than the mesh of screen face **11**, and of a sufficient size to pass the liquid detergent therethrough and into shaker reservoir **23**. The remaining cuttings stream, substantially free of detergent and other liquid, falls from screen top **22** and into a discharge chamber **24**.

The chamber **24** includes a receptacle to receive and retain the flow of cleaned cuttings for further cleaning. The receptacle **24** in the instant arrangement is communicated with seawater drawn from the immediate area by conduit **26** and pump **27**, or from an alternate source of water. Ater further cleaning by contact with seawater, the cleaned cuttings are discharged into a downcomer **44**.

The member comprises in its simplest form an elongated tubular conduit that extends downwardly beneath the water's surface terminating short of the sea floor. Cuttings deposited at the upper end thereby are directed toward the floor where they tend to settle without the concern of prompting a water polluting situation on the surface.

Solution, including detergent separated from the drilling cuttings within separator **21**, is received in reservoir **23**. The latter includes an outlet communicated with the inlet of skimmer tank **28** by a valved connecting line **29**. Skimmer tank **28** embodies a first compartment **31** into which the detergent is fed and into which additional detergent can be added if such addition is required for reconstituting the material.

A second compartment **32** is communicated with the first compartment across a transverse panel **33**. Compartment **32** is provided with an outlet to receive detergent in valved line **34**, which in turn is communicated with the suction of detergent pump **36**. The discharge of the pump **36** is communicated with one or more hydrocyclone units **37** and **38** or similar fluid separating units, by way of line **41**. The function of the units **37** and **38** is to provide a final separation of detergent from any remaining materials in the flow stream.

Hydrocyclone units **37** and **38** function to centrifuge detergent from any remaining mud, water, and/or other fluidized or particulated components. The separated and cleaned fine solids pass upwardly into manifold **42** and are carried by line **43** into conduit **44**. The detergent material, essentially free of solids, discharges into wash chamber **16** to again contact the incoming mud and cuttings flow.

Open ended downcomer conduit **44**, as noted herein is disposed in the body of water, normally depending from the offshore drilling platform. The conduit **44** preferably is positioned with its lower open end spaced from the floor at the offshore location, or provided in the alternate with openings formed about the lower end. The conduit or caisson upper end is open to the atmosphere and disposed in alignment with the discharge opening of cuttings collector **24**. Thus, in the course of the process, substantially clean cuttings are fed into the caisson upper end. The clean cuttings thus enter the water and flow downwardly by gravity through the caisson, to be deposited at the ocean floor.

Use of the method described above for treating drilling cuttings results in a cleaner operation as well as a more economical one. The method serves to maintain a nonpolluting condition at the offshore production or drilling site and also permits maximum recovery of both drilling mud and washing detergent for the subsequent reuse of both items.

References

(1) Verdin, S.M.; U.S. Patent 3,554,290; January 12, 1971.
(2) Duffy, J.; U.S. Patent 3,576,738; April 27, 1971; assigned to The Signal Companies.
(3) Teague, L.P.; U.S. Patent 3,693,733; September 26, 1972; assigned to Texaco, Inc.

# OIL (REFINERY WASTE)

In the operation of a refinery, a petroleum crude oil is processed to form many products. In the course of such operations, large quantities of water are used for many purposes. When used as a purifying medium, the water becomes contaminated with the contaminants removed from the petroleum, and particularly with particles of oil. Formerly, the water used in the refinery for either purification or processing purposes, when contaminated, was disposed of by discharging it into a stream or some other available body of water.

With the increase in the number and size of refineries and the number of processing steps utilizing water in the refining operation, the amount of contaminants has reached such proportions as to pollute streams, lakes or other bodies of water. The water pollution has reached the extent that the fish, marine life and animals in the vicinity suffer and the obnoxious odors of refinery waste materials are a nuisance to the surrounding population.

The matter of disposing of refinery wastewater is a major problem from the standpoint of conservation. In order to prevent the pollution of water, laws and regulations have been passed by the Federal and State governments regulating the discharging of water containing contaminants. These laws and regulations pertain to both salt and fresh water, and compliance has proven to be an expensive problem for oil refiners.

In many areas, processing water is not available in sufficient quantity to permit its disposal after use. In such cases, it is essential to reuse the water rather than to dispose of it and continuously use fresh water. Additionally, in view of the vast quantities of water utilized in the refining processes, large quantities of oil are inadvertently discarded along with the water.

## Removal from Water

A process developed by A.M. Kivari (1) involves separating hydrocarbons from an aqueous waste by pressurizing the waste with a gas so that the gas is dissolved in the waste. The pressure on the waste is thereafter reduced whereby some dissolved gas is freed from solution. The freed gas carries the more volatile hydrocarbons from the waste, and the evolved gas and the volatilized hydrocarbons are separated from the waste.

The pressure on the waste is thereafter further reduced and the waste is held in a substantially quiescent state so that additional gas is freed from solution and aids in floating remaining liquid and solid hydrocarbons in the waste to the surface of the waste. These floating hydrocarbons are then removed from the upper part of the waste. In those aqueous wastes containing settleable solids, the solids are allowed to settle to the bottom of the waste and are continuously removed therefrom.

A process developed by R.J. Austin et al (2) relates to an improved system for treating oil-containing wastewater by an improved bioflotation technique and it pertains more particularly to the treatment of petroleum refinery wastewater so that it may be discharged into lakes or streams without undue contamination thereof.

Petroleum wastewaters from conventional American Petroleum Institute (API) separators contain organic matter and/or organisms which require biochemical oxidation and the biochemical oxygen demand (BOD) of such waste may be of the order of about 30 to 200 or more parts per million (ppm); in lakes and streams the BOD is usually less than 1 ppm and effluent water discharged thereto should have a BOD of less than 15 ppm and preferably less than about 6 ppm. Such secondary wastes contain oil in dissolved or solutized form to the extent of 100 to 200 ppm; it is desirable that the oil content be reduced to not more than about 30 ppm and to as great an extent as possible.

The phenolics content of such waste may range from 1 to 15 ppm or more and it is desirable that the phenolics be reduced to not substantially more than about 0.3 ppm and preferably to a much lower figure. The threshold odor number (TON) of the secondary waste may be of the order of 20,000 to 50,000 or more and it should be reduced to not more than about 6,000 or lower. The extent of required purification will, of course, depend upon each particular situation. The general appearance of the wastewater is also an important factor, it being desirable that the dark colored and almost black waste with an oily caste be converted into a relatively clear water of lighter and brighter color.

The process is preferably applied to the petroleum refinery waste discharged from the usual API separator. The acidity of this wastewater is first adjusted to a level in the pH range of about 6½ to 8½, e.g., about 7 and if necessary, its temperature to a range of about 40° to 100°F., preferably about 90°F. If the wastewater contains excessive amounts of sulfides, particularly $H_2S$, it is desirable to strip out $H_2S$ by passing air upwardly through the water from distributors or diffusers at the base of a preliminary stripping zone; in most cases, however, this stripping step is not required since beaters may accomplish the required stripping.

The pH adjusted stream at the defined temperature is then preferably passed through a preliminary settling zone where it is held for about 0.2 to 4 hours or more to enable removal by sedimentation and flotation of substances which can thus be removed, thereby minimizing the load on the subsequent bioflotation treatment. The bottom sediment accumulates so slowly that it may be periodically removed at intervals of about a year or more but any oil or scum should be skimmed from the top of the water leaving the preliminary settler before it is subjected to the bioflotation treatment.

In the bioflotation treatment, the presettled wastewater at the defined acidity and temperature is passed in series through a plurality of aerobic zones with holding times in each zone in the range of 0.3 to 3 hours, preferably about 1 hour, and with an average flow rate through the zones in the range of 0.1 to 10 feet per minute, preferably 0.5 to 5 or 1 to 2 feet per minute. Beaters are installed at the inlet surface of each aerobic zone for beating or triturating air into the water whereby a myriad of fine bubbles are formed, dispersed and impelled into the water so that it assumes a somewhat milky appearance.

Downwardly, inclined baffles are mounted adjacent the downstream side of the beaters, preferably at an angle of 20° to 50°, the upper edge of the baffle being above or at the approximate level of the upper part of the brush and the lower end of the baffle extending about 2 inches to a foot or more beneath the surface of the water. The purpose of the baffle is to improve aeration efficiency (note Pasveer's "Research on Activated Sludge," *Sewage and Industrial Wastes,* November 1953, pp. 1253-1258, and December 1953, pp. 1397-1404) but it serves an added function of downwardly directing the liquid containing entrained air bubbles and thus substantially limiting any turbulence to the inlet end of each aerobic zone so that the liquid in the downstream portion of each zone may remain relatively quiescent. The combined action of the beater and the baffle causes the gas bubbles to be directed to the bottom of the tank so that oxidation and subsequent flotation is effected throughout substantially the entire mass of the liquid.

After the initial biological growth period which may require several weeks, particles which may be called sludge but are better described as slime appear and grow, a relatively small part tending to be internally recycled at the inlet end of each aerobic zone and a larger part accumulating in bulk in the downstream portion of the zone. This biological slime functions in a manner analogous to sludge in accelerating both chemical and biological oxidation so that most of the oxygen content of the entrained bubbles is effectively utilized. The stone particles apparently provide nuclei or absorptive floc for coalescing oil and entrained organic matter and at the same time they tend to occlude the nitrogen and unused portion of the oxygen, the occluded gas tending to make the particles sufficiently buoyant so that they float to the surface of the liquid in the downstream quiescent section of the aerobic zones.

For best results it is essential to provide a holding period of at least about 0.3 hour in order to effectively utilize the entrained oxygen and it is also essential to remove the slime which is floated to the surface because if this slime becomes deaerated it will sink back into the liquid and/or accumulate on the bottom of the zone, thereby decreasing the effectiveness of the treatment.

By limiting the holding period to approximately 1 hour, one can effectively utilize the oxygen content and at the same time recover substantially all of the resulting oily slime from the surface of the water by a skimmer. The technique thus defined enables a large part of the impurities to be removed without necessity of biological or chemical oxidation thus enormously minimizing the overall oxygen requirements of the treating system.

The number of aerobic zones employed will depend on the nature of the wastewater and the requirements for final effluent purity. It is preferred to employ at least three zones in series and in one commercial design five such zones are utilized. Each zone is preferably from 15 to 300, preferably about 50 to 200 feet in length with a water depth of 3 to 10 feet and with a width to provide the beater length necessary for incorporating the required amount of oxygen and flotation gas in the zone, 1 foot of beater length usually being adequate to incorporate 6 to 10 lbs./day of oxygen and 24 to 40 lbs. of inert gases into the water. The inlet to each zone is preferably from near the bottom of the preceding zone after slime is skimmed therefrom.

Total beater length which may be divided up into a plurality of stages is about 1 foot for each 5 to 100 gallons per minute of wastewater to be treated depending upon the extent of treatment required.

A process developed by J.W. Fella (3) involves removing soluble and entrained insoluble oil from oil-laden water obtained during a delayed coking operation and in particular it relates to a method which permits the use of the oil-laden water in the hydraulic decoking cycle of the coking operation. In addition, it relates to a method of reducing the oil content of wastewater to a level which permits the water to be easily disposed of in the refinery effluent water.

Oil-laden wastewater presents a real problem in the disposal of wastewater from a refinery which is either located on inland waters where such waters are used for drinking purposes or located on inland bays or river outlets where oysters and other fish foods are grown for commercial purposes and thus are protected by fish and game laws. In addition, at a number of refineries the lack of adequate fresh water supply has become a problem. As industries continue to use increasing amounts of fresh water, all means of conserving the fresh water by reemploying usable contaminated water in refinery operations must be observed.

Previously, the delayed coking operation has resulted in a substantial amount of oil-laden wastewater which cannot be reused and must be disposed of with the refinery effluent wastewater. The water used in the hydraulic decoking cycle of the coking operation must have a very low oil content since any significant amount of oil will cause emulsion of the high pressure centrifugal pumps used for pumping the water under pressure to the jets during the hydraulic decoking.

The emulsion substantially reduces the effectiveness of hydraulic decoking. Previously water used in the delayed coking operation has been disposed of after becoming oil-laden.

In the process substantial savings of fresh water makeup can be obtained in the delayed coking operation by a technique that removes the oil from oil-laden water which was formerly discharged from the operation. This technique is to use the oil-laden water as cooling water during the hot coke bed cooling cycle. When the oil-laden water is withdrawn from the hot coke bed, the water contains less than 100 ppm of oil. At this oil concentration the water is suitable for use in the hydraulic decoking cycle without causing emulsion.

Figure 100 is a flow sheet illustrating the operation of the process. The drawing shows a partial view of the delayed coking unit. Means for heating and introducing the oil to the coking drum and means for carrying out the coking cycle are not shown. Neither are means shown for hydraulic decoking of the coke drum and the removal of the discharged coke.

After coking drum 11 has been filled with coke during the coking cycle of the delayed coking operation, the temperature of the bed is about 900°F. and must be cooled to below 212°F. before the hydraulic decoking operation can proceed. The cooling of the hot coke bed is effected by introducing water fron tank 12 through lines 13 and 14 and valve 15 by pump 16. The water entering coke drum 11 is vaporized by the hot coke and water vapors laden with oil are withdrawn through line 17 and passed to blow down drum 18 where the oil and water vapors are condensed.

The introduction of water into the hot coke bed is continued until the temperature of the coke in coke drum 11 is less than 212°F., at which time the interstices in the coke drum are filled with water. After this the water is withdrawn through line 19 and valve 20 to tank 12 where it can be reused for coke bed cooling or during the hydraulic decoking cycle.

FIGURE 100: DIAGRAM OF PROCESS FOR USE OF OIL-LADEN WATER TO COOL HOT COKE

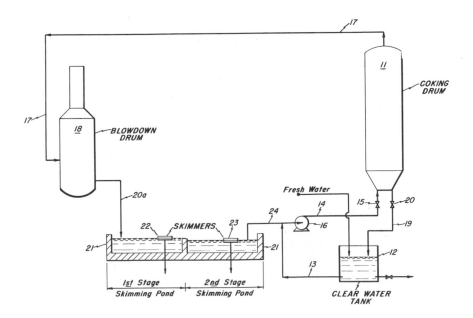

Source: J.W. Fella; U.S. Patent 3,146,185; August 25, 1964

The condensed water and oil vapors in blowdown drum **18** are passed by line **20a** to the first stage of skimming pond **21** where the bulk of the oil is removed from the surface by skimmer **22**. The oil then passes into the second stage of the skimming pond **21** where further oil is removed by skimmer **23**. Even after more than a reasonable amount of settling and skimming, and skimmed water from the second stage of the skimming pond contains up to about 2,000 ppm of oil.

Since water containing this amount of oil will emulsify in high pressure centrifugal pumps and in the coke bed during the hydraulic decoking, previously this water has not been used further in the coking process. Consequently, the water separated in the second stage of the skimming pond was passed to the API separator for further processing before discharge to the refinery wastewater effluent.

Substantial amounts of water were introduced and discharged during a complete cycle of the delayed coking operation. If the oil-laden water from the second stage of skimming pond **21** is passed by line **24** to the suction side of pump **16** and introduced to coking drum **11** during the hot coke cooling cycle, the water withdrawn from the cooled coke in coking drum **11** to tank **12** contains less than 100 ppm of oil.

Thus the oil-laden water which has previously been discharged from the delayed coking unit is freed of the bulk of the soluble and entrained insoluble oil and conserved for further use in the delayed coking operation. Any emulsion that may be formed when using the oil-laden water in the cooling step will not interfere with the cooling of the bed and will be broken when brought into contact with the hot coke.

It can be seen that a unique method of removing oil from water is achieved. Contrary to a filtering step where a bed of absorbent material would have to be activated and discarded or regenerated after saturation, the oil removed from the water by the hot coke during the cooling cycle adheres to the coke and is removed along with the coke during the decoking cycle of the coking operation. In effect, with each new bed of hot coke a new water-treating bed is available. Contrary to what might have been expected, the quality of the coke obtained from the coking unit was not altered.

In the operation of a refinery on an East Coast inland bay which contains oysters, clams and fishes protected by fish and game laws, it has been necessary to use elaborate means to to assure that the wastewater being discharged to the bay is of sufficiently low oil content so as not to impart odor or taste to the fish. As the refinery expanded, means for conserving fresh water supply and reducing the load on the wastewater cleanup plant were also needed.

The practice of the process resulted in a solution to this problem. Skimmed water separated from the second stage of the skimming pond contained about 1,600 ppm of oil. This water was introduced into a hot coke bed of a delayed coking unit during a cooling step and used to cool a bed of coke about 50 feet high. The water which was withdrawn from the cooled bed of coke contained less than 100 ppm of oil and was suitable for use in a subsequent cooling cycle or for a hydraulic decoking cycle.

A process developed by N.H. Wolf (4) utilizes an apparatus for removing oil particles from water comprising an absorption tower containing a diffusing medium, a solvent having a higher specific gravity than water and being insoluble in water, and a porous material which is permeable to water and impermeable to oil and solvent. A method for removing oil particles from water comprising the steps of passing water containing oil particles through a particulate diffusing medium containing particles such as sand, gravel or Raschig rings, and high specific-gravity solvent, the oil particles adhering to the solvent by a process of absorption, and the oil-free water discharging through a water-permeable porous mass.

Electrically operated sensing means is located near the discharge of the oil-free water to verify that adequate separation is achieved and to guarantee efficient recovery of the solvent.

A process developed by B.R. Carlstedt (5) involves the separation of oil from an oil-water mixture comprising introducing an oil-water mixture to a surface region of a layer of grainy oil resistant material housed in a tank, flowing an aqueous phase substantially vertically down through the layer of the material, maintaining a ground water zero-pressure level in the layer and finally collecting a substantially pure oil phase at a place in the layer situated at a distance from the region where the mixture has been introduced.

Figure 101 shows a suitable form of apparatus for the conduct of the process. The drawing shows a cross section of an elongated tank 1 containing a bed which consists of an upper stratum of gravel, an intermediate stratum of sand, and a thin bottom stratum of gravel. The bottom stratum is drained by communication with a water-collecting basin or well 3 having an overflow 5. At one of the longitudinal sides of the tank close to the tank wall an oil-collecting channel 2 is arranged in the upper gravel stratum. This channel has a wall in contact with the gravel layer and provided with vertical slots, and moreover has an overflow 6 at a level somewhat higher than the level of the water overflow 5.

**FIGURE 101: APPARATUS FOR THE SEPARATION OF OIL FROM AN OIL–WATER MIXTURE**

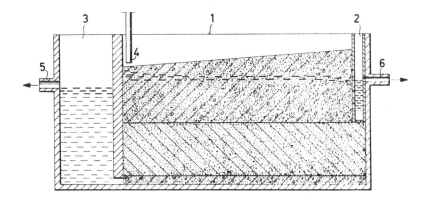

Source:  B.R. Carlstedt; U.S. Patent 3,574,096; April 6, 1971

The apparatus is operated in the following way. The oil-water mixture to be separated is introduced into the surface of the bed through the conduit 4, the flow of the mixture being maintained calm and continuous so as to prevent this oil-water mixture from spreading too much laterally. The bed surface, therefore, is somewhat inclined down towards the region where the mixture is introduced.

In the capillary zone in the gravel stratum above the ground water level the oil phase then spreads laterally towards the oil-collecting channel, while the aqueous phase is flowing straight through the gravel-sand-gravel strata and is finally discharged via the overflow 5. The height of the ground-water level as defined by the overflow 5 is determined mainly by the oil-water mixture added and by the flow resistance of the water in the gravel and sand strata.

Oil that has moved laterally towards the oil-collecting channel will enter the channel through the vertical slots and collect on the liquid surface in the channel, to thus form a layer increasing in thickness. When the top of the oil layer reaches the overflow 6, oil will flow over into a separate collecting vessel (not shown). Laboratory tests for demonstrating the oil-separating effect have been performed with an apparatus of this type where the bed had a surface of 200 cm.² (5 x 40 cm.) and consisted of one 35 cm. top statum of gravel (grain size 3 to 8 mm.), one 30 cm. intermediate stratum of sand (grain size 0.5 to 1 mm.) and one 5 cm. bottom stratum of the same gravel as in the upper stratum. The difference in height between the oil overflow and water overflow was 2 to 4 cm.

The oil-water mixture that was tested had a content of 37.5 cc of colored fuel oil per liter of water. Samples of the discharged water which were taken after testing through the water storage basin of water volume corresponding to 3 to 5 times the volume of the basin showed a content of 20 mg. of fuel oil in dissolved state per liter of water, i.e., a separation of 99.95% of the oil from the water had been achieved. In this connection it should also be pointed out that the fuel oil employed in these test-runs had a maximum solubility in water amounting to 25 mg./l.

A process developed by D.K. Beavon (6) involves treating oily water to remove oils and particulate matter therefrom.

Figure 102 is a flow diagram of the process. As shown there, the first phase of the process involves separation of solids, normally oil-wet solids, from an oil-water mixture. In this filter phase the oil-water system is passed through a filter media **10** which will retain the solids contained in the oil-water system. For a typical oil-water system the filter media **10** is normally contained in a pressure vessel **12** to allow forced flow of the oil-water system therethrough. Although forced flow is desirable under modern processing conditions, it is to be understood that gravitation separation is also possible.

## FIGURE 102: FILTRATION SCHEME FOR SEPARATION OF OIL FROM WATER

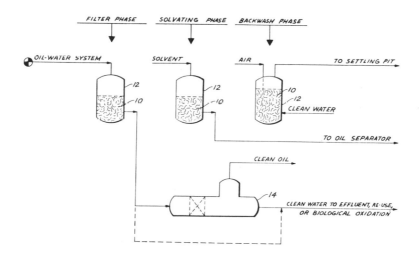

Source: D.K. Beavon; U.S. Patent 3,574,329; April 13, 1971

The effluent from the oil-water system filtration phase is either clarified water of a mixture of water and oil which will readily separate into a water phase and oil phase in separator **14**. The oil is withdrawn from the top as clean oil and water withdrawn from the base and passed directly to effluent or a biological oxidation system.

The oil-water system treated in the filter phase may be obtained directly from any process source, such as petroleum refinery streams, oil purification streams, vegetable and oil processing streams, rolling mill streams from steel mills and like processes where there is formed an oil-water system containing particulate solids, usually oil-wet solids. The solids may vary from ordinary earth residues to more sophisticated salts, corrosive entrainments and processing by-products. Being generally oil-wet, they retard fine separation of oil from water.

Filtration removes them and allows water in an oil-water system to be readily reconstituted for reuse or disposal without fear of pollution. As indicated, either clarified water or an oil-water mixture free of solids is the effluent from this phase. If a mixture of oil and water is obtained, it will separate into sharply defined water and oil layers as the solids, which impede separation, have been effectively removed.

The nature of the filter media 10 is not narrowly critical. Generally, it is composed of particles of a size which will yield a bed destined to retain the particulate solids, resolve emulsions and coalesce oil which allowing water, with or without oil, to flow therethrough. For most oil-water systems a filter bed of a media having a particle size from 0.1 to 5 mm. in average diameter may be expected to adequately entrain essentially all of the particulate solid matter contained in the process stream.

Finer sized filter media may be used for extremely fine particulate matter in a process stream. The filter media particle size may be conveniently varied by building layers of increasing particle size to generally provide, in a direction opposed or concurrent with the flow of the oil-water system, to promote gradient filtration.

The nature of the media which comprises the filter bed is not narrowly critical, although it should be resistant to the solvating action of the oil extract and used in the subsequent filter bed regeneration. Examples of media, which may be used are sand, particulate or crushed coal, finely divided polymeric materials, finely divided carbon, such as coke, glass beads, sintered metal and the like, of a size which will provide a free path for the flow of water, with or without oil, while resisting the flow of particulate matter contained in the oil-water system. The use of sand or a combination of sand and particulate coal is preferred, with a combination of sand and particulate coal particularly preferred for treatment of a highly loaded system as it will naturally separate into coal and sand layers which offer a combination of coarse and fine filter medias.

There will, of course, be some resistance to the flow which will increase with particulate solids entrainment. There will therefore be some pressure differential between outlet pressure and inlet pressure. For a newly formed or regenerated bed it has been observed that substantial retention of particulate solids contained in an oil-water system may be achieved at a pressure drop through the filter bed as low as about 1 psig, but that with operation, pressure, which must be applied to maintain desirable transport, will increase with bed loading. Generally, for efficient operation throughput of water or oil and water should be in the order of 2 to 10 gpm/ft.² of bed surface area and in the instance of downward flow as aided by gravity preferably from 2 to 5 gpm/ft.² of bed surface area.

Actual pressure of operation will depend on the nature of the oil-water system being processed with particular emphasis on the end use of the water, as this determines the net lower operating pressure for the system. Normally, however, when pressure drop through the filter media exceeds about 20 psi, it has been observed that the bed has reached maximum solids entrainment and must be regenerated for reuse.

With reference to the oil solvating phase of the drawing, this is accomplished generally by discontinuing input, allowing free fluids to drain, then stripping entrained oil from the filter, without disrupting entrained solids, using a stripping media, normally steam or an organic solvent, for the oil. This is accomplished by duplicating the direction flow of the oily process water. Where steam, the preferred oil extractant, is used, oil will be removed without disturbing the collected solids by the general combined actions involving steam distillation of the more volatile oils and steam heating of oils to reduce viscosity and thereby cause flow and preferential displacement of oil. The solids remaining are oil-free and water wet.

Steaming is preferably carried out at temperatures of from 212° to 350°F. at as low a pressure as can be convenient to the system.

When an organic solvent for the oil is used, it should not display a solvating action towards the entrained solids and should have a sufficiently high vapor pressure to allow the bed to free itself of vapors prior to removal of oil-free particulate solids. Normal petroleum distillates, such as petroleum naphtha, may be effectively used to remove oil without disturbing solids entrained in the bed. As with steam stripping, the oil solvent is passed through the filter media in the same direction of flow as the oily water.

When the bed has been stripped of oil, the entrained oil-free solids are removed by backwashing with water. With reference to the backwash phase of the drawing, this is accomplished generally by passing water through the filter media in a direction countercurrent to the normal flow of the oil-water system. As illustrated in the drawing, this is shown as an upward flow accompanied by aeration for the situation when the oil-water system has been passed downwardly through the filter bed.

The rate of water flow is not narrowly critical but should be sufficient to agitate the filter media particles and release therefrom the oil-free particulate solids. Generally, water flow may vary from 3 to 15 gpm/ft.$^2$ of bed surface area. As it is important to clean the filter bed thoroughly, the bed is preferably agitated with air while flooding with water, either before or during backwashing. The resulting mixture of backwash water and solids, and air, if present, are then removed to a settling pond where solids are recovered by sedimentation. Clean sedimentation occurs because oil is not present. After backwashing, the filter bed is ready for reuse to further filter an oil-water system.

Since the sequence of filtering, steaming and backwashing may be repeated indefinitely, and since it may be desired to operate the filter continuously, it is preferred to operate two or more filters in parallel, one or more filters being used to filter oily solids from the oil-water process stream while one or more filters are being regenerated.

The recovered water, either directly from the filter or after subsequent gravity separation, is normally returned to the system for reuse or oxidized biologically in a surface aerated pond before disposal to waste.

It is also within the scope of the process to aid filtering of oily particulate matter from an oil-water system by the addition of a coagulating agent to the oil-water system. Coagulating agents are those which serve to agglomerate oil particles and oily solids and are normally polyelectrolytes, such as acrylamide polymers. Because the filter media, however, serves a prime function for solids retention, the amount of coagulating agent required will be substantially less than employed in prior art systems.

The particular advantages offered by the practice of the process are water clarification with reduction of entrained oil to values of 10 ppm or less and essentially complete removal of solids. The practice of this process also affords production of dry oil, free of emulsions, and the production of solids substantially free of oil content suitable for disposal without further processing. The system may be further constructed and operated at costs substantially less than the cost of constructing and operating prior separation systems.

It has been found that, using a sand media as a filter and filtering in a downward direction to a bed, as illustrated in the drawing, clear water, containing less than 10 ppm free oil and essentially no solids, has been consistently obtained from an oily feed of a crude contaminated brine from a crude oil desalter containing oily condensates from a petroleum refinery.

The filtered water, when oxidized biologically in a surface aerated pond, was found to support fish life without difficulty. When the sand media was steamed and backwashed, the solids obtained were essentially free of oil and settled rapidly in water for disposal by burial without causing pollution.

A process developed by J.R. Bilhartz et al (7) involves treating waste materials containing oil, water, oil and water emulsions, and oil-coated solids, particularly slop oil from an integrated petroleum refinery, to separate oil, water and solids which can be utilized or disposed of without environmental contamination. The waste material is subjected to ultrasonic treatment at subcavitation power levels and permitted to settle, for example at a power level of about 2 to 10 w./bbl. of oil and with hourly alternate treatment for 5 to 30 minutes and settling for 30 to 55-minute periods for an 8-hour treating cycle, followed by a 16-hour settling period, to provide a 24-hour total time cycle.

A clarified oil phase is recovered as an upper phase; a lower phase is removed and subjected to ultrasonic treatment at cavitation power levels, for example, in a continuous flow and at a power level of 1 to 10 kwh/bbl. of fluid treated. The cavitated product is then separated to recover an upper free oil phase and a lower water and solids phase; and the water and solids are then separated for use and/or disposal.

The lower phase of sonically treated products from either the subcavitation treatment or the cavitation treatment or both may be further separated to recover an intermediate emulsion phase and this emulsion phase may be recycled to the subcavitation treatment while the remaining water and solids phase is subjected to cavitation treatment.

Alternatively, an intermediate emulsion phase from the first subcavitation treatment may be separated and subjected to a second subcavitation treatment prior to subjecting the bottoms product to the cavitation treatment. The water and solids from the process may be separated by filtering or centrifuging and, if desired, the solids material recovered may be washed with a solvent to remove any residual oil therefrom.

A process developed by J.F. Grutsch et al (8) is one in which contaminants in refinery wastewater are removed by passing the wastewater through a biological treating zone, adding a coagulating agent to the effluent from this treating zone to coagulate into solids contaminants remaining in the effluent, and then separating the solids from the water. Separation can be achieved by either air flotation and/or filtration. Preferably, the effluent from the biological treating zone contains unflocculated biological solids which reinforce the effect of the coagulating agent.

A process developed by L.C. Waterman (9) involves the clarification of oil-contaminated water from a desalter or other source, the water being clarified by gas flotation effected in a closed vessel having an upper inclined wall guiding the oily waste material to a collection zone of the vessel without the use of mechanical surface skimmers. The flotation is usually effected at a pressure other than atmospheric. The oily waste material and the separated gas may be further treated for recovery of valuable products, as by being returned to the desalter or to refinery equipment.

As shown in the flow diagram in Figure 103, a salty crude oil of relatively low water content is pressured by a pump 10 and delivered to a treater, exemplified as a desalter 12, through a pipe 13 that may incorporate a mixer 14 of any suitable type. If desired, water may be delivered to the pipe 13 by a pump 15. A demulsifying chemical may likewise be delivered to the pipe 13 by a pump 16, all as well-known in the art of dehydrating and/or desalting crude oils. The desalter 13 separates the mixture into a desalted crude oil, exiting through a pipe 18, and an oil-contaminated water that is bled from the desalter through a pipe 19 equipped with a valve 20.

Following known practices, the desalted oil is pressured by a pump 21 and enters the refinery equipment 22 through a pipe 23. Various petroleum fractions or products are produced in the refinery equipment 22 and are removed therefrom through product lines a, b, c, d, etc., all with the production of a waste gas. Such gas, usually along with waste gases from other equipment, is pumped by a pump 24 through a pipe 25 and a valve 26 to a refinery gas-recovery system 27 for recovery of valuable components from the gas.

Oil-contaminated water withdrawn from the bottom of the desalter 12 through the valve

20 enters a pump 31 for delivery through a pipe 32 and a pressure-reduction valve 33 to the clarifier of the process, indicated generally by the numeral 35. Alternatively or in addition, an oil-contaminated water from any other source may be delivered to the pump 31 through a valve 36 for treatment and separation in the clarifier 35.

FIGURE 103: APPARATUS FOR CLARIFYING OIL-CONTAMINATED WATER BY
FLOTATION IN A CLOSED SYSTEM

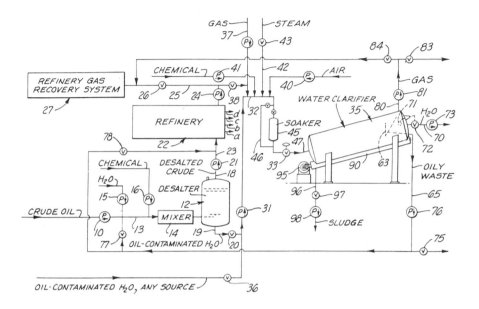

Source: L.C. Waterman; U.S. Patent 3,625,882; December 7, 1971

It is desirable that the oil-contaminated water should have dispersed gas therein at the time of release into the clarifier 35. Some or all of this gas may be present in dissolved form in the oil-contaminated water from the desalter 12 if the latter is operating under significant superatmospheric pressure. At least a portion of this dissolved gas is converted to gas in a dispersed state upon pressure reduction in the valve 33.

Usually however, the process operates best when such gas is augmented by supplementary gas from other sources. In this connection gas from any source 37 may be pressured by a pump if necessary and introduced into the pipe 32. It is often desirable to use $CO_2$ as such a gas in view of the high solubility thereof in water, but any fixed or stable gas can be used.

In many instances the supplementary gas may desirably be a refinery gas. For example, a portion of the gas pressured by the pump 24 and flow-controlled by a valve 38 can be forced into the pipe 32 to mingle with the oil-contaminated water therein. Likewise atmospheric air may be pressured by a pump 40 and delivered to the pipe 32 as a substitute for or in addition to the gas supply from the sources 27 or 37. Some or all of any such supplementary gas delivered under pressure to the pipe 32 may dissolve in the liquids therein at the existing pressure but usually some of the supplementary gas will exist in entrained state beyond the point of its introduction.

A small amount of a chemical agent can be metered into the pipe 32 through a pump 41.

Chemicals acting to destabilize the oil droplets dispersed in the water or that will facilitate phase separation by gravity are well-known in the art. Likewise steam may be introduced into the pipe **32** through a steam line **42** as controlled by a valve **43**. Such steam may be used to heat the oil-contaminated water advancing to the clarifier **35** or destabilize the dispersed oil droplets therein. The condensate will join the aqueous phase of the advancing stream.

The steam or supplementary gas supplied to the pipe **32** under pressure will readily mingle with the oil-contaminated water therein as a result of turbulence in the pipe **32** resulting from flow therein. No special mixing device is required but such can be incorporated in the pipe **32** if desired. However, it is often desirable to subject the oil-contaminated water and its dissolved or entrained gas to a soaking period before the stream enters the clarifier **35**. For this purpose a soaker **45** can be interposed in the pipe **32** and may represent a pressure-type vessel, usually upright, providing a residence time of 1 to 5 minutes or so. If a soaking period is not desired the soaker **45** may be bypassed by the line **46** and the valving shown.

As the stream passes through the valve **33** its pressure is significantly reduced and much or all of the dissolved gas is liberated in a pipe **47** that extends into the clarifier and terminates in some type of spreader or distributor.

Separated water, now substantially free of dispersed oil, can be withdrawn from the body of water through a pipe **70**. The interface can be maintained at a constant level by a level-control float **71** operatively connected to a valve **72** in the pipe **70** controlling this valve in a way to maintain the interface at a constant level. In instances where a subatmospheric pressure is maintained in the vessel, flow of the clarified water through the pipe **70** is induced by a pump **73**.

The oily waste material exiting through the pipe **65** may be discarded to waste through a valve **75** and a pump **76** may be used to induce this flow if the pressure in the vessel **50** is subatmospheric. Preferably however this oily waste material is further processed to recover valuable components therefrom. By opening a valve **77** all or a portion of this oily waste material can be pumped into the pipe **13** and mixed with the crude oil before the latter reaches the desalter **12**. In the desalter the oily waste material can be largely separated into an oil component which will exit through pipe **18** with the desalted crude oil and a water component which will exit through the pipe **19** as a part of the oil-contaminated water reaching the clarifier **35**.

The separated gas is withdrawn from the gas zone through a pipe **80** with the flow being assisted by an aspirator or pump **81** if the interior of the vessel **50** is at subatmospheric pressure. This gas can be discharged to waste by opening a valve **83**, but is desirably returned to the refinery gas-recovery system **27** by opening a valve **84** which conducts the gas to the pipe **25**. This procedure permits recovery of valuable components from the gas and avoids any discharge of the gas into the atmosphere. Atmospheric pollution is eliminated by the desirable recycle of the gas to the refinery gas-recovery system **27**.

Some oil-contaminated waters from one or more desalters or from other sources carry heavier materials that do not readily associate with and be carried upward by the gas particles rising in the body of water. These heavier materials may be solids or heavier sludge-forming materials that settle in the water zone to the bottom of the vessel. The settled material is usually in the form of a sludge that accumulates in the bottom of the vessel with little or no tendency to flow along the bottom wall thereof toward the lowest portion of the vessel. This sludge can be periodically washed from the vessel during shutdown of the equipment.

In those instances where continuous or intermittent removal thereof is desired without shutdown the lower wall of the vessel may be dished downward to form a longitudinal trough **90**. Positioned there is a helically bladed screw which is turned by a motor **95** to advance the sludge along the trough **90** toward a pipe **96** through which the sludge exits from the vessel. The flow of the sludge can be controlled by a valve **97**, and the flow may be assisted by a pump **98**, particularly when subatmospheric pressure conditions are maintained in the vessel **50**.

A process developed by J.F. Grutsch et al (10) is one in which wastewater, particularly refinery wastewater, is treated in a series of lagoons in which thrive microorganisms that feed on contaminants in the wastewater. The system is characterized by mobile aerator apparatus moving across the lagoons' surfaces. These apparatus introduce air into the lagoons and churn surface water, and they have means which direct at least some of the churning surface water towards the lagoons' bottoms. The underwater turbulence scours the lagoons' bottoms and provides a way of controlling sludge deposits.

### References

(1) Kivari, A.M.; U.S. Patent 2,876,863; March 10, 1959; assigned to Process Engineers, Inc.
(2) Austin, R.J., Grutsch, J.F. and Mallatt, R.C.; U.S. Patent 2,948,677; August 9, 1960; assigned to Standard Oil Co.
(3) Fella, J.W.; U.S. Patent 3,146,185; August 25, 1964; assigned to Standard Oil Co.
(4) Wolf, N.H.; U.S. Patent 3,554,906; January 12, 1971.
(5) Carlstedt, B.R.; U.S. Patent 3,574,096; April 6, 1971.
(6) Beavon, D.K.; U.S. Patent 3,574,329; April 13, 1971.
(7) Bilhartz, J.R. and Nellis, A.G., Jr.; U.S. Patent 3,594,314; July 20, 1971.
(8) Grutsch, J.R. and Mallatt, R.C.; U.S. Patent 3,617,539; November 2, 1971; assigned to Standard Oil Co.
(9) Waterman, L.C.; U.S. Patent 3,625,882; December 7, 1971.
(10) Grutsch, J.F. and Mallatt, R.C.; U.S. Patent 3,675,779; July 11, 1972; assigned to Standard Oil Co.

## OIL (TRANSPORT SPILLS)

Spills sometimes occur when oil is transferred from tankers to ships for refueling or other purposes, when it is discharged from tankers into refinery storage tanks on shore, when valves are opened or transfer lines break accidentally during these operations, and when there are leaks from tanks on ships or ashore. Water contaminated with petroleum or fuel oil may initially form an oil slick of approximately three to four inches in depth which if not promptly removed spreads rapidly in the course of three or four days over the surface of the water to approximately 0.01 inch in depth.

In the past, removal of this oil has been attempted by the application of oil absorbing materials such as hay or straw to the oil surface while traveling through the slick in boats and distributing the hay over the entire surface of the slick. The oil adheres to the floating straw or hay, and the hay is collected by suitable means such as pitchforks and removed from the area. Such a method is not at all to be desired because it is inefficient, costly, time consuming and requires large amounts of the hay where the slick covers a large area. Also, it does not stop the slick from spreading during the removal operation.

Surface active agents have also been employed to reduce surface tension over the oil slick area, either the anionic types such as alkyl aryl sulfonates, and other sulfonate soaps or any of the numerous nonionic surfactants known to the trade, in an attempt to disperse the oil into the water so that it disappears from the surface. This method does not result in actual removal of the oil, but merely breaks it up in finer particles, contaminating the water and thus creating a serious threat to marine fauna and flora. Also, such method is rather costly as far as the materials used are concerned.

### Removal from Water

The control of oil spills has been reviewed by workers at Texas A & M University (1). After consideration of preventive methods, containment is considered and pneumatic and mechanical barriers are discussed. Finally, removal is described as performed by: [1] mechanical skimmers, [2] use of vortex principle, [3] absorption processes and [4] chemical treatment using: (a) dispersants, (b) floating absorbents, (c) sinking agents, (d) gelling agents and (e) burning agents.

*Treatment on Shipboard:* A process developed by A.W. Kingsbury et al (2) utilizes a special form of apparatus for separating viscous oil from ships' ballast water. As the fuel oil tanks on sea going vessels become empty, they are customarily filled with seawater as ballast. The oil remaining in the tanks mixes with the seawater and is later discharged overboard during a deballasting operation prior to refueling. In order that seas and harbors not be contaminated with oil, it is desirable that the oil content of the ballast water be reduced before deballasting.

Many ships burn Bunker "C" fuel oil, which is difficult to separate from ballast water. Bunker "C" is the name commonly given to No. 6 fuel oil as defined by the Bureau of Standards in Commercial Standard CS 12–48. The specific gravity of Bunker "C" fuel oil is not constant and varies from 0.93 to 1.06. Thus the Bunker "C" may be heavier, lighter or the same density as water and so is not consistently separable by gravimetric methods. In addition, density of the water itself may vary by reason of temperature variation or salinity differences.

Knitted metallic mesh fabric or fine mesh screens have been used for separating oils free from suspended matter from clear water. The fuel oil used in ships may contain residues, and in addition the water used for ballast purposes may be picked up from rivers in which the water contains dirt and other suspended particles. A major problem in the separation of oil from water by wire mesh screens has been the difficulty in cleaning the screens. A deficiency of the wire mesh screens has been that they are not satisfactory in coalescing high viscosity oils. Bunker "C" being a high viscosity oil, having a kinematic viscosity of between 45 and 300 seconds, Saybolt Furol, has not been separable by known coalescing methods on a practical basis.

Ideally, the screen material for use in separating oil from water has a low pressure drop, does not clog readily and is easily cleanable. Apparatus utilizing such screens desirably has means for continuously removing the oil collected from the screens. It is an object of this process to provide an apparatus for reducing the high viscosity oil content of oil-laden water. It is a further object to provide an apparatus to continuously separate high viscosity oils from ships' ballast water, which apparatus is easily cleaned.

It is a further object to provide an apparatus to limit the residual oil being discharged in oil-laden ballast water to small size droplets which are widely dispersed, have less tendency to coalesce and are more susceptible to bacterial degradation in seawater than large droplets. It is another object of this process to provide a woven mesh screen which coalesces oil from oil-laden water, has a low pressure drop and is readily cleanable.

These objects may be achieved by passing a mixture of Bunker "C" oil and water through a series of open weave wire mesh screens. Surprisingly, the Bunker "C" oil is largely separated from the water by such operation. In passing through the screens, the oil coalesces and collects in a cohesive mass on the downstream side of the screen. The wire mesh screens are oleophilic in that they retain the oil on their surfaces.

A simple drawoff device may be used to remove the coalesced oil from the screen continuously or intermittently. It has been found that by using relatively open weave mesh screens, sufficient coalescing action can be obtained to provide satisfactory removal of viscous oil from water. While the open mesh screens tend to clog, they are easily cleaned by a simple process of backwashing.

Figure 104 shows a suitable form of apparatus for the conduct of the process. The separator consists of a shell 2, which is essentially cylindrical in shape, and contains heads 4 and 6 closing its ends. Head 4 contains an inlet conduit 8 through which oil-laden water is introduced. The shell is divided into chambers **10, 12, 14** and **16** by liquid pervious septa **18, 20** and **22.**

Each septum extends across the separator shell intersecting the liquid flow path which extends from inlet **8** to outlet **24** in head **6.** The septa may contain one or more open mesh screens and may be supported in the shell by any well-known means, such as by flanges **26.** Drawoff conduits **28** are mounted in the shell adjacent to the downstream face of each septum. Valves **30** control the flow from the drawoff conduits **28** through conduits **32** into the collection headers **34.**

**FIGURE 104:  OIL-WATER SEPARATOR FOR SHIPBOARD USE IN BALLAST WATER TREATMENT**

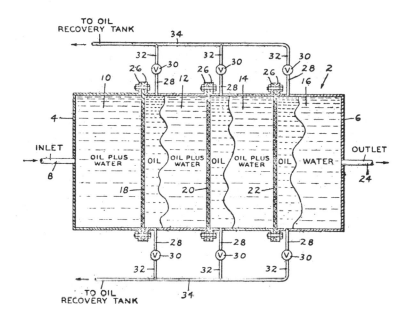

Source:  A.W. Kingsbury and W.S. Young; U.S. Patent 3,231,091; January 25, 1966

In operation, the oil-laden water enters chamber **10** of shell **2** through inlet **8** and passes through septum **18.** A portion of the oil contained in the water is coalesced on the downstream face of septum **18.** The slightly purified water passes across chamber **12** and through septum **20.** Additional oil is coalesced on the downstream face of septum **20** and builds up in chamber **14.** The additionally purified water passes across chamber **14** and through septum **22** where a final stage of oil coalesces on the face of septum **22.** The purified water passes through chamber **16** and leaves the separator through outlet **24.**

The oil which builds up on the downstream faces of the septa **18, 20** and **22** may be withdrawn through drawoff conduits **28.** Valves **30** may be used to control the flow of oil so that the oil may be withdrawn continuously or intermittently. The withdrawn oil passes through conduits **32** into collection headers **34** abd is discharged into an oil recovery tank, not shown.

Screen weaves which have been used with good results include: [1] plain weave, where each chute wire passes over and under successive rows of warp wire, and each warp wire passes over and under successive rows of chute wire; [2] twill, which is similar to the plain weave, except that each chute wire successively passes over and under two warp wires, and each warp wire successively passes over and under two chute wires; [3] plain Dutch weave, which has a similar interlacing as plain weave except that the warp wires are heavier and that the chute wires are lighter, and driven close together and crimped at each pass; and [4] twill Dutch weave, which is similar to Dutch weave, except that the warp wires are usually the same size as the chute wires.

A process developed by W.S. Young (3) also relates to the separation of fuel oil from ballast water. Since the usual deballasting operation is rapid, a flow rate of 600 gal./min. would not be unusually high, it is important that the separation system be able to operate efficiently at high throughput rates. Moreover, in view of the space limitations imposed by the necessities of maritime design and efficiency, a satisfactory separation system should be compact.

It has been found that when a ballast filled fuel oil tank is pumped out during a deballasting operation, the residual fuel oil is not evenly mixed with the ballast water. On the contrary, the oil concentration will vary from 0 to 100% and the bulk of the oil will come in relatively substantial slugs of almost 100% oil. Except for these slugs of oil, which occur largely near the beginning and ending of the deballasting operation, the pumps are handling relatively clean seawater. The concentration of the oil in the pumped fluid can readily be measured and then the treatment accorded to the pumped fluid can be selected to meet the oil concentration present.

One scheme for the operation of the process is shown in Figure 105. A ballast pump **10** draws fluid from the ship ballast tanks through a pipe **11**. The fluid discharged from pump **10** is supplied through a pipe **12** to a three-way motor operated diverting valve **13**. Valve **13** is operated by an electric motor **14** and, depending on the valve setting, may close off pipe **12**, may connect pipe **12** to a pipe **15** or may connect pipe **12** to a pipe **16**.

## FIGURE 105:  ALTERNATIVE DESIGN FOR BALLAST WATER OIL REMOVAL

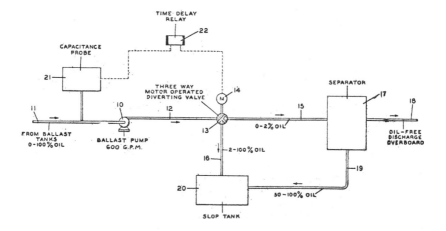

Source:  W.S. Young; U.S. Patent 3,253,711; May 31, 1966

Pipe 15 delivers fluid to an oil-water separator 17 which may be of any suitable type. If the residual oil in the ballast tanks is Bunker "C" which is the name commonly given to No. 6 fuel oil as defined by the Bureau of Standards in Commercial Standard CS 12–48, the separator 17 is preferably of the type described and claimed in U.S. Patent 3,231,091. The oil-free effluent of separator 17 is discharged overboard through a pipe 18. The oil separated in separator 17 is supplied through a pipe 19 to a slop tank 20. Pipe 16 delivers fluid directly to slop tank 20.

Valve 13 will be set to deliver fluid to pipe 15 and thus to separator 17 when the oil content of the fluid from the ballast tanks is less than a predetermined amount. This amount will be dependent on the capacity and efficiency of separator 17 and on the permissible oil content of the water discharged overboard.

In the case of Bunker "C" residual oil and a separator of the type described in the aforementioned patent (U.S. Patent 3,231,091), the valve 13 will usually be set to supply fluid to separator 17 when the oil content of that fluid is between 0 and 2%. Where the oil content of the fluid from the ballast tanks exceeds the predetermined amount, e.g., 2%, valve 13 will be set to deliver the fluid from pipe 12 to slop tank 20.

As mentioned above, oil in the ballast fluid tends to come in slugs of almost 100% oil, and in the absence of such slugs the ballast fluid is relatively clean seawater which can readily be handled by separator 17 to achieve the desired freedom from oil content for the water discharged overside. Because the oil tends to come in slugs, the presence of such slugs in pipe 11 may conveniently be detected by a capacitance probe 21 which measures continuously the dielectric constant of the fluid in pipe 11.

Since the dielectric constant of oil is many times less than that of seawater, a slug of oil will be easily detected. Detection of a slug of oil in pipe 11 operates a time delay relay 22 which in turn operates motor 14 to control valve 13. Time delay relay 22 is provided to insure that the probe 21 has detected a substantial slug or solid stream of oil and has not been affected by a small slug of oil passed in pipe 11.

Time delay relay 22 similarly provides a time delay on release when detection shifts from oil to water to insure that valve 13 is not operated by a small slug of water in an otherwise solid stream of oil. It has been found that a suitable time delay when oil is detected is of the order of about 1 second while a suitable time delay when water is detected is of the order of 5 to 10 seconds.

In operation, when the fluid in pipe 11 has an oil content less than the predetermined level, e.g., less than about 2%, relay 22 will be in a condition thereof in which valve 13 passes fluid from pipe 12 to pipe 15. When the oil content of the fluid in pipe 11 exceeds the predetermined level, capacitance probe 21 will produce an output which, after the set time delay, e.g., 1 second, will cause relay 22 to shift to another condition thereof in which valve 13 connects pipe 12 to pipe 16. Valve 13 will be returned to its initial condition when the oil content of the fluid in pipe 11 drops below the predetermined level and after the set time delay, e.g., 5 to 10 seconds.

Since the oil content of the ballast water tends largely to be either well below 2% or close to 100%, great sensitivity in capacitance probe 21 is not required. Hence probe 21 may be selected from among many available types. However, the probe 21 may, if desired, be made very sensitive to afford a control operation sensitive to small changes in oil content of the ballast water.

The measurement of the oil content of the ballast water need not be effected by capacitance means, although measurement of the dielectric constant is deemed to be the most suitable measurement scheme for this purpose. Other measurement means, e.g., optical inspection, may be used. The mixture of fuel oil and seawater in slop tank 20 may be disposed of in any desired way. For example, the slop tank may be pumped out into suitable disposal means on land or may be discharged overboard at a later time far out at sea. Or the slop tank mixture may be subjected to a separating operation at a later time when the high throughput requirement of the rapid deballasting operation is not present. In a typical installation the ballast tanks might contain 500,000 gallons of seawater and 3,000 gallons of Bunker "C" oil. For this ballast, the slop tank **20** might be expected to receive about 10,000 gallons of seawater and substantially the entire 3,000 gallons of oil.

A process developed by E.G. Greenman (4) utilizes a separator in the form of an impregnated fiber body for separating an oil-contaminant from water. The separator utilizes a strenghtened fiber body and a fluorocarbon which is oleophobic but readily wetted by water. In the process of oil removal the contaminated water is directed to such a fiber body in the form of a paper sheet and the water passes through while the oil is retained.

The process permits relatively large quantities of contaminated ballast water to be handled relatively quickly and under conditions present in the treatment of waters to protect rivers and ocean harbors.

A process developed by J.J. Sheehy et al (5) is one in which means for separating oil from water are provided in a system for handling tank washings from oil tankers to assure that water passing overboard will not have present therein oil in excess of a predetermined concentration.

According to the process in a tanker the residue from washing the cargo tanks or the oily ballast water retained on board is sent to a collecting tank. Pumping means consisting of first a high capacity pump and alternatively, a second low capacity pump, remove the water which has settled to the bottom of the tank. The high capacity pump (or sometimes only the low capacity pump if slower pumping is required) may be used for direct discharge of such settled water overboard from the collecting tank during a process commonly referred to as decanting.

The high or low capacity pump is deactivated by an oil/water interface sensor which substantially reduces the risk of discharging contaminated water during decanting. If the high capacity pump is used first, ordinarily, the low capacity pump subsequently is operated to direct the remaining collected liquid to a separator which separates the oil from the water before the latter is permitted to pass overboard. An oil/water detector and associated conduit means cooperate to shunt mixtures of oil/water back to the settling tank when the separated water contains above a preselected concentration of oil.

*Agglomeration or Freezing on Surface:* A process developed by E.R. De Lew (6) involves subjecting an oil slick on the water to an agent which will congeal the oil to a substantially stiff mass or gel that will stop the oil from spreading and that can be mechanically handled by physical means, such as fine mesh nets or screens which will hold the congealed oil but allow water to drain through. Perforated clam shell buckets can also be employed for removal of the stiffened oil.

In greater detail, any suitable congealing agent can be employed, for example, molten materials which congeal when in contact with the cold oil layer at ambient temperature on the water. Of materials of this type, molten wax, such as paraffin or slack wax, or molten soap solutions which are solid at room temperature are satisfactory. Materials of this type entrap the oil slick when applied to the water and congeal therewith, thus enabling mechanical removal.

Since these materials generally require initial heating to bring them to a molten state before application to the water, a more desirable and consequently preferred type of congealing agent is any suitable material which is liquid at atmospheric temperature, but which will form a relatively stiff water-in-oil gel or emulsion when applied to the oil slick in contact with the water. The term water-in-oil is employed in contradistinction to an oil-in-water emulsion produced by dispersing and emulsifying agents.

The most desirable gel or emulsion forming agents for the particular purpose are soaps of the natural fatty acids contained in wool grease. The natural wave motion or turbulence of the water generally provides sufficient agitation for converting into a relatively stiff water-in-oil gel or emulsion the oil slick to which these lanolin soaps have been added. To insure sufficient agitation and mixing of the liquid congealing agent with the oil and water, particularly with viscous oils such as lubricating oils and heavy fuel oils, it is desirable to apply the congealing agent over the oil slick by means of a nonatomizing spray with sufficient pressure to penetrate the oil film.

Lanolin liquefied by heating also can be used provided vigorous mechanical agitation, such as paddling, is applied to the oil surface after the lanolin is applied thereto. However, its gelling action is much slower than that of its soaps.

Generally speaking, all liquid fatty acid soaps or liquid solutions of such soaps, which are derived from divalent metals, such as calcium, strontium, barium, magnesium, zinc and manganese, as well as certain trivalent metals, such as aluminum and iron, are characterized by their ability to form water-in-oil emulsions. This applies also to divalent and trivalent metal soaps of the so-called naphthenic acids, which are not derived from fats, but from petroleum and have a more complex structure than the fatty acids.

A process developed by E.L. Cole et al (7) involves cleaning up marine oil spills by freezing the surface layer of oil, preferably with particles of Dry Ice or the like, to enable the layer to be screened off the surface as a cake.

A process developed by C.O. Bunn (8) involves recovering oil leakage from the surface of a body of water, comprising the steps of mixing discrete core particles of wood material, polyethylene and finely ground coal to form particles having a high affinity for oil. The particles have a high degree of integrity, thereby permitting storage of the same at points relatively remote from the point of leakage and airlifting of the same to the oil spot. The oil-saturated particles are processed after collection for further use as a fuel source, with the Btu content being as great or greater than bituminous coal.

*Dispersion:* A process developed by W.H. McNeely (9) is one in which an oil slick on a body of water is dispersed by dividing the oil slick and concentrating the oil on the bow wave created by a boat propelled through the oil slick. At the same time a mixture of water and chemical dispersant is sprayed in high pressure jets which are swept across the bow wave in a cyclic oscillating motion substantially perpendicular to the length of the boat, thus producing a zig-zag spray pattern on the oil slick due to the forward motion to the boat.

A near constant angular speed in the oscillatory motion of the jets automatically applies a greater concentration of dispersant adjacent the boat, where the oil is heaviest on the bow wave. The high dilution of the dispersant with environment water increases emulsification and turbulence for increased efficiency.

*Contaminant on Surface:* A process developed by F. McCormick (10) involves containment and deflection of inorganic and organic aqueous surface pollutants such as an oil slick, flotsam debris and jellyfish. This method comprises the generation of an air or bubble barrier which permits the passage therethrough of surface vessels and large fish but halts the movement of floating surface pollutants by the creation of a flexible continuous band of surface turbulence.

The system in one specific application is designed to protect harbor and beach areas and fishing grounds from contamination with oil from oil tankers. The system can also be employed to recover oil from sunken or leaking tankers at sea by containment and collection of the oil released therefrom within the circumference of the bubble barrier wall created in a geometric pattern about the location of the stricken vessel.

A process developed by J.P. Latimer (11) utilizes a slick confining boom adapted to confine and prevent the spread of oil on the surface of water. Oil which has inadvertently

escaped and which lies on the surface of water is often driven by winds and tides into undesirable areas and may be carried ashore where it can do considerable damage to wildlife
and property and where a serious pollution situation is created. The problem has been
compounded by the advent of offshore oil well drilling and the development of present day
super tankers whereby very large quantities of oil may escape due to accidental leakage.

It has accordingly been necessary to develop means by which the slick can be confined so
as to prevent spread of the oil and to enable the oil to be collected and disposed of. Slick
confining booms for this purpose may be of relatively great length ranging from a few hundred feet to several thousand feet or more in length.

Various designs have been employed in the construction of slick confining booms to prevent the spread of oil spilled on the surface of water. The booms as manufactured,
although differing in materials and details of construction, have consisted essentially of a
device of necessary length that floats on the surface of a body of water and has a skirt
that extends downwardly from a few inches to a few feet, and a vertically upward projecting portion that extends from a few inches to a foot or more upward from the surface of
the water. Some form of buoyancy is incorporated in these devices so that at least in
smooth water they float with the skirt immersed to the design depth.

When such booms are used to retain an oil spill, the ends are normally attached to fixed
or stationary structures in such a manner that the boom lies across the path of spreading
oil. Currents and winds create tension forces along the boom, which are conveyed to the
structures to which the ends of the boom are attached. The tension forces along the boom
may be borne by the material in the skirt and the above-water portion, or by ropes or
other strength members contiguously attached to them.

It should be noted that if such booms are exposed to wind or current, a sufficient length
must be used to permit the boom to trail in an arc or catenary between fixed end points.
Otherwise, the lateral tension along the boom becomes excessive and it is pulled apart.

In the process, a boom is connected with buoyancy means for retaining the boom in operative position in a body of water. A foot extends downwardly at an angle therefrom to
define a space between the boom, the foot and the surface of the body of water. Means
are provided for reducing the pressure within the space to facilitate accumulation of a
large quantity of oil adjacent to the body. The means for reducing pressure in the space
extends lengthwise of the boom so that oil can be collected substantially throughout the
entire length of the boom.

Such a device is shown in Figure 106. The key to the reference numerals is as follows:

| | |
|---|---|
| 10 | Boom assembly. |
| 12 | Plastic foam buoyancy means for boom. |
| 14 | Skirt. |
| 16 | Upward extending portion of boom. |
| 20 | Foot. |
| 22 | Upward directed portion of foot. |
| 24 | Plastic foam buoyancy means for foot. |
| 28 | Air space. |
| 30 | Tubular air pipe. |
| 34 | Holes spaced in tubular air pipe. |
| 40 | Oil pipe. |
| 42 | Slot in oil pipe. |
| 44 | Buoyant support means for oil pipe. |
| 46 | Flexible connectors. |
| 50 | Flexible electrical elements for when oil heating is needed. |

FIGURE 106: APPARATUS FOR CONFINING AND COLLECTING OIL FROM SLICK

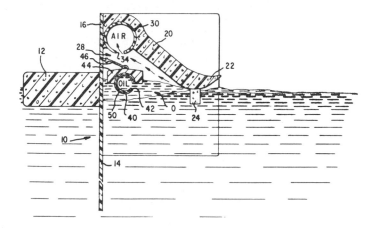

Source: J.P. Latimer; U.S. Patent 3,565,254; February 23, 1971

A process developed by H.J. Fitzgerald et al (12) involves the use of an apparatus for collecting oil from the surface of a body of water having two V-shaped assemblies of flexible inflated floats, one arranged 5 to 25 feet leewardly of the other on the same central axis. The wind and current drive the oil into the open end of the assemblies and cause it to be funneled rearwardly to their apices. Each V-shaped assembly is provided with a depending skirt of impermeable sheet material, the lower edges of the skirts at either side of the inner assembly being interconnected by shock cords and the lower edges of the skirts on the outer assembly being connected to the inner assembly by netting.

*Skimming from Surface:* A process developed by J.E. Woolley (13) is a process for separating oil from water, particularly oil floating on the surface in a dock or at sea, in which an oil/water mixture is pumped continuously into one end of an elongated container in which the mixture separates into an upper oil layer and a lower water layer, and water is drained continuously from the lower water layer.

The operation is continued until oil starts to appear from the water outlet, thus indicating that the container is substantially full of oil. The container is then emptied of oil, and the operation commenced again. The container may be an elongated flexible container or envelope capable of floating in water when containing a sufficient amount of air and/or oil, alternatively the container may be rendered buoyant by any suitable means.

A process developed by D. Bucchioni et al (14) for removing oily and other floating wastes from water surfaces comprises flowing at least the upper water layer through a canal and under at least one floodgate immersed a selected distance in the water and extending transversely of the direction of flow. The cross section of the canal is markedly increased under the floodgate, for example, by increasing the depth of the canal, in order to decrease the velocity of the water and promote decantation. Wastes collecting in front of the floodgate are led off with a portion of the water to at least one decantation tank where the wastes are separated from the water.

A process developed by H.J. Fitzgerald (15) for skimming an oil film from the surface of a large body of water uses an apparatus including a towed funnel assembly with a flexible cover and side skirts of impermeable sheet material with floats to keep the leading edge of the cover spaced above the surface of the water so that the oil film will pass beneath it, with the remaining portions of the cover supported on the floating oil, a bottom panel of netting to hold the side skirts in downwardly projecting position to confine the oil laterally, while permitting the water beneath it to escape freely, and a sump at the apex of the funnel to receive the oil for transfer to a storage vessel.

A device developed by J.M. Valdespino (16) is a portable inflatable apparatus for confining and collecting oil on the surface of water, separating the oil from the water and containing such oil until collected, without the use of mechanical parts.

A process developed by W.L. Bulkley et al (17) involves removing floating material, particularly oily material, from a liquid. At least one pair of spaced, revolving pickup members which dip into the liquid are used to recover the floating material. This material adheres to the members as they come into contact with the liquid, and means adjacent these members remove and collect the material adhering to them.

The characterizing feature of this apparatus is that the surface of one member is smooth and oleophilic, and the surface of the other member is porous and deformable. The member having a smooth, oleophilic surface is in advance of the member having the porous, deformable surface, so that the smooth surfaced member contacts the floating material before the porous surfaced member.

A device developed by H.J. Fitzgerald et al (18) for removing a film of oil from a large body of water consists of a pair of generally similar funnel assemblies, one positioned behind and in the wake of the other with a harness for towing the same along their common central axis, each funnel assembly having an impermeable cover with spaced floats to support its wide leading edge above the water to capture the oil with the rest of the cover being supported on the floating oil, depending skirts at the tapered trailing edges of the cover to funnel the oil inwardly toward its apex, an enclosing sack of reinforcing netting covering the top and bottom of the apparatus, a sump to receive the oil from the apex of the rearward assembly, and a pump to transfer the oil from the sump to a storage barge.

A device developed by R.L. Yahnke (19) is one in which a rotating cylinder covered with a layer of porous polyurethane absorbs oil flowing on a body of water, and a roller squeezes absorbed oil from the layer into a collecting trough.

FIGURE 107: SQUEEZABLE ROTATING CYLINDER FOR OIL COLLECTION FROM WATER

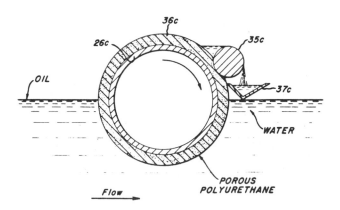

Source: R.L. Yahnke; U.S. Patent 3,578,585; May 11, 1971

Figure 107 shows the regenerable absorption filter means **36c** which forms a sheath or cover about the moving mass comprising the closed cylinder **26c**. The regenerating means including roller **35c** and the lower adjacent trough **37c** function as a wringer to remove oil, some water and other impurities from the filter means for collection and transportation from the system by trough **37c**.

The moving means is a closed rotatable cylinder **26c** having the filter means **36c** forming a sheath about the cylinder wall. As the moving means rotates, the filter means passes through the oily water flowing in the direction shown. The oil preferentially wetting the surfaces of the interstices of the filter means **36c** is carried along with some water to the regenerating means where the water, oil and other impurities are removed from the filter means and the system by the regenerating means **35c** and **37c**. Means **36c** is polyurethane.

An apparatus developed by R.E. Hunter (20) functions to separate oil from a water surface by distributing many small buoyant bodies of oil absorbent material upon such surface, continuously lifting such bodies from the surface, treating the bodies to remove the absorbed oil, and again distributing the bodies upon the surface for reuse.

The apparatus preferably includes booms for gathering the distributed bodies of absorbent material toward a conveyor which lifts the bodies upwardly. The booms are articulated and include floats so that the booms rise and fall with any wave action of the water, such as would exist in the unprotected waters of the open sea. The apparatus also preferably includes a system for compressing the oil from the bodies, and thereafter dropping the bodies onto the water surface for recycling.

A device developed by M.H. Van Stavern et al (21) is a rotatable drum type of oil skimmer which is continually rotated to pick up a film of oil and water on the surface of the drum, having a supplemental or transfer drum located substantially above the oil slick and well out of contact therewith, so as to come into contact with film on the pickup drum and receive a portion of the film, product oil being recovered from both drums.

A process developed by R.G. Will et al (22) is one in which oil-contaminated water is purified using apparatus having a revolving polyurethane foam belt which is mounted on an incline relative to the horizontal. The contaminated water moves past and through the revolving belt or the belt is mounted on a boat which moves the belt through the water.

In either case, as the belt and water move relative to each other, the oil-contaminated water filters through the belt and is purified. The belt is then squeezed twice, first gently to remove water, and then vigorously to remove oil. The belt may include inner and outer abutting sections which are reinforced by a network of threads. The outer section which first contacts the contaminated water has a larger pore structure than the inner section.

A process developed by I. Ginsburgh et al (23) is one in which oil and debris are removed from the surface of water using a revolving, partially submerged, endless brush belt. The brush belt has outwardly projecting bristles which ensnare the debris and pick up oil. Polypropylene bristles are preferred. The oil and debris are removed from the belt before the belt is reimmersed in the water. Alternatively, a brush-type drum could be used in place of the endless brush belt.

A process developed by E.L. Cole et al (24) involves separation and recovery of oil from oil slicks on the surface of water by continuously and selectively picking up the surface oil in a relatively thick blanket of a bulk fabric composed of highly oleophilic fibers expressing the oil from the blanket and leaving the fibers in an open condition highly receptive to additional oil.

A device developed by A.R. Budris et al (25) is a watercraft which comprises a body member of the catamaran-type having spaced-apart twin hulls between which is disposed an oil recovery system which can recover oil (for instance in the form of an oil slick) which has been properly directed by the hulls of the craft.

A process developed by R.H. Burroughs et al (26) is one in which oil is removed from contaminated water by means of a fibrous structure of low denier polyolefin fibers attached to a pumping system. The polyolefin fiber structure can absorb many times its own weight of oil while absorbing little or no water. The oil is then easily removed from the fibrous structure by pumping.

An apparatus developed by C. Bezemer et al (27) for removing oil slicks from water consists of an elongated absorbent porous body positioned along the water surface by base means which include squeezers for removing oil from the body, a container for collecting the oil and drive means for moving the body past the squeezers.

*Burning from Surface:*  A process developed by P.R. Tully et al (28) provides an improved method for the elimination of water and land borne spills by burning. Broadly, certain particulate solids are applied to the spill and the resulting system is thereafter fired. Such treated spills are more easily ignited and the combustion thereof is more complete than experienced with untreated spills.

When certain conditions pertaining to the type and amount of treating agent applied to the spill are met even further benefits accrue. The benefits reside in improved physical character of the burned residue which is more amenable to physical removal thereof from the water or land mass than the burned residuum of untreated spills.

A process developed by A. Molin et al (29) involves destroying drifting oil layers on the surface of water basins by sustained combustion in a zone contiguous to the oil layer and in relative motion with respect thereto. In the method for thus combating drifting oil, a plurality of jets of combustion sustaining gas, in particular compressed air, are blown against the oil layer in the zone for sustaining combustion therein.

In the combating means a hollow element is connected to a source of pressure gas, in particular compressed air, and kept afloat at the surface of the water with longitudinally spaced discharge openings on the element blowing the pressure gas against the oil layer in the zone for sustaining combustion therein.

*Recovery from Aquifers:*  A process developed by D.N. Dietz (30) is one in which taste-spolling components of oil spilled in a drinking water catchment area are removed from the aquifer by supplying liquid such as water or tasteless oil to the spill area and recovering the liquid together with the undesired components via an auxiliary well.

### References

(1) Texas A & M University, "Control of Oil Spills," *Report COM-72-10810;* Springfield, Va., National Tech. Information Service (March 1972).
(2) Kingsbury, A.W. and Young, W.S.; U.S. Patent 3,231,091; January 25, 1966; assigned to Pfaudler Permutit Inc.
(3) Young, W.S.; U.S. Patent 3,253,711; May 31, 1966; assigned to Pfaudler Permutit Inc.
(4) Greenman, E.G.; U.S. Patent 3,494,863; February 10, 1970; assigned to Kimberly-Clark Corp.
(5) Sheehy, J.J. and Sait, P.A.; U.S. Patent 3,565,252; February 23, 1971; assigned to Esso Research and Engineering Co.
(6) De Lew, E.R.; U.S. Patent 3,198,731; August 3, 1965; assigned to Yosemite Chemical Co.
(7) Cole, E.L. and Hess, H.V.; U.S. Patent 3,614,873; October 26, 1971; assigned to Texaco Inc.
(8) Bunn, C.O.; U.S. Patent 3,651,948; March 28, 1972; assigned to Col-Mont Corporation.
(9) McNeely, W.H.; U.S. Patent 3,532,622; October 6, 1970.
(10) McCormick, F.; U.S. Patent 3,491,023; January 20, 1970; assigned to Submersible Systems, Inc.
(11) Latimer, J.P.; U.S. Patent 3,565,254; February 23, 1971; assigned to Deepsea Ventures, Inc.
(12) Fitzgerald, H.J. and Koepf, E.H.; U.S. Patent 3,590,584; July 6, 1971; assigned to Ocean Pollution Control, Inc.
(13) Woolley, J.E.; U.S. Patent 3,508,652; April 28, 1970; assigned to The Dunlop Company Limited, England.
(14) Bucchioni, D. and De Toffoli, M.F.; U.S. Patent 3,517,812; June 30, 1970.
(15) Fitzgerald, H.J.; U.S. Patent 3,523,611; August 11, 1970; assigned to Ocean Pollution Control, Inc.
(16) Valdespino, J.M.; U.S. Patent 3,532,219; October 6, 1970; assigned to Water Pollution Controls, Inc.

(17) Bulkley, W.L., Ries, H.E., Jr. and Will, R.G.; U.S. Patent 3,539,508; November 10, 1970; assigned to Standard Oil Company.
(18) Fitzgerald, H.J. and Koepf, E.H.; U.S. Patent 3,557,960; January 26, 1971; assigned to Ocean Pollution Control, Inc.
(19) Yahnke, R.L.; U.S. Patent 3,578,585; May 11, 1971; assigned to Standard Oil Company.
(20) Hunter, R.E.; U.S. Patent 3,581,899; June 1, 1971; assigned to Ocean Design Engineering Corp.
(21) Van Stavern, M.H., Jones, W.T., Cossey, H.F. and Clark, W.J.; U.S. Patent 3,612,277; October 12, 1971; assigned to Texaco Inc.
(22) Will, R.G. and Grutsch, J.F.; U.S. Patent 3,617,552; November 2, 1971; assigned to Standard Oil Company.
(23) Ginsburgh, I. and Will, R.G.; U.S. Patent 3,617,555; November 2, 1971; assigned to Standard Oil Company.
(24) Cole, E.L. and Hess, H.V.; U.S. Patent 3,617,556; November 2, 1971; assigned to Texaco Inc.
(25) Budris, A.R., McGowan, F.J., Evans, L.M., Wayne, T.J., Lithen, E.E. and Darcy, C.B.; U.S. Patent 3,646,901; March 7, 1972; assigned to Worthington Corporation.
(26) Burroughs, R.H. and Cox, P.R., Jr.; U.S. Patent 3,667,608; June 6, 1972; assigned to Hercules Incorporated.
(27) Bezemer, C., Tadema, H.J. and Houbolt, J.J.H.C.; U.S. Patent 3,700,593; October 24, 1972; assigned to Shell Oil Company.
(28) Tully, P.R., Fletcher, W.J. and Cochrane, H.; U.S. Patent 3,556,698; January 19, 1971; assigned to Cabot Corporation.
(29) Molin, A. and Carlsson, O.; U.S. Patent 3,586,469; June 22, 1971; assigned to Atlas Copco Aktiebolag, Sweden.
(30) Dietz, D.N.; U.S. Patent 3,628,607; December 21, 1971; assigned to Shell Oil Company.

# OIL (VEGETABLE)

## Removal from Water

A process developed by P. Bradford (1) applies to the removal of soapy materials from vegetable oil refinery wastewaters. Considerable difficulty has been experienced in the vegetable oil refining industry in separating insoluble soaps from the condensate of the steam deodorizers. It is common practice in the refining of vegetable oils to subject the oil to a steam deodorization operation during which the oil is held in a large enclosed vessel and a vacuum drawn on its surface while steam is bubbled therethrough.

The volatile materials, which include principally free fatty acids, are drawn off with the steam by the vacuum and later condensed by spraying with a cooling liquid in barometric condensers. In those refineries which are located adjacent the ocean, seawater may be used for condensing the vapors in the barometric condensers. Seawater contains large quantities of calcium and magnesium ions which react with the free fatty acids to form insoluble soaps.

The soaps are widely dispersed and form insoluble flocs, which tend neither to float nor settle and will remain suspended in the liquid for long periods of time. The seawater from the condensers of the steam deodorizers is collected in a basin known as a hot well. It has proven difficult to economically separate these insoluble soaps from the hot well liquor, presumably because of their low concentration and because of their tendency to remain suspended.

Flotation separation of the precipitated soaps of the hot well liquors of the vegetable oil refinery industry by direct introduction of air into the liquor has generally proven ineffective. It has been proposed that the hot well liquors be aerated and the aerated liquor held for a period of time at a substantial pressure to promote intimate mixing of the air and liquor.

It is thought that by the latter practice greater quantities of the air will become dissolved in the aerated liquor with the result that when the pressure is released on the waste liquor there will be large quantities of air available to coat the precipitated soaps. This latter scheme may or may not be effective for removing insoluble soaps from the hot well liquor depending apparently upon the concentration of the soap and the willingness of the one treating the material to provide ample air. The cost of such treatment may be prohibitive.

The process involves introducing an aerated second wastewater which carries fine solid particles or a precipitate into the first waste. The treated waste and the introduced aerated stream are intimately mixed and then passed to a quiescent zone where the floc and the precipitate of the two aqueous waste bodies may be floated to the surface of the combined liquors.

It has been found that in order to remove the insoluble soaps present in hot well seawater that there must be introduced into the seawater a second aqueous stream containing both dissolved air and a fine precipitate or fine particles. It may be that the precipitate of the introduced aerated stream provide nuclei for the conglomeration of the widely dispersed insoluble soaps found in the seawater, but in any event, in order to separate effectively the floc of the hot well seawater, the latter water must be intimately mixed with the aerated stream and this is best accomplished by conducting the aqueous bodies upwardly through a zone. It has also been found that by utilization of such a zone and by directing the two mixed streams upwardly therein, the floc and precipitate rise readily to the surface and may be separated in an adjacent quiescent zone.

In a preferred embodiment the aerated precipitate containing stream is a treated floor waste material of the refinery. This latter waste material contains large quantities of soluble soaps, fats, and some protein materials. The soluble soaps are first placed in an insoluble form prior to the aeration of the floor waste waters by mixing with that water a portion of the hot well seawater. The rest of the seawater, and by far the major portion of it, is introduced into the aforementioned confined zone together with the aerated floor waste. This scheme has proven very effective for the treatment of both hot well seawater and the floor waste.

A process developed by G.L. Kovacs (2) involves treating the oil-bearing cooling water from industrial plants such as cold rolling mills with a view to separating the oil and water in such a manner that the oil is suitable for further processing into reusable or saleable form rather than fit only for disposal, and the water is in an acceptable state for discharging into adjacent sewers or water courses.

In the process of cold rolling steel strip to light gauges, palm oil is often employed as an aid in rolling. Other natural oils, animal fats and artificially compounded oil mixtures are also employed, especially in the rolling of heavier gauges. Some mills use the oils undiluted and apply it with sprays, others spray a mixture or emulsion of water and oil. Some mills use recirculating systems where a large quantity of the oil/water emulsion is flooding the work; in others a smaller quantity of oil is applied on the strip or on the rolls to be used only once, then discarded.

According to conventional practice, the effluent cooling liquid in the rolling process, containing spent rolling oil, cooling water, iron and impurities, is allowed to settle in sumps, pits or settling tanks. A black colored semisolid scum is skimmed off, heated in settling tanks, allowed to stand for periods ranging up to several months to remove water dried by evaporating the remaining water, and sold as a by-product known, if it originated from pure palm oil, as "Refuse Palm Oil."

From the bottom of settling sumps or tanks a heavy sedimentary solid layer containing considerable quantities of oil is cleaned out periodically and discarded as waste. These sludges and scums usually contain considerable amounts of finely abraded iron from the rolling process. The watery portion remaining after the removal of these scums and sludges, which is the bulk of the effluent, is usually handled in one of the following ways: (a) directly discharged untreated, (b) further settled in skimming tanks prior to discharge and (c) lime treated and clarified, prior to discharge.

The practice of discharging the watery portion directly to waste is still practiced extensively. As stream pollution has become worse, the purity requirements for bulk liquid discharging have risen correspondingly. As a result, the practice of further settling the watery portion of the effluent in skimming tanks prior to discharge is used in many places. Because the settling and skimming equipment necessary requires considerable investment, attempts have been made to offset some of the costs involved by reusing the recovered oil. However, failure of the recovered oil to function satisfactorily has discouraged further development in this direction.

In some areas, a high degree of stream pollution control is imperative, and several processes have been developed to treat the watery portion of the effluent which have satisfactorily met these high standards. Such processes, however, are costly to operate and consume considerable space and chemicals. They further result in the rolling oil being recovered in a form which has so far precluded all possibility of reuse or sale as a by-product. Thus, no economy is available to offset the depollution costs, and the recovered oils themselves become a disposal problem.

Canadian Patent 597,986, "Process for Recovering Rolling Oil," issued May 17, 1960 describes a process which has been in use at a number of plants whereby the fresh sludges removed from the effluent may be rendered into reuseable rolling oils. However, the water from which these sludges are removed still is not satisfactory for immediate discharge, because the standards set by government pollution control agencies (usually less than 15 ppm oil) are not met.

Unlike any of the above approaches to the problem, the process consists of a method for treating the fresh effluent immediately, rather than permitting the effluents to stand in settling tanks so that scums and sludges may be formed.

The object is to provide a process for treating the effluent cooling liquid from the rolling operation, with two other waste materials available from the mill which results in the recovery of the rolling oil in a form suitable for further treatment, for example by the process taught in Canadian Patent 597,986, and which results in water containing less than 15 parts per million oil, thus meeting the antipollution requirements.

The process involves forming in the effluent cooling liquid an insoluble precipitate which coagulates and concentrates the oil, placing the total effluent under pressure and dissolving pressurized air therein, releasing the pressure on the effluent liquid so that the air returns to the gaseous state and removes the coagulum and thus the oil upwardly by flotation, and skimming off the resultant layer of scum. By "insoluble precipitate" is meant a material which is insoluble in the effluent cooling liquid. It is mandatory that this material be capable of dissolution in some subsequent process for oil recovery.

The precipitate is that formed by mixing an alkaline waste referred to as washer effluent with an acid waste known as pickle liquor. In steel mills there are "cleaning lines" which are alkaline washers for steel strip which produce the alkaline waste used in this process. The alkaline waste usually contains free caustic soda, soap, and oil in an emulsified form which will not settle. This material is herein referred to as washer effluent.

Waste pickle liquor, which is usually available in all steel mills and which normally constitutes a disposal problem is the second waste used to treat the rolling mill effluent. Waste pickle liquor consists primarily of sulfuric acid ($H_2SO_4$) and ferrous sulfate ($FeSO_4$). It is important to form the precipitate in situ, that is, the first waste should be fully dispersed in the mill effluent before the second is added, in order to insure the scrubbing out of as much oil as possible.

Either the washer effluent or the waste pickle liquor may be the first to be added to the effluent, and usually the choice of the one over the other results in little practical difference. However, it is advisable to avoid having the pH of the mixture remain in the neutral or alkaline range for any length of time, because the oil content of the effluent tends in such circumstances to become saponified. This problem is avoided if the pH of the effluent is in the acid range. For this reason the waste pickle liquor is ideal because it contains sulfuric acid which is useful to keep the pH of the final mixture in the acid range.

Figure 108 shows the arrangement of apparatus involved in the conduct of the process. The waste pickle liquor is metered by a metering pump **10** into a surge tank **11** into which the tandem mill effluent to be purified is fed. The surge tank **11** is violently agitated by the incoming effluent and thereby complete dispersion of the waste pickle liquor throughout the effluent is assured. A pressurizing pump **12** draws the effluent containing the waste pickle liquor out of the surge tank **11** and pressurizes it to within the range of 40 to 80 psig.

**FIGURE 108: PROCESS FOR TREATING COLD ROLLING MILL EFFLUENT CONTAINING OIL EMULSIFIED IN WATER**

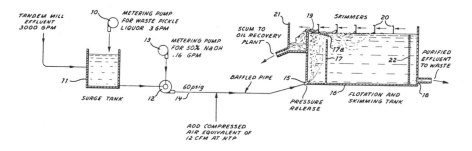

Source: G.L. Kovacs; U.S. Patent 3,301,779; January 31, 1967

A second metering pump **13** feeds the washer effluent into the suction of the pressurizing pump **12**, and the pressurizing pump effects the complete mixing of both additives in the effluent. On issuing from the pump **12** the effluent should have a pH of 5 to 6. The cleaning line waste and the pickle liquor combine in the effluent to form an insoluble precipitate which scrubs out substantially all of the oil in the effluent.

In the above step it is advisable to avoid cleaning line wastes, or substitutes therefor, which contain silicates because although the water is purified to the necessary degree, the presence of the silicates in the reclaimed oil is detrimental. It should also be pointed out that the use of lime, $Ca(OH)_2$ as a substitute for washer effluent is undesirable due to the deleterious effect it has in later processing and on the quality of the recovery oil.

After mixing the waste pickle liquor and the cleaning line waste in the mill effluent, some indication as to the resultant pH of the effluent can be had from its color. If the effluent is neutral or alkaline, it retains a yellow-brown color. If the pH is maintained on the acid side, the effluent is colorless and has a sparkling clear appearance.

Whether the precipitate is formed so as to be voluminous or finely divided, no difference in the final result is noticeable. The formation of the precipitate in situ has the important advantage of effecting rapid and complete coagulation of the oil since the precipitate when formed is already completely dispersed throughout the effluent, which insures that the scrubbing out of the oil particles will be complete.

The coagulum at this point may either float or sink, the latter case requiring that air be used to float it. Although in the past the air that is normally dissolved in water at atmospheric pressure or dissolved under pressure in a recirculated side stream has, in many processes, been successfully used to give adequate flotation, in the process this has proven unsatisfactory because the bulk of the coagulum escapes flotation. Thus, this method contemplates that the total effluent be placed under a pressure of from 40 to 80 psig, and that pressurized air at the same pressure be admitted thereto so that it can dissolve in the effluent.

Dissolving air in less than the total effluent decreases the efficiency of the separation of air from the effluent stream. The pressurized or compressed air is preferably admitted to the effluent while the latter is flowing through a pipe 14 having upper, semicircular baffles (not shown) which produce great turbulence.

The step of releasing the pressure on the effluent to cause the air to remove the precipitate and thus the oil upwardly by flotation is carried out by releasing the effluent, saturated with air, through a restrictive orifice 15 into one end of an open-top skimming tank 16 exposed to the atmosphere. In the drawing the skimming tank 16 is shown schematically. A vertical baffle 17 within the skimming tank directs the incoming, treated effluent upwardly toward the surface of the effluent in the skimming tank.

The incoming effluent passes above the baffle and then moves toward a drainage point 18 from which it is discharged to waste. The provision of the baffle tends to prevent mixing and agitation due to the incoming effluent that could interfere with rapid flotation of the floc. It has been found that a skimmed gravity settler, mostly of conventional design, will perform satisfactorily as a skimming tank.

A beach 17a is located above the baffle and angles upwardly toward the left end wall of the skimming tank. A space is left between the beach and the skimming tank through which undissolved air entering the tank can be vented upwardly.

If the coagulum were not present, the air would come out of solution as a fine milky fog which would rise slowly. However, the action of this material is such that the greater part of the air is associated with it when the pressure is released. The coagulum then floats very rapidly and strongly. It has been found that the bulk of it is floated very early in its passage through the skimming tank, and that most of the scum 19 formed will surface in the first one-third of the length of the skimming tank. It is thus advantageous to skim off the scum layer in the direction opposite to that of effluent flow in the skimming tank, so that the scum is not dragged the full length of the skimmer.

In the drawing are shown skimming paddles 20 which move slowly in the direction opposite to the direction of effluent flow in the tank. In this way there is afforded an opportunity for particles, which may have been redeposited through agitation by the skimming action, to be floated again. The paddles drag the scum up over the beach and into a scum collection trough 21 located at the end of the skimming tank.

The fact that flotation of the precipitate occurs rapidly after the pressure is released permits high effluent feedrates with satisfactory results. A further baffle 22 is located at the downstream end of the skimming tank, and acts as a weir for regulating the depth of effluent in the tank, so that the effluent level can be maintained at the beach.

*Example 1:* This method was applied to the effluent from a 5 stand tandem high speed cold mill, producing cold reduced steel strip of 0.008" thickness for tin-plate manufacture. The effluent contained the following impurities: 800 ppm rolling oil, 40 ppm fine iron powder and 75 ppm lubricating oils.

The balance of the effluent was cooling water used to cool the steel passing through the mill. In addition to the above impurities the effluent contained varying quantities of emulsifying agents, and some emulsification of the oil and iron impurities had likely taken place. The effluent was at an average temperature of 80°F. The effluent was treated as follows:

[1] The total effluent was pumped at 3,000 gpm to surge tanks.

[2] A metering pump added to the surge tanks at a rate of 3 gpm waste pickle liquor containing approximately 12% by weight $FeSO_4$ and 8% by weight $H_2SO_4$.

[3] A pressurizing pump drew effluent from the surge tanks and pumped it through a pipeline into a baffled pipe at a pressure of 60 psig.

[4] A second metering pump delivered to the suction of the pressurizing pump 6.0 gpm of a washer effluent. Precipitate was formed.

[5] Compressed air was admitted to the baffled pipe at a pressure in excess of 60 psig and a rate of about 12 cfm as measured at normal temperature and pressure (NTP). The effluent became substantially saturated with air.

[6] The air-saturated, precipitate-laden effluent was then released into a settling tank at atmospheric pressure, and the precipitate was floated and removed by skimming.

The skimmed-off material was treated as described in Canadian Patent 597,986 to produce rolling oil which was subsequently reused. The effluent from the settling tank had a sparkling, clear, colorless appearance, a pH of 5 to 6, and an oil content less than 15 ppm.

*Example 2:* The method was applied to the effluent from a second tandem mill. The cooling water for the mill was derived from clarified river water. The composition of the effluent was as follows: 600 ppm rolling oil, 35 ppm iron powder and 76 ppm lubricating oil.

The remainder of the effluent was water, and the whole was at an average temperature of 75°F. The effluent was treated as follows:

[1] The total effluent was pumped at 3,000 gpm to surge tanks.

[2] A metering pump added to the surge tanks at a rate of 3.5 gpm waste pickle liquor containing 8% by weight $FeSO_4$.

[3] A pressurizing pump drew effluent from the surge tanks and pumped it through a pipeline into a baffled pipe at a pressure of 45 psig.

[4] A second metering pump delivered to the suction of the pressurizing pump 0.45 gpm of a solution of an alkaline washing compound used in electrolytic washers. Complete mixing was effected by the pressurizing pump and precipitate was formed.

[5] Compressed air was admitted to the baffled pipe at a pressure in excess of 45 psig and at a rate of about 12 cfm as measured at normal temperature and pressure (NTP). The effluent became substantially saturated with air.

[6] The air-saturated, precipitate-laden effluent was then released into a settling tank at atmospheric pressure, and the coagulum was floated and removed by skimming.

The skimmed-off material was treated as described in Canadian Patent 597,986 to produce rolling oil which was subsequently reused. The effluent from the settling tank had a sparkling, clear, colorless appearance, a pH of 5 to 6, and an oil content of less than 15 ppm.

### References

(1) Bradford, P.; U.S. Patent 2,925,383; February 16, 1960; assigned to Swift & Co.
(2) Kovacs, G.L.; U.S. Patent 3,301,779; January 31, 1967; assigned to New Canadian Processes Limited, Canada.

## ORGANIC VAPORS

The reader is also referred to the section of this handbook on Solvents, Removal from Air.

**Removal from Air**

A process developed by D.H. Haigh et al (1) is one in which alkylstyrene polymer particles are used as an absorbent for organic vapors and are particularly effective for removing organic vapors from air.

**References**

(1)  Haigh, D.H., Hall, R.H. and Lange, C.E.; U.S. Patent 3,686,827; August 29, 1972; assigned to The Dow Chemical Company.

## OXYDEHYDROGENATION PROCESS EFFLUENTS

In a recently developed butene dehydrogenation process known as oxidative dehydrogenation, an oxygen-containing gas is fed to the catalytic reaction zone containing a catalyst such as stannic phosphate along with the butenes feed and steam, and a substantial portion of the hydrogen produced by dehydrogenation is combusted to water vapor.

This not only removes the inhibiting effect of the hydrogen on further dehydrogenation, but also supplies heat to this endothermic reaction resulting in high conversions and per-pass yields of butadiene at relatively good selectivity. By this method, additional steam is produced which is recovered from the process effluent as condensate. Also, moderate concentrations of oxygenated hydrocarbons are generated which similarly appear in the condensed steam and/or in the hydrocarbon effluent.

It has been found that as much as 4 mol percent of the olefin feed may be converted to oxygenated hydrocarbons such as carboxylic acids, aldehydes, ketones, etc., especially acetic and propionic acids and acetaldehyde; the nature and quantity of these compounds depending on the conditions under which dehydrogenation is effected. Under normal plant operating conditions, these oxygenated by-products will be ultimately vented to the atmosphere and/or discharged with wastewater from the process, depending upon the separation and recovery processes employed and their operating conditions.

However, it has been found that these by-products are toxic and result in damage to property, particularly crops and foliage and are probably contributors to photochemical smog and haze, especially when vented as aerosols. It is not only desirable to eliminate or at least reduce this source of air and water pollution, but such control is essential in many locations.

**Removal of Carbonyl Compounds from Water**

A process developed by T. Hutson, Jr. and R.E. Riter (1) is one in which excess purge water containing oxygenated hydrocarbons resulting from hydrocarbon dehydrogenation processes is rendered nontoxic by the conversion of the oxygenated hydrocarbons to water and carbon oxides in the presence of microorganisms, e.g., saprophytic bacteria, protozoa, yeast and fungi, under aerobic conditions. Preferably the purge water prior to treatment with microorganisms is reboiled or stripped with a hot gas to reduce the concentration of oxygenated hydrocarbons.

A process developed by R.A. Hinton and J.E. Cottle (2) is one in which water containing oxygenated hydrocarbons resulting from hydrocarbon oxidative dehydrogenation processes is rendered substantially nontoxic by stripping with steam to remove the oxygenated hydrocarbons including carbonyls. It may then be returned to the oxidative dehydrogenation process to suppress the formation of additional oxygenated hydrocarbons and a portion of the steam stripped water converted to steam for use as stripping medium.

### References

(1) Hutson, T.,Jr. and Ritter, R.E.; U.S. Patent 3,646,239; February 29, 1972; assigned to Phillips Petroleum Company.
(2) Hinton, R.A. and Cottle, J.E.; U.S. Patent 3,679,764; July 25, 1972; assigned to Phillips Petroleum Company.

# PAINT AND PAINTING EFFLUENTS

## Removal of Paint Particles from Air

A process developed by O.M. Arnold and R.H. Harbin (1) involves the recovery of overspray paint in liquid or semiliquid form. One of the important problems in the use of sprayed paints is the collection of the overspray paint in an economical and efficient manner. The quantity of overspray paint may represent a substantial portion of the original material. For example, in some paint operations 70% of the paint results in overspray. Irrespective of whether this overspray paint is to be reprocessed so it can be used again, it must be collected and prevented from affecting other operations or environments.

In general, relatively efficient air washing systems have been used in which the paint particles are captured in a water solution. This solution ordinarily contains surface active agents provided to kill, coagulate, and float the paint material so as to prepare it for easy disposition, for example, by scooping the solid components out of one section of the spray booth. Such a method works quite satisfactorily with many types of paint materials provided efficient formulations of paint killing materials are used. Thus, lacquer type materials and most paints can be handled effectively by such processes. However, other types of paints are not readily handled in this manner, and merely form gummy masses which adhere to the surfaces of the spray booth to the moving parts of the air washer.

One example of such a paint is the material referred to as chassis-black paint which is one used to paint, for example, the understructures of automobiles. There are many different formulations of chassis-black paint, but ordinarily it is formed of gilsonite or asphalt dissolved or dispersed in a petroleum base solvent such as mineral spirits. A low boiling solvent of the aliphatic type is generally used so that the parts will dry quickly.

The base components are asphalt type materials including gilsonite, also called asphaltite, which is a natural form of bitumen. Sometimes a small quantity of carbon black pigment is introduced to increase the blackness and luster of the dried paint. The asphalt material is readily soluble in the mineral spirit solvent so that substantial quantities can be dispersed in a stable form. Other types of chassis-black paint include rosins or other low cost resin materials.

In part, this material is difficult to handle because of the characteristics of the bituminous material and in part because of the absence of, or limited quantity of, suspended pigment which in other types of paint assists in forming the solid phase. In addition, these bituminous based paints readily wet metal surfaces and accordingly form a coating on the exposed surfaces of the spray booth so that it is difficult to keep the spray booth in clean operating condition.

Moreover, paint of this type is ordinarily relatively cheap and is used in applications where a large volume of overspray is produced, and it is thus rendered even more difficult to successfully dispose of the overspray.

The process provides an improved method for handling overspray paint materials, particularly those including bituminous type materials, in which the overspray is retained in a liquid state in the spray booth so that it can be handled hydraulically, eliminating the difficult and expensive cleaning operations ordinarily required when paint of this type is being used. In this method, instead of converting the paints to solid conglomerate, the paints are treated by reagents that cause the paint to form into a concentrated viscous or semiviscous liquid state which will not stick to the sides of the paint spray booth, and which can be collected by flotation with the use of suitable hydraulic equipment.

In order to accomplish this, a solvent of higher boiling point, one which will not be vaporized under the usual operating conditions of the spray booth is substituted for the volatile vehicle of the paint. This higher boiling solvent becomes a substitute solvent or dispersing media for the original vehicle and liquid components carrying the paint pigment, resin, and other materials.

In addition to being a higher boiling material with lower vapor pressure than the solvent originally in the paint, this substitute solvent must have the ability to dissolve or dispense the paint selectively in itself; it should be immiscible with the substrate media such as a water solution, although it may be temporarily dispersed by mechanical means in the aqueous substrate; and it should be of lower specific gravity than the water solution in order to assist in the flotation and rapid separation of the paint dispersed in this solvent.

In operation, the initial volatile solvent in the paint is vaporized in part as the material is sprayed and most of the remainder of the most volatile solvent materials is removed in the air washing portion of the spray booth which may include, for example, air washers of the type described in U.S. Patent 2,889,005.

A process developed by L.C. Hardison (2) utilizes a multistage cleaning system which includes the treating of the air stream so as to permit it to be returned to the processing zone, as for example, a paint spray booth in a clean, warm and purified state.

Various systems have been utilized for handling and treating an air stream containing particulates and volatiles from a paint spray booth, or from operations providing an equivalent type of problem. For example, there has been used a method of water washing the exhaust stream from a spray booth to remove entrained paint particles. Such a system results in a disposal problem for the paint sludge and further causes saturation of the air stream so that it cannot satisfactorily be reused in the spray booth.

For proper paint spraying and drying conditions, there is a need for a dry air environment within the spraying zone. Also, there is a particular cost advantage which can be obtained by eliminating the heat requirements for drying and preheating all of the air that is being introduced into the spray booth. The water wash system or any other system which precludes reuse of the air and necessarily results in the total discharge of the air stream from the system is thus costly and undesirable.

Briefly this improved purifying system includes, at least one moving metal screen type of belt for collecting particulates, a burn-off zone for the continuous oxidation and removal of the deposited material on the belt and at least one fume elimination zone usable for the further incineration and purification of combustion gases from the burn-off zone, and or for use in treating the air stream to be returned to the processing zone which evolves the particulates and combustible materials.

Figure 109 illustrates the essentials of this process. There is indicated a spray booth zone 1 which is generally an elongated type of room designed to accommodate a continuous conveyor means, such as 2, for carrying material that is to be painted as it moves continuously through the zone. The drawing indicates auto bodies 3 being carried by the conveyor means 2 and such car bodies being subjected to enameling or lacquering from operators stationed in the spray booth. For simplicity, the drawing shows merely half of the spray booth zone and the accompanying treating stages in combination therewith.

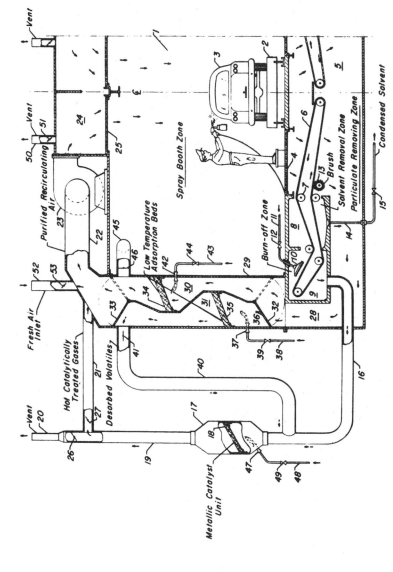

FIGURE 109: APPARATUS FOR PAINT REMOVAL FROM AUTOMOBILE SPRAY BOOTH EXIT GASES

Source: L.C. Hardison; U.S. Patent 3,395,972; August 6, 1968

Actually the total unit may be considered as being symmetrical about a vertical center line extending through the middle of the auto body 3. The floor of the spray booth 1 is provided with a continuous grill or grid means 4 which permits the down flow of a heating and/or ventilating air stream through the booth whereby entrained volatiles and particulates will be carried into a lower particulate removing zone 5 that is housing a continuously moving mesh type filter belt 6. The latter passes over a plurality of spaced crown rollers 7 within zone 5, a solvent removal zone 8 and a burn-off zone 9. The latter is equipped with a plurality of burners 10 being supplied by a fuel-air mixture from line 11 having control valve 12.

Preferably, as indicated hereinbefore, the burners 10 are of the infrared type suitable for providing controlled high energy heating to a confined relatively small area of the surface mesh type filter belt 6 within the burn-off zone 9. The impinging temperature should of course be adequate to effect the complete burn-off of the deposited material on the belt surface and permit the belt to be returned to the filtering or particulate removing zone 5 in a cleaned state. A brush means 13 may be positioned at the belt outlet section of the burn-off zone or at the solvent removal zone 8 such that oxidized ash-like particles will be completely removed from the openings in the mesh of belt 6 prior to its reuse for filtering the air stream descending through grid 4.

The temperature from the burner means 10 in zone 9 will of course provide some transfer of heat into the adjacent zone 8, which first receives the filter belt 6 from zone 5. Thus, there will be some heating and removal of the volatile components entrained with the deposited paint particulates in zone 8. At the same time, any condensation of volatile solvent materials that may be collected in zone 8 can be removed from the floor by way of line 14 having control valve 15. However, in a preferred operation, the burner means 10 shall be operated in a rich manner to minimize the presence of oxygen and to preclude any upstream burning of paint or other deposited particulates on the moving belt 6.

The gases from the burn-off zone 9 are passed by vent means 16 through an oxidizing catalyst zone 17 having a catalyst bed 18. The latter may comprise a permeable unit containing subdivided particles of an active oxidation catalyst or a mat-like unit of crimped alloy ribbon that is coated with a noble metal, and particularly a platinum group metal, whereby there will be oxidation and conversion of combustible materials in the gas stream to provide harmless, odorless, oxidation products which are primarily carbon dioxide and water.

For example, one desirable form of catalytic fume incinerating means may comprise the all metal catalyst unit such as described in U.S. Patent 2,658,742. The treated oxidized gas stream from catalyst oxidation zone 17 may be discharged into the atmosphere by way of duct 19 and stack 20 or in part recirculated and reused within the spray booth zone by passage through duct 21 and 22, fan means 23 and air distributing plenum 24 which in turn releases the gases downwardly through grill means 25 into the spray booth zone 1. Suitable valve means 26 and 27 can be adjusted to accommodate the desired gas flow through the recirculation fan and to the spray booth zone or to the atmosphere by way of stack 20.

The air stream from zone 5 is shown to flow by way of passageway 28 into an adsorption section 29 which in turn accommodates two separate adsorption zones 30 and 31. The air flow to one or the other of the zones is controlled by a lower damper means 32 and an upper damper 33. As shown for one case, air flow with entrained volatile components carries by way of zone 30 into the transfer duct 22 after first passing through the adsorption bed 34. When the bed 34 becomes heavily saturated with volatile materials then it is subjected to desorption while air flow is routed (by means of the movement of damper means 32 and 33 into the dashed lined positions) such that adsorption is effected by bed 35 in zone 31.

As previously noted, the beds 34 and 35 for completing the purification of the air stream, may utilize activated carbon particles as the adsorption media since such material is particularly adapted to provide high adsorption activity for the volatile components being encountered from paint and lacquer spraying operations. During the desorption cycle, as indicated in the drawing for zone 31, there may be heated air provided by a bleed opening 36 in damper 32 and a burner means 37 for fuel being supplied by line 38 and control valve 39.

A low velocity hot gas stream will thus pass through the bed 35 at a desorption temperature and at a rate sufficient to substantially and completely effect the removal of adsorbed volatiles. The desorption stream is carried by way of duct 40 and control valve 41 into duct 16 at a point upstream from the catalyst zone 17. Thus the desorbed volatiles can be catalytically oxidized and removed from the air stream to permit discharge into the atmosphere or be recirculated into the air stream by way of duct 21 with control valve 27.

In the alternative operation where the adsorption bed 34 is undergoing desorption, then the main air flow is through bed 35 and a desorbing heated air stream is passed upwardly through bed 34 by means of burner 42 which receives fuel by way of line 43 and control valve 44. The desorbed volatiles carry downstream from bed 34 through duct 45 which, although not shown on the drawing, may be made connective with duct 40 to in turn pass the volatiles into contact with the catalyst unit 18 in zone 17.

During this cycle, the dampers 32 and 33 will be in the dash line positions and valve 46 at the inlet of duct 45 opened to accommodate the desorption stream and the entrained desorbed volatiles. Generally, the desorbing gas stream used during the reactivation cycle for each of the adsorption beds 34 and 35 will be at a temperature of the order of about 500°F. or more, so as to provide an effective removal of all of the adsorbed components.

The streams passing by way of lines 40 and 45 into the oxidation zone 17 may be generally at a temperature sufficient to maintain catalytic oxidation within bed 18, particularly where the latter utilizes an active catalyst coating. However, where additional heat may be required to sustain complete incineration of the combustible entrained components, then a burner 47, being supplied fuel by line 48 and control valve 49, will assist in adding heat to the catalyst oxidation zone and insure complete conversion of the combustible materials as they pass through the bed.

In an operation where the oxidized gases leaving the zone 17 are maintained in the system, by recirculation through ducts 21 and 22 into fan 23, there may be provision for a continuous slip-stream removal of a portion of the combustion gases in order to preclude any build up of carbon dioxide and, at the same time, provision to continuously introduce a small portion of fresh air into the system. Venting may be effected from above the plenum by duct 50 with adjustable valve 51 and fresh air introduced by duct 52 with adjustable valve 53 at a point connective with duct 22 just upstream from the fan 23.

In a normal operation, the air stream through the spray booth zone 1 will be maintained at a relatively constant temperature of about 75°F., or any other desired room temperature. The filtering and adsorption steps will have no effect on the air stream inasmuch as it is continuously circulating through only the moving filter belt 6 and the adsorption beds 34 and 35 which in turn will operate at room temperature in their respective contact zones.

As indicated briefly hereinbefore, in an operation where there are substantially no obnoxious volatiles in the filtered stream, then it may be returned directly to the processing section. In other words, the absorption beds may be bypassed or eliminated from the system. The combustion products stream of oxidation zone 17 which can be passed by way of duct 21 into the recirculating air system will be at a high temperature of about 1,000°F. or more and thus be of advantage for some heating in cool seasons. During the summer months, such stream may be discharged to the atmosphere to prevent any heat build up in the system.

A process developed by J. Saubesty (3) is one in which the paint room comprises along its upper portion one or a plurality of air inlet orifices. It is equipped with an elongated sealed sedimentation tank connected on the one hand to an air suction device and on the other hand to an air duct provided with a device for forming wash-liquid sheets which comprises a single elongated passage extending in the horizontal direction, formed in the bottom of the paint room between two inclined walls thereof and located substantially in a vertical plane containing the longitudinal center line of the room. A wash-liquid supply system is so disposed along and near the upper end of each one of the inclined walls that the liquid will stream down the greater part of their top surface.

Figure 110 shows such an arrangement applied to an automobile plant. The upper portion of the paint room **1** constitutes a suction housing **2** connected to an air inlet duct **3**. A distributor grid **4** is disposed between the housing **2** and the room proper to ensure a uniform distribution of the ventilation air throughout the horizontal cross-sectional area of the room. The articles to be painted, for instance the bodies **5** of automotive vehicles in the example illustrated, are caused to travel through the room **1** on suitable carriages **6**. The paint spray or application means are not shown.

The floor **7** for the operators of the paint plant is perforated to permit the passage of air therethrough. The bottom of the room **1** consists of a pair of plane surfaces **8** and **9** inclined downwards at different angles from their point of junction with the lateral walls of the room.

Substantially at the vertical longitudinal median plane of the paint room, these two inclined planes have superposed and adequately shaped edges **10** and **11** leaving a vertical passage **12** therebetween.

A water supply device is provided in the form of water distributing or sprinkling rails **13** and **14**. These distributor rails **13** and **14** are so disposed that the water issuing therefrom will stream over the greater part and preferably throughout the top surface of the inclined surfaces **8** and **9**, the adequately shaped edges **10** and **11** thereof causing the water to form regular sheets flowing therefrom.

The passage **12** opens into a vertical air duct **15** provided with baffle means **16** also provided with suitably shaped edge **17**, the arrangement being such that the water streaming from the upper edge **10** onto the lower inclined plane **8** flows over the edge **11** thereof onto the first baffle member **16**, then over the edge **17** thereof and then onto the next baffle, if any, whereby the air sucked from the room into this duct is caused to flow through several sheets of water.

Of course, the baffle system may be replaced by water-atomizing or sprinkling rails mounted in the duct **15**. The air duct **15** opens into the sealed sedimentation tank **18** equipped at its upper portion with a plurality of fans **19** forcing the air to the outside so as to create a certain vacuum in the tank. The tank **18** and duct **15** are advantageously of elongated configuration and substantially of same length as the paint room, the fans **19** being disposed at spaced intervals along one of the longitudinal walls of the tank.

As shown in the lower portion of the figure which is a vertical section on a smaller scale, the tank comprises at one end a compartment **20** separated from the main body of the tank by a vertical filtering partition **21** having its upper edge substantially level with the liquid **22** in the tank. Secured above this partition is an inclined plane **23** having one end immersed in the liquid **22**, as shown, and its other end overlying a paint receiving vat **24** disposed in the compartment **20**.

An inspection door or lid **25** is mounted on the top wall of compartment **20**, above the vat **24**, so that an operator may use a suitable doctor blade for transferring the settled paint sludge **26** into the vat **24** and can remove the sludge from this vat without penetrating into the compartment. Finally, a pump **27** draws the wash liquid from the compartment **20** and forces it again through the sprinkling rails through a return line **28**. This pump is also adapted to supply liquid to the water sprinkling rails disposed in the duct **15** in case these rails are used instead of the baffle means illustrated. The paint plant illustrated and described operates as follows.

Due to the vacuum created in the tank **18**, the air sucked into this tank is caused to follow the path shown by the arrows and flows through the duct **3** into the housing **2**, then vertically across the paint room **1** proper and the space surrounding the body **5** to be painted, whereby the paint particles in suspension which are not deposited on the body **5** are entrained.

The air stream loaded with paint particles will thus flow through the perforated floor **7**, along the inclined planes **8** and **9** sprinkled with water, then through the water sheets formed in the longitudinal passage **12** and between the baffle means **16**, so as to be freed of the paint particles, the scrubbed or cleansed air penetrating into the tank **18** and being subsequently

FIGURE 110:  ALTERNATIVE SCHEME FOR AUTOMOBILE PAINT SPRAY BOOTH
FUME HANDLING

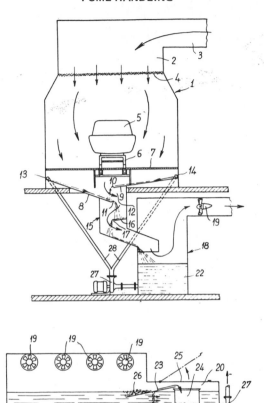

Source:  J. Saubesty; U.S. Patent 3,516,230; June 23, 1970;

extracted therefrom by the air fans **19**.  The water loaded with paint particles is also caused
to flow through this tank **18**.  A sedimentation thus takes place and the particles are thus
caused to float as the water slowly penetrates into the compartment **20** through the filtering
partition **21** where it is sucked by the pump **27** and forced back into the paint room through
the return line **28**.

It is only necessary to open the door **25** for periodically transferring by means of a suitable
scraper or doctor the sedimentation or flotation sludge from the inclined plane **23** into the
vat **24**.  When this vat is filled with paint, it can be either emptied or removed from the plant
without having to penetrate into this plant.  Of course, the liquid level **22** in the tank remains
constant because the water sucked and forced by the pump **27** returns very rapidly to the
tank.  The paint plant built according to this process is characterized by many advantageous
features, some of which are listed below.

(a)  The surface area of the paint room is reduced.

(b)  The paint room construction is lighter in weight and cheaper.

(c) The air circulation is better balanced due to the single central suction system.

(d) The operators' work is facilitated due to the reduction of noise levels, as the fan mounting is remote from, and at a lower level than, the paint rooms, instead of being disposed laterally thereto.

(e) All maintenance, cleaning and repair works are simplified due to the easier access through the door 25, independently of the paint room.

### Removal of Paint Particles from Water

A process developed by J.F. Wallace and J.G. Ransom (4) applies to a painting plant in which articles are coated with a water borne paint and then water rinsed. The rinse water is conveyed back to the coating tank, the contents of which are conveyed under pressure to a reverse osmosis unit to separate water and low concentrate solutions of pigments and resins, the high concentrate solution of resins and pigments being conveyed to a filter bank, the output from which is passed back to the reverse osmosis unit or under pressure to a second reverse osmosis unit to separate still further relatively high and low concentration of pigments and resins, the circulation of the effluent continuing until separation is substantially complete.

Another process developed by J.F. Wallace and J.G. Ransom (5) is a process for the treatment of solutions and mixtures, which become diluted by rinse water in the course of treatment, such as a water borne paint, used to coat an article. The contaminated rinse water is conveyed back to a treatment tank from a rinsing station and the thus diluted contents of the treatment tank are passed as an effluent under pressure through a reverse osmosis unit to separate the effluent into parts of high and low concentration of the original contents of the treatment tank.

A process developed by W.H. Gardner and N.F. Stanley (6) involves the removal of waste solids from industrial and municipal wastewater in flotation apparatus, the improvement comprising blending into the wastewater, prior to flotation, a sufficient amount of flotation agent and flocculant to effect rapid flotation of the waste solids from the wastewater, passing the blended mixture thus formed into the flotation apparatus, removing the waste solids from the flotation apparatus as scum, and removing the remaining wastewater from the flotation apparatus as effluent. Microballoon agents and flocculant polymers are utilized. The following is one specific example of the application of this process.

*Example:* An industrial wastewater from a paint pigment plant at Glens Falls, N.Y. was obtained. This particular waste is a plant effluent which is acid (pH $\approx$ 3.0) and contains significant amounts of inorganic paint pigment solids. This waste effluent was treated by adding Bakelite microballoons as the flotation agent and then adding Hercofloc as the flocculant in the proportion of about 0.1 g. microballoons and about 0.001 g. Hercofloc to 1,000 cc of the waste effluent. Complete visual clarification was obtained within about 30 seconds upon subjecting the mixture to flotation.

### References

(1) Arnold, O.M. and Harbin, R.H.; U.S. Patent 2,982,723; May 2, 1961; assigned to Ajem Laboratories, Inc.
(2) Hardison, L.C.; U.S. Patent 3,395,972; August 6, 1968; assigned to Universal Oil Products Company.
(3) Saubesty, J.; U.S. Patent 3,516,230; June 23, 1970; assigned to Regie Nationale des Usines Renault, France.
(4) Wallace, J.F. and Ransom, J.G.; U.S. Patent 3,528,901; September 15, 1970; assigned to Pressed Steel Fisher Limited, England.
(5) Wallace, J.F. and Ransom, J.G.; U.S. Patent 3,556,970; January 19, 1971; assigned to Pressed Steel Fisher Limited, England.
(6) Gardner, W.H. and Stanley, N.F.; U.S. Patent 3,637,490; January 25, 1972; assigned to Hercules Inc.

## PAPER MILL EFFLUENTS

For details of pulp and paper mill pollution problems and their solutions, the reader of this handbook is referred to the review volume by H.R. Jones (1).

### References

(1)  Jones, H.R., *Pollution Control and Chemical Recovery in the Pulp and Paper Industry,* Park Ridge, N.J., Noyes Data Corp. (1973).

## PARTICULATES

The removal of particulates will not be discussed in detail here for two reasons:

[1]  This handbook is directed now to the removal of specific air and water contaminants rather than to broad classes of material such as particulates.

[2]  The topic of particulate removal has been covered in recent literature. The reader is particularly referred to H.R. Jones, *Fine Dust and Particulates Removal,* Park Ridge, N.J., Noyes Data Corp. (1972).

## PESTICIDES

Pesticides can cause poisoning by ingestion, absorption through the intact skin, or inhalation. In cases of accidental occupational poisonings, it has usually been impossible to determine if the exposure was predominantly respiratory or dermal. Of the 111 accidental deaths caused by pesticides in 1961 in the United States, 5 deaths were attributed to respiratory exposure.

Of all the pesticides, the chlorinated hydrocarbon and organophosphorus insecticides are of major concern because of their health hazard. The acute toxicity of the organophosphates, on the average, is somewhat greater than that of the chlorinated hydrocarbons. However, the latter group is considerably more persistent because of their greater stability.

Some members of the chlorinated hydrocarbon group, especially DDT, dieldrin, and BHC, have been found as residues in human fat tissue in all parts of the world. The mean storage level of DDT in the body fat of the general population in the United States in 1961 to 1962 was reported to be 12.6 ppm. The dieldrin and BHC storage levels have been reported in the United States as 0.15 and 0.2 ppm, respectively, according to Finkelstein (1).

Acute poisonings of commercial and domestic animals have usually been accidental and involved the more toxic organophosphorous insecticides. Animals also store the chlorinated hydrocarbon residues in fat tissue, and as with humans, the significance of this storage is not completely known. When ingested, as little as 7 to 8 ppm of DDT residue on hay will result in 3 ppm being excreted in cow's milk, and butter made from such milk will contain 65 ppm.

Fowl, fish, and many forms of wildlife have been adversely affected by pesticides, especially the chlorinated hydrocarbons. Birds are affected by DDT resulting in thin-shelled eggs and a decrease in hatchability. Wildlife in general have been affected in various parts of the country. Herbicides may cause damage to other than the target plants if the dosage is too great. Some insecticides have produced undesirable flavors in plants used as food. Translocation of DDT and other insecticides into crops from the soil has been observed, but apparently this does not result in a high residue level. The primary source of pesticides in the air is the process of application. Even under the most ideal conditions, some amount will remain in the air following the application. However, under certain meteorological conditions, the pesticide spray or dust does not settle and can drift some distance from the area of application. Many episodes have occurred in which these drifting pesticide clouds have caused inhalation poisonings as well as toxic residues on crop lands.

## Removal of Pesticides from Air

The abatement of pesticidal contamination of the ambient air is a complex problem but it is being attempted. It appears that the contamination arising from the production processes can be controlled. Although incidences of occupational poisonings have been reported, proper protective measures are available. Precautions similar to those used in general chemical industries are taken to prevent the dusts and fumes from leaving the production plant into the outside environment.

Bag packers, barrel fillers, blenders, mixing tanks, and grinding operations are generally completely enclosed or hooded and the air is vented through baghouses or cyclone separators. Similar control procedures are used when liquids are involved; liquid scrubbers are used however, instead of baghouses (2)(3).

Although Tabor (4) has monitored the air near a formulating plant and found air levels similar to that observed in earlier agricultural samplings in the same area, too little air monitoring data are available at the present time to properly evaluate the production air control measures.

The control of chemical drift as a source of pesticide air contamination has been studied extensively. Akesson and Yates (5) have reviewed the literature, including their own research on drift control. They considered that three factors affect the control of a given application: [1] the distribution equipment; [2] the physical state of the pesticide; and [3] the microclimatology of the area. Although the emphasis has been placed on control of aerial applications, applications made with ground equipment can also result in drift. However, greater operator control is possible with a ground unit, since it generally has a lower discharge rate than aerial equipment.

The physical state of the pesticide is quite important in drift. The drift potential from pesticide dust is very high because of particle size. Dust materials are generally screened to incorporate only particles ranging from 1 to $25\mu$ in size; on the average, 80 to 90% of the particles in the formulation are under $25\mu$. Spray droplets of $50\mu$ in size show less drift than dusts of smaller particle size. Therefore, the use of dusts has been decreasing in recent years and the Federal Government has banned the use of 2,4-D dust (5).

MacCollom, (6) in a study of a Vermont apple orchard where Tedion dusting for apple insect control had been the standard practice for the previous 10 years, found that drift could be a problem even under ideal weather conditions. He noted that under conditions of a windspeed of 1.3 mph, a temperature of 81°F., and relative humidity of 40%, drift occurred up to 300 feet. He suggested that a buffer zone of at least 300 feet be used in future applications where forage areas are adjacent to the sprayed area.

Van Middelem (7) cited the work of Yoe, who found that spray droplets ranging from 10 to $50\mu$ in size can drift several miles from the area of application, whereas $100\mu$ sized particles usually do not produce a drift hazard unless the winds are high. Akesson and Yates (5) in their review have cited the work of Brooks (8) on the relationship between particle size and anticipated drift pattern.

Evaporating the spray droplet in spraying operations has been considered as a drift control measure. Water is the most frequently used diluent because of its availability, its low cost, and its nonphytotoxic effects on agricultural crops. However, water droplets may experience considerable evaporation and reduction in size while airborne in the spray, resulting in a sufficiently smaller droplet size and therefore more drift than had been anticipated. As a control measure, oils which are relatively nonevaporative are frequently used as the pesticide carrier in sprays. They are especially used as carriers for low volume (1 to 3 gal./acre) applications of very fine sprays for forests and range lands. (5)

Improvement in delivery equipment and techniques is another partial solution to the abatement problem. Equipment has been developed for producing invert emulsions (water-in-oil emulsions) of which greater than 90% is in the water phase. In contrast to conventional oil-in-water emulsions, preparations containing a high water content are quite viscous.

When using the invert emulsions, sprays consisting of large droplets can be delivered aerially, minimizing both drift and evaporation. Trials indicated that smaller quantities of pesticide could be used in the invert emulsion with equivalent results in terms of insect kill or herbicidal efficiency. At the same time, the accuracy of delivery was improved so that the invert emulsions could be applied under more adverse meteorological conditions than conventional sprays. (9)

Various spray nozzles have been designed and used with varied pesticide formulations having different viscosity, density, and surface tension in attempts to control drift during application. Additional factors such as the angle of the nozzle with airstream or the use of screens or discs at the nozzle also contribute to the characteristic of the spray. These factors are discussed in detail by Akesson and Yates. (5)

Meteorological conditions are extremely important parameters that are considered in the application of pesticides and the control of potential drifts. Wind direction and velocity, humidity and temperature at ground and higher levels, and the amount of sunshine or rain are all interrelated factors that are considered. Because of the importance of such meteorological information, the U.S. Weather Bureau provides this specialized data as part of their service.

Dusting and spraying advisories are sent out on a teletypewriter circuit 24 hours a day emphasizing local weather conditions for aerial and ground applications for various agricultural chemicals. In addition, these advisories include information relating certain insect and other pest activities with weather conditions, so that pesticides can be applied at the proper time to produce maximum pest control. (10)

Volatilization of pesticides into the air from soil, water, plants, and other treated surfaces is known to occur. However, the control of this is complicated by the fact that the extent to which volatilization occurs is not known. Some control over volatilization from soil can be effected by the use of cover crops. It has been observed that two to three times more insecticidal residues were recovered from alfalfa covered plots than from fallow ones. (11)

### Removal of Pesticides from Water

A process developed by F.F. Sako et al (12) permits the removal of insecticides such as DDT, Aldrin and the like from water. Figure 111 shows a suitable arrangement of apparatus for the conduct of such a process. Effluent is introduced via line 10 into the apparatus. It is conducted into an activated sludge chamber or basin 12. The outlet of basin 12 is connected by line 14 to settling basin 16. Sludge that settles out in basin 16 is returned to sewage feed line 10 by line 17.

The water that leaves the settling basin 16 is relatively pure, except that it may contain high molecular weight soluble organics, which the bacterial and chemical actions of the sewage plant cannot entirely remove. The concentration of such organics, such as DDT, in the water leaving the settling basin, may be 5 or more parts per million. This water leaving the sewage system is conducted by a line 18 to a pump 20, which delivers the water to an inlet line 22 and then to an inlet control valve 24, leading to the fractionating tower T of the process.

FIGURE 111:  PROCESS EQUIPMENT FOR REMOVING CHLORINATED INSECTICIDES
FROM WATER

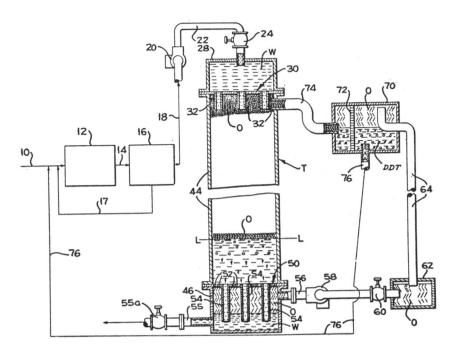

Source:  F.F. Sako and J.A. Abbott; U.S. Patent 3,247,104; April 19, 1966

In the apparatus being described, the fractionating tower T is formed in three sections.
There is an upper, or water admission section 28, into which the water W is conducted
under control of the inlet valve 24.  The lower portion of the upper section 28 of the
fractionating tower is bounded by a water admission plate 30, from which depends water
admission pipes, tubes or downcomers 32.  The number of downcomers 32 depends upon
the diameter and capacity of the tower.  The lower end of each tube 32 is restricted somewhat, forming a nozzle, which nozzle is surrounded by a baffle.  The lower end of each
baffle is closed and the water is released through a number of ports that pierce the baffle.

There is also a main, fractionating, or treatment section of the tower T, formed as a column
44.  The column 44 is relatively long, in order to provide ample time for the soluble organics to form a film of enriched solution on the surface of the oil droplets in the column.
A lower section 46 completes the tower T.  Section 46 is a combined oil admission and
water removal section.  The upper wall of the lower section 46 is in the form of an oil
dispersing, or droplet forming plate 50, which is formed with a number of nozzles or bores
52, through which oil O is dispersed from a body of oil lying beneath the oil dispersing
plate 50.

Depending from the oil dispersing plate 50 are a number of downcomers 54 for conducting
the water W through the body of oil O, and out of the main or fractionating section 44
of the tower.  In the form illustrated, the oil in the system is lighter than the water so
that there is a body of water W beneath the body of oil O in the lower section 46 of the
tower.  The water is withdrawn from the lower section of the tower by a water effluent
line 55, under control of a valve 55a.

Oil is admitted to the lower section **46** of the tower **T** by an oil inlet line **56**, connected to a pump **58**. The amount of oil admitted is under control of a valve **60**, the oil being withdrawn by the pump from a sump unit **62**. Oil enters the sump **62** by means of a line **64** that extends downwardly from a combined oil coalescing and phase separator unit **70**. In the separator unit **70**, an oil breaker **72** in the form of a honeycomb member is provided, although such a breaker is not indispensable to the practice of the process. The surface enriched droplets of oil are conducted from the upper portion of the fractionating section **44** of the tower **T** by a line **74** from which line the oil droplets enter the phase separator unit **70**.

In the phase separator **70**, the surface enriched oil droplets are broken and coalesced into a body of oil **O** and a body of organic enriched water solution. The honeycomb breaker **72** assists in coalescing of the oil droplets. When pesticides, such as DDT, are contained in the water, the water phase in separator **70** is a solution of DDT that is more concentrated than the solution entering the tower **T** for treatment.

The enriched solution of DDT or the like is returned to feed line **10** by way of line **76**. Experience with sewage disposal plants has demonstrated that bacteria and other active agents found in sewage disposal systems will attack and break down relatively enriched solutions of DDT and other high molecular weight organics, whereas very dilute solutions of the same materials are not effectively treated. This recycling has been proven to reduce the concentration of the active agents in the effluent from that present when no enriched solution is introduced into the system.

### Removal of Pesticides from Solid Wastes

The problem of pesticides in solid wastes has been reviewed by Floyd (13). As the production and use of chemical pesticides continue to increase so do pesticide wastes. Many of these wastes are accumulated in the environment and present hazards in many forms. The magnitude of this problem is presented as it relates to solid waste management.

Some of the important factors that influence the production of pesticides are listed. Current studies on processing and disposal techniques (composting, incineration, waste management, ultimate disposal, recycling and reuse) for the pesticide wastes and containers are discussed, and new avenues of approach and needed research efforts are suggested for mitigation of the problem.

### References

(1) Finkelstein, H., "Air Pollution Aspects of Pesticides," *Report PB 188,091*; Springfield, Va., National Tech. Information Service (Sept. 1969).
(2) "Air Pollution Control in Connection with DDT Production," Informative Report No. 6, *J. Air Pollution Control Assoc.* 14, 49 (1964).
(3) Danielson, J.A., *Air Pollution Engineering Manual,* U.S. Dept. of Health, Education and Welfare, Public Health Service, National Center for Air Pollution Control (1967).
(4) Tabor, E.C., "Pesticides in Urban Atmospheres," *J. Air Pollution Control Assoc.* 15, 415 (1966).
(5) Akesson, N.B. and Yates, W.E., "Problems Relating to Application of Agricultural Chemicals and Resulting Drift Residues," *Ann. Rev. Entomol.* 9, 285 (1964).
(6) MacCollom, G.B., "Orchard Air Dusting and Drift Residues on Adjacent Hayfields," *J. Econ. Entomol.* 55, 999 (1962).
(7) Van Middelem, C.H., "Fate and Persistence of Organic Pesticides in the Environment," in *Organic Pesticides in the Environment,* Advan. Chem. Ser. 60 (1966).
(8) Brooks, F.A., "The Drifting of Poisonous Dusts Applied by Airplanes and Land Rigs," *J. Agr. Eng.* 28, 233 (1947).
(9) Gill, R.E., "A New Spraying Process," *J. Roy. Aeron. Soc.* 69, 864 (1965).
(10) Scotton, J.W., *Atmospheric Transport of Pesticide Aerosols,* U.S. Dept. of Health, Education, and Welfare, Public Health Service, Washington, D.C. (1965).
(11) Lichtenstein, E.P., "Persistence and Degradation of Pesticides in the Environment," in *Scientific Aspects of Pest Control,* Natl. Acad. Sci.-Natl. Res. Council, Publ. 1402 (1966).
(12) Sako, F.F. and Abbott, J.A.; U.S. Patent 3,247,104; April 19, 1966; assigned to FMC Corp.
(13) Floyd, E.P., *Occurrence and Significance of Pesticides in Solid Wastes,* Public Health Service, Environmental Health Service, Bureau of Solid Waste Management, Washington, D.C. (1970).

## PHENOLS

McKee and Wolf (1) have listed the following industries as characteristic sources of phenolic pollutants:

| | |
|---|---|
| Gas Works (production) | Explosives |
| Wood Distillation | Coal Tar Distilling |
| Oil Refineries | Mine Flotation Wastes |
| Sheep and Cattle Dip | Insecticides |
| Chemical Plants | Resin Manufacture |
| Photographic Developers | Coke Ovens |

In addition, aircraft maintenance (2)(3), foundry operations (4), Orlon manufacture (5), caustic air scrubbers in paper processing plants (6), rubber reclamation plants (7)(8), nitrogen works (9), fiberboard factories, plastics factories, glass production, and stocking factories (10) have been reported as contributing phenols to wastewaters. Table 15 summarizes the levels of phenol found in wastes of various industries.

**TABLE 15: SUMMARY OF PHENOL CONCENTRATION REPORTED IN INDUSTRIAL WASTEWATERS**

| Industrial Source | Phenol Concentration (mg./l.) | Reference No. |
|---|---|---|
| Coke Ovens: | | |
| Weak ammonia liquor, without dephenolization | 3,350 – 3,900 | 11 |
| | 1,400 – 2,500 | 12 |
| | 2,500 – 3,600 | 13 |
| | 3,000 – 10,000 | 14 |
| | 580 – 2,100 | 15 |
| | 600 – 800 | 16 |
| Weak ammonia liquor, after dephenolization | 28 – 332 | 17 |
| | 10 | 14 |
| | 10 – 30 | 12 |
| | 4.5 – 100 | 9 |
| Wash oil still wastes | 30 – 150 | 18 |
| Oil Refineries: | | |
| Sour water | 80 – 185 (140 ave.) | 18 |
| General waste stream | 40 – 80 | 19 |
| Post-stripping | 80 | 20 |
| General (catalytic cracker) | 40 – 50 | 21 |
| Mineral oil wastewater | 100 | 9 |
| Petrochemical: | | |
| General petrochemical | 50 – 600 | 22 |
| Benzene refineries | 210 | 9 |
| Nitrogen works | 250 | 9 |
| Tar distilling plants | 300 | 9 |
| Aircraft maintenance | 200 – 400 | 2,3 |
| Other: | | |
| Rubber reclamation | 3 – 10 | 7,8 |
| Orlon manufacturing | 100 – 150 | 5 |
| Plastics factory | 600 – 2,000 | 23 |
| Fiberboard factory | 150 | 23 |
| Wood carbonizing | 500 | 23 |
| Phenolic resin production | 1,600 | 24 |
| Stocking factory | 6,000 | 10 |

Source:  Report PB 204,521

Although described in the technical literature simply as phenols, this waste category may include a wide range of similar chemical compounds. In terms of pollution control, reported concentrations of phenol are thus the result of a standard analytical methodology which measures a general group of similar compounds rather than being based upon specific identification of the single compound, phenol(hydroxybenzene).

### Removal of Phenolic Materials from Air

In a process developed by F.E. Warner and A.P. Rice (25) for removing phenolic air pollutants in the production of glass fiber products, the polluted air is passed sequentially through at least two low energy contacting zones in each of which it is contacted with descending scrubbing liquor, each succeeding contacting zone having at its bottom individual liquor collecting means.

In passing from one contacting zone to the next zone the air stream undergoes acceleration in a transfer passage in which no interzone flow of liquor occurs, and then impinges on baffle means disposed in its path as it enters the next zone at a level above the pool of scrubbing liquor collected at the bottom of the next zone in the liquor collecting means.

In the large scale manufacture of glass fiber products such as thermal insulation materials, a random deposition of the glass fiber onto a conveyor is carried out in the presence of a spray of an organic resin binder. The particular orientations of the glass fiber spinners and the organic resin sprays are such that the glass fibers depositing on the conveyor are substantially coated with the organic resin binder which is subsequently cured by heat.

The whole process is conducted in a flow of air which carries the glass fibers and the resin spray droplets onto the moving conveyor. This air is well above ambient temperature due to the fact that the spinners which produce the fibers by extrusion of molten glass have to be kept red hot. The hot air, laden with unused organic binder material, passes through the conveyor mesh and the associated ducting and fan system, and is finally discharged to the external atmosphere. Also there is a discharge of fumes evolved in the resin-curing oven.

Many different organic resin formulations may be employed in such a process. For the more common products accounting for the major part of output in large factories, resin binders based on phenol-formaldehyde condensation products, usually but not exclusively in the so-called resole form, are still the preferred materials. The materials have many advantages among which is the compatibility of resoles with water in which they are soluble. On the other hand; the use of phenol-formaldehyde systems seriously aggravates the problem of atmospheric pollution because of the toxicity and the acid odor of the fumes, which are perceptible to smell at very low concentration.

A variety of methods have been tried to remove the phenolic pollutants from the discharge air streams. Hitherto, none of the methods has proved acceptable either on efficiency of collection of the pollutants or on economic feasibility on the large scale. Consideration of the resin binder formulation, its method of application and the general mechanics of the manufacturing process as a whole shows that the pollution of the outgoing air streams is due to two coexisting factors.

The first factor is the presence of particulate material in the form of liquid droplets ranging in size from some scores of microns diameter down to submicron aerosol particles, and the second factor is the presence of vapors, in particular phenol vapor and to a lesser extent formaldehyde vapor. Failure to understand properly the true nature of the pollution in the air streams is believed to account for the unsatisfactory results obtained previously in trying to reduce the degree of pollution to a harmless and undetectable level.

It will be apparent that the time factors appropriate to the efficient removal of particulate material are generally much shorter than those required by absorption from the vapor phase into a liquid system. For particulate removal a high energy scrubbing system is required to obtain effective performance when particle size is very small. This implies high air velocities and short contact times, so that vapor absorption is negligible. Furthermore, the high energy input results in a rise in temperature and thus in the equilibrium vapor pressure over the aqueous scrubber liquors, so establishing a lower concentration of the volatile pollutants in the final liquor discharge.

On the other hand, the conventional low energy scrubbing systems are less effective in the removal of very small particulate pollution, and in order to provide a sufficiently long vapor-liquid contact time, the equipment has to be very large and expensive. Furthermore, running costs are high due to pressure drop and to the high liquid-to-vapor ratios necessary to maintain low phenolic concentrations in the scrubber liquors.

It is possible to overcome the last objection by using sodium carbonate or hydroxide as the scrubbing liquor, thus retaining the pollutants in solution as the respective sodium phenates etc. This is only justifiable when the actual quanties of phenolic substances removed from the air stream are such that recovery on a commercial basis is feasible. Such conditions do not apply in the case we are considering specifically, namely the production of glass fiber products. It is an object of the process to achieve a technique for the removal of phenolic air pollutants in the production of glass fiber products which takes full account of the two factors in the polluting content.

### Removal of Phenolic Materials from Water

Treatment technology for phenolic wastes is widely reported in the literature. Methods are available for reduction of all levels of initial phenol concentration, and frequently there is a region of overlap between methods. Both chemical-physical and biological treatment are in successful full-scale industrial use, and high efficiencies of treatment are reported. However, any particular industrial system may encounter difficulties in employing a specific treatment method, depending upon the overall composition of the waste stream.

A case in point involves attempting biological treatment on a phenolic waste containing high levels of heavy metals, prior to metal removal. Phenolic wastes often contain large quantities of other waste constituents which require special treatment procedures. This is particularly true for refinery and coke oven wastes. Oil and cyanides are commonly present in large quantities in phenolic waste waters, and their removal, with attendant economical disadvantages, prior to phenol removal may be required. In general, however, phenol removal is successfully practiced by industry.

Treatment methods for phenol can be best discussed in terms of the phenol concentration in the waste to be treated. It is apparent from Table 15 that industrial wastes may contain an extremely wide range of phenolic material. Therefore, treatment technology is discussed below under three headings: concentrated, intermediate and dilute phenolic wastewaters.

*Concentrated Wastewaters:*   High concentrations of phenol in a waste stream make recovery an attractive economic consideration. Phenol recovery value from coke plant ammonia liquor is estimated at 20 cents per ton of coal processed (14). Wurm (9) has reviewed the more common and successful methods employed for phenol recovery from concentrated wastes. All processes are based upon extractive recovery into an immiscible organic solvent. Efficiencies are extremely high, with recoveries of 98 to 99% reported.

However, even such high percent efficiencies can leave significant residual phenol in the treated effluent of a concentrated phenolic waste stream. The problem of solvent loss due to slight solubility in the wastewater versus efficiency of phenol removal must be balanced economically, as solvent costs can be high. In addition, any solvent dissolved in the extracted water represents increased organic matter in the waste. Fisher reports that the Koppers light oil extraction process reduces 1,500 to 2,000 mg./l. phenol to 10 to 30 milligrams per liter (12). Other processes are available and are summarized in a 1957 article (26).

However, Wurm (9) has presented those in most common use and of highest removal efficiency. All extractive phenol processes appear capable of high recovery, but residual phenol may be too high to meet effluent standards. Ross (27) reports that phenol solutions of 7% (7,000 mg./l.) could be incinerated for fuel costs of 1.5 cents/1,000 gal. Capital costs for the incinerator were estimated at $10 to $15,000.

Besselievre (28) reports the following: an extraction system at the Steel Company of Canada, treating 0.15 MGD of phenolic wastes cost $500,000; an oil refinery employing a light oil extraction system coupled with an electrostatic field invested $200,000 in their phenol treatment unit; and a Podbielniak phenol extractor (Chemizon process) treating 0.2 MGD would cost between $500,000 and $600,000 installed.

Cost data compiled in 1957 for the Barrett Phenol Recovery Process (a multistage extraction process) are reported by Heller, A.N., Clark, E.W. and Reiter, W. (26) and are summarized below, assuming an initial phenol concentration of 5,000 mg./l. and an effluent concentration of 5 mg./l.

| Flow, (mgd) | Phenol Recovery Value per Year ($) | Capital Costs ($) |
|---|---|---|
| 0.01 | 5,500 | 90,000 |
| 0.10 | 40,000 | 200,000 |
| 0.17 | 70,000 | 260,000 |
| 0.50 | 200,000 | 670,000 |

Capital costs decrease for lower removals. Assuming an inflow of 5,000 mg./l., and allowing an effluent concentration of 50 mg./l. lowers the capital costs of $150,000 for a 0.1 thousand gallons per day plant and $70,000 for a 0.01 MGD plant. Increasing effluent phenol concentration to 500 mg./l. (90% removal) reduces the 0.01 MGD costs to $60,000 (26). Reduced recovery value is, however, associated with decreased recovery.

*Intermediate Level Treatment:* An arbitrary classification for intermediate phenol concentration might be 5 to 500 mg./l., since this encompasses the levels resulting from recovery process residuals up to levels that are not sufficiently high to warrant phenol recovery (from an economic viewpoint). In the absence of high concentration of toxic substances or in the event of their successful prior removal, biological treatment is widely employed for treatment of wastewaters containing intermediate phenol levels. The biological processes include lagoons, oxidation ditches, trickling filters and activated sludge. Benger (20) claims that the activated sludge process is preferable because of its high treatment efficiency and ease of control.

Phenol concentrations from 50 to 500 mg./l. are generally considered suitable for treatment by biological processes, but much higher levels have been successfully subjected to biological treatment. The exact concentration of phenol that can be directly treated biologically will depend to some extent upon other contaminants in the waste, and whether or not these contaminants are toxic to the microorganisms employed in the biological process. The accumulative effect of toxic constituents in phenolic wastes has been discussed by Reid and Libby, wherein chromium accumulation gradually decreased the biological activity in a phenol treatment system (3).

Installation costs for a Bethlehem Steel activated sludge system was reported as being $310,000 (1962 prices) for a 0.11 MGD plant (11). Costs included supplemental aeration equipment. Benger (20) reports installation costs of $800,000 and 1965 operating costs of 22 cents per 1,000 gallons for a 1.92 MGD activated sludge system treating phenolic wastes.

Sample and Rea (29) report capital costs approximately $1,000,000 for an air flotation-holding basin–trickling filter combination capable of handling a phenolic waste stream totaling 6.5 MGD. These values seem to correspond closely to conventional biological unit operation costs.

Culp and Culp (34) report capital costs of $1.25 million for the Lake Tahoe municipal sewage activated sludge plant of 7.5 MGD design capacity. Capital and operating costs of activated sludge and trickling filtration processes, as a function of design flow, are presented by Smith (30).

Other treatment methods have been employed for phenol removal at intermediate phenol concentrations. Several articles in 1951 and 1952 dealt with chemical oxidation of phenols (8)(31)(32). Chamberlin and Griffin (32) report that coagulation with alum and iron salts at various pH value removes only 10 to 20% of phenols at initial levels of 100 to 125 mg./l. Only 62.4% removal was accomplished at 125 mg./l. with permanganate oxidation, and at a cost of $1.83/1,000 gals.

However, the authors suggest that chlorination plus lime addition would result in 100% removal. With chlorination, extremely high concentrations of chlorine must be added to effect complete phenol removal (31)(32).

Chlorination must be carried to completion and at pH less than 7, or toxic chlorophenols are produced (31). Using chlorination, residual phenol was reported as 0.00, reduced from ~5 mg./l. for rubber reclamation wastes (8).

Ozone and chlorine dioxide have also been reported as successful oxidants for phenol removal (8)(19)(23)(31). The relationships between oxidant dose and phenol removals have been published (31). Although ozone is capable of efficient phenol removal, McPhee and Smith (19) indicate that operating and capital costs would be extremely high and that ozone should be considered only as a low level or polishing treatment. Besselievre (28) reports that 1.5 to 2.5 parts ozone to 1 part phenol is effective for phenol removals down to 3 parts per billion.

Schutt and Loftus (33) have reported on a wastewater hydrocarbon stripper which reduced initial phenol levels of 5 to 6 mg./l. to less than 1 mg./l. Stripped gases were combusted rather than discharged to the atmosphere. Capital costs were estimated at $25,000 for a 0.18 MGD wastewater flow and $81,000 for 1.1 MGD wastewater flow. Operating costs for the fully automated units were estimated at 9.3 and 7.0 cents/1,000 gal. respectively. Costs were expected to drop to 5 to 6 cents/1,000 gal. for 1.3 to 2.0 MGD plants.

*Dilute Wastes Treatment:* Biological treatment generally is capable of reduction of phenol down to the 1 to 2 mg./l. level, except for very high influent concentrations. Removal below the 1 to 0.1 mg./l. level can be achieved by other processes. Normally, chemical or physical-chemical methods replace the biological processes for treatment of these dilute phenolic wastes.

McPhee and Smith (19) report that biologically treated refinery waste with a phenol residual of 0.16 to 0.35 mg./l. was reduced to 0.003 mg./l. with ozone treatment. The ozone treatment was reportedly so effective that an activated carbon polishing unit designed to follow the ozonator was not required. Culp and Culp (34) report capital and operating costs for general ozone treatment at $2.80 per thousand gallons per milligrams per liter of ozone dosage required, for a 10 MGD plant. For levels at which McPhee and Smith (19) applied ozone, costs would be of the order of $80 to $100 for a 10 MGD plant. Their plant capacity was approximately 1 MGD capacity.

Activated carbon use is also well established for removal of trace organics as a tertiary treatment process (34). It was reported (19) that a good grade of activated carbon should have a capacity for removal of 0.09 mg. of phenol for every 20 mg. of activated carbon. Designing for phenol removal from 0.020 down to 0.005 mg./l., Ross (27) summarized capital and operating costs for granular activated carbon treatment of secondary municipal treatment effluents; costs for activated carbon treatment of like wastes should be similar (30).

*Proprietary Processes:* A process developed by R.E. Dickey and W.A. Weaverling (35) permits substantially complete recovery of purified cresylic acids from aqueous caustic refinery waste streams containing cresols, mercaptans and neutral oils. In the process the caustic solution is contacted directly with the hot combustion gases to partially neutralize the solution to a pH of about 11. The partially neutralized aqueous caustic cresylate is simultaneously oxidized by the excess oxygen introduced during the contacting with the hot submerged combustion gases, thereby converting mercaptans and thiols to disulfides.

The contacting with the hot submerged combustion gases and the free oxygen may be continued until the mercaptan number is reduced to not more than about 30 and is preferably extended to obtain a substantially zero mercaptan number. Such conversion to disulfides at the defined pH range can be effected in a relatively short time, of the order of 2 to 10 hours, after the solution reaches the boiling point. By using an oxidation catalyst such as nickel cresylate, nickel hydroxide, nickel sulfide, etc., or other known types of oxidation catalysts, the time required for substantially complete conversion of the mercaptans to disulfides may be further decreased, thereby permitting removal of the organic sulfur before the carbonation springs the cresols.

The hot submerged combustion gases generated within the caustic extract solution raises the temperature of the solution to its boiling point and steam is driven off. The action of the steam generated in the solution serves to distil the neutral oils from the solution during the conversion of the mercaptans to disulfides and the partial acidifying of the solution by carbonation. When the desired extent of conversion to disulfides has been effected, the passage of the combustion gases into the solution may be stopped and two layers allowed to separate by settling.

The supernatant organic layer contains the disulfides and can be withdrawn from the system. Unseparated disulfides may be scrubbed from the solution by means of an organic solvent such as a hydrocarbon oil. The passage of hot combustion gases through the residual solution from which the disulfides have been removed is then resumed for completing the carbonation of the solution and the liberation of the cresylic acids.

Water is continuously added to the caustic solution during the carbonation period during which springing is effected in order to maintain the concentration of the solution approximately constant. Such water may comprise condensate recovered from the system. The return or introduction of water is necessary throughout the carbonation period, since evaporation proceeds more rapidly than carbonation. When the cresylic acids have been liberated (as determined by sampling) the passage of the submerged combustion gases into the solution is again stopped and the two layers allowed to separate by settling.

The liberated cresols which accumulate in an upper immiscible phase are removed before final concentration. The carbonation period may require between about 8 and 10 hours, and ultimate concentration to a solid salt residue comprising sodium carbonate requires an additional 2 or 3 hours. The solid or crystalline residue which remains upon completion of the evaporation is substantially uncontaminated by cresylic acids and is removed manually from the contacting vessel.

Previously the disposal of the contaminated crystalline salts resulting from the ultimate concentration of the aqueous caustic solution has been difficult in view of the fact that sulfides and cresylic acids are considered by public health authorities to be particularly undesirable for discard to public waters. However, by this procedure stream pollution is alleviated because the salt liquor or solid salts remaining after the cresols and neutral oils have been removed are not obnoxious contaminants, the chemical compounds which remain being present in natural waters. In fact, the resulting carbonate salts may be used in water treatment systems. Figure 112 shows the essential features of the process. The treatment of a crude caustic cresylate solution obtained by extraction of heater oils and containing about 18.6 weight percent sodium hydroxide, 2.2 weight percent sodium sulfide, 0.4 weight percent sodium chloride and 18.7 liquid volume percent cresols along with solutized neutral oil will be described with reference to the figure. In this example, the thiol content of the waste caustic solution was relatively low and no step of separating initially formed disulfides was performed prior to the springing of the cresols. This crude caustic cresylate solution is charged to a contacting vessel 10 by line 11.

A preheated submerged combustion burner 12 is lowered into the vessel 10 via manhole 13 and hot combustion gases are discharged from the burner directly into the caustic extract solution. The construction of the submerged combustion burner 12 consists of three essential parts, a mixing chamber which will produce a homogeneous gas-air mixture, a flame arrester which may comprise a velocity tube through which the gaseous-air mixture flows at a rate greater than the rate of flame propagation and a combustion chamber containing a refractory surface which becomes incandescent and acts as an ignition point for the gas-air mixture. A submerged combustion apparatus of this general type is described in *Industrial and Engineering Chemistry* (September 1933), page 984 et seq.

An unglazed porcelain or zirconia tube is satisfactory for the refractory surface and has been found to glow brilliantly in the combustion zone and keep the flame continuously ignited. Adequate insulation between the combustion tube and an outer metal jacket is important. This insulating material may be diatomaceous earth or other finely divided refractory substance. A burner constructed in this manner remains ignited throughout the 10 to 12 hour carbonation and evaporation periods.

Before submerging the burner 12 into the solution, it is preheated for about 10 minutes in air to bring the combustion zone refractory liner to a cherry red. The quantity of gas and air supplied by valved lines 14 and 15 to the burner is measured by flow meter means 16 and 17 adjusted to a heat input of about 900 Btu/hr./gal. of solution charged to the contactor 10. An excess of air of between 10 and 20%, preferably about 12 to 15%, is supplied so that both free oxygen and combustion gases may be introduced into the contactor 10.

As the heating progresses the temperature of the aqueous caustic solution is increased to its boiling point and neutral oils together with some water vapor are withdrawn overhead from the contactor 10 via line 18 and condenser 19 to separator 20. From this separator flue gases are withdrawn via line 21, neutral oils via line 22, and water via line 23. Ordinarily the condensate water from separator 20 may be recycled to the concentrator 10 via line 11 except during the ultimate concentration step when it may be discarded by valved line 24.

The submerged combustion gases discharged into the solution comprise nitrogen, carbon dioxide, steam and free oxygen together with a small amount of carbon monoxide. The temperature is maintained at the boiling point of the caustic cresylate solution and at atmospheric or slightly superatmospheric pressure for a period sufficient to carbonate the solution to about pH 9, i.e., to a pH low enough to effect springing or liberation of cresylic acid as a separate phase.

During the liberation or springing step, it is necessary to return condensate from separator 20 via lines 23 and 11 or to add water via line 26 throughout the carbonation period of between about 8 and 10 hours because evaporation of water from the solution proceeds more rapidly than carbonation. Sufficient water should be maintained in the solution to prevent the boiling temperature from exceeding about 300°F. and to facilitate the

## FIGURE 112: APPARATUS FOR CRESYLIC ACID RECOVERY FROM CAUSTIC REFINERY WASTE STREAM

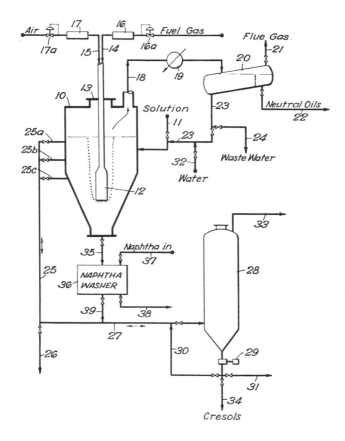

Source: R.E. Dickey and W.A. Weaverling; U.S. Patent 2,686,105; August 10, 1954

subsequent separation of the cresol phase. Dilute caustic cresylate solution, introduced by line **11** as condensate is collected, may comprise the source of added water and is advantageous in this case since it is desirable that the cresol content be high in the springing step. Although agitation is violent and the hot gases from the burner **12** are finely dispersed, the submerged combustion conversion offers no difficulty with foaming. When the cresylic acids have been liberated, the passage of the combustion gases into the solution is interrupted and the liberated cresols accumulate in an upper immiscible phase which can be withdrawn to the desired level via lines **25a, 25b** or **25c** before final concentration of the salts in treater **10**.

The withdrawn cresols are transferred via lines **25** and line **27** into the separator **28**. In this separator the small amount of aqueous caustic solution withdrawn with the liberated cresylic acids is allowed to separate by settling into two layers; the lower layer comprising the caustic solution is withdrawn via pump **29** and returned via lines **30,27** and **25** to the convertor **10** for further and ultimate concentration. Ultimate concentration of solid residue is obtained in this case containing 0.2 weight percent sodium chloride, 0.2 weight percent sodium sulfate and 34.9 weight percent sodium carbonate contaminated by a very minor amount of cresols, such solid residue being in contact with a mother liquid containing 1.2 weight percent sodium chloride, 1.0 weight present sodium sulfate and 19 weight percent sodium carbonate with a very small amount of cresols. The cresylic acids remaining in separator 28 can then be distilled under vacuum from vessel 28 or can be withdrawn via pump 29 and valved line 31 for storage or further purification and/or processing.

A process developed by V.A. Hann and S.J. Niegowski (36) involves the treatment of liquid effluents containing phenolic type compounds so as to render such effluents unobjectionable, in so far as the phenolic substances are concerned, and thereby to facilitate the safe disposal thereof.

Certain liquid effluents, such as those resulting from coke oven operations, are found to contain phenolic type substances such as phenol, cresols, xylenols, and the like. Phenolic type substances are objectionable in a disposable waste because of their toxicity to animal and plant life, and particularly because of the odor and taste intensity imparted by such substances to the water into which such a waste is discharged, especially when water containing such compounds is chlorinated at water treatment plants.

A specific example of this type of waste is the effluent from the ammonia stills in coke oven operations. The ammonia still effluents, even where the oven liquors are passed through a dephenolizer, contain a residual concentration of phenolic type compounds which hereinafter will sometimes be referred to as phenols.

It has been proposed to subject these phenolic wastes to a chlorine oxidation treatment in order to remove the phenols. However, effective chlorination requires a large excess of chlorine, it being estimated that approximately 5,000 to 6,000 ppm is required to insure oxidation of the phenols; and this would raise the chloride content of the waste considerably. Since a large excess has to be removed before final discharge because, when a waste having a high residual concentration of chlorine is admixed with other phenolic waste materials (which also are produced normally in coking operations) very objectionable forms of chlorinated phenols are produced.

Furthermore, chlorine preferentially attacks ammonia so that, in the treatment of coke oven ammonia still wastes, more oxidant is required to oxidize the phenols than would otherwise be required. Beyond this, chlorinated phenols are compounds of very high toxicity and they introduce a taste and odor intensity of about a hundred or more times that of phenol alone. In addition, such chlorination treatments are expensive and require very close laboratory control.

Accordingly, it is an object of this process to avoid objections of the kind just noted and to provide an economical method for the treatment of phenolic wastes, particularly coking plant wastes, so as to render them unobjectionable. It is another object to provide such a process which is simple to operate and which requires a minimum of laboratory control or other manual attention. These objects and advantages are accomplished by oxidizing the phenolic type compounds contained in an effluent by passing ozone into such a liquid effluent at alkaline pH's.

The ozone treatment can be carried out in any equipment capable of being employed to effect the dispersion of gases in liquids. Preferably the ozone is introduced into the liquid through orifices placed near the bottom of a treating tank, basin, tower, or the like, and allowed to percolate to the surface. The treatment can be carried out either as a batch or continuous process. It is possible to employ either ozone or ozone diluted with other gases such as ozonized air or ozonized oxygen and such substances can be produced by any of the well-known methods.

It has been found that, in spite of the known instability of the ozone in alkaline solutions and particularly within the higher pH ranges, it is possible to remove phenols from liquid wastes with ozone even if the waste is highly alkaline and pH's over 10.5 have been found to be especially suitable.

The phenols are thereby oxidized to unobjectionable forms and such treated wastes can be safely discharged without fear of phenolic pollution. Further it has been found that when the ozonation of the phenols is carried out in effluents on the alkaline side which contain other oxidizable impurities, the phenols will be preferentially oxidized if the initial pH is maintained at a range between about 10.5 and 12.5, with an optimum generally in the neighborhood between 11.5 and 12.5 depending upon the individual waste under treatment.

A process developed by W.T. Dickens and A.D. Evans (37) utilizes an apparatus adapted to destroy the phenolic components of aqueous solutions by biological oxidation. The disposal of aqueous phenolic wastes constitutes a serious industrial problem, particularly in that the concentration of phenolic components contained in effluent waters discharged into rivers and streams must in many cases be reduced to the order of a few parts per million. It is known that phenolic compounds can be destroyed by biological oxidation in the presence of certain strains of aerobic bacteria, but the methods and apparatuses available for carrying out such reactions are not readily adaptable to the handling of industrial phenolic wastes.

Such oxidations can be carried out by passing the aqueous phenolic wastes through rock filled, tricklebed filters, but for a number of reasons, including their massive size, tricklebed filters are not an attractive apparatus to be employed in industrial installations. Aqueous phenolic wastes also can be oxidized in activated sludge beds, but the sensitivity of the activated sludge process to chemical and thermal shock renders its use hazardous in industrial installations. It is an object of this process to provide an improved apparatus for effecting the biological oxidation of aqueous phenolic wastes.

Decomposition of the phenolic wastes is started by inoculating the aqueous wastes with a suitable strain of aerobic bacteria. At the same time air is introduced into the apparatus through an inlet pipe and thoroughly aerates the aqueous wastes which flow downwardly through the apparatus in a tortuous cascade path. The air which is introduced is maintained at a predetermined temperature which favors the growth of the particular strain of aerobic bacteria introduced into the aqueous wastes. The phenolic content of the circulating liquid is followed by analysis and, when the phenolic concentration has fallen to a predetermined level, fresh feed of aqueous phenolic wastes is introduced into the apparatus. At this time dephenolized liquid is withdrawn from the apparatus.

A process developed by G. Stoneburner (38) involves removing phenol from wastewaters used in a fluid catalytic cracking process. In the fluid catalytic cracking process oxygenated compounds, such as phenolic compounds are produced as by-products to the main cracking reaction. In some fluid catalytic cracking operations water is used as a quench and water also enters the process in form of steam, which is condensed. Due to the solubility of the phenolic compounds in water these quench and condensate waters collect phenols and other phenolic compounds produced as a by-product of the operation. In addition, the glands of high temperature and high velocity pumps require cooling water.

In the refinery this water is commonly called gland water. The gland water becomes saturated with the water-soluble phenolic compounds that inadvertently leak into such water. Cooling and wastewaters from refinery operations can also become contaminated with water-soluble phenolic compounds from other sources in the refinery. These phenolic-containing wastewaters present a real problem in the wastewater disposal of a refinery which is either located on inland waters where such waters are used for recreational or drinking purposes or where they are located on inland bays or river outlets which are used for recreation or where oysters and other fish foods are grown for commercial purposes and thus protected by fish and game laws.

One of the methods which can be used to remove the phenols from wastewater is by chemical reaction. However, this is expensive and essentially prohibitive on a very large scale. One refinery uses the high phenolic water in the crude desalting operation. During this operation the crude will absorb up to about 90% of the phenolic compounds. However, the amount of phenolic compounds still remaining in the crude desalter effluent water is substantially higher than is permissible to discharge into the effluent wastewater stream from the refinery. Phenolic compounds are particularly objectionable in drinking water, as well as in waters containing oysters, clams and other fishes, since relatively small amounts will yield disagreeable odors and taste.

It has been discovered that if the high phenolic-containing wastewater from refinery operations is used as the cooling water to cool a hot bed of coke contained in the coking drum of a delayed coking unit, the phenolic compounds will be removed by some manner when in contact with the coke so as to reduce their amount in the wastewater sufficient to enable the wastewater to be discharged directly to the refinery effluent or require a very small amount of treatment before such discharge. Thus a dual purpose is served. Since fresh water is normally used for cooling the hot coke bed in the coking operation, fresh water is conserved at refineries which are located in areas where fresh water supply is limited.

A process developed by I. Herrick, M.F. Adams and E.M. Huffaker (39) involves selectively and substantially quantitatively removing water-soluble compounds from aqueous streams, which compounds include ortho-hydroxyphenols by the addition of MgO-containing compounds to aqueous streams containing such phenols.

A process developed by W.E. Walles (40) is one in which a phenolic substance is removed from water in which it is dispersed by contacting the water bearing such compound with an insoluble polymer comprising recurring morpholinone groups.

A process developed by N.N. Li, R.P. Cahn and A.L. Shrier (41) is based on the discovery that a liquid membrane such as that disclosed in U.S. Patent 3,454,489 may be utilized to separate organic compounds such as phenol from wastewaters. The liquid membrane coats emulsion size droplets which contain a material more basic than or acidic than the organic compound; in the event the organic compound is phenol, a strongly basic material would be used such as sodium hydroxide, ammonium hydroxide, lime, soda ash, potash, etc.

The organic compound, such as phenol, will permeate through the particular surfactant membranes utilized; these membranes are liquid surfactant membranes formed from a solution of oil-soluble surfactants in an organic compound which is immiscible with the aqueous stream being treated and also immiscible with the aqueous solution contained therein in the form of droplets. Preferably, the solubility of the membrane materials in the aqueous phases should be very low to prevent losses and contamination.

A process developed by C.S. Wang and J.P. Easterly, Jr. (42) is one in which carbocyclic oxyaromatic and heterocyclic contaminants such as phenols, pyridinols, benzofuran and derivatives thereof in water are destroyed by dissolving excess alkali in the contaminated water and contacting the alkaline solution with sufficient chlorine to reduce the pH below seven.

A process developed by M. Adegeest (43) involves the removal of impurities from wastewaters resulting from the manufacture of phenol-formaldehyde resins. This process involves the steps of adding at least one phenol to such wastewater until the phenol-formaldehyde molar ratio is brought to from 1:1.02 to 1:1.12; adding reagents as needed to adjust the total acid normality of the wastewater to from 0.015 to 0.10 N; heating the solution for a prolonged interval at 80° to 85°C., and subsequently removing excess water by heating the liquid mixture to a temperature between its boiling point and 5°C. below its boiling point.

At the end of the prolonged heating mentioned, liquid resin is removed from a lower zone in the heated solution to cool storage, volatilized constituents are removed from an upper zone, and water containing reagents added to adjust the normality is removed to an evaporating zone where it is subjected to temperatures sufficiently high to evaporate excess water, the liquid resin remaining being removed from the heated area to cool storage.

A process developed by T.R. Morrow (44) involves the processing of refinery waters to re-cover phenolic constituents produced in catalytic cracking by adsorption with light process naphthas which are then passed to catalytic reforming.

A process developed by R.A. Wiley (45) involves removing and concentrating acidic organic material from a water stream, such as a sour phenolic steam condensate stream from an oil refinery. The water stream is dispersed within an organic liquid solvent for removing substantial amounts of acidic organic material, such as phenol, mercaptans and thiophenols, etc., from the water. The enriched organic solvent phase is separated from the purified water stream phase.

Next, the enriched solvent is intimately contacted with substantially stoichiometric amounts of an immiscible concentrated caustic solution forming a three phase mixture in a second dispersion. This three phase liquid mixture is separated into a regenerated solvent phase, a second liquid phase of the alkali metal salts of extracted acidic organic material, and a third phase of excess caustic solution. Preferably, both phase separations are undertaken in the presence of an electric field.

The regenerated solvent is recycled into contact with the water stream; the high purity alkali metal salts of extracted acidic organic material are passed to some suitable utilization, and the excess caustic solution is recycled for regenerating further amounts of the enriched solvent. Only small amounts of caustic need to be added to maintain a circulating inven-tory of the caustic solution.

*Summary:* Phenol removals to very low residual levels is technologically feasible and in practice. Removals to a few parts per billion are accomplished by multiple treatment sequences, since increasing efficiency for any one process, e.g., 99.0 to 99.9%, usually is associated with an exponential increase in cost. Processes are available for 99% or greater removals for the 500 to 10,000 mg./l. phenol level. Such processes are typically product recovery oriented.

Biological processes have been shown capable of up to 99% phenol removal in the 10 to 500 mg./l. range. Specialized biological systems are capable of treating even higher con-centrations. Although not favorable for direct treatment of concentrated or intermediate levels phenolic waste, chemical oxidation processes such as chlorination and ozonation, and carbon adsorption demonstrate favorable economics when employed as polishing pro-cesses after biological systems. Costs associated with treatment processes generally are of the order of equivalent water and wastewater treatment processes under municipal man-agement.

### References

(1)  McKee, J.E. and Wolf, H.W., *Water Quality Criteria,* 2nd Ed., California State Water Quality Control Board, Publication No. 3-A (1963).
(2)  Reid, W., Daigh, R. and Wortman, R.L., "Phenolic Wastes from Aircraft Maintenance," *Jour. Wat. Poll. Control Fed.* 32, 353 to 391 (1960).
(3)  Reid, W. and Libby, R.W. "Phenolic Waste Treatment Studies," *Proc. 12th Purdue Industrial Waste Conf.,* pp. 250 to 258, (1957).
(4)  Barzler, R.P., Giffels, D.J. and Willoughby, E., "Pollution Control in Foundry Operations," in *Industrial Pollution Control Handbook,* Herbert F. Lund, ed., New York, McGraw-Hill Book. Co., (1971).
(5)  Schesinger, A., Dul, E.F. and Fridy, T.A., Jr., "Pollution Control in Textile Mills," in *Industrial Pollution Control Handbook,* Herbert F. Lund, ed., New York, McGraw-Hill Book Co. (1971).
(6)  Ross, R.D., *Industrial Waste Disposal,* New York, Reinhold Book Corporation (1968).
(7)  Parsons, W.A., *Chemical Treatment of Sewage and Industrial Wastes,* National Lime Association, Washington (1965).
(8)  Sechrist, W.D. and Chamberlin, N.S., "Chlorination of Phenol Bearing Rubber Wastes," *Proc. 6th Purdue Industrial Waste Conf.* pp. 396 to 412 (1951).
(9)  Wurm, H.J., "The Treatment of Phenolic Wastes," *Proc. 23rd Purdue Industrial Waste Conf.,* pp. 1054 to 1073 (1969).

(10) Ide, T., Formal Discussion to Biczysko and Suschka, "Investigations on Phenolic Wastes Treatment in an Oxidation Ditch," in *Advances in Water Pollution Research, Munich Conference*, 2, pp. 285 to 295, New York, Pergamon Press (1967).

(11) Kostenbader, P.D. and Flecksteiner, J.W., "Biological Oxidation of Coke Plant Weak Liquor," *Jour. Wat. Poll. Control Fed.* 41, 199 to 207 (1969).

(12) Fisher, C.W., "Coke and Gas," in *Chemical Technology Volume 2, Industrial Wastewater Control*, F. Fred Gurnham, ed., New York, Academic Press (1965).

(13) Carbone, W.E., Hall, R.N., Kaiser, H.R. and Bazell, C.G., "Commercial Dephenolization of Ammonical Liquors with Centrifugal Extractors," *Proc. 5th Ontario Indust. Waste Conf.*, pp. 42 to 58 (1958)

(14) Resource Engineering Associates, *State of the Art Review on Product Recovery*, U.S. Department Interior, Washington (1969).

(15) Biczysko, J. and Suschka, J., "Investigations on Phenolic Wastes Treatment in an Oxidation Ditch," in *Advances in Water Pollution Research, Munich Conference*, 2, pp. 285 to 295, New York, Pergamon Press (1967).

(16) Clough, G.F.S., "Biological Oxidation of Phenolic Waste Liquor," *Chem. Proc. Eng.* 42, (1), 11 to 14 (1961).

(17) Lesperance, T.W., "Biological Treatment of Phenols," *Proc. 8th Ontario Industrial Waste Conference*, pp. 59 to 66 (1961).

(18) Graves, B.S., "Biological Oxidation of Phenols in a Trickling Filter," *Proc. 14th Purdue Industrial Waste Conf.*, pp. 1 to 6, (1959).

(19) McPhee, W.T. and Smith, A.R., "From Refinery Waste to Pure Water," *Proc. 16th Purdue Industrial Waste Conf.*, pp. 311 to 326 (1961).

(20) Benger, M., "The Disposal of Liquid and Solid Effluents from Oil Refineries," *Proc. 21st Purdue Industrial Waste Conf.*, pp. 759 to 767 (1966).

(21) Steck, W., "The Treatment of Refinery Wastewater with Particular Consideration of Phenolic Streams," *Proc. 21st Purdue Industrial Waste Conf.*, pp. 783 to 790 (1966).

(22) Dickenson, B.W. and Laffey, W.T., "Pilot Plant Studies of Phenol Waste from Petrochemical Operations," *Proc. 14th Purdue Industrial Waste Conf.*, pp. 780 to 799 (1959).

(23) Noack, W., Formal Discussion to Biczysko and Suschka, "Investigations on Phenolic Wastes Treatment in an Oxidation Ditch," in *Advances in Water Pollution Research, Munich Conference* 2, 285 to 295, New York, Pergamon Press (1967).

(24) "The Cost of Clean Water," Vol. III, *Industrial Waste Profile No. 10, Plastics Materials and Resins*, U.S. Dept. Interior, Washington (1967).

(25) Warner, F.E. and Rice, A.P.; U.S. Patent 3,528,220; September 15, 1970; assigned to Fibreglass Limited, England.

(26) Heller, A.N., Clark, E.W. and Reiter, W., "Some Factors in the Selection of a Phenol Recovery Process," *Proc. 12th Purdue Industrial Waste Conf.*, pp. 103 to 122 (1957).

(27) Ross, R.D., "Pollution Waste Control," in *Industrial Pollution Control Handbook*, Herbert F. Lund, ed., New York, McGraw-Hill Book Co. (1971).

(28) Besselievre, E.B., *The Treatment of Industrial Wastes*, New York, McGraw-Hill Book Co., (1969).

(29) Sample, G.E. and Rea, R.D., "Floats Away Refinery Wastes, Oil Phenols Reduced 98%," *Chem. Proc. 32*, (12), 41042 (1969).

(30) Smith, R., "Cost of Conventional and Advanced Treatment of Wastewater," *Jour. Wat. Poll. Control Fed.* 40, 1546 to 1547 (1968).

(31) Cleary, E.J. and Kinney, J.E., "Findings from a Cooperative Study of Phenol Waste Treatment," *Proc. 6th Purdue Industrial Waste Conf.*, pp. 158 to 170 (1951).

(32) Chamberlin, N.S. and Griffin, A.E., "Chemical Oxidation of Phenolic Wastes with Chlorine," *Sew. Ind. Wastes* 24, 750 to 760 (1952).

(33) Schutt, H.C. and Loftus, J., "Wastewater Conditioned by Carrier Gas," *Oil and Gas Jour.* 64 (32), 70 to 72 (1966).

(34) Culp, R.L. and Culp, G.L., *Advance Wastewater Treatment*, New York, Van Nostrand Reinhold Co. (1971)

(35) Dickey, R.E. and Weaverling, W.A.; U.S. Patent 2,686,105; August 10, 1954; assigned to Standard Oil Company.

(36) Hann, V.A. and Niegowski, S.J.; U.S. Patent 2,703,312; March 1, 1955; assigned to The Welsbach Corporation.

(37) Dickens, W.T. and Evans, A.D.; U.S. Patent 2,865,617; December 23, 1958; assigned to Monsanto Chemical Company.

(38) Stoneburner, G.; U.S. Patent 3,284,337; November 8, 1966; assigned to Standard Oil Company.

(39) Herrick, I., Adams, M.F. and Huffaker, E.M.; U.S. Patent 3,350,259; October 31, 1967; assigned to Northwest Magnesite Company.

(40) Walles, W.E.; U.S. Patent 3,492,223; January 27, 1970; assigned to The Dow Chemical Company.

(41) Li, N.N., Cahn, R.P. and Shrier, A.L.; U.S. Patent 3,617,546; November 2, 1971; assigned to Esso Research and Engineering Company.

(42) Wang, C.S., Easterly, J.P., Jr.; U.S. Patent 3,617,581; November 2, 1971; assigned to The Dow Chemical Company.
(43) Adegeest, M.; U.S. Patent 3,655,047; April 11, 1972; assigned to Corodex, N.V., Netherlands.
(44) Morrow, T.R.; U.S. Patent 3,671,422; June 20, 1972; assigned to Mobil Oil Corporation.
(45) Wiley, R.A.; U.S. Patent 3,673,070; June 27, 1972; assigned to Petrolite Corporation.

# PHOSGENE

## Removal from Air

A process developed by H. Richert and E. Zirngiebl (1) involves decomposing phosgene to produce hydrochloric acid and carbon dioxide. It has been proposed in U.S. Patent 2,832,670 to decompose phosgene by intimately contacting it with activated carbon and water. Water must be used in such proportions to the quantity of phosgene to be decomposed that the concentration of the hydrochloric acid does not exceed 10%. In other words, at least 7.5 parts of water must be used per part of phosgene and the low concentration acid produced must be neutralized and discarded with 9 times its weight of water. It is stated in the patent that a high concentration of hydrochloric acid prevents the complete hydrolysis of phosgene.

In the production of isocyanates by phosgenating amines, large quantities of waste gases which contain principally hydrogen chloride, phosgene and inert gases are formed. It is necessary to decompose the phosgene because of its highly poisonous property. In the old process for decomposing phosgene, large amounts of wastewater and low concentration acid are produced and seriously add to the expense of the process. Moreover, the dilute hydrochloric acid produced is valueless. This process accomplishes the decomposition of phosgene by contacting it with water in the presence of activated carbon in the vapor phase at a temperature above the dew point of the mixture of water and phosgene.

This process involves working up waste gases which contain hydrogen chloride and which may also contain other impurities in addition to phosgene and inert gases, and comprises adiabatically absorbing the hydrogen chloride and hydrolyzing the phosgene on active carbon, wherein, possibly after an active carbon absorption, the hydrogen chloride is first of all removed from the waste gas by adiabatic absorption with water and/or dilute hydrochloric acid, concentrated hydrochloric acid being formed.

Thereafter, the gas mixture still containing substantially phosgene and steam (possible after adding more steam) is conducted at temperatures above the dew point of the mixture over active carbon, whereby a molar ratio of water:phosgene of at least 1:1, preferably of between 5:1 and 20:1 is adjusted, the hydrogen chloride thereby formed is condensed and the hydrochloric acid is recovered by separation from the residual constituents of the gas mixture.

The essentials of the process are shown in the flow diagram in Figure 113. Through a pipe 1, the system is charged with crude gas, which contains volatile organic and inorganic impurities in addition to HCl and phosgene. Of two initial purification towers A and B which are filled with active carbon, one tower is in operation and the other is switched over for regeneration. When tower A is used as initial purifier, the gas enters by way of the opened valve a by way of a pipe 10. The tower B which is under regeneration is charged through a valve b which is only slightly open and by way of a pipe 3, since the gas stream for the purification represents only a fraction of the total stream and is heated by suitable means (e.g., steam heating) to about 150° to 250°C.

The gas charged with the desorbed impurities leaves the tower by way of a valve b' in a pipe 4, which opens into a pipe 6. This latter pipe leads to a gas condenser C, in which most of the desorbate is condensed, while the gas is supplied through a pipe 8 to a fan D which returns it through a pipe 9 into the pipe 1.

FIGURE 113:  BLOCK FLOW DIAGRAM OF PROCESS FOR DECOMPOSITION OF
PHOSGENE IN WASTE GASES

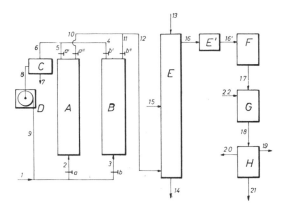

Source:  H. Richert and E. Zirngiebl; U.S. Patent 3,314,753; April 18, 1967

The separate impurities are drawn off through pipe **7**.  With the switching over of the containers from **B** to **A**, the valves are appropriately adjusted.  Thus the crude gas flows through
pipe **3** into the tower **B** and leaves it through pipe **11**, while the regeneration gases flow
through the pipe **5** into the pipe **6**.  The initially purified hydrogen chloride gases flowing
in a pipe **12** into adiabatic absorption tower **E** are sprinkled with water through pipe **13**.

A substantially 30% hydrochloric acid leaves the tower through a pipe **14**.  The inert gases
and the phosgene (30 to 80% of the phosgene crude gas content) leaves the tower **E** with
the steam formed through a pipe **16** and are brought in a preheater **E'** to a temperature
above the dew point (95° to 120°C.).  Through a pipe **16'**, this gas mixture now enters an
active carbon reactor **F**, in which the reaction is completed in accordance with the equation

$$COCl_2 + H_2O \xrightarrow{\text{active carbon}} CO_2 + 2HCl$$

The final temperature being adjusted with the phosgene hydrolysis on active carbon depends on the ratio between steam and phosgene and can be between 120° to about 400°C.
or even higher.  The dephosgenated gas containing hydrogen chloride is now condensed in
a gas condenser **G**, which it reaches by way of a pipe **17**, and flows through pipe **18** into
a receiver **H**.  The acid which is formed can be drawn off through a pipe **20** and either used
as such or again fed to the tower **E** by means of a pump (not shown) at a position suitable for this purpose, e.g., pipe **15**.

Any possibly settling impurities can be drawn off through a pipe **21**.  The cold gas, which
consists mainly of $CO_2$ and inert gases, is supplied by way of a pipe **19** and a waste gas
scrubber to a chimney.  With high phosgene contents, such a quantity of water is injected
through a pipe **22** for complete precipitation in the condenser **G** of the hydrogen chloride
formed by the hydrolysis that the acid discharging from the receiver **H** by way of a pipe
**20** is about 30%.

If the pure acid in the gas condenser **G** shall not be recovered, the initial purification can
be excluded, this being achieved by direct passage from pipe **1** to pipe **12**.  In this case,
very volatile substances which are in the crude gas and which are insoluble in hydrochloric

acid are deposited in the receiver H and can be extracted therefrom. The acid forming at H can be worked up directly in the tower E. If the proportions of phosgene in the crude gas is low and the acid thereby formed in H is very dilute, it can be admixed with the water in the pipe 13. Therefore in no case is any wastewater formed.

A process developed by R.W. Beech and J.T. Polley (2) for removing phosgene from gas streams especially those containing relatively small quantities of phosgene, comprises contacting the gas streams with crystalline alumina in the presence of sufficient water to dissolve the HCl formed by the resulting conversion of phosgene.

### References

(1)  Richert, H. and Zirngiebl, E.; U.S. Patent 3,314,753; April 18, 1967; assigned to Farbenfabriken
     Bayer AG, Germany.
(2)  Beech, R.W. and Polley, J.T.; U.S. Patent 3,411,867; November 19, 1968; assigned to Allied Chemical
     Corporation.

## PHOSPHATES

Phosphates occur in rivers and wastewaters from a variety of sources such as the following: detergent laden wastes where phosphate builders were present in the detergent formulations; agricultural runoff from lands fertilized with phosphate-containing fertilizers; and miscellaneous industrial and municipal wastewaters.

### Removal of Phosphates from Water

A review of available processes for phosphate removal from water has been prepared by Cecil (1).

A process developed by L.E. Lancy (2) involves removing and neutralizing phosphate type waste or carry-over on workpieces from a metal finishing bath. The carry-over usually consists of large concentrations of iron or zinc phosphates or both and some free phosphoric acid which has been used for phosphatizing, pickling, or metal surface preparation.

An aqueous chemical treatment wash solution is used having a pH of less than about 8 and is applied to surfaces of the workpieces during their movement; the solution as thus contaminated is circulated in a system having a treatment solution reservoir or tank, and is subjected to the introduction of hydrated lime, slaked lime, or powdered limestone, either immediately before its introduction into the reservoir or at the time of its introduction, in an amount determined to be sufficient to precipitate and settle-out iron and zinc as well as calcium phosphates in the reservoir.

Thereafter, the reconditioned solution is moved from the treatment reservoir back to a workpiece treating tank to provide a continuous washing off of the surfaces of the metal workpieces. Care is taken to assure that the solution as returned to the treating tank is, for all practical purposes, free of dissolved calcium compounds. An alternative is to employ a caustic soda addition to the solution for precipitating the metal phosphates and to employ a small quantity of calcium ion added as a secondary treatment to the solution for the purpose of removing the minor constituent of the carry-over, namely, free phosphoric acid.

A process developed by J. Block (3) is one in which phosphate ions in wastewaters can be removed from the water by first adding an ion which, when added in sufficient amounts, would form an insoluble phosphate precipitate, but here, being added in relatively small amounts, does not per se form such precipitate; then adding an anionic surfactant to form a phosphate-containing precipitate, and floating the precipitate to the surface with bubbles. The precipitate which has been floated to the surface can then be removed in the resultant froth, leaving the remainder of the solution relatively free of phosphate ions. Figure 115 shows a laboratory scale apparatus suitable for the conduct of this process. The solution which contained excess $PO_4^{\equiv}$ was put into container 1, to which was connected an air line 2, which in turn was connected to a compressed gas cylinder. A precipitating agent 5 was added at the top of the container 1, followed by the addition of anionic surfactant 6, also through the top of the container. As air was released into the solution via the air line 2, bubbles 4 floated to the top of the solution and deposited a foam 7, which contained the precipitated phosphate complex, which was easily removed. A glass tube 3 then removed the foam to a collector 8 which collected the phosphate-containing foam.

**FIGURE 115: APPARATUS FOR REMOVAL OF PHOSPHATE ION BY FLOTATION WITH AN ANIONIC SURFACTANT**

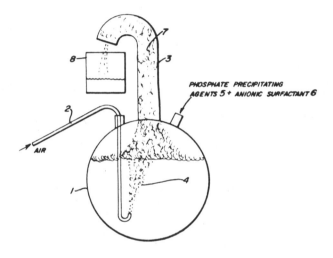

Source: J. Block; U.S. Patent 3,583,909; June 8, 1971

A process developed by D.F. Bishop and J.B. Stamberg (4) is one in which nitrogen and phosphorus are removed from wastewaters by a process including the steps of biological nitrification, chemical precipitation of phosphorus and biological denitrification. Buffering capacity of the water is substantially reduced and in some cases nearly eliminated by reaction of acid, produced in the nitrification step, with bicarbonate ion contained in the wastewater. A precipitate containing phosphate in high concentration is recovered at low chemical cost.

A process developed by R.B. Hudson, R.E. Mesmer and R.A. Rose (5) involves recovering phosphorus values from plant effluents by precipitating the phosphorus values with lime. The sludge obtained which is difficult to handle is improved by treating the sludge with phosphoric acid so as to convert phosphorus values in the sludge from basic calcium phosphate to crystalline calcium phosphates which are readily filterable and useful as an animal feed supplement.

A process developed by G.V. Levin and G.J. Topol (6) is an activated sludge sewage treatment process in which phosphates are removed from phosphate enriched sludge by aerating the phosphate enriched sludge with an oxygen-containing gas. During aeration, the organisms in the sludge, after consuming the available food substrate, go into endogenous respiration, consuming much of their own cellular material.

Thus, the aeration serves to reduce the volume of sludge as well as to cause the organisms in the sludge to release phosphate.  A phosphate enriched supernatant liquor is formed on settling.  The sludge, having a reduced phosphate content, is separated from the phosphate enriched supernatant liquor and at least a portion thereof is recycled for mixing with influent sewage material in an activated sludge sewage treatment process.  The resultant mixed liquor is aerated to reduce the BOD content and to cause the organisms present to take up phosphate and phosphate enriched sludge is separated from the mixed liquor to provide a substantially phosphate-free effluent.

Figure 116 is a block flow diagram of the process.  A raw sewage influent stream 1 is passed through conventional screening and grit removing units and is optionally subjected to primary settling in a tank 2 from which primary sludge is removed in line 3.  The primary settled sewage is mixed with recycled, activated sludge hereinafter described to form a mixed liquor and is passed by line 4 to the aeration tank 5.

In the aeration tank, the mixed liquor is aerated at a rate sufficient to maintain at least about 0.3 mg. of dissolved oxygen per liter of mixed liquor for a period of at least 10 minutes.  During aeration, the bacteria present take up phosphate and consume organic matter present in the sewage.  A high degree of BOD removal is obtained during aeration.

After aeration, the mixed liquor is fed into a secondary settling tank 6.  In the secondary settling tank 6, phosphate enriched sludge settles and thereby separates from the liquor. The sludge contains a substantial portion of the phosphate present in the sewage.  The substantially phosphate-free effluent is discharged for disposal in a conventional manner by line 7.

The phosphate enriched sludge is removed from the settling tank 6 by line 8.  A portion of the sludge may be delivered to waste and the remainder is passed to the phosphate stripper 9.  In the phosphate stripper 9, the phosphate enriched sludge is aerated, preferably at a rate of from 2 to 20 cubic feet of air per gallon of sludge for from 3 to 24 hours. This causes the organisms in the sludge to release the phosphate which they have taken up in the aeration tank 5.  The phosphate leaks out of the sludge into a liquid phase.  The aeration also causes the organisms to undergo endogenous respiration thereby consuming much of their own cellular material.  This results in a reduction of the amount of sludge which must be handled and subsequently disposed of.  The aeration should be controlled to insure that it does not reduce the amount of viable sludge below the amount which is required for recycle to form the mixed liquor.

FIGURE 116:  PROCESS FOR AEROBIC REMOVAL OF PHOSPHATE FROM
ACTIVATED SLUDGE

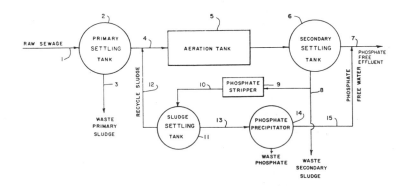

Source:  G.V. Levin and G.J. Topol; U.S. Patent 3,654,146; April 4, 1972

The amount of sludge required for recycle depends on the strength of the sewage, the phosphate content of the sewage and the amount of viable organisms in the sludge. If aeration is carried out for too long, there are not sufficient viable organisms left in the recycle sludge to effectively consume the available nutrients and to take up the phosphates from the sewage. The sludge having a reduced phosphate content is separated from the liquid phase by passing the aerated mixture from the phosphate stripper 9 by line 10 to the settling tank 11. After settling, the sludge is withdrawn from the settling tank 11 and passed by line 12 for mixing with the raw sewage which is being fed to the aeration tank 5.

A phosphate enriched liquid phase is produced in the phosphate stripper 9. This liquid phase is withdrawn as a supernatant liquor from the settling tank 11 and is passed by line 13 to the phosphate precipitator 14. A phosphate precipitant, such as lime, is mixed with the phosphate enriched supernatant liquor in the phosphate precipitator 14 to precipitate phosphate. The phosphate precipitate may be combined with any waste phosphate enriched sludge removed from the secondary settling tank 6 and converted into a fertilizer or otherwise disposed of by conventional methods.

A phosphate-free supernatant liquor is withdrawn from the phosphate precipitator 14 and passed by line 15 to line 7 wherein it is combined with the phosphate-free effluent from the secondary settling tank 6. This process not only reduces the phosphate content of the phosphate enriched sludge, but it also reduces the amount of waste sludge which must be disposed of to less than 50% of the amount which would be produced in a conventional activated sludge sewage treatment process.

A process developed by H.G. Flock, Jr. and E.G. Rausch (7) is one in which phosphate is removed from municipal and industrial wastewater by treating the water with a synergistic admixture of a water-soluble high molecular weight nonionic polymer, preferably polyacrylamide, and a water-soluble salt containing ferric ions, preferably ferric chloride.

A process developed by M.A. Kuehner (8) is one in which aluminum surfaces with good corrosion resistance and paint adhesion properties, together with a waste liquid substantially free of objectionable ions are produced when the aluminum is treated with a solution of phosphate, molybdate and fluoride, and when the waste stream is rendered basic with lime (pH 11) and then neutralized with sulfuric acid (pH 7).

### References

(1) Cecil, L.K., "Evaluation of Processes for Removal of Phosphorus from Wastewater," *Report PB 211,191,* Springfield, Va., Nat. Tech. Inf. Serv. (Feb. 1972).
(2) Lancy, L.E.; U.S. Patent 3,562,015; February 9, 1971; assigned to Lancy Laboratories, Inc.
(3) Block, J.; U.S. Patent 3,583,909; June 8, 1971; assigned to W.R. Grace & Co.
(4) Bishop, D.F. and Stamberg, J.B.; U.S. Patent 3,617,540; November 2, 1971; assigned to U.S. Secretary of the Interior.
(5) Hudson, R.B., Mesmer, R.E. and Rose, R.A.; U.S. Patent 3,650,686; March 21, 1972; assigned to Monsanto Company.
(6) Levin, G.V. and Topol, G.J.; U.S. Patent 3,654,146; April 4, 1972; assigned to Biospherics Inc.
(7) Flock, H.G., Jr. and Rausch, E.G.; U.S. Patent 3,655,552; April 11, 1972; assigned to Calgon Corp.
(8) Kuehner, M.A.; U.S. Patent 3,697,332; October 10, 1972; assigned to Amchem Products, Inc.

# PHOSPHORIC ACID

## Removal from Air

In the production of phosphoric acid by the thermal process, the major pollutant is phosphoric acid mist, according to Athanassiadis (1). There are several pollution control devices used which are discussed below; one of the most important factors influencing emission rates is maintenance of the collection equipment.

*Scrubbers:* The use of packed and open tower scrubbers is a simple and low-cost abatement technique. Collection efficiencies vary from 40 to 95% for gas velocities of 2 to 7 feet per second (2). Improved efficiency has been achieved by installing wire mesh mist eliminators after the scrubber. Venturi scrubbers are also used and have higher collection efficiency, ranging from 78 to 98% for particle sizes of from 0.5 to 1$\mu$, respectively.

*Cyclones:* Cyclones have comparatively low collection efficiencies since they are effective only in the particle size range of 10$\mu$ and above. In some plants they are complemented with wire mesh mist eliminators or used in series with venturi scrubbers. In this case a collection efficiency of up to 99.9% of the acid mist can be achieved (2).

*Fiber Mist Eliminators:* Plants using glass fiber mist eliminators operating with gas velocities ranging from 0.4 to 13 ft./sec. provide collection efficiencies of 96 to 99.9%.

*High Energy Wire Mesh Contactors:* High energy wire mesh contactors can operate at high vapor velocities of 20 to 30 ft./sec. and provide collection efficiencies that exceed 99.9% (2).

*Electrostatic Precipitators:* Electrostatic precipitators show high collection efficiencies of 98 to 99%, but also high maintenance costs. The factors affecting their performance are the rate of gas flow or temperature and the electrical conditions during their operation.

In a study (3) made by the International Minerals and Chemicals Corporation, it was found that the optimum type of equipment (in the farm chemicals industry) for the control of phosphorus pentoxide ($P_2O_5$) losses in the fumes is a modified single-stage venturi scrubber followed by an impingement basin. The data reported on emissions from seven plants, together with abatement devices used, are shown in Table 3.

### TABLE 3: EMISSIONS AND ABATEMENT DEVICES USED IN SEVEN PHOSPHATE FERTILIZER PLANTS (3)

|  | Average | Range |
|---|---|---|
| Grade manufacturing: 5-10-10 to 10-20-10 |  |  |
| production rate, tons/hr. | 21 | 15 – 32 |
| Emission rates |  |  |
| Solids, lb./hr. | 36 | 13 – 138 |
| Fume, lb./hr. | 110 | 31 – 241 |
| Number of Cyclones* | 3 | 2 – 4 |

*Scrubbers used in five out of the seven plants in conjunction with cyclones.

### References

(1) Athanassiadis, Y.C., "Air Pollution Aspects of Phosphoric Acid & Its Compounds," *Report PB 188,073,* Springfield, Va., National Tech. Information Service (Sept. 1969).
(2) Striplin, M.M., "Development of Processes & Equipment for Production of Phosphoric Acid," *Chemical Engineering Report No. 2,* Tennessee Valley Authority (1948).
(3) "Industry Answers the Challange," *Farm Chemicals,* pp. 21, 24, 26, 28 (June 1967).

## PHOSPHORUS

### Removal of Phosphorus from Water

A process developed by F. Muller, K.H. Stendenbach and H.H. Weizenkorn (1) involves the continual workup of wastewater having phosphorus sludge therein, such as that obtained in the electrothermal production of phosphorus. The wastewater is first filtered in a filtration zone, the resulting filter cake is predried in a preliminary drying zone, the predried filter cake is conveyed through at least two additional drying zones maintained at temperatures between about 100° and 380°C., gaseous and vaporous matter issuing from the additional drying zones is delivered to a condensation zone, the phosphorus and water are condensed therein and separately removed therefrom.

Figure 114 is a flow diagram showing the essential features of the process. Phosphorus sludge-containing wastewater, obtained in the electrothermal production of phosphorus, is pumped through conduit 2 into filtration apparatus 1 and filtered therein. The resulting filter cake travels through downpipe 3 to be delivered to intermediary tank 5, which has a discharge means connected to it, and is conveyed then through a cell lock 6 to drier 7, wherein the bulk of the water contained in the filter cake is expelled together with a quantity of phosphorus, which corresponds to its vapor pressure at substantially 100°C.

The residue, which is now anhydrous, travels through cell lock 9 to distilling apparatus 8 to be completely freed therein from phosphorus, at internal temperatures above the boiling point of yellow phosphorus. Powdery and uninflammable material is drawn off through cell lock 11.

Each of apparatus parts 7 and 8 is maintained under nitrogen travelling through line 10. The vaporous matter issuing from drier 7 and distillation apparatus 8, respectively, is delivered through lines 12 and 13 to injection condenser 14 and precipitated therein with the use of a portion of filtrate flowing through the conduit 4 for filtrate obtained near the centrifugal filter 1. Water from conduit 4 is supplied through nozzles 18.

## FIGURE 114: APPARATUS FOR USE IN THE WORK-UP OF WASTEWATER CONTAINING PHOSPHORUS SLUDGE

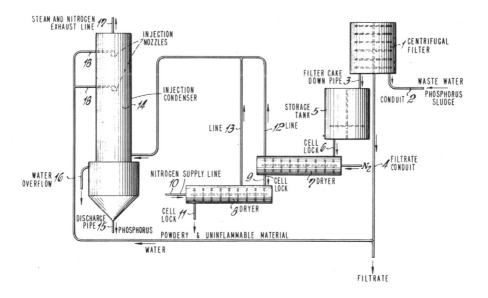

Source:   F. Muller, K.H. Stendenbach and H.H. Weizenkorn; U.S. Patent 3,684,461; August 15, 1972

Nitrogen and steam escape near head of the condenser via line **17** and a water layer and a phosphorus layer deposit in the condenser's bottom portion. The phosphorus is drawn off at discharge line **15** and water overflows at line **16**.

**References**

(1) Muller, F., Stendenbach, K.H. and Weizenkorn, H.H.; U.S. Patent 3,684,461; August 15, 1972.

## PICKLING CHEMICALS

The problem of the disposal of waste acid pickling liquor is serious. In some cases the quantity to be disposed of may be as much as a million gallons per day or more. The strength of such an acid waste will usually vary more or less periodically but it may at times contain as much as 1% of acid or more. Such quantities of acid waste cannot ordinarily be discharged into a stream for disposal without prior treatment.

The problem of disposal of waste pickle liquor is thus one of the major waste disposal problems of the heavy metal industries. The most common method of disposal has been by ponding, and by discharging of the waste liquor into rivers, lakes or tide waters whenever possible. Lime, limestone, or scrap iron can also be used to neutralize the excess acid, the resulting suspension being ponded so that solar evaporation or ground seepage would dissipate the water.

Other methods have endeavored to develop a process whereby usuable by-products could be manufactured and recovered, to compensate for the cost of treatment. These processes involve recovery of various iron compounds, the recovery of sulfuric acid or of ammonium sulfate or gypsum. These processes have a disadvantage of comparatively high cost of installation, equipment, maintenance and operation, due to the corrosive nature of the liquor. Moreover, in many instances, there has been a lack of an adequate market for the by-products.

### Removal of Sulfuric Acid Pickle Liquor from Water

In addition to the material cited at this point in the handbook, the reader is also referred to the section on "Iron; Removal of Soluble Iron from Water" for additional text and references on the treatment of sulfuric acid pickle liquor.

A process developed by J.M. Kahn and E. Kominek, Jr. (1) relates to the treatment of waste pickle liquor commonly produced by the steel industry in the acid pickling of steel. In this process the acid liquor is neutralized in a first step with a magnesium compound whereby a solution of a magnesium salt of the acid is obtained. This solution is then treated with dolomitic lime whereby the calcium hydrate or oxide of the lime will precipitate the magnesium of the solution as magnesium hydroxide, commonly called magnesium hydrate.

The magnesium hydrate of the dolomitic lime is left in solid form and obtained with the precipitated magnesium hydrate. A suitable part of the thus obtained magnesium hydrate is returned to the step wherein the acid is neutralized while the remaining magnesium hydrate is available for any desired use.

The cyclic process for the treatment of waste pickle liquor containing free sulfuric acid, ferric sulfate and ferrous sulfate, thus comprises mixing magnesium hydroxide with the liquor to neutralize free acid and precipitate all of the ferric iron as ferric hydroxide and a minor portion of the ferrous iron as ferrous hydroxide, separating the resulting precipitate from the solution, adding calcium hydroxide to the liquor to precipitate the balance of the iron as ferrous hydroxide thereby forming a mixed solution of calcium sulfate and magnesium sulfate, the liquor being kept sufficiently dilute to prevent precipitation of calcium sulfate therefrom, separating the solution from the ferrous hydroxide, adding dolomitic lime to the solution to precipitate magnesium hydroxide and utilizing a portion of the magnesium hydroxide for the treatment of additional quantities of waste liquor.

A process developed by P. Simpson, G.F.G. Clough and K.H. Todhunter (2) involves treating waste acid liquor with ammonia liquor for the recovery of ferrous hydroxide and ammonium salts therefrom. It comprises the steps of adding the ammonia liquor simultaneously with the waste acid liquor to a reactor operating under vigorous agitation at a temperature in excess of 50°C., removing reacted liquor from the reactor at the same rate as that of the addition thereto, separating the ferrous hydroxide precipitate from the reacted liquor so removed, precipitating by oxidation substantially all the iron remaining in the reacted liquor and evaporating the supernatant liquor from which the precipitates have been removed, whereby to initiate the formation of substantially iron-free ammonium salt crystals.

Figure 117 is a flow diagram of the process. Ammonia liquor, which may be an aqueous solution of synthetic ammonia or the effluent from the gas scrubbing and ammonia liquor purifying plant of a coke oven, is collected in a liquor storage vessel 11; the scrubbing of the gas and the purification of the ammonia liquor, where this is coke oven effluent, previously having been carried out in the manner known in the art.

Waste sulfuric acid pickle liquor from steel pickling baths is collected in a pickle liquor storage vessel 12. The ammonia liquor from the vessel 11 and pickle liquor from the vessel 12 are fed through conduits 13, 14, respectively, to the top of a reactor 15, which is a totally enclosed vessel to which heat is applied by suitable means (not shown) so that the heat within the reactor 15 does not exceed the boiling point of the liquor therein. Suitable valves 16,17 are provided in the respective conduits 13 and 14 so that the flow in these conduits can be adjusted to provide substantially stoichiometric proportions of the two liquors.

Normally, the reactor will operate at a temperature not greater than 95°C., but it may be arranged to operate under pressure, in which case the temperature may be increased to as much as 150°C., so long as the pressure in the reactor 15 is sufficient to prevent ebullition at the chosen operating temperature. The heat for the reaction may be provided by injected steam, by an external furnace or by heat exchange means within the reactor.

The ammonia liquor fed into the reactor is substantially pure, except that, in the case of coke oven effluent, it may contain up to 0.3% pyridine, and it is fed into the reactor 15 in such a proportion that the amount of ammonia added thereto balances with the sulfate ions in the pickle liquor to form ammonium sulfate in the reaction therein.

In order to obtain the desired properties of the ferrous hydroxide, the mixture of liquors in the reactor 15 is vigorously agitated as by an agitator 18. Any pyridine contained in the ammonia liquor is liberated in the reactor 15 and passed through the conduit 19 to a condenser 20. The reacted suspension from the reactor 15 is passed through a conduit 21 to a centrifuge 22 wherein the solid matter in the form of ferrous hydroxide precipitate is removed and passed to storage through a conduit 23.

Liquor from the centrifuge 22 is passed through a conduit 24 to an oxidizer 25, which is provided with agitation means 26, which may be of the known turbine and sparger type. Here, the liquor is aerated to oxidize to the ferric state the iron still remaining in solution. A substantially complete removal of iron is effected, the small amount remaining, which comprises approximately 0.2 part per million, being complexed with the small amount of organic matter present in the waste pickle liquor.

The oxidized liquid from the oxidizer 25 is then passed through a conduit 27 to a centrifuge 28 wherein the solid matter is removed and passed to storage through a conduit 29. Ammonia liquor from the centrifuge 28 is passed through a conduit 30 to an evaporator and crystallizer 31 wherein the ammonium sulfate crystals are formed in the known way.

### FIGURE 117: FLOW DIAGRAM OF PROCESS FOR AMMONIA TREATMENT OF WASTE PICKLE LIQUOR

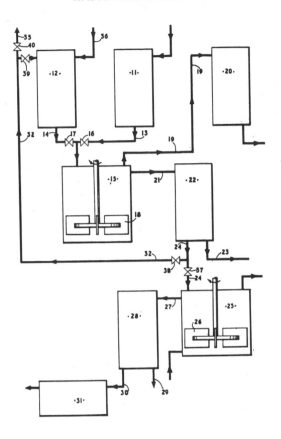

Source:  P. Simpson, G.F.G. Clough and K.H. Todhunter; U.S. Patent 3,167,390;
        January 26, 1965

Some of the treated liquor from the centrifuge **22** may be by-passed from the conduit **24** through a by-pass conduit **32** to the pickle liquor storage vessel **12** for the purpose of reducing the iron concentration in the waste pickle liquor therein; the reduction in concentration improving the settling properties of the precipitate. Treated liquor may also be recycled through the conduit **35** to the conduit **36** connecting the pickle bath (not shown) with the waste pickle liquor storage vessel **12** to prevent the crystallization of ferrous sulfate in the conduits **36** and **14**. The by-pass of liquor is controlled by adjustment of the valves **37, 38, 39** and **40**.

In a process developed by L.W. Heise and M. Johnson (3) lime produced as a carbide residue is first mixed with water or with recirculated treated pickle liquor to form a slurry. The slurry is continually fed into a neutralizing tank into which the pickle liquor is also fed at a rate to give the mixture a pH of about 7.5. The mixture is agitated to insure complete neutralization of the pickle liquor and aerated to oxidize from about 2 to 5% of the ferrous salts in the liquor so as to give the mixture the proper filtering characteristics. The mixture continually overflows from the neutralizing tank into a vacuum filter where the same is filtered.

The filtered cake containing the insoluble salts is scraped from the filter and conveyed away and dumped. The filtrate is removed to a storage tank and recirculated to the mixing chamber for make up of the slurry. In this process the make up of the slurry, neutralization and filtration are continually occurring so that no time or labor is lost in the handling or transporting of the liquor as is necessary in the conventional batch process.

In addition a given volume of pickle liquor will be completely treated in about 5 to 15 minutes as compared to the usual 1 to 2 hours for treatment in the former processes. This time saving is brought about through the use of a more effective agitation and neutralization operation and by the fact that only 2 to 5% of the iron need be oxidized rather than the 50% required in the ordinary process.

Figure 118 is a flow diagram of the process. The lime employed to make up the slurry is contained in a hopper 1 and may take the form of slaked lime produced either by the roasting of limestone or as a residual by-product of the carbide process of making acetylene. However it is contemplated that any oxide or hydroxide of an alkaline earth metal may be employed in place of lime as the neutralizing agent. The lime is conveyed from hopper 1 to a mixing tank 2 by a bucket conveyor 3 or the like.

To begin the process, water is mixed with the lime in tank 2 to form an aqueous suspension or slurry of lime. Mixing of the lime and water is accelerated by an agitator 4. After the process is initiated the treated pickle liquor filtrate is employed in place of the water for the purpose of making the slurry. The lime and water or filtrate are continually added to tank 2 and the resulting lime slurry will therefore continually flow from tank 2 and pass through line 5 to the neutralizing tank 6. The rate at which the lime and filtrate are added to tank 2 determines the rate of flow of the slurry into tank 6.

The waste pickle liquor to be treated passes through the line 7 from a storage lagoon into neutralizing tank 6. Line 5 conducting the lime slurry and line 7 conducting the pickle liquor terminate adjacent to each other at the bottom of tank 6 so that the slurry and liquor will tend to mix together at the moment of their entry into the tank.

FIGURE 118: PROCESS FOR LIME NEUTRALIZATION OF WASTE PICKLE LIQUOR

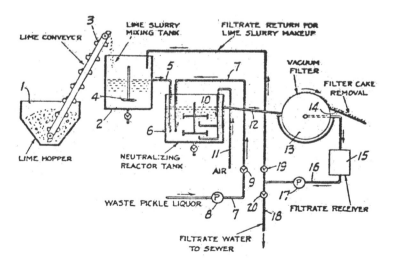

Source:  L.W. Heise and M. Johnson; U.S. Patent 2,692,229; October 19, 1954

The pickle liquor is conveyed within line **7** by a suitable pump **8** and the rate of discharge of the liquor from line **7** into tank **6** is controlled by valve **9**. The waste pickle liquor and the lime slurry are thoroughly intermixed in tank **6** by means of a paddle-type agitator **10** so as to insure complete reaction between the ferrous sulfate and/or the sulfuric acid with the lime in accordance with the following equations.

$$FeSO_4 + Ca(OH)_2 \rightarrow Fe(OH)_2 + CaSO_4$$

$$H_2SO_4 + Ca(OH)_2 \rightarrow CaSO_4 + 2H_2O$$

Under the conditions stated whereby the lime and pickle liquor are introduced in contiguous streams and are constantly mixed by agitator **10** these reactions proceed to substantial completion. A portion of the ferrous compounds in the liquor are oxidized to the ferric state in the neutralizing tank **6** so as to obtain maximum filterability of the liquor. This oxidation is accomplished by introducing air into tank **6** through line **11**. It is preferred to introduce the air beneath the circular path of the rotating paddles of agitator **10** so that the air will be thrown outwardly by the paddles and tend to atomize or disperse in finely divided particles.

The atomization causes the air to be distributed equally throughout the entire mass and the effectiveness of oxidation is thereby increased. As shown in the figure, the agitator **10** is provided with two series of paddles, and the air is introduced beneath each series. The oxidation is brought about in accordance with the following equation.

$$4Fe(OH)_2 + O_2 \rightarrow 2Fe_2O_3 + 4H_2O$$

Previously it was thought that incomplete oxidation of the ferrous compounds resulted in a iron precipitate which was colloidal or gelatinous in nature and very difficult to separate from the solution. Therefore, the universal tendency was to oxidize at least 50 to 100% of the ferrous material so as to avoid the poor filterability accompanying any lesser degree of oxidation.

However, an important feature of this process resides in the discovery that excellent filterability can be obtained by merely oxidizing from about 2 to 5% of the ferrous material. The filterability of the precipitate at this percentage of oxidation is equally as good as that obtained at substantially complete oxidation of the ferrous compounds.

The pH of the solution in tank **6** is controlled by regulating the flow of pickle liquor into the tank by means of valve **9** in line **7** or by regulating the flow of slurry through line **5**. The pH control may be performed by manual operation of the necessary valve controls or it may be accomplished automatically by making the flow through lines **5** and **7** responsive to the readings of a pH meter or measuring device. The liquor properly oxidized and having the required pH overflows from tank **6** and passes through line **12** to a suitable filter **13** or separator. As shown, the filter **13** may be a vacuum disc or drum-type filter.

The insoluble salts containing substantially all of the iron in the original pickle liquor are deposited on filter **13** as a filter cake and removed therefrom by scraper **14**. The filter cake is conveyed to a suitable dumping ground or it may undergo subsequent processing to recover valuable ingredients. The filtrate, which consists essentially of a saturated solution of calcium sulfate, passes from filter **13** to a vacuum filtrate receiver **15**. The filtrate may then either be recirculated through line **16** by pump **17** to the mixing tank **2** to be employed to make this lime slurry or it may be discharged from line **16** through **18** to the sewage system.

The rate of flow of the filtrate through line **16** to the mixing tank **2** is controlled by valve **19** and the flow to the sewerage system is regulated by valve **20**. The recirculation of the filtrate is desirable for it eliminates the expense of using fresh water to make up the slurry, and the filtrate also serves as a source of calcium for the slurry. This latter advantage reduces the amount of lime necessary to neutralize a given volume of pickle liquor.

A process developed by E. Wainer (4) provides for complete recovery of usable sulfuric acid from the pickle liquor, together with saleable by-products such as Portland cement, pig iron, and pure carbon dioxide. In general the process comprises the following sequence of steps. Waste pickle liquor from the pickle line is pumped to a precipitation tank. Ground silica is added to the liquor in the tank. The molten blast furnace slag is granulated by direct quenching of the hot slag into the pickle liquor-silica mixture in the tank.

Ground limestone is then added to the tank, with stirring, and the gases developing as a result of the ensuing reaction (consisting primarily of wet carbon dioxide) are first dried and then collected in a suitable chamber for the preparation of solid carbon dioxide. The slurry in the precipitation tank (formed as the result of the addition of calcium carbonate and after the elimination of the carbon dioxide represents a mixture of hydrated calcium sulfate, hydrated oxides of iron, silica and blast furnace slag) is passed to a filter press wherein it is filtered and washed, the wash water is reserved as process water. The filter cake is dried in a waste heat dryer from which it emerges as relatively hard porous pieces or pellets, at a temperature between 800° and 900°C.

The hot dry filter cake is fed directly to a shaft located directly above the hearth of a submerged arc furnace wherein a hearth temperature of about 1800°C. is maintained, in order to insure that the contents in the hearth remain molten. Air is blown into the porous charge in the lower portions of the stack. As a consequence, all of the sulfur trioxide bound in the calcium sulfate is evolved from the dried pellets, this evolution being accelerated by the catalytic action of the iron oxides in the same. The evolved sulfur trioxide is passed to a collector and then through an electrostatic precipitator and the process water originally used and available as the filtrate from the original filtration is used to absorb this sulfur trioxide after it leaves the electrostatic precipitator so that the sulfuric acid is reconstituted.

The fused mass is then flowed into the upper end of a second heated chamber containing hot coke and as a result the iron oxide in the melt is reduced to metallic iron of blast furnace quality which collects at the base of the chamber, along with a slag consisting primarily of a mixture of the oxides of calcium, alumina and silica which floats on the pool of iron. The hot iron is tapped to a hot ladle and may be used as a feed for the open hearth or it may be cast into pig iron. The molten slag is tapped and the slag is blown to facilitate subsequent grinding. The blown slag is then ground in continuous closed circuit grinders to produce the particle size required by federal specifications for Portland cement.

As a consequence of these operations substantially all of the sulfuric acid originally utilized in the steel pickling operation is returned in usable form to the pickling line, the process water is retained in tanks for reuse if desired, or may be dumped to the sewer without fear of causing pollution, but more important, Portland cement, molten iron suitable for open hearth feed and relatively pure carbon dioxide, are also recovered so that the operation may be carried out profitably.

A process developed by M.C.-S. Chang (5) involves neutralizing acidic sulfate, iron value-containing waste pickle liquor with a basic, iron value-containing blast furnace flue dust. In somewhat more detail, the process comprises subjecting a dust laden blast furnace effluent gas to a particle classification and separation operation to remove from the gas a first, relatively coarser and chemically relatively inactive dust fraction, then removing from the effluent gas a second finer and chemically active dust fraction by means of a water-scrubbing operation.

This will form an aqueous dispersion of the second dust fraction. The aqueous dispersion is concentrated to form a slurry having a greater neutralizing capacity as compared to a slurry containing an equal quantity of unclassified dust. Mix the slurry with the waste pickle liquor maintained at a temperature from 100° to 160°F. to neutralize the acid values therein, thus forming a precipitate comprising recoverable iron values and basic sulfates. The process continues by separating the precipitate and sintering it to produce a basic, low sulfur, reducible, iron value-containing blast furnace feed product.

A process developed by J.A. Gross and B.G. Mandelik (6) is one in which pickle liquor is added to a circulating body of strong acid solution. The mixed liquid stream is then suddenly concentrated by a type of flash evaporation of water. This evaporation is accomplished by projecting the liquid stream transverse to a highly accelerated stream of hot combustion gas or other heated gas. As a result, the liquid is dispersed into very fine droplets by the action of the hot gas, and sudden evaporation of water takes place.

The cooled and moisture laden gas stream is then separated from the residual concentrated acid solution, which now contains the precipitated crystals of ferrous sulfate monohydrate. The solid crystals are readily settled to a slurry and overhead strong acid solution containing fine seed crystals is recycled as the aforementioned circulating body of strong acid solution. The slurry is then filtered to recover product crystalline ferrous sulfate monohydrate, while the filtrate consisting of crude strong sulfuric acid solution is also recovered as the other major product of the process.

A process developed by B.I. Karsay (7) is one in which spent pickle liquor comprising an aqueous solution of ferrous sulfate and sulfuric acid is regenerated for use in pickling by concentrating the sulfuric acid contained in the spent pickle liquor and separating ferrous sulfate therefrom, by adding crystalline ferrous sulfate monohydrate to the spent pickle liquor in a quantity which is substantially larger than corresponds to its solubility in the spent pickle liquor.

A quantity of ferrous sulfate monohydrate is used such that the resultant slurry will contain 20 to 50%, preferably 25 to 35% by weight ferrous sulfate based on the total weight of slurry. The temperature of the slurry is maintained within the range of 130°F. to the freezing point of the slurry preferably within the range of 50° to 0°F. for a sufficient length of time for the ferrous sulfate monohydrate to extract water from the spent pickle liquor and combine with it to form ferrous sulfate heptahydrate.

Simultaneously with the concentrating of the sulfuric acid in the pickle liquor as a result of the extraction of water therefrom, solubility of ferrous sulfate in the pickle liquor decreases with the result that the ferrous sulfate heptahydrate precipitate in the slurry is greater in quantity based on the ferrous sulfate than the amount of ferrous sulfate monohydrate added to the spent pickle liquor. The ferrous sulfate heptahydrate crystals are separated from the liquid leaving a more concentrated pickle liquor, i.e., a stronger sulfuric acid solution than is initially present.

The recovered heptahydrate may be reconverted to monohydrate by dehydration and recycled to the water extraction step for the regeneration of a new amount of pickle liquor. Thus, the process is self-sustaining, as after the initial start-up, it furnishes its own water extracting agent, and no ferrous sulfate monohydrate is required from external sources to maintain the operation.

Moreover, an excess of ferrous sulfate monohydrate is produced, equivalent to the amount which precipitates out of the pickle liquor or concentration. This material may be withdrawn from the process and utilized in the sulfuric acid manufacture. The concentration of the sulfuric acid in the pickle liquor may be increased to about 25 to 35% by water extraction with ferrous sulfate monohydrate.

A further concentration of the acid to the desired final strength may be accomplished by fortification with a concentrated sulfuric acid, e.g., 66°Bé. sulfuric acid, which at the same time, supplies the make-up acid. On addition of the 66°Bé. acid, additional ferrous sulfate precipitates out of the pickle liquor, because the solubility of the salt further decreases with increasing acid strength.

However, this portion of the recovered total ferrous sulfate is monohydrate in contrast to the first portion, which was heptahydrate, and may directly be, after separation from the liquid, recycled to the water extraction step. The filtrate of this second separation is the regenerated pickle liquor, which, after dilution to the proper strength with water, may be returned to the pickling operation.

A process developed by L.E. Lancy (8) is one in which ferrous metal-containing work-pieces which have been subjected to a pickling or other treatment in an acid bath have potentially polluting iron salt waste or carry-over solution on their surfaces fully reacted and neutralized by a calcium containing aqueous chemical solution, without forming un-desirable calcium deposits and while enabling efficient precipitation of iron compounds.

The workpiece may be subsequently subjected to a rinse water wash without fear of con-taminating a stream if the wash water is to be discharged therein. The aqueous treating solution is continuously reconditioned and reused and substantially fully removes, reacts with or neutralizes the waste carry-over on a workpiece that is being moved continuously into an aqueous washing bath.

## Removal of Hydrochloric Acid Pickle Liquor from Water

A process developed by C.B. Myers (9) is one in which spent hydrochloric acid pickle liq-uors, especially those liquors which contain iron, may be disposed of in an efficient and economical manner by reacting these liquors with a waste material, which waste material is formed primarily from the discharge of an ammonia-soda plant. In this manner the iron is substantially completely removed from the liquor and therefore does not enter the sur-rounding watershed.

The reader of this handbook is also referred to the section on "Hydrochloric Acid, Recovery from Water" for additional data and references on the recovery of hydrochloric acid from pickle liquor.

## Removal of $HNO_3$–HF Pickle Liquor from Water

A process developed by K.E. Gunnarsson (10) involves treating industrial wastewater con-taminated with metal ions, nitrate ions and fluoride ions which comprises the steps of con-verting the fluoride ions of the wastewater to insoluble fluorides, neutralizing the waste-water with an ammoniacal neutralizing agent and adding excess ammonia to a pH of at least 8, filtering, adding an alkali hydroxide to the filtrate, distilling off the content of am-monia and finally removing any solid phase from the distillation residue.

The pickling solutions used for the pickling of stainless steel consist of mixtures including substantially nitric acid and hydrofluoric acid. The acid concentration normally is between 10 and 15% nitric acid and 4% hydrofluoric acid. During the pickling process metallic oxides and metals are dissolved, while forming metal nitrates and metal fluorides, water and nitrous gases (nitric oxides), which in their turn volatilize. Thereby the concentrations of free acids decrease and are restored by the addition of the new acid.

The content of metal ions, nitrate ions and fluoride ions in the pickling bath gradually in-creases and at a certain content of metal ions the pickling effect is reduced appreciably. The worn out bath is poured off and replaced by a new pickling solution. The worn out pickling bath is treated according to the process as follows.

The hydrofluoric acid is converted to insoluble calcium fluoride by the addition of equiv-alent amounts of lime (CaO). This conversion gives rise to a heat development, which is of benefit to the entire reaction process. However, no apparent volume increase (dilution) takes place, as it is the case, for example, when using calcium hydroxide [$Ca(OH)_2$].

Ammonia is added until neutralization (pH = 7) is achieved, and thereafter in excess (pH $\geq$ 8) for the complete formation of metal amines. The deposit consisting, for ex-ample, of iron hydroxides, chromium hydroxides, manganese hydroxides and other com-pounds of metallic elements, for example, molybdenum and calcium fluoride, is filtered or separated from the filtrate, which in addition to $NH_4NO_3$ and $NH_4OH$ substantially contains amino compounds of nickel and to a smaller extent amino compounds of copper and cobalt which are easily soluble. The deposit may be added, for example, to ore con-centrates or to oxides and finally be converted to metals.

Alkali hydroxides such as NaOH, KOH and/or hydroxides of the alkaline earth metals, for example $Ca(OH)_2$, but advantageously and in this example KOH, are added to the filtrate in order thereby to bring about the conditions required for an easy separation of $NH_4OH$ and for the replacement of $NH_4$-ions by K-ions. The ammonia is distilled off by water vapor. Ammonia as distillate is returned to the process as in the second step.

The hydroxides of the metals nickel, copper and cobalt which were precipitated during the addition of potassium hydroxide and during the distilling-off of ammonia, are separated by filtering from the potassium nitrate solution. The deposit consisting substantially of nickel hydroxide is dried and annealed. The hydroxide is thereby converted to oxide, which can be introduced directly into the metallurgical process or be converted to metal. The hydrate consisting of pure potassium nitrate solution is utilized in the production of fertilizers. Hereby the costs of the method are reduced considerably, because a full financial compensation is obtained for the potassium used in the method.

References

 (1) Kahn, J.M. and Kominek, E., Jr.; U.S. Patent 2,433,458; December 30, 1947; assigned to Infilco Inc.
 (2) Simpson, P., Clough, G.F.G. and Todhunter, K.H.; U.S. Patent 3,167,390; January 26, 1965; as-
       signed to Simon-Carves Limited, England.
 (3) Heise, L.W. and Johnson, M.; U.S. Patent 2,692,229; October 19, 1954; assigned to A.O. Smith Corp.
 (4) Wainer, E.; U.S. Patent 3,203,758; August 31, 1965; assigned to Horizons Inc.
 (5) Chang, M.C.-S.; U.S. Patent 3,205,064; September 7, 1965; assigned to Crucible Steel Co. of America.
 (6) Gross, J.A. and Mandelik, B.G.; U.S. Patent 3,211,538; October 12, 1965; assigned to Chemical
       Construction Corporation.
 (7) Karsay, B.I.; U.S. Patent 3,340,015; September 5, 1967; assigned to Allied Chemical Corporation.
 (8) Lancy, L.E.; U.S. Patent 3,562,016; February 9, 1971; assigned to Lancy Laboratories, Inc.
 (9) Myers, C.B.; U.S. Patent 3,468,797; September 23, 1969; assigned to Diamond Shamrock Corp.
(10) Gunnarsson, K.E.; U.S. Patent 3,647,686; March 7, 1972; assigned to Nyby Bruks AB, Sweden

## PLASTIC WASTES

The treatment of plastic wastes is a complex problem both technologically and economically. Plastics themselves cover a wide range of chemical compositions and they are often associated with other materials such as plasticizers, fillers, laminate materials, etc. The use of waste plastic materials for the production of process chemicals has been reviewed by M.E. Banks, W.D. Lusk and R.S. Ottinger (1). The disposal of polyethylene plastic waste has been studied by K. Gutfreund (2). Problems in the processing of plastics in urban refuse have been reviewed by J.L. Holman, J.B. Stephenson and J.W. Jensen (3).

### Plastic Waste Degradation

A material developed by J.P. Pijst (4) is a plastic material for protecting a vegetative plant, the material being air and water permeable and sensitive to light so as to be gradually disintegrated thereby. More specifically, the product preferably consists of a sheet of foamed plastic material containing open cells. The plastic material may be an adduct of a multifunctional isocyanate with a multifunctional alcohol of a polyether. More particularly, the adduct may be of toluene diisocyanate with di-, tri- or higher functional alcohols of a polyether.

A material developed by T.H. Shepherd (5) is a polybutene-1 film for use as an agricultural mulch. This film degrades and substantially disintegrates of its own volition within a growing season.

A process developed by G.C. Newland, R.M. Schulken, Jr. and R.C. Harris (6) is one whereby polyolefin films for use as agricultural ground cover can be made heat absorbent to assist in plant development while preventing weed growth, and thereafter heat reflective to give increased crop production, and thereafter friable, by the incorporating into the film composition certain fugitive heat absorptive colorants, light reflective opaquing pigments, and prooxidants.

A device which has been developed by F.W. Stager and R.N. Minor (7) is a container which deteriorates after a prolonged exposure to moisture and oxygen in an atmospheric environment with the aid of biological agents. The container material has organic components that are nutrients to fungi and various bacteria which bring about the decomposition. The material is composed of an organic filler material, such as peat, and a phenolic resin impregnated with a metallic salt of a fatty acid. The decomposition process transforms the container material into a rich organic substance which enhances soil characteristics.

A product developed by B.F. Anderson (8) is a plastic film for use as an agricultural mulch. The film is a blend of an ethylene polymer, a substantially amorphous ethylene/propylene copolymer, a natural rubber and a paraffin wax. The films are sufficiently tough and flexible to allow handling and laying out on a planted soil surface, yet they readily rupture under the pressure of a growing seedling to allow plant penetration. The films degrade during a growing season to the point that they can be readily disposed of by plowing and disking into the ground.

A composition developed by J.W. Henry (9) comprises (a) a copolymer of ethylene and 0.1 to 18% by weight of carbon monoxide, the percentage range being the proportion of carbon monoxide in the copolymer, and (b) an ultraviolet light degradable amount of an organometallic compound or salt of a common metal, e.g., ferrocene, cobalt acetyl acetonate, etc.

Film produced from such materials upon exposure to ultraviolet rays emitted by the sun are capable of accelerated deterioration thereby causing such materials to decompose forming dust or flakes which may become embedded in the soil or carried away by rain water under those circumstances where they are thrown away out of doors so as to be exposed to natural environmental conditions.

## Plastic Waste Incineration

A device developed by J.A. Boyd and D.E. Boyd (10) provides an improved incineration system for thermoplastic materials wherein a combustion chamber is provided with an inlet or charging port. A moving stream of air is generated outside the incineration combustion chamber and is directed toward and through the charging port into the interior of the combustion chamber. The thermoplastic materials to be incinerated are introduced into the air stream at a location remote from the charging port and are conveyed along and supported by the air stream and are thus introduced through the charging port into the combustion chamber where they are burned.

By using the air stream for supporting and conveying the thermoplastic materials, such materials are maintained out of contact with the elevated temperature portions of the incinerator itself and hence will not melt and stick to surfaces of the incinerator. Instead, the thermoplastic materials will be carried into the central portion of the combustion chamber where they are to be burned, and the air stream which carries the thermoplastic materials additionally creates turbulent flow conditions within the combustion chamber to promote and facilitate burning of the materials.

A process developed by R.F. Stockman (11) involves incinerating a meltable plastic material by means of first reducing the plastic to a liquid, heating the liquid to vaporization, and then burning the fumes that are given off from the surface of the liquid before they may be exhausted to the surrounding atmosphere. Figure 119 shows a suitable form of apparatus for the conduct of the process. In the figure, the numeral 10 refers to a housing or a container for the incineration of plastic material or the like. The housing has substantially vertical side walls 12 and an apertured top 14 with an inlet duct 16 which extends

FIGURE 119: PLASTIC WASTE INCINERATION

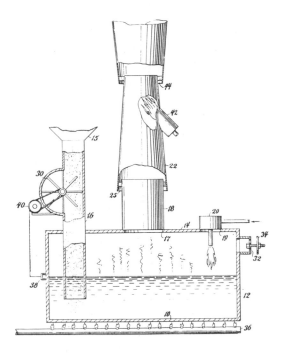

Source:  R.F. Stockman; U.S. Patent 3,572,265; March 23, 1971

therethrough and terminates above the floor of housing **10** for adding solid plastic material
into the container while an outlet port **17** at the top of the housing exhausts the fumes
given off therefrom.  The port **17** is provided with an outlet duct **18** of cylindrical design
which terminates a short distance above the housing **14** and is centrally positioned within
the housing in a manner that permits a reasonable freedom from the congestion of fumes
given off from the molten plastic.

The stack **22** exhausting from the container **10** may be of any approved construction, however, one of modified venturi configuration performs best in mixing the fumes with air for
combustion and then subjecting the mixture to a secondary process of combustion.  At the
lower end of the stack are provided air supply ports **25** whereby the upward flow of fumes
from the container **10** will induce the flow of air therethrough.  As the air mixes with the
fumes it is compressed slightly by passing through the converging section of the venturi
shaped stack.

Adjacent its point of greatest restriction a burner **42** including an independent source of
gas and air is injected into the stack and ignited whereby the fumes from the plastic together with the air supplied through ports **24** are then subjected to combustion.  As the
gases ignite in the stack they expand and draw in still more air through the ports **44**
whereby combustion is nearly complete in the remaining section of the stack before the
spent gases are exhausted to the atmosphere.  The top **14** of the container includes a third
opening **19** for a start up burner **20** which is inserted therethrough to initiate combustion
of the plastic fumes.  An air supply port **32** in the upper portion of the housing **12** includes
an adjustable valve **34** that may be opened a predetermined degree to partially burn the

fumes being given off from the surface of the liquefied plastic and produce sufficient heat in the combustion process to insure that the mass of plastic material in the housing is maintained in a liquid state. In operation a mass of plastic material is loaded into the hopper 15 leading to container 10 through the inlet duct 16. The size of the individual pieces of plastic will determine the size of duct 16, thus for a mass of particle sized plastic an inlet duct 16 of smaller diameter is adequate while for larger chunk sized pieces of solid plastic an inlet of larger diameter is required.

A feeding device such as a star wheel 30 may be required to feed the plastic evenly and to preclude the back flow of excess quantities of vapor or fumes given off from the molten plastic which lies within the inlet tube 16. A temporary source of heat such as a burner 36 is applied to the container 10 adequate to heat the contents thereof and thus reduce the solid plastic to a liquid state. As the plastic is heated it liquefies and its rate of vaporization increases so that it increasingly gives off fumes and vapors which are in turn ignited by the burner or other ignition means in the opening 20. Fumes given off from the incomplete combustion of fumes in the housing 10 are exhausted through the outlet 18 to the stack 22.

The fumes from the incompletely burned plastic are then mixed with air entering through ports 25, mixed in the throat of the venturi and subjected to the incinerating action of the secondary burner 42. Such incineration causes the inspiration of additional air through ports 44 so that combustion is complete before the gases are exhausted to the atmosphere. The step of injecting air and combustible fuel can be located at as many points along the stack as may be required to provide for the complete elimination of combustible products in the exhaust gas.

Heat from the plastic burning in the container 10 is sufficient to melt additional plastic material being added thereto so that auxiliary or temporary source of heat such as the burner 36 is not required for continuous operation. A liquid level sensor 38 on the container 10 may be made to control operation of the motor 40 and star wheel feeder 30 so that the level of plastic supplied through tube 16 to container 10 is maintained above the lower end opening to the inlet tube 16 so that little vapor from the molten plastic is allowed to escape up the tube 16 to the atmosphere. The star wheel 30 provides an additional barrier in tube 16 preventing still further the escape of vapor from the container.

The amount of combustion air permitted to enter the container 10 through supply valve 34 is controlled carefully to provide heat for the combustion of plastic within the container 10 adequate only for the continuous reduction to a molten state of the solid plastic being added through inlet 16, it being understood that level of the plastic within the container is continuously controlled by the liquid level controller 20 so that it remains substantially constant at a point lying between the open bottom end of tube 16 and the top of the container.

Therefore the surface are of the body of molten plastic remains substantially constant, its temperature too is almost constant, and the air for its combustion is allowed to vary little so that an almost constant rate of combustion is permitted. Inasmuch as wide fluctuations in the amount of fuel available for burning are substantially eliminated, combustion is nearly complete and little smoke or obnoxious exhaust gas escapes to the atmosphere.

### Plastic Waste Recycling

A process developed by G.P. Monet (12) is a process for recovering from waste nylon the salt-forming components from which it is made. The normal operation of a nylon plant results in the unavoidable collection of waste fiber and polymer at various points in the process. This material is generally not suitable for remelting and extrusion into high grade fibers. It is therefore desirable that the salt-forming components of nylon be recovered from this material for the production of new polymer. In the prior art, adipic acid and hexamethylenediamine have been recovered from nylon polymer by repeated hydrolysis of the polymer in sulfuric acid, recovering the adipic acid by crystallization after each hydrolysis step and the diamine by distillation after neutralizing the acid. In this process, the hydrolysis and crystallization of the adipic acid have to be carried out in a stepwise fashion, requiring frequent filtrations and a great deal of time for optimum recovery of the adipic acid. Such a process, then, is not efficient for handling the greatly increased volume of waste which has resulted from the continually increasing production of nylon fiber.

Accordingly, it is desired to provide an improved process for the recovery from nylon of its salt-forming components. Another object is to provide a process comprising a continuous and rapid crystallization of adipic acid. A further object is to provide continuous and rapid crystallization of adipic acid in the form of crystals of such size as to promote rapid filtration of the crystals.

These objects are accomplished in the process by continuously hydrolyzing nylon waste with an aqueous mineral acid of from about 30 to 70% concentration, continuously feeding the resulting hydrolysate to a crystallization zone, continuously crystallizing the acid component and removing the crystals from the hydrolysate, continuously adding calcium hydroxide to the resulting mother liquor to substantially neutralize it and liberate at least about 95%, but less than 100%, of the diamine component, and separating this component from the mother liquor.

A process developed by R.T. Hurley (13) involves treating the waste and scrap material resulting from the fabrication of polyurethane-type resins, whereby a useful liquid resin is recovered. Polyurethane resins are made from a variety of intermediates, including polyisocyanates and polyesters. The polyisocyanates generally used are the diisocyanates, while the polyesters are generally esters of polyfunctional alcohols with polycarboxylic acids. If a cellular or foam-like structure of the resulting polyurethane resin is desired, it is customary to add to the diisocyanate-polyester mixture an activating mixture which includes water or an equivalent hydrolyzing agent. In the resulting hydrolysis of the diisocyanate, carbon dioxide is evolved concurrently with the polycondensation of the principal reactants, whereby the desired resin foam is obtained.

In the practice of this process there is generated an appreciable quantity of scrap, in the form of waste and trimmings, the market value of which is comparatively low. It has been proposed in U.S. Patent 2,729,618 to reclaim such scrap by making it into useful products, heat, pressure and shaping being used to reform the material. It has also been proposed in U.S. Patent 2,998,395 to burn the scrap, which produces a liquid which can be employed as a substituent for fresh polyester in a further reaction with a polyisocyanate. The product of this burning process, although it is quite satisfactory, even superior for certain uses, is not as desirable for other uses.

It is an object of this process to improve the method of recovering scrap polyurethane resins, whether they be solid or foam, and particularly to improve the quality over that of the product of the decomposition process described in U.S. Patent 2,998,395. According to the process polyurethane scrap is subjected to thermal decomposition by methods not involving oxidation. The product thus produced is superior to that produced by combustion or other oxidation processes and it may be used alone or in combination with fresh polyester in the typical reactions with polyisocyanates.

The process is carried out at temperatures between about $200°$ to $400°C$. Other temperatures, outside this range can be used but they frequently require too long a time if carried out below $200°C$. and result in an inferior product if carried out above $400°C$. The preferred range is between $260°$ and $350°C$. and in actual practice the optimum is reached between $290°$ to $310°C$.

The reaction is preferably carried out in an enclosed vessel under a nitrogen or other inert gas atmosphere. The nitrogen atmosphere prevents the oxidation of the materials, improves the color, and increases the yield of the product. It is desirable to use pressure on the material, particularly when foam scrap is being recovered. The foam is filled with cells which contain air and the air is expelled from them by the use of pressure.

A satisfactory method of applying pressure is to pass the scrap to a pair of rollers at the entrance to the reaction chamber, the air being squeezed out and the material, largely free of its air, discharged within. It is a simple matter to provide sufficient sealing around such rollers and to maintain an adequate predominance of nitrogen within the chamber to prevent excessive oxidation.

A satisfactory method of heating the scrap to decomposition temperature is to pass it between heated rollers under resilient pressure. By this means uniform heat is applied to the parts of scrap and the resulting product is homogeneous. The product is a liquid which issues from between the hot rollers and is collected by any suitable means as it falls. Gases of decomposition are given off and these are allowed to escape with the current of nitrogen and may be condensed to recover the useful ingredients which have been entrained. All types of polyurethane resins can be decomposed in this way and will produce a product capable of being reacted with polyisocyanates to make new useful compounds.

A process developed by R. Lotz, G. Wick and C. Neuhaus (14) involves decomposing scrap polyethylene terephthalate into a pure dimethyl terephthalate under conditions which do not require elevated pressures or high pressure apparatus. It is known that polyethylene terephthalate, in the form of waste or scrap material, can be decomposed or degraded into dimethyl terephthalate by reaction with methanol at high temperatures and under high pressures. The known reactions proceed with relatively good conversions and purity of the dimethyl terephthalate product.

However, carrying out the reaction under high pressures is quite difficult and requires expensive apparatus. Even a slight leakage or loss of pressure in this apparatus requires an interruption of the decomposition reaction. It is necessary to shut down and restart the pressurized decomposition apparatus, and this procedure always entails substantial material losses of the polyester and methanol as well as a considerable reduction in the rate of throughput and in the quality of the final product.

An attempt has been made to decompose molten polyethylene terephthalate into dimethyl terephthalate without use of pressure by conducting methanol vapors over or through the molten material. In such known reactions, however, there is still much to be desired in terms of reaction speed and the yield of dimethyl terephthalate, and it is quite difficult to prevent excessive decomposition and the formation of undesirable impurities.

This process for the pressureless regeneration of dimethyl terephthalate from polyethylene terephthalate can be achieved by first treating the polyethylene terephthalate with steam at a temperature of about 200° to 450°C., and then reducing this steam-treated polyethylene terephthalate in the form of a brittle solid product to a powder having a mean particle size of about 0.0005 to 0.002 mm.

Subsequently, this fine powder is atomized together with at least one gaseous substance selected from the group consisting of an inert gas and methanol vapor to form an aerosol, the aerosol being conducted in turbulent flow through a reaction zone maintained at a temperature of about 250° to 300°C. in the presence of excess methanol vapors. The desired dimethyl terephthalate is then separated from the effluent reaction product of the reaction zone.

In particular, it is preferred to first cool the effluent reaction products to a temperature of about 165° to 175°C. with separation of the solid and liquid components at this temperature, and the remaining components are further cooled to form a liquid which can be subjected to fractional crystallization for the recovery of dimethyl terephthalate. It is especially desirable to carry out the process in a continuous manner so that unreacted solid and condensed liquid components separated at about 165° to 175°C. can be continuously recycled to the reaction zone after first reducing the solids to a mean particle size of approximately 0.001 mm.

A process developed by K.T. Barkey, E.B. Lefferts and D.C. May (15) for recovering polyester scrap for reuse involves initially degrading the polyester with a lower alkyl alcohol and subsequently recovering glycol, dicarboxylic diester, and alcohol from the resulting reaction mixture by distillation. It has been discovered that substantially higher yields of desired components result when a small amount of a phosphorus-containing material that is capable of reacting with the ester exchange catalysts in the scrap is blended into the reaction mixture at the end of the alcoholysis step of the process.

A process developed by E.R. Hittel and F.W. Rennie (16) provides for the recovery of polyester base material from films, including photographic films having at least one macromolecular organic polymer coating. The process comprises (a) treating pieces or flakes of such films with a caustic alkali solution to form a slurry of flakes; (b) feeding the slurry into a classification column and allowing flakes to move downwardly to a rising column of aqueous liquid to separate the polyester material from the coating material; (c) removing flakes settling to the bottom of the column, and (d) recovering the polymeric coating material from the top of the column. The flakes are then dried and used as a source of polyester material.

A process developed by P.J. Boeke (17) is one in which recovery of waste plastic materials in industrial manufacturing operations is enhanced by applying a coating material such as polyvinylchloride to a substrate such as polyethylene in a predetermined geometrical pattern. Subsequent manufacturing operations affect only the coated portion of the substrate. The noncoated substrate outside the processing zone is readily recovered in condition for reuse.

References

(1) Banks, M.E., Lusk, W.D. and Ottinger, R.S. *New Chemical Concepts for Utilization of Waste Plastics,* Washington, D.C., U.S. Environmental Protection Agency (1971).
(2) Gutfreund, K., *Feasibility Study of the Disposal of Polyethylene Plastic Waste,* Washington, D.C., U.S. Environmental Protection Agency (1971).
(3) Holman, J.L., Stephenson, J.B. and Jensen, J.W., "Processing the Plastics from Urban Refuse," *Report PB 208,014,* Springfield, Va., National Tech. Information Service (Feb. 1972).
(4) Pijst, J.P.; U.S. Patent 3,590,527; July 6, 1971; assigned to N.V. Hollandsche Draad-en Kabelfabriek, Netherlands.
(5) Shepherd, T.H.; U.S. Patent 3,590,528; July 6, 1971; assigned to Princeton Chemical Research Inc.
(6) Newland, G.C., Schulken, R.M., Jr., and Harris, R.C.; U.S. Patent 3,592,792; July 13, 1971; assigned to Eastman Kodak Co.
(7) Stager, F.W. and Minor, R.N.; U.S. Patent 3,647,111; March 7, 1972; assigned to Biocor Corp.
(8) Anderson, B.F.; U.S. Patent 3,673,134; June 27, 1972; assigned to E.I. du Pont de Nemours and Co.
(9) Henry, J.W.; U.S. Patent 3,676,401; July 11, 1972; assigned to Eastman Kodak Company.
(10) Boyd, J.A. and Boyd, D.E.; U.S. Patent 3,490,395; January 20, 1970; assigned to Washington Incinerator Sales & Service, Inc.
(11) Stockman, R.F.; U.S. Patent 3,572,265; March 23, 1971; assigned to The Air Preheater Co., Inc.
(12) Monet, G.P.; U.S. Patent 3,069,465; December 18, 1962; assigned to E.I. du Pont de Nemours and Company.
(13) Hurley, R.T.; U.S. Patent 3,143,515; August 4, 1964; assigned to Reeves Brothers, Inc.
(14) Lotz, R., Wick, G. and Neuhaus, C.; U.S. Patent 3,321,510; May 23, 1967; assigned to Vereinigte Glanzstoff-Fabriken AG, Germany.
(15) Barkey, K.T., Lefferts, E.B. and May, D.C.; U.S. Patent 3,488,298; January 6, 1970; assigned to Eastman Kodak Company.
(16) Hittel, E.R. and Rennie, F.W.; U.S. Patent 3,652,466; March 28, 1972; assigned to E.I. du Pont de Nemours and Company.
(17) Boeke, P.J.; U.S. Patent 3,551,242; December 29, 1970; assigned to Phillips Petroleum Co.

## PLATING CHEMICALS

### Recovery from Water

A process developed by D.C. McKissick (1) is a process for economically recovering electroplating solutions, which are usually discarded, from an electroplating production line. The process provides for one or more static rinses where the plated material is washed free of the plating solution. The rinse solution is then filtered and treated in a cation exchanger where impurities are removed from the rinse and the purified solution returned to the plating tank for immediate use or to a holding tank for future use.

### References

(1) McKissick, D.C.; U.S. Patent 3,681,212; August 1, 1972; assigned to American Standard Inc.

## PLATINUM

### Removal from Air

A process developed by H. Rudorfer (1) involves recovering noble metals, particularly platinum metals, which are lost, e.g., by evaporation or mechanical separation during exothermic chemical reactions performed on catalysts consisting of such metals or alloys thereof. In order to recover the largest possible amount of the catalytic noble metal which is evaporated or which is lost in the form of dust it has been proposed to use very expensive collecting substances and collecting devices and to arrange the same at various points of the reaction apparatus.

In some cases the recovered amounts of platinum, rhodium or other alloying constituents of such noble metal catalysts were unsatisfactory and in other cases the manipulation and processing required for separating and isolating the metals were complicated and the collecting substances used were expensive themselves and in danger of being lost. After time consuming experiments aiming at an optimum recovery of platinum metals with the aid of least expensive auxiliary substances which can be processed more easily this method has been developed.

This process for recovering platinum and platinum-rhodium alloys from gases containing same comprises passing the gases at a temperature from 600° to 900°C. through a gas-permeable layer of solids, at least the surface of which is magnesium oxide, and dissolving the oxide in an acid solvent in which the platinum and platinum-rhodium alloys are substantially insoluble.

A process developed by H. Wimmer, A. Wagner, R. Staudigl and H. Rudorfer (2) constitutes an improvement in the process for the recovery of precious metals which are lost during exothermic chemical reactions performed on catalysts consisting of such metals or alloys thereof.

In a known process, heat-resisting acid-insoluble materials are employed in chemical devices for such metals. The recovery capability of such materials is increased by an activation treatment prior to their employment in collection devices. The materials are subjected to chemical treatment wherein the chemicals attack the surface of the collection substances and increase substantially the surface area thereof. Following the treatment the substances are employed in the collection devices in the known manner.

### References

(1) Rudorfer, H.; U.S. Patent 2,920,953; January 12, 1960; assigned to Oesterreichische Stickstoffwerke AG, Austria.
(2) Wimmer, H., Wagner, A., Staudigl, R. and Rudorfer, H.; U.S. Patent 3,515,541; June 2, 1970; assigned to Osterreichische Stickstoffwerke AG, Austria.

## PROTEINS

### Removal from Water

A process developed by R.A. Grant (1) involves the use of a particulate ion exchange material for the purification of waste effluents, such as washings obtained from slaughter houses, which contain protein or fat, or both. The use of the material can provide effluent with a sufficiently low contamination level for it to be readily disposed of, or even reused for further cleaning purposes. By suitable elution of the material, the protein or fat can be released and isolated for use, for example, as animal food. The ion exchange material can be regenerated for reuse.

### References

(1) Grant, R.A.; U.S. Patent 3,697,419; October 10, 1972; assigned to Tasman Vaccine Laboratory, Limited, New Zealand.

## RADIOACTIVE MATERIALS

### Removal from Air

Radioactive materials used in industry are a definite hazard today and will become an increasing rather than a diminishing hazard in the future. In industry, the maximum permissible dose of direct, whole body radiation of persons from all radioactive materials, airborne or nonairborne, is 5,000 millirem per year. There is greater likelihood that this limit will be reduced than that it will be increased. Airborne radiological hazards can result from routine or accidental venting of radioactive mists, dusts, metallurgical fumes, and gases and from spillages of liquids or solids.

Presently existing governmental regulation of the rate of venting airborne, radioactive materials consists primarily of specific limitations based upon individual chemical compounds or upon concentrations of radioactivity from single vents. No concepts have been promulgated concerning methods of controlling total radioactive air pollution from all sources in an entire area. Whether it will be either desirable or necessary to find a solution or solutions to these problems is an unanswered question.

The characteristics of radioactive, gaseous or airborne, particulate wastes vary widely depending upon the nature of the operation from which they originate. In gaseous form they may range from rare gases, such as argon (A 41) from air cooled reactors, to highly corrosive gases, such as hydrogen fluoride from chemical and metallurgical processes. Particulate matter or aerosols may be organic or inorganic and range in size from less than 0.05 to 20 microns. The smaller particles originate from metallurgical fumes caused by oxidation or vaporization. The larger particles may be acid mist droplets, which are low in specific gravity and remain suspended in air or gas streams for longer periods.

Solid, radioactive wastes are of two general classes, combustible and noncombustible. Typical combustible solid wastes are paper, clothes, filters and wood. Noncombustible, solid wastes may include nonrecoverable scrap, evaporator bottoms, contaminated process equipment, floor sweepings, and broken glassware. If inadequate provisions are made for proper handling and disposal of these wastes, a distinct nuisance, and, under certain circumstances, even a hazard, could result.

Liquid, radioactive wastes are evolved in all nuclear energy operations, from laboratory research to full-scale production. Liquid wastes with relatively small concentrations of radioactivity originate in laboratory operations where relatively small quantities of radioactive materials are involved. Other sources are the processing of uranium ore and feed material; the normal operation of essentially all reactors, particularly water-cooled types; and the routine chemical processing of reactor fuels. High activity liquid wastes are produced by the chemical processing of reactor fuels. Removal of radioactive suspended particles, vapors, and gases from hot (radioactive) exhaust systems before discharge to the atmosphere is a serious problem confronting all nuclear energy and radiochemistry installations.

Removal is necessary in order to prevent dangerous contamination of the immediate and neighboring areas. Air pollution brought about through discharge of radioactive stack gas wastes from ventilation systems is only partially avoided by filter devices, no matter how efficient they may be, if the discharge contains radioactive gases. In systems using filter media such as paper, cloth, glass fiber, and so forth, activity eventually builds up in the filter media through dust loading; the same situation applies to electrical precipitators.

Another problem in the control of airborne, radioactive waste is the low dust loading of exhaust streams. The dust concentration of ambient air is usually about 1 grain per 1,000 cubic feet. At installations handling radioactive material, owing to precleaning of the entering air, aerosols may have concentrations as small as $10^{-2}$ to $10^{-3}$ grain per 1,000 cubic feet. In contrast, loadings of some industrial gases may reach several hundred grains per cubic foot, though values of 20 grains or less per cubic foot are more common.

An outstanding feature to consider with air cleaning requirements for many nuclear operations is the extremely small permissible concentrations of various radioisotopes in the atmosphere. Often, removal efficiencies of about 99.9% or greater for particles less than 1 micron in diameter are necessary. This high removal efficiency limits the selection of control equipment for radioactive applications.

*Hooding and Ventilation Requirements:* Hooding for radiochemical process must prevent radioactive contaminants, such as dust and fumes, from escaping into the work area and must deliver them to suitable control devices. Radioactive sources require proper shielding to prevent the escape of radiation and are not considered in this section. The materials used for construction for hoods depend upon the type and quantities of radioactivity and the nature of the process. Stainless steel, masonite, transite, or sheet steel, surfaced with a washable or strippable paint, can be used. Where it is necessary in a process to handle material that may cause dusts or fumes to form, a completely enclosed hood should be used, equipped with a glove box or dry box. Any tools used for manipulation should not be removed from the hood.

The recommended airflow for toxic material across the face of a hood is 150 fpm (Manufacturing Chemists' Association, 1954). Turbulence of air entering a hood can be reduced by the addition of picture frame airfoils to the edges. Hoods should not be located where drafts will affect their operation. When more than one hood is located in a room, fan motors should be operated by a single switch. The fan should freely discharge to the atmosphere and be connected to the outlet side of any control device, the motors being located outside the air ducts to prevent their contamination. Hood and ducts should be equipped with manometers to indicate that they are operating under a negative pressure.

*Air Pollution Control Equipment:* Reduction at the source has been defined as the design of processes so as to minimize the initial release of particulate matter at its source. The principle is not new; it is applied, for example, in the ceramics industry where dry powders are wetted and mixed as a slurry to minimize the production of dust. But its application to radioactive aerosols is particularly worthwhile since it (a) provides a cleaner effluent, (b) reduces radiation hazards involved in the maintenance of air cleaning equipment or those resulting from the buildup of dust activity, (c) permits the use of simpler and less expensive air cleaning equipment, and (d) becomes a part of the process once reduction has been established. In general, preventing the formation of highly toxic aerosols is preferable to cleaning by secondary equipment.

The design or redesign of processes for reduction at the source should be based upon a study of the quantity and physical characteristics of the contaminant, and the manner in which it is released. Examples of this concept are installation of glass fiber filters on the inlet of ventilating or cooling air to minimize the irradiation of ambient dust particles, and treatment of ducts to minimize corrosion and flaking. The most satisfactory control of particulate contamination with air cleaning equipment results from using combinations of the various collectors. These installations should be designed to terminate with the most efficient separator possible, the nature of the gases being considered. To reduce maintenance less efficient cleaners capable of holding or disposing of most of the weight load should be placed before the final stage. It is good practice to arrange the equipment in order of increasing efficiency.

A typical example of such an arrangement is a wet collector such as a centrifugal scrubber to cool the gases and remove most of the larger particles, an efficient dry filter such as a glass fiber filter to remove most of the remaining particulate matter, and a highly efficient paper filter to perform the final cleaning. If the gases are moist, as in this example, the paper filter should be preceded by a preheater to dry the gases. An air cleaning installation for highly toxic aerosols should fulfill the following requirements.

(a) It should discharge innocuous air.

(b) The equipment should require only occasional replacement and should be designed for easy maintenance. Frequent replacement or cleaning entails excessive exposure to radiation and the danger of redispersing the collected material.

(c) The particulate matter should be separated in a form allowing easy disposal. The use of wet collectors, for example, poses the additional problem of disposing of volumes of contaminated liquid. Wet collection does, however, reduce considerably the danger of redispersion.

(d) Initial and maintenance costs, as well as operating costs, should be as low as possible while fulfilling the preceding three conditions. In this respect, pressure drop is generally an important consideration.

One type of commercially available dust collector that meets the requirements of filtering airborne, radioactive particles from ventilation exhaust streams is a bag filter employing what is called reverse jet cleaning. This type of baghouse has an efficiency as high as the conventional cloth bag or cloth screen collector and is particularly adapted to an installation where the grain loading of the effluent is low. The bag material is a hard wool felt of the pressed type, about $\frac{1}{16}$" thick, or a cloth woven of glass fibers.

The gas flow is likely to be around 10 to 40 cfm per square foot of bag area when the pressure drop is maintained at usual values such as 2 to 7 inches water column. The conventional cloth bag or cloth screen collectors, which are cleaned periodically by automatic shaking devices, may allow a puff of dust to escape after the shaking operation. The problem of maintenance in this instance presents a contamination and radiation hazard. For this reason, the reverse jet baghouse is generally preferred. Another method of treating contaminated exhaust air before discharge to the atmosphere involves the use of wet collectors of various types. These collectors are relatively effective on gases.

Investigation covering changing of water supply or recirculating has shown the latter procedure useful for considerable periods of time without apparent adverse effect. Evaporation is compensated for by fresh supply. Insoluble radioactive salts, soluble salts, and other radioactive particles that may form a solution, suspension, or sludge in the reservoir result in fairly high radioactivity of the scrubbing media. Precautions must be taken during maintenance to avoid carryover of the scrubbing media since the radioactive contamination of entrained liquid would be transferred to the preheater or filter, resulting in high radiation levels at those points.

Some important disadvantages of wet collectors make them less attractive than other types of collectors. Wet collectors present the difficult problem of separating the radioactive, solid material from the water in which it is suspended. Maintenance and corrosion are serious problems. Considerable quantities of water are required, and, if the radioactive solids are not separated from the water, this in turn leads to a final storage and disposal problem. Radioactive, airborne particles, when given an electrical charge, can be collected on grounded surfaces. The fact that the particles are radioactive has very little to do with their behavior in an electrical precipitator. Experiments conducted with precipitators using the alpha-emitter polonium and the beta-emitter sulfur 35 indicate that neither material behaves in a way different from nonradioactive material.

Water flushed type, single stage, industrial precipitators, and air-conditioning type, two-stage precipitators are used for separating radioactive dusts and fumes from gases at atomic energy plants and laboratories. A small electrical precipitator of the water flushed type with a design capacity of 200 cfm was installed to test efficiency of collecting and removing particulate radioactivity from the off-gas system of an isotope recovery operation. This precipitator consists of 23 vertical collecting pipes with an ionizing wire centered in each pipe. The inside surfaces of the pipes serve as collecting walls.

For wet operation, the collecting walls are water flushed by means of spray nozzles installed at the top of each pipe. This water is recycled continuously at a rate of 35 gpm over the collecting walls while high voltage is applied to the electrodes. This unit reportedly collects more than 99.99% of the particulate radioactivity in the off-gas at 50 to 55 kilovolts when the concentration of radioactivity as solids is greater than $5.0 \times 10^{-4}$ microcuries per cubic centimeter of off-gas. Based upon tests made at the Oak Ridge National Laboratory, the following evaluation of precipitators used in radioactive applications may be made.

(a) Electrical precipitators are not intended to collect the ultrafine particles which may be discharged from radiochemistry installations.

(b) With uneven airflow, the air velocity through some of the collector cells may be sufficiently above velocity limits to blow off collected wastes which would then be discharged to the atmosphere.

(c) Efficient operation depends a great deal on the regularity with which the unit is cleaned. At best the electrical precipitator is only approximately 90% efficient. This may be demonstrated by the fact that dense clouds of tobacco smoke fed into the precipitator will escape from it in concentrations great enough so that the escaping smoke can be seen. The blue color of tobacco smoke is evidence that most of its particles have a diameter less than the wavelength of light, which is roughly 0.5 micron.

(d) For absolute efficiency an after filter of the Cambridge or MSA Ultra-Aire type is necessary to catch the dirt should the precipitator short circuit.

(e) Difficulty may be experienced if the dustload builds up faster than it can be removed, eventually becoming so heavy that arcing occurs between the dirt bridges resulting in a fire hazard.

(f) Devices such as the single-stage industrial precipitator and the air-conditioning type two-stage precipitator accomplish only one phase of the problem. The final disposal of radioactive wastes collected and accumulated during operation and maintenance still remains.

Glass fiber or glass fiber paper is often used as a filter medium and is effective in the operation of radiochemistry hoods, canopies and gloved boxes. One of the most efficient lightweight, inorganic filters developed to date is made with a continuous, pleated sheet of microglass fiber paper. The pleats of the glass paper are separated by a corrugated material (paper, glass paper, aluminum foil, plastic, or asbestos paper) for easy passage of air to the deep pleats of the filter paper.

The assembly of the filter paper and corrugated separators is sealed in a frame of wood, cadmium plated steel, stainless steel, or aluminum. This construction permits a large area of filter paper to be presented to the airstream of a correspondingly low resistance. Glass fiber, from which filters are made, withstands temperatures up to 1000°F. It is noncombustible and has extremely low thermal conductivity and low heat capacity.

The fibers are noncellular, are like minute rods of glass, and do not absorb moisture; however, water can enter the interstices. The material is relatively nonsettling, noncorrosive, and durable. It is resistant to acid fumes and vapors, except hydrogen fluoride. The installation and replacement costs of glass fiber filters are low. Final disposal of used filters may be accomplished by incinerating at over 1000°F. with provisions for decontaminating the stack gases. This melts the glass fibers, reducing the physical mass to the size of a glass bead. Thus, glass fiber filters provide, in part, a very good answer to the problem of control and final disposal of radioactive contaminants.

A highly efficient paper filter medium can be used with adequate effectiveness on incoming ventilating air and as a final cleaner in many instances. This type filter is composed of asbestos cellulose paper. A more recently developed filter has a glass fiber web. It is designed and manufactured in corrugated form to increase the available filter area and loading capacity and to reduce initial resistance. The filter units are tested at rated capacity with standard U.S. Army Chemical Corps test equipment for resistance and initial penetration and are unconditionally guaranteed to be at least 99.95% effective against 0.3 micron diameter dioctyl phthalate particles. This filter performs as well as, or better than, the earlier paper types and under temperatures up to 1000°F.

Airborne, radioactive wastes are only part of the control and disposal problem of nuclear energy and radiochemistry installations. Solid and liquid, radioactive wastes are subject to the same limitations on disposal to the environment. The methods of disposing of the final waste from the collection systems present additional problems, as follows. (a) Incineration results in stack gas and particle discharge which is a cycle of the entire problem repeated over again. (b) Direct burial results in redispersal and ground contamination with associated problems related to the ground water table. (c) High dust or particle loading capacity results in high radioactivity of the collecting media.

(d) Vapors, acid fumes and unfilterable gases may cause rapid deterioration and disintegration of filter media resulting in a maintenance and health hazard problem. (e) Mechanical replacement costs are high because of the remote handling involved. (f) An auxiliary unit for emergency or maintenance shutdown must be available to prevent the possibility of reverse flow of the air stream out of hot equipment into controlled rooms and areas.

The overall aspects of air pollution by radioactive substances have been reviewed by S. Miner (1).

### Removal from Water

A process developed by J.W. Loeding and A.A. Jonke (2) relates to a method of treating aqueous nuclear wastes whereby disposal problems are vastly simplified. An earlier method provides for the disposal of aqueous nuclear wastes in which the radioactive material is obtained in a form which is easily stored and yet available to obtain potentially valuable isotopes therefrom. This is accomplished by reducing the waste solution to dryness and simultaneously calcining the solids contained therein to the oxides by introducing the solution into a heated fluidized bed formed of the oxides. The method suggests the use of air and of steam as fluidizing gases.

A problem arising in connection with operation of the described process resides in the fact that ruthenium is volatilized at the temperatures at which it is desirable to carry out the process. Since ruthenium is a radioactive fission product, the condensate obtained is radioactive.

This radioactive solution then must be treated to remove the ruthenium therefrom before it can be disposed of. Obviously it is desirable to carry out the operation so that substantially all of the radioactivity remains in the solids. It is the object of this process to carry out the evaporation and calcination of radioactive nuclear wastes in such a way that ruthenium is retained in the solids. It has now been found that this and other advantages can be obtained by the use of carbon monoxide, preferably in combination with nitrogen, as the fluidizing gas. Figure 120 shows a suitable form of apparatus for the conduct of the process. In the figure a reaction vessel 10 contains reaction chamber 11 and disengaging chamber 12. Reaction chamber 11 is separated from a gas chamber 13 by a porous plate 14. Product delivery conduit 15 leads from reaction chamber 11 to a product delivery reservoir 16. A gas inlet pipe 17, passing through a heater 18 leads into gas chamber 13.

Disengaging chamber 12 contains two banks of filters 19 separated by baffle 20. Off-gas conduits 21 join to form conduit 22 and lead the gas passing through the filters 19 to a condenser 23. Gases leaving the condenser 23 will be exhausted to the atmosphere through a stack (not shown). One bank of filters may be used at a time while a flow of gas is passed in reverse direction through the other bank to remove the dust collected thereby.

Feed to the reaction vessel 10 is by way of a plurality of pneumatic atomizing spray nozzles 24 located above but near to the porous plate 14 and located in a single horizontal plane. The spray nozzles include an air inlet 25, a feed inlet 26, and a mixing chamber 27. A charge hopper 28 is connected to the upper portion of the reaction chamber 11 by conduit 29. Electric heaters 30 are disposed about the reaction vessel 10. Appropriate shielding (not shown) is provided to protect operating personnel. To operate the device the reaction chamber 11 is filled to a point above the spray nozzles 24 with granular particles between 20 and 200 mesh in size. A heated gas is passed into gas chamber 13 through air inlet 17 and heater 18.

FIGURE 120: APPARATUS FOR REDUCING AQUEOUS RADIOACTIVE NUCLEAR
WASTES TO SOLID FORM

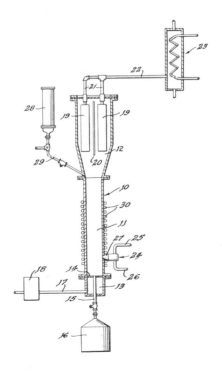

Source: J.W. Loeding and A.A. Jonke; U.S. Patent 2,977,194; March 28, 1961

The gas passes through porous plate 14 and up through the particles in the reaction chamber 11, causing the mass of particles to behave very much as a vigorously boiling liquid. This fluidized bed is heated to the desired operating temperature by heaters 30. Feed solution is then introduced into the bed below the surface thereof. The solution is evaporated and the salts contained therein are calcined to the oxide of the elements contained in the solution by contact with the heated particles in the bed. As the oxide is produced it is withdrawn through product delivery conduit 15 and retained in reservoir 16.

The wastes will then be stored till short-lived fission products have decayed. The potentially valuable strontium-90 and cesium-137 can then be recovered by leaching or dissolution and precipitation. The gas, carrying the vaporizable portion of the feed, rises and passes through filters 19 which remove entrained solids from the gas. The gas is then condensed in condenser 23 and the condensate collected. The noncondensable gases are then evacuated to the atmosphere through a stack (not shown). If necessary the condensate may be treated by such methods as ion exchange to remove any radioactive materials which have been volatilized.

A process developed by B.M. Johnson, Jr. and G.B. Barton (3) for processing radioactive nonacidic waste solutions, derived from the processing of neutron-irradiated uranium, for disposal, comprises adding to the solution a water-soluble compound containing an ion selected from the group consisting of phosphate, borate, and silicate ions and mixtures thereof; spraying the solution thus obtained with steam into a space heated at between 325° and 400°C. whereby water evaporates and a powder is formed, heating the powder to melting temperature of between 800° and 1000°C. whereby calcination takes place; and separating water vapor and gaseous substances developed from the calcined product.

Figure 121 is a flow sheet of the process. Referring to this figure in detail, the reference numeral 1 designates a dehydration and calcination tower, a calciner which is equipped at its top with a fluid spray nozzle 2. An inlet pipe 3 for steam and an inlet pipe 4 for the waste solution to be treated lead into the spray nozzle 2. The tower has a dehydration zone 5, a calcination zone 6 and a cooling zone 7. To create these zones, independent heating means (not shown) arranged around zones 5 and 6 and cooling means (also not shown) arranged around zone 7 are operated.

At the bottom of the tower there is an outlet pipe 8 which leads to a cyclone 9, and the cyclone, in turn, is connected in series with a glass bag filter 10. Discharge pipes 11 and 12 of the cyclone and the glass filter respectively, lead to a common pipe 13 through which the solids deposited at the bottoms of the cyclone 9 and the glass bag filter 10 are withdrawn and charged to a packaging unit (not shown).

At the top of filter 10 there is arranged a pipe 14 which is connected with a heat exchanger and condensor 15 located near the bottom of an evaporator 16. An outlet pipe 17 for the condensate is arranged in the bottom of heat exchanger 15; it leads to a holdup tank 18. Tubes 19 and 20 connect the holdup tank 18 with the evaporator 16 and tube 19 connects it also with a scrubber 21. The scrubber is linked to a separator 22, and the latter is connected with the evaporator 16 through pipes 23 and 20. Pipe 24 connects condensor 15 and scrubber 21. A filter 25 is arranged above and connected with the separator 22. The top of the evaporator 16 is equipped with an exhaust line 26 which leads to a vacuum condenser (not shown) where the steam and any other vapors escaping the evaporator are condensed. A discharge line 27 connects the bottom of the evaporator with inlet pipe 4.

The waste solution is introduced in the form of a spray at the top of the calciner where it passes through zones 5, 6 and 7 for dehydration, calcination and cooling, respectively. The powder suspended in the gas leaving the calciner at its bottom through pipe 8 is separated roughly from the gas in cyclone 9, the solid being discharged through pipe 11. The gas which still contains some fine powder particles is withdrawn at the top of the cyclone and introduced into the top of glass bag filter 10 where further gas solid separation is accomplished.

# FIGURE 121: ALTERNATIVE PROCESS FOR REDUCING LIQUID RADIOACTIVE WASTES TO SOLID FORM

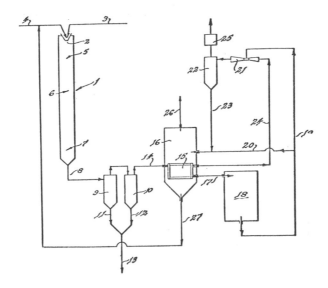

Source:  B.M. Johnson, Jr. and G.B. Barton; U.S. Patent 3,008,904; November 14, 1961

The gaseous component is then subjected to a cooling and condensation step by heat exchange in condensator **15**; noncondensed fractions leave the evaporator at **26**. The cooled condensed liquid is withdrawn through pipe **17** and passed into holdup tank **18**. Part of the condensate accumulated in tank **18**, preferably about 90% of it, is cycled as cooling medium into evaporator **16**, while the rest is introduced into scrubber **21** and utilized for washing the gas. The scrubbed gas leaving the separator is filtered in **25**, while the scrubbing liquor is recycled, via pipes **23** and **20**, into evaporator **16**. Liquid condensed in evaporator **16** is withdrawn at **27** and recycled into the feed line **4** and back into the calciner. Gas noncondensable in the heat exchanger **15** leaves through pipe **24** whence it passes through scrubber **21**, etc.

While phosphate and borate anions are the preferred ions for the purpose of converting the calcined waste solution to a glassy water-insoluble material, silicate and mixtures of any of the three can also be used to advantage. The disadvantage of silicate is that a higher processing temperature is needed for the material to melt. A mixture of borax and sodium silicate, in an equivalence ratio of 2:1 was found to reduce the solubility of the final product of 1.3% of the initial weight as measured in 100 parts of boiling water for one hour; the melting point of the product thus obtained was 850°C.

For a waste solution derived from the extraction of uranium, plutonium, etc. with tributyl phosphate, the addition was preferably a mixture of phosphate and borate. All phosphates, borates and silicates that are water-soluble are suitable, including the free acids.

Sugar is preferably added in a quantity ranging from 200 to 400 g./l. The wall temperature of the reactor should be maintained at at least 700°C. in order to obtain decomposition of the sugar to carbon. The process is usable for all kinds of waste solutions; it has been found particularly advantageous for waste solutions that are derived from extraction processes, e.g., from methyl isobutyl ketone (hexone) or with tributyl phosphate.

A process developed by M.D. Nelson, Jr. (4) provides for the safe and complete containment of high level radioactive wastes. The processing of radioactive matter produces large amounts of radioactive wastes. The principal state of these radioactive wastes is liquid. Although the liquid wastes contain nuclides in a small concentration, the level of radio-activity may be very high. These wastes are usually stored as a means of disposal. The storage of wastes containing nuclides which have a long life requires absolute confinement for extended periods of time, generally between 15 and 20 years.

However, in certain cases the period of storage may be centuries in duration. The most common means of storage for the disposal of high level atomic wastes is the method of complete containment. The method of complete containment of high level radioactive wastes consists in storing the wastes in absolute confinement until all or most of the radio-activity has decayed. The absolute confinement can be obtained by storing the radioactive wastes in leakproof shielded containers.

The leakproof shielded containers usually are provided by underground storage tanks having enclosing walls of concrete approximately 8 ft. in thickness with an inner fluidproof liner of stainless or mild steel. These tanks are surrounded by a second storage facility, spaced from the first, with means for monitoring the intermediate space to detect the escape of radioactive matter or nuclides. Further, expensive high speed fluid transferring equipment must be available at all times to transfer the high level radioactive wastes from one storage facility to another in case of a leak.

The above complete containment method requires expensive storage facilities with continuous supervision and maintenance. It is known to reduce the cost of storage by reducing the amount of the stored radioactive liquid wastes such as by evaporation or other means of concentration of the liquid wastes. However, even the storage of concentrated liquids is very excessively expensive. There is always the possibility with this method of complete containment that as a result of an earthquake or other catastrophe, the contents of such facilities may escape confinement and migrate to inhabited areas and thereby contaminate them to a high level of radioactivity.

It would be desirable to confine the radioactive wastes within the earth in conventional cement, especially where the wastes are solids. However, because of the danger of formation fluids leaching the conventional cement, and the loss of fluids from the cement before and after the initial setting period, the cement-contained radioactive wastes are not disposed within the earth, but rather are cased in steel barrels or other small impervious containers. This method is expensive because of the large number of containers which must be disposed usually by dumping at sea.

Radioactive liquid wastes can be confined in certain impervious geological structures, such as in voids present within an impervious formation. However, such storage facilities require a specific type of formation which is free from migrating fluids which can leach the wastes and then escape confinement. In particular, migrational formation water must be absent. Further the storage facilities may be ruptured by earthquakes or the like, thereby permitting the radioactive wastes to escape containment.

It is therefore an object of this process to provide an economical method and composition for the complete containment of radioactive materials to insure the absolute protection of the public health for the period of time required for the radioactivity of the radioactive materials to decay to a safe level.

In accordance with this process there is provided a radioactive containment composition comprising a mixture of an inverted emulsion cement and the radioactive wastes. Further, there is provided a method of placing such composition into a subsurface storage area for the complete containment of the radioactive wastes. Additionally, there is provided a system for storing the disposed composition and a nonmigrational, cement-like radioactive material from which nuclides cannot escape.

A process developed by B.A. Podberesky and H.T. Loeser (5) relates to the processing of radioactive liquid wastes and more particularly, the removal of radioactive ions from a liquid. One of the most important problems in the field of nuclear technology is the processing of liquid wastes containing radioactive isotopes of iron, cobalt, chromium, tantalum, sulfur and other metallic and nonmetallic elements. The methods in operation to process liquid wastes involve at least one of the following techniques: evaporation; crystallization; demineralization; fixation; and centrifugation.

These methods suffer from the disadvantage that the highly radioactive residues remaining after processing are in a form which is difficult for handling and storage. Furthermore, the methods are costly since they utilize relatively expensive material and equipment. Demineralization with ion exchange resins is somewhat limited in its application to radioactive liquid wastes since the regeneration of the exhausted resin produces highly radioactive liquids which themselves must be processed or stored in shielded tanks.

In addition, replacement of the resin involves elaborate and expensive packaging of the exhausted resin for burial. Still another problem in using ion exchange resin for processing radioactive liquid wastes is the direct effects of radiation on the resin resulting in a decrease in the exchange capacity and degradation of the physical and chemical characteristics of the resin. Thus, a demand exists in the nuclear industry for a more efficient and economical method of removing radioactive ions from liquid wastes.

This process involves contacting a liquid containing radioactive nonmetallic ions with electrodes including at least one anode and one cathode, the anode being fabricated from a metal which is reduced and goes into solution thereby combining with the nonmetallic ions to form an insoluble precipitate.

A process developed by D.A. Shock, J.D. Sudbury and P.L. Gant (6) involves disposal of radioactive waste solutions and slurries in surface and subsurface earthern reservoirs and the like by means of fixing, in an economical manner, and such that discarded wastes will not present any hazards. In another aspect, it relates to method of disposal of radioactive waste in fractured impermeable formations.

More specifically, this process comprises mixing a radioactive waste solution or slurry with clay minerals to form a slurry similar to drilling mud, converting the slurry to a high lime mud having a pH of above 10 by the addition of lime and caustic. The above procedure will be found sufficient in many cases to attain solidification; however, in others it will be found necessary for the gel to be heated to above 500°F. by self absorption of the heat evolved by the radioactive emission, and/or heat from an outside source. It must be pointed out that, generally, the additional heat is for the purpose of shortening the time for solidification not to cause solidification per se.

In another aspect, the process is directed to the disposal of radioactive waste in impermeable rock formations in horizontal fractures provided therein, by injecting a solidifiable material containing radioactive waste into the fracture and thereafter injecting additional solidifiable material free of radioactive waste to seal the radioactive waste in the impermeable rock formation.

In a more specific aspect of this method, injection of a solidifiable material free from radioactive waste into the fracture is commenced, thereafter radioactive waste is introduced to the flowing solidifiable material and finally the introduction of radioactive waste is terminated and flow of the solidifiable material into the fracture is continued for a period of time to seal the radioactive waste in the impermeable rock formation.

**References**

(1) Miner, S., "Air Pollution Aspects of Radioactive Substances," *Report PB 188,092,* Springfield, Va., National Tech. Information Service (Sept. 1969).
(2) Loeding, J.W. and Jonke, A.A.; U.S. Patent 2,977,194; March 28, 1961; assigned to U.S. Atomic Energy Commission.

(3) Johnson, B.M., Jr. and Barton, G.B.; U.S. Patent 3,008,904; November 14, 1961; assigned to the U.S. Atomic Energy Commission.

(4) Nelson, M.D., Jr.; U.S. Patent 3,262,274; July 26, 1966; assigned to Mobil Oil Corporation.

(5) Podberesky, B.A. and Loeser, H.T.; U.S. Patent 3,320,175; May 16, 1967; assigned to General Dynamics Corporation.

(6) Shock, D.A., Sudbury, J.D. and Gant, P.L.; U.S. Patent 3,274,784; September 27, 1966; assigned to Continental Oil Company.

# RARE EARTHS

A process has been developed by H.L. Recht and M. Ghassemi (1) for chemically removing trivalent rare earth ions from wastewater by treating the trivalent rare earth ion-containing water with a carbonate salt, such as an alkali metal carbonate, to form an insoluble rare earth carbonate precipitate and then separating the precipitate from the water. The rare earth ions may be regenerated in the form of a soluble rare earth salt, for example, by treating the separated rare earth carbonate precipitate with acid, such as hydrochloric or sulfuric acid.

### References

(1) Recht, H.L. and Ghassemi, M.; U.S. Patent 3,692,671; September 19, 1972; assigned to North American Rockwell Corporation.

# ROLLING MILL DUST AND FUMES

## Removal from Air

A process developed by T. Eklund (1) relates to apparatus for the collecting and transporting away of dust, fumes and gases in rolling mills. As a part of the general tendency to obtain better and more hygienic working conditions in the industry, arrangements of the abovementioned type have come into use. It has, however, proved difficult to obtain an effective and suitable exhaust of dust, fumes and gases without the use of exhaust hoods of such dimensions and such placement that they jeopardize the attending and supervising of the mills.

A primary object of the process is to eliminate the abovementioned difficulties and to make possible a more widespread use of such exhaust devices. The principal form is characterized in that the suction boxes are located just above the path of the rolled stock and provided with a suction slot facing the path. Further, a throttling valve is arranged in the branch duct from each suction box, the valves being adapted to be automatically adjusted by means of impulses from the driving means of the mill in such a manner that the throttling valve located at the discharge side of the roll stand always is open, while the other valve simultaneously is closed.

By arranging the exhaustion point close to the points where the development of fumes primarily takes place the fumes can effectively be removed by using relatively small quantities of air and in this manner the suction boxes can be given dimensions so that they do not disturb the running process. Because in reversing type mills the exhaust need be employed at one side at a time the required fan may be relatively small and this also works well for the dust separator used in the exhaust line. Therefore the plant as a whole is simple and inexpensive while simultaneously taking but little space and working effectively. According to a suitable form of the process the suction boxes are combined with the strippers of the roll stand and are connected to the exhaust ducts by means of flexible tubes disposed at the pivot center of the strippers.

The exhaust system may also include a dust separating device of known design. Figure 122 shows a suitable form of apparatus for the conduct of this process. In the figure, 1 designates a housing for a mill and 2 and 3 designate respectively two suction boxes arranged at each side of a roll stand 30. The suction boxes, by means of branch ducts 4 and 5 and a main duct 6, are connected to an exhaust fan 20 which may include a dust separator 25.

The suction boxes are arranged adjacent the path of the rolled stock 21 and are provided with a suction slot 2a and 3a respectively, facing the path. In this form the suction boxes are combined with roll strippers 10 and 11 of the mill and are connected to an exhaust system including exhaust ducts 4 and 5 by means of flexible tubes 2b and 3b respectively, disposed at the pivot center 22 of the strippers.

FIGURE 122:  APPARATUS FOR THE COLLECTING AND TRANSPORTING AWAY OF
DUST, FUMES AND GASES IN ROLLING MILLS

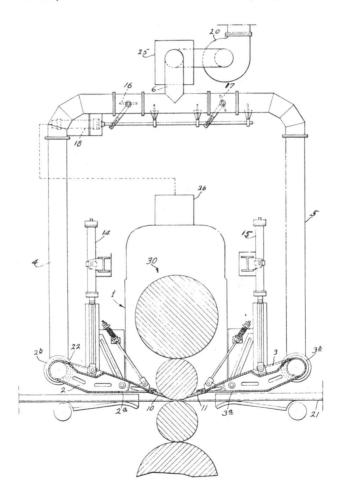

Source:  T. Eklund; U.S. Patent 3,204,393; September 7, 1965

The numbers **14** and **15** stand for two lifting means, in this instance pneumatic, for raising the strippers together with the suction boxes united thereto and for replacing the same in their proper position. In order to gain an effective exhaust and still only use a limited quantity of air, the exhaust is carried out only by means of that suction box which at the moment is situated at the discharge side of the mill.

The alternate connection of the suction boxes is performed by valve means, in this instance comprising the valves **16** and **17** inserted in the branch ducts. The valves are arranged to be automatically adjusted by means of impulses from the driving means **26** of the mill, which impulses govern valve actuating means which in the illustrated form comprise a pneumatic adjustment device **18**. The blades of the valves are mutually displaced 90° so that one valve is closed while the other is open.

### References

(1) Eklund, T.; U.S. Patent 3,204,393; September 7, 1965; assigned to Aktiebolaget Svenska Fläktfabriken, Sweden.

## ROOFING FACTORY WASTES

### Removal from Air

A process developed by H.B. Johnson (1) is one in which the solid wastes of combustible and noncombustible roofing material are comminuted in an attrition mill, fed to a surge bin, and then used as fuel in a solid waste and fume incinerator which also receives and ecologically incinerates the exhaust fumes from a saturator.

A fume incinerator also receives saturator exhaust fumes along with effluent gas from the oxidizer of the factory, and ecologically incinerates the same with a minimum consumption of conventional fuel. The fume ducts from the saturator exhausts to the incinerators include a cross duct and switching valves for selectively controlling the flow of the saturator exhaust fumes from one or both saturators to one or both incinerators for optimum operational efficiency of the system in accordance with current production of the roofing factory. Granules resulting from the solid scrap incineration are washed and recovered for reuse.

Figure 123 is a flow diagram showing the essential elements of this process. Saturators **10** and **12**, and oxidizer **14** constitute components of a roofing factory which also includes a felt section **16**. A by-product of the factory is scrap, including tabs which may be termed solid waste that contains combustible material used in making the roofing products. The scrap from the factory is loaded on a platform **18** for movement by a conveyor **20** to an attrition mill **22**. The tabs **24** and **26** are also moved by conveyors **28** and **30** onto the conveyor **20** which, in turn, carries them to the mill **22**. The output of mill **22** goes to a surge or storage bin **32** on a conveyor **34**, and from the bin **32** to a fuel input of incinerator **36** by a suitable conveyor **38**.

Fume exhaust blowers or fans **40** and **42** are associated with the saturators **10** and **12**, respectively, and are connected by ducts **44** and **46** to incinerators **48** and **36**, respectively. A cross duct **50** having a damper or valve **52**, is connected to the ducts **44** and **46** for selectively separating or putting the latter ducts in communication with each other.

The ducts **44** and **46** are also provided with switching valves **54** and **56**, and **58** and **60** at opposite sides of the cross duct **52** for selectively controlling the flow of exhaust fumes from either one or both saturators **10, 12**, to either one or both of the incinerators **36, 48**. The effluent gas from the oxidizer **14** flows to the incinerator **46** through a duct **62**. The effluent of the incinerator **48** goes to a suitable stack or chimney by way of an air heater **64** from which the heated air flows by way of a duct **66** to the felt section **16** of

## FIGURE 123: ROOFING FACTORY FUME AND SOLID WASTE DISPOSAL SYSTEM

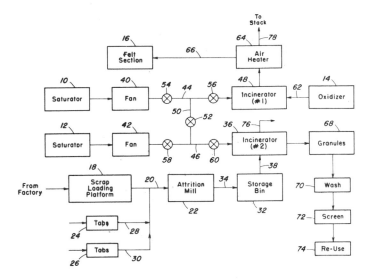

Source: H.B. Johnson; U.S. Patent 3,662,695; May 16, 1972

the factory, providing auxiliary heat for the vapor absorption system of such felt section 16. Residual noncombustible granules produced as a by-product of the incinerator 36, are collected at station 68, then washed and screened at stations 70 and 72, respectively, for reuse from a station 74 where they are finally collected for such purpose.

Effluent from the incinerator 38 go directly to an individual stack from effluent outlet 76. Such effluent as well as that of incinerator 48 going to its stack through outlet 78, is ecologically clean due to complete combustion of the exhaust fumes, and the oxidizer effluent gas, as well as all of the combustible material contained in the solid wastes that are fed to the incinerators 36 and 42. The residual granules of incinerator 36 are also ecologically clean even before washing by virtue of the operation of the complete incinerator 36.

Briefly, the illustrated system efficiently functions to dispose of the solid wastes in the form of various roofing scrap, and incinerates the saturator exhaust fumes, as well as the oxidizer effluent gas. The system takes full advantage of the fuel value of the roofing scrap and also incorporates heat recovery to minimize the consumption of conventional fuel. The resulting stack effluent is ecologically acceptable and the noncombustible solids are reclaimed.

Furthermore, the system is adjustable to handle any waste material that is available as a result of the particular roofing product being made at the time by the factory. Thus, such waste is taken care of efficiently and ecologically by the system regardless of such particular roofing product. Changeover to another type of roofing product requires merely the adjustment of the duct switching valves, as well as adjustment of conventional fuel and air (or oxygen) supplied to the incinerators to bring about complete combustion, and the resulting clean stack effluent.

### References

(1) Johnson, H.B.; U.S. Patent 3,662,695; May 16, 1972; assigned to GAF Corporation.

# RUBBER

## Rubber Processing Effluent from Air

A process developed by J.J. Cunningham, Jr. (1) for removing dust and smoke from tire buffing comprises a cyclone dust collector, a spray chamber and an absolute filter in combination. There is preferably a bypass from a high pressure point in the cyclone dust collector back to a hood adjoining the tire buffing for lengthening the time of a dust and smoke cycle.

## Rubber Recycling

The topic of rubber recycling and reuse has been discussed in some detail by R.J. Pettigrew et al (2).

A process developed by C.E. Scott (3) is one whereby waste vulcanized rubber is depolymerized in a liquid hydrocarbon medium by means of agitation, heat, free radical initiators and molecular oxygen. The resulting solution of rubber modified hydrocarbon can then be utilized in the manufacture of end-use products, thereby disposing of the waste rubber without pollution of the ecological environment. The rubber modified hydrocarbon can, for instance, be vulcanized to produce a moisture barrier or insulation coating. It can also be mixed with asphalt to provide rubberized asphaltic compositions. In still another application, the rubber modified hydrocarbon can be thermally decomposed to produce carbon black.

### References

(1) Cunningham, J.J., Jr.; U.S. Patent 3,579,314; May 18, 1971; assigned to Quinn Brothers, Inc.
(2) Pettigrew, R.J. and Roninger, F.H., *Rubber Reuse and Solid Waste Management,* Washington, D.C., U.S. Environmental Protection Agency (1971).
(3) Scott, C.E.; U.S. Patent 3,700,615; October 24, 1972; assigned to Cities Service Company.

# SELENIUM

Selenium poisoning of humans and animals from ingestion of foods containing toxic amounts of selenium has been and still is a problem of great concern in the United States. Many plants used as food by humans and animals can accumulate high concentrations of selenium from the soil. The soils of the Midwest are particularly high in selenium. Extensive studies have been made of the people, animals, food products and soils of the high seleniferous areas. Selenium has also been found to be an essential nutrient for animals and may be necessary for humans.

Selenium compounds, although toxic, have seldom presented a serious problem in industry. Inhalation of the dust, fumes, or vapors of selenium compounds can irritate the eyes, nose, and throat, causing lacrimation, palpebral edema, conjunctivitis, sneezing, nasal congestion, anosmia and coughing. A study made in Boston, Mass. showed that there was approximately 0.001 $\mu$g./m.$^3$ of selenium in samples of rain, snow, and air analyzed. The sources of atmospheric selenium are believed to be terrestrial, such as fuels and ores used by industry, or possibly the burning of trash, particularly paper.

## Removal of Selenium from Air

No study has been made of the methods for control of selenium and its compounds. However, based on the properties and on the methods of recovery and purification of selenium, wet scrubbers and high voltage electrostatic precipitators should be effective, according to Stahl (1). The common selenium atmospheric pollutants are probably selenium dioxide (or selenious acid in moist environments) and, to a lesser extent, hydrogen selenide.

These compounds are all soluble in water, the former a solid and the latter a gas at ambient temperatures. Selenium metal, except for the red allotrope form, is insoluble in water. However, it is soluble in strong bases, forming selenium compounds, and it will burn in air, producing selenium dioxide. In the recovery and purification processes of selenium, (2) wet scrubbers and electrostatic precipitators are used to collect selenium dioxide fumes given off during the smelting or roasting, as well as to collect red selenium. Furthermore, selenium emissions from trash burning are controlled by the use of wet scrubbers (3).

The chief industrial control methods are the use of a good ventilation system and personal protective equipment, including safety goggles and a respirator when working in a low air-borne concentration and a supplied air respirator in heavily contaminated areas (4). Selenium-containing liquid wastes are disposed of by washing them down the drain, white solid wastes are buried.

A process developed by A. Yomiyama and S. Yonekawa (5) involves the recovery of selenium or selenium compounds used in the production of unsaturated aldehydes or nitriles from unsaturated hydrocarbons.

### Removal of Selenium from Water

An intensive search of the water pollution literature, the industrial waste literature, and the technical literature of the industries which use selenium (paint, pigment and dye producers, glass manufacturers and insecticide industries) yielded essentially no information on levels of selenium in industrial wastewaters, treatment methods for selenium wastes or costs associated with removal of selenium from industrial wastewaters. Available information relating to the general subject of selenium wastewaters and potential methods of treatment is discussed by J.W. Patterson and R.A. Minear (6).

H. Johnson (7) has reported that selenium is present in almost all conceivable types of paper, and it might be concluded from this information that pulp and paper mill wastes could contain selenium. Johnson's primary concern was with the release of selenium as an air pollutant, upon incineration of paper products. However, he did report selenium wastewater levels as follows. For incinerator fly ash quench water, soluble selenium was measured at 5 to 23 $\mu$g./l., while incinerator residue quench water contained 3 $\mu$g./l. of soluble selenium. The incinerator handled a solid waste containing approximately 55 to 69% paper.

K.D. Linstedt, C.P. Houck and J.T. O'Connor (8) measured 2.5 $\mu$g./l. of selenium in the effluent from a secondary sewage treatment plant. These workers investigated the efficiencies of removal of selenium, as $SeO_3^{-2}$, by several advanced wastewater treatment processes and reported the results shown in Table 16.

### TABLE 16: REMOVAL OF SELENIUM BY ADVANCED WASTEWATER TREATMENT PROCESSES (BENCH-SCALE)

| Process | Removal (percent) |
|---|---|
| Lime coagulation–settling | 16.2 |
| Cation exchange | 0.9 |
| Cation plus anion exchange | 99.7 |
| Process Sequence* | |
|   1st. Sand filtration | 9.5 |
|   2nd. Activated carbon | 43.2 |
|   3rd. Cation exchange | 44.7 |
|   4th. Anion exchange | 99.9 |

*Cumulative removal after indicated process.

Source: Report PB 204.521

The authors concluded that efficient removal of selenium could be achieved with a strong acid-weak base ion exchange system. R.M. Ahlgren has presented capital and operating costs for such a weak base system, as a function of dissolved solids content of the waste-water (9). The nonselective nature of ion exchange resins is a drawback when attempting to remove a low level contaminant in the presence of significant quantities of other ions. The simultaneous presence of other ions rapidly increases ion exchange costs.

According to the work of Linstedt and his coworkers, however, other commonly employed industrial metal treatment processes, e.g., lime coagulation, settling, and sand filtration, were ineffective in recovering selenium, at least as the negatively charged anion. According to J.D. Hem (10), selenium is most likely to occur in aqueous solution in anionic form, and presumably would be treated for removal in that form. In that event, ion exchange appears to be the most effective technique reported.

**References**

(1) Stahl, Q.R., "Air Pollution Aspects of Selenium and Its Compounds," *Report PB 188,077;* Spring-field, Va., National Tech. Information Service (Sept. 1969).
(2) Lansche, A.M., "Selenium and Tellurium – A Materials Survey," *U.S. Bur. Mines Inform. Circ. 8340,* Washington, D.C. (1967).
(3) West, P.W., Louisiana State University, Baton Rouge, La. in *Report PB 188,091.*
(4) "Selenium and Its Compounds," *National Safety News* 93, 5 (1966).
(5) Yomiyama, A. and Yonekawa, S.; U.S. Patent 3,052,522; September 4, 1962; assigned to Asahi Kasei Kogyo K.K., Japan
(6) Patterson, J.W. and Minear, R.A., "Wastewater Treatment Technology," *Report PB 204,521;* Spring-field, Va., Natl. Tech. Information Service (Aug. 1971).
(7) Johnson, H., "Determination of Selenium in Solid Waste," *Envir. Sci. Tech.,* 4, 850 to 853 (1970).
(8) Linstedt, K.D., Houck, C.P. and O'Connor, J.T., "Trace Element Removals in Advanced Wastewater Treatment Processes," *Jour. Wat. Poll. Control Fed.,* 43, 1507 to 1513 (1971).
(9) Ahlgren, R.M., "Membrane vs. Resinous Ion Exchange Demineralization," *Indust. Water Engrs.,* 8, 12 to 14 (1969).
(10) Hem, J.D., "Study and Interpertation of the Chemical Characteristics of Natural Water," *Geological Survey Water - Supply Paper 1473,* U.S. Dept. Interior, Washington, D.C., (1959).

# SILVER

McKee and Wolf (1) have reported that silver, as the solid metal, is used in the jewelry, silverware, metal alloy, and food and beverage processing industries. Little soluble silver waste would be expected to result from most uses of the solid metal. The only appreci-ably soluble common silver salt, silver nitrate, is used in the following industries: porcelain, photographic, electroplating, ink manufacture, and has also found application as an anti-septic (1). The two major sources of soluble silver wastes are the photographic and elec-troplating industries.

## Removal of Silver from Water

Basic removal methods for silver in wastewaters are discussed under four categories: (A) precipitation, (B) ion exchange, (C) reductive exchange and (D) electrolytic recovery. More than one of these types of processes may be combined in order to achieve high levels of silver removal, since recovery of silver from waste streams is more lucrative than for most other metals. A recent study on product recovery places a net recovery value of $1.60 to $9.00/1,000 gal. of dilute plating rinse waters containing from 50 to 250 mg./l. of silver (2).

*Precipitation:* Silver is frequently removed from wastewaters by precipitation as silver chloride. In contrast to most metal chlorides, which are relatively soluble, silver chloride is an extremely insoluble compound. The salt will dissolve in water to a maximum con-centration of approximately 1.4 mg./l. silver ion (3). Slight excess of chloride ion will re-duce this value but greater excess increases the solubility of silver through the formation of soluble silver chloride complexes (3)(4). Thus, silver can be selectively recovered as silver chloride from a mixed metal waste stream without initial wastestream segregation or concurrent precipitation of other metals. In the event that treatment conditions are alkaline, thus resulting in precipitation of hydroxides of other metals along with the silver chloride, acid washing of the precipitated sludge removes contaminant metal ions, leaving the insoluble silver chloride.

Plating wastes generally contain silver in the form of silver cyanide, which interferes with precipitation of silver chloride. Therefore, cyanide removal is necessary prior to precipitation of silver as the chloride. Oxidation of the cyanide with chlorine releases chloride ions into solution, which in turn react to form silver chloride directly (4). In the event that cyanide is present in great excess over silver, high chloride concentrations resulting from cyanide oxidation may greatly reduce the effectiveness of this silver removal process (4).

Bench-scale studies by C.A. Walker, W. Zabban, W. Southworth and E.P. Heslin (4) were performed in the course of designing a silver waste treatment process. Treatment consisted simply of addition of chlorine in the ratio of 3.5 mg./mg. of cyanide, and a 30 minute reaction time. Silver reductions of 90 to 99% were achieved. Modification of the batch treatment process to the sequence described below resulted in even higher removals.

    (a) Chlorination and 10 minute reaction

    (b) Adjust pH to 6.5 to complete oxidation of cyanide

    (c) Addition of ferric chloride

    (d) Skimming and sedimentation

    (e) Lime addition to pH 8

    (f) Settling of floc and decanting supernatant

Improved removal resulted, reducing the residual silver ion in the effluent to less than 0.1 milligram per liter. High removal of other heavy metals was achieved. The treatment plant resulting from the above studies handled silver recovery in batch processes of 10,000 gallons. Operating data from the full-scale recovery system is presented by Eichenlaub and Cox (5). The high residual levels in the silver recovery system effluent would not be reflected in the plant discharge, as the silver wastes were diluted 40-fold by other plant waste streams. Cost data were not given for the process.

Other silver precipitation processes are mentioned in the literature. One, a patented process of Eastman Kodak Co. (6), precipitates silver from waste photographic solutions containing high levels of organic acids by adding magnesium sulfate and lime. The silver likely precipitates as a mixed sulfate-oxide and is recovered from the sludge.

Another Eastman Kodak patent (7) is one in which waste photographic processing solutions which contain silver thiosulfate complex, such as exhausted thiosulfate fixing solutions, are chlorinated under alkaline conditions to precipitate a silver-bearing sludge from which the silver can be separated and to destroy thiosulfate ion and thereby reduce the oxygen demand of the solution. Chlorination can be effected by introduction of chlorine gas or by use of hypochlorite solution. Since thiosulfate ion accounts for a major part of the oxygen-consuming material which is discharged in photographic processing, a significant source of water pollution is substantially eliminated by this method.

Use of sulfide for precipitating silver from photographic solutions as the extremely insoluble silver sulfide is reported by Fusco (8), but no operating data or costs are provided. From a solubility standpoint, this would appear to be an attractive method, although solids separation and addition of sulfide to the wastewater are considered disadvantages (8). In a review of methods of silver recovery in the photographic film processing industry, Schreiber (9) has pointed out that sulfide use is one of the oldest methods for silver precipitation.

Use of hydrosulfite for precipitation yields both free silver and the sulfide, as well as a more compact precipitate with good settling characteristics. Heat is required however, and chemical costs are high.

*Ion Exchange:*  A patent issued to the Permutit Co. (10) describes the recovery of silver from dilute photographic washwaters by passage through basic ion exchange resins. Recovery is by elution of silver salts from the resin, or by direct resin incineration and recovery of pure silver.  Linsted, Houck and O'Connor (11) achieved 85.8% removal of trace levels of silver from an extremely dilute secondary sewage effluent by cation exchange, and 91.7% removal by combined cation-anion exchange.

Heidorn and Keller (12) report the use of a series cation-anion exchanger combination for recovery of silver cyanide from plating rinsewaters.  The recovered sodium silver cyanide may be returned to the plating bath, or converted to silver chloride by the precipitation process discussed above.

Schreiber (9) indicates that problems of recovering silver from the resin may make ion exchange uneconomical for photographic waste recovery.  Arden (13) suggests using ion exchange in conjunction with electrolytic recovery for thiosulfate developer baths.  In that case, fixing solution silver is recovered by plating out the silver.  Washout silver waste is accumulated by ion exchange, and the resin regenerant containing the concentrated silver is passed through an electrolytic process, thereby removing silver and functioning as developer bath makeup water.

*Reductive Exchange:*  This method involves precipitation of silver onto another metal. Usually, zinc or iron is used because they are relatively inexpensive.  The net result is a substitution of iron or zinc ion for silver in solution.  This process had been in operation at the Oneida Silverware plant prior to installation of a cyanide chlorination-silver chloride recovery system (14).  Replacement of the reductive exchange process was related to the necessity to install cyanide treatment facilities.

Schreiber (9) indicates that by using zinc or steel wool under proper conditions to maintain sufficient metal surface exposed, 95% removal of silver can be accomplished with column operation.  Silver ions are replaced with an equivalent quantities of zinc or iron ions however.

*Electrolytic Recovery:*  This process has been mentioned in conjunction with photographic developer solutions (15), although without any data reported.  As with any electrolytic process, high wastewater ionic concentrations are necessary for the method to function effectively.  If sufficient metallic ions are in solution such an operation is feasible, however, most silver wastewaters do not appear to contain sufficient concentrations of silver ion alone.

Schreiber (9) supports this position indirectly in reporting normal operating levels for various electrolytic units in service to the photographic industry.  Many units will operate at silver concentrations of 500 mg./l. and some as low as 100 mg./l.  Usually operation begins at 5,000 mg./l. however, and reduces bath silver concentrations to 500 mg./l.  He reports that a silver recovery unit can cost as little as $450.

A process developed by W.R. Luck, Jr. (15) utilizes an apparatus for silver recovery which consists generally of a container whose outer wall is a cylindrical cathode and having an anode suspended centrally thereof.  Electrical current is passed between the anode and the cathode through the liquid causing silver or other metallic material in solution in the fluid passed through the container to be deposited on the cathode by electrolysis.  An impeller is disposed centrally and interiorly of the bottom of the container which is driven by an external motor through a magnetic drive. Fluid is admitted centrally of the upper wall of the container and discharged through the upper wall of the container outwardly of the center thereof. The impeller draws fluid through the central passage which is at lower pressure and the fluid is discharged through the outer passage which is at higher pressure, thereby eliminating the necessity of a pump to circulate the fluid therethrough. Although the device is primarily designed for recovery of silver from photographic fixer liquid, it may be employed to remove metal from other liquid.

*Summary:* The value of silver makes recovery from process streams attractive. Electrolytic recovery, long practiced in photographic processing is operative only down to solution concentrations of 100 to 500 mg./l. Precipitation with chloride ion can remove silver to the milligram per liter level. However, coprecipitation with other metal hydroxides under alkaline conditions improves silver removal to less than 0.1 mg./l.

Silver recovery is possible from mixed precipitates by redissolving the metal precipitates under acidic conditions. Ion exchange for low silver concentrations appears feasible for separated waste streams, where total ionic strength is not appreciably greater than the silver salt concentration. Very low residual silver concentrations are possible with ion exchange. Sulfide precipitation of silver has been used on a small scale for many years, but large scale operation appears to pose problems, particularly with respect to solids separation and handling.

### References

(1) McKee, J.E. and Wolf, H.W., "Water Quality Criteria," 2nd. Ed., *California State Water Quality Control Board Publication No. 3A,* (1963).
(2) Resource Engineering Associates, *State of the Art Review on Product Recovery,* Department of the Interior, Washington, D.C., (1969).
(3) Butler, J.N., *Ionic Equilibrium,* Reading, Mass., Addison-Wesley (1964).
(4) Walker, C.A., Zabban, W., Southworth, W. and Heslin, E.P., "Disposal of Electroplating Wastes by Oneida, Ltd. II, Development of Treatment Processes," *Sew. Ind. Wastes* 26, 849 to 853 (1954).
(5) Eichenlaub, P.W. and Cox, J., "Disposal of Electroplating Wastes by Oneida, Ltd. V. Plant Operation," *Sew. Ind. Wastes,* 26, 1130 to 1135 (1954).
(6) Pool, S.C.; U.S. Patent 2,507,175; May 9, 1950; assigned to Eastman Kodak Co.
(7) Hendrickson, T.N. and Dagon, T.J.; U.S. Patent 3,594,157; July 20, 1971; assigned to Eastman Kodak Company.
(8) Fusco, R., "Recovery of Silver from Photographic Fixing Baths and Other Sources," *Chim. Industr.* 24, 356 (1942).
(9) Schreiber, M.L., "Present Status of Silver Recovery in Motion Picture Laboratories," *Jour. Soc. Motion Pict. Tech. Engrs.* 74, 505 to 513 (1965).
(10) Permutit Co., Ltd., *Improvements Relating to the Recovery of Silver from Dilute Solutions,* B.P. 626,081; July 8, 1949.
(11) Linstedt, K.D., Houck, C.P. and O'Connor, J.T., "Trace Element Removals in Advanced Wastewater Treatment Processes," *Jour. Wat. Poll. Control, Fed.,* 43, 1507 to 1513 (1971).
(12) Heidorn, R.F. and Keller, H.W., "Methods for Disposal and Treatment of Plating Room Solutions," *Proc. 13th Purdue Industrial Waste Conf.,* pp. 418 to 427 (1958).
(13) Arden, T.V., *Water Purification by Ion Exchange,* New York, Plenum Press, (1968).
(14) Culotta, J.M. and Swanton, F.M., "Case Histories of Plating Waste Recovery Systems," *Plating* 57, 251 to 255 (1970).
(15) Eastman Kodak Co., "Disposal of Photographic-Processing Waste," *Kodak Publication No. J–28,* Rochester, New York (1969).
(16) Luck, W.R., Jr.; U.S. Patent 3,694,341; September 26, 1972.

# SODIUM CARBONATE (TRONA)

## Removal from Air

A process developed by C.J. Howard and E.B. Port (1) is one in which dust and fines issuing from trona processing systems in the manufacture of sodium carbonate are removed by scrubbing the dust ladened gases with an aqueous scrubbing solution under conditions sufficient to remove the fines from the gases and preferably effect particle growth of the dust and fines in the scrubbing solution. The solids of larger particle size are separated and returned to the trona processing system.

### References

(1)  Howard, C.J. and Port, E.B.; U.S. Patent 3,634,999; January 18, 1972; assigned to Allied Chemical Corporation.

## SODIUM MONOXIDE

### Removal from Air

A process developed by G.L. Fairs and W. Plaut (1) is of particular use in removing solid particles such as, for example, sodium monoxide fume. Thus when diaphragms are being replaced in cells for the electrolysis of fused mixtures of sodium chloride and calcium chloride for the manufacture of sodium and chlorine a fume is obtained containing fine particles of sodium monoxide together with a little calcium chloride and sodium chloride and a little chlorine.

Again a fume containing fine particles consisting almost entirely of sodium oxide is obtained when sodium residues are being disposed of in burner bays. Such fumes are very objectionable and numerous attempts have been made to try to remove such fume effectively. Thus the fume has been passed through a filter composed of slag wool fibers but it was found that such a filter quickly blocked with collected solids. This filter was also attacked by small quantities of chlorine which were present in the fume.

Again a fiber filter composed of very fine untreated glass fibers (wettable fibers) will not effectively remove fumes of sodium and potassium oxide particles entrained in air since high pressure drops are required and in any case the filter soon blocks. Moreover such filters cannot be irrigated with a liquid such as water to remove deposited solids as it very quickly logs and becomes useless as a filter medium. However, when using a nonwettable fiber filter, suitably one composed of Terylene polyester fibers or one composed of glass fibers having an adherent silicone surface, which is continuously irrigated with water it was found that these filters were completely effective in removing sodium monoxide fumes.

Moreover when using such filters the process of removing the particles may be operated at relatively low pressure drops. While such filters can be used to remove extremely fine liquid and solid particles of extremely fine particle size, for instance of size less than $5\mu$, they may also be used for the removal of particles of considerably greater size, for instance $25\mu$ or more.

The following is one specific example of the application of this process. Glass fibers of diameter in the range 5 to $40\mu$ were packed and compressed in a mold to a density of 160 kg./m.$^3$ to form a layer 5 cm. deep. The glass fibers were then heat treated and provided with an adherent silicone surface. The filter was in the form of a flat cylindrical mat 37.5 cm. in diameter.

From a burner bay disposing of sodium residues air was withdrawn containing sodium oxide particles having a top particle size of about $9\mu$, 50% by weight being less than $5\mu$. The concentration of the particles varied between 0.25 and 0.8 g./m.$^3$ of air. This air was passed through the filter which was continuously irrigated with water, the irrigation rate being of the order of 5.4 l./hr. square meter of the filter area. The throughput of air passing through the irrigated filter varied between 280 m.$^3$/hr. square meter of the filter area at 40 cm. water gauge pressure drop and 205 m.$^3$/hr. square meter of filter area at 35 cm. water gauge pressure drop.

Water containing dissolved sodium hydroxide was removed at a point in the apparatus below the filter. At no time during a run of 700 hours was any fume visible in the exit gases after passing through the filter and any solids present in the exit gas were in quantities which were certainly less than 0.001 g./m.$^3$.

### References

(1) Fairs, G.L. and Plaut, W.; U.S. Patent 3,135,592; June 2, 1964; assigned to Imperial Chemical Industries, Limited, England.

## SOLVENTS

Air pollutants discharged by industrial concerns, often contaminate the atmosphere and, in areas of industrial concentration, to even a greater degree than automobiles. Generally it can be said that the source of industrial pollutants are incomplete combustion products from heat generating processes and the discharge of combustible solvents which have been evaporated. One of the most common applications of the latter is found in painting processes where paints are dried by baking or the like to leave a thin film of pigment on the article being painted. Evaporated solvents, particularly hydrocarbons, represent one of the most undesirable air pollutants. When subjected to elevated temperatures and sunlight, these pollutants create an eyestinging haze commonly referred to as smog.

To counter the problem of air pollution a number of attempts have been made in the prior art to prevent potentially polluting particles from being discharged into the atmosphere. In industrial applications evaporated solvents may be subjected to a sufficient heat to incinerate them. By incineration, organic pollutants, which are by far the majority of all pollutants, are chemically changed into water ($H_2O$) and carbon dioxide ($CO_2$).

Previous incineration systems employed combustion chambers into which the exhaust gases containing the evaporated solvents are discharged. There they are subjected to high temperature flames, most commonly fed by natural gas, and thoroughly mixed therewith. To incinerate the majority of the pollutants the exhaust gases are retained in the heated zone for a minimum length of time. The required time is considered to be at least 0.3 second.

The volume of air that is exhausted from the ovens is very large. For example, in a paint drying oven it takes approximately 10,000 scf of air for each gallon of a common paint solvent that must be evaporated. In large operations the amount of solvent that must be evaporated often reaches several gallons a minute. This air volume must be heated to such temperature as will assure a nearly complete incineration of the evaporated solvent or waste particles in the exhaust air. The incineration temperature for exhaust air including hydrocarbons is a minimum of about 1200°F.

As a safety factor systems are designed to operate at 1500°F. The required thermal energy to raise the temperature of the exhaust gases to between 1200° and 1500°F. is substantial and, for a system wherein no more than about one-half gallon of solvents is evaporated a minute, it is in the neighborhood of 4,680,000 Btu an hour. For larger systems the heat requirement is correspondingly greater.

The economic cost is substantial, particularly when considering the continuous operation of these systems. For the most part the thermal energy contained in the hot gases is discharged into the atmosphere and, therefore, economically lost. Although it has been suggested to use the hot gases in an economic manner as, for example, by introducing them into heat exchangers, this has on the whole not significantly reduced the high cost of incinerating air pollutants. A main reason is the relatively high temperature of the gases which makes them difficult to handle and a relatively large pressure drop in the combustion chamber of the incinerator.

In the combustion chamber the exhaust gases are thoroughly mixed into an already existing gas flame by injecting them into the flame in, for example, a tangential manner or by
mechanically agitating the flame and the exhaust gases. The great amount of turbulence
causes pressure drops across the combustion chamber which are normally as high as several inches of $H_2O$. To introduce these hot gases into a heat exchanger a blower is generally necessary. The temperature of the gas and its volume make the blower very large and
expensive to construct. Moreover, it is subjected to such high temperatures that failure of
portions of the blower, commonly its blades, is frequent.

Thus, the high cost of building and maintaining the blower, together with its frequent
failures, has made it more attractive for industry to simply discharge the hot gases into the
atmosphere through relatively tall stacks. The lack of a satisfactory, inexpensive incineration system has therefore increased the cost of maintaining or regaining a nonpolluted air
in our metropolitan centers. What is perhaps more serious, its high cost has prevented
many from installing an incineration system. These industries continue to discharge the
health impairing pollutants into the atmosphere.

### Removal of Solvent Vapors from Air

A process developed by J.J. Moon (1) involves recovering solvent from vented gases accumulated from several pieces of equipment in an olefin polymerization plant. Commercial
operations of olefin polymerization and other chemical processes which employ a hydrocarbon process diluent, such as a solvent or liquid carrier for polymer and/or catalyst, have
shown that appreciable losses of this hydrocarbon solvent can be expected, particularly
during initial operation. Losses occur primarily in vented off-gas streams which are accumulated from the various pressure relief lines connected to process equipment, such as reactors, filters, heat exchangers and the like.

Off-gases that are vented in this manner can be saturated or substantially saturated with
the process solvent, and in continuous operation of large installations, the amount of such
pressure relief off-gas over a period of time can be large enough to involve considerable
cost in the loss of solvent. Even so, treatment of off-gases cannot ordinarily be justified
economically because of wide variations in the flow of vented gases, and the usual approach is to burn such gases continuously at a remotely located flare.

In the process a hydrocarbon solvent having from 3 to 12 carbon atoms per molecule
(such as that customarily employed as a process diluent in olefin polymerization of the
type described in U.S. Patent 2,825,721) is recovered from plant off-gas by contacting the
gas with an absorption oil, thereby absorbing solvent from the gas, stripping the absorption
oil to remove the solvent therefrom and recirculating the absorption oil to the contacting
step.

A substantially constant amount of off-gas is passed from the off-gas stream through the
absorption zone so that the absorption column is operated under steady conditions and
the substantially solvent-free off-gas removed from the absorption zone is returned to the
off-gas stream and disposed of in the usual manner, generally by burning at a remotely
located flare.

A process developed by W.R. Evans, Jr. (2) involves solvent recovery from cellulose acetate
drying operations. During many operations such as the formation of synthetic fibers and
films, e.g., cellulose ester fibers, sheet, and films or the application of organic coatings,
considerable amounts of organic solvents are evaporated into the atmosphere or other
gaseous media. In order to make such processes economically profitable, it is necessary
that such evaporated solvents be recovered from the air or other gaseous media.

In this process the condensation of the vaporized solvent is achieved by bringing the coolant
into direct contact with the gaseous medium. This may be advantageously accomplished
by spraying the refrigerated coolant into the gaseous medium to condense and carry off
or entrain the vaporized solvent as a coolant-solvent admixture.

When this spray condensation process is used, the coolant preferably comprises the organic solvent composition in the liquid state. This avoids the need to completely separate the condensed solvent from coolant prior to the rerefrigeration of the coolant since the condensed solvent forms part of the coolant. If desired, the sprayed coolant may contain a liquid absorbent for the vaporized solvent such as light mineral oil.

A process developed by J. Sanders (3) involves recovering solvents such as white spirit, trichloroethylene and perchloroethylene. It comprises delivering the solvent laden vapor into a recovery apparatus for flow upwardly through a diffuser including a carbon bed while an evacuating fan is connected to the apparatus for removing substantially only air from the location above the carbon bed after the solvent has been absorbed by the carbon bed.

A process developed by D.M. Wilkinson (4) involves a paint drying system in which the vapors of the volatile solvents are burned to eliminate pollution of the atmosphere and the heat generated by such combustion is employed to preheat the incoming fresh air thereby reducing the fuel required to dry the paint. The continuous painting or coating of strip material such as sheet metal requires the use of high efficiency continuous oven dryers which apply heat in various stages to evaporate the volatile solvents and to bake the paint. In the normal case the paint used contains considerable quantities of volatile solvents which are evaporated in large quantities and constitute a serious disposal problem.

In the past, it has been common practice to vent these volatile solvents directly into the atmosphere. As a result, municipal regulation have been adopted in certain areas prohibiting the dumping of these solvent effluents which otherwise constituted a serious source of air pollution. Operators of such coating plants have therefore been forced to adopt some system for the removal of the volatile effluents down to a predetermined maximum level.

Unlike sooty carbon deposits, volatile effluents are difficult to filter, and, although they may be recovered by a distillation or condensation process, the equipment required for such processes is costly to manufacture and difficult to install in existing plants and may be subject to variations in the efficiency of effluent extraction due for example, to different concentrations of effluents encountered in the waste air stream.

An alternative method of effluent removal is simply to oxidize the effluents by burning them in the waste air stream. Such oxidization is again subject to variations due to variations in the effluent content of the waste air system and although the equipment required is very much cheaper than the distillation or condensation equipment referred to above, there is no effluent recovery to offset such cost and the process becomes increasingly wasteful.

Accordingly, it is an objective of this process to provide a paint drying and baking system incorporating high efficiency ovens for effective and rapid drying and baking of the paint on a continuously moving strip and further incorporating effluent oxidizing means which is regenerative in nature and assists in maintaining the efficiency of the ovens.

A process developed by H.A. Price and D.A. Price (5) provides a waste gas incineration system for preventing the discharge into the atmosphere of oxidizable waste particles in exhaust gases of ovens. Exhaust gases are introduced into a combustion chamber of an incinerator where they are mixed with a combustible gas and ignited.

The gases are retained in the combustion chamber sufficiently long to assure substantially complete incineration of all waste particles. An impeller withdraws the waste gases from the chamber. Ports are provided for mixing the gases with ambient air to reduce the temperature of the air and gas mixture. The air-gas mixture is transported to locations where heat or thermal energy is required.

A process developed by N.J. Handman (6) for preventing hot noxious vapors, comprising aerosols of organic materials emanating from an oven in a manufacturing process, from

**References**

(1) Moon, J.J.; U.S. Patent 3,131,228; April 28, 1964; assigned to Phillips Petroleum Company.
(2) Evans, W.R., Jr.; U.S. Patent 3,232,029; February 1, 1966; assigned to Celanese Corporation of America.
(3) Sanders, J.; U.S. Patent 3,368,325; February 13, 1968.
(4) Wilkinson, D.M.; U.S. Patent 3,437,321; April 8, 1969; assigned to B & K Machinery International Limited, Canada.
(5) Price, H.A. and Price, D.A.; U.S. Patent 3,472,498; October 14, 1969; assigned to Gas Processors, Inc.
(6) Handman, N.J.; U.S. Patent 3,618,301; November 9, 1971; assigned to Clermont Engineering Co.

# STARCH

## Removal from Water

A process developed by J.T. Gayhardt (1) involves the disposal of wastes derived from starch compositions employed as adhesives in certain industries, with particular reference to the corrugated board and box industries. In many industries it is conventional practice to employ various starch compositions for adhesive purposes, and among these may be mentioned the corrugated board and box industries where starch compositions are employed for sealing or laminating two or more pieces or piles of corrugated board or box components together during the course of their manufacture.

These components may be paper, cardboard, plywood, etc. The waste starch residues that result from the manufacturing operations in these industries pose serious problems of waste control and disposal due to the physical and chemical characteristics of the starch adhesive refuse materials. Normally this starch refuse material is directed into a pit or tank from the corrugators or other starch composition utilizing apparatus, together with an indeterminate amount of flushing water, grease, oil and/or other liquid debris. On standing, this material becomes compacted and doughy or sludge-like in consistency with concomitant microbial and bacterial action, resulting in odors of decay rendering its removal problem even more difficult.

Since many municipalities have imposed severe restrictions upon the dumping of such malodorous and otherwise objectionable waste materials into streams, it is frequently impossible to dispose of this waste residue in such fashion. Moreover, even where the plant may be in a relatively isolated location with perhaps few, if any, legal restrictions upon the method of disposal of the waste, nevertheless to attempt to dispose of the waste in such manner still gives rise to serious problems.

Heretofore, removal of this waste residue was effected periodically such as approximately every month by means of manual excavation from the pit or tank, transfer of the refuse to an isolated area and burial of same. This, however, entails considerable expense in terms of labor costs, the procurement of permits from municipal authorities and general inconvenience to all concerned. It is therefore an object of this process to provide for handling waste materials resulting from the use of various starch formulations as adhesives in industries in a fashion that will result not only in substantial savings in labor and other costs, but also with a minimum of nuisance to neighboring persons and facilities.

The process comprises mixing the residues with a treating composition comprising cyclohexylamine, an alkyl phenyl polyethylene glycol ether and tetrakis (2-hydroxy-propyl) ethylene diamine in order to disperse components of the sludge-like residues therein and running the resulting mixture to waste in the normal sewage system of an industrial plant. A suitable form of apparatus for the conduct of the process is shown in Figure 125. The apparatus consists of a treatment tank or well 1 which may be of any suitable material, such as concrete, steel or brick.

FIGURE 125:  APPARATUS FOR DISPOSAL OF WASTE STARCH-CONTAINING
RESIDUES

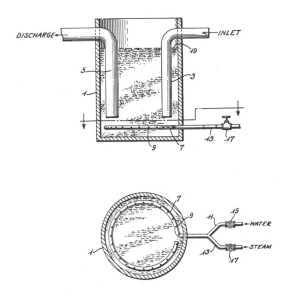

Source:  J.T. Gayhardt; U.S. Patent 3,037,931; June 5, 1962

For a rated capacity of handling 50 pounds per day of waste residue it is found that a tank
3 ft. deep by 2 ft. in diameter is adequate.  For other capacities of course the dimensions
and ratios of treating agent will be modified accordingly.  An inlet pipe 3 is provided for
introducing waste residue from the corrugating or other starch adhesive composition utilization machine in the plant proper.  This inlet pipe preferably extends downwardly to
about 6 in. from the bottom of the tank 1, thereby helping to trap oils, grease and other
liquid debris from the plant and minimizing the necessity for agitation.

A discharge pipe 5 is provided for removing the treated material from the tank and discharging it to the sewer or other waste.  This pipe also extends to within about 6 in. of
the bottom of tank 1.  Disposed in the lowermost part of the tank is a ring shaped pipe 7
having numerous perforations 9 through which, if desired, various fluids may be introduced
into the tank 1.  Pipe 7 communicates with branches 11 and 13 (see the plan view at the
base of the figure), by means of which fluids such as water and live steam, respectively,
may be introduced into the tank 1.  Control valves 15 and 17 are provided for controlling
the rate of input of fluids thus introduced.

In operation the tank will become filled with waste starch adhesive composition up to about
the level of 19 whereupon, after having added thereto the chemical treating agent as set
forth above, and with or without agitation by fluids via pipe 7, the suspended or solubilized
waste materials will be discharged by pipe 5 and a pump (not shown) to suitable disposal
such as a sewer or the like.  The chemical treating agent may be added continuously or
intermittently, as desired, directly to the tank 1.  By way of illustration of starch adhesive
compositions to the waste residues of which the process is particularly applicable, there
may be mentioned a normal starch adhesive composition for the manufacture of corrugated
cardboard boxes consisting of a mixture of pearl starch (1.82 lbs./gal.), caustic soda

(0.045 lb./gal.) and 5 mol percent borax solution (0.039 lb. borax/gal.) For more rugged uses, such as where the boxes are intended for overseas shipments and the like, one may make use of a weatherproof starch adhesive composition consisting of a mixture of a starch-resorcinol reaction product (0.43 lb./gal.), formaldehyde (0.195 lb./gal. in the form of a 37% solution), and caustic soda (0.0285 lb./gal.).

At the end of the manufacturing shift, the corrugators and refuse troughs are flushed down in the normal fashion with the refuse starch and other liquid debris flowing into the treatment tank which allows the water to run off and the starch to settle after which from about 450 to 600 ml. of additive or treating agent are placed in the tank. The actual weight of starch adhesive refuse from three corrugators will amount to about 40 to 60 pounds. This procedure is repeated at suitable intervals, such as every 24 hours, with ultimate overflow from the pit to the drainage system, carrying suspended and dissolved starch refuse through the sewage system of the plant.

### References

(1) Gayhardt, J.T.; U.S. Patent 3,037,931; June 5, 1962.

## STEEL MILL CONVERTER EMISSIONS

### Removal from Air

A process developed by J.D. Marino (1) involves cleaning exhaust gases from oxygen steel-making furnaces. In an oxygen steelmaking process, a stream of oxygen contacts molten iron in a furnace and oxidizes various impurities from the iron. The exhaust gases from the furnace are extremely hot and dirty and must be cooled and cleaned before they discharge into the atmosphere. The gases may be cleaned either in a precipitator or by wet-washing.

One difficulty encountered in usual wet-washing systems is that heavy particles recovered from the gases tend to plug parts of the equipment and otherwise interfere with its proper operation. An object of this process is to provide an improved method and apparatus, operable in conjunction with an otherwise conventional wet-washing system, for handling and removing heavier particles from the water and preventing their interference with operation of the equipment. The heavier particles are retained in a critically designed settling column ahead of the disposal equipment for the lighter particles.

A process developed by H. Hoff (2) provides a dust removal system for eliminating dust from the hot gases emerging from the steel-making converter wherein a stack delivers partially cool exhaust gas to a gas-washing column which may be bypassed by a conduit opening upwardly in this column into the downwardly turned end of a pipe carrying the washed gases to an electrostatic filter.

A temperature responsive control system detects the gas temperature at the entrance to the electrostatic filter and operates a valve member cooperating with the upwardly turned end of the bypass conduit to regulate the amount of hot exhaust gas mixed with the washed gas to maintain the gas temperature at the entrance to the electrostatic filter substantially constant.

A process developed by J.A. Finney, Jr. and R. Jablin (3) involves cleaning and recovering heat from particle laden hot waste gases from a basic oxygen converter. The gases are passed serially through a refractory heat accumulator, a waste heat boiler, a bag filter means, fan and stack in that order, the cleaned gases then being discharged to atmosphere. Between converter blows outside air is drawn through the above apparatus in the same flow direction whereby to maintain a sustained heat release in the waste heat boiler.

Figure 126 shows a suitable form of apparatus for the conduct of this process. There is indicated in a diagrammatic manner a converter **1**, which is adapted to receive a downwardly injected oxygen lance, to in turn provide a rapid high temperature carbon removal from molten metal in the converter in the manner of the typical basic-oxygen process. Generally, the oxygen lance operation or blow will last from 18 to 30 minutes during which period there will be a virtual eruption of flame, hot gases and sparks from the top of the converter into an enclosing hood **2**.

As the carbon is being burned out of the steel in the converter, there will be large quantities of carbon dioxide, carbon monoxide and nitrogen, as well as considerable amounts of metal vapors, particularly of iron, silica and manganese, together with entrained particulates in the hot gaseous stream. The actual temperature of the gas stream is extremely high and will generally be of the order of 2,500° to 3,000°F. or higher.

At the top of the hood **2**, there is indicated an explosion type of hinged door or hood cap **3** which, in the operation, will generally remain closed inasmuch as all of the hot gases and entrained particles will carry through the multiple stages of the system to be discharged as a cleaned stream. The upper part of the hood **2** is connected with duct **4** to the upper end portion of a heat accumulator chamber **5** indicated as containing a checkerwork of solid refractory material **6**. The heat accumulator zone **5** serves to operate in the manner of a heat sink by virtue of the heat absorbing material **6** which is placed throughout the height of chamber **5** in a manner effecting a direct contact between the gas stream and the heat absorbing surfaces.

It is of course not intended to limit the present heat accumulator design and heat absorption operation to the use of any one particular type of refractory material. The refractory material **6** may comprise typical checkerwork patterns of fire brick which in turn may be formed of suitable types of fire clay, high silica content brick, carborundum, slag brick, etc. Also suitable types of heat absorbing metallic forms of brick work may be to advantage in the heat accumulator. A preferred design of chamber **5** and arrangement of packing **6** will provide means (not shown) for effecting the periodic removal of settled particulates from within the interior of the chamber and on the heat absorbing material.

The lower end of the heat accumulator chamber **5** is connected by means of duct work **7** to the lower portion of a waste heat recovery unit **8**, normally having a plurality of tube banks housed therein to provide for an indirect heat exchange contact with a fluid medium to be heated in the tube bank means. The figure indicates diagrammatically that feed water is initially introduced to the economizer section **9** of the boiler unit **8** by way of line **10** and valve **11** while steam is being withdrawn by way of line **12** from a superheater section **13**.

The resulting cooled waste gas stream will thus leave the waste boiler unit **8** by way of duct **14** at a temperature below that of the steam being produced in the unit. Duct **14** is shown being directly connective with a particle removal section **15**. As is also shown in the figure, for safety purposes, air inlet means may be provided for the introduction of a cooling or tempering air stream by way of duct **16** and control damper **17** into duct **14** such that there may be adequate temperature control and cooling of the gas stream being introduced into the separation zone **15**. For automatic control, there is indicated diagrammatically, at the inlet to the particle collection zone **15**, a thermocouple or other temperature sensitive member **23** which connects to a temperature controller **24**.

The latter connects to and operates a suitable motor means **25** that adjusts damper **17**. Thus, where for some unusual reason, the gases leaving the heat recovery unit are too hot for entering the separator means **15** there can be automatic air tempering. The separation zone **15** may house suitable bag filtering means to insure the discharge of a cleaned stream to the atmosphere although, as noted hereinbefore, other types of particle separating equipment may be used. Hopper means **18** is indicated diagrammatically at the lower end of the particle separation zone **15** so as to provide means for feeding separated particulates into suitable conveyor or transporting means.

## FIGURE 126:  PLANT FOR CLEANING STEEL CONVERTER EXIT GASES

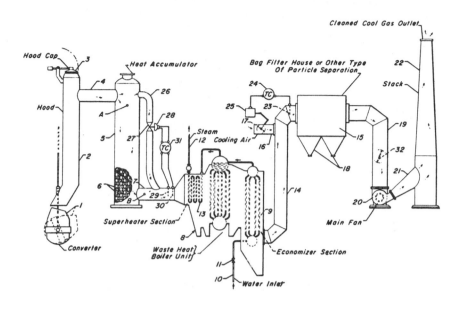

Source:  J.A. Finney, Jr. and R.Jablin; U.S. Patent 3,395,512; August 6, 1968

At the cleaned gas outlet end of the separation zone **15**, a duct **19** carries the waste gas stream through a suction fan **20** which in turn connects by way of duct **21** into the lower end of a stack **22** whereby the cleaned and cooled gas stream will be discharged into the atmosphere.

As hereinbefore noted, a preferred gas cooling and cleaning system provides a bypass duct, such as **26** with control valve **27**, around the heat accumulator **5**.  The control valve **27** can in turn have suitably controlled motor means **28** so that the valve can be opened where there is need of hot gas flow to increase temperature to the waste heat boiler **8** or, alternatively, when after a blow period there may be need of bringing in cooler air from the duct **26** to effect a tempering of hot gases from the accumulator **5**.  There are shown diagrammatically, high and low temperature sensitive means **29** and **30**, at the inlet to the waste heat boiler, which connect to a temperature controller means **31**.  The latter in turn connects to the motor means **28** of control valve **27** in duct **26** to effect desired adjustments of hot gas tempering to the boiler **8**.

In the initial period of a blow from the converter, the temperature control means **31** may be set to let some hot gases through duct **26** because of gases from the heat sink being too low in temperature.  During most of the converter blow period or cycle, the system operates to have the hot converter gas stream pass up through the hood **2** into the top of the heat accumulator **5** and provide a temperature that may well build up to the order of 2000°F. or more at point A, being indicated as the inlet end of the heat absorbing checkerwork material **6**.

Varying quantities of heat absorption material may of course be utilized in the accumulator zone **5**; however, in a properly designed and sized unit, sufficient quantities will be used to provide a first-stage temperature reduction for the waste gas stream so that point B at the discharge end of the accumulator **5**, will be of the order of 800° to 1000°F.

Also, in the waste heat water boiler 8 sufficient tube bank means and economizer means is provided to effectively supply useful steam energy and to reduce the waste gas temperature to the order of 250°F. or less. Thus, generally there will be no need for cooling air introduction through duct 16 into the particle separator 15, except where special low level cooling may be required to accommodate a particular type of particle removal equipment embodied therein or where there is a temporary misoperation of the heat recovery section. Where bag filtering means using synthetic fibers is utilized, the inlet air stream with entrained particulates should be of the order of not more than about 250°F. With glass fibers, temperatures of 500°F. may be accommodated.

On the other hand, where mechanical separator means are used, or where combinations of, for example, centrifugal and electrostatic separator means are utilized, then a slightly higher temperature may be permissible. Preferably, bag or cloth filter means are embodied in a final stage for the cleaning of the waste gas stream in order that there may be a highly efficient particle removal operation carried out, with a very minimum of particulate discharge into the surrounding atmosphere.

During the periods between converter blows, i.e., while the converter is being recharged with molten metal, the system is designed to maintain a continuous operation, with outside air being drawn by means of the suction fan 20 into converter hood 2 and heat accumulator 5 and thence through the rest of the system. This operation provides for the gradual cooling down of the principal portion of the heating material 6 by reason of the outside air flow coming down through the height of accumulator 5. In other words, there is a cyclic cooling at point A, in the inlet end of heat accumulator 5 whereby the temperature of solid heat absorbing material at such point gradually varies from the order of 2000° to the order of 600° to 800°F.

Generally, the temperature at point B will remain substantially constant at 800° to 1000°F. so that the inlet temperature into the waste boiler section 8 likewise remains relatively constant. This operation provides the continuous cyclic heating and cooling of the refractory material 6, with a temperature gradient occurring throughout the height of the heat accumulator zone 5. The inlet end having a wide temperature fluctuation and the outlet end having a very little variation. However, this system, with the use of automatic temperature controller means 31, can also bring cool air through duct 26 at a time right after a blow period when the hot gases from the accumulator 5 may be too hot to give an even temperature to the boiler 8.

This improved system, utilizing an initial flow through the heat accumulator zone, necessarily requires the use of a heat accumulator unit for each converter unit. Thus, where there are two or more converters in a bank arrangement in order to have one converter being charged while another is being blown with oxygen; then additional heat accumulator zones should be furnished so as to provide means for the cyclic heating and cooling of each heat accumulator section along with the accompanying operation of a companion converter.

On the other hand, two or more heat accumulator units may be connected with a single waste heat boiler means and single particle separator means such that there may be heat recovery and particle separation from a waste gas stream being combined from two or more converter units. In large installations there may also be duct work means provided to connect any one accumulator section with different converter hoods whereby one hood may be relined while another is being utilized for a converter blow.

Still further it should be pointed out that the equipment arrangement shown in the figure may be readily varied and that it is not in any way limiting. For example, various types of horizontal heat accumulator arrangements may be provided in lieu of the vertical chamber design, and there may be different types and arrangements of placing the heat absorbing material in the unit. In addition, various more elaborate temperature and valve control systems may be incorporated into the overall system to effect desired temperature controlling to the waste heat boiler unit 8 and to the bag filter zone 15 through the regulation of one or more dampers or valves in each of the ducts **16** and **26**. Adjustable damper means **32** may also be provided within duct **19** where it is desirable to control gas flow rates to the downstream treatment zones. Alternatively, damper means can be provided within ducts **7** or **14** to regulate overall gas flow rates, although the damper in duct **19** is generally preferable because of being located in a cooler and cleaner zone. This system has also been primarily directed for use with a basic oxygen converter, however, such system might well be utilized with an electric arc furnace or any processing unit which is similarly operated in an intermittent manner.

A process developed by H. Hoff (4) involves the treatment of dust laden combustible exhaust gas of a steelmaking converter in which the $O_2/CO$ reaction is carried out to completion in a gas collection tube having a length of 5 to 20 meters and oriented to sustain an air factor n of 0.05 to $\leqslant n \leqslant 0.5$. The tube opens upwardly into a dust-separating column and directs the gas against an impingement body above which a nozzle is provided to direct washwater into the gap between the impingement baffle and the wall of the column. Thereafter, the gases pass through a dust-collecting unit.

### References

(1) Marino, J.D.; U.S. Patent 3,315,443; April 25, 1967; assigned to United States Steel Corporation.
(2) Hoff, H.; U.S. Patent 3,372,528; March 12, 1968; assigned to Firma Gottfried Bischoff KG, Germany.
(3) Finney, J.A., Jr. and Jablin, R.; U.S. Patent 3,395,512; August 6, 1968; assigned to Universal Oil Products Co. and Bethlehem Steel Corp.
(4) Hoff, H.; U.S. Patent 3,497,194; February 24, 1970; assigned to Firma Gottfried Bischoff KG, Germany.

# STRONTIUM

## Removal from Water

A process developed by D.W. Rhodes, J.R. McHenry and L.L. Ames, Jr. (1) involves the removal from solutions of strontium ions including those of the radioactive isotope $Sr^{90}$ or radiostrontium, particularly when such ions are present in very small or trace amounts not economically removable by known methods.

$Sr^{90}$, one of the products of all known nuclear fission reactions, combines two properties which make it highly dangerous to public health: a close chemical relationship to calcium giving rise to a bone-seeking tendency whereby it becomes lodged within the bones of human being and animals, and an unusually long life, its half-life being about twenty years. Even trace amounts of this radioactive isotope or nuclide are considered a menace to the well being of communities if they find their way into the public water supply through ground seepage, or into the milk supply since, due to its chemical kinship to calcium, it can be carried through the calcium in the milk of dairy cows which have grazed on herbage containing it.

Once ingested by a human being or animal $Sr^{90}$ finds its way into the predominantly calcium structure of the bones where due to its long life it remains for all practical purposes, permanently. No means of dislodging it is known, nor is there much expectation among scientists that such means will be found within the foreseeable future, if ever.

Since, as previously stated, $Sr^{90}$ is formed in all known fission reactions, all nuclear reactors, which are now becoming quite numerous throughout the world, produce, either directly or indirectly, solutions containing ions of this isotope, which require treatment to remove it. Certain reactors, such as those where the fissionable material is in solution or in a slurry with a liquid moderator, produce the ions directly; other types of reactors are designed to confine their fission products within metal covered or canned fuel rods while the reactor is in operation, but after the fuel rods become spent it is necessary to dissolve them in some solvent, usually an aqueous solution of nitric acid, in order to recover the unaffected original fissionable material, the plutonium which has been produced by the neutron radiation, both of which are very valuable, and the fission products, many of which have economic value. In any event, at some stage in the operation of all kinds of nuclear reactors a solution containing $Sr^{90}$ ions is encountered. There are a number of known methods by which the fission products, including $Sr^{90}$, may be removed in a gross sense, but this process is addressed not to this stage but to the removal of the trace or residual or waste solutions after the conventional, large scale methods have done their utmost.

These residual concentrations are often of a magnitude that cannot be detected by conventional chemical methods such as weighing, titrating, spectroscopy and the like, and can only be detected by radiation counting devices. This method could be used to remove strontium from solutions on a large scale but it is probably less efficient and more expensive than other methods for this purpose. On the other hand, it will surpass known methods in its ability or economy in removing trace concentrations of strontium from large quantities of such residual or waste solutions and it is therefore important from a public health standpoint. This method will remove all strontium ions whether of mass number 90 or some other mass number, but its main utility is, of course, due to the removal of the isotope $Sr^{90}$ which carries the dangerous radioactivity described above.

The suggestion has been advanced of constructing large, liquid tight storage tanks to store the residual solutions described above until all radioactivity had spent itself. This, while superficially plausible, is not possible either economically or physically; the costs of such a program are economically unbearable, and the hazards of earthquakes, lightning, corrosion, defective steel plates, defective workmanship and the like in the construction and erection of the tanks make complete physical security of the dangerous solutions unattainable. If the erection of nuclear reactors is not to be brought to a complete halt some method must be found whereby large volumes of the residual solutions from their operation may be safely discharged into rivers, watercourses or disposal pits in the ground stripped of practically all radioactivity including that attributable to $Sr^{90}$.

Containment tanks being impractical, it has been suggested that the residual solutions be dumped into pits in the ground where the soil is calcareous. It was reasoned that since the health problem referred to is due originally to an ion exchange ability of the phosphate and carbonate anions of the calcium salts of the bones, whereby calcium atoms are displaced by those of strontium, this unfortunate circumstance could be turned to advantage by discharging solutions containing strontium into soils containing such salts.

Soils containing calcium carbonate are to be found in many places throughout the world; in fact, all sweet soils are characterized by the presence of sufficient calcium, usually in the carbonate form, to prevent their having an acidic reaction. Calcium phosphate is present in significant amounts only in isolated localities, but in either case it appeared probable that the atomic displacement reaction described above would take place in the soils in the same manner as it die in the bones. While not wholly incorrect, this conjecture proved to lead to rather disappointing results; the carbonate and phosphate anions while possessing sufficient ion exchange capacity to create the health problem within men and animals referred to, did not have enough of it to make for efficient removal of strontium either in the soil or in laboratory tests where the pure salts were employed as ion exchange materials.

This seemingly paradoxical statement is, of course, but another consequence of the huge difference in orders of magnitude inherent in the equivalence of mass and energy in the nuclear field, whereby amounts of matter once looked upon as inconsequential become highly significant when translated into terms of energy. It is, accordingly, an object of the process to provide a method whereby strontium ions may be removed from solutions where they are present in trace or residual concentrations in a practical, economic manner. This process is based on the discovery that while either calcium carbonate or calcium phosphate alone make but indifferent ion exchange materials for the removal of $Sr^{90}$ ions from solutions, calcium and other alkaline earth phosphates, as well as other metal phosphates, in the process of being created through the reaction of carbonates or other salts with phosphate ions, make highly efficient ion exchange materials for this purpose. A typical, but of course not the only reaction whereby a metal phosphate is created is the following.

$$5CaCO_3 + NaOH + 3Na_3PO_4 \longrightarrow Ca_5(PO_4)_3(OH) + 5Na_2CO_3$$

Chemists are familiar with a number of similar reactions whereby phosphate salts are created. This process is based upon the discovery that if strontium is present during the course of these main reactions it will be found to be removed from the solution of the reaction, even if present in only trace, or residual amounts.

Combinations of man-made structures and discharge into soil are possible, for example, the residual solutions containing $Sr^{90}$ and added phosphate ions can first be made to flow through a calcite column and then discharged into a pit in a calcareous soil, or into a bed of calcite and later discharged into the soil, or any combination of these. All that is required is a vessel filled with calcite; the water to be purified has sodium phosphate added to it and is then poured into the vessel, shaken and poured off after settling. This removes the $Sr^{90}$ and any unreacted phosphate remaining in the water is, for all practical purposes, harmless. Alternately a column of calcite can be used through which the water with added phosphate ions flows slowly.

**References**

(1)  Rhodes, D.W., McHenry, J.R. and Ames, L.L., Jr.; U.S. Patent 3,032,497; May 1, 1962; assigned to
     U.S. Atomic Energy Commission.

## SULFIDES

Certain chemical processes result in the production of noxious effluents containing quantities of sulfides in solution, usually spent alkaline wash liquors such as those obtained in petroleum refineries. The disposal of such untreated effluents presents considerable problems since strict limits are commonly applied to the maximum permissible sulfide concentration in effluents which are to be discharged into rivers or other waterways. The maximum permissible sulfide concentration may be less than 1 part per million.

For this reason it is often necessary to eliminate the sulfide from the effluents. Elimination may be achieved by means of an oxidation process whereby sulfides are converted to thiosulfates, which, in effluents, are less objectionable constituents. However, the thiosulfate is still capable of taking up oxygen and this is a disadvantage when the treated effluent is to be passed to a biological oxidation plant since the consumption of oxygen in that plant is considerably increased. Thiosulfate, by reason of its oxygen demands, is also an undesirable contaminant if passed direct to a river.

### Removal from Water

A process developed by M. Benger (1) involves the conversion of sulfides in aqueous solutions to sulfates by oxidation in the presence of a small quantity of certain catalysts.
In the process the treatment of aqueous solutions containing sulfides, particularly petroleum refinery effluent, comprises contacting the aqueous solution with oxygen or a gas mixture containing free oxygen, preferably air, at elevated temperature in the presence of a small quantity of a catalyst consisting essentially of a compound of copper or an iron group metal.

Cupric, ferrous and ferric compounds are particularly suitable and chlorides are preferred. The use of iron salts may be preferred in installations largely comprising iron or steel pipework since the use of copper salts in such installations may under certain conditions bring about corrosion.

The catalysts are preferably used in the form of aqueous solutions and preferably the pH values of the catalyst solutions are adjusted so that they are slightly alkaline, for example, between 7 and 9 or higher. This may be achieved, for example, by treatment with caustic alkali solution.

The sulfide-containing solution is preferably contacted with catalyst under moderately elevated conditions of temperature and pressure, for example, a temperature within the range of 95° to 135°C., preferably above 100°C., suitably within the range 105° to 130°C., and a pressure sufficient to maintain liquid phase conditions, preferably 0 to 50 psig. Sufficient oxygen-containing gas is contacted with the solution to provide an excess of free oxygen over the stoichiometric quantity required to oxidize the sulfide in the solution, for example, between 1 and 5 times, preferably 3 to 4 times, the stoichiometric quantity.

The reaction time for complete oxidation of sulfide to sulfate decreases with increasing severity of process conditions, i.e., with increasing temperature, pressure and oxygen flow rate. Preferably the catalyst is contacted with solution at a rate sufficient to bring about substantially complete oxidation of sulfide to sulfate, this being readily determinable by experiment. Effluent waters may contain up to 15% weight of sulfide or more but generally speaking the sulfide content will be less than 5% weight. Generally speaking, sufficient catalyst to provide between 10 and 1,000 parts of metal per million parts of solution, preferably 30 to 100 ppm will be adequate.

A process developed by P. Urban and R.H. Rosenwald (2) is one in which a hydrocarbon charge stock containing sulfurous and nitrogenous contaminants is converted and elemental sulfur and ammonia is simultaneously recovered by the following steps.

(a) Contacting in a hydrocarbon conversion zone the hydrocarbon charge stock, hydrogen and an aqueous recycle stream containing $(NH_4)_2S_2O_3$ with a hydrocarbon conversion catalyst at conversion conditions sufficient to form an effluent stream containing substantially sulfur-free hydrocarbons, hydrogen, $NH_3$, $H_2S$, and $H_2O$;

(b) Cooling and separating the effluent stream from step (a) to form a hydrogen-rich gaseous stream, a hydrocarbon-rich liquid product stream, and an aqueous waste stream containing $NH_4HS$;

(c) Catalytically treating the aqueous waste stream from step (b) with oxygen at oxidizing conditions effective to produce an effluent stream containing elemental sulfur or ammonium polysulfide, $NH_4OH$ and $(NH_4)_2S_2O_3$;

(d) Separating sulfur and ammonia from the effluent stream from step (c) to produce the aqueous recycle stream containing $(NH_4)_2S_2O_3$; and

(e) Passing this last aqueous stream to step (a).

Key feature of the resulting process is the continuous recycle of the treated water back to the hydrocarbon conversion process with consequential abatement of water pollution problems and substantial reduction of requirements for makeup water.

**References**

(1) Benger, M.; U.S. Patent 3,186,942; June 1, 1965; assigned to The British Petroleum Company, Limited, England.

(2) Urban, P. and Rosenwald, R.H.; U.S. Patent 3,530,063; September 22, 1970; assigned to Universal Oil Products Company.

# SULFUR

## Removal of Elemental Sulfur from Air

A process developed by M.W. MacAfee (1) involves the recovery of elemental sulfur from a gas stream in which the sulfur may exist as a vapor or an entrained liquid or solid. Sulfur vapor may be recovered from hot gases containing it by scrubbing it in water, but the resulting product is not suitable for market and requires dewatering and other operations before it can be sold. Consequently, there is an advantage to absorbing the sulfur from a gas stream in molten sulfur, but heretofore customary practices along this line have not been entirely successful and tend to produce a sulfur aerosol which passes through the system and escapes.

Sulfur is particularly difficult to recover from highly heated gases, say those heated above the boiling point of sulfur. Sulfur boils at approximately 445°C. or 832°F. at atmospheric pressure and it would be expected that sulfur would condense and drop out of a gas stream when cooled to such a temperature. In practice, however, this does not occur, and so long as the temperature of the gas stream stays about 507°F., the sulfur apparently continues to behave as a gas and no condensation appears to result, at least within a reasonable time.

If a gas stream containing sulfur vapor at a temperature near the boiling point of sulfur is brought into contact with molten sulfur, some condensation results and a small amount of sulfur is absorbed by the molten sulfur from the gas stream. However, the bulk of the sulfur in the gas stream, in some cases as much as 70%, appears to be converted into an aerosol or mist which is not absorbed by the molten sulfur and tends to pass through a scrubber through which the molten sulfur is circulated irrespective of the circulation rate.

In accordance with this process, sulfur vapor is recovered from a gas stream by cooling the gas stream to a temperature below the boiling point of sulfur but not below about 507°F. by indirect heat exchange and thereafter the cooled gas is brought into direct contact with molten sulfur, preferably in a scrubbing tower. For optimum results the gas stream and the molten sulfur stream should flow concurrently with each other through the scrubber so that the hottest gas comes into contact with the coolest molten sulfur.

After the contact between the gas and the molten sulfur, some sulfur may remain entrained in the gas stream, but the great bulk of this can be removed by swirling the gas to throw entrained sulfur out of it by centrifugal action. Small residual amounts of sulfur, if any, may be removed thereafter by passing the gas stream through a porous solid medium such as a tower packed with Raschig rings or beri saddles, or by electrostatic precipitation.

Sulfur, if heated somewhat above its melting point, tends to become so viscous that it will not flow. Consequently, it is important to control the temperature of the circulated sulfur within a close range, say 250° to 300°F. Control is facilitated by bringing the circulating sulfur into indirect heat exchange with another liquid, preferably superheated water. In normal operation, the molten sulfur from the scrubber contains excess heat and the heat transfer will be from the sulfur to the other liquid, say superheated water. However, during starting-up operations, the transfer of heat may be in the other direction. In either case, the use of superheated water is desirable because it may be maintained at a substantially constant temperature by regulating the pressure of a closed system through which the superheated water circulates.

References

(1)   MacAfee, M.W.; U.S. Patent 2,780,307; February 5, 1957; assigned to Pacific Foundry Company, Ltd.


SULFUR DIOXIDE

Sulfur dioxide has become a major pollutant of the atmosphere, particularly in urban areas. The presence of sulfur dioxide in the atmosphere is due primarily to the combustion of fossil fuels, i.e., coal and oil, which contain sulfur. Electric power plants constitute a major source of sulfur dioxide pollution of the atmosphere. Various processes have been suggested for removal of sulfur dioxide from flue gas, although none has gained a general industry acceptance to date. These processes may be grouped generally as wet processes and dry processes. Wet processes are those which employ an absorbent solution, usually aqueous, for removal of sulfur dioxide from a gas stream.

A flue gas desulfurization process has several requirements. First, it must be capable of removing most of the sulfur dioxide content of the flue gas, preferably 90% or more of the $SO_2$ present, under widely varying load conditions. Second, it should not create any air or water pollution problems. Third, the process should be easy to operate and maintain. The process should have a low net cost.

In many instances this would require the production of a salable by-product. The process should be capable of incorporation into existing power plants if it is to achieve maximum application. This requirement favors wet processes, which operate at a low temperature and therefore can be placed after the conventional air preheater in which incoming air for combustion is heated by the hot flue gas. Dry processes usually require a much higher operating temperature, and therefore must be inserted ahead of the preheater and integrated with the power plant.

### Removal of $SO_2$ from Air

Techniques for the removal of $SO_2$ from air have been reviewed by A.V. Slack (1) and by the staff of the U.S. Department of Health, Education and Welfare (2).

### Recovery as Sulfur Dioxide

A process developed by R.R. Cantrell and F.P. Wiley (3) involves removing and recovering $SO_2$ from gaseous mixtures such as power plant flue gases by absorbing the $SO_2$ in dimethyl sulfoxide (DMSO), and subsequently desorbing the $SO_2$ from the DMSO/$SO_2$ mixture by contacting the mixture with alumina, or other suitable adsorbent.

Figure 127 is a flow diagram showing the essential features of the process. In the figure it is seen that flue gas enters the wet scrubber 9 through conduit 10, and passes along the line, 11, to the sorption column 12, where the gas passes upwardly through a packing such as Berl saddles 13. The gas is met by a downward flow of DMSO projected into the column 12 through the distributing nozzle, 14. Scrubbed flue gas leaves column 12 through conduit 15 and is vented to the atmosphere. Liquid DMSO containing the sulfur dioxide washed out of the flue gas exits from the column 12 through line 16, and passes through heat exchanger 17.

The solution leaves the heat exchanger 17 through conduit 18 and may receive supplemental heating in heat exchanger 19, then passes by way of distributing nozzle 21 into a desorption column 20, where the DMSO/$SO_2$ solution passes downwardly through a packing 22 of activated alumina. In the column, 20, which is preferably operated under vacuum, the sulfur dioxide is removed from its solvent DMSO and is absorbed in the solid phase activated alumina.

In the column 20, which is preferably operated under vacuum, the sulfur dioxide is removed from its solvent DMSO and is adsorbed in the solid phase activated alumina. The DMSO falls to the bottom of column 20, exits through line 23 and returns to the sorption column 12 after passing through a pump 24 and the heat exchanger 17. Supplemental cooling may be necessary and is supplied in the cooler 25. The $SO_2$ which is released continuously from the alumina when column 20 is under vacuum, is drawn into a compressor 26 through conduit 27, the gas is compressed to approximately 55 psi and is delivered as a liquid to $SO_2$ tank 28 through line 29. Assuming that the total gas flow from the boilers of a porous station is 3,170 pounds of gas per minute, that the water scrubbed flue gas temperature is 86°F., and that the gas contains about 0.7% of sulfur dioxide, a flow of 5,978 pounds of

FIGURE 127:  PROCESS FOR SO$_2$ RECOVERY FROM FLUE GASES BY ABSORPTION
IN DIMETHYL SULFOXIDE

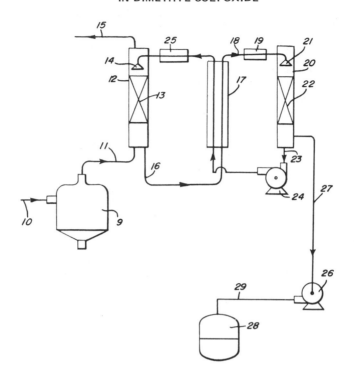

Source:   R.R. Cantrell and F.P. Wiley; U.S. Patent 3,538,681; November 10, 1970

DMSO/minute through the sorption apparatus will pick up 22½ pounds of sulfur dioxide.
Exact values and efficiencies are, of course, dependent on the specific recovery system.
Generally however, DMSO temperature as it enters the sorption tower should be about
30° ± 5°C. for, in this range, viscosity is favorable for pumped operation, yet the vapor pressure is reasonably low.

The weight of DMSO which should be brought into contact with a given weight of flue
gas to achieve a 99% SO$_2$ recovery depends on absorber construction, the efficiency of the
packing in developing extended surface area, and to a certain extent on the composition
of the combustion effluent.  Nevertheless, a ratio of DMSO pumped to pounds of flue gas
passing into the sorption tower which will permit 99% SO$_2$ removal will be found.  Its
value usually lies between equal weights of DMSO and flue gas and 2 weight units DMSO
to 1 weight flue gas.

A process developed by D.M. Little and G.R. Hettick (4) is one in which stack gases resulting from combustion of sulfur-containing fuels are treated in combination of steps to
remove SO$_2$ and CO$_2$ therefrom.  The gases are cooled, as by direct water quench to a
suitable treatment temperature, say 150°F., then intimately contacted with an absorbent
for SO$_2$ and CO$_2$, specifically a sulfolane, unabsorbed gases are vented to atmosphere or
otherwise recovered.  The SO$_2$-rich sulfolane stream also containing CO$_2$ is stripped as by
heating to recover a lean sulfolane bottoms, which can be recycled, a side stream of SO$_2$
and an overhead stream containing the CO$_2$.

For best operation, the cooled gases are compressed to a pressure of about 1 psig prior to the absorption step. Gases containing as little as 0.2 or less $SO_2$ mol percent can be treated to recover completely therefrom, the $SO_2$ contained therein. The operation can be continuous and can be effected in a simple apparatus.

A process developed by W.A. McRae, D.L. Brown and S.A. McGriff (5) is a cyclic process for the removal and recovery of sulfur dioxide from waste stack gases to lessen atmospheric pollution. The process involves: (a) electrolytically converting a salt solution into an acid and base; (b) employing the base to absorb the sulfur dioxide from the waste gas; (c) neutralizing the resulting spent base with the electrolytically produced acid to reform the original salt solution and to release the absorbed sulfur dioxide gas, and (d) recycling the salt solution to the electrolytic cell and recovering sulfur dioxide gas.

A process developed by T. Owaki (6) comprises treating exhaust gas containing sulfurous and/or sulfuric acid gas with a solution or suspension of the hydroxide or carbonate of calcium, to produce the corresponding sulfite and/or sulfate. Waste sulfuric acid is then added to the liquid so obtained to convert the sulfite to the corresponding sulfate simultaneously with the evolution of sulfurous acid gas. The liquid thus obtained is degassed by blowing air thereinto, to recover the sulfate and sulfurous acid gas from the exhaust gas and consequentially eliminate a public nuisance otherwise caused thereby.

Figure 128 shows the essential features of the process. Numeral **1** represents a dryer for $SO_2$ gas as a starting material, **2** a converter in which $SO_2$ is converted to $SO_3$, and **3** an absorber in which aqueous sulfuric acid is produced. The exhaust gas withdrawn at the top of the absorber still contains sulfurous and sulfuric acid gas left unabsorbed, which shouldn't be discharged into the atmosphere without any treatment to remove the acid gases.

The exhaust gas is then introduced through pipe **4** to an acid gas remover **5** which may be of any type so far as it is structurally such that no stoppage will be caused therein. The remover is connected through pipe **6** to a tank for absorbent **7** which holds lime milk or a suspension of calcium carbonate as an absorbent. The tank **7** is also connected through pipe **9** and pump **8** to the remover **5** so that the absorbent may be circulated between the remover and tank.

Numeral **10** shows an exhaust pipe provided on the top of the remover. The absorbent tank **7** is provided with pipe **11** through which the absorbent is supplied to the tank to which is added a certain calcium compound in suitable amounts according to the amounts of $SO_2$ and $SO_3$ being absorbed in the absorbent so that the circulating absorbent may be kept in a fixed range of composition. Numeral **12** represents a closed-type reactor in which are carried out the reactions in the 2nd step of the process.

The reactor **12** is provided with a suitable agitator therein, and with pipe **14** leading to the tank **7**, pipe **15** for supplying sulfuric acid and pipe **16** for discharging sulfurous acid gas. It is further provided with pipe **17** for withdrawing slurried plaster thereunder. Numeral **19** represents a degassing tower in which is recovered sulfurous acid gas from slurried plaster; the acid gas being dissolved in the plaster from the reactor **12**. The recovered acid gas is passed through pipe **21** to pipe **16** and then introduced into pipe **18** together with that discharged from the reactor, the pipe **18** being connected to the apparatus for producing sulfuric acid by contact process. The slurried plaster so degassed is withdrawn through pipe **22** at the bottom of the degassing tower.

In the practice of this process in combination with the apparatus for producing contact process sulfuric acid, sulfurous acid gas as a starting material is passed through the dryer **1** to the converter **2** to be converted to sulfuric acid gas which is then introduced to the absorber **3** to be dissolved into sulfuric acid. The remaining undissolved gas, which is an exhaust gas in this case, is passed through pipe **4** to the acid gas remover **5**. The exhaust gas is contacted with the absorber by passing upwards within the remover so that sulfurous and sulfuric acid gas still contained in the exhaust gas may completely be absorbed or

FIGURE 128: PROCESS FOR SO₂ RECOVERY FROM SULFURIC ACID PLANT
OFF-GAS BY ABSORPTION IN LIMESTONE SUSPENSION

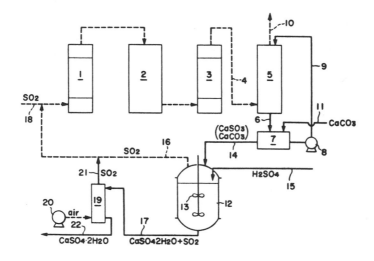

Source:  T. Owaki; U.S. Patent 3,556,722; January 19, 1971

dissolved, thereby rendering the exhaust gas harmless and discharging it into the atmosphere via pipe 10. During the aforesaid operation, the absorbent or absorbing liquid is introduced to the remover 5 at the top via pump 8 and pipe 9, passed downwards through the tower, introduced to the absorbent tank 7 in which the absorbent is replenished with fresh absorbent, and then recycled to the acid gas remover 5.

When calcium sulfite formed in the absorbent has been increased in concentration to a predetermined value, a part of the absorbent is passed through pipe 14 to the reactor 12 and then allowed to contact with sulfuric acid introduced via pipe 15 thereby to carry out the reactions in the 2nd step previously mentioned. The sulfurous acid gas produced by these reactions is supplied via 16 to pipe 18 and then used in the preparation of contact process sulfuric acid, while the slurry containing the resulting plaster separated is withdrawn through pipe 17 from the reactor 12.

The slurried plaster in which sulfurous acid gas is dissolved, is introduced via pipe 17 to a degassing tower where the plaster is degassed using the air. The sulfurous acid gas liberated from the plaster is supplied to pipe 21, and then passed through pipe 16 to pipe 18 together with the sulfurous acid gas withdrawn from the reactor 12, the pipe 18 leading to the apparatus for producing contact process sulfuric acid.

A process developed by I.S. Shah (7) permits one to efficiently and economically absorb and recover sulfur dioxide from a waste gas stream such as the tail gas from a sulfuric acid facility, or flue gas from the combustion of a sulfur-containing fuel. The process thus eliminates air pollution, and produces a useful sulfur-containing product from the recovered sulfur dioxide. The waste gas stream is scrubbed with a recirculating aqueous slurry containing magnesium oxide and magnesium sulfite. A small quantity of magnesium sulfate may be present if formed as a result of oxidation of magnesium sulfite or absorption of sulfur trioxide.

Another process developed by I.S. Shan (8) is one in which sulfur dioxide is recovered from a waste gas stream, such as the flue gas from a power boiler which burns sulfur-containing coal or other fuel, by scrubbing the waste gas stream with an aqueous solution of magnesium bisulfite-magnesium sulfite. Sulfur dioxide is absorbed into the aqueous scrubbing solution by reacting with magnesium sulfite to form magnesium bisulfite in solution. The resulting solution is divided into two portions. Magnesium hydroxide is added to the first portion at a controlled rate, to convert magnesium bisulfite to magnesium sulfite.

The first portion is then recycled for further waste gas scrubbing. The magnesium sulfite and magnesium bisulfite content of the second solution portion are processed at elevated temperature to produce magnesium oxide and a gas stream of high sulfur dioxide content, usually about 10% by volume. The magnesium oxide is slaked with water to form magnesium hydroxide which is added to the first solution portion, and the gas stream of high sulfur dioxide content is utilized to produce a sulfur-containing product.

A process developed by J.D. Terrana and L.A. Miller (9) involves the recovery of sulfur dioxide from $SO_2$-containing gases in a system involving contacting the gas with an aqueous sulfite solution to produce a solution of the corresponding bisulfite which is a precursor of $SO_2$ and subsequently separating the bisulfite and $SO_2$ partial pressure lowering materials to obtain purer bisulfite which is then decomposed to regenerate sulfur dioxide. Examples of sulfites include potassium, cesium and rubidium sulfites.

In the process separation of the bisulfite is accomplished by vaporization of water from the bisulfite-containing solution, e.g., in a flash chamber to supersaturate the solution and precipitate the bisulfite, e.g., in an amount equivalent to the amount of the sulfur dioxide removed from the waste gases. Conditions are chosen to precipitate crystals having a size suitable for vacuum drum filtering. The crystals are advantageously redissolved prior to heating to regenerate the sulfur dioxide and a portion of this solution is used to wash the crystals on the drum filter and improve the purity thereof.

A process developed by R.A. Meyers, J.S. Land, C.C. Shih and J.L. Lewis (10) is one in which sulfur dioxide contained in flue gases from a coal-burning, thermal generating plant are sorbed on finely divided cellulose such as shredded paper at temperatures in the range of 215° to 300°F. After the cellulose has been saturated with the sulfur dioxide, it is desorbed by flue gas at about 350° to 450°F. The desorbed sulfur dioxide may be then conveniently forwarded to a plant for processing to sulfuric acid. Spent cellulose may be forwarded to the plant for burning or it may be further digested to carbohydrates.

A process developed by J.F. Villiers-Fisher and A. Warshaw (11) is one in which waste gases containing sulfur dioxide, which may also contain sulfur trioxide, are scrubbed with an aqueous slurry containing solid reactant particles of an oxide or carbonate of calcium, magnesium or barium, together with an additive containing an acidic radical which solubilizes the calcium, magnesium or barium ion, so that the aqueous slurry absorbs sulfur dioxide from the waste gas.

The dissolved sulfur dioxide forms the soluble sulfite radical in solution, which reacts with the dissolved calcium, magnesium or barium ion to precipitate a solid sulfite and regenerate the additive in solution. The solid sulfite or calcium, magnesium or barium is separated from the aqueous liquid phase which is recycled for further scrubbing, together with makeup solid reactant particles. The scrubbed waste gas is discharged to atmosphere.

A process developed by J.F. Villiers-Fisher is one in which waste gases containing sulfur dioxide and entrained solids such as fly ash are scrubbed with an aqueous magnesium oxide-sulfite slurry, which dissolves sulfur dioxide to form further magnesium sulfite and also scrubs entrained solids from the gas stream into the aqueous slurry. Magnesium oxide is regenerated by calcining a portion of the slurry or the solids component of a slurry portion, which generates a dust laden off-gas rich in sulfur dioxide.

Buildup of entrained solids such as fly ash in the system is effectively prevented by scrubbing the dust laden off-gas with an aqueous solution, to which a portion of the solids component of the slurry or regenerated magnesium oxide may be added.

The scrubbing step solubilizes magnesium compounds as magnesium bisulfite, due to high sulfur dioxide concentration, and the resulting aqueous magnesium bisulfite solution is filtered to remove residual solids such as fly ash. Magnesium oxide is added to the residual liquid phase to precipitate solid magnesium sulfite, which is filtered from the final residual liquid phase. A portion of this final liquid phase may be discarded to prevent buildup of alkali sulfates or the like, and the balance of the final liquid phase is recycled to the off-gas scrubbing step. The precipitated solid magnesium sulfite is recycled to the calcining step.

A process developed by S.L. Torrence (13) is a process for removing sulfur dioxide and sulfur trioxide from a gaseous stream whereby sulfur dioxide is oxidized to sulfur trioxide and the sulfur trioxide is adsorbed onto an activated carbon adsorbent having elemental sulfur adsorbed thereon; thereafter the adsorbent is regenerated producing sulfur dioxide and activated carbon adsorbent with a minimal loss of adsorbent due to burn-off; elemental sulfur is adsorbed onto the regenerated adsorbent and the adsorbent is then recycled to contact the gaseous stream and the sulfur dioxide produced during regeneration is recovered.

A process developed by Y. Eguchi (14) involves removing sulfur oxides from waste gases with employment of activated carbon as the adsorbent. The chemical exhaustion of activated carbon and the lowering of its ability for adsorbing sulfur oxides in the regeneration step can be remarkably prevented by regenerating the sulfur oxides-adsorbing activated carbon with a desorbent containing carbon monoxide gas and/or hydrogen gas in a concentration at least about 40% at a temperature from about 230° to 450°C. The most advantageous results can be attained by employing vanadium oxide-supporting activated carbon as the adsorbent.

Figure 129 shows a suitable arrangement of apparatus for the conduct of this process. Waste gases containing sulfur oxides 1 are conducted into the dust collection chamber 2 wherein dusts such as fly ash and rubble in the waste gases are removed. The dust-free waste gases are cooled by the cooler 3 to a temperature lower than about 200°C., preferably between the dew point of the waste gases to be treated and about 180°C. and then are conducted into the adsorbent tower 4 charged with activated carbon, advantageously with the vanadium oxide-supporting activated carbon, wherein substantially all the sulfur oxides in the waste gases are adsorbed on the activated carbon.

The sulfur oxides-free waste gases are exhausted via the smokestack 5 into the atmosphere. The activated carbon is moved counter to the flow of waste gases in the adsorbent tower 4 and the sulfur oxides-adsorbing activated carbon is transferred from the bottom of the adsorbent tower to the regenerating tower 6.

Carbon monoxide gas and/or hydrogen gas produced in the generator 7 is combined with the circulating desorbent 9. The resulting mixture is heated by the heating apparatus 10, e.g., heat exchanger to a temperature between about 250° and 350°C., and then conducted into regenerating tower 6 at a flow velocity of about 0.01 cm./sec. and about 10 cm./sec. as superficial flow velocity, whereby the sulfur oxides adsorbed on the activated carbon are desorbed as sulfur dioxide into the gas passing through the regenerating tower and the activated carbon is regenerated.

The bulk of the gas passing through the regenerating tower is allowed to circulate through the blower 11 and the heating apparatus to the regenerating tower for reuse. The rest of the gas is conducted through the blower 12 into the condenser 13 where the gas is cooled to about 4°C. to separate a diluted sulfuric acid solution, and subsequently into another condenser 14 where the gas is cooled to a temperature lower than about −10°C. to recover liquid sulfur dioxide.

The treated gas still contains carbon monoxide and/or hydrogen, and may be allowed to circulate through the regenerator 7 or directly to the heating apparatus for reemployment as the desorbent in the regenerating tower, or may be employed for another use such as

FIGURE 129:  PROCESS FOR SO$_2$ RECOVERY BY ADSORPTION ON ACTIVATED
CARBON

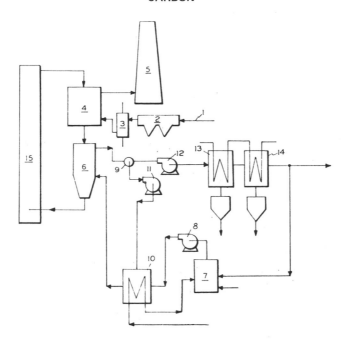

Source:  Y. Eguchi; U.S. Patent 3,667,910; June 6, 1972


fuel gas.  The regenerated activated carbon is transferred continuously or at intervals, from
the regenerating tower to the adsorbent tower for reuse through the transferring system
**15**.  This process may be advantageously applied to any kind of waste gases containing sulfur oxides, such as flue gases from thermal power stations, waste gases from chemical factories, waste gases from smelting furnaces and the like.

A process developed by A.B. Welty, Jr. (15) is one in which buildup of sulfate ions in a
flue gas desulfurization system employing an aqueous ammoniacal absorbent is prevented
by reducing a portion of the sulfate to sulfur dioxide.  Sulfur dioxide is removed from flue
gas by absorption in an aqueous solution of ammonium sulfite or ammonia.  Some oxidation of tetravalent sulfur to the hexavalent state takes place.

At least a portion of the absorber effluent solution is regenerated by acidification with ammonium bisulfate to liberate sulfur dioxide and to form an aqueous ammonium sulfate-
ammonium bisulfate slurry.  This slurry is introduced into a decomposition zone where a
major portion of the salt content is decomposed into ammonium bisulfate and ammonia,
and a minor portion into a gas mixture comprising ammonia, nitrogen, sulfur dioxide and
steam.  The ammonium bisulfate thus produced furnishes the bisulfate used for liberation
of sulfur dioxide.  The gas mixture formed in decomposition is used to ammoniate a second portion of the absorber effluent solution and thereby produce fresh absorbent solution.

Figure 130 is a flow sheet of this process.  A flue gas stream containing about 0.2 to 0.3%
by volume of sulfur dioxide plus small amounts of sulfur trioxide is introduced via line
**10** into sulfur dioxide absorber **11**.  The flue gas stream is contacted in absorber **11** with

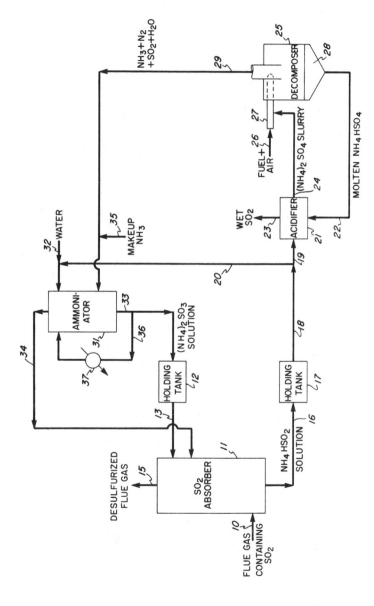

FIGURE 130: PROCESS FOR FLUE GAS DESULFURIZATION WITH AMMONIA

Source:   A.B. Welty, Jr.; U.S. Patent 3,676,059; July 11, 1972

fresh aqueous absorbent solution containing ammonium sulfite as its principal solute. This aqueous solution is conveyed from holding tank 12 through feed line 13 to scrubber 11. All of the absorbent solution is shown as entering the top of the scrubber, although the solution may be supplied at two or more vertically spaced locations. The solution rate is substantially proportional to the rate of flow of sulfur dioxide in the flue gas.

Desulfurized flue gas, having about 10% or less of the original sulfur dioxide content, is removed from the absorber through overhead line 15. An aqueous absorber effluent solution comprising ammonium bisulfite with small amounts of ammonium bisulfate present, is removed from the base of the absorber through line 16. The ammonium bisulfate is due to the small amount of oxidation of sulfur from the tetravalent to the hexavalent state which takes place in the sulfur dioxide scrubber 11. This absorber effluent solution flows from line 16 into holding tank 17.

An absorber effluent solution stream 18 is withdrawn from holding tank 17. This absorber effluent solution stream is divided into two portions. The first and smaller portion, which is stream 19, is treated to liberate sulfur dioxide. The second and larger portion, stream 20, is used to prepare fresh absorbent solution.

The absorbent effluent solution stream 19 is introduced into acidifier 21, where it is reacted with excess ammonium bisulfate introduced in the molten state through recycle line 22. The sensible heat of the molten ammonium bisulfate stream maintains the acidifier at its desired operating temperature of about 200° to 225°F. Wet sulfur dioxide, i.e., a mixture of sulfur dioxide and water vapor, is liberated in acidifier 21 and withdrawn through overhead line 23. The water vapor may be separated from the sulfur dioxide by known means and the sulfur dioxide converted to either sulfur or sulfuric acid.

An aqueous slurry of ammonium sulfate and ammonium bisulfate, the former present in greater amount, is formed as a by-product in the acidifier 21. This slurry is withdrawn from the acidifier through line 24 and is introduced into a decomposer 25. A preferred mode of heating the decomposer is by means of hot combustion gases. In the form shown a mixture of fuel and air is introduced through line 26 into a refractory lined combustion tube 27, where the fuel is burned. The ammonium sulfate-ammonium bisulfate slurry in line 24 is introduced into the hot combustion gas stream in combustion tube 27 which is formed by combustion of the fuel therein.

The water in the slurry is evaporated and the ammonium sulfate is decomposed into ammonium bisulfate, ammonia, nitrogen, sulfur dioxide and water vapor. The major portion of the ammonium sulfate is decomposed into ammonium bisulfate and ammonia. The ammonium bisulfate is collected in molten form in a pool 28 at the bottom of decomposer 25. This ammonium bisulfate is returned in molten form to the acidifier 21 through line 22. The mixture of combustion gas and gaseous decomposition products is removed from decomposer 25 through overhead line 29.

The gas mixture is introduced into the ammonia absorber 31. The gas mixer contacts the second portion of the absorber effluent solution, which is conveyed to the ammonia absorber or ammoniator 31 via line 20. Makeup water is added as required through line 32 and makeup ammonia is added through line 35. The ammonia and sulfur dioxide in the gas mixture react with the ammonium bisulfite to form fresh absorbent solution, containing ammonium sulfite as the principal solute.

This fresh absorbent solution is returned to holding tank 12 through line 33. The desired temperature is maintained in ammoniator 31 by recirculating part of the effluent solution through a pumparound circuit which includes recirculation line 36 and cooler 37. Gases which are not absorbed, i.e., nitrogen and carbon monoxide, are removed from the absorber 31 through overhead line 34. These gases are returned to the sulfur dioxide absorber 11 in order to remove any ammonia or sulfur dioxide which may be present.

### Recovery as Sulfur Trioxide

A process developed by G.C. Blytas (16) involves removing sulfur dioxide from hot gas that contains sulfur dioxide and oxygen by contacting the gas with a molten salt system including potassium sulfate and potassium pyrosulfate and a third salt that depresses the melting point of the system, circulating the salt system between a gas contacting zone at a maximum temperature of 550°C. to a regenerating zone where the temperature of the system is raised to at least 600°C. to drive off sulfur trioxide.

Figure 131 is a flow sheet of this process. A sulfur dioxide scrubbing zone 1 is illustrated where the hot oxygen and sulfur dioxide-containing gas is introduced into the lower portion of zone 1 through line 2 and countercurrently contacted therein with a descending stream of molten salt system containing potassium sulfate and potassium pyrosulfate introduced through line 3.

As a result of the countercurrent contact, a substantially sulfur dioxide-free gas stream passes through line 4 to a stack, while a molten salt system stream enriched in potassium pyrosulfate passes through line 5. The stream in line 5 contains more potassium pyrosulfate than the stream entering line 3, which contains m   ore potassium sulfate than the stream discharging through line 5; and it will contain potassium sulfate, potassium pyrosulfate and preferably a melting point depressant such as potassium metaphosphate.

The molten salt system enriched in potassium pyrosulfate passes through line 4 into the upper portion of a regenerating zone 6 where the temperature of the molten salt system is raised and it is contacted with an upwardly flowing stream of inert gas introduced at 7. As a result of the hot countercurrent contact with inert gas, a regenerated molten salt system is produced which contains a substantially higher proportion of potassium sulfate than the feed to zone 6, and it passes from the lower portion of regenerating vessel 6 through line 3 which passes the system into the upper portion of contacting vessel 1.

A gas stream containing large quantities of SO$_3$ in inert gas passes from the upper portion of regenerating vessel 6 through line 8 and it is introduced into the lower portion of absorbing vessel 10 where it is contacted with a descending stream of water or dilute sulfuric acid introduced through line 11 which scrubs substantially all of the SO$_3$ from the inert gas to produce a concentrated sulfuric acid stream passing from the bottom of absorber 10 through line 12. The substantially SO$_3$-free inert gas stream passes from the upper portion of absorber 10 through line 7 to the lower portion of regeneration vessel 6.

## FIGURE 131: REMOVING SULFUR DIOXIDE FROM HOT GAS

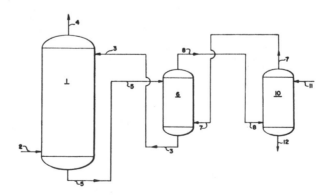

Source: G.C. Blytas; U.S. Patent 3,552,921; January 5, 1971

In a typical form of the process, a typical stack gas will enter contacting zone 1 at about 400°C. The lean melt entering the upper portion of contacting zone 1 will be about 370°C. and it will be a slurry containing about 30% solids, which will be a slurry of potassium sulfate in a liquid potassium pyrosulfate-potassium metaphosphate melt. The slurry contains 10% potassium metaphosphate, 23% potassium sulfate and 67% potassium pyrosulfate as it enters the sulfur dioxide absorber.

The fat melt discharging from the bottom of contacting zone 1 through line 5 will normally contain about 9.25% potassium metaphosphate, 5.5% potassium sulfate, 85.25% potassium pyrosulfate and it will comprise a slurry containing about 20% solids. Each pass through contacting zone 1 will remove about 6.4% of the weight of slurry of sulfur dioxide. The contacting zone is externally cooled sufficiently so that the fat melt leaves the contacting zone at about 345°C. The regeneration zone 6 functions at about 650°C. and ordinary heat exchange equipment is employed to heat the stream in line 5 to the regeneration temperature and to cool the stream in line 3 to the temperature at which it is introduced into the absorber.

A process developed by A.B. Welty, Jr., A.K.S. Raman and C.M. Lathrop (17) is one in which sulfur dioxide is removed from flue gas by contacting the flue gas with a vanadium pentoxide absorbent, then passing air in contact with the absorbent, causing desorption of $SO_3$. The reaction may be carried out in a rotating reactor containing a plurality of beds of absorbent.

A process developed by S. Hori, T. Inoue, S. Yamamoto, K. Tatara, M. Kitagawa, M. Watanabe, Y. Okada and N. Negishi (18) utilizes an apparatus for eliminating sulfur oxides from combustions exhaust gases comprising a plurality of compartments each containing activated carbon and means for washing the latter with water.

A pair of inlet ducts, a pair of outlet ducts, and inlet and outlet flow control means associated with each compartment are provided so that the latter can be placed in communication with either of the inlet ducts and either of the outlet ducts, respectively, or isolated from all of the ducts, the inlet ducts branching from a common main duct for leading exhaust gas to the apparatus. One of the outlet ducts communicates with one of the inlet ducts at a position downstream of the branching point of the main duct and the other of the outlet ducts leads gas out of the apparatus. Dilute $H_2SO_4$ is the product of the washing process.

### Recovery as Elemental Sulfur

A process developed by I.S. Shah (19) is one in which a waste gas stream containing sulfur dioxide is scrubbed with aqueous sodium sulfite solution to dissolve sulfur dioxide and form sodium bisulfite in solution. A portion of the resulting solution is reacted with sodium carbonate to convert bisulfite to sulfite, and the sodium sulfite is reacted with a carbon-containing reducing medium to form a sodium sulfide smelt or solid particles of sodium sulfide, which is dissolved in water and reacted with a gas stream containing carbon dioxide, to form a gaseous stream containing hydrogen sulfide and regenerate sodium carbonate for recycle. The hydrogen sulfide is recovered in the form of a valuable sulfur-containing product such as elemental sulfur.

Figure 132 is a flow sheet of this process. The waste gas stream 1 containing sulfur dioxide and entrained fly ash is derived from a waste or flue gas source such as a steam power plant or the like. The stream will typically contain in the range of 0.01 to 1% sulfur dioxide content, together with entrained fly ash and soot, balance carbon dioxide and nitrogen, and the stream will typically be derived at an elevated temperature generally in the range of 100° to 300°C.

Stream 1 is passed into the upper end of waste gas scrubber 2, which may be any suitable device for gas-liquid contact and scrubbing for sulfur dioxide removal. The unit 2 accomplishes waste gas scrubbing and sulfur dioxide absorption by the provision of an internal inverted frusto-conical baffle 3, so that a venturi-type scrubbing action is attained. Aqueous scrubbing solution stream 4 containing dissolved sodium sulfite and also containing a minor proportion of residual unconverted dissolved sodium bisulfite in most cases, is passed into unit 2 above and adjacent to the upper end of the baffle, and flows downwards on the upper surface of the baffle for projection into the highly accelerated high velocity waste gas stream at the lower central opening of the baffle.

The aqueous scrubbing liquor is thus projected and finely dispersed into the waste gas stream within unit 2 at a temperature typically in the range of 50° to 90°C., and sulfur dioxide absorption together with fly ash entrainment into the liquid phase take place in the lower part of unit 2. The scrubbed and cooled waste gas stream 5, now at a temperature typically in the range of 50° to 90°C. is discharged from unit 2 below the baffle and is passed to atmosphere or further utilization as desired. In most cases the waste gas stream will be saturated with water vapor and will contain less than 0.01% sulfur dioxide content by volume.

The absorption of sulfur dioxide into the aqueous scrubbing liquor within unit 2 also results in the conversion of at least a portion of the dissolved sodium sulfite to sodium bisulfite. The resulting aqueous liquor stream 6 discharged from the unit now contains dissolved sodium bisulfite, residual sodium sulfite and entrained fly ash. A portion of stream 6 is recycled for further waste gas scrubbing via stream 7, which is combined with recycle sodium carbonate solution or solids stream 8.

FIGURE 132: APPARATUS FOR SO$_2$ REMOVAL FROM GASES BY SCRUBBING WITH Na$_2$SO$_3$ SOLUTION TO LIBERATE H$_2$S AND PRODUCE ELEMENTAL SULFUR

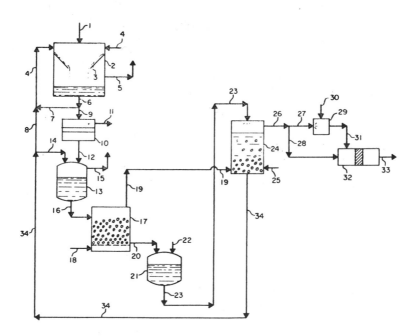

Source: I.S. Shah; U.S. Patent 3,542,511; November 24, 1970

The addition of stream **8** to stream **7** serves to convert sodium bisulfite in stream **7** to sodium sulfite by reaction with sodium carbonate. Combination of streams **7** and **8** thus results in the formation of aqueous scrubbing liquid stream **4**, which is utilized as described supra. The balance of stream **6** passes via stream **9** to solids filter or centrifuge **10**, which is any suitable device for separating entrained solids from a liquid stream. The separated solid fly ash stream **11** removed from unit **10** is passed to waste disposal or other utilization.

The filtered and solids-free liquid stream **12** discharged from unit **10** now contains dissolved $Na_2SO_3$ and $Na_2HSO_4$, and the dissolved $Na_2SO_3$ in stream **12** is converted to $NaHSO_3$ by reaction of stream **12** with added $Na_2CO_3$ within reactor unit **13**. The $Na_2CO_3$ in solid or aqueous solution form is passed via stream **14** into unit **13**, together with stream **12**, and the resultant reaction of sodium bisulfite with sodium carbonate serves to form sodium sulfite and evolve gaseous carbon dioxide, which is removed via stream **15**. Suitable heating means, not shown, may also be provided in unit **13** in order to heat and concentrate the solution by evaporation of water vapor, which is also removed via stream **15**.

The resulting concentrated sodium sulfite solution stream **16** discharged from unit **13** is now passed to a suitable furnace or the like for elevated temperature reaction with a reducing gas and formation of a sodium sulfide smelt. Stream **16** is passed into fluid bed reactor **17**, which is heated to a temperature generally in the range of 500° to 900°C., and a hot reducing gas stream **18** typically consisting of preheated carbon monoxide or methane or the like is passed into unit **17** below the fluid bed. As an alternative, stream **18** may consist of a hot combustion gas, in which case pulverized coal or other solid carbonaceous particles would be admitted into unit **17** together with stream **16**. In any case, a hot off-gas stream **19** containing carbon dioxide is discharged from unit **17**, together with a solids stream **20** consisting of a smelt or solid granules of sodium sulfide.

Stream **20** is now passed into solids dissolving tank **21**, which is any suitable unit for dissolving the solids stream **20** in water stream **22**. Unit **21** may be provided with a suitable internal stirrer or other agitation and dissolving means, not shown. The resulting aqueous $Na_2S$ solution stream **23** formed in unit **21** is passed to carbonator **24** which is any suitable carbonation tower or gas-liquid contact mixer. The hot off-gas stream **19** containing $CO_2$ is passed into the lower end of unit **24** and rises through the body of liquor maintained at 40° to 90°C. Also, $CO_2$ stream **25** derived from external sources, such as streams **5** or **15**, may be passed into unit **24**.

The rising gaseous carbon dioxide within unit **24** reacts with the dissolved sodium sulfide component in the liquid phase, to form dissolved sodium carbonate and liberate gaseous hydrogen sulfide. The resulting off-gas stream **26** discharged from the upper end of unit **24** is rich in hydrogen sulfide, and stream **26** is processed in this form of the process to form elemental sulfur. Stream **26** is divided into portion streams **27** and **28**. Stream **27** is passed to combustor **29** and reacted with air stream **30** in an exothermic combustion reaction which burns hydrogen sulfide to form sulfur dioxide.

The resulting process gas stream **31** discharged from unit **29** now contains sulfur dioxide and streams **31** and **28** are passed into catalytic reactor **32** for the reaction of sulfur dioxide with hydrogen sulfide in accordance with the Clause reaction, to yield elemental sulfur. The resulting process gas stream **33** discharged from unit **32** is rich in elemental sulfur vapor, which may be recovered by cooling of stream **33** to produce condensed liquid or by other means known to the art such as by scrubbing stream **33** with liquid elemental sulfur or a suitable liquid solvent for elemental sulfur. In any case, elemental sulfur may be produced from stream **33** as a valuable commercial product.

Returning to unit **24**, liquid stream **34** consisting essentially of an aqueous sodium carbonate solution is withdrawn from the bottom of unit **24**, and stream **34** is employed as a source of sodium carbonate for the conversion of sodium bisulfite to sodium sulfite. Stream **34** may be concentrated in means not shown, to produce a highly concentrated solution or even produce solid sodium carbonate, prior to recycle.

In any case, stream **34** is divided into streams **8** and **14**, which are utilized as described.

A process developed by S.L. Torrence (20) for removing sulfur oxides from gas streams is one whereby the sulfur oxides are adsorbed onto a carbonaceous adsorbent. Sulfur dioxide is recovered from the spent adsorbent by contacting the adsorbent with an oxide-reducing gas, i.e., hydrogen sulfide, whereby the adsorbed sulfur oxides having been adsorbed as sulfuric acid and sulfur trioxide are reduced to sulfur dioxide and elemental sulfur. The elemental sulfur is removed from the adsorbent by contacting with a sulfur-reducing gas and thereafter recycling the adsorbent.

A process developed by J.M. Potts (21) for the recovery of sulfur dioxide from stack gas consists of scrubbing the gas with potassium polyphosphate solution to form a slurry of crystallized potassium pyrosulfite. The potassium pyrosulfite is regenerated for further pickup of sulfur dioxide by (a) heating the slurry directly; (b) separating the potassium pyrosulfite and treating it in a reflux stripper; (c) heating the potassium pyrosulfite in solid form, or (d) heating solid potassium pyrosulfite to drive off one-third of the sulfur dioxide and reducing the remaining two-thirds to form hydrogen sulfide, the sulfur dioxide and hydrogen sulfide being reacted to form elemental sulfur.

### Recovery as Hydrogen Sulfide

A process developed by P.A. Lefrancois and K.M. Barclay (22) is one in which compounds of sulfur and of nitrogen and fly ash are removed from gases by absorption at between 1500° and 2000°F. under atmospheric to 5 atmospheres pressure in a molten carbonate salt of sodium, potassium and/or lithium, and the decontaminated combustion gas stream is cooled and vented to the atmosphere.

To describe the process in somewhat more detail, a metal carbonate melt of sodium, potassium, or lithium or mixtures thereof and a contaminated industrial sulfur-containing gas are contacted under conditions such that a pressure of from about atmospheric to 5 atmospheres is maintained with a pressure drop of less than 2 psi, preferably not more than 1 psi, during contact and a critical temperature within the range of from 1500° to 2000°F. is maintained in the contacting zone so that contaminants of the gas are absorbed by the melt and the formation of a metal sulfite is substantially avoided.

A preferred method of contacting the gas and melt is to direct the gas by means of a baffle or the like to impinge on the surface of a moving film of the melt disposed on a tray in a contacting zone. More preferably a plurality of baffles are employed over each tray to provide multipoint impingement of the gas on the surface of the melt and a series of trays with downcomers employed to successively pass the melt onto each tray in series and finally into the liquid outlet of the contacting zone from which a portion of the liquid melt can be withdrawn for purification and/or recovery of contaminating elements and the remaining portion can be subjected to heating, e.g., by oxidation of any contained or introduced carbonaceous material to maintain the molten state prior to recycle of the melt to the absorption zone.

Alternatively the remaining portion can be directly recycled to the first tray of the series for further contact with gas in a continuous operation wherein indirect heating is provided. The contaminants concerned herein are those constituents, whether liquid, gaseous or solid, whose presence is objectionable when emitted to the atmosphere. Among these are sulfur dioxide, hydrogen sulfide, sulfur trioxide, the oxides of nitrogen and carbon, fly ash, soot, and combinations thereof as contained in industrial off-gases.

The materials may be contained in any vented gas such as carbon dioxide, nitrogen, oxygen, air, water vapor or combinations thereof as found in off-gases from power plants, petroleum refining, chemical plants, pulp and paper, soap, detergent, and textile manufacturers, automobile and steel plants and leather processing plants. The carbonate melt is continuously withdrawn and recycled to the contacting zone with a portion being removed for regeneration and a like portion of fresh carbonate feed being added to the recycle stream to supplement the amount of carbonate passed to regeneration. The preferred regeneration procedure is described in U.S. Patent 3,567,377, although any other convenient method for regeneration of the carbonate melt can be employed. Generally, the process comprises diluting the melt with an aqueous solution of the corresponding metal bicarbonate and, if necessary, filtering solids from the resulting solution.

The resulting solution is then passed to a two-stage carbonation zone where the first stage effects conversion of the metal carbonate to bicarbonate precipitate with carbon dioxide to provide a more viscous, less alkaline solution. This solution is then passed to the second stage where it is separately contacted with carbon dioxide for removal of sulfur values as hydrogen sulfide. The process continues by converting at least a portion of the metal bicarbonate to the corresponding metal carbonate and concentrating and heating the resulting carbonate to provide a suitable recycle of carbonate melt to the process.

A process developed by E. Gorin and P.M. Yavorsky (23) accomplishes removal of sulfur dioxide from hot flue gas by passing the flue gas in contact with potassium formate, sodium formate, or ammonium formate, in either a molten state or in aqueous solution, at a temperature above 140°F., whereby the sulfur dioxide and the formate react to form principally thiosulfate. At temperatures above 475°F., secondary reactions of the thiosulfate formed occur, and increase in rate as the temperature is increased. Among such secondary reactions is the following.

$$K_2S_2O_3 + 4HCOOK \rightarrow 2H_2S + 3K_2CO_3 + CO_2$$

Since simply replacing the $SO_2$ by $H_2S$ is undesirable in most instances, the $H_2S$ is usually sent to a plant for the production of elemental sulfur.

### Recovery as Pulp Cooking Liquor

A process developed by J.L. Clement (24) utilizes a system for the absorption of $SO_2$ from gases of combustion where the gases are passed in series through a plurality of direct contact zones. The absorption liquid is made up of a solution of magnesium and sulfur which is sprayed into the gas, and the makeup water in the solution is selectively added into the last stage zone for optimum $SO_2$ absorption efficiency of the entire system. When used in the chemical recovery system of a pulp and paper installation, both the magnesium compounds and the $SO_2$ absorbed in the magnesium may be reused in the chemical process.

### Recovery as Ammonium Sulfate Fertilizer

A process developed by J. Jonakin and A.L. Plumley (25) involves the addition of magnesium-containing additives such as dolomite to fossil-fuel fired equipment followed by the wet scrubbing of the flue gases resulting in the reaction of the sulfur oxides with the additive. The process further involves the addition of ammonium hydroxide to the solution removed from the scrubber to convert the soluble magnesium sulfate to soluble ammonium sulfate and precipitate magnesium hydroxide. The ammonium sulfate solution is removed from the magnesium hydroxide and either used for fertilizer or further reacted with calcium oxide to precipitate calcium sulfate and form ammonium hydroxide which is recycled for reaction with the scrubber effluent.

### Disposal as Waste Calcium Sulfate

A process developed by K.-A.G. Gustavsson (26) involves cleansing flue gas from primarily $SO_2$, and includes a container adapted to accommodate a quantity of treatment liquid and provided with a gas inlet so arranged that the gas is caused to flow substantially at right angles down towards the surface of the liquid in the container; and a treatment column directed essentially perpendicularly to the surface of the liquid and through which the gas flows essentially linearly away from the surface of the liquid while entraining droplets of liquid therefrom, and wherein the container below the liquid level communicates with an equalizing tank provided with means for maintaining the level of the liquid therein constant.

Among the alkalis which can be used as treatment or absorption agents are primarily sodium hydroxide (caustic soda) and calcium hydroxide (hydrated lime). The caustic soda is obviously a more superior absorption agent, but since it is much too expensive aqueous calcium hydroxide is generally used.

The hydrated lime may conveniently be metered direct in powder form to the equalizing tank, which may thus simultaneously serve as a dissolver, by means of a screw conveyor or other mechanical devices. The strong circulation of water effects the mixing and dissolving of the substances. Metering of the lime can also be controlled by instruments for controlling the pH value of the return liquid from the top of the treatment column so that an alkali of pH 7 to 11 is maintained there. The lime solution thus becomes supersaturated in remaining parts of the system and conventional iron plate can be used as construction material without too much risk for corrosion.

The only waste product is the sludge formed by the consumed hydrated lime and solid constituents and sulfur compounds washed from the flue gas. The sludge can be tapped off intermittently through a pocket under the liquid bath. However, it is desirable to obtain a higher concentration than that obtained by self sedimentation and accordingly, a fully automatic operating sludge thickener can be connected in the return circuit for the return liquid so that the sludge is discharged in the form of a concentrated paste while the liquid freed from sludge is passed back to the liquid bath. This liquid may suitably be returned so that it flushes the walls of the container free from any sludge which might have settled thereon.

### References

(1)  Slack, A.V., *Sulfur Dioxide Removal from Waste Gases,* Park Ridge, N.J., Noyes Data Corp. (1971).
(2)  U.S. Department of Health, Education and Welfare, *Control Techniques for Sulfur Oxide Air Pollutants,* Washington, D.C., National Air Pollution Control Administration (Jan. 1969).
(3)  Cantrell, R.R. and Wiley, F.P.; U.S. Patent 3,538,681; November 10, 1970; assigned to W.R. Grace & Co.
(4)  Little, D.M. and Hettick, G.R.; U.S. Patent 3,553,936; January 12, 1971; assigned to Phillips Petroleum Company.
(5)  McRae, W.A., Brown, D.L. and McGriff, S.G.; U.S. Patent 3,554,895; January 12, 1971; assigned to Ionics, Inc.
(6)  Owaki, T.; U.S. Patent 3,556,722; January 19, 1971; assigned to Furukawa Mining Co., Ltd., Japan.
(7)  Shah, I.S.; U.S. Patent 3,577,219; May 4, 1971; assigned to Chemical Construction Corporation.
(8)  Shah, I.S.; U.S. Patent 3,617,212; November 2, 1971; assigned to Chemical Construction Corporation.
(9)  Terrana, J.D. and Miller, L.A.; U.S. Patent 3,627,464; December 14, 1971; assigned to Wellman-Lord, Inc.
(10)  Meyers, R.A., Land, J.S., Shih, C.C. and Lewis, J.L.; U.S. Patent 3,629,996; December 28, 1971; assigned to TRW, Inc.
(11)  Villiers-Fisher, J.F. and Warshaw, A.; U.S. Patent 3,632,306; January 4, 1972; assigned to Chemical Construction Corporation.
(12)  Villiers-Fisher, J.F.; U.S. Patent 3,650,692; March 21, 1972; assigned to Chemical Construction Corporation.
(13)  Torrence, S.L.; U.S. Patent 3,667,908; June 6, 1972; assigned to Westvaco Corporation.
(14)  Eguchi, Y.; U.S. Patent 3,667,910; June 6, 1972; assigned to Takeda Chemical Industries, Ltd., Japan.
(15)  Welty, A.B., Jr.; U.S. Patent 3,676,059; July 11, 1972.
(16)  Blytas, G.C.; U.S. Patent 3,552,921; January 5, 1971; assigned to Shell Oil Company.
(17)  Welty, A.B., Jr., Raman, A.K.S. and Lathrop, C.M.; U.S. Patent 3,615,196; October 26, 1971; assigned to Esso Research and Engineering Company.
(18)  Hori, S., Inoue, T., Yamamoto, S., Tatara, K., Kitagawa, M., Watanabe, M., Okada, Y. and Negishi, N.; U.S. Patent 3,686,832; August 29, 1972; assigned to Kogyo Kaihatsu Kenkyusho, Japan.
(19)  Shah, I.S.; U.S. Patent 3,542,511; November 24, 1970; assigned to Chemical Construction Corp.
(20)  Torrence, S.L.; U.S. Patent 3,563,704; February 16, 1971; assigned to Westvaco Corporation.
(21)  Potts, J.M.; U.S. Patent 3,630,672; December 28, 1971; assigned to Tennesee Valley Authority.
(22)  Lefrancois, P.A. and Barclay, K.M.; U.S. Patent 3,671,185; June 20, 1972; assigned to Pullman, Inc.
(23)  Gorin, E. and Yavorsky, P.M.; U.S. Patent 3,687,615; August 29, 1972; assigned to Consolidation Coal Company.

(24) Clement, J.L.; U.S. Patent 3,615,165; October 26, 1971; assigned to The Babcock & Wilcox Co.
(25) Jonakin, J. and Plumley, A.L.; U.S. Patent 3,637,347; January 25, 1972; assigned to Combustion Engineering, Inc.
(26) Gustavsson, K.-A.G.; U.S. Patent 3,640,053; February 8, 1972; assigned to Aktienbolaget Bahco Ventilation, Sweden.

# SULFURIC ACID

## Removal from Air

A process developed by R.G. Hartig and J.R. Archer (1) is based on the discovery that chemical mists such as sulfur trioxide can be eliminated by passage of mist-bearing gas through a packing of inert materials having a critical interstitial pore size, which can be controlled within limits through control of the volume of liquid used to wet the packing.

In utilizing the principle herein referred to, the mists and gases are passed through a confined space such as a tower. This confined space is filled in its entirety or in part, depending upon use conditions, with a packing inert with respect to the gases and mists such as Orlon filaments or woven Orlon cloth, asbestos, sawdust, coke, rock, ceramic material and the like. When using cloth or filaments and asbestos, interstitial space must be provided so as to be equivalent to that provided by a size graded bed of coke, rock, sawdust, ceramic material, or the like, having particles of a size to pass through standard screens in the range between about 10 mesh, 1,650 microns and about 80 mesh, 175 microns (Tyler Standard screens).

Packing in the confined space may be wetted with liquid which is capable of absorbing the nuisance gases, i.e., water, dilute acids, aqueous solutions such as sodium carbonate, sodium bicarbonate and the like. The greater the volume of liquid passing through the confined space and over the packing, up to the point of flooding the combined space, the greater the effectiveness of the absorption. A factor limiting the volume of liquid flowing through the confined space is the back pressure on the gas as the fluid reduces the interstitial space free to pass the gases.

Reasonable pressure drops for commercial operations are of the order of about 10 to 30 inches of water although pressures outside this range can be utilized. For example, a sawdust bed 10 feet square and 7 inches thick will filter sulfur trioxide mist from 7,000 cfm of gas if a pressure drop of 19 inches of water is maintained; whereas, it will only filter about 5,500 cfm of gas if the back pressure is increased to 25 inches of water.

A process developed by C.L. Leonard, C.R. Schneider and J.K. Sheppard (2) involves efficient recovery of mist from the exhaust gases from the concentration of sulfuric acid. Sulfuric acid mist is removed from a gaseous mixture containing the same by passing the mist containing gaseous mixture upward through porous plates covered with a free level of sulfuric acid of 60 to 72% concentration.

A process developed by F. Schytil and H. Krollmann (3) involves the collection of the liquid contained in a mist and more particularly such collection from mists in which the particle size of substantial quantities of the mist droplets is extremely small, for example, below 50$\mu$ in diameter.

Various processes are known for the collection of liquids contained in mists. The process with the highest efficiency previously known is the electrostatic precipitation. This process gives efficiencies up to 99% but requires very high investment costs and is applicable only to a limited degree to the precipitation of highly corrosive components. For example, disturbances occur in electrostatic precipitation when mists of sulfuric acid are precipitated whose droplets consist of an acid with higher strength than about 80%. Mechanical means have been used for some time for the collection of mists, such as, filters consisting of porous material, for example, ceramics, fabrics, felts or packings of granular material (coke, glass globules, etc.). In the operation of such filters the following principles have been applied.

(a) Pure sieving action, that is ordinary droplets larger than the pore size of the filter are accumulated on the side of the filter facing the gas stream passed therethrough. The thickness of the filter is of no importance in this case, but it is important to use low gas velocities as otherwise the collected liquid would be pressed in causing the pores to clog. In filters operated according to this procedure, it has been proposed to employ filter materials whose surfaces are not wetted by the liquid being collected.

(b) In the use of filters having a larger pore diameter than the mist droplets to be collected, it has been proposed to pass the gas through the filter so slowly that sufficient time is permitted for the transverse Brownian movement to bring the particles to the surface of the wetted pores where they are collected by the liquid film covering the pore surface.

Filters operating according to these principles show a decrease in efficiency with increase in velocity so that the investment costs and the cost of operation increase disproportionately with increase in desired efficiencies, especially when mists treated contain particles of small size, for example, below $50\mu$ in diameter. Actually mists with much smaller droplets occur, such as from 0.1 to $5\mu$ in diameter, so that these mists cannot be precipitated according to the abovementioned principles in a commercially feasible way with a higher efficiency than about 70 to 80%.

For example, commercially operated coke filters operate usually with an efficiency of 60 to 65%, even with coarse mists, although theoretically any desired degree of efficiency can be obtained by decreasing the particle size of the coke and the gas velocity. Any increases in efficiency obtained in this way is paid for with such an increase in cost of investment and operation that the cost of the electrostatic precipitation is quickly arrived at or even surpassed. Therefore, even with other methods such as laboratory procedures, where commercial considerations play only a minor role, only efficiencies of about 94 to 95% have been achieved.

(c) Inertia phenoma are used in another method of collecting mists. If a stream of mist containing gas is deflected from a straight path, the heavy mist particles are thrown out by centrifugal force onto a surface much more rapidly than lighter particles and are collected on such surface. However, this gives satisfactory results only if the inertia of the individual mist particles is much greater than the force of the gas resistance, i.e., only relatively large droplets, such as above $50\mu$ in diameter, have sufficient inertia. Actually the efficiency with mist droplets in the neighborhood of $50\mu$ is relatively poor and it is only with droplets of about $200\mu$ and above that really satisfactory results are obtained. Cyclones, perforated plates and baffle plates operate according to this principle.

As the abovementioned types of procedures give satisfactory results only with coarse mists, proposals have been made to carry out the operations which give rise to mist formation in such a way that either the mist formation is suppressed as much as possible or the formation of large mist droplets is favored. However, these expedients are in many cases not convenient and it has therefore also been proposed to subject fine mists to sonic or ultrasonic waves to effect coalescence of the fine mist droplets to coarser droplets. However, the high energy consumption and the almost unbearable noise connected with this measure prohibited the commercial use of this proposal.

The process for collecting the sulfuric acid contained in a sulfuric acid mist containing a substantial quantity of mist particles of a size between 0.01 and $3\mu$ comprises passing the mist through a wettable porous ceramic filter of bonded granules having a pore diameter larger than the smallest particles of the mist to be collected at a velocity which is sufficient to cause a pressure drop of between 100 and 500 mm. of water.

The process requires substantially less investment cost than an electrofilter (approximately one-tenth of the investment cost of an electrofilter). Even in operation with highly corrosive components, no disturbances have been observed after a run of several years.

A process developed by W. Plaut and G.L. Fairs (4) involves sulfuric acid mist removal using a fiber filter made up of a class of fibers which, for convenience, are termed nonwettable fibers. An outstanding and most surprising increase in filtration efficiency is obtained; these are fibers whereon the mist is deposited in separate and independent droplets not connected (as in the case of the wettable fibers hitherto used) by a bridging film of liquid. The following is a specific example of the application of this process.

*Example:* In a plant for the manufacture of sulfuric acid the gases leaving the contact chamber were first cooled, then absorbed in strong sulfuric acid and subsequently passed through alkali-containing absorption towers. The sulfuric acid content of the mist-containing exit gas varied between 0.05 and 0.1 gram $H_2SO_4$ per cubic meter of gas and the mist particles were all of size less than $2\mu$, 10% of them by weight being less than $1\mu$.

Glass fiber of diameters in the range of 5 to $50\mu$ was treated with the Silicone Fluid M441 and was packed and compressed to a density of 10 pounds per cubic feet (160 kg./m³) to form a layer 5 cm. deep and the filter was held between confining gauzes made of resin coated stainless steel. The surface area of the filter presented to the gas stream was approximately 0.46 square meter. The mist-containing exit gas was passed downwardly through the filter at the rate of 300 to 350 m³/hr. per square meter of filter surface and the pressure drop was 19 cm. water gauge.

While the filter was in continuous operation for over 900 hours there was no visible fume in the exit gas and a weak acid varying in strength between about 2½ and 10% $H_2SO_4$ was collected by drainage from the filter. The sulfuric acid content of the exit gas as measured by means of an electrostatic sampler was less than 0.0007 to 0.0008 g./m³. By way of comparison the example was repeated except that the mist-containing gas was passed through a similar filter of untreated glass fiber but in this case a persistency visible gas left the filter, the sulfuric acid content of the tail gases being 0.007 to 0.012 g./m³.

### Removal from Water

A process developed by A.A. Spinola (5) relates to the neutralization of acid wastewaters and, in particular, to a method utilizing as a neutralizing agent, the solids removed from the effluent gas leaving a cement kiln and apparatus for carrying out the method.

Another process developed by A.A. Spinola (6) is one in which acid wastewaters, such as mine drainage water and spent pickle liquor, are neutralized by the addition thereto of solids collected from the gases discharged from a kiln used in making cement clinker to produce a sludge product including neutralized acid water. The process is low in cost, highly effective, rapid in action and results in a small volume of sludge which can easily be handled. It involves only the addition to the waters to be neutralized, of an appropriate amount of cement kiln flue dust and agitating the mixture or simply permitting it to stand until neutralization is complete.

A process developed by J.J. Birch (7) for the treatment of acid mine water comprises first adding alkaline reactant such as lime in powdered form, carrying the mixture of water and powdered reactant through a rotating bed of limestone in the presence of air, thus adding calcium carbonate particles to the water, combining the water thus treated with raw acid mine water, aerating the mixture and allowing a flocculant precipitate or floc of ferric hydroxide and alumina to separate out.

A process developed by J.C. Rhodes (8) involves treating mine drainage water, either in situ or any near adjacent impounding basins such as abandoned quarries, strip mine excavations, dams and the like, as to precipitate out the iron content thereof and to neutralize and precipitate the sulfuric acid content thereof, with the result that the water will thereby be nearly neutral if not indeed neutral and free or substantially free of iron and sulfate (or sulfur) ions. By this process other metals, particularly toxic metals, as for instance, copper are similarly removed. The process comprises subjecting mine drainage water and the like to the action of an excess of metallic iron and an excess of air in intimate contact with the water and for a sufficient length of time.

The added metallic iron reacts with the sulfuric acid to form ferrous sulfate which oxidizes to ferric sulfate. Concurrently, namely, as this reaction gradually reduces and finally exhausts the sulfuric acid content of the water, the ferric sulfate is hydrolyzed into insoluble basic ferric sulfates whose composition may vary somewhat according to the extent of hydrolysis of the ferric sulfate, but all insoluble in the absence of sulfuric acid.

By such absorption of the sulfuric acid by the added metallic iron and the oxidation of the ferrous sulfate to ferric sulfate by the added oxygen and the hydrolysis of the ferric sulfate to the insoluble basic iron sulfates, the original sulfuric acid content of the water (as well as the sulfuric acid) produced by the hydrolysis and the original iron content of the water are converted to insoluble basic iron sulfates, with the result that the decanted water (or the water from which such insoluble basic iron sulfates have been filtered) may then be safely released into streams or rivers or may be used directly as a source of domestic or industrial water supply.

A process developed by J.G. Selmeczi (9) involves the removal of anions from water such as acid mine drainage water, by providing a body of weakly basic anion exchange resin in substantially its free base form, introducing carbon dioxide into a stream of water to be purified, passing the water through the resin with removal of the anions and regulating the carbon dioxide flow into the stream to control the removal of anions from the water, the anions being removed in order of decreasing exchange potential. Additionally, ferrous salts will precipitate out as ferrous carbonate.

A process developed by J.W. Dixon (10) involves ion exchange treatment of coal mine acid drainings to produce a mine discharge which will not pollute streams and rivers. Figure 133 is a flow diagram which illustrates the process in somewhat more detail.

### FIGURE 133:  PROCESS FOR ION EXCHANGE TREATMENT OF COAL MINE ACID DRAINAGE

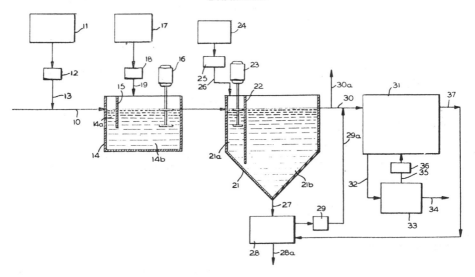

Source:  J.W. Dixon; U.S. Patent 3,403,099; September 24, 1968

Influent coal mine acid drainings enter the treating apparatus through inlet line 10 to a mixing tank 14 containing a baffle 15 and mixing device 16. A cationic polymer of suitable viscosity is contained in tank 11 and controlled additions thereof are metered through a metering device 12 and through line 13 into the drainings passing through line 10. The mixture of cationic polymer and the drainings enter the baffled zone 14a of mixing tank 14. A base material such as sodium carbonate is contained in tank 17 and controlled additions thereof are metered through a metering device 18 and through line 19 into zone 14b of tank 14. The total mixture is agitated by mixing device 16.

The mixture of drainings, cationic polymer and base material are passed through line 20 to a settling tank 21 containing a baffle 22 and enter zone 21a thereof. An anion exchange polymer is contained in tank 24 and controlled additions thereof are metered through a metering device 25 and through line 26. Zone 21a is provided with a mixing device 23. The mixture in zone 21a contains a precipitate which begins to settle towards drain 27. The clarified water passes into zone 21b. Concentrated floc material passes through drain 27 into a filter means 28 to be dewatered. The filtrate from dewatered floc is pumped through pump 29 through line 29a to line 30. The dewatered floc is removed through line 28a and passed to land fill or disposed of by other conventional means.

The supernatant liquid flows from the top of tank 21 through line 30. The effluent is in a treated state that it may be discharged into streams and rivers through line 30a without causing pollution. However, if a potable water is desired, portions of the effluent in line 30 may be passed to an aerated filter means 31 containing sand and charcoal or coke. Passage therethrough removes residual iron and reduces the sulfate content. Potable water is obtained through line 32 and stored in tank 33 for removal when desired through line 34. Backwashing of the aerated filter means is accomplished by pumping water with pump 36 from tank 33 through line 35 into filter means 31. Backwash drain effluent is removed through line 37 and passed to filter 28 for removal of floc picked up by the backwashing operation.

A process developed by M. Deul and E.A. Mihok (11) is one in which dilute acid wastes, such as mine drainage waters, are neutralized by reaction with an extremely finely divided limestone slurry after which the neutralized effluent is aerated to strip carbon dioxide and oxidize ferrous iron. After clarification, the product stream is suitable for disposal in surface waters.

**References**

1) Hartig, R.G. and Archer, J.R.; U.S. Patent 2,901,061; August 25, 1959; assigned to International Minerals & Chemical Corporation.
(2) Leonard, C.L., Schneider, C.R. and Sheppard, J.K.; U.S. Patent 2,906,372; September 29, 1959; assigned to Union Carbide Corporation.
(3) Schytil, F. and Krollmann, H.; U.S. Patent 2,947,383; August 2, 1960; assigned to Metallgesellschaft, AG, Germany.
(4) Plaut, W. and Fairs, G.L.; U.S. Patent 3,107,986; October 22, 1963; assigned to Imperial Chemical Industries Limited, England.
(5) Spinola, A.A.; U.S. Patent 3,541,008; November 17, 1970; assigned to United States Steel Corp.
(6) Spinola, A.A.; U.S. Patent 3,511,777; May 12, 1970; assigned to United States Steel Corporation.
(7) Birch, J.J.; U.S. Patent 3,516,931; June 23, 1970; assigned to Barnes & Tucker Company.
(8) Rhodes, J.C.; U.S. Patent 3,347,787; October 17, 1967; October 17, 1967.
(9) Selmeczi, J.G.; U.S. Patent 3,420,773; January 7, 1969; assigned to Ionics, Incorporated.
(10) Dixon, J.W.; U.S. Patent 3,403,099; September 24, 1968.
(11) Deul, M. and Mihok, E.A.; U.S. Patent 3,617,560; November 2, 1971; assigned to U.S. Secretary of the Interior.

# TANTALUM

## Recovery from Scrap

A process developed by W.M. Baldwin, E.O. Fuchs, D.J. Sharp and J.H. Swisher (1) is one in which scrap tantalum bodies having impurities therein are reclaimed as substantially pure tantalum metal by a process which includes immersing the bodies in a molten metal such as calcium or magnesium at elevated temperatures, and under nonoxidizing conditions for a prolonged period of time sufficient to cause entrapped impurities within the body to diffuse out and react with the molten metal.

In the manufacture of tantalum capacitors, the properties of the tantalum metal are very critical to production. These properties are strongly dependent on the concentration of interstitial impurities within the metal, itself. Presently, there are sources available which supply tantalum metal commercially that meet the critical property requirements to manufacture suitable capacitors. However, these sources are limited. Moreover, tantalum is commonly considered as one of the critical metals whose source is being depleted in the environment. This is attested to in an article by B.D. Wakefield, "U.S. Heads for the Mineral Poorhouse," *Iron Age,* June 4, 1970, pp. 71 to 76. Because of this trend, it is becoming necessary to salvage scrap tantalum by-products to meet production demands.

In the processing of scrap tantalum products, e.g., rejected tantalum capacitors, for recovery into a suitable tantalum metal to fabricate new capacitors, there are a number of essential operations to be performed. These operations have proven to be quite complex. For example, a rejected tantalum capacitor includes a tantalum metal body which has a first overlying anodic film of tantalum oxide and a second overlying counterelectrode film consisting of layers of manganese oxide, graphite and silver. The composite body also has soldered metal leads extending axially therefrom and is further encapsulated with an epoxy based composition.

In addition, excess oxygen is interstitially entrapped within the tantalum body because oxygen readily diffuses through the encapsulator and into the open pores of the tantalum body when the capacitor is in use. The diffused oxygen affects the electrical leakage properties of the capacitor to the point where the capacitor is no longer serviceable. To recover the tantalum, obviously, the encapsulator, and metal finishing must be separated from the tantalum body.

Further, the overlying films which are now impaired by delineation and other imperfections no longer possess the necessary electrical properties required in a newly formed capacitor, and they have to be removed. Moreover, in order to produce a suitable high purity tantalum metal which can be used in the fabrication of a new tantalum capacitor, it is necessary to reduce the content of entrapped oxygen in the tantalum body to a minimal level.

The encapsulator, metal finishings and overlying films are generally removed in a step-by-step operation employing physical and chemical techniques that are regarded as well-known in the art. However, it has proven very difficult to remove the entrapped oxygen from the tantalum body to a level which would be suitable for manufacturing new tantalum capacitors.

A known technique for removing the entrapped oxygen includes reacting carbon powder with the powder of a tantalum metal body to be processed. This reaction causes the entrapped oxygen to react with the carbon if the carbon is stoichiometrically mixed with the tantalum powder. While this process is effective for producing satisfactory high purity tantalum powders, extreme control of the amount of carbon employed must be maintained during this reaction. Too little carbon will not be sufficient to react all the oxygen in the tantalum, and too much carbon will result in excess carbon mixed with the tantalum metal.

Carbon cannot be removed from the metal unless it is converted to a volatile constituent and carbon is another impurity that is undesirable in the tantalum metal. Moreover, the tantalum body must be ground into a powder before it can be mixed with the carbon. Because of these complexities this technique demands extreme control precautions and these precautions negate many of the advantages in recovering acceptable tantalum for manufacturing capacitors. For instance in U.S. Patent 3,415,639 issued December 10, 1968, a process is disclosed for reclaiming a purified tantalum powder, suitable for making anodes, from a tantalum oxide source which also has a high content of entrapped oxygen. The tantalum oxide is first treated with highly reactive metals, such as calcium or magnesium, to reduce the oxide. These metals react very rapidly to reduce the oxide to tantalum and the chemical thermodynamics of their reactions are well-known in the art. The powder is then further treated with carbon in the previously described manner to remove the trapped oxygen therefrom. Needless to note, the disadvantages in this process are the same as previously mentioned.

The process is based on the discovery that a rejected tantalum body having entrapped oxygen within can be treated by a relatively simple procedure to remove most of the entrapped oxygen. Furthermore, extensive quality control measures need not be taken. Briefly, the tantalum body is reacted with a high purity molten metal selected from the group consisting of calcium, magnesium and mixtures thereof at elevated temperatures in a nonoxidizing atmosphere for a prolonged period of time exceeding the ordinary reaction period.

The entrapped oxygen within the tantalum body diffuses to the surface of the body in the presence of a nonoxidizing atmosphere and also reacts with the molten metal until substantially complete deoxidation of the tantalum body occurs. A reaction film is formed on the surface of the body, but this film does not impede prolonged reaction of the diffused oxygen because the film is discontinuous, therefore quite porous, permitting the oxygen to diffuse therethrough. Furthermore, it is noted that during a prolonged period the molten metal does not react with the tantalum nor does it diffuse therein in the presence of a nonoxidizing atmosphere even though it is in intimate contact with the body.

Therefore, excess amounts of molten metal may be employed without danger of contaminating the deoxidized tantalum body. By excess, it is meant that amounts of molten metal greater than the stoichiometric amount required to react with the diffused oxygen may be employed. The oxide of the molten metal, which is formed as a result of the reaction between the metal and the diffused oxygen and any excess molten metal are easily dissolved from the surface of the deoxidized tantalum metal body with a suitable inorganic solvent which is inert to the tantalum. With this improved technique the tantalum body need not be pulverized during the deoxidation, unless it is desired to do so.

**References**

(1) Baldwin, W.M., Fuchs, E.O., Sharp, D.J. and Swisher, J.H.; U.S. Patent 3,697,255; October 10, 1972; assigned to Western Electric Company, Inc.

## TELLURIUM HEXAFLUORIDE

### Removal from Air

A process developed by D.R. Vissers, M.J. Steindler and J.T. Holmes (1) is one in which tellurium hexafluoride and fluorine are removed from a gas by passing the gas through a fluidized bed of activated alumina to remove the greater portion of fluorine and thereafter passing the effluent from the fluidized bed through a packed bed of activated alumina to remove the tellurium hexafluoride.

A continuous fluoride volatility process has been studied at the Argonne National Laboratory to separate uranium and plutonium from spent nuclear fuel elements. During the process, in which fuel elements are oxidized and later fluorinated to produce volatile fluorides, tellurium hexafluoride as well as other fission product fluorides must be separated for storage and eventual disposal. Because the gas stream which contains the tellurium hexafluoride will also contain large amounts of elemental fluorine, systems designed to eliminate the tellurium hexafluoride must also be able to handle the fluorine present. Wasted disposal and storage are two problems which confront the designers of nuclear fuel reprocessing plants. While moderately large volumes of liquid waste can be tolerated because there are relatively few nuclear reactors operating in the United States, a better solution must be found in order to accommodate the much greater volumes of waste which will result from the operation of the next generation of nuclear reactors. To that end sorption of fission product gases or other radioactive gases onto solids which have a relatively small volume and are easily stored for long periods of time is one answer to the waste disposal problem.

### References

(1) Vissers, D.R., Steindler, M.J. and Holmes, J.T.; U.S. Patent 3,491,513; January 27, 1970; assigned to U.S. Atomic Energy Commission.

## TETRABROMOMETHANE

### Removal from Air

A process developed by R. Dietrich (1) involves the removal of gaseous tetrabromomethane from gases which contain only a relatively slight amount. This process serves primarily for air purification. Technically, it is connected with the manufacture of copying materials in which tetrabromomethane is used as a radical former. It is particularly advantageous for the purification of exhaust air from photocopying devices wherein solid or dissolved tetrabromomethane is processed.

Processes are known for manufacturing copies and stencils using formation of free radicals from halogen compounds to develop colored or polymer reaction products. Preferred radical formers are short chain polyhalogen compounds particularly tetrabromomethane. An explicit description of such a process is given, e.g., in U.S. Patent 3,042,519.

The high volatility and toxicity of the tetrabromomethane is a hindrance to the practical use of this process which is to produce higher quality copies. At any one place during the manufacture of the copying material or its processing, tetrabromomethane must be passed on to or into the copying material before exposure to light. In this way a certain part of the tetrabromomethane is constantly being evaporated in the surrounding gas phase. Due to the toxicity of the tetrabromomethane this quantity must be kept low. If this is not directly successful, then care must be taken that the operator does not inhale any air containing tetrabromomethane.

It is known, for example, from the French Patent 1,321,064 that the tetrabromomethane can be brought into the coating shortly before the copying material is exposed to light. In this manner, the copying procedure can be carried out in a closed copying device from which the tetrabromomethane cannot leak. Within a closed device, excess tetrabromomethane can be driven out by heating the copying material. So far, no solution has been found for the problem of disposing of the gaseous tetrabromomethane which is closed inside the device.

Attempts to remove gaseous tetrabromomethane from other gases by adsorption to solids having a large surface area or by absorption in suitable difficultly volatile solvents have not been successful up to now. The required final low concentration of less than about 0.3 part per million parts by volume of air has not been attained with these solvents.

Concentrations lower than 0.3 ppm are definitely harmless, concentrations only very slightly higher can be smelled and are unpleasantly felt, and concentrations which are much higher are injurious to health. The range of concentration of tetrabromomethane in the air which occurs at different stages of the process in question is very wide. It ranges from a few parts per million when the tetramomethane evaporates from individual sheets of coated copying material, to some tenths of a percent by volume after many copies have been produced in compact copying devices. Accordingly, a process for the removal of gaseous tetrabromomethane should preferably have a wide sphere of operation. However, none of the processes hitherto examined have been entirely successful. Therefore, it is an object of the process to provide for the removal of gaseous tetrabromomethane from gases which operates effectively both with high and low concentrations, and which further provides purification of contaminated air to a level of less than about 0.3 ppm of tetrabromomethane.

The decontaminating process is characterized by the fact that the gas containing the tetrabromomethane is washed with a liquid which contains at least one aliphatic or araliphatic amine or an amine closed to form a ring and/or an amino alcohol with up to 20 carbons atoms in such a way that, in a conventional manner, an extensive surface of contact between the washing liquid and the gas phase is provided.

The process is based on a very fast acting chemical reaction between tetrabromomethane and the amine group, the mechanism and reaction products of which are not yet exactly known. The fact that the reaction concerned is a chemical reaction means that the washing liquid cannot be regenerated, as is possible in an adsorption process. The process is not limited to air purification, but can also be applied to other inert gases which contain tetrabromomethane.

#### References

(1)  Dietrich, R.; U.S. Patent 3,437,429; April 8, 1969; assigned to Kalle AG, Germany.

## TEXTILE INDUSTRY EFFLUENTS

The general problem of pollution control in the textile industry has been reviewed in some detail in a book by H.R. Jones (1).

#### References

(1)  Jones, H.R., *Pollution Control in the Textile Industry,* Park Ridge, N.J., Noyes Data Corp. (1973).

## THIOSULFATES

### Removal from Water

A process developed by C.J. Armstrong and A. Cronig (1) involves a method and apparatus for the control of the contamination level of photo washwater supporting an environment for ionic substitution of simple thiosulfate salts in the water with complex argento thiosulfates on the photographic emulsion surface. A water purifier is placed in closed loop with the wash tank to control the level of contamination.

A process developed by P. Urban (2) is one in which a water-soluble inorganic thiosulfate compound is reduced to the corresponding sulfide compound by contacting an aqueous solution of thiosulfate compound and hydrogen with a catalyst, comprising a catalytically effective amount of cobalt sulfide combined with a porous carrier material, at reduction conditions. Principal utility of this treatment procedure is associated with the cleanup or regeneration of aqueous streams containing undesired thiosulfate compounds so that they can be reused in the process which originally produced them or discharged them into a suitable sewer without causing a pollution problem.

Key feature of the disclosed method is the use of a unique catalyst which has extraordinary activity for converting thiosulfate to sulfide in an aqueous solution when hydrogen is utilized as the reducing agent.

The reader is also referred to the section of this handbook on "Silver, Removal from Water" for a discussion of the treatment of silver thiosulfate solutions by chlorination whereby the silver is precipitated as recoverable sludge and the thiosulfate ion is destroyed.

### References

(1)  Armstrong, C.J. and Cronig, A.; U.S. Patent 3,531,284; September 29, 1970; assigned to Itek Corp.
(2)  Urban, P.; U.S. Patent 3,709,660; January 9, 1973; assigned to Universal Oil Products Company.

# TIN

### Removal from Water

A process developed by R.C. Williamson (1) relates to the removal of tin and fluoride from aqueous solutions of the same.  Certain industrial operations generate aqueous waste that contains both tin and fluoride in solution.  One of these is electrolytic tinplating from halogen tin lines.  In such operations, wash solutions and other aqueous wastes are produced from rinsing sprays, the scrubbing of effluent fumes, the liquid effluent from tin recovery operations, and from other sources.

Both tin and the fluoride must be removed from or substantially reduced in such waste solutions, for two reasons.  In the first place the loss of large quantities of recoverable tin is economically unacceptable and in the second place the fluoride is a pollutant that cannot be pumped into streams and rivers in large quantity.

In the past, it has been conventional practice to remove and recover tin from aqueous solutions of this type by treating the solutions with sodium carbonate so as to precipitate the tetravalent tin solute as stannic hydroxide.  However, such treatment was ineffective to precipitate or otherwise remove the fluoride from the solution.  A separate treatment had to be provided for fluoride removal.  Even if a single reagent could have been found that would precipitate both the tin and the fluoride, the recoverable stannic hydroxide sludge would be greatly diluted by precipitated fluoride and would be of correspondingly decreased market value.  Accordingly, it is an object of the process to provide methods for the removal both of tin and of fluoride from aqueous solutions of the same, while at the same time recovering a tin concentrate of desirably high tin content.

The process is based on the discovery that both tin and fluoride can be removed from aqueous solution by precipitation with calcium hydroxide and that, moreover, the precipitation can be carried out stepwise in such a way that in at least one early precipitation stage a greater percentage of the original tin solute can be precipitated at fairly low pH compared to that of the original fluoride solute, while during a later precipitation stage at fairly high pH, a greater proportion of the original fluoride solute can be precipitated than of the original tin solute.  In other words, the process can be used to produce first a tin-rich and second a fluoride-rich precipitate.  The tin-rich precipitate is valuable for recovery of its tin values. The fluoride precipitate assures that the liquid can be discharged without giving rise to objectionably high levels of fluoride contamination.

The process is best adapted for the treatment of large volumes of water in which the tin and fluoride are in relatively low concentration.  For relatively small volumes of highly concentrated tin solute, it is preferable to rely on tin recovery by precipitation with sodium carbonate if fluoride is present only at such a low level as not to give rise to undesirable fluoride contamination.

Figure 134 is a flow diagram of this process. As shown, the wastewater containing tin and fluoride in solution is introduced to the cycle through a feed conduit **1** and proceeds into a first mixing chamber **3**. In this mixing chamber a rotor **5** vigorously agitates the feed water in admixture with calcium hydroxide supplied through a lime conduit **7** and air is supplied through an air conduit **9**. Relatively heavy particles, such as sand and the like introduced with the lime, may settle out in the mixing chamber, and for this purpose a valved cleanout conduit **11** is provided for removing and discarding such material.

Wastewater from the mixing chamber continuously overflows over a horizontal weir **13** into a relatively large settling chamber **15**. This settling chamber may be of any of a variety of constructions, either circular or rectilinear or the like, and in the figure is provided at its end remote from weir **13** with a launder **17**. Liquid that has been treated with lime and air overflows the weir and moves slowly toward the launder, a relatively large proportion of stannic hydroxide precipitate and a relatively small proportion of calcium fluoride precipitate, plus any undissolved calcium hydroxide, settling out in the settling chamber and collecting on the bottom of the settling chamber where it moves by gravity or may be propelled toward a sump **19** for the collection of first stage precipitate.

This first stage precipitate may then be removed either continuously or batchwise through a valve controlled discharge conduit **21** in the form of a tin-rich sludge. The sludge flows through the conduit to a centrifugal separator **23** in which at least most of the water is removed from the sludge. The relatively dry sludge is discharged either continuously or batchwise through a discharge conduit **25**, while the water from the separator is returned to feed conduit **1** through a return conduit **27**.

**FIGURE 134: PROCESS FOR SIMULTANEOUS REMOVAL OF TIN AND FLUORIDES FROM AQUEOUS SOLUTION**

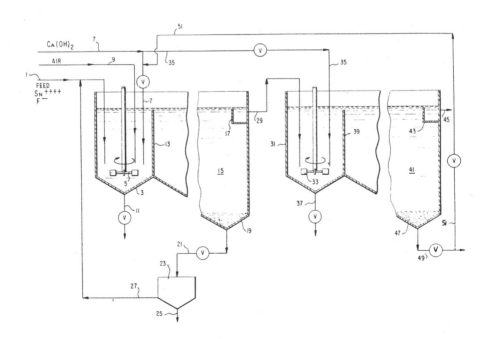

Source: R.C. Williamson; U.S. Patent 3,284,350; November 8, 1966

This water will contain some fine precipitate not removed in the centrifugal separator, as well as various products or reagents of the reaction that must be recycled either for economical tin recovery or for precipitation to avoid pollution. Accordingly, very little water leaves the cycle through discharge conduit 25. Substantially all of the water, therefore, proceeds from the first stage at the left of the figure to the second precipitation stage at the right of the figure. Specifically, water with most of the dissolved tin removed leaves launder 17 through discharge conduit 29 and proceeds to a second stage mixing chamber 31, in which a rotor 33 agitates the feed liquid into intimate association with further calcium hydroxide introduced into the second stage mixing chamber through a valved lime conduit 35 that is a branch of conduit 7. As before, coarse solids introduced into the mixing chamber can be cleaned out through a valved cleanout conduit 37.

As described in the first stage precipitation, so also in the second stage precipitation the liquid at the top of mixing chamber 31 continuously overflows over a weir 39 into a second stage settling chamber 41 provided with a launder 43 at its end opposite weir 39. Water is continuously removed from the launder through a discharge conduit 45. This water is sufficiently free from dissolved fluoride and other contaminants to be sewered or otherwise disposed of in streams, rivers, settling ponds and the like.

A second stage precipitate collects in the sump 47 of the second stage settling chamber and may be continuously or batchwise removed through a valved discharge conduit 49. This sludge that precipitates in the second stage is fluoride-rich as distinguished from the tin-rich precipitate of the first stage of precipitation. This is to say that in the first stage, a greater percentage of the original tin is precipitated than of the original fluoride, while in the second stage a greater percentage of the original fluoride is precipitated than of the original tin. Preferably, a greater absolute quantity of tin is precipitated in the first stage than in the second stage and a greater absolute quantity of fluoride is precipitated in the second stage than in the first stage.

The solids withdrawn through discharge conduit 49 will also contain a minor proportion of tin. In certain cases this low tin sludge may nevertheless be marketable for its tin content. Also, it is desirable to recycle a portion of this second stage sludge to the first stage precipitation, through a valve return conduit 51. This second stage sludge will contain some undissolved calcium hydroxide and some recoverable stannic hydroxide and the recirculation of at least a portion of the second stage sludge performs the useful function of reducing calcium hydroxide costs as well as recovering tin and reducing the quantity of dumped solids. The recirculation of second stage precipitate to the first stage precipitation also furthers the separation of stannic hydroxide and calcium fluoride. The following is a specific example of the conduct of this process.

*Example:* Wastewater at a temperature of 25°C. is introduced at a flow rate of 100 gallons per minute through conduit 1. This wastewater has a pH of 3.7, a total tin concentration of 0.13 ounce per gallon, and a total fluoride concentration of 0.173 ounce per gallon. It is admixed with 75 cubic feet per minute of air introduced through conduit 9 and 0.13 ounce of lime per gallon of feed is introduced in the form of a calcium hydroxide slurry through conduit 7. The treated feed has an average dwell time of 20 minutes in mixing chamber 3 and an average dwell time of 40 minutes in settling chamber 15.

The pH in the settling chamber is 5.6. The sludge is removed through conduit 21 at a rate of 1.4 pounds per minute on a dry solids basis and contains 69.2% of the original tin, as stannic hydroxide, and 12.3% of the original fluoride content as calcium fluoride. The sludge itself is 40% tin, that is 63% stannic hydroxide.

The calcium fluoride precipitated in the first stage is 19% of the sludge by weight, while the undissolved calcium hydroxide is 9% of the sludge by weight. The remaining 10% of the sludge is made up of about half sand and other insolubles from the lime slurry, or 5% of the total weight of the sludge, while the remaining 5% of the total is a group of minor constituents such as iron, cyanides, etc. The liquid effluent from the first stage leaves through conduit 29 and has about the same flow rate as originally, that is, about 100 gallons per minute. It contains 0.04 ounce per gallon of tin and 0.152 ounce per gallon of fluoride. It is introduced into second stage mixing chamber **31** and is there admixed by agitation from rotor **33** with a slurry of calcium hydroxide at the rate of 0.26 ounce per gallon of calcium hydroxide based on the flow rate of the influent wastewater. This addition of calcium hydroxide brings the pH to 11.95 and reduces the tin concentration remaining in solution to 0.01 ounce per gallon and the fluoride concentration to 0.011 ounce per gallon.

By this time, 92.3% of the original total tin and 93.8% of the original total fluoride have been precipitated. The second stage precipitate thus contains 23.1% of the original tin and 81.5% of the original fluoride. The tin content of this second stage precipitate is 5% by weight, which is to say that the second stage precipitate contains about 8% stannic hydroxide, with the balance principally calcium fluoride and calcium hydroxide. To achieve this precipitate, the liquid in the second stage has an average dwell time in the mixing section of 20 minutes and an average dwell time in the settling section of 200 minutes. Precipitate is removed through valved conduit **49** at a rate of 3.7 pounds per minute on a dry solids basis. 20% of the precipitate is recycled through conduit **51**.

References

(1) Williamson, R.C.; U.S. Patent 3,284,350; November 8, 1966; assigned to National Steel Corp.

## TITANIUM

### Removal of $TiO_2$ from Air

A process developed by W.P. Vosseller (1) involves removing aerosols and certain obnoxious gaseous components from the effluent gases discharged from kilns used to calcine $TiO_2$ hydrate material. The calcination of titanium dioxide hydrate as normally carried out in the industry is performed in inclined rotating kilns, the titanium dioxide hydrate to be calcined being introduced into the upper end of the kiln and traveling slowly down the length of the kiln to the lower end from which it is discharged as calcined titanium dioxide pigment.

During its passage through the kiln the titanium dioxide hydrate is calcined by a combustible fuel burning at the lower end of the kiln. During calcination gases are formed in the kiln in part, by combustion of the fuel and in part by the evolution of water vapor and sulfur oxides from the titanium dioxide during its heat treatment. A typical composition of calciner exhaust gases (before treatment) is given below.

### Analysis of Effluent Gases of a $TiO_2$ Calciner*

| Materials | Percent by Volume |
|---|---|
| $N_2$ | 54.0 |
| $H_2O$ | 35.0 |
| $O_2$ | 7.0 |
| $CO_2$ | 4.0 |
| $SO_3 + SO_2$ | 0.3 |
| $TiO_2$ (grains/dry scf) | 0.45 |

*Calculated from material balance.

Heretofore it has been the practice to cool these effluent kiln gases, by passing them through a succession of spray towers and/or electrostatic precipitators, so as to remove the sulfur oxide components and in particular $H_2SO_4$ before exhausting the gases to the atmosphere. However, the removal of the liquid aerosols, including liquid $H_2SO_4$ and water, formed as a result of cooling the kiln gases has been quite incomplete. In carrying out the method of this process the effluent gases from a calciner are piped or otherwise conveyed into a multistage filtering system which, in essence, comprises two filtering areas or chambers arranged in sequence each provided with a filter arranged so that the gases will pass therethrough; and with means for spraying a liquid onto the upstream side of the first filter to wash the aerosols and gaseous components therethrough in a direction concurrent with the gas flow. Each filtering area or chamber is provided also with an outlet on the downstream side of its respective filter for draining off the liquids and/or solids burdened liquids which collect on the downstream side thereof.

The filters used may be in any suitable form, i.e., substantially flat pads, filter candles or a combination thereof, and are formed of a liquid repelling material and in particular hydrophobic fibers such as silicone treated glass wool or preferably a synthetic, hydrophobic polyester fiber such as Dacron (polyethylene terephthalate), Terylene or the like. While the size of the fibers is not critical it is preferred to use fibers having a diameter less than about 30 microns.

### Removal of TiCl$_4$ from Air

A process developed by E.O. Kleinfelder and H. Valdsaar (2) involves removing small amounts of titanium tetrachloride and other chloride impurities from waste gas produced during the chlorination of a titaniferous ore by first acid scrubbing the gas with sulfuric acid of 75 to 95 weight percent concentration and then scrubbing with water to produce a clear gas that may be vented to the atmosphere without fuming.

A process developed by S.F. Brzozowski (3) is one in which waste gas from the chlorination of titaniferous ores is treated with steam before being vented and burned. During extended periods of titanium tetrachloride production, it has been found that a gelatinous substance accumulates in the waste gas blower. Such accumulation eventually blocks the blower and requires that it be shut down. When such blockage occurs, the entire chlorination process also must be shut down.

When the gelatinous-type material found in the blower was analyzed, it was found that it was composed chiefly of the hydrolysis products of the residual metal halides, e.g., titanium, silicon, etc., left in the waste gas stream. It has now been discovered that formation of the aforementioned gelatinous substance, which is extremely sticky, can be avoided by treating the waste gas stream before the scrubbing step with steam. The overall disposal process is shown in the flow diagram in Figure 135.

FIGURE 135:  PROCESS FOR TREATMENT OF WASTE GAS FROM THE
CHLORINATION OF TITANIUM ORE

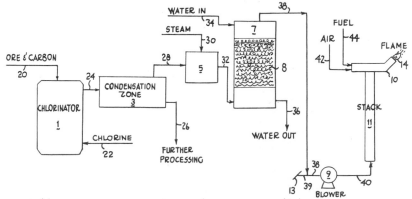

Source:  S.F. Brzozowski; U.S. Patent 3,615,163; October 26, 1971

Referring to the figure, ore, carbon and chlorine are introduced into chlorinator **1** as indicated by lines **20** and **22** respectively. Product vapors from the chlorinator are removed from chlorinator **1** and forwarded to condensation zone **3** as indicated by line **24**. The product from condensation zone **3** is removed therefrom, as indicated by line **26**, and subjected to further processing, e.g., chemical purification and rectification, in a known manner. Uncondensed vapors (the waste gas stream) are removed from the condensation zone as indicated by line **28** and are introduced into vessel **5** to which steam is also introduced as indicated by line **30**. Vessel **5** can be of any suitable acid resistant vessel capable of handling the volumes of waste gas and steam introduced therein.

Steam is introduced into the vessel in a manner which results in intimate contact between the waste gas stream and steam. Typically, several steam lines, preferably with sparging nozzles, are connected to the vessel so as to insure intimate waste gas-steam contact. The resulting vapors are removed from the vessel, as indicated by line **32**, and introduced into scrubber **7** which, as shown, is a countercurrent packed water scrubber. As shown, the vapors from the vessel enter the bottom of the scrubber and pass upwards through packing **8** countercurrent to a downwardly flowing stream of water charged to the scrubber from a source not shown through line **34**. Scrubbed waste gas vapors are removed from the scrubber, as indicated by line **38**. Water is removed from the scrubber as indicated by line **36**.

The amount of steam with which the waste gases is treated can vary, but typically will be that amount which is necessary to avoid formation of the gelatinous material in the blower following scrubbing. Usually this amount will be that amount which is sufficient to hydrolyze any residual hydrolyzable titanium tetrachloride contained in the waste gas stream. Preferably, sufficient steam is used to hydrolyze all of the residual hydrolyzable metal halide components of the waste gas stream, e.g., the chlorides of silicon, titanium, chromium, tin, etc. Generally, because of the minimal cost of plant steam, which is usually available in the immediate plant area, a stoichiometric excess (based on residual $TiCl_4$) of steam, e.g., 1 to 10 times the theoretical stoichiometric is utilized.

The exact manner by which steam treatment of the titanium tetrachloride waste gas stream in the manner described herein prevents formation of the gelatinous residue previously described is not exactly known. It is believed, however, that the steam, by virtue of the existence of fine water vapor particles present therein, effects more intimate contact (and therefore more complete hydrolysis) with the finely divided metal halide particles suspended in the waste gas than is obtained by the water scrubber. It is also believed that the hydrolysis product of, for example, titanium tetrachloride with steam as opposed to water, is different, e.g., less sticky and gelatinous.

The type of steam (temperature and pressure) used can be any conveniently available plant steam line. Typically, the steam pressure will vary from 0 to 100 psig, more typically from 10 to 50 psig, the temperatures of which will vary respectively from 212° to 330°F., more typically from 240° to about 280°F. Usually, steam having temperatures below 350°F., more usually below 300°F. is used. The aforementioned properties are those of saturated steam.

In carrying out the process it is unnecessary to heat the waste gas steam, either before, during or after steam treatment (but before flaring) to temperatures of 200°C. (392°F.) or more, in accordance with Japanese Patent 7,414, for the reason that the hydrolysis products produced in vessel **5** are removed from the system by means of water scrubber **7**. The process is, thus, easily adaptable to existing facilities and more economic in the use of readily available plant steam and low pressure equipment.

The water used in scrubber **7** is typically untreated plant water, e.g., river or well water, which is used at their respective prevailing temperatures. The amount of water used in the scrubber is that amount which is sufficient to obtain intimate contact between it and the waste gas stream and effectively remove all the noxious and otherwise detrimental components of the stream, e.g., chlorine and the hydrolysis products of the residual metal halide components of the stream. Typically, the ratio of water to waste gas will be from about 15 to 20 pounds of water per pound of gas. The type of packing used in scrubber **7** can be of any conventional packing, such as Raschig rings, Berl saddles, glass, crushed stone, stoneware rings, etc. Although a packed scrubber is shown in the figure, any type of water scrubber which provides intimate water-gas contact can be used. As the steam treated waste gas passes through scrubber **7**, the steam introduced into the waste gas in vessel **5** is condensed and flushed out together with the residual metal halide hydrolysis products by the scrubbing water.

The treated waste gas, now consisting principally of carbon dioxide and carbon monoxide, is removed from the scrubber, as indicated by line **38**, and forwarded to blower **9** wherein the waste is admixed with air in quantities that is sufficient to prevent the formation of $CO/O_2$ explosive mixtures. Air is introduced into line **38** through damper **13** and line **39** which is openly connected to line **38**. Typically, sufficient air is aspirated into the waste gas stream so that the carbon monoxide content thereof is maintained below 14 volume percent.

The operation of the blower creates a vacuum on the intake side of the blower, which expedites the movement of waste gas from the scrubber to the blower. After admixture with air, the waste gas stream is forwarded from blower **9** to vent stack **11**, as indicated by line **40**, and burned in flare **10** as flame **14**. Combustion of the combustible portion of the waste gas stream is accomplished by introducing air and a fuel, usually natural gas, into flare **10** from sources not shown through lines **42** and **44** respectively and igniting the fuel. The flame produced by the flare is typically operated at at least 1600°F., usually between 1600° and 1900°F., to insure complete combustion.

A process developed by D.N. Low (4) is one in which an anhydrous gas stream composed predominantly of inert gases and containing appreciable quantities of HCl and $TiCl_4$ is to be contacted with an aqueous base liquid in an amount in excess of that required to saturate the stream with water to thereby absorb the HCl in the form of an HCl solution. It has been found that an improvement in such a method can be achieved leading to elimination or at least substantial reduction in the opacity of the gas upon being vented to the atmosphere.

The improvement involves evaporating an initial and restricted quantity of water into the stream of anhydrous gas at least one-twentieth second prior to the regular scrubbing operation in which the gas is contacted with a saturation amount of water. The initial quantity of water evaporated into the stream must be less than that required to saturate the stream at the existing temperature and pressure conditions but greater than that stoichiometrically required to react with the $TiCl_4$ in the stream to form $TiO_2$.

### References

(1)  Vosseller, W.P.; U.S. Patent 3,250,059; May 10, 1966; assigned to National Lead Company.
(2)  Kleinfelder, E.O. and Valdsaar, H.; U.S. Patent 3,564,817; February 23, 1971; assigned to E.I. du
       Pont de Nemours and Company.
(3)  Brzozowski, S.F.; U.S. Patent 3,615,163; October 26, 1971; assigned to PPG Industries, Inc.
(4)  Low, D.N.; U.S. Patent 3,690,041; September 12, 1972; assigned to E.I. du Pont de Nemours and Co.

# TRIARYLPHOSPHATES

### Separation from Water

A process developed by G.E. Schmauch and J.L. Stolz (1) is one in which water contaminated with oil heavier than water and with water-soluble components, such as an air compressor condensate in the form of an emulsion containing such contaminants, is treated for recovery of the oil and for the removal of such water-soluble components.

The successive steps of the operation by which this is accomplished follow:

(a) Deaerating the contaminated condensate emulsion, as needed, in a preseparating zone for release of excessive amounts of incorporated air;

(b) Adding a chemical deemulsifying agent to the relatively air-free emulsion;

(c) Mildly agitating the mixture in a mixing and settling zone by injection of gaseous material, such as air, in an amount sufficient only to promote mixing and to accelerate the resultant reaction, while permitting concomitant partial settling of the heavier oil-containing fraction;

(d) Coalescing and filtering the deemulsified mixture in a coalescing zone to remove any heavy sludge;

(e) Separating the remaining liquid in a settling zone into a lower layer containing the oil component and an upper layer containing the water-soluble components;

(f) Separately withdrawing the heavier recoverable oil fraction and the lighter, oil-free, contaminated water fraction from the settling zone; and

(g) Adsorbing the water-soluble components from the lighter, contaminated water fraction in an adsorbing zone.

In a typical operation, a condensate emulsion containing triarylphosphate in an amount between about 0.1 and 0.5 weight percent is treated with about 0.2 weight percent of calcium chloride as the deemulsifier.

A process developed by J.M. Binder (2) treats oil contaminated water in a very similar manner to that previously described. The steps in this modified process are as follows:

(a) Deaerating the contaminated condensate emulsion, as needed, in a preseparating zone for release of excessive amounts of incorporated air;

(b) Maintaining a surge supply of the relatively air-free emulsion in a holdup zone;

(c) Adding a chemical deemulsifying agent to the degassed fluid condensate;

(d) Distributing the deemulsifying agent throughout the condensate to accelerate the resultant reaction;

(e) Coalescing and filtering the deemulsified mixture in a coalescing zone to to remove any sludge; and

(f) Separating the remaining liquid in a settling zone into a lower layer comprising recoverable oil and an upper layer comprising water contaminated by the water-soluble components.

The sludge is separately withdrawn from the coalescing zone and may be recycled, if desired. The contaminated water fraction is treated in an absorbing zone containing a body of adsorbent material, so as to remove the water-soluble components. In a typical operation, a condensate emulsion containing triarylphosphate in an amount between about 0.1 and 0.5 weight percent is treated with about 0.2 weight percent of calcium chloride as the deemulsifier.

**References**

(1) Schmauch, G.E. and Stolz, J.L.; U.S. Patent 3,363,399; January 16, 1968; assigned to Air Products and Chemicals, Inc.
(2) Binder, J.M.; U.S. Patent 3,420,039; January 7, 1969; assigned to Air Products and Chemicals, Inc.

## URANIUM

### Removal from Water

A process developed by J.J. Odom, Jr., T.P. Shumaker and D.B. Griffin (1) involves the removal of uranium from wastewater systems by treatment with a phenolic aldehyde resin solubilized by alkali to effect removal of uranium salts. The conventional processing of uranium compounds from naturally occurring uranium minerals such as pitchblende, uranite, carnotite and others is designed to obtain uranium compounds having very low solubility in water. These uranium compounds are not completely insoluble, however, and since large amounts of water are used in their processing, it is not surprising to find uranium compounds in industrial wastewater systems.

In fact, the industrial processing of uranium has thus resulted in contamination of wastewater streams with small but significant amounts of both soluble uranium compounds and ultrafine particles of uranium. An important step in the processing of uranium compounds is the conversion of naturally occurring common uranium minerals like pitchblende, uranite, carnotite, and others into their respective soluble derivatives. These soluble derivatives are separated from the remainder of the raw ore and uranium is precipitated as an insoluble salt. This separation phase of the processing is frequently preceded by a roasting process. One of the methods employed in ore separation is the so-called alkaline leach.

To attain maximum effectiveness, the leaching operation, which is usually a carbonate leaching, is performed under carefully controlled conditions of carbonate concentration, temperature and pressure. Recovery of the uranium compounds from the carbonate leach liquors is accomplished by one of two procedures. In the first procedure, the leach liquor is acidified to a low pH and boiled to remove carbon dioxide. The uranium salts are then precipitated by raising the pH to 7 or above with sodium hydroxide or ammonia.

In the second procedure, sodium hydroxide is added directly to the carbonate solution to raise the pH above 11, whereupon reasonably complete precipitation of uranium occurs. Where precipitation is effected by addition of sodium hydroxide, sodium polyuranate salts are obtained, while the use of ammonia as a precipitant results in the production of complex salts whose structure is not clearly defined.

The compounds obtained by either of these commonly used uranium processing procedures have a low solubility in water, but are partially soluble, thus raising wastewater contamination problems. The presence of even small quantities of uranium compounds in water systems is most undesirable because of the high chemical toxicity of these compounds which manifests itself in kidney damage and in acute necrotic arterial lesions.

There are also compelling economic reasons for the recovery of residual quantities of uranium salts present in wastewater systems, and particularly from wastewater systems from uranium processing plants. Typical samples taken from the wastewater exiting uranium processing operations have been found to contain approximately 0.4 g. of uranium per liter of water. The following is one specific example of the operation of this process.

*Example:* An industrial, ammoniacal uranium waste effluent having a pH of 10 and a uranium content of 0.4 g./l. (expressed as uranium ion and determined by the method described in Scott's *Standard Methods of Analysis,* Fifth edition, volume 1, pages 1017 to 1022) was treated with a solution of 0.8 g./l. of phenol-formaldehyde resin. A flocculant precipitate was formed almost immediately. The effluent was subsequently filtered through a normal plate and frame filter press. The uranium content of the effluent was reduced to 0.02 g./l.

The preparation of the phenol-formaldehyde resin used may be described in further detail as follows. A clean kettle is charged with 150 pounds of water, 15 pounds of methanol and 150 pounds of phenol. The kettle is cooled with water to prevent heat exotherm during the addition of 200 pounds of 44% formaldehyde.

Then 85 pounds of 50% caustic sodium hydroxide solution are added to the reaction vessel. The addition is controlled to prevent the temperature of the reaction mixture from rising above 50°C. After all the caustic soda is added, temperature is allowed to rise to 96°C., and held there until the mixture reaches an extrapolated viscosity of F to G (Gardner). Thereafter the batch is cooled rapidly to 20° to 25°C. The reaction product exhibits the following properties: nonvolatiles, 40 to 42%; specific gravity, 1.17 to 1.19; viscosity, D to M (Gardner); pH, 10.5 to 12.5; and free formaldehyde, 0.5% maximum.

**References**

(1) Odom, J.J., Jr., Shumaker, T.P. and Griffin, D.B.; U.S. Patent 3,634,230; January 11, 1972; assigned to Reichhold Chemicals, Inc.

## VANADIUM

The major sources of vanadium emissions are the metallurgical processes producing vanadium metal and concentrates; the alloy industry; the chemical industry; power plants and utilities using vanadium-rich residual oils and; to a lesser extent, the coal and oil refining industries. Vanadium production is concentrated in the states of Colorado, Utah, Idaho and New Mexico, while the highest concentration of industries producing vanadium chemicals is found in New Jersey and New York. Domestic vanadium consumption has more than doubled since 1960, and the domestic mine production of ores and concentrates increased from 1,482 short tons of vanadium in 1945 to 5,226 short tons in 1965.

In communities in the United States in which vanadium concentrations were measured, the average values (quarterly composites) ranged from below detection (0.003 $\mu$g./m.$^3$) to 0.30 (1964), 0.39 (1966), and 0.90 (1967) $\mu$g./m.$^3$. Vanadium is toxic to humans and animals, expecially its pentavalent compounds. Exposure of humans through inhalation of relatively low concentrations (less than 1,000 $\mu$g./m.$^3$) has been found to result in inhibition of the synthesis of cholesterol and other lipids, cysteine, and other amino acids, and hemoglobin. Low concentrations also act as strong catalysts on serotinin and adrenaline. Chronic exposure to environmental air concentrations of vanadium has been statistically associated with the incidence of cardiovascular diseases and certain cancers.

Human exposure to high concentrations of vanadium (greater than 1,000 $\mu$g./m.$^3$) results in a variety of clinically observable adverse effects whose severity increases with increasing concentrations. These effects include irritation of the gastronintestinal and respiratory tracts, anorexia, coughing (from slight to paroxysmal), hemoptysis, destruction of epithelium in the lungs and kidneys, pneumonia, bronchitis and bronchopneumonia, tuberculosis, and effects on the nervous system ranging from melancholia to hysteria.

### Removal of Vanadium from Air

No information has been found on abatement of air pollution by vanadium emitted from its production or product sources, according to Athanassiadis (1). When additives such as magnesium oxide are used in oil fired burners, a chain of reactions occurs resulting in the reduction in the amount of fine particulates and amounts of vanadium escaping to the atmosphere.

The portion of total particulates in the less than 10$\mu$ range may be reduced from 60 to 40% in the case of high vanadium content oils. This is significant, since particulate size distribution is an important parameter of abatement efficiency. Centrifugal collectors are preferred over electrostatic precipitators because they reduce the difficulties associated with the acid character of the ash, especially where bag filters and scrubbers are used. However, for any given particle size distribution, the efficiency of various centrifuges may vary as much as from 50 to 65% and from 70 to 85%, around the particle size range of

5 and 10µ, respectively (2). The use of efficient fly ash control equipment in modern coal fired power plants may considerably reduce the emission of particulates containing vanadium. The control equipment most used are cyclones and electrostatic precipitators. When additives are used, which result in the formation of larger ash particles, cyclones are more efficient and economical to use than electrostatic precipitators (3).

### References

(1)  Athanassiadis, Y.C., "Air Pollution Aspects of Vanadium & Its Compounds," *Report PB 188,093;* Springfield, Va., National Tech. Information Service (September 1969).
(2)  Burdock, J.L., "Fly Ash Collection from Oil-Fired Boilers," Paper presented at the 10th meeting of the New England Air Pollution Control Association, Hartford, Connecticut (1966).
(3)  Gerstle, R.W., et al, "Air Pollutant Emissions from Coal-Fired Power Plants: Report No. 2, *J. Air Pollution Control Assoc.* 15 (2), 59 (1965).

# VEGETABLE PROCESSING INDUSTRY EFFLUENTS

A review of pollution control problems and techniques in the fruit and vegetable processing industry has been published by H.R. Jones (1).

### References

(1)  Jones, H.R., *Waste Disposal Control in the Fruit and Vegetable Industry,* Park Ridge, N.J., Noyes Data Corp. (1973).

# VIRUSES

### Removal from Water

The general topic of virus removal from water and wastewater has been discussed by W.A. Drewry (1). Conventional wastewater treatment processes, including conventional chlorination practices, have proved to be less than satisfactory for virus removal (2) (3) (4) (5) (6) (7) (8) (9) (10) (11). Even absorption on activated carbon columns has not proved 100% effective for virus removal from wastewater (12). These observations along with recognition of the increasing contamination of water supply sources and recreational waters by the expanding population and the widespread use of private and public swimming pools, etc., have increased considerably the possibilities for virus transmission by the water route.

A process developed by S.P. Felix (13) utilizes a special filtering apparatus for the removal of virus from contaminated water. Since viruses are difficult to detect, their removal cannot be determined with certainty. Moreover, chlorination is not always effective in destroying viruses as contrasted with bacteria. One effective technique for virus removal from water involves chemical flocculation which results in the formation of a metal-virus complex which aggregates to form a separable precipitate.

The formation of the metal-virus complex appears to be the result of a metal cation-protein ion reaction. The use of iron or aluminum salts, such as iron sulfate or aluminum sulfate, for the formation of the metal virus salts has been found to provide satisfactory results. The virus is not destroyed by the flocculant but is concentrated in the floc.

In accordance with prior methods of removing the virus by flocculation, after the metal-virus complex was formed and aggregated to form a precipitate, the precipitate was detained in a coagulating basin to allow the coagulant to form aggregates of a sufficient size so that they would settle out by gravity.

In this settling method, the size of the coagulating and settling tank must be large enough to slow the liquid velocity so that settling can take place. Thus, in the case of the public water supplies where large volumes of water must be handled, coagulating and settling tanks of extremely large sizes are required. Moreover, detention times of approximately three hours are normally involved in these settling methods.

It is an object of this process to provide a filtering process and apparatus involving the use of a filter, such as the diatomaceous earth type, for the removal of virus from water. The method makes use of the fact that the metal-virus complex formation is an ionic reaction which occurs instantaneously. Broadly speaking, the process involves admixture of the raw water and the flocculating agent in a detention tank for a very short time, only sufficient to insure uniform admixture of the contents and passing the liquid from the detention tank through a diatomaceous earth filter which removes the floc containing the virus. The detention time need only be sufficient to permit the metal-virus complex to aggregate to a size large enough to be filterable by the diatomaceous earth filter, which can remove particles of very small sizes (though it would not remove the unmodified virus).

**References**

(1) Drewry, W.A., "Virus Removal — Water and Wastewater," *Report PB 210,857,* Springfield, Va., National Tech. Information Service (May 1972).
(2) Burns, R.W. and Sproul, O.J., "The Virucidal Effects of Chlorine in Wastewater," *New England Water Works Journal* 31, 26 (1967).
(3) Bloom, H.H., et al, "Identification of Enteroviruses in Sewage," *Jour. Infectious Diseases,* 105, 61, (1959).
(4) Clark, N.A., et al, "Removal of Enteric Viruses from Sewage by Activated Sludge Treatment," *American Journal of Public Health,* 51, 1118, (1961).
(5) Mack, W.N., et al, "Enterovirus Removal by Activated Sludge Treatment," *Journal Water Pollution Control Federation,* 34, 1133 (1962).
(6) Bush, A.F. and Isherwood, J.D., "Virus Removal in Sewage Treatment," *Jour. San. Engr. Div.,* Proc. ASCE, 92, SA-1, 99 (1966).
(7) Malherbe, H.H. and Strickland–Cholmley, M., "Quantitative Studies on Virus Survival in Sewage Purification," *Transmission of Viruses by the Water Route,* New York, Interscience Publishers, (1967).
(8) Kelly, S. and Sanderson, W.W., " The Effect of Sewage Treatment on Viruses," *Sewage and Industrial Wastes,* 31, 683 (1959).
(9) Merrell, J.C., Jr., and Katko, A., "Reclaimed Wastewater for Santee Recreational Lakes," *Journal Water Pollution Control Federation,* 38, 1310 (1966).
(10) Kelly, S., Sanderson, W.W., and Neidl, C., "Removal of Enteroviruses from Sewage by Activated Sludge," *Journal Water Pollution Control Federation,* 33, 1056 (1061).
(11) Shuval, H.I., et al, "The Inactivation of Enteroviruses in Sewage by Chlorination," *Advances in Water Pollution Research,* Vol. 3, New York, MacMillan Co. (1967).
(12) Larochelle, L.R., "Virus Removal from Wastewater by Adsorption on Activated Carbon," Unpublished Master's Thesis, University of Maine (1966).
(13) Felix, S.P.; U.S. Patent 3,214,369; October 26, 1965; assigned to DeLaval Turbine Inc.

# ZINC

The primary sources of zinc compounds in the atmosphere are the zinc, lead, and copper-smelting industries, secondary-processing operations which recover zinc from scrap, brass-alloy manufacturing and reclaiming, and galvanizing processing. Average annual production and consumption of zinc in the United States have increased steadily during this century and it is predicted that this trend will continue.

As the emission of zinc into the atmosphere in most of these operations represents an economic loss of the zinc material, control procedures are normally employed to prevent emission to the atmosphere. In those industries where zinc is a by-product, control procedures are normally employed to prevent emission to the atmosphere.

In those industries where zinc is a by-product, control procedures for zinc are not as effective and greater quantities of zinc therefore escape into the environment. It is not possible to assess fully the role of zinc and its compounds as air pollutants. Despite the fact that specific effects attributed to certain compounds of zinc have been noted, the common association of zinc with other metals, and the frequent presence of toxic contaminants (such as cadmium) in zinc materials, raise questions which have yet to be answered concerning the synergistic effects of these metals.

The most common effects of zinc poisoning in humans are nonfatal metal-fume fever, caused by inhalation of zinc oxide fumes and illnesses arising from the ingestion of acidic foods prepared in zinc-galvanized containers. Zinc chloride fumes, though only moderately toxic, have produced fatalities in one instance of highly concentrated inhalation. Zinc stearate has been mentioned as a possible cause of pneumonitis. Zinc salts, particularly zinc chloride, produce dermatitis upon contact with the skin.

Accidental poisoning of cattle and horses has occurred from inhalation of a combination of lead and zinc-contaminated air. Zinc oxide concentrations of 400 to 600 $\mu$g./m.$^3$ are toxic to rats, producing damage to lung and liver, with death resulting in approximately 10% of the cases. Although dogs and cats were found to tolerate high concentrations (up to 1,000,000 $\mu$g./day) of zinc oxide for long periods, evidence of glycosuria and damage to the pancreas became apparent. Concentrations of 40,000 to 50,000 $\mu$g./m.$^3$ of zinc ammonium sulfate produced no appreciable effects on cats.

Some evidence exists of damage to plants from high concentrations of zinc in association with other metals. No information was found on damage to materials from zinc or its compounds in the atmosphere. Measurements of the 24 hour atmospheric concentrations of zinc in primarily urban areas of the United States reveal an average annual value of 0.67 $\mu$g./m.$^3$ for the period 1960 to 1964; the highest value recorded during that period was 58.00 $\mu$g./m.$^3$, measured in 1963 at East St. Louis, Ill., according to Athanassiadis (1).

Industries discharging waste streams which carry significant quantities of zinc, according to J.W. Patterson et al (2) include zinc and brass metal works; zinc and brass plating; silver and stainless steel tableware manufacturing; viscose rayon yarn and fiber production; groundwood pulp production and newsprint paper production. In addition, recirculating cooling water systems employing Cathanodic Treatment contain zinc which is discharged during blowdown (3).

Rock (3) estimates that 10,000 tons of waste zinc per year are discharged from viscose rayon production, and that 5 tons of zinc may be used per day in a large groundwood pulp mill. Similar quantities were estimated for a typical brass company (4).

The primary source of zinc in wastewaters from plating and metal processing industries is the solution adhering to the metal product after removal from pickling or plating baths. The metal is washed free of this solution, referred to as dragout and the contaminants are thus transferred to the rinse water. The pickling process consists of immersing the metal (zinc or brass) in a strong acid bath to remove oxides from the metal surface. Finished metals are brightened by submergence in a bright dip bath containing strong chromate concentrations in addition to acid.

Plating solutions typically contain 5,000 to 34,000 mg./l. of zinc. The concentration of zinc in the rinse water will be a function of the bath zinc concentration, drainage time over the bath, and the volume of rinse water used. Zinc and brass plating solutions generally contain cyanide as zinc and copper-cyanide complexes.

Waste concentrations of zinc range from less than 1 to more than 1,000 milligrams per liter in various waste streams reported in the literature. Average values, however, seem to be between 10 and 100 milligrams per liter. Table 17 summarizes values reported for various zinc bearing wastewaters.

## TABLE 17: CONCENTRATIONS OF ZINC IN PROCESS WASTEWATERS

| Process | Zinc Concentration, (mg./l.) | | Reference |
|---|---|---|---|
| | Range | Average | |
| Metal Processing: | | | |
| Pickle baths | 4.3 – 41.4 | | 5 |
| Bright dip wastes | 0.2 – 37.0 | | 5 |
| Brass mill wastes | 40 – 1,463 | | 5 |
| Brass mill wastes | 8 – 10 | | 4 |
| Pickle bath | 0.5 – 37 | | 4 |
| Plating: | | | |
| General | 2.4 – 13.8 | 8.2 | 6 |
| General | 55 – 120 | | 7 |
| General | 15 – 20 | 15 | 8 |
| Zinc | 70 – 150 | | 5 |
| Brass | 11 – 55 | | 5 |
| Zinc | 70 – 350 | | 9 |
| Brass | 10 – 60 | | 9 |
| General | 7.0 – 215 | 46.3 | 5 |
| Silver Plating: | | | |
| Silver bearing wastes | 0 – 25 | 9 | 5 |
| Acid waste | 5 – 220 | 65 | 5 |
| Alkaline | 0.5 – 5.1 | 2.2 | 5 |
| Rayon Wastes: | | | |
| General | 250 – 1,000 | | 10 |
| General | 20* | | 11 |

* After process recovery of zinc by ion exchange.

Source: Report PB 204,521

### Removal of Zinc from Air

*Primary Zinc–Smelting Operations:* Following is a description of the system of air pollution control equipment used at a major zinc-smelting operation, as described by Y.C. Athanassiadis (1). Sinter machine exhaust gases are preconditioned to lower the electrical resistance of the gases. They are then processed in electrostatic rod curtain type precipitators, further treated in cyclone scrubbers and then vented to the atmosphere through a 402 foot chimney. Plant dusts and fumes (other than from the sinter machine) are collected in a 275,000 cfm bag collector which discharges to a 168 foot chimney.

Sinter machines, in the agglomeration of the roaster calcines, produce large quantities of zinc fume. As zinc fume has high electrical resistance, it must be preconditioned prior to processing by the electrostatic precipitators. This conditioning is achieved by treating the fumes with large quantities of finely atomized water in conditioning chambers. Water at a pressure of 500 psi and a flow of 350 gpm is used to achieve the required atomization.

Following the preconditioning, the gases and fumes pass to rod curtain type electrostatic precipitators, operating with a collection efficiency of 90%, with an associated dust loading in the exit gases of 0.05 grains per cubic foot.

Two concrete cyclone scrubbers (25 feet in diameter by 50 feet high) further treat sintering gases before they are vented to the atmosphere. Their collection efficiency is approximately 50%, so that the overall removal of fume from the sintering gases is of the order of 95%, with a concentration in stack gases of approximately 0.02 grains per cubic foot STP. For the control of dust and fume other than from the sinter machines, a central cloth bag collector system is used with a capacity of 275,000 cfm. The dust-loading entering the collector is about 4 grains per cubic foot and the collector's efficiency is 99.9%.

The dust has approximately the same composition as sinter and is returned to the sinter circuits. The collector is a 56 section Dracco with 4,480 bags (56 sections by 80 bags each, the bags measuring 8 inches by 10 feet). The discharge duct is 11 feet in diameter and leads to the inlet of a No. 19 SCLD Buffalo exhauster discharging at a 168 foot masonry chimney. The exhauster is operated at 225,000 cfm (total at high speed operation). The above described system is in use in the Josephtown zinc-smelting plant. No comparative data exist for the major primary zinc-smelting operations identified elsewhere.

The quantities of zinc dust emitted to the atmosphere by a copper-smelting plant before installation of a filtering system have been published (12). After installation of an efficient filtering system more than 35 tons of dust were recovered per day with a zinc content of approximately 30%.

*Secondary Zinc Melting Operations:* When hoods of the proper type are used, they have proved to be very effective in all three types of furnaces (Belgian, distillation and muffle). The hoods are usually placed directly above the furnace or over the charging doors of the sweat chambers (muffle furnaces). Baghouses and precipitators are the most commonly and efficiently used control devices on the abovementioned types of furnaces (12).

*Galvanizing Operations:* In job shops it is necessary to use high canopy or room type hoods. The amount of ventilation volume necessary for use with high canopy hoods as well as the size of the collector required increases considerably with the height of the hood. Low canopy hoods can be used for a galvanizing kettle when head room is not required. Due to the fact that particles range from submicron to 2 microns in size, the only efficient control devices are baghouses or high efficiency electrostatic precipitators (used in the presence of oil mists).

Cotton bags have proved effective as collectors in most galvanizing operations. The tendency of fumes to coagulate makes it difficult to clean the bags with mechanical shakers. Because of high velocities (above 2 fpm) which also make cleaning of the bags extremely difficult and because of the large exhaust volumes required, large sized baghouses are needed.

*Zinc–Alloy Sweating Operations:* The types of hooding and ventilation mentioned in a foregoing section on brass-smelting operations are also used with good results in zinc-alloy sweating operations. In low temperature operations, auxiliary hooding is usually necessary, varying with the type of sweating operation. Normally, an inlet velocity of 100 to 200 feet per minute is sufficient to prevent emissions to the atmosphere. A baghouse together with an afterburner is used to collect the dust and fumes. The maximum recommended baghouse filter velocity is 3 fpm.

*Secondary Brass Melting Operations:* The collection of zinc oxide fumes in brass melting operations is a problem both economically and technically. Most zinc oxide particles are of submicron size and can be handled by precipitators; however, zinc oxide particles emitted from brass furnaces are found in the 0.3 to $1.0\mu$ range where even the most efficient precipitators operate at low efficiency ratios. Zinc oxide fumes have been found to represent, on the average, 59% of the stack emissions from brass furnaces. The absolute amount of dust and fume from these sources varies from 0.022 to 0.771 grains per standard cubic foot at stack conditions.

The usual controls for the various types of furnaces are slag (crushed glass) and flux (borax, soda ash, etc.) covers. However, it is necessary to control the thickness of slag covers in order to minimize emissions. In addition, if the alloy contains more than 7% zinc, these covers do not sufficiently suppress emissions. On the other hand, flux covers usually destroy the furnace walls. Of all the available types of air pollution control equipment, only baghouses have proved sufficiently economical and efficient for secondary brass smelting operations. Electric precipitators have not proven entirely satisfactory for the control of zinc oxide fumes because of the following: the high resistivity of the fumes; the unavailability of small high voltage precipitators suitable for the average size of the operations in question; and the fact that at the particle size range below 0.5 micron the efficiency of electrostatic precipitators drops considerably. Scrubbers (dynamic) or washers (mechanical) have proved (a) to be ineffective in the submicron range; (b) to consume much power; (c) to be subject to mechanical wear; and (d) usually to necessitate separation of the fumes and other particulate matter. Collectors (centrifugal) are also inefficient in the submicron range. In one instance, a tested cyclone (wet) had to be replaced by baghouses in the end.

*Open-Hearth and Electric-Arc Furnaces, Steel Operations:* Metal fumes and oxides from open-hearth and electric-arc furnaces for making steel alloys amount to 99 and 75% (by weight) of the total emissions respectively. The corresponding percentages of zinc are 15 and 37. In spite of this, no air pollution control devices have been used or recommended. The alleged reason is that the absolute amounts of particulate emission have been found to be 0.00278 grain per standard cubic foot or 0.14 pound per hour (for a furnace rated at 1,000 pounds capacity). This has been considered low.

*Effectiveness:* A measure of the effectiveness of pollutant control devices employed in the zinc and related industries can be seen from Table 18, which presents data on the reduction of emissions into the atmosphere of Los Angeles County during the period 1950 to 1960 (13).

TABLE 18

| Metal | Melted or Refined (tons/month) | Reduction in Air Pollution Due to Controls (tons/month) | Net Emissions to the Atmosphere (tons/month) |
|---|---|---|---|
| Zinc (except zinc oxide production and galvanizing) | 3,000 | 24 | 2 |
| Galvanizing (zinc) | 600 | 12 | 4 |
| Copper and Bronze | 1,500 | 4 | 1 |
| Red Brass | 2,000 | 3 | 1 |
| Semired Brass | 2,000 | 18 | 2 |
| Yellow Brass | 1,200 | 33 | 2 |
| Totals (rounded) | 10,300 | 124 | 12 |

## Removal of Zinc from Water

Treatment processes employed for wastewater zinc removal may involve either destruction or recovery. The destruction process is essentially chemical precipitation, with disposal of the resultant sludge. Recovery processes include ion exchange and evaporative recovery. Barnes (6) has pointed out that recovery of plating wastes frequently proves to be more economical on an overall basis than conventional destructive treatment.

*Chemical Precipitation:* The precipitation process most frequently involves adjustment of pH to achieve alkaline conditions and precipitation of zinc hydroxide. Lime addition has been the most widely accepted method for pH adjustment, in spite of the concurrent precipitation of calcium sulfate in the presence of high sulfate levels in pickling bath wastes. The precipitation of calcium sulfate along with the zinc hydroxide increases the total amount of sludge to be disposed of. Banerjee and Banerjee (14) report the use of hydrogen sulfide at pH 2 for complete removal of zinc from an electrolytic nickel recovery system. However, this approach is not mentioned to any extent in the literature encountered.

In the case of cyanide and/or chromate in the zinc waste stream (both are frequently encountered in zinc and brass plating wastes), the waste must undergo oxidative treatment for cyanide destruction, followed by reductive treatment of hexavalent chromium before precipitation. Zinc is then removed by precipitation along with other metal species present. For a metal bearing waste containing 55 to 120 mg./l. of zinc, Nyquist and Carroll (7) reported treated effluent concentrations ranging up to 5.4 mg./l. over one month of monitoring, but with an average value less than 1 mg./l. The waste flow varied from 2,750 to 3,000 gal./hr. and required a lime dose of 56 lb./1,000 gal. The theoretical minimum dosage, based upon the stoichiometry of the chemical reaction involved in treating the waste, was 52 lb./1,000 gal.

The use of a Dorr Clariflocculator reduced zinc levels to 0.5 to 2.5 mg./l. in a 1.5 to 2.0 MGD waste stream. After installation of sand filters, the effluent zinc was reduced to 0.1 to 0.5 mg./l. Process pH was 8.5 after lime addition (15). Sharda and Manwannan (11) have reported that lime treatment reduces 20 mg./l. zinc to less than 1 mg./l. by precipitation of zinc hydroxide. Rock (3) reported 95% or greater removal of zinc from rayon acid wastes upon addition of sodium hydroxide. This treatment yielded a residual zinc concentration of 3.5 mg./l. for the initial 70 mg./l. zinc contained in the waste stream.

For a tableware plant with three waste streams, one of 5,000 gal./day with an average zinc concentration of 9 mg./l.; one of 46,000 gal./day with zinc concentration of 65 mg./l.; and one of 160 gal./day with a zinc concentration of 2.2 mg./l., lime treatment (after cyanide oxidation and hexavalent chromium reduction) followed by sand filters produced a combined effluent zinc content of 0.02 to 0.23 mg./l. This represents a net reduction of zinc exceeding 99%.

Lancy (16) claims that his integrated (and patented) treatment system is the only way to guarantee effluent zinc levels less than 1.0 mg./l. in plating and pickling systems, due to difficulty in settling metal hydroxide precipitates. Others have also reported that the gelatinous zinc hydroxide sludges formed in the precipitation process are difficult to settle or filter (3)(10). Stone (15) reported, however, that a 10% solids sludge was obtained prior to vacuum dewatering. No mention was made of polyelectrolyte use. Rock (3) reported that a viscous rayon manufacturing zinc waste sludge initially capable of settling only to 0.5% solids after 18 hours could be thickened to 5% solids by constantly recycling the solids. This was accomplished without settling, dewatering or the use of polymers.

A process reported by DuPont (17) was developed primarily for small zinc and cadmium-cyanide plating operations. A proprietory formulation called Kastone (composed of a catalyst, hydrogen peroxide and stabilizers) oxidizes the cyanide and promotes the formation of metal oxides instead of hydroxides. The former are readily removed by filtration apparatus which most small plating plants already have in operation. No removal levels are given, but the company claims the method offers a great advantage in metal removal.

Costs for zinc removal systems cannot usually be isolated from the cost of treatment for other constituents contained in the particular waste. Neutralization must be accomplished even in waste streams containing no zinc; cyanide must be oxidized and chromium reduced. Costs directly related to zinc treatment result from the lime requirements, clarifier costs in the absence of other metal hydroxides and sludge dewatering and disposal costs per ton of zinc hydroxide produced. Culotta and Swanton (18) point out that in the past precipitation of metal hydroxides was relatively inexpensive, but that costs have increased considerably in achieving more effective treatment.

Nyquist and Carroll (7) reported waste treatment plant installation and treatment costs for the Conneaut, Ohio General Electric plant. Capital costs (1957 prices) for the 30,000 gallons per day plant were $276,000, or $9,200/1,000 gallons per day treatment capacity. These costs are itemized below.

| | |
|---|---|
| Treatment building | $121,300 |
| Meter building | 3,400 |
| Utilities | 11,600 |
| Installed equipment* | 107,200 |
| Design costs | 28,000 |
| Miscellaneous | 4,500 |
| | $276,000 |

*Largest single equipment expenditure was for a vacuum filter unit at $30,000.

Operating costs were itemized as follows.

|  | Per 1,000 Gallons |
|---|---|
| Mixing | 0.060 |
| Neutralization | 0.360 |
| Filtration | 0.913 |
| Sludge removal | 0.280 |
| Manpower | 2.135 |
| Miscellaneous | 0.045 |
|  | 3.793 |

Thus, total operating costs were $3.79/1,000 gallons for that waste stream, with almost two-thirds of this cost being due to manpower requirements. Sludge disposal costs, also included in the total, represented $0.28/1,000 gallons of wastewater treated. Zinc was reduced to an average effluent of less than 1 mg./l. from a raw waste level of 55 to 120 milligrams per liter (7). The waste stream also contained copper at 204 to 385 mg./l., reduced concurrently with zinc to an average of 0.5 mg./l. Lancy (16) estimates that in general, chemical costs for zinc waste treatment are $0.05 to $0.07/1,000 ft.$^2$ of zinc plated surface, or $0.70 to $1.00/1,000 lbs. of zinc plated small parts.

Ross (19) estimated operating costs for lime treatment of 35,000 gal./day of a zinc-nickel waste, containing 5.5 mg./l. of nickel and 3.5 mg./l. of zinc, to be $0.34/1,000 gal. Lime quantity was estimated at 185 lbs./day and sludge disposal amounted to approximately $0.03/1,000 gal. of wastewater treated.

*Recovery Processes:* Unless other process modifications are necessary, or extremely high levels of zinc are present and minimal purification of recovered material is required, zinc recovery is generally not economical. Precipitation can be considered as a recovery process in situations when zinc is essentially the only ion precipitated and the precipitant is regenerated by acid dissolution. This situation has been reported in viscose rayon manufacturing (20). Recovery processes include the following: (a) ion exchange (which may also function as a polishing unit for final effluent); (b) evaporative recovery; or (c) simple process modification such as countercurrent rinsing.

Kantawala and Tomlinson (21) reported on laboratory studies which indicated that at flow rates of 2.5 gal./hr./ft.$^2$ ion exchange was not as effective as lime precipitation for zinc removal. Ross (19) compared the treatment of zinc and nickel wastes by ion exchange versus lime treatment, and found ion exchange to be more than twice as costly as lime treatment for a waste flow of 35,000 gallons per day.

Precipitation costs were $0.34/1,000 gal. Recovery value from the ion exchange process was credited for water reuse only. Zinc recovery by ion exchange is practiced in viscose rayon manufacturing (10)(11), where the acid regenerant can be reused in the manufacturing process directly without purification. A residual effluent zinc concentration of 20 milligrams per liter was reported in one case after ion exchange recovery, necessitating further waste treatment (11).

The major source of zinc bearing wastewaters is from the metal processing and plating operations and specifically from pickling or plating dragout into rinse water, although periodic dumping of the concentrated plating bath solution may occur as a result of contaminant buildup. A major problem in treating dilute wastewater is the large volumes involved and attendant capital and operation costs. Direct recycle or evaporative recycle of pickling and plating bath rinse water is made feasible by countercurrent rinse flow.

Lancy (16) claims rinse water volumes can be reduced to one-third to one-fourth of normal volumes by the use of cascade or countercurrent rinse practices. This reduction in volume yields higher concentrations in the rinse water. Barnes (6) was able to reduce zinc cyanide plating rinse requirements from 310 to 12 gal./hr. for two units by employing countercurrent rinse.

The remaining rinse waters were of sufficient zinc concentration to allow evaporative concentration and recovery of the plating dragout. Final rinse dragout required no further treatment, due to the low zinc levels present. Lancy (16) points out however, that complete elimination of dragout by evaporative recovery does result in impurity buildup in the plating baths, and subsequent need for periodic dumping of the bath solutions.

Culotta and Swanton (18) reported on treatment of a zinc and brass plating operation by evaporative recovery. Plant modifications reduced process water from 3,000 to 50 gal./hr. and resulted in process chemical savings of $18,000/yr. plus equal savings in chlorine and other treatment chemicals. Capital and operating costs associated with such a closed loop system were reported to be approximately the same as for destructive recovery.

*Summary:* Zinc removal costs often cannot be isolated from overall waste treatment costs for industrial waste streams, which frequently contains mixtures of other heavy metals and chemicals also requiring treatment. In the case where zinc removal is the only consideration and recovery is not warranted, removal by precipitation can be accomplished by standard pH adjustment, lime addition, precipitation and flocculation, and sedimentation, employing standard waste treatment equipment.

Operational data for existing chemical precipitation units indicate that levels below 1 mg./l. of zinc are readily obtainable with lime precipitation, although assurance of consistent removal of precipitated zinc from the effluent stream may require sand filtration. Approximate costs for this general type of treatment have been published (22).

Other treatment alternatives exist and may be more economical. Among these is process modification for evaporative or ion exchange recovery, although existing process layout and current treatment facilities may cause the costs to vary widely with industrial sites and nature of product. Ion exchange recovery seems feasible only for large scale operations and with high zinc concentration in the wastewaters. Evaporative recovery, in concert with countercurrent rinsing to reduce waste volumes, has proven successful on both an economic and zinc removal efficiency basis.

### References

(1)  Athanassiadis, Y.C., "Air Pollution Aspects of Lime and Its Compounds," *Report PB 188,072;* Springfield, Va., National Tech. Information Services (September 1969).
(2)  Patterson, J.W. and Minear, R.A., "Wastewater Treatment Technology," *Report PB 204,521;* Springfield, Va., National Tech. Information Service (August 1971).
(3)  Rock, D.M., "Hydroxide Precipitation and Recovery of Certain Metallic Ions from Wastewaters," presented at Annual Meeting, American Institute for Chemical Engineers, Chicago, Illinois, (November to December 1970).
(4)  McGarvey, F.X., Tenhoor, R.E. and Nevers, R.P., "Brass and Copper Industry: Cation Exchangers for Metals Concentration from Pickle Rinse Waters," *Ind. Engr. Chem.,* 44, 534 to 541 (1952).
(5)  Nemerow, N.L., *Theories and Practices of Industrial Waste Treatment,* Reading, Mass., Addison-Wesley Publishing Co. (1963).
(6)  Barnes, G.E., "Disposal and Recovery of Electroplating Wastes," *Jour. Wat. Control Fed.,* 40, 1459 to 1470 (1968).
(7)  Nyquist, O.W. and Carroll, H.R., "Design and Treatment of Metal Processing Wastewaters," *Sew. Ind. Wastes,* 31, 941 to 948 (1959).
(8)  Pinkerton, H.L., "Waste Disposal," in *Electroplating Engineering Handbook,* 2nd. ed., A. Kenneth Graham, ed. in chief, New York, Reinhold Publishing Co., (1962).

(9) "State of the Art: Review on Product Recovery," *Water Pollution Control Research Series,* U.S. Department Interior, Washington, D.C. (November 1970).

(10) McGarvey, F.X., "The Application of Ion Exchange Resins to Metallurgical Waste Problems," *Proc. 17th. Purdue Industrial Waste Conf.,* pp. 289 to 304 (1952).

(11) Sharda, C.P. and Namwannan, K., "Viscose Rayon Factory Wastes and Their Treatment," *Technolgy, Sindri,* 3, 58 to 60 (1966); *Water Poll. Abstr.* 41:No. 1698 (1968).

(12) Robertson, D.J., Filtration of Copper Smelter Gases at Hudson Bay Mining and Smelting Company, Limited, *Can. Mining Met. Bull.* 53, 326 (1960).

(13) *Air Pollution Control District, County of Los Angeles, Technical Progress Report - Control of Stationary Sources,* vol. 1. (April 1960).

(14) Banerjee, N.G. and Banerjee, T., "Recovery of Nickel and Zinc from Refinery Waste Liquors: Part 1 - Recovery of Nickel by Electrodeposition," *J. Sci. Industr. Res.* 11B, 77 to 78 (1952).

(15) Stone, E.H.F., "Treatment of Nonferrous Metal Process Waste at Kynoch Works, Birmingham, England," *Proc. 25th. Purdue Industrial Waste Conf.,* pp. 848 to 865 (1967).

(16) Lancy, L.E., "An Economic Study of Metal Finishing Waste Treatment," *Plating,* 54, 157 to 161 (1967).

(17) "New Process Detoxifies Cyanide Wastes," *Env. Sci. Technol.,* 5, 496 to 497 (1971).

(18) Culotta, J.W. and Swanton, W.F., "The Role of Evaporation in the Economics of Waste Treatment for Plating Operations," *Plating,* 55, 957 to 961 (1968).

(19) Ross, R.D., *Industrial Waste Disposal,* New York, Reinhold Book Company (1968).

(20) Saxena, K.I. and Chakraborty, R.N., "Viscose Rayon Wastes and Recovery of Zinc Therefrom," *Technology, Sindri,* 3, 29 to 33, (1964); *Wat. Pollut. Abstr.,* 41:No. 1699 (1968).

(21) Kantawala, D. and Tomlinson, H.D., "Comparative Study of Recovery of Zinc and Nickel by Ion Exchange Media and Chemical Precipitation," *Water Sew. Works,* III, R281 to R286 (1964).

(22) "Cost of Wastewater Treatment Processes," *Robert A. Taft Research Center, Report No. TWRC-6,* U.S. Department Interior, Washington, D.C. (1968).

# NOTICE

Nothing contained in this Review shall be construed to constitute a permission or recommendation to practice any invention covered by any patent without a license from the patent owners. Further, neither the author nor the publisher assumes any liability with respect to the use of, or for damages resulting from the use of, any information, apparatus, method or process described in this Review.

# WASTEWATER CLEANUP
# EQUIPMENT 1973

## Second Edition

Water pollution is becoming more of a problem with every passing year. Plants engaged in all types of manufacture are being more and more carefully watched by federal, state, and municipal governments to prevent them from pouring their untreated effluents into the nation's waterways, as they used to do. The sewage treatment plants of many municipalities are becoming too small for the burgeoning population, and many communities once served by individual septic tanks are having to build sewers and treatment plants.

Water pollution will be solved primarily by application of techniques, processes, and devices already known or in existence today, supplemented by modifications of these known methods based on advanced technology. This book gives you basic technical information and specifications pertaining to commercial equipment currently available from equipment manufacturers. Altogether the products of 94 companies are represented.

This second edition of "Wastewater Cleanup Equipment" supplies technical data, diagrams, pictures, specifications and other information on commercial equipment useful in water pollution control and sewage treatment. The data appearing in this book were selected by the publisher from each manufacturer's literature at no cost to, nor influence from, the manufacturers of the equipment.

It is expected that vast sums will be spent in the United States during the remaining portion of this decade for control and abatement of water pollution. Much of the expenditure will be for the type of equipment described in this book.

Today's environmental control is taken to mean a specialized technology employing specialized equipment designed to process the discarded and excreted wastes of human metabolism and human activity of any sort.

Next to air, water is the most abundant and utilized commodity necessary for the maintenance of human life. The average consumption of water per person in residential communities in the United States is between 40 and 100 gallons in one day. In highly industrialized communities the average consumption pro head can be as high as 250 gallons per day.

The reuse of wastewater after cleanup is not only becoming a cogent necessity, but it is also becoming more attractive economically. The degree of purity required for industrial water use is in many cases greater or vastly different from that acceptable for potable water.

Special equipment for cleanup of wastewater is therefore an absolute must, and this book is offered with the intention of providing real help in the selection of the proper equipment.

The descriptions and illustrations given by the original equipment manufacturer include one or more of the following:

1. Diagrams of commercial equipment with descriptions of components.

2. A technical description of the apparatus and the processes involved in its use.

3. Specifications of the apparatus, including dimensions, capacities, etc.

4. Examples of practical applications.

5. Graphs relating to the various parameters involved.

Arrangement is alphabetically by manufacturer. A detailed subject index by type of equipment is included, as well as a company name cross reference index.

ISBN 0-8155-0487-X          $36          372 pages

# POLLUTION CONTROL AND CHEMICAL RECOVERY IN THE PULP AND PAPER INDUSTRY 1973

## by H. R. Jones

### *Pollution Technology Review No. 3*

The pulp and paper industry has serious problems to solve, both in the air pollution and water pollution areas.

Many effluent wastes from paper mills are biodegradable, but treatment costs are increasing, effluent discharge requirements are becoming more stringent, and urbanization increasingly limits the availability of land. Economic control procedures applied by the industry, therefore, form an essential part of the pollution control procedures.

In this book are condensed vital data from government sources of information that are scattered and difficult to pull together. Important processes are interpreted and explained by examples from 54 U.S. patents. One should have to go no further than this condensed information to establish a sound background for action towards combating pollution in the pulp and paper industry.

A partial and condensed table of contents follows. Chapter headings are given, followed by examples of important subtitles.

ISBN 0-8155-0479-9      $36      337 pages

# POLLUTION CONTROL
# IN THE TEXTILE INDUSTRY 1973

## by H. R. Jones

*Pollution Technology Review No. 2*

The major pollution problems confronted by the textile industry are water pollution problems. There is some air pollution by chemicals and lint, but it is of minor concern. Wastes in textile mill effluents may be divided into the following categories:

1. Naturally occurring dust and dirt, salts, oils and fats on cotton and wool.
2. Chemicals added and removed during the many different process operations.
3. Fibers removed by chemical and mechanical action during processing.

Solutions to these problems are afforded by this book which is based on various government surveys and reports with practical examples detailed from late U.S. patents. A partial and condensed table of contents follows:

ISBN 0-8155-0470-5                    $36                    324 pages

# WASTE DISPOSAL CONTROL
# IN THE FRUIT AND VEGETABLE INDUSTRY 1973

### by H. R. Jones

### *Pollution Technology Review No. 1*

The fruit and vegetable processing industry is certainly a major industry. As pointed out in a recent report by the National Canners Association ca. 170,000 persons are employed in ca. 1,800 plants of the canned and frozen fruits industry in the United States.

It is estimated that in the U.S. alone (incl. Hawaii) this industry annually:

**Processes 26 million tons of raw product.**
**Discharges 83 billion gallons of wastewater.**
**Generates 800 million pounds of biochemical oxygen demand (BOD),**
**And 392 million pounds of suspended solids,**
**Together with 8 million tons of solid residue.**

While these huge wastes are biodegradable, they add enormously to the garbage waste and disposal problem.

This book, largely based on authoritative government reports and surveys, attempts to clarify the ways and means open to the food processor who must keep his polluting wastes down to a minimum.

In the following partial list of contents chapter headings are given, together with the more important subtitles.

ISBN 0-8155-0466-7     $36     261 pages

# ENVIRONMENTAL SCIENCE TECHNOLOGY INFORMATION RESOURCES 1973

*Edited by Dr. Sidney B. Tuwiner in conjunction with*
*The Chemical International Information Center*
*(Chemists' Club, New York)*

Environmental science is a new science in many of its aspects. Being a subject of high actuality, it has led to a proliferation of many new publications, stemming from a pressing need for continual updating of the technology and its literature, codes regulating discharges and emissions, etc.

Environmental science is interdisciplinary, involving sociology, law and economics, as well as technology. Some of the problems arising from this fact are discussed here and information retrieval solutions are discussed.

The individual papers give a close look at the many sources of environmental information from the several viewpoints of librarian, editor, and specialist in government sources or industry associations, enabling the user to obtain a broad perspective on environmental information.

**Section I** includes the proceedings and panel discussions of the Symposium on Environmental Science Technology Information Resources sponsored by the Chemical International Information Center, and held at the Chemists' Club, New York, N.Y. on April 28, 1972.

**Introduction** — Dr. Sidney B. Tuwiner
**Resources and Industrial Cooperation on Environmental Control at the American Petroleum Institute** — E. H. Brenner, American Petroleum Institute
**The Environmental Protection Agency and its R & D Program** — William J. Lacy, Chief, Applied Science & Technology, Office of Research & Monitoring, EPA
**Government Sources of Information on Environmental Control** — Marshall Sittig, Director of Special Projects, Princeton University
**Sorting It Out - Data vs. Information** — Steven S. Ross, Editor, New Engineer
**Environmental Control Resources of a Large Technical Library** — Kirk Cabeen, Director of the Engineering Societies Library

**Section II** includes selected papers presented at the National Environmental Information Symposium, sponsored by the EPA, held at Cincinnati, Ohio, September 24-27, 1972.

**Technical Information Programs in the Environmental Protection Agency** — A. C. Trakowski, Deputy Asst. Administrator for Program Operations, Office of Research & Monitoring, EPA
**The Environmental Science Information Center** — James E. Caskey, Director ESIC (Environmental Science Information Center)
**Federal Environmental Data Centers and Systems** — Arnold R. Hull, Associate Director for Climatology & Environmental Data Service, NOAA, U.S. Dept. of Commerce
**Scientific and Technical Information Centers Concerned with the Biological Sciences** — William B. Cottrell, Director, Nuclear Safety Information Center, Oak Ridge National Library
**Secondary Technical and Scientific Journals** — Bernard D. Rosenthal, President, Pollution Abstracts, Inc.
**Environmental Litigation as a Source of Environmental Information** — Victor John Yannacone, Jr., Attorney
**Scientific and Technical Primary Publications Carrying Environmental Information** — D. H. Michael Bowen, Managing Editor, Environmental Science & Technology, American Chemical Society
**Applications of Socioeconomic Information to Environmental Research and Planning** — William B. DeVille, Director of Program Development, Gulf South Research Institute
**Socioeconomic Aspects of Environmental Problems: Secondary Information Sources** — James G. Kollegger, President, Environment Information Center, Inc.

**Section III** is a bibliography of basic governmental, institutional, and organizational documents assembled by the United Nations Conference on the Human Environment, held at Stockholm, Sweden, June 5-16, 1972.

**Introduction**
**Basic Documents Received from States Invited to the Conference**
**Basic Documents Prepared Within the United Nations System**
**Basic Documents Received from Other Sources**

ISBN 0-8155-0467-5          $18          218 pages

# PRACTICE OF DESALINATION 1973

## R. Bakish — Editor

This down-to-earth practical book is the first publication developed under the auspices of the St. Croix Corrosion Installation West Indies Laboratory, College of Science and Engineering, Fairleigh Dickinson University.

This particular volume is based upon a series of practical instructional papers presented as a special topics course by various authorities at St. Croix, U.S. Virgin Islands, during December 1971. This course was sponsored jointly by Fairleigh Dickinson University and the Association of Carribean Desalination Plants Owners and Operators.

The book therefore reflects the present state of the art of the American desalination industry in both its underlying principles and their applications which should remain valid for some time to come. A complete table of contents follows:

Profusely illustrated, this book is a valuable guide for professional engineers, technicians, and operators; also for business and municipal managements.

ISBN 0-8155-0478-0                $18                273 pages